Robert Schlögl (Ed.)
Chemical Energy Storage

Also of Interest

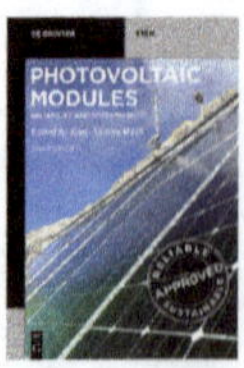

Photovoltaic Modules.
Reliability and Sustainability
Weiß (Ed.), 2021
ISBN 978-3-11-068554-1, e-ISBN 978-3-11-068555-8

Energy and Sustainable Development
Warren, 2021
ISBN 978-1-5015-1973-4, e-ISBN 978-1-5015-1977-2

Chemistry and Energy.
From Conventional to Renewable
Benvenuto, 2022
ISBN 978-3-11-066226-9, e-ISBN 978-3-11-066227-6

Electrochemical Energy Storage.
Physics and Chemistry of Batteries
Job, 2020
ISBN 978-3-11-048437-3, e-ISBN 978-3-11-048442-7

Chemical Energy Storage

Edited by
Robert Schlögl

2nd Edition

DE GRUYTER

Editor
Prof. Dr. Robert Schlögl
Fritz-Haber-Institut
der Max-Planck-Gesellschaft
Abteilung Anorganische Chemie
Faradayweg 4-6
14195 Berlin
Germany

ISBN 978-3-11-060843-4
e-ISBN (PDF) 978-3-11-060845-8
e-ISBN (EPUB) 978-3-11-060859-5

Library of Congress Control Number: 2021940760

Bibliographic information published by the Deutsche Nationalbibliothek
The Deutsche Nationalbibliothek lists this publication in the Deutsche Nationalbibliografie; detailed bibliographic data are available on the Internet at http://dnb.dnb.de.

Cover image: ollo/E+/Getty Images
Typesetting: Integra Software Services Pvt. Ltd.
Printing and binding: CPI books GmbH, Leck

www.degruyter.com

Contents

3 Thermochemical

4 Biobased

List of Contributing Authors

Dr. Atte Aho
Process Chemistry Centre
Åbo Akademi University
Biskopsgatan 8
20500 Turku/Åbo
Finland
atte.aho@abo.fi
Chapter 19

Prof. Dr. Markus Antonietti
Department of Colloid Chemistry
Max Planck Institute of Colloids and Interfaces
Forschungscampus Golm
14424 Potsdam
Germany
Markus.Antonietti@mpikg.mpg.de
Chapter 20

Dr. Sebastian Arndt
Department of Chemistry
Technische Universität Berlin
Straße des 17. Juni 124
10623 Berlin
Germany
arndt@chem.tu-berlin.de
Chapter 10

Prof. Dr. Malte Behrens
Kiel University
Institute of Inorganic Chemistry
Max-Eyth-Strasse 2
24118 Kiel
Germany
mbehrens@ac.uni-kiel.de
Chapter 14

Dr.-Ing. Cornelius M. Berger
Ruhr University Bochum
Laboratory of Industrial Chemistry
Universitätsstr. 150
44801 Bochum
Germany
cornelius.berger@gmx.de
Chapter 14

Dr. Eckhard Bill
Department of Inorganic Spectroscopy
Max Planck Institute for Chemical Energy Conversion
Stiftstraße 34-36
45470 Mülheim an der Ruhr
Germany
eckhard.bill@cec.mpg.de
Chapter 16

Dr. James Birrell
Department of Inorganic Spectroscopy
Max Planck Institute for Chemical Energy Conversion
Stiftstraße 34–36
45470 Mülheim an der Ruhr
Germany
James.birrell@cec.mpg.de
Chapter 8

Dr.-Ing. Armin Brandner
Green Chemicals & Fuels Product Line
Air Liquide
Olof-Palme-Str. 35
60439 Frankfurt am Main, Germany
armin.brandner@airliquide.com
Chapter 18

Prof. Dr. Gabriele Centi
Dipartimento ChiBioFarAm (Chimica Industriale),
Chimica ed Ingegneria Industriale
Università degli Studi di Messina
Viale Ferdinando Stagno d'Alcontres 31
98166 Messina
Italy
centi@unime.it
Chapter 15

Prof. Dr. Peter Claus
Technische Universität Darmstadt
Technische Chemie II
Ernst-Berl-Institut für Technische und Makromolekulare Chemie
Petersenstraße 20

https://doi.org/10.1515/9783110608458-203

64287 Darmstadt
Germany
claus@ct.chemie.tu-darmstadt.de
Chapter 18

Prof. Dr. Nicholas Cox
Australian National University
Research School of Chemistry
Canberra, Australia
nick.cox@anu.edu.au
Chapter 8

Prof. Dr. Serena DeBeer
Department of Inorganic Spectroscopy
Max Planck Institute for
Chemical Energy Conversion
Stiftstraße 34-36
45470 Mülheim an der Ruhr
Germany
serena.debeer@cec.mpg.de
Chapter 16

Dr. Nikolai DeMartini
Process Chemistry Centre
Åbo Akademi University
Biskopsgatan 8
20500 Turku/Åbo
Finland
nmartini@abo.fi
Chapter 19

Dr. Karl Doblhofer
Department of Physical Chemistry
Fritz-Haber-Institut der
Max-Planck-Gesellschaft
Faradayweg 4-6
14195 Berlin
Germany
acsek@fhi-berlin.mpg.de
Chapter 5

Dr. Dmitry Doronkin
Karlsruhe Institute of Technology (KIT)
Institute for Chemical Technology and
Polymer Chemistry (ITCP)
and
Institute of Catalysis Research and
Technology (IKTF)
Kaiserstraße 12
76128 Karlsruhe
Germany
dmitry.doronkin@kit.edu
Chapter 12

Prof. Dr. Hans-Joachim Freund
Department of Chemical Physics
Fritz-Haber-Institut der
Max-Planck-Gesellschaft
Faradayweg 4-6
14195 Berlin
Germany
freund@fhi-berlin.mpg.de
Chapter 13

Dr. Maurice van Gastel
Max-Planck-Institut für Kohlenforschung
Department for Molecular and Theoretical
Spectroscopy
Kaiser-Wilhelm-Platz 1
45470 Mülheim an der Ruhr
Germany
maurice.van-gastel@kofo.mpg.de
Chapter 16

Prof. Dr. Dr. h.c. Carl Friedrich Gethmann
Lebenswissenschaftliche Fakultät
Universität Siegen
WeidenauerStr. 167
57076 Siegen
Germany
carl.gethmann@uni-siegen.de
Chapter 2

Prof. Dr. Jan-Dierk Grunwaldt
Karlsruhe Institute of Technology (KIT)
Institute for Chemical Technology and
Polymer Chemistry (ITCP)
and
Institute of Catalysis Research and
Technology (IKTF)
Kaiserstraße 12
76128 Karlsruhe
Germany
grunwaldt@kit.edu
Chapter 12

Prof. Dr. Mikko Hupa
Process Chemistry Centre
Åbo Akademi University
Biskopsgatan 8
20500 Turku/Åbo
Finland
mikko.hupa@abo.fi
Chapter 19

Dr. Georg Kamp
Forschungszentrum Jülich GmbH
Wilhelm-Johnen-Straße
52425 Jülich
Germany
g.kamps@fz-juelich.de
Chapter 2

Dr. Edward Kunkes
BASF Corp.
Iselin
New Jersey
08830 USA
edward.kunkes@basf.com
Chapter 14

Dr. Roh Pin Lee
Institute for Energy Process Engineering and Chemical Engineering
TU Bergakademie Freiberg
Germany
Roh-Pin.Lee@iec.tu-freiberg.de
Chapter 21

Dr. Jan van de Loosdrecht
Council for Scientific and Industrial Research
Meiring Naudé Road
Pretoria 0001
South Africa
Loosdrecht1@outlook.com
Chapter 14

Prof. Dr. Dr. h.c. Wolfgang Lubitz
Max Planck Institute for
Chemical Energy Conversion
Stiftstraße 34-36
45470 Mülheim an der Ruhr
Germany
wolfgang.lubitz@cec.mpg.de
Chapter 8

Prof. Dr. Joachim Maier
Max Planck Institute for
Solid State Research
Heisenbergstraße 1
70569 Stuttgart
Germany
s.weiglein@fkf.mpg.de
Chapter 9

Prof. Dr. Dietrich Menzel
Physik Department E20
Fakultät für Physik
Technische Universität München
James-Franck-Straße 1
85748 Garching
Germany
dietrich.menzel@ph.tum.de
Chapter 13

Prof. Dr.-Ing. Bernd Meyer
Institute for Energy Process Engineering and Chemical Engineering
TU Bergakademie Freiberg
Germany
Bernd.Meyer@iec.tu-freiberg.de
Chapter 21

Prof. Dr. Martin Muhler
Ruhr University Bochum
Laboratory of Industrial Chemistry
Universitätsstraße 150
44801 Bochum
Germany
muhler@techem.rub.de
Chapter 14

Prof. Dr. Dmitry Yu. Murzin
Process Chemistry Centre
Åbo Akademi University
Biskopsgatan 8
20500 Turku/Åbo
Finland
dmitry.murzin@abo.fi
Chapter 19

Prof. Dr. Frank Neese
Department of Molecular Theory and Spectroscopy
Max-Planck-Institut für Kohlenforschung

Kaiser-Wilhelm-Platz 1
45470 Mülheim an der Ruhr
Germany
frank.neese@kofo.mpg.de
Chapter 16

Prof. Dr. (Hans) J.W. Niemantsverdriet
Syngaschem BV
SynCat@DIFFER
De Zaale 20
5612 Eindhoven
The Netherlands
jwn@syngaschem.com
Chapter 14

Dr. Niklas Nilius
Department of Chemical Physics
Fritz-Haber-Institut der
Max-Planck-Gesellschaft
Faradayweg 4-6
14195 Berlin
Germany
nilius@fhi-berlin.mpg.de
Chapter 13

Dr. Klaus Friedel Ortega
Kiel University
Institute of Inorganic Chemistry
Max-Eyth-Strasse 2
24118 Kiel
Germany
kfriedel@ac.uni-kiel.de
Chapter 14

Prof. Dr. Regina Palkovits
RWTH Aachen University
Institut für Technische und
Makromolekulare Chemie (ITMC)
Worringerweg 1
52074 Aachen
Germany
palkovits@itmc.rwth-aachen.de
Chapter 17

Dr. Dimitrios A. Pantazis
Max-Planck-Institut für Kohlenforschung
Department of Molecular Theory and
Spectroscopy
Kaiser-Wilhelm-Platz 1
45470 Mülheim an der Ruhr
Germany
dimitrios.pantazis@kofo.mpg.de
Chapter 16

Prof. Siglinda Perathoner
Università degli Studi di Messina
Dipartimento ChiBiofarAm (Chimica
Industriale)
Viale Ferdinando Stagno d'Alcontres 31
98166 Messina
Italy
perathon@unime.it
Chapter 15

Dr. Taras Petrenko
High Performance Computing Center
Stuttgart (HLRS)
Nobelstr. 19
70569 Stuttgart
Germany
taras.petrenko@hlrs.de
Chapter 16

Dr. Mauricio Prieto
Fritz-Haber-Institut der Max-Planck-
Gesellschaft,
Faradayweg 4-6, 14195 Berlin
Germany
prieto@fhi-berlin.mpg.de
Chapter 13

Prof. Dr. Drs.h.c. Ortwin Renn
IASS Potsdam
Institute for Advanced Sustainability Studies
e.V.
Berliner Straße 130
14467 Potsdam
Germany
ortwin.renn@iass-potsdam.de
Chapter 1

Prof. Dr. Marcus Rose
Technische Universität Darmstadt
Technische Chemie II
Ernst-Berl-Institut für Technische und
Makromolekulare Chemie

Petersenstraße 20
64287 Darmstadt
Germany
rose@tc2.tu-darmstadt.de
Chapter 18

Prof. Dr. Jan Rossmeisl
University of Copenhagen
Department of Chemistry
Universitetsparken 5
2100 København Ø
Denmark
jan.rossmeisl@chem.ku.dk
Chapter 6

Dr. Dominik Samuelis
Heraeus Battery Technology GmbH
Heraeusstr. 12-14
63450 Hanau
Germany
dominik.samuelis@heraeus.com
Chapter 9

Prof. Dr. Robert Schlögl
Department of Inorganic Chemistry
Fritz-Haber-Institut der
Max-Planck-Gesellschaft
Faradayweg 4-6
14195 Berlin
Germany
acsek@fhi-berlin.mpg.de
Chapter 4

Dr. Thomas Schmidt
Fritz-Haber-Institut der Max-Planck-Gesellschaft,
Faradayweg 4-6, 14195 Berlin
Germany
schmidtt@fhi-berlin.mpg.de
Chapter 13

Prof. Dr. Reinhard Schomäcker
Department of Chemistry
Technical University Berlin
Straße des 17. Juni 124
10623 Berlin
Germany
schomaecker@tu-berlin.de
Chapter 10

Prof. Dr. Ferdi Schüth
Department of Heterogeneous Catalysis
Max-Planck-Institut für
Kohlenforschung
Kaiser-Wilhelm-Platz 1
45470 Mülheim
Germany
schueth@kofo.mpg.de
Chapter 3

Dr. Shamil Shaikhutdinov
Department of Chemical Physics
Fritz-Haber-Institut der
Max-Planck-Gesellschaft
Faradayweg 4-6
14195 Berlin
Germany
shaikhutdinov@fhi-berlin.mpg.de
Chapter 13

Dr. Martin Sterrer
Department of Chemical Physics
Fritz-Haber-Institut der
Max-Planck-Gesellschaft
Faradayweg 4-6
14195 Berlin
Germany
sterrer@fhi-berlin.mpg.de
Chapter 13

Prof. Dr. Peter Strasser
The Electrochemical Energy, Catalysis,
and Materials Science Laboratory
Department of Chemistry
Chemical Engineering Division
Straße des 17. Juni 124
Technical University Berlin
10623 Berlin
Germany
pstrasser@tu-berlin.de
Chapter 7

Dr. Annette Trunschke
Department of Inorganic Chemistry

Fritz-Haber-Institut der Max-Planck-Gesellschaft
Faradayweg 4-6
14195 Berlin
Germany
trunschke@fhi-berlin.mpg.de
Chapter 11

Dr. William R. H. Wright
Institut für Technische und Makromolekulare Chemie (ITMC)
RWTH Aachen University
Worringerweg 1
52074 Aachen
Germany
right@itmc.rwth-aachen.de
Chapter 17

Dr. Shengfa Ye
Max-Planck-Institut für Kohlenforschung
Department for Molecular and Theoretical Spectroscopy
Kaiser-Wilhelm-Platz 1
45470 Mülheim an der Ruhr
Germany
shenfa.ye@kofo.mpg.de
Chapter 16

1 General

Ortwin Renn

1 Energy and Society: Toward a Socio-ecological Foundation for Research on Energy Transformations

1.1 Introduction

At first glance, energy research appears to be a topic primarily for the engineering and natural sciences, which can be enriched, at best, by social science studies. However, the study of the physical consequences of energy production, transportation, and consumption cannot only be related to the effects of human interventions on nature and society, but must start with the investigation of the social and cultural causes that have triggered these interventions in the first place. Humans make decisions based on mental models of what they would like to accomplish. These models are framed by social and cultural aspirations, values, and norms [1–4]. Based on this broader perspective, three areas of overlap can be identified that link the biophysical with the cultural world [59, 5, 6]:

- Basic human needs and demands with respect to resources from the natural environment (in particular energy)
- The consequences of human interventions for natural cycles, processes, and structures, including biological changes in and among humans (such as new lifestyles, new health risks)
- The feedback of these interventions on cultural self-image, social structures, and social processes (cultural identity)

Undoubtedly, all three dimensions are closely and inevitably associated with insights from the natural sciences. Without sufficient knowledge of the structure and dynamics of natural systems, the estimation of anthropogenic causes and consequences of energy supply systems remains speculative. Interactions arise from the energetic, material or communicative exchanges between humans and the natural environment. However, the natural science perspective is not sufficient to understand the interactions between society and energy.

Traditional energy research is largely based on the ideal of a value-free exploration of the physical resources, their efficient use, and their sustainability to meet human needs. This concept of energy research is problematic for two reasons [7, 8]: First, the choice of what people need is already culturally influenced. The postulated assumptions about correlations between the use of the physical resources and their

Ortwin Renn, IASS Potsdam, Institute for Advanced Sustainability Studies e.V., (IASS) Berliner Straße 130, 14467 Potsdam, Germany, e-mail: ortwin.renn@iass-potsdam.de

https://doi.org/10.1515/9783110608458-001

utilization for human purposes are determined by preconceived cultural frames, which require their own explanation and questioning. Secondly, the supposed chain of knowledge and action falls short [9]. Knowledge alone does not change behavior. Although knowledge is a prerequisite for motivating action, it does not replace the necessity of having normative priorities when weighing the consequences for the biophysical environment against other positive or negative impacts on the economy and society [10]. At the same time, the balancing of these processes itself is dependent on social preferences. For example, the political scientist Aaron Wildavsky concluded that the U.S. environmental legislation and regulatory practices systematically hide the true environmental impact when damage is distributed widely and evenly, while pollution that affects socially prominent groups is targeted by the media and triggers more regulatory action [60].

In conclusion, energy research has to combine both components: the consequences of human behavior on the natural and socio-cultural environment as well as the reflexive perception, assessment, and evaluation of human behavior when experiencing transformations in energy supply systems [11]. Such an integrative approach includes five key objectives [5, 12, 13]:

- To gain systematic understanding of the processes of knowledge generation and of value formation regarding human interventions into nature and society, and, from these findings, contribute to a better understanding of the *human–environment–nature relationship* and the cultural patterns for selecting options to meet energy demands.
- To gain better knowledge of processes and procedures that shape or enlighten the *social discourse(s)* about the right balance between different options of meeting energy service demand and, consequently, about ethically justifiable degrees of interventions into the natural environment on the basis of comprehensible and politically legitimate criteria.
- To investigate institutional processes and organizational structures that review, revise, and regulate *individual and collective energy-related decisions.*
- To identify patterns of technology development and transformations with the goal of identifying and understanding *potential impacts of energy systems and their organizational environment.*
- To investigate *obstacles and barriers,* but also opportunities and incentives, which affect energy transitions at both individual and collective levels, and to systematically explore and develop constructive suggestions for *resolving conflicts* about priorities for designing and legitimizing energy policies.

The following sections will demonstrate how a socio-ecological approach to energy systems analysis can be designed to meet these five objectives. Given the limitation of space and the main interest of the readers of this volume to get practical advice on how social science can assist transformations toward climate-neutral energy supply, the main focus of the remaining sections will be placed on the fifth objective. In which way can social science insights contribute to overcoming obstacles and barriers to a successful transformation of energy systems? We will argue that citizen empowerment, deliberative reasoning, and targeted communication are a key to facilitate energy transformations in democratic societies.

1.2 Characteristics of a Socio-ecological Approach to Energy Research

The main proposition of the social–ecological approach is that knowledge can only advance in the area of energy research by reflecting on the interaction between the social processes of knowledge generation and their interpretation based on language, research tools, and cultural contexts. This task of establishing relationships between human observers and nonhuman environments by means of interventions thus stands at the beginning of any socio-ecological research process. Social–ecological risk research is self-reflexive and self-critical of its knowledge claims. This allows one to look beyond the confines of relativism and allows selected interactions between nature, environment, and culture to be intersubjectively validated [12, 14–16].

The second essential characteristic of the social–ecological approach to understand energy transformations is the combination of knowledge and values [17, 18]. In addition to cultural selection and interpretation regarding energy selection and policy making, knowledge needs to be supplemented by a culture of values and moral principles in order to derive meaningful and responsible decisions. Decision-making includes two components: assessment of likely consequences of each decision option (the knowledge component), and evaluation of these consequences according to personal or collective desirability and ethical acceptability (the value component).

Classic risk research has been adopting a value-free approach. Research questions are normally phrased as "What if . . .?" By not abiding to one scenario, many researchers felt that they were able to represent the plurality of societal values. This approach, however, blends out the fact that each of these "if–then" questions represents but one of many possible scenarios and that a selection must be made. Furthermore, even the most scientific scenario construction will be limited to potential consequences that researchers think will or should matter to society [19].

The perspective of social ecology can no longer rely on neutrality. Valuations not only creep into the factual process of consideration of what could be researched but are already incorporated into the design of assessment processes [20]. The catalytic function of science, part of the five requirements mentioned in the section above, is based on normative foundations, such as that all those who could be affected should also be considered in the analysis or that the factual findings of appraisal processes should be made available to all parties [21, 22]. That these normative standards must be made explicit should be considered as self-evident.

The third component of social–ecological research includes the methodological approach, often referred to in the literature as transdisciplinary [23–27]. This term includes, despite all the differences in respective positions, the following key aspects:
- Methodological approach based on using multiple methods employed by different disciplines or methods that bridge different disciplines
- A problem-oriented, not phenomenon-oriented approach
- A close connection between theoretical analysis and practical implications
- Direct involvement of individuals and groups affected by the respective problem in the research process (recursive participation)

The most controversial feature of social–ecological research in the literature is the participatory approach that relates to the last point in the list of requirements for transdisciplinary research [28–30]. Especially when it comes to problem selection, to localizing knowledge bases, to developing and evaluating options for socio-ecological approaches, it is recommended that those who are affected by the problem directly and indirectly should participate in the research project. The definition of the problem, the question of which values to include, and the selection of methods used to consider and evaluate solutions, all presuppose normative assumptions that need to be aligned to the preferences and values of those who will be affected. These requirements cannot be derived legitimately from research itself, but must result from social discourse. In this respect, transdisciplinary research is also conditional on the involvement of affected persons and groups in the process of knowledge discovery and evaluation. Participation is not a panacea to resolving all conflicts. The danger of the participatory approach lies in the indiscriminate blending of factual knowledge, normative orientation, values, and interests [31, 32]. This mix should, however, not be seen as a problem but rather as an inevitable research condition that can be managed by choosing the appropriate designs and ensuring professional support [33].

For socio-ecological research, a division into three categories of knowledge has proven useful [34–36]:
- *Knowledge for orientation*: What normative assumptions will we accept as benchmarks for our own actions, and what does that imply for the specific energy problem that we want to resolve?
- *Knowledge for system understanding*: What is needed tom understand energy as a cross-cutting system that affects the natural, social, and cultural environment? What are the connections between the variables that define the problem and its context? What are the factors that determine the present situation? What actions are possible to facilitate the required transitions?
- *Knowledge for transformation*: What measures are appropriate to achieve the desired objectives given the knowledge for orientation (goals) and the knowledge for understanding (causal web)? How can we reach these objectives under conditions of effectiveness, efficiency, and acceptability?

The combination of systemic, orientative, and transformational knowledge is directed toward a *rapprochement* between the models representing reality (knowledge) and the real manifestations of reality in form of empirical observations. Using some vague notion of desirability as a selection mechanism of knowledge should be avoided and an organized skepticism must be strictly adhered to, examining the validity and reliability of research results [37].

However, when it comes to framing problems, interpreting results, and judging the acceptability of solutions, subjective values are essential for creating important benchmarks for the research process. Aspects such as selecting problems to be investigated, deciding which dimensions are classified as socially relevant, as well as balancing the arguments for and against a course of action require subjective input. The big challenge is to shape the necessary integrative research process, so that, on the one hand, the analytical separation of states of knowledge and values is maintained but, on the other hand, a procedural "locking in" of analysis and evaluation is ascertained. For the actual research process, this means there must be differing participatory principles and procedures in order to assess the implicit requirements of the three different categories of knowledge: (i) Knowledge for orientation requires value input from all affected populations; (ii) knowledge for systems understanding requires input from all who are able to contribute to the analysis of risks and opportunities of each energy option; and (iii) knowledge for transition requires the input of all those who are needed to translate insights into collective action [27].

1.3 From Public Acceptance to Societal Discourse

Integrating the three types of knowledge (system, orientation, transformation) requires both: a strong scientific analytic input providing the evidence for what is likely to be expected when pursuing option (a) versus option (b) and a refined deliberative discourse process that ensures the inclusion of knowledge, interests, and values by stakeholders and the affected public(s). This combination has been coined an analytic-deliberative discourse by the US Academy of Sciences [30, 38, 39]. Such a combination of science and public participation needs to be enlightened by a social science analysis of what matters to people and why. The following section summarizes the research results of what drives public acceptance and approval of energy policies and measures.

1.3.1 Public Acceptance: Drivers and Motivations

Acceptance for new energy policies or for siting of energy facilities does not necessarily require a positive attitude or endorsement of the planned projects [40, 41]. Acceptance can be divided into three levels of approval: (i) tolerance; (ii) positive attitude; and (iii) active engagement (or involvement). In the first case, citizens do not take action against government measures or planning, even if they oppose them mentally. This tolerance often originates in the fact that people do not consider the issue significant enough to deal with, or they believe they cannot make a difference anyway. Most planning, especially when it comes to on-site facilities, is not embraced with enthusiasm. For the political implementation of planning, it is usually sufficient if the people affected by it tolerate the associated measures and facilities in their living environment. However, it is of central importance for the dynamics of the discussion and for the formation of political will that there are individuals and groups with a positive attitude as well as committed citizens who support the implementation of the planned policies or infrastructural changes. From this point of view, a positive attitude and active commitment are conducive to the concrete implementation of energy transition measures, but in terms of acceptance, it is sufficient if the planned measure is tolerated. In order for policies or infrastructure changes to gain acceptance in this sense, four prerequisites must be met.

- *Orientation and insight:* If there is insight into the necessity of the policy or measure and if people support the goals and means targeted by the measures, then acceptance is more likely to follow. However, in order for people to get an idea of what to expect from energy transition policies, they need information about the reasons and arguments in favor of the respective policies or measures. This includes whether alternatives are available and, if so, why they were not chosen. Citizens demand transparent, that is, comprehensible, reasoning when it comes to justifying decisions.
- *Self-efficacy and personal agency:* People tend to reject interventions in their living environment if they associate them with the idea that their scope for freedom and their sovereignty over their living habits could be negatively affected. For example, people perceive an intervention in their usage habits (such as shutting down power-hungry machines through smart grid programs when the demand for electricity is too high) as an impermissible impairment of their sovereignty. The more a measure gives the impression of restricting freedom, the more likely it is that it will not be accepted.
- *Positive risk–benefit balance:* Acceptance is more likely the more the planned policies or measures benefit oneself or the groups and individuals one particularly values. Higher acceptance can also be expected if measures are subjectively judged as a contribution to the general welfare. With all the information, it is important for most people to know whether they themselves or others close to them will experience a positive risk–benefit balance with the realization of the measure. Without information about benefits and risks, it is also difficult to assess the desirability of the goals intended by the measure. In the risk–benefit balance, both components: risk and benefit are of particular importance. As a rule, people try to avoid cognitive dissonance, that is, contradictory arguments

and opposites. Therefore, empirical studies show that most people associate a high risk with a low risk and vice versa a high benefit with a low risk [42, 43]. Thus, as soon as an energy source is considered risky or dangerous by the public, the benefits are automatically considered less relevant for one's own judgment. Therefore, risk perception is an important factor in determining the acceptability of a policy or a measure.
- *Identity:* The more one can identify with a measure, also emotionally, the greater the willingness to accept it. In the context of energy transitions, information is significant that helps people to grasp the significance of the measure for their own ideas about the future and to check the fit of the measure in the self-image and external image of their own social and cultural environment. In the context of the energy transition, for example, new operator models and ownership options should be mentioned (such as cooperatives, issuing share certificates, and profit sharing), which can create identification through an emotional attachment to ownership or usage rights.

If one wants to influence these four aspects in favor of a greater acceptance of energy policies and measures, then it is at least necessary that the information and communication programs are related to all four aspects. Only if one can make clear how the planned projects improve the benefits for oneself and others and to what extent positive identification is made possible, can higher acceptance be expected.

However, the effectiveness of communication in influencing acceptance is limited. Especially when it comes to projects that impose burdens on local residents and where the general benefits are socially controversial, it is almost impossible to achieve a change in acceptance through information and communication alone, even if they are offered in the form of a dialogue [44]. In addition, large-scale changes tend to be seen as a foreign body and less as an enrichment of the local environment. Here, the communicator has bad cards.

In this respect, due to the lack of effectiveness of communication alone, it is appropriate to give the affected people greater opportunities for participation, so that they can decide for themselves on the basis of different variants to what extent the four acceptance criteria are met. The perspective of participation changes the political decision-making process. Communication is aimed at making the people affected aware of a decision once it has been legally made in favor of a particular option, in the hope that they will also accept this view or at least tolerate it. In contrast, the perspective of participation assumes open decision-making processes and leaves it up to the citizens involved in the process, within the legal boundaries, to create new options and evaluate existing ones on the basis of their own ideas and assessments. At the moment when those affected by decisions become decision-makers, identity is created by the process alone [45, 46].

1.3.2 Benefits of a Participatory Approach to Energy Policy Making

There are many reasons for involving citizens more in the decision-making process for complex decisions on energy policies, such as those needed to implement climate protection. Some of these reasons are listed below [47–49]:

- By including locally affected parts of the population, the knowledge base can be expanded. In addition to the systematic knowledge of experts and the process knowledge of decision-makers, experiential knowledge, and indigenous knowledge on the ground can be of particular importance for many decision-making problems. These non-scientific sources of relevant knowledge can be an important enrichment and occasionally a correction of expert knowledge.
- Citizen participation can provide the relevant decision makers with important information about the distribution of preferences and values of the affected populations. Since decisions are based on consequential knowledge and judgments about the desirability of the expected consequences, it is often indispensable for decision-makers to explicitly ascertain the perception of the desirability of the consequences and (also) take it up as a basis for their own decision. Particularly in the case of complex problems such as the energy transition, there are often conflicting goals, for example, between economic interests and ecological concerns. Here, it is essential for socially acceptable policies to learn how affected social groups and individuals distribute relative weights on the target dimensions (trade-offs).
- Citizen participation can serve as a tool to fairly negotiate interests and values. The literature on game theory, mediation, arbitration, and negotiation processes is eloquent testimony to this conflict-mediating function of citizen participation [50]. This insight is inspired by the proposition that those who lay claim to public resources in competition with others (such as in the 2019 Coal Commission) should best find their own solution for fair distribution in a "fair" process (procedural equity).
- Citizen participation can be seen as a contest of arguments, with the aim of placing collective decisions on a normatively secure footing on the basis of justifications [30, 51]. The goal of such a participation process is the discursive contestation of justified positions under the framework conditions of an examination of the arguments presented in each case based strictly on logic and consistent derivation (claim to validity and proof of validity). The main issue here is the reasonableness of normative settings for all those affected (beyond the circle of those participating in the discourse). Especially in the case of energy transition, it is often a question of the reasonableness of burdens for certain groups of the population (such as employees in coal-fired power plants, electricity consumers, car owners).

- Citizen participation has an impact on the shape and design of one's own living environment [52]. In this function, the people concerned are given the opportunity to bring about changes in their own lifeworld in the form of voluntary commitments or attributions of responsibility (e.g., in the siting of wind or solar farms).

However, active participation of citizens in issues of energy transition is subject to conditions. It requires two things: legitimation through the process and an open, transparent discussion with the affected population groups. In this context, an open dialogue must not be limited to informing those affected but must also include the rights of those affected to participate in the decision-making process. Without such feedback, any dialogue will ultimately fail and end up in frustration. Participation and openness about options are thus necessary conditions for a successful participation process [53].

1.3.3 Conditions for Successful Participation

The concrete design of participation processes depends on the purpose and topic to be discussed. However, there are some overarching conditions that need to be adhered to independent of the specific format or mandate of the participatory exercise. The most important conditions are [44]:

- *Fairness:* The persons participating in a participation procedure should be selected at an early stage according to fair, comprehensible criteria and should be able to claim equal rights and obligations within the procedure. Participants should be aware of the responsibility of their function.
- *Competence*: The expertise required in each case to assess the consequences and side-effects of decision options must be made available to the participants. Common rules for conducting discussions are essential. Furthermore, a successful translation of scientific input into comprehensible language suited for dialog formats and participatory decision-making needs to be ensured and prepared in advance of the participation process.
- *Legitimacy:* Those with political responsibility should give the participants in the process the benefit of the doubt and open up room for designing creative solutions. This also includes the binding assurance that the results of participation procedures will be examined constructively and sympathetically and that deviations from the results achieved will be justified in detail.
- *Efficiency*: It is necessary to ensure that the recommendations can be linked to the political decision-making process. Participants in the process should have a clear mandate that they can fulfill within a limited period of time. The expected positive effect and the effort involved in participation must be in an acceptable ratio for all those involved.

- *Neutrality of the moderator:* The competence of the moderator to conduct the discussion in an objective, impartial and constructive manner must be guaranteed.
- *Willingness to learn:* Participation processes are about exchanging arguments according to defined rules. A willingness to engage in mutual learning processes should be assumed.
- *Open-endedness*: The process must be open-ended and offer potential for new win-win solutions or at least involve burdens that are acceptable to all participants.
- *Public resonance:* It must be ensured that the procedure and the results have an appropriate resonance among the general public. Both the process and the outcome of the process should be made transparent and presented in a form that can be understood by all interested parties.
- *Connectivity:* The procedure should be designed in such a way that the results can be introduced into the political decision-making process and implemented there in compliance with all legal provisions.

If the principles listed above are observed in the implementation of participation procedures, the procedure can help to ensure that the decision-making process about energy policies and the resulting outcome are accepted by a large number of citizens and that they have the momentum to be effective in shaping energy policies toward a sustainable energy future.

1.3.4 Synthesis in Form of an Analytic-Deliberative Process

The US-National Research Council has addressed the issue of participatory decision making for complex topics and recommended a combination of (i) analytic rigor based on scientific evidence and comprehensive peer review and (ii) deliberative argumentation among a broad representation of stakeholders and representatives of the various publics [38, 39]. The concept suggests a policymaking process based on the inclusion of experts, stakeholders, and the general public [54, 55]. Such process promises the inclusion of the best scientific knowledge and incorporates the plurality of cognitive, ethical, and aesthetical claims in society. Furthermore, it provides the input to combining system, orientation, and transformation knowledge.

In order to integrate scientific evidence and public deliberation, it is suggested that analysts and policymakers work together by using participatory modeling as a tool [56]. This way, scientific knowledge (e.g., computer modeling and simulations) can be integrated with stakeholder/lay knowledge (e.g., what needs to be modeled) in order to create scenarios and policy options that policymakers can apply for designing transition pathways toward a sustainable energy future. By combining social and technical elements, participatory modeling has the potential to contribute to better knowledge exchange, collective decision-making, negotiation, and integration, thereby supporting reciprocal understanding, creativity, and social learning [57].

1.4 Conclusion

The paper advocated a socio-ecological approach to investigate the complex associations between energy technology, policymaking, and public acceptance. This approach is based on three requirements:

- Integration of systemic, orientative, and transformational knowledge as a prerequisite for creating a common understanding of a problem and exploring a solution space for resolving the problem.
- Procedural integration of knowledge generation and selection of individual and collective evaluation processes without causing an impermissible mixing of perception and interest during the assessment and evaluation phases of the process (analytic component).
- Inclusion of stakeholders and affected citizens in co-designing their environment and becoming empowered to contribute to public policymaking and transformation practices (deliberative component).

The socio-ecological approach is interdisciplinary, problem-oriented, practical and participatory. Within the transdisciplinary realm, an integration of systemic, orientative, and transformational knowledge provides the basis for a rational and problem-oriented approach to the process of designing, implementing, and monitoring energy policies. Particularly important is the combination of expertise and deliberation based on the model of the analytic-deliberative discourse advocated by the US National Academy of Science [39].

Ambitious energy transition goals can hence only be implemented politically if societal structures, decision-making processes, and developments are considered from a perspective of scientific analysis in combination with public deliberation. In order to take these processes of social change sufficiently into account, it is necessary to involve the population at an early stage in problem definition, problem analysis, and decision-making. Only if citizens are involved in the processes of implementing energy policies from the outset will it be possible to perceive this transformation as a task for society as a whole and to implement it successfully.

Participation processes are used to jointly assess the consequences and implications associated with each of the possible courses of action and, on this basis, to formulate recommendations for representative bodies and/or for fellow citizens. Accordingly, democracy does not need a substitute for its representative bodies but rather needs a functional enrichment through discursive forms of direct citizen participation, which in turn benefit the representative bodies in their own quest for legitimacy. This enrichment of democracy is essential, especially in the case of ambitious projects such as those in energy transition and climate policy.

For participation to succeed, it requires competent planning and professional implementation of the participation process. There is no general recipe book for this [30, 44, 46, 58]. However, it has been shown that, especially at the beginning of the process, the central cornerstones must be made transparent for all participants. It must be clear in what form and with what commitment citizens can contribute to the success of the transition. At the same time, however, it must also be clear where the limits of the process lie. Communicating this is a major challenge, since a clear position should be taken in the run-up to participation. However, it also prevents the risk of later disappointment if less influence can be exerted by citizens than hoped. A realistic and as concrete as possible idea of the possibilities and limits of the process is therefore of central importance for all participants in order to be able to successfully shape energy policies as a central component of a necessary overall social transformation process.

References

[1] Adams M. The reflexive self and culture. Br J Sociol. 2003;54:221–38.
[2] Bell M. An Invitation to Environmental Sociology, 4th ed. Thousand Oaks: Sage, 2012.
[3] Bostrom A, Fischhoff B, Morgan MG. Characterizing mental models of hazardous processes: A methodology and an application to Radon. J Curr Issues. 1992;48:85–100.
[4] Pearce BJ, Ejderyan O. Joint problem framing. A reflexive practice: Honing a transdisciplinary skill. Sustainability Sci. 2019;15(3):683–98.
[5] Renn O. Towards a socio-ecological foundation for environmental risk research. In: Lockie S, Sonnenfeld DA, Fisher DR, (editors). Routledge International Handbook of Social and Environmental Change. London: Routledge; 2014, 207–20.
[6] Von Wehrden H, Guimarães MH, Bina O, et al. Interdisciplinary and transdisciplinary research: Finding the common ground of multi-faceted concepts. Sustainability Sci. 2019;14: 875–88. DOI: https://doi.org/10.1007/s11625-018-0594-x.
[7] Meyer H. Quo vadis? Perspektiven der sozialwissenschaftlichen Umweltforschung. In: Müller-Rommel F, Meyer H, editors. Studium der Umweltwissenschaften – Sozialwissenschaften. Heidelberg & Berlin: Springer; 2001, 153–68.
[8] Nowotny H, Scott P, Gibbons M. "Mode 2" revisited: The new production of knowledge. Minerva. 2003;41:179–94.
[9] SAPEA, Science Advise for Policy by European Academies (2019): Making sense of science for policy under conditions of uncertainty and complexity. Report to the EU Commission. Berlin: SAPEA, https://doi.org/10.26356/MASOS.
[10] Renn O. Decision making under uncertainty: Trade-offs between environmental health and other risks. In: Nriagu J, editor. Encyclopedia of Environmental Health. New York: Elsevier; 2019a, Vol. 2, 10–16. DOI: https://dx.doi.org/10.1016/B978-0-12-409548-9.10675-X.
[11] Rip A. A co-evolutionary approach to reflexive governance – and its ironies. In: Voß JP, Bauknecht D, Kamp R, editors. Reflexive Governance for Sustainable Development. Cheltenham: Edward Elgar; 2006, 82–100.
[12] Becker E, Jahn T, Hummel D, Stiess I, Wehling P. Sustainability: A cross-disciplinary concept for social-ecological transformations. In: Klein JT, Grossenbacher-Mansuy W, Häberli R, Bill A, Scholz RW, Welti M, editors. Transdisciplinarity: Joint Problem Solving among Science, Technology, and Society. Basel: Birkhäuser; 2001, 147–52.

[13] Dunlap RE, Lutzenhiser LA, Rosa EA. Understanding environmental problems: A sociological perspective. In: Bürgenmeier B, editor. Economy, Environment, and Technology: A Socio-Economic Approach. Armonk & New York: Sharpe; 1994, 27–49.
[14] Despres C, Brais N, Avellan S. Collaborative planning for retrofitting suburbs: transdisciplinarity and intersubjectivity in action. Futures. 2004;36:471–86.
[15] Dunlap RE, Catton WR. Struggling with human exemptionalism: The rise, decline and revitalization of environmental sociology. Am Sociol. 1994;25:5–30.
[16] Gallopin GC, Funtowicz S, O'Connor M, Ravetz JR. Science for the 21st century: From social contract to the scientific core. Int J Soc Sci. 2001;168:219–29.
[17] IRGC, International Risk Governance Council. White Paper on Risk Governance. Towards an Integrative Approach. Geneva: International Risk Governance Council; 2005.
[18] Ramadier T. Transdisciplinarity and its challenges: the case of urban studies. Futures. 2015;36:423–39.
[19] Vecchione E. Science for the environment: Examining the allocation of the burden of uncertainty. Eur J Risk Regul. 2011;2(2):227–39.
[20] Stirling A. Precaution, foresight and sustainability: Reflection and reflexivity in the governance of science and technology. In: Voß JP, Bauknecht D, Kamp R, editors. Reflexive Governance for Sustainable Development. Cheltenham: Edward Elgar; 2006, 225–72.
[21] Redclift M. Dance with the wolves? Sustainability and the social sciences. In: Becker E, Jahn T, editors. Sustainability and the Social Sciences. London: Zed Books; 1999, 267–73.
[22] Scholz R. The normative dimension in transdisciplinarity, transition management und transformation sciences: New roles of sciences of universities in sustainable transitioning. Sustainability. 2017;9:1–31.
[23] Bergmann M, Jahn T, Knobloch T, Krohn W, Pohl C, Schramm E. Methods for Transdisciplinary Research: A Primer for Practice. Frankfurt am Main: Campus; 2012.
[24] Defila R, Di Giulio A. Integrating knowledge: Challenges raised by the "Inventory of Synthesis". Futures. 2015;65:123–35.
[25] Hirsch Hadorn G, Biber-Klemm S, Grossenbacher-Mansuy W, Hoffmann-Riem H, Joye D, Pohl C, Wiesmann U, Zemp E. The Emergence of transdisciplinarity as a form of research. In: Hirsch Hadorn G, Hoffmann-Riem H, Biber-Klemm S, Grossenbacher-Mansuy W, Joye D, Pohl C, Wiesmann U, Zemp E, editors. Handbook of transdisciplinary research. Berlin: Springer; 2008, 19–39.
[26] Lang D, Wiek A, Bergmann M, Stauffacher M, Martens P, Moll P, Swilling M, Thomas CJ. Transdisciplinary research in sustainability science: Practice, principles, and challenges. Sustainability Sci. 2012;7(Supplement 1):25–43.
[27] Renn O. Die Rolle transdisziplinärer Wissenschaft bei konfliktgeladenen Transformationsprozessen. GAIA. 2019b;28(1):44–51.
[28] Mobjörk M. Consulting versus participatory transdisciplinarity: A refined classification of transdisciplinary research. Futures. 2010;42:866–73.
[29] Scholz RW. Environmental Literacy in Science and Society. Cambridge, U.K.: Cambridge University Press; 2011.
[30] Renn, Schweizer P-J. Inclusive governance for energy policy making: Conceptual foundations, applications, and lessons learned. In: Renn O, Ulmer F, Deckert A, editors. The Role of Public Participation in Energy Transitions. London: Elsevier Academic Press; 2020, 39–79.
[31] Gethmann CF. Participatory technology assessment: Some critical questions. In: Decker M, editors. Interdisciplinarity in Technology Assessment: Implementation and its Chances and Limits. Heidelberg & Berlin: Springer; 2001, 3–14.
[32] Warren ME. Can participatory democracy produce better selves? Psychological dimensions of Habermas discursive model of democracy. Polit Psychol. 1993;14:209–34.

[33] IRGC, International Risk Governance Council. Involving Stakeholders in the Risk Governance Process. Lausanne: International Risk Governance Council; 2020.
[34] Daschkeit A. Von der naturwissenschaftlichen Umweltforschung zur Nachhaltigkeitswissenschaft? GAIA. 2006;15(1):37–43.
[35] Pohl C, Hirsch Hadorn G. Gestaltungsprinzipien für die transdisziplinäre Forschung – Ein Beitrag des td-net. Munich: oekom; 2006.
[36] Rosendahl J, Zanella MA, Rist S, Weigelt J. Scientists' situated knowledge: Strong objectivity in transdisciplinarity. Futures. 2015;65:17–27. DOI: http://doi.org/10.1016/j.futures.2014.10.011#.
[37] Rosa EA. Metatheoretical Foundations for Post-Normal Risk. J Risk Res. 1998;1(1):15–44.
[38] Stern PC, Fineberg V. Understanding Risk: Informing Decisions in a Democratic Society. National Research Council, Committee on Risk Characterization. Washington, D.C.: National Academies Press; 1996.
[39] US-National Research Council. Public Participation in Environmental Assessment and Decision Making. Washington, D.C.: National Academies Press; 2008.
[40] Becker S, Renn O. Akzeptanzbedingungen politischer Maßnahmen für die Verkehrswende: Das Fallbeispiel Berliner Mobilitätsgesetz. In: Fraune C, Knodt M, Gölz S, Langer K, editors. Akzeptanz und politische Partizipation in der Energietransformation. Wiesbaden: Springer VS; 2019, 109–30.
[41] Hildebrand J, Renn O. Akzeptanz in der Energiewende. In: Radtke J, Canzler W, editors. Energiewende. Wiesbaden: Springer VS; 2019, 261–82.
[42] Slovic P. Perceived risk, trust and democracy. Risk Analysis. 1993;13:675–82.
[43] Siegrist M, Arvai J. Risk perception: Reflections on 40 years of research. Risk Analysis. 2020;40(11):2191–206. DOI: https://doi.org/10.1111/risa.13599.
[44] Benighaus C, Renn O. Teil A Grundlagen (17–102). In: Benighaus C, Wachinger G, Renn O, editors. Bürgerbeteiligung. Konzepte und Lösungswege für die Praxis. Berlin: Wolfgang Metzner; 2016.
[45] Bohman J. Public Deliberation: Pluralism, Complexity and Democracy. Cambridge Mass: MIT Press; 1996.
[46] OECD. Innovative Citizen Participation and New Democratic Institutions: Catching the Deliberative Wave. Paris: OECD Publishing; 2020. DOI: https://doi.org/10.1787/339306da-en.
[47] Radtke J. Bürgerenergie in Deutschland. Partizipation zwischen Gemeinwohl und Rendite. Wiesbaden: Springer VS; 2016.
[48] Radtke J, Renn O. Partizipation und bürgerschaftliches Engagement in der Energiewende. In: Radtke J, Canzler W, editors. Energiewende. Wiesbaden: Springer VS; 2019, 283–316.
[49] Dryzek J, Bächtiger A, Chambers S, Joshua Cohen J, et al. The crisis of democracy and the science of deliberation. Science. 2019;363(6432):1144–46. DOI: 10.1126/science.aaw2694.
[50] Curato N, Dryzek JS, Ercan SA, Hendriks CM, Niemeyer S. Twelve key findings in deliberative democracy research. Daedalus. 2017;146(3):28–38.
[51] Webler T, Tuler S. Fairness and competence in citizen participation: Theoretical reflections from a Case Study. Adm Soc. 2000;32(5):566–95.
[52] Schweizer P-J. Partizipation bei der Energiewende und beim Ausbau der Stromnetze: Philosophische Fundierung. In: Schippl J, Grunwald A, Renn O, editors. Die Energiewende verstehen – orientieren – gestalten. Baden-Baden: Nomos; 2017, 341–50.
[53] Fung A. Putting the public back into governance: The challenges of citizen participation and its future. Public Adm Rev. 2015;75(4):513–22.
[54] Rauschmayer F, Wittmer H. Evaluating deliberative and analytical methods for the resolution of environmental conflicts. Land Use Policy. 2006;23:108–22.
[55] Renn O. Risk Governance. In: Coping with Uncertainty in a Complex World. London: Earthscan; 2008.

[56] Mendoza GA, Prabhu R. Participatory modeling and analysis for sustainable forest management: Overview of soft system dynamics models and applications. For Policy Econ. 2006;9(2):179–96.
[57] Renn O, Klinke A. Risk perception and its impacts on risk governance. Envi Sci. 2016. DOI: https://doi.org/10.1093/acrefore/9780199389414.013.2.
[58] Nanz P, Fritsche M. Handbuch Bürgerbeteiligung: Verfahren und Akteure, Chancen und Grenzen. Bonn: Bundeszentrale für politische Bildung; 2012.
[59] Matthies, E. & Homburg, A. (2001). Umweltpsychologie (pp. 95–124). In: F. Müller-Rommel & H. Meyer (eds.): *Studium der Umweltwissenschaften. Sozialwissenschaften*. Heidelberg & Berlin: Springer, here 5–6.
[60] Wildavsky, A., & Dake, K. Theories of risk perception: Who fears what and why? Daedalus, 119, 4 (1990), 41–60.

Carl Friedrich Gethmann, Georg Kamp

2 Energy Ethics

Contribution to Chemical Energy Conversion, 2nd. Ed.

2.1 Introduction

The availability of energy is the basis of every human way of life that wants to rise above a subsistence economy, a life dependent on the vicissitudes of nature and always at risk. The mechanization of energy production, supply, and use forms the basis of modern civilizations and is deeply interwoven with them. Cheap, always reliable, and abundantly available energy supplies have made modern lifestyles possible in the first place. For production, for their transportation needs, for home lighting and air conditioning, and for their leisure activities, the average European consistently used over 4,200 W of primary energy in 2014. Americans consistently claimed 7,700 W, while the global average was 2,370 W.[1] The importance of energy for our lifestyles and the efficiency of the technologies used to provide it become clear when one looks at comparative figures. Horsepower, the reference value introduced by James Watt for the generally comprehensible representation of the performance capabilities of the steam engines he developed, is given as 735 W; without technical support, the human body is capable of a continuous power output of about 60 W. The technical exploitation of fossil resources, especially coal, gas, and oil, and their mass utilization in increasingly complex, industrialized processes form the basis of the historical development that has led to the formation of the energy-saturated lifestyle in the Western world. This is also clearly reflected in the average values that emerge for regions where this development has been delayed or has hardly set in so far: in 2014, Asia and South America had an average of around 1,800 W of primary energy available per inhabitant, and across the African continent, people had on average of around 500 W to their disposition.

The differences in the technically produced availability of energy go hand in hand with unequal distributions of benefits and burdens. For example, the developed industrialized nations can import energy sources in enormous quantities and use them efficiently in relatively low-risk and low-emission processes, while in the countries of origin, they are often extracted at the cost of high risks to people and

1 Data from the United Nations Conference on Trade and Development (UNCTAD): Online database: UNCTADstat (https://unctadstat.unctad.org/EN/; reference year 2014), own calculations.

Carl Friedrich Gethmann, Lebenswissenschaftliche Fakultät Universität Siegen, Weidenauer Str. 167, 57076 Siegen, Germany, e-mail: carl.gethmann@uni-siegen.de
Georg Kamp, Forschungszentrum Jülich GmbH 52425 Jülich, Germany

https://doi.org/10.1515/9783110608458-002

the environment and under circumstances that are in some cases questionable in terms of human rights. In addition, they use economic and military power, which they have gained largely in part because of their energy wealth, to secure the future inflow of energy and, under certain circumstances, inhibit development in the exporting countries.

An unequal distribution of benefits and burdens also exists with regard to the emissions associated with the provision and use of energy, especially the so-called greenhouse gases, which lead to global climate change. The amount of CO_2 emissions per capita in the Unites States is about 10 times than that of India (1.57:14.95 tons in 2016).[2] It is precisely the greenhouse gases that have accumulated in the atmosphere over decades that are decisive for climate change, so that the unequal distribution of benefits and burdens is multiplied once again in the historical dimension. This is exacerbated by the fact that particularly dramatic developments in climate change are expected precisely for regions that have historically made only a minor contribution to CO_2 emissions and do so today.[3] Last but not least, a considerable amount of energy will again be required to protect against the predicted consequences of expected climate change. The production of essential building materials such as steel, aluminum, and concrete is particularly energy-intensive, as is the treatment of drinking water, the artificial air conditioning of the living environment, or the preservative cooling of food. Accordingly, the energy-poor regions in Central Africa and the Indian subcontinent are considered particularly vulnerable.[4] Also, with regard to the temporal dimension, it can be assumed that the polluters are not the affected and the affected are not the polluters. Since the consequences of climate change will only occur gradually as greenhouse gases accumulate in the atmosphere, it will be the members of future generations who will be exposed to these consequences.

Unequal distributions of benefits and burdens have always raised the question of the justice of distribution – where distributions are perceived as unjust, indignation is the natural reaction and manifested conflicts are often the result. How to deal with such conflicts of justice is anything but trivial – and it cannot always be assumed from the outset that they can be resolved amicably. This is particularly true where – as in the case of the production, transport, and use of energy sources and the climate change associated with emissions – the actions of one person are accompanied by consequences for others who live under different natural conditions, operate under different economic conditions, and are accustomed to different cultural customs. And this is especially true where the consumption of non-renewable resources, infrastructural interventions in the living environment, and long-term changes in climatic conditions mean that members of future generations are affected by the

2 https://de.statista.com/statistik/daten/studie/167877/umfrage/co-emissionen-nach-laendern-je-einwohner/ (accessed 01-Feb-2021).
3 https://gain.nd.edu/our-work/country-index/ (accessed 01-Feb-2021).
4 https://gain.nd.edu/our-work/country-index/ (accessed 01-Feb-2021).

energy decisions made today. In the case of conflict constellations of such a nature and scope, it is particularly important to warn against unthinkingly seeking orientation in the moral convictions of one's own social environment. The strategies for conflict resolution that have been established and proven in small social groups regularly fail when they are transferred to the interaction conditions of anonymous large groups. Thus, procedures according to which goods are distributed within a circle of friends or common facilities are shared in an apartment building cannot be transferred to large cities and nations. It would be just as misguided as to use the price signal to control distributions in the narrower social environment as to trust mutual affection and social control in a nationwide distribution of public goods.[5] When different moral intuitions collide in conflicts that cross cultural boundaries, this is not infrequently accompanied by an aggravation of the conflicts and a decrease in the willingness to cooperate. Here, then, ethics, as a rational discipline of conflict management reflecting on the given conditions of action, is called upon to examine the existing strategies and, in critical further development, to recommend those measures that seem suitable for working toward a more equitable distribution of benefits and burdens on a global scale.

2.2 The Relevance of Energy Availability for Individual Life and Society

Whenever the way of life wants to rise above the pure subsistence economy, above the proverbial "living hand-to-mouth", energy was and is the lever that increases the efficiency of action by the decisive moment. But the sober eye also sees the impact that the use of this lever has always had on fellow human beings and the natural environment. It is the increasing ability to use this lever that has enabled mankind to emancipate itself from the dependence on contingent natural processes and to take forward-looking and planned measures to safeguard its own subsistence. An essential measure for the efficiency, quasi for the length of this "lever", is the ratio between invested time and own labor power on the one hand and the yield of available energy on the other hand. According to calculations by Nordhaus [66], Stone Age man around 9,300 BCE

5 This problem of inadmissible transfer does not contradict the thesis that ethically justified sentences claim universal validity (cf. Kant [12], Sidgwick [67], Hare [39]). Rather, the very idea of universalizability owes itself to the observation that in the formation of anonymous large-scale societies, the orientation to the proven small-group morality fails. The classical formulation of the problem leads back to F. von Hayek's historical–analytical reconstruction of forms of social organization: "The greatest change which man has still only partially digested came with the transition from the face-to-face society to what Sir Karl Popper has appropriately called the abstract society: a society in which no longer the known needs of known people but only abstract rules and impersonal signals guide actions toward strangers. This made a specialization possible far beyond the range any one man can survey" [40].

had to invest about 50 h of labor to obtain a thousand lumen-hours of light (this corresponds to about one hour's light output from a conventional 75-watt incandescent lamp). Still at the beginning of the nineteenth century, when in the light of the whale oil lamps the industrialization begins, nearly 5 h had to be spent for an appropriate light yield. Development took a significant leap with the expansion of a central gas supply in the large cities of the late nineteenth century. But a member of western industrialized nations today needs only about 0.00012 working hours – calculated before the spread of LED lamps. The relation can – cum grano salis – be exemplary for the progress in many fields of use of energy offers, which have been made accessible by technical and organizational developments. Just as "artificial" light makes human life independent of the natural change of light and extends the time in which humans can be productive, educate themselves, or teach their children, the controlled availability of heat and power allows a development that makes human life increasingly independent of the vicissitudes of nature and increasingly increases the productivity and range of human activity.

Over thousands of years and into modern times, almost all available energy has been directly or indirectly tied to arable, pasture, and forest land. Even around 1800, 20–25% of the total energy consumption was accounted for by food for humans and draft animals, the remaining 50–60% by collected or cut firewood. In earlier eras, and in some economies even into modern times, part of the arable land is also devoted to the maintenance of labor forced from slaves.[6] Only about 1–2% of the energy is provided by wind and waterpower, mainly for conversion into mechanical energy in mills and pumps [1].[7] To a minor extent and depending on availability, animal and vegetable fats have also been used. Though an energy supply based so predominantly on the use of biomass provides a source that is essentially reliable and sustainable, the incident solar energy, which ultimately forms the output for this, can only be used to a small extent (only approximately 1%) through the process of photosynthesis. The availability of the biomass is always endangered by droughts, storms, fungal infestation, and so on, the production and preparation, the transport, and the storage are considerably costly. In pre-modern Europe of the late eighteenth and early nineteenth centuries, 60–70% of the total economic output was directly or indirectly related to the availability of energy, and the land consumption was considerable. Depending on the region, an area of one to three hectares per inhabitant was required for the supply [1]. The majority of the population was therefore tied to agricultural work; even in normal times, the citizen of the early nineteenth century spent two-thirds of his earned income and more on food; the always-uncertain supply prevented long-term planning; malnutrition and undernourishment

6 On the value and economic performance of slaves in the United States, see https://www.measuringworth.com/slavery.php (accessed 01-Feb-2021).

7 The figure for human food refers only to the proportion of supplied energy that was converted (based on numerical estimates) into physical work.

were part of the everyday experience of life for large sections of the population. Even in the mid-nineteenth century, the simplest means of transportation were available only to a minority, and even for the wealthiest citizens, their homes were not frost-free in winter [2].

It is true that the onset of industrialization by no means immediately led to an improvement in living conditions. However, industrialization set in motion a development of prosperity that created the basis for the modern world of work and life. The intensive development, extraction, and use of fossil fuels played a major role in this development – due to their mass availability, which allows for a plannable and inexpensive use, and not least due to the enormous energy density of coal, gas, and oil [3], the pressure on the resources previously used decreased. The newly developed energy sources relieve and strengthen human labor, they end the widespread exploitation of slaves and animals, the energy-related land use shrinks considerably, and the overused forest stock can recover. Instead, settlements in which people live and farm together are growing: Whereas previously, the size of settlements was essentially limited by the area needed in the surrounding area to supply the population with energy, now the large cities can emerge, in which the population is increasingly concentrated. The short distances and high population density then make it worthwhile to develop and provide for the city dwellers the services and public goods that largely determine modern life – an efficient supply of consumer goods, electricity and gas networks, traffic networks for individual and mass transport, hospitals and, last but not least, education systems that create the basis for further progress.

But the problems associated with the growing consumption of fossil energy soon became apparent. As early as the nineteenth century, industrialized cities experienced an increased incidence of rickets and bronchial diseases caused by exhaust fumes from industrial chimneys and thousands of domestic fireplaces. Up to now, just – as for instance for [4] – the air or the atmosphere had been regarded as the example of a substance which belonged to the "most necessary of all things", but over which there could not be conflicts of use. No one could commit an injustice by "using it lavishly and enjoying its blessings". First locally, via the smog over the cities, later globally, via the complex interactions of the so-called greenhouse gases that are released during combustion processes and are identified as one of the main causes of climate change, the intensive use of fossil energies contributes to considerable and, in part, completely new ethical challenges. A not-inconsiderable role is played here by the fact that in the last 200 years or so, the world population – and thus the number of potential users of the resource "atmosphere" – has grown to a considerable extent at the same time as prosperity. Particularly due to the progress achieved in many areas of life, the world population has grown from approximately 1.6 billion people to 7.6 billion people since the beginning of the twentieth century alone. The projections of the UN vary between a leveling off at approximately

9.6 billion by the end of the century and a further increase in the world population to 13.2 by 2100 and a further increase beyond that.[8]

The further course of development will depend, to no small extent, on the availability of energy. While the development of population figures is stagnating almost everywhere in the world, enormous increases are expected in particular for individual countries in Africa. There is widespread agreement among demographers that population development in these countries also depends, to a large extent, on the level of education of the female population in particular [5]. The reliable and affordable availability of technically generated energy, which reduces the cost of energy supply in line with the development outlined by Nordhaus, can make an important contribution to creating the space needed for education and development. In order to escape poverty and the life risks of subsistence farming, energy must also be the lever with the help of which billions of people who will be added can become producers who can provide for themselves and their dependents with what they produce and acquire.

Martin Heidegger has pointed out that the use of what is available at any time as a matter of course ("ready-to-hand", German: "zuhanden") is, so to speak, absorbed in the everyday contexts of action and eludes perceptibility. Only when the self-evident contexts of action are disturbed, because the means are not available or their use is accompanied by undesirable consequences, they become "conspicuous" [6, 7]. For example, modern man in Western civilizations has come to take for granted the now given availability of energy at all times. On the other hand, burdens such as noise, particulate matter, or land consumption associated with the intensive use of energy are often directly experienced by the individual. Other burdens, such as the risks of nuclear energy or climate change, are regularly the subject of media coverage and public debate. The burdens of large-scale energy supply thus play a major role in determining the perception of the citizen of many developed countries, while the benefits, taken for granted, are consumed on a daily basis without separate awareness. Conversely, the attention of people who must continue to obtain modest units of energy at great daily effort is bound precisely by that effort and by the burdens that can be directly experienced. So, for example, over 2.5 billion people worldwide are still dependent on the traditional use of biomass (firewood, manure, and charcoal) for cooking. Estimately, about 780,000 people die each year from indoor air pollution caused by the emissions of incomplete combustion in the existing primitive furnaces, many more suffer from serious health damages [8–10].

But where it is a question of an appropriate weighing of benefits and burdens of production, provision, use, and its consequences, and of their fairer distribution, this asymmetry of attention has to be taken into account – ultimately, benefits and

8 Department of Economic and Social Affairs, 2017 Revision, https://www.un.org/development/desa/publications/world-population-prospects-the-2017-revision.html (accessed 01-Feb-2021).

burdens are always the result of a perception and assessment of individual experiences against the yardstick of individual claims and preferences formed in the respective circumstances of life. In view of the asymmetry of attention to be assumed, it is therefore obvious to use not only the proclamations of the actors for an appropriate reconstruction of the conflict constellations to be judged ethically, but also their claims and revealed preferences documented by their daily behavior (so-called revealed preferences). The preferences that the individual expresses and documents in his behavior are always co-determined by the respective life circumstances under which the actor has formed them and by the possibilities that he perceives in his life situation. And for the ethical debate, one will not already assume the just distribution of these possibilities [11].

Energy is deeply woven into the fabric of daily life, and the quantity and quality of its availability is a major determinant of development potential. Where an efficient energy supply is not available, development is hardly possible. Efficient and reliable energy supply systems are usually highly complex and interconnected, but they make the availability of energy at all times appear to be an unreflected matter of course. The high degree of organization that makes these supply systems possible in the first place makes remote consequences, such as those arising from the procurement of energy sources and from emissions, almost invisible to the user of modern supply systems. Therefore, the fair balancing of claims that is sought cannot be about the factual claims formulated on the basis of the respective life situation. Rather, the effort to achieve a viable balance that will meet everyone's approval in the long term, regardless of changing circumstances, and that can endure beyond the present day, must be based on enlightened, well-informed claims.

2.3 Energy Availability, Conflicts of Claims, and Ethics

Conflicts over energy are by no means a modern phenomenon where the power of disposition over primeval land and timber resources, over human and animal labor always also means the power of disposition over energy, territorial disputes, and campaigns of conquest must also be considered under this aspect. However, in modern times, conflicts about energy arise above all from the increasing competition for the use of available energy sources and raw materials as well as for the sinks that are claimed for the emissions. And to a large extent, markets and diplomacy became the instruments of conflict resolution. But due to very different starting situations, often the market power and political power are very uneven distributed, and questions of a fair distribution of benefits and burdens arise. These often comprise a temporal component: To what extent, for example, is there an obligation to leave a share of certain resources to future generations? Are present generations ethically obliged to change their energy supply in such a way that the expected consequences of climate change do not occur or are at least mitigated? And at what price? There is a clear conflict between the interest of catching up development in regions still largely characterized by subsistence farming and the interest of future generations: If subsistence and gradual prosperity cannot be secured or can only be achieved to a limited extent without climate-damaging emissions, is the present generation obliged to forego or slow down development for the benefit of future generations? Or are developed nations, which are considered the main contributors to climate change due to the emissions accumulated in their development trajectory, obligated to provide special support here?

In how far the claim of individuals to live their lives on a certain level of prosperity can be ethically justified is disputed in philosophical debates. The answer depends essentially on the assumed background theories. Pure considerations of justice, for instance, turn out to be completely indifferent. Justly distributed resources, just energy availability and just distribution of emission burdens and other risks, for instance, can be imagined on the level of a pre-industrial, agrarian-based energy economy as well as on the level of lifestyle found in industrial societies today. The same is true for ethical concepts, which are oriented toward universalizability in accordance with Immanuel Kant. The ethical test of action-determining principles ("maxims") he proposes, according to which they must at the same time be suitable as principles of a general legislation [12], is indifferent to the level of prosperity of the relevant general public. Utilitarian theories, as developed by Jeremy Bentham [13] and John Stuart Mill [14] are based on the assumption that humans ought to maximize the sum of available utility, that is, bring about "the greatest amount of good for the greatest number". In contrast, an ascetic tradition – also often supported by spiritual–religious convictions – rejects any striving for material prosperity in favor of a higher common good, in favor of one's own "inner" perfection or in order to re-establish harmony with nature.

In the self-understanding of philosophical ethics, as it has established itself with roots in antiquity since the modern Enlightenment, it is, however, in no case a matter of designating certain ends as generally desirable and recommending this or that form of life to people. The considerations of philosophical ethics are rather directed to strategies for the accomplishment of conflicts under renunciation of means of power or force. Conflict constellations can be described purely structurally – among them conflicts about just distributions are a significant part, but only a part of ethically relevant conflicts. A conflict can arise, for example, because actor A is striving for a state of affairs that actor B is trying to avoid. Or a conflict may exist because the creation of the state A strives for is accompanied by consequences and side effects that make it impossible for B to implement his own plans of action. Or a conflict may exist because A claims means for the realization of his ends which B would also need for the realization of his ends. Conflicts can accordingly be understood as the incompatibility of plans of action of different actors – irrespective of whether the plans aim at the appropriation of consumer goods or the support of people or animals in need of help, at short-term pleasure or long-term precaution. Conflicts can be appropriately managed by at least one actor abandoning his or her plan of action or modifying it in such a way as to remove the incompatibility – whether by changing his or her ends or by choosing different means to achieve the ends. Where ascetic approaches, insofar as they are ethically motivated, seek to cope with conflicts by renouncing claims, theories of justice, the Kantian test procedure, and the utilitarian approach all require the consideration of the entitlements of others in making one's plans for action and the application of non-violent strategies to make plans for action of conflicting actors compatible with each other. Thus, they are essentially discursive procedures that are directed at negotiating claims and exploring options for the compatibility of claims [15]. Since claims are first and foremost justified in such discursive procedures, none of the participants in the discourse can presume their claims as justified without there being a justification, accepted procedurally correct and well founded by all participants in the discourse. Accordingly, no one has privileged access or a privileged role in the discourse. All parties to the conflict are equally entitled to participate in the discourse, and the enlightened, well-informed claims of all are equally subject to the discourse.

Only sometimes – and rather only in elementary school examples – the solution of a justice conflict consists in establishing an equal distribution of benefits and burdens. In fact, equality is neither sufficient nor necessary for justice, nor should it be striven for per se as a long-term goal or guiding idea of development: An equal distribution among unequals can be extremely unjust, and a just distribution that responds to the particularities of the recipients can be extremely unequal.[9] A closer look at the reality of life as well as at political history then also shows that an equal distribution can often only be enforced and maintained by massive coercion.

In philosophical ethics, justice is usually understood as justified inequality – whoever can point to his special performance in the production of a good, whoever can comprehensibly demonstrate his special need, whoever can show that he can perform special services with the help of this good, perhaps even services that then benefit everyone, should certainly be able to claim more of a good without committing an injustice [16]. The object of ethical consideration is not only the question of

9 It is merely a common misunderstanding that Marx's and Engels' idea of a communist society seeks to establish justice through absolute equality. In their programmatic writings, they are unambiguous: the prominent distribution principle "From each according to his ability, to each according to his need" [41], for example, seeks to define abilities and needs as the yardstick for the social distribution of goods instead of class affiliation and ownership of capital and means of production. It is a pre-supposition of this principle that abilities and needs are unequally distributed. Cf. also F. Engels' Preface to Marx' "The Poverty of Philosophy" [41]: "Justice and equality of rights are the cornerstones on which the bourgeois of the eighteenth and nineteenth centuries would like to erect his social edifice over the ruins of feudal injustice, inequality and privilege". For a more detailed discussion, cf. Buchanan [42].

which of these principles and others is more appropriate in this or that case. On the contrary, this can usually be left to the decision rules of morality and law as they are socially established and culturally practiced. If, however, the conflict constellations exceed the range of the established conflict management routines, for instance with regard to the complexity of the decision situations, due to the cultural disparity of the groups involved or due to specific peculiarities, then it belongs to the tasks of philosophy to structure the discourse. Here, a professionalized philosophical ethics can offer support by sharpening the terminology, by a critical examination of arguments with regard to their soundness and mutual compatibility, and by an examination and further development of the standards to be used for attempts of justification. The unequal distribution of benefits and burdens of energy supply and its consequences for man and the environment pose such a task for philosophy: As indicated before, energy availability has a significant influence on the living conditions and thus on the effectiveness of action as well as on the level of aspiration of people. And striving for justice in these cases is not so much about the fair distribution of goods, but of risks and opportunities for the shaping of life. Therefore, the questions of justice have to be dealt with in accordance with a comprehensive ethical question. The global reach of trade in energy carriers and materials for energy infrastructure, which extends far beyond the borders of states, cultures, and economic communities, as well as the global climate impact of their use, evokes basic ethical considerations, since conflicts here transcends cultures and therefore cultural based expectations of how conflicts cannot be applied [17]. Last but not least, in view of the scope of the problem, it will also have to be examined whether and to what extent obligations exist toward future generations.

Where energy is available in abundance and at low cost and increases the efficiency of action like a lever, options tend to increase and competition to decrease. As the scope for action increases, new opportunities arise to make one's own plans of action compatible with those of others without giving up one's own demands. And because more goods can be made available, more and more actors have the opportunity to realize their plans of action.

In the reverse, people set their purposes in each case against the background of growing possibilities and make ever more far-reaching plans of action. This creates new scarcities for energy sources, for raw materials needed for the construction of energy-technical plants, and for available road and settlement space, if more and more people use the growing mobility potentials to achieve their purposes. Where the procurement of energy is accompanied by externalities, by emissions, sinks become scarce, and under certain circumstances, others are burdened with the consequences and the risks of large-scale technical energy supply. To make demands compatible, a strategy emerges on the basis of this constellation to improve the conditions for cheap and abundant availability of energy for all through technical and organizational development, choosing strategies that minimize the associated burdens on people and their environment. In particular, efforts will be made, for example, to decouple the provision of energy from emissions.

On the other hand, it must be considered particularly problematic if a society ensures the functionality of its supply systems at the expense of others by keeping others dependent, perpetuating exploitative systems or outsourcing the risks associated with emissions. Last but not least, the leverage that abundant and easily available energy gives to actors can also be used to secure one's own supply at the expense of others, precisely by using the technical and energetic potential to settle conflicts by means of warlike violence.

If the discursive processes and the resulting decisions are to be relevant to life, they must be geared to the actual circumstances. For questions of energy supply, in view of the worldwide networking of supply systems and global climate effects, this means taking into account the demands of all people in their sometimes extremely different life situations as well as the presumed demands of future generations. To what extent the consideration of the claims of others and participation in such a discourse is in the interest of all actors involved, and to what extent the willingness to show such consideration can be won through mere argumentation, is ultimately an empirical question. This cannot be assumed for all constellations, and it cannot be assumed from the outset that the ethical approach to discursive conflict resolution is promising tout court. In any case, conflict constellations of high complexity require organizational forms that enable the representative representation of the claims of larger groups. In the current world situation, it is above all, but not only, the autonomous nation-states that appear here as actors in international and global forums. Even within these groups, the claims of individuals will be disparate – so an aggregate must be formed. Ideally, the formation of this aggregate already takes place through the free discourse of all members of the group; if necessary, aggregate claims of subgroups are to be determined. Future generations that by reasons of principle cannot participate in the discourse are to be assigned appropriate claims – likewise based on a discourse of the relevant present groups – as will be shown below.

2.4 Supranational and Internal Conflicts

Management of internal conflicts, the peaceful reconciliation of claims between citizens in the countless questions of individual lifestyles, and the maintenance of a willingness to cooperate on all sides in a large collective are only some of the numerous tasks and functions of state organizations.[10] Ideally, the state is legitimized

10 For detailed technical, ethical, and political considerations of the challenges and opportunities of energy supply in a global perspective and from an interdisciplinary point of view cf. Gethmann et al. [8]. Chapters 4 and 7 are based on this study.

as a “cooperative enterprise” by serving the mutual advantage of its citizens wherever possible [18] and enabling them to reap the fruits of peaceful cooperation without requiring and spending substantial resources on it. Where the rule of law is established and institutionalized, state agencies themselves are subject to the law, and conflicts between state representatives are manageable as efficient as conflicts between citizens. The state protects the citizen against attacks by others on property and life, thereby creating the planning security required for investment in the future and sustainable action. Ideally, supranational organizations and their relationship with the individual states will also be designed precisely in accordance with these requirements. However, since it is hard to imagine anything other than tiered responsibilities and multilevel systems, the coupling is also associated with challenges that have ethical relevance and can entail considerable burdens for the establishment of equitable conditions.

In order to maintain the willingness of individual citizens or like-minded citizens to cooperate, who unite to achieve common goals, internal conflict regulation can have a binding effect that is precisely contrary to supranational cooperation interests. This becomes particularly clear when, for example, a volume of emissions defined by its accumulated effects on the climate is defined as a budget that is to be distributed among the global community as a whole. Measures that are adopted at the national level to ensure internal cooperation and serve to reconcile the interests of actors who pursue incompatible goals due to different expectations of prosperity, different risk appetites, or different conditions for action are often at the expense of third parties. Whether it is regional economic interests that allow one nation to hold on to the recycling of the locally cheap and abundantly available coal, or whether it is the rejection of risks associated with the capture and storage of CO_2 or the continued operation of largely CO_2-neutral nuclear power plants – compromises that are found on this basis might be an effective conflict resolution at the national level but at the same time might be unfair in view to third parties insofar they do not exploit available potentials for emission reduction and requiring an unnecessary share of the overall available residual budget. This is especially the case if there was little or no scope for others to switch to other energy options – for example because of the level of social or technical development or because of other regional circumstances. The reduction of the budget by one side would then either mean that they would have to compensate for the use by others by reducing energy consumption or would have to accept the risks of climate change to which others would be exposed. The examples chosen above make it clear that this is above all a question of the fair distribution of risks: the low technical and economic risks associated with the maintenance of coal-fired power plants, for example, and the considerable use of CO_2 quotas, are offset by the risks to which others are exposed as a result of the remaining options. Anyone who fails to exploit the savings potential associated with the operation of nuclear power plants or other low-emission technologies will expose others to a certain amount of risks that goes along with climate change. Given the scientific assessment of the respective risks, this would not be in accordance with the rule of risk allocation.

In western industrial nations – precisely because the question of who is allowed to impose what risks on whom is the subject of social compensation – decisions on the energy infrastructure of such scope are taken by representatives of society and often with the involvement of the affected part of society. In particular, the binding effect of such decisions, which are the result of intensive direct citizen participation, may be regarded as particularly high. A representative of the state would have hardly any room for maneuver in international negotiations to even relativize the decisions thus found. However, as the arguments about the risks of nuclear power or about installations such as power lines or wind turbines, which are intended to make energy supply with renewable energy sources possible show that the requests for participation are generally not raised with a view to the available emissions budget, but rather with a view to quasi local risks, which are then passed on to inhabitants of other states as "externalities".

Such externalities arise not only because of the global interconnectedness due to an atmosphere that is common to all. Such externalities are as well formed by the creation of common markets and global trade: waste is shipped to countries where few benefit from cash payments, while large sections of society, which are not involved in decision-making and do not have sufficient political power, have to bear the health risks. On international markets, raw materials are traded that help the consumer to spend a comfortable and healthy life while people of the region where they are mined accept considerable risks for themselves and the environment. More often than not, the cash flows allow for perpetuating a balance of power that stabilizes the exploitative conditions.

While in countries which are already largely developed as "cooperative enterprises for the benefit of all" or which owe their existence to this idea, the discovery of profitable mineral resources finds good prerequisites for promoting general prosperity (e.g., Norway), such a discovery can easily lead to exploitative conditions in societies which are already extremely unequal and not protected by robust constitutions under the rule of law. In the literature, this phenomenon is referred to as the resource trap or resource curse [19, 20]. It is associated with dictatorship and oppression, with corruption and instability, with a high debt risk, lasting political instability, and persistent violence in so-called failed states [21]. Supranational institutions alone cannot provide the necessary balance here. At the same time, individuals are overwhelmed by the diversity of information and the complexity of the interrelationships. It is therefore the responsibility of the individual states in particular not only to take up expressions of interest and to orient themselves toward the willingness of citizens to accept them but also to promote acceptable decisions supported by rational argumentation in an international vote. Not least because of the potential offered by the new media to organized groups to assert their special interests, the state, as a cooperative enterprise, is called upon to the advantage of all to review the generalizability of the incentives given to it by citizens and domestic organizations.

Where justice is to be established and conditions are to meet with general acceptance, the standards and the measures have always had to and must be gained from deficient conditions in a joint effort. Justice is an unstable equilibrium, and the organizational basis of discourse on which it must be established cannot be taken for granted. Often, it must first be created, and where it exists, it requires constant formation. The goal to be striven for under these conditions is not to be seen in the idea of an absolutely just state, but in a state which is the more just compared to the existing one. But there, where favorable conditions for this development have already been created – for instance by the creation of efficient organizations – the yields of cooperation should be visible and advertised for themselves. Especially, the efficient development and provision of energy, which is only possible in complexly networked and routinely cooperating collectives and at the same time creates collective prosperity, can be an essential factor in this, provided that it is as far-reaching as possible decoupled from the unwelcome consequences of its use. And it is precisely the increasing globalization of supply networks and the need to strive for an optimized use of local conditions and potentials on a global scale under the conditions of climate change that offers options under which the support of the disadvantaged can become a "cooperative venture for mutual advantage".

2.5 Obligations to Future Generations

In modern, industrialized societies energy availability goes along consequences that not only reach over the boundaries of societies, but also over the boundaries of generations. Non-renewable resources used today in large amounts will not be available for future generations. According to climate research models, the effect of greenhouse gas emissions today will only gradually and with intergenerational delay lead to changes that make life more difficult in certain regions. And radioactive waste will be a legacy that might be a danger for thousands of years.[11] This raises the by no means trivial question of whether people living now are obliged at all to take account of the possible needs and requirements of future generations in their actions, even if they are not in direct interaction with them. A further question to be answered is, what they are committed to and how they can effectively fulfil obligations across temporal

11 For ethical considerations on the use of renewable and non-renewable resources cf. de Haan/Kamp et al. [43]; for ethical considerations on radioactive waste cf. the author's contribution in Streffer et al. [44], Chapter 3, for ethical considerations on climate change see Gethmann/Kamp [45]. This chapter and the next are derived from these sources, and are largely taken from the latter.

distances – the effects of their actions would then have to be justifiable in view of the action requirements and needs of members of future generations.

The idea that in their actions people do not only have obligations toward those they face here and now is by no means new to ethics.[12] Furthermore, options for action with distant consequences have been discussed for a long time in other contexts as well, such as economic and technical ones [22]. The problems of national debt, population growth, disposal of radioactive waste, and so on could be pointed out. In the debates on how to deal with these challenges, it is regularly assumed that there is an obligation, even toward more distant generations.

The question is aggravated if one includes the trivial insight that the circumstances of action of the members of distant generations are more or less unknown to us living now ("prognosis problem"). The scope of this insight can be estimated if one imagines that the inhabitants of the British Empire of 1900 or China of 1950 wanted to take precautions for the welfare of their descendants living today. Even if one tries the thought experiment that mankind of 1900 had the means to stop a process that foreseeably makes parts of the earth uninhabitable, the picture is not clear. What has been considered uninhabitable in 1900 will not necessarily continue to be considered uninhabitable after 100 years of technical and social development. Perhaps, it would have meant comparatively less effort for future generations to restore habitability than it has cost previous generations to maintain habitability – all the more so if the resources invested in the conservation process had instead been invested in research and education or growth-enhancing measures.[13]

With regard to the prognosis problem, the ethical assessment of intergenerational obligations is not only subject to the problem of uncertainty and requires complex risk/opportunity considerations. It is also permeated by uncertainties, by problem areas in which the actors not only do not have reliable answers but also are not even in a position to ask appropriate questions that are relevant to future environments (so-called unknown unknowns).[14] It should be borne in mind that

---

12 I. Kant already points out in his historical–philosophical draft of a "world society of eternal peace" that "human nature is so constituted that we cannot be indifferent to the most remote epoch our race may come to, if only we may expect it with certainty" [26]. He thereby describes cross-generational precaution as part of a development process of humanity in which "the earlier generations appear to carry through their toilsome labor only for the sake of the later" (ibid. Third Thesis) and the later beneficiaries of development are those who have advanced the earlier ones.

13 Well-founded estimates of how, for example, the level of prosperity in western societies has developed over the past 100 years are vague and difficult. Indicator-based estimates range from four to five times (e.g., DeLong [46]) to twenty times (e.g., Roberts [47]) the level of prosperity at that time.

14 "If we ask ourselves a question and can come up with an exact answer, that is a known known. If we ask ourselves a question and cannot come up with a very precise answer, that is a known unknown. An unknown unknown is when we have not really thought to ask the question in the first place" [48].

human efforts to gain prognostic knowledge are subject to systematic biases that are difficult to overcome [23, 24]. For example, causal developments are generally easier to predict than those driven by creative processes. This often makes it easy to foresee the problems arising from natural and social developments, while possible solutions cannot be anticipated for logical reasons. We simply cannot know what we will know tomorrow – if that was the case, we would already know it today.[15] It should also be borne in mind that not only acts of execution, but also acts of omission are subjects of moral qualification. Since the conceptual relationship between the execution and the omission of acts is disjunctive (tertium non datur), it is not possible to be morally indifferent about waiting.

Situations when an obligation to take the needs of others into account become an issue usually when those in which such a consideration is explicitly demanded or a failure to obey such an obligation is lamented. The moral intuition that is formed and consolidated in the resulting conflict constellations is therefore "presentist" in its lifeworld roots. It is therefore by no means trivial to discuss the question of whether *actors are only committed to those interacting with them at the same time*.

If obligations arise from discursively generated agreements to overcome existing conflicts, and if conflicts are defined as constellations of actions in which actors pursue incompatible purposes, then a positive answer to this question seems to be obvious – at least prima facie. Therefore, it is often attempted to derive obligations toward future generations from anticipated claims of future generations, from human dignity or from universal rights, that is, rights to which all human beings are entitled.[16] The argument is that this would also include people who will be living in the future into a trans-temporal collective of formally equal holders of entitlements. However, the idea is connected with a logical–conceptual problem. In a finely resolved form, reconstructed according to the logical grammar of this argument, the problematic details become visible:

(1) For all x: If x is a human being, then x has rights.
(2) For all x: If x is a member of a future generation, then x is a human being.
(3) So: For all x: If x is a member of a future generation, then x has rights.

The phrasing "For all x" renders the premises in the first two sentences universalistic – and accordingly the consequence in the third sentence is universalistic, too.

15 Cf. the so-called Popper's Paradox: "if there is such a thing as growing human knowledge, then we cannot anticipate today what we shall know only tomorrow" [49].

16 As an example, cf. Broome [28]: "If future generations have a right to an unpolluted environment, we violate their right by leaving greenhouse gas in the air". He concludes that future generations are to be compensated for the harm we impose on them. For an overview of the debate on obligations to future generations and further examples, see the contributions by Gosseries and Meyer [50].

Universal formulas, however, are always more or less general. Thus, the predicator "is a human being" is more general than the predicator "is member of a future generation". Therefore, despite the assumed universalism, it does not follow from the premises that x has rights, especially not that x has rights *today* – but merely the if-then-clause that x has rights *if* he is a member of a future generation. The "if-then" only provides information about a logical connection (in the language of logicians: the subjunction) and not about temporal relations. But of course, someone who is a member of a future generation is not a member of this *future* generation *today* – after all, he or she will only be member of a future generation in the *future*. Thus, as the future King of England does not wear a future crown today (he will wear one in the future), the members of future generations do not have future rights today (they will have them in the future). This applies regardless of how human rights and their origin are understood, whether as the result of fair agreements, as quasi-natural rights, or conferred by divine grace. And the negative diagnosis remains unchanged – mutatis mutandis – if one wants to speak of claims or needs of (members of) future generations, of universal human dignity, of universal values or something else instead of rights [25].

Consequently, if future generations do not have any rights now, then they cannot commit us now either – only the present can do that among themselves. A look at everyday practice shows that it is not mandatory to assume that claims are made explicitly, but only that someone presents a justification for the fact that a certain entitlement is to be recognized – for a valid justification and the practical recognition of the entitlement, the explicit making of claims by the persons concerned is neither sufficient nor necessary. Moreover, where claims are made explicitly by the present and the recognition of entitlements is demanded, an intersubjectively accessible and comprehensible justification is generally expected. The raising of claims is just the cause and starting point of justification efforts, and only if the efforts lead to success, then is it justified to claim entitlements or to attribute them to others.

From many contexts, the practice is also familiar that people mutually commit themselves in favor of third parties to whom they ascribe claims without the latter being able to express them. As an example, people make agreements or contracts in favor of third parties such as unborn life, comatose patients, animals, or even landscapes and grant them explicit rights. A mutual obligation of the contemporaries for the benefit of future generations would therefore not be a special ethical situation, rather it is in line with the usual practices of life.[17] And as in the other contexts, contemporaries can effectively control and adjust their mutual agreements to the benefit of the future by registering deviant behavior, demanding justifications, and sanctioning unjustified failures. The question that remains to be answered, however,

---

17 In contrast, Gardiner [51] assumes not only that actors but also generations act in a fundamentally self-referential manner and speaks of a "tyranny of the contemporary".

is why the members of an earlier generation should establish a practice by which they impose on themselves and others consideration for the presumed interests of future generations, a practice that restricts their own scope for freedom and possibly demands expenditure and restrictions that will – prima facie – hardly pay off for themselves. In the context of ethics, the question does not aim at an explanation, but at a justification. Thus, Immanuel Kant's view of man's nature, according to which "human nature is so constituted that we cannot be indifferent to the most remote epoch our race may come to" [26] could explain a practice of mutual obligation in favor of future generations. However, the practice would not yet be justified by this, just as the equally held view [27] that human beings by nature had a tendency to act in a present-oriented manner and only for their own benefit would justify a practice to develop a shortsighted practice at the expense of future generations.

The path toward justification is taken by people expressing interests – whether on the basis of their natural inclinations or rational considerations – that collide with the interests of others and raise conflicts that are resolved by agreement. If successful, action plans are coordinated in such a way that the action plans of one are taken into account in the planning of the other [15]. For the purposes in question, the ethical debate on whether the justification of such decisions must be based on fundamental principles or whether, for example, the increased benefit of the actors through cooperation alone is decisive is of no significance. The question remains why a today's actor should have an interest in taking into account the concerns of future generations, to include them in their planning and to require others to take their future-oriented plans into account while planning. In fact, it can be justified rationally on the basis of pure subjective egoistic motives and the well-understood self-interest of people living today, that they should demand and enter into such mutual obligations for the benefit of future generations, even those farther away. The justification is based on game-theoretic findings and considerations of institutional theory, but at the end it agrees with Kant's "transcendental" ethical justification, according to which the actor in performing an action A should not erode the practice, which first of all creates the prerequisite for the performance of A and thus is a condition of the possibility of his situational planning [12]: Game-theoretic analyses show that numerous forms of our cooperation for mutual advantage can only work in the "shadow of the future".[18] The argument is based on an understanding of conventions and organizations (treaties, morals, institutions, states, etc.) as socially

---

18 Axelrod [52]; Scheffler [53] argues in a similar way – though without foundations in game theory. Avner de-Shalit [54] with his concept of the "transgenerational community" also assumes such an aspect, albeit on a communitarian basis: According to this, the individual cannot see himself as part of a community that defines itself through a canon of common rules and norms, but does understand itself as "transgenerational community extending into the future", "hence recognizing obligations to future generations". Just as the past of the community is constitutive for its cultural self-understanding, so too is its future Heidegger ([7], § 65: "Temporality as the ontological meaning

established agreements that form communities as a "cooperative venture for mutual advantage" [18].

In modern civilizations, a complex structure of norms and customs has developed in which people regularly cooperate with each other without even knowing each other, but which nevertheless form a reliable framework also for longer-term action and life planning, for example, to provide for one's own future. The continued existence of these norms and practices is based on recognition by the members and must be confirmed by regular cooperation and rule-compliant behavior. The community members must invest in the continued existence of the social framework conditions that are the prerequisite for their long-term planning, if only by refraining from non-cooperative behavior and one-sided exploitation of the common institutions for their own benefit, which could ruin them. If, however, the actors would have to fear that the agreements would be broken at the time t and would not continue to exist, it would be better for them, after rationally weighing up the opportunities and risks, to pre-empt the expected exploitation of their own investments at t with their own refusal to cooperate at an earlier time t–1. A rational counterpart, however, would anticipate this and refuse to cooperate at t–2 and so on. Under such conditions, the social and organizational framework would not exist, that is indispensable for any reliable longer-term planning and that makes the decision to make provisions for later times at the expense of immediate consumption a rational one. For being rationally motivated for such a long-term planning, it is not necessary to give in to the illusion that the established framework conditions continue to exist permanently. But long-term planning will not occur in a society if its members will not be able to act under the assumption that the established framework conditions continue to exist for an indefinite, open-ended period. The present situation in modern civilizations, which enables actors to provide for their and their relatives' lives beyond the immediate future, is only what it is because their planning takes place in the shadow of a future that exceeds the actor's own lifetime and in which the framework, conventions, organizations, and institutions can be expected to remain constant. It is therefore in the self-interest of the members of a community to act in a way, that preserves this framework and secures the continuity of the shadow of the future at any time, so that it will further be rational to make provisions and to invest in a future, that outlast the actor's own lifetime. The assumption of an indefinite, open-ended continuation of the framework of one's own long-term planning presupposes that a shadow of the future spreads over the planning and acting of the following generations. It therefore lies in the well-understood self-interest for the members of each generation to consider an open succession of following generations and for an in principle permanent continuation of the conventions and institutions that shape

of care") as well emphasizes the fundamental-ontological significance of past and (open) future for individual existence.

their own civilization. Therefore, it is in the well-understood self-interest of contemporaries to demand and enter into mutual obligations that include presumed or attributed interests of future generations into the mutual obligation contemporaries for the benefit of future generations.

With the conclusion that provision for future generations is in the well-understood self-interest of the present, it is not yet clear how one should act in accordance to this interest and especially how goods and bads are to be distributed between members of generation in distance to each other. Assuming, for example, a constant economic growth and assuming that future generations tend to have more technical possibilities, higher productivity and more goods, it seems appropriate, according to a widespread intuition, that the present – as the in comparison relatively "poorer" generation – receive (or keep for themselves) a higher share of the goods to be distributed.

In economic theory, this corresponds to concept of discounting. The discount rate to be applied depends on assumptions about the development of prosperity. Whether a lump sum consideration without a closer look into the specific peculiarities of the goods to be distributed intergenerationally is appropriately a matter of controversy in economics. Not all goods seem to be arbitrarily reproducible or substitutable. Whether and to what extent drinking water or an atmosphere in a beneficial composition may be available in the future depends on assumptions about technical innovation processes.[19] Broome [28] argues that ethical aspects also play a significant role in the appropriate approach to discount rates. In particular, the pure discounting of the economic exchange value of goods should be distinguished from the discounting of the presumed effect that a distribution may have on the well-being of the individuals involved.[20]

In general, it is far from clear what exactly should count as the object of distribution and thus of the justice debate in general [29]. Some argue, for instance, in favor of a more "holistic" approach, in which the appropriateness of distributions is not only to be determined by budgets of carbon, but should be the sum of the overall sum of climate-related costs [30]. Where, as in relation to generations in some distance to each other, costs and value can only be estimated or even only hypothetically assumed, one will rather talk about more or less probable benefits, that is, about opportunities, which of course are then confronted with a probable damage, a risk. Especially in the effort to achieve just relations across generational boundaries, focusing on risks and opportunities is helpful. In order to use opportunities and to avoid risks, certain framework conditions, skills, and competences are usually required as technical know-how and education, but also political and economic

19 Hampicke [55], for example, shows that with regard to perishable, non-renewable goods, an equal distribution over all relevant time periods has the highest utility integral.

20 On this basis, Broome [28] derives a duty of justice that imposes on every actor to avoid emissions as thoroughly as possible and to compensate for released emissions with a corresponding carbon offset.

freedom. On this basis, Amartya Sen and Martha Nussbaum [11] have developed the so-called capability approach. Capabilities can be described as the practically effective ability to freely define purposes and the means to pursue them.

A purely macro-economic view, in which goods are to be distributed across time and generations over "humanity" and the recipients are constructed as "generations" or "cohorts", may find useful applications in institutional planning but in any case is questionable from the ethical point of view with its focus on the individual [31]. Rather than "discounting", actors will look for strategies that allow them to do what is necessary to create or maintain appropriate living conditions for future individuals, to ensure the continued existence of society and to maintain the prerequisites for their own long-term planning in a systematic and reliable manner. Where the scarcity of resources, the changing intervention in natural processes or the lack of sustainability of the necessary framework for the provision of capabilities become risks for long-term survival, it is accordingly rational to include their conservation or the controlled provision of suitable substitutes in the objects of planning. However, with an increasing temporal scope of action and due to the increasing prognostic uncertainties, the certainty with which the desired effects can be achieved in a planned manner is rapidly diminishing. It is therefore quite rational and justifiable for the members of a generation G_0 to mutually commit themselves with first preference to the immediately following generation G_1, with second preference to the following generation G_2, with third preference to the generation G_3 and so on (whereby for illustrative purposes it is assumed that the generations follow each other instead of continuously overlapping in their life span). There are no recognizable reasons that would justify the end of the obligation with any generation G_n – and thus, it would violate a moral prohibition of arbitrariness to demand obligations toward a generation of G_{n-1}, but not to consider the immediately following generations anymore. In this respect, it is rational and appropriate to everyday intuition to mutually commit tout court in favor of "future generations", but nevertheless to gradate the obligations along the time axis according to their degree of commitment [32].

2.6 Obligation, Responsibility, and Acceptability

The justification from the "shadow of the future", presented with a view to complex civilizations of Western custom, applies in a very similar form where conventions, rites, morals and beliefs are deeply integrated into the practice of life, for example, in tribal cultures. The self-evidence of its regular practice becomes questionable when it is perceived as finite – when, for example, a rise in the sea as a result of climate change threatens the island's habitat and thus the continuation of the cultural community. Here, too, the assumed continuity is the condition of the possibility of the current practice. The action and life plans drawn up under these conditions – like all action plans and life plans – must be taken into account in the event of conflict, for example, when the lifestyle in industrialized countries associated with high emissions changes the habitat of oceanic cultures via the mechanisms of climate change. However, no fundamental right to the continuation of a culture can be derived that is exempt from all weighing up – the ethical demand to consider the action plans of others in one's own making plans and to strive for coordination and for mutual benefit in the event of conflict exists for all sides. However, a suitable formal framework for such coordination efforts is needed so that the results are not determined by majority or power relations, but by a fair process acceptable to all sides.

The justification structure for perceived obligations toward future generations clearly reveals the conflict constellations that arise if a community is not able to any long-term planning at all. If, due to immediate need or crisis circumstances, the members of a community are not capable of long-term planning and have to provide for themselves and their relatives directly, the elimination of the need in generation G_0 deserves priority – because it provides this community first and foremost with the prerequisites for a long-term planning perspective.[21] In fact, the great importance attached to the issues of sustainability and intergenerational justice in public debates and political agenda-setting, especially in parts of Europe and America, should not be seen as an indicator of an existing need for the protection and rescue of civilization and its environment – rather, such indicators should be taken from the data with which public and private agencies attempt to monitor the status of the development of the environment and the development of civilization.[22] The increased attention paid to the long-term effects of our actions and to cross-generational provisions can also be seen as an indicator of the high degree of stability and planning security that the countries concerned have achieved.

As these considerations illustrate, commitments made with a view to long-term interests for the benefit of future generations can easily collide with other interests, such as short-term interests, or with other commitments made to contemporaries. Both individually and socially, there is therefore an interest in self-commitment: to prevent short-term interests from repeatedly being prioritized at the expense of long-term interests, it is rational to invest in measures that prevent too shortsighted planning. The spectrum ranges from publicly announced objectives that will put the author's reputation at risk in case he fails to achieve them ("naming and shaming"), to explicit contractual provisions with a reliable threat of sanctions. Another instrument that can also increase the effectiveness of planning, decision-making, and action can be made available by division of labor, specialization, and the delegation of commitment.

21 Contrasts between nations that plan for the long term and those that depend on immediate solutions to existential needs are impressively described, for example, in Deaton [56].

22 Cf. World Bank [57] and the overviews on the website "Our World in Data" https://ourworldindata.org/; (accessed 01-Feb-2021).

Here, a distinction is to be made between the *obligation* for an action and the *responsibility* for an action. Actors have a direct and undivided obligation, but they bear responsibility by delegation of a collective commitment. Responsibility results from obligation, quasi by "moral division of labor". Following the example of H.L.A. Hart ([33]: "Postscript"), various forms of responsibility must be kept apart, the distinction between role-responsibility and liability-responsibility being particularly important. In the first sense, an actor is given the decision-making power which he is supposed to exercise in the presumed sense of others. The guardian, for example, is responsible for deciding in the sense of an assumed welfare of the children "entrusted" to him, the captain is responsible for the safety of the passengers, and the deputy is responsible for deciding in the spirit and sense of an office holder. To delegate or accept responsibility for future generations and others, who cannot participate in common discourses, in this sense seems inappropriate: Responsibility for the welfare of others is delegated if these are unable or only partially able to decide or act for themselves. It would be inappropriate to apply this understanding in those cases. Furthermore, being aware of the cultural differences, the abovementioned asymmetry of attention and – with regard to future generations – the prognosis problem, it must seem like hubris to want to act like a deputy on their behalf and in accordance with their interests.

An appropriate understanding of "responsibility" is therefore based in the concept of liability-responsibility: decisions should be made and actions planned and executed in such a way that they are discursively justifiable to everyone – also and especially to those who have an interest in the long-term continuation of a lifeworld according to their ideas and needs. Even in the case of actions with long-term effects, the current circumstances of action and decision situations as well as the current state of knowledge must serve as the bases for justification. As such, knowledge is always incomplete and provisional, justifications take the form of rational risk/opportunity considerations. Thereby, the probability assumptions used in the risk/opportunity assessment as well as the applied decision-rules are to be justified. Thus, not only the uncertainty and incompleteness of knowledge are to be taken into account, but also the possibility that present actors presumably are not even in a position to model an adequate decision situation appropriate to people from other cultural conditions and future conditions. Under decision constellations like this, the option that leaves greater scope for decision for all parties affected by the measure is always preferable, whether they are involved in the implementation of the measure and thus have the possibility of adjustment and readjustment, or whether they are only passively affected by the measure. Moreover, the responsibility does not only extend to the content of the decisions, but also to the fact that the decisions have been reached by appropriate means: Negligence is reproachable, also in the context of responsibility for future generations. This must be countered not only by continuous research into the relevant interrelationships, but also by appropriate organizational precautions in decision-making. This requires a multifaceted advisory infrastructure which ensures the most rational possible examination of the organizations of special-purpose vehicles and the assessment and evaluation of the consequences of the action considered (including omission), but also the uncertainties. Last but not least, there is a need for an organization of decision-making bodies in which the comparison of controversial views leads to the inclusion of as many aspects as possible and thus to the optimization of the results, but which also, despite the complex discourse situation, leads to effective decision-making within a reasonable period of time. Science claims to make its findings invariant to the positions and interests of the discourse parties and invariant to the respective situational circumstances. In order to safeguard this claim as best as possible, organizational precautions have been developed to neutralize the biases of the individual researchers as far as possible. The structured cooperation of the scientists in the Intergovernmental Panel on Climate Change represents this in a highly elaborated form. Scientists provide the informational basis for decision-making discourses and are explicit about what we know and what we do not know about the effects of climate change, but are not – at least not in their role as scientists – representatives of positions in the decision making. It can also support the process by, for example, identifying and structuring the often complex preference situations and acceptance conditions in affected groups, and it can support decision-makers in systematizing and organizing complex and often inconsistent sets of objectives [34]. Decisions must be taken – in accordance with the mandate to assume responsibility – in a way that they can count as acceptable for everyone, including members of future generations. This kind of justifiability is, of course, a fiction. Factual trustees of this as-if responsibility remain those who delegate the responsibility: the delegation of responsibility does not release the actors who mutually commit themselves to consider the projected interests of future generations from their obligations. The moral division of labor owes itself to only rational considerations and is justified by the expected greater impact alone. The obligation of the individual actor therefore also implies that the work of the decision-making bodies must be provided with the necessary resources and that their impact is critically examined.

2.7 Distribution Postulates

As far as justice is at stake in such deliberations, a basic question leads to the principles that can be invoked as justification for distributions. As already mentioned, a seemingly obvious answer – often suggested by textbook examples and in educational contexts – is not possible here: equal distribution. The principle of equal distribution (it must be distinguished from the principle of equality of rights and from the principle of equal opportunities), however obvious it may seem at first, loses its power of persuasion as soon it is applied to larger collectives which, oriented toward the different interests and the different capacities of individuals, base their supply of resources on a division of labor and thus an unequal distribution of competences and responsibilities. In order to achieve an equal distribution, a considerable authoritarian intervention in the freedom of the individual would be required in order to create and maintain the necessary preconditions. The same applies mutatis mutandis to any effort, to level different inclinations, motivations and requirements as well as the different capabilities and prerequisites to use a given framework for one's own purposes and to enforce one's own interests against the interests of others. And with regard to future generations, any effort of equal distributions is doomed to fail from the start, since we do not even know how many generations there are to be taken into account. Would the present generation – with a certain understanding of to the shadow of the future argument – assume an infinite sequence of generations, the share of resources it would be allowed to keep for itself would be infinitely small. But, in any case, it would be inappropriate to take only x generations into account and keep any possible generation x+n out of the equation. In any case, the principle of equal distribution would lead to inappropriate consequences.

Nevertheless, an understanding of justice as equality – as equality of citizens before the law and of equal rights and opportunities as market actors – offers an ethically and pragmatically justifiable yardstick for the quality of existing organizational forms and for decisions on the further development of organizations. But equality must not be the sole yardstick here either. Those who adhere in principle to the ideal of equality will have to balance out social equality and the advantages that a free development of individual talents and potentials can have for the whole of society.

If one wants to realize distributions that are sensitive to the different concerns and prerequisites of the participants and have the aim of balancing interests, then the common formation of framework conditions in a free discourse of all participants will be the desired ideal [35]. Such a discourse will either directly address the solution of concrete distribution tasks ("material justice discourse") or rules, procedures, and possibly an institutional framework through which just distributions are to be ensured or better ensured than before ("procedural justice discourse"). In the factual realizations of such ideal discourses, it is to be expected in any case that the parties will invoke principles for the defense and examination of distribution or procedural proposals that are at the center of justice-theoretical considerations since the antiquity and that determine theoretical and practical debates to this day. They have their regular place in local and global debates on questions of equitable supply of resources of all kind, including energy, as well as in questions of the equitable distribution of family rights and duties, a fair tax system and the equitable distribution of public goods. They are all mentioned in a much quoted passage by John Stuart Mills – exemplarily related to the fair distribution of a collectively generated labor income:

> Some [...] consider it unjust that the produce of the labour of the community should be shared on any other principle than that of exact equality; others think it just that those should receive most whose wants are greatest; while others hold that those who work harder, or who produce more, or whose services are more valuable to the community, may justly claim a larger quota in the division of produce. And the sense of natural justice may be plausibly appealed to in behalf of every one of these options. [14]

Mill already indicates the finding of a detailed philosophical examination of these principles, which is widely accepted today. Which distribution principles are proposed in a distribution constellation and which can be enforced in the discourse is not decided in advance – for an equitable distribution of resources, no direct general rules can be established ex ante, since the justification depends, among other things, on the subjective perception of the distribution constellation, the subjective evaluation of options and on the individual and collective weighing of needs in discourses. In the context of a fair energy supply and the associated question of a fair distribution of risks and consequential burdens, they are all familiar in one form or another and have already been addressed in part above. Is it rather understood occasionally as a singular distribution or as part of a whole sequence of expected distributions? Is, maybe, the distribution understood as compensation of previous distributions? Can future distributions be included in the current assessment of needs? Is there a state of emergency that forces actors into certain minimum requirements? Questions like these and many more are open for negotiation in an ideal discourse, and since there are no defining standards or unquestionable criteria to answer them with unanimous consent, no valid statements can be made ex ante about the outcome of the forthcoming discourses. The resulting distribution will be not least a matter of factual morality, co-determined by traditions and well-rehearsed practices, and less a question of rational examination and ethical rules. The assessments of needs and how they are weighed against others, for example, are not subject to any moral "jurisdiction". In the same way, there is no such thing as an appropriate yardstick for assessing risks or the willingness to live in the vicinity of technical installations. And practical outcomes often even succeed without requiring consistency and general validity of the solutions achieved. In the opposite, what people are willing to accept and are willing to take into account is very much dependent on what is socially established and what they are used to [36]. But while in most areas a distribution of energy resources has prevailed and is widely accepted that is based on the willingness of the demanders to pay, in a globally perspective the demand for a just energy supply is very much based on the intuition of an affordable and secure energy supply for everyone in the manner of a human right.[23] Those, however, who make concrete proposals and try to justify them specified with the one or

---

23 For example, the UN Secretary-General heads a "Sustainable Energy for All Initiative", the aim of which is to ensure universal access to modern energy services http://www.un.org/sustainabledevelopment/energy/; (accessed 01-Feb-2021).

other distribution principle, regularly face counter-arguments that give preference to other distribution principles, that – to cite Mill again – "the sense of natural justice may be plausibly appealed to", trained by tradition and well-rehearsed customs.[24] In any case, those who propose a distribution or are unwilling to accept a completed distribution and do not simply want to impose their position with power and force will have to justify their claims. Such justification cannot, however, be based on arguments that are not already universal or universalizable. Statements of the kind "A shall get something (more)" or "I shall get something (more)" merely formulate a claim and are no more an argument as statements of the kind "I need more", "I have the older rights", or also "We have been suppressed for centuries, now it's our turn". An argument that wants to have prospect of examination and consideration on the participants in discourse though must be of the form "Everyone who has property E shall get ...". "E" thereby may stand for a particular ability or productivity, the degree of need, inherited class or family privileges, the historical suffering of a group, or any other potentially claiming quality. Anyone who claims a certain right for him- or herself or another person or group characterized on the basis of certain properties must then, for reasons of pragmatic consistency, grant this right to anyone else who possesses this characteristic. In moral communities, entitlement rights are thus "always already" claimed and justified with universal arguments – ultimately because otherwise the discourse community refuses to accept them.[25] Whether and to what extent then universalized principles such as "everyone according to his need", "everyone according to his possibilities", "everyone according to his abilities", "everyone according to his commitment", "everyone according to his merit", and "everyone according to his achievements" are regarded as decisive by the parties to the discourse, and which principle flows into the decision-making, is not predetermined by any external circumstances. Here, the sciences and humanities can offer support services and provide valuable advisory services, for example, with references to historical distribution effects, statistically collected or theoretically derived estimates of expected consequences of any kind of distribution, or simply with linguistically sharpened differentiation services. In no way, however, can the question of which distribution principles are to be applied be answered by pure reasoning or scientific analyses "in abstracto". According to the well-known "suum cuique", any consideration will have to take individual interests into account, as they can be ideally expressed in a free and open discourse. Therefore, the requirement for universalizability is rather to be understood as a formal principle, not directed toward distribution outcomes, but toward the distribution procedure, as a requirement for its design: design distribution

24 Public debates are full of examples of more or less quickly withdrawn proposals, from flat tax to income-based tiered pricing for access to public goods, income-related fines for traffic offences and much more.

25 This is a variant of Hare's (esp. [58]) universalizability argument, reformulated for distributive justice and relativized for moral communities.

procedures in a way so that everyone can assert his or her claims and that everyone's individual prerequisites are taken into account! Therefore, in the modern era, the philosophical debate on justice has increasingly turned away from the so-called material questions, that is, questions directed at the concrete distribution, and toward the more formal questions of the design of framework conditions and procedures that lead to just distributions.

From a historical point of view, questions about criteria, structures, and procedures have increasingly been included in the debates on ethics, particularly with the emergence of anonymous large societies. Promising proposals and recommendations have been developed on these questions and their ability to consent can be effectively determined by the actual participation of those involved ("political philosophy"). Thus, the thinkers of the so-called Scottish Enlightenment discuss questions of justice on the basis of the phenomenon of the spontaneous self-organization of interacting communities and the allocation principles of free markets, Montesquieu proposes to protect the individual's freedom of action and expression by division of power and mutual control of governmental branches, and Kant's categorical imperative formulates the principle of rule of law against the absolute claim to power of the state ("Act only according to that maxim whereby you can at the same time will that it should become a universal law"). Where the behavioral expectations that people have of each other become insecure and volatile, a debate about the (further development of) procedures for the (re-)establishment of relationships that everyone (sufficiently) finds fair can, on the one hand, have the prospect of general agreement and, on the other, have a regulative effect. In this sense, Rawls [18, 37] and the discussion that followed him then endeavored to show criteria for the assessment of social orders that above all emphasize their suitability to establish conditions that are justly presented. In the local context, these can be required to ensure adequate access to an affordable and reliable energy supply. It cannot be expected that all scarcity problems will be eliminated – but it can be expected that all rivalry arising from scarcity conditions will be overcome in the most efficient and conflict-free way possible. Ideally, a pareto-optimal distribution is realized. There would then be no alternative to the implemented distribution that would place one participant in a position he prefers without having to place another participant in a position he does not prefer.

Especially with regard to ethical considerations that will helpful in transcending cultural-historical borders on a global scale, ethics is thus confronted with basic challenges, and only a few approaches, if any, have been developed to counter them.[26] In principle, the question would have to be answered as to what proposals and recommendations for "universally valid" and at the same time practically implementable rules would look like if, on the one hand, the implementability is not

26 Cf. the overviews of the diversity of approaches in Moellendorf/Widdows [59] and the thematic compilations in Stückelberger [60].

independent of what the addressees are prepared to accept on the basis of their respective moral presettings, and, on the other hand, it is not the factual, but often arbitrary agreement ("acceptance") on the basis of given moral presettings that serves as the yardstick, but the general acceptability, that is rationally founded in that it includes all aspects.

The orientation toward general acceptability aims at generalizability in a multiple sense:

Non-arbitrarity

As is expressed by the most basic legal principle, same cases should always be treated equally and unequally. In jurisprudence, this rule is referred to as the prohibition of arbitrariness. Only if same cases are treated equally, only if action (or action planning) A and action (or action planning) B, which corresponds to A in all relevant points, consequences included, are judged equally, actors can establish stable mutual behavioral expectations. If within a group, resources are allocated today according to principle p_1, tomorrow, in an otherwise identical situation, according to principle p_2 and the next day in an again similar situation according to principle p_3, those who support this allocations and do not give a justification that refers to relevant differences in the cases will not be seen as a reliable interaction partner. If today's distribution of joint earnings from work favors A while yesterday's distribution favored B without there being any justification for the unequal treatment, if the behavior that has been applauded today is punished tomorrow, then actors cannot plan social interaction for the future and cannot anticipate the other's action in the future. In case of doubt, therefore, it seems rational to refuse to cooperate further on or to consume today what one might otherwise have traded tomorrow in expectation of a greater benefit. Non-arbitrarity, based on generalizable justifications, therefore is essential for the establishment and preservation of a social framework that allows for plannable cooperation. It is, what Kant in his "Groundwork of the Metaphysic of Morals" [12] identifies as a "condition of possibility" for stable social relations and not the least for moral ("Sittlichkeit") itself. Generalizability thereby aims at the assessment of entire classes (sets) or types of identical cases and considers individual subjects of assessment in the light of these similarities. In this light, individual distributions for which there would have been a more favorable alternative for certain actors concerned may appear acceptable, even if they find it difficult to accept the alternatives spontaneously and only a reference to the overriding rationality of equal treatment is needed in order to achieve reflected acceptance.

Robustness

General statements of the type: "whenever ..., then ...", logically analyzed as "for all cases x: if x, then ...", do not only claim validity for previously observed applications, but also for future applications. In particular, if the connection established with such statements is to become the basis for future action planning, the cases will be differentiated and specified until the statements are as robust as possible against the manifold variance of the circumstances. If, after countless successful cases, the means M does not always prove to be successful when actor A wants to achieve a certain purpose, then the if-clause may have to be supplemented by further conditions ("if x, then ..., unless ...", resp. "for all cases x, for all cases y: if x and y, then... "). The same also applies to agreements or rules designed to ensure future coordination of the action plans of various actors with as little disruption as possible. The fundamental fallibility of scientific statements corresponds to the concept of incomplete contracts: All eventualities cannot be anticipated.

Resilience

The target systems on which people plan are never complete and consistent, they change in the course of life, but also in different life situations. Actors sometimes subordinate goal a to goal b, sometimes goal b to goal a, be it that the framework conditions change (e.g., slippage of prosperity into poverty), that roles or perspectives change (e.g., between that of beneficiary of public goods and that of the taxpayer), be it that social moods fluctuate.[27] Agreements should be agreeable for each of the participants among all the changes that arise.

Inclusion

Agreements should be acceptable to each of the participants and should not be at the expense of third parties. Those whose action plans are affected should be able to give their approval, whether directly affected or only by distant consequences in time or space.

In this context, it is worth mentioning, that generalizability and acceptability are not characteristics that can be attributed once and for all to an agreement, convention, rule, or standard. As in the case of scientific statements, whose claim to general validity must always be put to the test and occasionally rejected, here too the reservation of fallibility always applies. In particular, the equality of cases cannot

27 Cf. the contributions in Lichtenstein/Slovic [61].

simply be stated – for example, by simply ostending to the cases. Correspondingly, the claim for an equivalent treatment could not simply be made. Textbook examples as they are often cited in ethical debates are deceptive here. A typical case is again the one of the passerby, witnessing how a child threatens to drown in a pond, who he could save while having to accept only small risks for himself, which is referred to in order to show that in any case, where there is a helpless person in need one could save with small risks for oneself, one it obliged to do so, regardless of where the person in need is, whether in close proximity or at a greater distance.[28] Even if one acknowledges – which is by no means trivial – the existence of an obligation to save the child in a situation as described, this would not imply an obligation of equal scope for those cases in which a fellow human being is not visibly threatened by drowning within reach, but a person in the distance, maybe on another continent, is threatened by starvation, who could be saved for the equivalent of only a dollar. In both cases, a life-threatening situation exists and the risk or the effort for the provider of assistance to remedy this situation may well be similar. In this respect, the cases are initially to be regarded as cases of the same type, so that, prima facie, by referring to the requirement of non-arbitrary nature from the given (or assumed) obligation toward the drowning child, an obligation could be derived toward all children in life-threatening situations, as long as they only could be saved with similarly assessed risks. But that the information about the situation in the distance is a mediated one, that other actors stand between the actor and the child threatened with fatal danger, that he has to make an effort to inform himself about sufficiently reliable help options that can only be implemented indirectly and with time delay, that he cannot react to immediate visual and acoustic stimuli, but if necessary, that his reaction to the situation in the distance requires deriving a state of emergency from words and numbers etc. makes the assumption of an equality of constellations, on which the conclusion of analogy is based, already problematic. Above all, however, the argument assumes that (i) generalizability would be a demand directed quasi from the outside to the actions of the individual and that (ii) the equality of cases, which any requirement of generalizability presupposes as given, is the subject of a statement, not of a (joint) determination. In both cases, this would be a normativistic fallacy.[29] The orientation toward generalizability and the joint development of an understanding

---

28 See Singer [62]. The analogy argument is often quoted and is – especially outside professional philosophical circles – often cited in order to promote financial transfers from developed regions to regions in need of development. Cf. Peter Singer's own online platform https://www.thelifeyoucansave.org; (accessed 01-Feb-2021).

29 A normativistic fallacy is the missed (but often superficially convincing) attempt to derive concrete requests or recommendations from general principles without due consideration being given to situational circumstances that stand in the way of the application of the principle to the given case. The definition of the normativistic fallacy goes back to Höffe [63], cf. for instance Gorke [64], Ch. 14.

of the same cases to be treated in the same way serves the organization of cooperation and the creation of a reliable environment for action, which provides security for individual planning. So, it is perhaps a praiseworthy and supportive project to promote equal treatment of cases and to fight politically for framework conditions which then also favor de facto equal treatment. The effort to demonstrate a general duty for everybody from the individual case, on the other hand, is as misguided as the effort to infer a general law of nature from a single observation. What is to be combined into a class of cases that require equal treatment and what is to be treated as unequal is rather the object and – if successful – the result of incessant negotiations and constant readjustment. And it is not even clear from the outset who in this negotiations bears the burden of proof: traditions and the rules included in them often form a kind of "default attitude" here, so that the burden of justification is attributed to those who want to follow deviating rules. Not only do traditional practices occasionally prove to be inconsistent, unsuitable for new situations or even disruptive for new challenges – for fundamental reasons the mere reference to existing practices can never be considered a rational answer when an existing practice is called into question. One argument that should not be underestimated, however, is the reference to the evolutionarily developed functionality of existing practices, even if further development is required in order to adapt the practices to new and changed framework conditions. In spite of all the restrictions on action that would result from the binding to the existing, established – fundamentally new, "tailor-made" solutions presuppose that the framework conditions for their design and implementation would have been fully grasped, or at least the essential sections would have been grasped. History gives clear indications that it is highly risky to make such prerequisites. With regard to the entitlement to participate in such distribution discourses and the obligation to comply with the discursive agreements reached there, it is, however, possible to formulate an equality rule at a "meta-level". A reliable conflict resolution will only be available if all those who are potentially able to make a claim are given the same opportunity to participate in discourse, and if, in addition, all those who participate in discourses are equally committed to a consensus of distribution. The individual, where this participation in the discourse of the individual takes place in an organizational framework through mandate holders and representatives, is not released from responsibility. It remains his or her duty to incentivize the incentivators to set fair, appropriate, and effective incentives [38].

An energy supply system (in general: a distribution of resources) that is justified in orientation to the principle of ethical universalism by a discourse equally accessible to all (its result may lie in equal or unequal distribution) is called "just".[30]

30 As altogether in this contribution here "justice" is used only in the sense of distributive justice (iustitia distributiva), not in the sense of exchange justice (iustitita commutativa) or justice before the law (iustitia legalis).

This shows that the prescriptive content of "justice" has two clearly distinguishable moments: first, a moment of equality, which, however, refers to the pre-discursive entitlements and obligations, such as equal access to discourse, equal treatment of all participants by the "rules of procedure", and so on. Second, there is a moment of justified inequality, which refers to the intra-discursively found distributions, that is, to the result of the distribution. Accordingly, two postulates can be formulated as an explication of the prescriptive content of "justice":

Equality Postulate

Act in such a way that everyone who makes relevant claims receives the same rights and obligations when participating in discourses!

Distribution Postulate

Distribute so that each allocation is justified by a discourse of equal participants!

These postulates refer to the scheme of distribution discourses and can therefore be described as "formal". The more "material" criteria of distributive justice are obtained by looking at the arguments put forward by the participants to justify their claims. By definition, actors try to create opportunities and avoid risks. Regardless of the question of whether there can be purely individual opportunities and risks at all, a fair distribution of opportunities and risks has to do with the question of what should be expected of others and what others should expect of us. For the decision of this question, individual or collective preferences of the respective individual or collective actors cannot therefore be (alone) decisive. This also applies when entire generations are assumed as actors. Ultimately, therefore, only procedures can be determined according to which it is decided whose claim is to be taken into account and whose "will" is to be realized in whole or in part.

For the distribution substratum "opportunities" and "risks", the demand for pragmatic consistency results first of all directly in the

Rule of Risk-Taking

Be prepared to take risks if you have already accepted or expected others to accept similar risks and therefore consider them acceptable!

Further rules, which then supplement such a general rule for concrete cases – derived from basic considerations – and develop it in a practical manner, are then possible further candidates for justification, to which the discourse community can commit itself:

Rule of Opportunity Sharing

Act in such a way that you let the risk takers share in the opportunities as much as possible!

Rule of Risk Allocation

Decide risk options in such a way that those who so far have benefited least from opportunities have the greatest relative advantage!

Risk Provisioning Rule

Act in such a way that you can compensate those who bear the risks of your opportunities as much as possible in the event of damage!

Such rules, however plausible they may be, immediately lead to considerable operational difficulties when considering interactions of a rapidly reached level of complexity, but especially when considering the interaction network of an entire society. Even if one makes optimal assumptions for the individual actors with regard to their insight and intentions for action, it is therefore quite unclear at the collective level how an equitable distribution of opportunities and risks can be thought out. Philosophers, economists, lawyers, and others have been working for several years on formal models that can make clear how distributional discourses function in relation to promising and risky actions.

References

[1] Kander A, Malanima P, Warde P. Power to the People. Energy in Europe over the Last Five Centuries. Princeton, NJ and Oxford, UK; 2013.

[2] Williams M. Deforestating the Earth. From Prehistory to Global Crisis. An Abridgment. Chicago, London, UK; 2006.

[3] Smil V. Power Density: A Key to Understanding Energy Sources and Uses. Cambridge, MS; 2015.

[4] Hume D. An Enquiry Concerning the Principles of Morals. Oxford, UK; 1751.

[5] DESA [Department of Economic and Social Affairs Population Division]. (2017): Changing population age structures and sustainable development. A Concise Report. United Nations, New York, NY.
[6] Gethmann CF. Dasein: Erkennen und Handeln. Heidegger im phänomenologischen Kontext. Berlin, Germany; 1993.
[7] Heidegger M. Sein und Zeit Freiburg Engl. In: translated by, Macquarrie J, Robinson E. Being and Time. Oxford, UK; 1927, 1962.
[8] Friedrich GC, Kamp G, Knodt M, Kröger W, Streffer C, Von Storch H, Ziesemer T. Global Energy Supply and Emissions. An Interdisciplinary View on Effects, Restrictions, Requirements and Options. Heidelberg, Germany/New York, NY; 2020.
[9] UNEP [UN Environment Programme]. The Asian Brown Cloud: Climate and Other Environmental Impacts, UNEP 2002. 2002; Nairobi, Kenya.
[10] UNEP [UN Environment Programme]. (2008) 2008 ECOREA; environmental review 2008. United Nations (ed) (1987) Report of the World Commission on environment and development, A/42/427.
[11] Nussbaum M. Creating Capabilities: The Human Development Approach. Cambridge, MS; 2011.
[12] Kant I. Grundlegung zur Metaphysik der Sitten (Engl.) Groundwork for the Metaphysics of Morals (trans. Wood AW) New Haven, CT 2002; original work published 1785.
[13] Bentham J. An Introduction to the Principles of Morals and Legislation. London; 1789.
[14] Mill JS. Utilitarianism. London, UK; 1863.
[15] Habermas J. A Theory of Communicative Action. Boston, MA; 1984.
[16] Hinsch W. Gerechtfertigte Ungleichheiten. Grundsätze sozialer Gerechtigkeit. Berlin (Germany), New York, NY; 2002.
[17] Bombaerts G, Jenkins K, Sanusi Y, Guoyu W, editors. Energy Justice Across Borders. Cham, Switzerland; 2020.
[18] Rawls J. A theory of justice. Cambridge, MA; 1971.
[19] Sachs JD, Warner A. The curse of natural resources. Eur Econ Rev. 2001;45(4–6):827–38.
[20] Wenar L. Blood Oil: Tyrants, Violence, and the Rules that Run the World. Oxford, UK; 2015.
[21] Acemoglu D, Robinson JA. Why Nations Fail: The Origins of Power, Prosperity, and Poverty. New York, NY; 2012.
[22] Gethmann CF. Ethical aspects of long-term responsibilities in research. In: Smelser NJ, Baltes P, editors. International Encyclopedia of the Social and Behavioral Sciences. Oxford, UK; 2001, Vol. 19, 13227–31.
[23] Tetlock PE. Expert Political Judgment: How Good is It? How Can we Know?. Princeton, NJ; 2006.
[24] Tetlock PE, Gardner D. Superforecasting: The Art and Science of Prediction. New York, NY; 2015.
[25] Düwell M, Bos G. Human rights and future people—possibilities of argumentation. J Hum Rights. 2016;15(2):231–50.
[26] Kant I. Idee zu einer allgemeinen Geschichte in weltbürgerlicher Absicht. [Idea for a universal history with a cosmopolitan purpose]. In: Reiss HS, editor. Kant. Cambridge Texts in the History of Political Thought. 2nd ed. Cambridge, UK; original work published 1784, 41–53.
[27] Gardiner SM. The pure intergenerational problem. Monist. 2003;86(3):481–500.
[28] Broome J. Climate Matters: Ethics in a Warming World. New York, NY; 2012.
[29] Sen A. Equality of what?. In: McMurrin S. Tanner Lectures on Human Values. Cambridge, MA; 1980, Vol. 1, 197–220.
[30] Caney S. Just emissions. Philos Public Aff. 2012;40(4):255–300.

[31] Hampicke U. Climate change economics and discounted utilitarianism. Ecol Econ. 2011;72: 45–52.

[32] Gethmann CF, Kamp G. Gradierung und Diskontierung bei der Langzeitverpflichtung. In: Birnbacher D, Brudermöller G, editor. Zukunftsverantwortung und Generationensolidarität. Würzburg, Germany; 2001, 137–53.

[33] Hart HLA. Postscript. In: Punishment and Responsibility: Essay in the Philosophy of Law. New York, NY; 1968, 210–37.

[34] Gethmann CF, Carrier M, Hanekamp G, Kaiser M, Kamp G, Lingner S, Thiele F. Interdisciplinary research and trans-disciplinary validity claims. Germany: Berlin; 2015.

[35] Habermas J. Diskursethik. Notizen zu einem Begründungsprogramm, Manuskript 1982. Veröffentlichungen in: ders., Moralbewußtsein und kommunikatives Handeln, Frankfurt, Germany 1983, 53–125; engl.: Discourse Ethics: Notes on a Program of Philosophical Justification. In: Moral Consciousness and Communicative Action. Cambridge, MS 1990, 1982, 43–115.

[36] Renn O. White Paper on Risk Governance: Toward an Integrative Framework. In: Renn O, editor. Global risk governance. Concept and practice using the IRGC framework. 2008, 3–73.

[37] Rawls J. The law of peoples. Cambridge, MA; 1999.

[38] Trapp RW. Klugheitsdilemmata und die Umweltproblematik. Paderborn, Germany; 1998.

[39] Hare RM. Moral thinking. Oxford, UK; 1981.

[40] Hayek F. Law, legislation and liberty, Vol. 3: The political order of a free people. UK: London; 1976.

[41] Engels F. Preface to the first German edition. In: Marx, editor The Poverty of Philosophy. Marx/Engels Internet Archive 1999. 1847.

[42] Buchanan AE. Marx and justice: the radical critique of liberalism. Law Philos. 1982;3(1): 147–53.

[43] De Haan G, Lerch A, Martignon L, Müller-Christ G, Nutzinger H-G. Nachhaltigkeit und Gerechtigkeit. Grundlagen und schulpraktische Konsequenzen. Heidelberg Germany/ New York NY; 2008.

[44] Streffer C, Gethmann CF, Kamp G, Kröger W, Rehbinder E, Renn O, Röhlig K-J. Radioactive Waste. Technical and Normative Aspects of its Disposal. Heidelberg, Germany/New York, NY; 2011.

[45] Gethmann CF, Kamp G (2020) Climate justice. In: Oxford research encyclopedia climate science. https://oxfordre.com/climatescience (accessed 01-Feb-2021)

[46] DeLong JB (2000). *Cornucopia: The pace of economic growth in the twentieth century [http://www.nber.org/papers/w7602]. NBER Working Paper No. 7602. Cambridge, MA: National Bureau of Economic Research. (accessed 01-Feb-2021)

[47] Roberts R. The Price of Everything: A parable of possibility and prosperity. Princeton, NJ; 2008.

[48] Silver N. The Signal and the Noise: Why so Many Predictions Fail—but Some don't. New York, NY; 2012.

[49] Popper K. The Poverty of Historicism. New York, NY; 1957.

[50] Gosseries A, Meyer LH. Intergenerational Justice. New York, NY; 2009.

[51] Gardiner SM. A Perfect Moral Storm: The Ethical Tragedy of Climate Change. New York, NY; 2011.

[52] Axelrod R The Evolution Of Cooperation. New York, NY, Rev ed., 2006; 1984.

[53] Scheffler S. Death & Afterlife. New York, NY; 2013.

[54] De-shalit A. Why Posterity Matters: Environmental Policies and Future Generations. London, UK; 1995.

[55] Hampicke U. Ökologische Ökonomie: Individuum und Natur in der Neoklassik. Opladen, Germany; 1992.

[56] Deaton A. The Great Escape: Health, Wealth, and the Origins of Inequality. Boston MA; 2013.

[57] World Bank. Global Monitoring Report 2015/2016: Development Goals in an Era of Demographic. Washington, DC; 2016.

[58] Hare RM. The Language of Morals. Oxford, UK; 1952.

[59] Moellendorf D, Widdows H. The Routledge Handbook of Global Ethics. New York, NY; 2015.

[60] Stückelberger C (2016) ed Global ethics applied, vols 1–4. Geneva 2, Switzerland [https://www.globethics.net/ (accessed 01-Feb-2021)]

[61] Lichtenstein S, Slovic P, editors. The Construction of Preference. Cambridge, UK; 2006.

[62] Singer P. The Life you can Save. New York, NY; 2009.

[63] Höffe O. Sittlich-politische Diskurse. Frankfurt, Germany; 1981.

[64] Gorke M. The Death of our Plant's Species. A Challenge to Ecology and Ethics. Washington, DC; 2003.

[65] Marx K (1875) Critique of the Gotha Programme. In: Marx/Engels selected works, vol. 3, Moscow 1970, 13–30

[66] Nordhaus WD. Do Real-Output and Real-Wage Measures Capture Reality? In: Timothy F. Bresnahan and Robert J. Gordon (eds): The Economics of New Goods, Chicago, IL 1996.

[67] Sidgwick H. Methods of ethics (7th ed. 1907). repr. 1981. Cambridge, MA: Hackett. (Original work published in 1874)

Ferdi Schüth

3 Energy Storage Strategies

3.1 Introduction

In the light of dwindling reserves of fossil fuels, which serve not only as a source of energy for our societies but also as the major form of storing energy, energy storage has become a pressing and modern topic. However, energy storage is probably as old as our civilization, although we mostly do not think about it in these terms. Collecting wood for maintaining a fire or to provide heat during the winter is a form of energy storage, and most of the features of modern energy storage are indeed present in firewood. It has a high energy density (about one-third of the gravimetric energy density of diesel fuel), it can be stored for practically an indefinite amount of time without losing its energy content, it is reasonably safe, and it can easily be converted to the useful forms of energy for which it is stored (i.e., heat and/or light).

In modern societies, energy storage is required for a whole range of different timescales and size scales (Figure 3.1). Capacitors in electronic devices store small amounts of energy for short times; batteries provide the energy for cell phones, laptops, or the propulsion of cars; gasoline, diesel fuel, and jet fuel provide the high storage density required for cars, trucks, or planes; and big subterranean caverns store grid-scale amounts of energy in the form of oil and natural gas. With more and more fluctuating energy from wind and solar power in the electricity grid and decreasing reserves of fossil fuels, energy storage will be a key component in any future energy infrastructure. In the following, the strategies for energy storage on different size scales and timescales will be discussed in more detail.

3.2 General Considerations

An important point for all further considerations is the question of whether the energy is initially obtained in a directly storable form, or whether it has to be converted to a storage form. Most renewable energies are not easily stored. Wind energy is converted directly to electricity; the same holds for photovoltaic, wave, and tidal energy. Solar thermal power plants also generate electrical energy, but timescales of hours can be bridged by integrated heat storage units. Of the renewable energies, only biomass, geothermal, and dam hydroelectric energy are easily stored. In contrast to renewable energies, all fossil forms of energy are initially produced as storable energy (i.e., oil,

Ferdi Schüth, Department of Heterogeneous Catalysis, Max-Planck-Institut für Kohlenforschung, Kaiser-Wilhelm-Platz 1, 45470 Mülheim, Germany, e-mail: schueth@kofo.mpg.de

https://doi.org/10.1515/9783110608458-003

gas, or coal). Uranium to run nuclear power plants can be stored for long time periods as well. From fossil and nuclear primary energy, useful forms of energy can be generated at will and if needed. This is the reason why our societies are predominantly relying on fossil energy, which is used as a kind of easily convertible "energy currency."

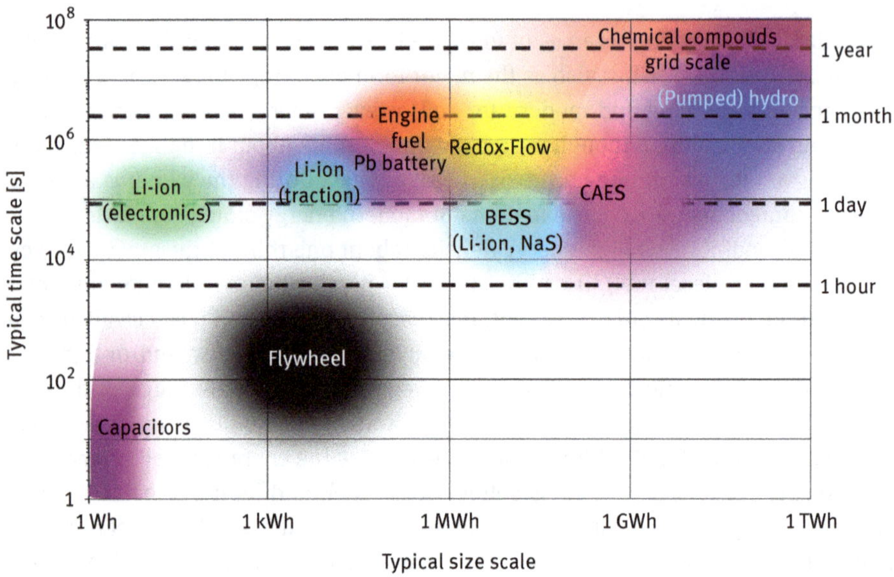

Figure 3.1: Typical time and size scales associated with different storage technologies.

On a crude scale, primary energy serves to supply one of three consumption sectors – that is, mobility, heat, and electrical energy, the latter of which, in turn, is converted to various forms of useful energy, such as mechanical energy, light, and heat. Very roughly, globally each of these sectors requires one-third of the primary energy, while electricity has a smaller fraction in the final energy consumption due to the conversion losses from primary energy to electrical energy. Energy storage has to meet completely different requirements for each of these consumption sectors, and the different storage concepts and technologies have to come together in a concerted manner to provide the basis of an energy system.

It should be noted here that, in many cases, energy storage is not without alternatives. Electricity has to be produced at any given time in exactly the same amount as it is being consumed. In times of overproduction – for instance, during periods of strong winds – electricity could be stored and then released in times of high demand and low production. Alternatively, however, one could install overcapacities, which would be shut down in case of overproduction. Excess energy would therefore not be stored. This would not make full use of the available energy resources (i.e., of wind or solar irradiation) but overall could be a more cost-efficient solution.

The higher the transfer capacities of electricity grids and the better the grids are connected, the lower the need for additional storage since exchange of electricity between different regions could help to balance supply and demand. Similarly, if most long-distance traffic would make use of trains, fuels for internal combustion engines or batteries of electric vehicles would not be needed to such an extent. Thus, as important as energy storage in various ways will be in future energy systems, the optimization of the whole system may rely on storage as well as on other technologies, with the balance between the approaches based on cost, practicability, societal acceptance, and other factors. It is expected that future energy systems will become increasingly electric, and the consumption sectors of mobility and electricity will merge to an appreciable extent, since more and more of the transportation sector will rely on electrical energy. Heat demand of buildings, especially of new buildings, will decrease due to improved insulation, and heat storage of solar thermal heat could become an important issue. The heating sector could become largely decoupled from the rest of the energy system, with the exception of cogeneration and purely electrical heating (heat pump or resistive) to meet the residual heat requirements of buildings. In the next section, therefore, the storage of heat energy will be discussed before we turn to the issue of storing electrical energy in future energy systems.

3.3 Heat (Cold) Storage

Heat and cold storage are among the oldest known forms of energy storage. Old tiled stoves store the heat for hours after the fire inside has burned down, and in ancient times, ice blocks were used and stored in ice cellars or ice pits to keep the cold from the winter for food preservation into the summer. Heat storage systems should meet a number of requirements (Table 3.1). Ideally, heat should be stored and released at constant temperature, which favors latent heat storage units. Such units also typically have higher storage densities than sensible heat storage systems. However, latent heat systems are often more complex since phase transitions or chemical reactions take place instead of simple heating up or cooling down. This is often associated with complex control problems, and thus for less demanding applications, simple sensible heat storage systems with water as the storage medium are often used.

There are two major emerging application fields for heat storage in future energy systems, storage on the scale of hundreds of MWh for solar thermal power plants and on the order of several MWh for domestic heating or district heating. If solar thermal plants should be able to provide base load power, then even in regions where there is almost never any cloud coverage, energy needs to be provided during the nighttime hours. Current state-of-the-art solar thermal plants have power on the order of 50 MW, which means that for bridging 8 h at night, 400 MWh of electrical

Table 3.1: Requirements for heat storage systems.

Property	Comment
Type of storage	Latent: constant *T*, but phase change, no losses; sensible: *T* varies, but no phase change, losses depend on insulation
Enthalpy effect	As high as possible for high storage density
Temperature range	Depends on application; 50–150 °C for domestic heating, 300–600 °C for solar thermal power plant
Cycle stability	As high as possible, more important for diurnal than for seasonal storage
Kinetics	As fast as possible
Hysteresis	Ideally absent
Losses	Ideally absent

energy needs to be generated. With typical efficiencies of steam turbines, heat storage capacities on the order of 1,000 MWh (thermal) are required. In the Andasol 50 MW solar thermal plant in Spain, this amount of heat is stored in 27,500 t of 60% $NaNO_3$/40% KNO_3 melt. During the day, part of the heat generated in the parabolic trough mirror field is used to power the steam turbine, while the remainder heats up the molten salt to approximately 390 °C. At night, the salt melt is passed through a heat exchanger and cooled to approximately 290 °C, and the extracted heat provides the energy for the steam turbine. This technology has proved to work in the field and could be deployed at a large scale in solar thermal power plants. It is rugged – provided that the salt does not solidify – and can be scaled so that it can be adapted to differently sized power plants. However, the cost of the material alone is substantial: at typical fertilizer prices (the salt is used for this purpose, prices fluctuate substantially), the cost is on the order of $ 10 million per plant. Alternatives thus appear to be attractive. Sensible heat storage is possible in other storage media as well. The use of huge concrete blocks with integrated heat exchangers is an option that is currently being explored. Another interesting system is Mg/MgH_2, which has very good heat storage properties and was extensively investigated in the 1980s by Boris Bogdanovic. The hydrogenation/dehydrogenation reaction has a thermal signature of 0.9 kWh kg^{-1}. Thus, for a 1,000 MWh thermal capacity, only somewhat over 1,000 t of magnesium instead of 27,500 t of salt melt is required, at a materials cost of $2–3 million. However, a more complex system is necessary, since the hydrogen also needs to be handled. An additional, dedicated hydrogen storage facility would probably render such a system more expensive than a molten salt-based heat storage facility. There could be situations, though, in which solar thermal plants are also used for the generation of hydrogen via thermochemical cycles so that a hydrogen infrastructure would be in place. Under such circumstances, hydride heat storage systems could be a competitive alternative, especially since the temperature level can be adjusted by the hydrogen pressure and/or the use of other metals or alloys. However, since solar thermal power plants have strongly fallen behind photovoltaic power plants with respect to cost of produced electricity, development of heat storage systems to be used in such plants is currently not pursued at high intensity.

Domestic heating in future homes will probably not require fossil fuel fired heating systems in most regions of the world due to much better insulation standards. Low-temperature heat can be stored in sensible or latent heat storage systems. The first installations are already in operation. In Munich, the complex "Ackermannbogen" has a 6,000 m^3 water-sensible heat storage, which is heated during the summer by solar panels. The stored energy is used to heat the 319 apartments until well into January; from then on, district heating provides the rest of the heating energy. This saves approximately 50% of the heating energy that would be required. In addition, the German parliament has a subterranean heat storage at a depth of approximately 300 m. Another water reservoir at 60 m depth is used to provide cooling during the summer months. Such concepts will become more and more widespread in modern buildings; the rest of the required heating energy will be provided by heat pumps and/or direct electrical heating.

While water is a cheap and simple solution, higher storage densities can be achieved by other technologies. Water stores 4.18 J K^{-1} g^{-1} in the temperature range up to 100 °C (i.e., assuming a useful temperature range of 50 K, approximately 200 kJ kg^{-1} can be stored), and water is one of the most efficient sensible heat storage media. The hydration and dehydration of zeolites, crystalline porous aluminosilicates, can be used in a similar temperature range. However, depending on the type of zeolite, the heats of hydration can be five times as high, and such a system would operate at constant temperature. Space requirements could be substantially reduced so that amounts on the order of 10–20 t, corresponding to one room in the cellar, would be sufficient for the heating of a well-insulated house during the whole winter. The costs of such systems are presently still not competitive, but with increasing energy prices, this may change in the future.

Other latent heat storage systems are also under development. Sodium acetate trihydrate with a melting point of 58 °C is the heat storage medium in hand warmers. When crystallization is induced, the heat of fusion is released. Such systems can be used not only on a small scale, such as in hand warmers, but also industrially. Low-temperature waste heat from industrial plants was used by the company LaTherm (Dortmund, Germany) to melt the salt in large containers. These were then transported to heat consumers, such as swimming pools, and there the heat generated during solidification was used to heat the water of the pools. However, after a few years of operation, the concept was not pursued further commercially and the company went out of business. Currently, such heat storage systems are still being explored in a number of research projects for heat storage on different size and time scales, but commercial solutions do not seem to exist.

3.4 Grid-Scale Storage of Electrical Energy

Electrical energy is the most versatile form of energy, but it is also the most difficult one to store. If only the grid without any storage options is considered, the supply and demand of electrical energy have to be exactly balanced at any moment to allow stable operation, both with respect to voltage and frequency. Small fluctuations can be balanced by various electrical measures available to the grid operators, and larger scale fluctuations by adapting the supply side (i.e., ramping up or down of power plants). In extreme situations, demand-side management may also be required (i.e., shutting down large consumers of electrical energy). While fluctuations can be managed to some extent by a highly interconnected and widely extended electricity grid, storage of energy can alleviate some of the problems associated with the strongly fluctuating energy supply from renewable sources. However, the cost of storage always has to be compared to the cost of installing backup capacity since this is an option to reduce the need for storage.

When storing electrical energy, there are always conversion processes involved, associated with energy losses. Cycle efficiency (i.e., the fraction of the electrical energy used to fill the storage that can be recovered in one cycle) is thus an important issue, and different storage technologies can have substantially different cycle efficiencies. Obviously, the cycle efficiency influences the operating cost of the storage facility, one important cost factor, while the other is in the capital expenditure for building the facility. Cost effectiveness of the overall system depends on both of these cost factors and on the frequency at which the storage system is used. For seasonal storage, investment costs should be low since otherwise the cost per kWh stored becomes prohibitively high. On the other hand, for short-term/high-frequency storage, one can afford higher capital expenditure since it is distributed over many operation cycles; however, cycle costs and losses should be small for such systems. The desirable features of storage for electrical energy are listed in Table 3.2.

Complex electricity grids consist of different levels. The transmission grid is used for long-distance, high-power electricity transport. It operates typically at voltage levels of around 400 kV. One level lower is the distribution grid, in Europe at 110 kV. Transformers convert this voltage to lower values for the intermediate voltage grid – operating at 10, 20, or 30 kV – which finally serves the local low-voltage grids to which the small-scale users of electricity are connected. Storage on the level of the transmission grid and on the sublevels requires different storage technologies, and thus they will be discussed separately.

Table 3.2: Requirements for storage of electrical energy.

Property	Comment
Cycle efficiency	High; benchmark small-scale capacitors close to 100%, grid scale pumped hydro 80%
Energy density	High; benchmark hydrogen 33.3 kWh kg^{-1}
Power density	High; often opposite to energy density
Cycle stability	High, problematic for various battery types
Self-discharge	Ideally zero, highly problematic for flywheels, problematic for some types of battery
Lifetime	As long as possible
Investment cost	Low but has to be considered together with cycle cost
Safety	High

3.4.1 Storage on the Transmission Grid Scale

3.4.1.1 (Pumped) Hydroelectric Power

A well-established method for grid-scale storage of electrical energy is hydroelectric power generated by hydroelectric dam power plants. Even if the upper reservoir is not filled by pumps from a lower reservoir, the water storage capacity of the reservoir can have an enormous balancing effect. In times of low electricity demand, the water supply to the turbines is shut off, while at high demand times, the turbines run at maximum power. The equilibrating effect of a dam power plant can be enhanced if pumps are added, by which at times of high electricity supply water is pumped from a lower reservoir into the upper reservoir, while the flow is inverted at times of high electricity demand. Pumped hydro is being used in many parts of the world as one of the most important means for stabilizing the grid, since reaction times on the order of minutes are possible, which allows fast reaction to changing demand and supply situations, and the pumps and turbines can be regulated over a wide range of power.

However, in densely populated areas or in regions with minimal height differences, the capacities for hydroelectric dam power are rather low, since mechanical energy generated from height differences does not have high storage density. Assuming a height difference of 300 m, which is not untypical, 1 m^3 of water has an energy content of

$$m \cdot g \cdot h = 1{,}000\,\mathrm{kg} \cdot 9.81\,\mathrm{m\,s^{-2}} \cdot 300\,\mathrm{m} = 2{,}943\,\mathrm{kJ}$$

(i.e., only about 0.8 kWh). In Germany, the total capacity of pumped hydro is approximately 40 GWh, which is very little compared to the electricity consumption of Germany (on the order of 550 TWh/a) or the total primary energy demand of Germany (approximately 3,700 TWh/a). Thus, if – hypothetically – one would shut off the primary energy supply to Germany, the amount of stored energy in water reservoirs would be sufficient to cover the energy demand for a little bit longer than 5 min. One additional problem has arisen over the last years; due to the increased fraction of solar electricity in the German energy grid and a reduced fraction of power plants operating more or less 24/7, the price spread between night-time and day-time electricity has almost vanished. In addition, legal boundary conditions have changed in 2008. This makes an investment in storage facilities, which should operate at reasonably high frequency (day–night), rather unattractive, so that many plans for the construction of new pumped hydro plants have been discontinued or at least put on hold. Other countries have much higher storage capacities in hydroelectric dams (although at present these are mostly not pumped, pumps can often be retrofitted). Austria, for instance, can store about 3 TWh in dams; Sweden, 33.8 TWh; and Norway, 81.7 TWh.

However, the balancing capacity of hydroelectric dams can only be exploited if there is a well-established grid with low energy losses over long distances. The only present-day method for low-loss/long-distance electricity transport is high-voltage direct current (HVDC) technology. To use hydroelectric storage to cope with more and more fluctuating energy demand and supply, a continental HVDC overlay grid structure seems indispensable. With such a grid structure available, hydroelectric dams, especially if combined with pumps, can provide an important storage element in future energy systems. The cycle efficiency of pumped hydro is on the order of 80%, and of all storage methods available nowadays, (pumped) hydroelectric energy is also the most cost-efficient method, both for short- and for long-term storage.

3.4.1.2 Compressed Air Energy Storage

Another grid-scale method for energy storage relies on air compressed in large subterranean caverns. The advantage over hydroelectric power is the availability of many possible sites and the low degree of interference with the landscape, since most of the system is underground.

In compressed air energy storage (CAES) storage systems, air is compressed into airtight underground caverns (several hundred thousand cubic meters volume) at times of high electrical energy supply. To recover the stored energy, the compressed air is used to charge a gas turbine, thus making the compressor of such turbines redundant. The cycle efficiency of the systems depends on the heat management: during compression, the air is heated. If this heat is lost, the efficiency of the system is below 50%. Full heat recovery (adiabatic operation) should lead to efficiencies around 70%, but currently, no such systems are installed. Plans for an adiabatic demonstration plant in Germany named ADELE were discontinued in 2015 due to uncertain market perspectives, with many similarities to the discontinued pumped storage units.

Two CAES plants are in operation at present, one in Huntorf in Germany (321 MW) and one in MacIntosh, Alabama, in the United States (110 MW). The plants have already been in operation for several decades, and this technology can be considered well proven. However, since capital expenditure for pumped hydro and CAES are estimated to be in the same range per kWh capacity, overall storage costs are higher for CAES due to the lower efficiencies, and the capacities are relatively low. The 300,000 m^3 storage caverns of the Huntorf plant can be used to drive the gas turbine for 2 h at full power, which corresponds to a storage capacity of approximately 500 MWh, considering that natural gas is also fed to drive the turbine. Nevertheless, CAES systems could find more widespread use if grid extension does not take place or proceeds at a slow pace and/or resistance against pumped hydro plants is high.

3.4.1.3 Storage in Chemical Compounds

Much higher energy densities than with pumped hydro or CAES can be achieved if the electrical energy is converted to a chemical compound. Such a compound could then either be reconverted to electrical energy or used in applications, where chemical compounds are indispensable – at least for the time being – such as for jet fuel. However, conversion to any chemical compound proceeds at low efficiency, and thus storage in the form of chemical compounds would probably only be used for long-term (seasonal) storage.

Five different compounds or groups of compounds are being discussed for grid-scale energy storage (i.e., hydrogen, methane, hydrocarbons, methanol, and ethanol). Ethanol has a special status in this list since it would probably not be produced from electrical energy, but directly by fermentation of biomass. Thus, it will not be discussed further here. The shortest route to a chemical compound from electricity is water electrolysis, which results in the formation of hydrogen. All other compounds are produced from hydrogen with a carbon source such as CO or CO_2. However, as long as fossil fuels are used to produce electricity, one should not hydrogenate CO_2 to produce methane, methanol, or liquid hydrocarbons via the Fischer–Tropsch process. Instead, the hydrogen should be directly reconverted to electrical energy and the power of a fossil-fuel power plant should be reduced because otherwise additional losses would occur.

For grid-scale storage, hydrogen appears to be the best compound. It has a number of advantages – that is, storage densities are very high, the production by electrolysis is an established technology, storage is possible in similar caverns as in CAES, it can be reconverted to electrical energy by fuel cells or turbines, and storage is essentially possible for indefinite periods of time (subterranean caverns are reported to loose less than 0.01% per year). However, hydrogen has one major disadvantage, the substantial energy losses during one cycle; electrolysis can be estimated to have an efficiency of 60%, transport and compression for storage may lead to another 10% losses (although some of this energy could possibly be recovered), and reconversion to electricity proceeds at most at 50% efficiency in fuel cells on the systems level (higher efficiencies could be possible in turbines, especially if cogeneration of heat is integrated). However, the overall cycle will most probably not be more efficient than 30%. The major advantages are the high storage density – compared to CAES, the capacity of the caverns could be increased by a factor of about 100 – and the possibility of long-term storage.

One should keep in mind that chemical compounds for energy storage are also accessible from other renewable, nonelectricity sources. Methane could be generated by fermentation from biomass, which is one of the most efficient ways for its energetic use; ethanol is accessible from sugar by fermentation, also a rather efficient pathway; and liquid hydrocarbons could be produced by various biomass-to-liquid processes. These pathways appear to be superior on the system scale, compared to hydrogenation of carbon compounds with hydrogen produced by electrolysis. Thus, hydrogen – in spite of all shortcomings – is probably the only feasible solution for directly storing electricity in the form of a chemical compound. However, conversion of hydrogen to other compounds could still be interesting for transportation purposes. If hydrogen is generated in sun-rich parts of the world, it needs to be transported to the demand centers. Transport of the hydrogen via pipelines or in liquefied forms are options but also chemical compounds which are easier to handle than hydrogen itself could be useful. The list of these compounds includes ammonia, liquid organic hydrogen carriers, and the chemicals discussed above, all associated with specific advantages and disadvantages. If hydrogen is converted to hydrocarbons by hydrogenation of CO_2 obtained via air capture, one may even consider not reconverting the hydrocarbons to hydrogen, but to continue using them as transportation fuels.

3.4.2 Storage on Distribution and Medium-Voltage Grid Scale

3.4.2.1 Battery Systems

While pumped hydro, CAES, and chemical compounds provide storage options on the level of the transmission grid, battery storage is more suitable for the distribution grid level. Such systems are often called battery energy storage system. Batteries would thus be deployed in connection with wind farms or solar farms to supply electricity at times when there is no wind or sunshine. Since typical farm sizes are in the range of several ten MW, and times on the order of 10 h shall be bridged, the size scale of such storage units is typically in the range of 100–1,000 MWh.

Battery storage has one important advantage: the cycle efficiency – depending on charge and discharge rates – can exceed 90%. However, necessary auxiliary components have to be taken into account as well, such as AC/DC converter systems, and this could reduce the cycle efficiency of the overall system. There is a multitude of different battery concepts, which will be discussed in more detail in later chapters of this book. For grid-scale storage, three main concepts are being pursued due to their durability and relatively low cost – although, at present, battery storage is much more expensive than the other methods of energy storage. For grid storage, the long-known lead acid battery is still an option. Weight – and thus low energy density, one of the disadvantages of the lead acid battery – is not such an important issue for stationary applications, and for large-scale installations, recycling of the lead and the acid is also not considered to be a major problem. Advantages of lead acid batteries are relatively low capital expenditure, ready availability of the materials, and a good safety standard. A disadvantage is the rather low cycle stability compared to other battery concepts. Sodium–sulfur (NaS) batteries are already in practical use in different application scenarios. The batteries are operated at about 300 °C since the sulfur and sodium need to be molten. This constitutes one of the biggest disadvantages: the energy consumption of the battery is high due to the requirement to keep the temperature constant. Thus, such batteries are only suitable for relatively high cycling frequencies on the order of hours to days. The major advantages are the high cycle and calendric life of such battery systems; however, the investment costs are relatively high compared to other batteries. NGK Insulators, Ltd. (Nagoya, Japan) produces such batteries, which are available with capacities on the order of 12 MWh per unit. Such units can be combined to bigger systems, with the biggest installation in 2020 reported to have a capacity of 300 MWh. Most current projects, however, seem to be realized using different types of lithium batteries, which have steep cost reduction curves and are thus gaining a competitive edge. At the end of 2020, Vistra power started operations of a 300 MW/1,200 MWh lithium-ion battery storage plant at Moss Landing Power Plant in California, with further expansions being planned. At an energy-to-power ratio of 4 (corresponding to the Moss Landing installation), a 2019 study from the Pacific Northwest National Laboratory found Li-ion technology the best option with respect to cost, performance, calendar and cycle life, and technological maturity.

Flow batteries are hybrids between a conventional battery and a fuel cell. The energy is provided by liquids at different electrochemical potentials, which can be passed along an electrode to deliver the power. The most well-known system is the vanadium redox-flow battery, in which in the charged state one electrolyte consists of V^{5+} solution, and the other one of V^{2+} solution. Discharge leads to reduction of V^{5+} to V^{4+} and oxidation of V^{2+} to V^{3+}; in charging, the process is reversed. Redox-flow batteries have a number of advantages: power (dependent on the electrode area) and capacity (dependent on electrolyte tank size) can be scaled independently, and there is basically no self-discharge. The cycle and calendric life of redox-flow systems are good, and investment costs are moderate, somewhere in between that of lead acid batteries and NaS batteries. However, the storage densities are rather low, so redox-flow installations have rather large footprints.

Battery technology is one of the most rapidly developing fields in energy storage, driven by the anticipated application in electromobility, so that in the future additional competitive technologies could become available for stationary applications as well.

3.4.2.2 Flywheel Storage

Intuitively, one would not expect flywheels to be suitable storage devices for larger amounts of electrical energy. However, advanced, very fast spinning types with carbon wheels can store substantial amounts of electrical energy. Commercial units with about 1 m diameter and 2 m height can store 25 kWh of electrical energy at a rotational speed of 16,000 rpm. Storage densities are substantially higher than for CAES and can reach almost the values of redox-flow battery systems. The most valuable property, however, are the high power ratings and very high ramp rates. They are often used in conjunction with battery systems for buffering power pulses. However, investment costs are very high, and self-discharge is rapid. Thus, such systems are only used for grid frequency regulation. A 20 MW plant with 200 independent flywheel modules is in commercial operation in the United States.

3.5 Energy Storage for Mobile Applications

The priorities in energy storage for mobile applications are quite different compared to those for stationary storage. For mobile applications, high energy density, both gravimetric and volumetric, is the most important criterion. This is the reason why liquid hydrocarbons have been so successful as the fuels for cars, ships, and planes. Diesel fuel has an energy density of approximately 12 kWh kg^{-1} and close to 10 kWh L^{-1}, a combination that is impossible to reach with any other method. Even with the efficiency of an internal combustion engine being only about 20%, all other technologies fall short of the system performance reached by conventional engines in combination with hydrocarbon fuels. For some transportation purposes, liquid hydrocarbons appear to be indispensable for the foreseeable future. It is difficult to imagine that jets the size of a Boeing 747 or an Airbus A380 could fly on any other fuel or on battery power. If fossil hydrocarbons become scarce, they should be reserved for such purposes, where no alternatives are available, and renewable resources, which can easily be converted to liquid hydrocarbons, should predominantly be used for this purpose. For regular cars, however, alternative options for onboard energy storage are within reach, even if there are some disadvantages associated with them; the different options shall briefly be discussed in the following.

3.5.1 Chemical Compounds

3.5.1.1 Synthetic Hydrocarbons

High-quality synthetic hydrocarbons can be synthesized via the Fischer–Tropsch process directly from synthesis gas (a mixture of CO, CO_2, and H_2) or indirectly from synthesis gas via methanol and a subsequent methanol-to-gasoline step. The synthesis gas can be produced from almost any carbon-containing material, such as natural gas, oil, coal, or biomass. However, depending on the source, the CO_2 footprint differs, with coal being the worst option and biomass being the best one. In spite of the high thermodynamic penalty due to the low concentration, in the future, even CO_2 capture from air with subsequent reverse water gas shift reaction could be used for the generation of syngas. Alternatively, hydrocarbons can also be produced from oil plants as hydrogenated vegetable oil; transesterification of vegetable oil with methanol leads to fatty acid methyl ester, which can be used as a substitute for diesel fuel. The energetic efficiency of liquid hydrocarbon production is moderate, but such fuels have the advantage that the current infrastructure can be retained. Moreover, they have the best combination of volumetric and gravimetric energy density. Liquid hydrocarbons will remain the most important storage compound for many years to come.

3.5.1.2 Ethanol

Ethanol, like liquid hydrocarbons, is used as fuel for internal combustion engines. Gasoline-driven motors, if suitably adapted as so-called Flexi-Fuel engines, can run on gasoline–ethanol mixtures in any ratio, and in some countries, such as in Brazil, most cars are equipped with Flexi-Fuel engines. Ethanol has a storage density of 7.5 kWh kg^{-1} and 5.8 kWh L^{-1} (i.e., substantially lower than diesel fuel or gasoline). Thus, larger tanks and more fuel are needed for the same driving range. At first sight, therefore, ethanol does not seem to have advantages. However, under optimum conditions, ethanol can be produced from sugarcane with only about one-fourth of the CO_2 footprint of fossil fuels at competitive costs, and this advantage makes ethanol a viable alternative to liquid hydrocarbons under specific conditions. However, a detailed life-cycle analysis is required to assess the advantage. For ethanol from wheat or corn grain, it has been shown that under certain conditions the CO_2 footprint is even bigger than that of fossil fuels.

3.5.1.3 Methanol

The energy density of methanol is even lower than that of ethanol (5.6 kWh kg^{-1}, 4.4 kWh L^{-1}). Like ethanol, it can be used in adapted internal combustion engines. In addition, conversion to electricity in direct methanol fuel cells is possible. However, such fuel cells have low power densities and low efficiencies, and thus it is not envisaged that methanol in combination with direct methanol fuel cells will find an application in cars.

In addition, there is no useful direct access to methanol from biomass, and the synthesis gas route must be used (i.e., gasification of biomass with subsequent methanol synthesis). Compared to the Fischer–Tropsch synthesis, the efficiency of methanol synthesis is higher, which is certainly an advantage. However, this is overridden by the necessity of additional infrastructure, the low energy density, and the toxicity of methanol. Balancing all factors against each other, methanol does not seem to be a promising energy storage compound for mobile applications or for energy storage on a grid scale.

3.5.1.4 Methane

Methane is already being used as fuel in some cars. It can be combusted in internal combustion engines after adaptation, but since it is a gas, an additional tank is required. While the gravimetric storage density of methane is very high (13.9 kWh kg^{-1}), the volumetric density is low since it is a gas at ambient temperatures. Even compressed methane does not reach the volumetric density of liquid hydrocarbons (0.0099 kWh L^{-1} at 1 bar). Methane is accessible from synthesis gas by the Sabatier process. However, it would probably be better to use anaerobic fermentation of biomass to produce biogas, which could then be purified to methane. There is well-established methane infrastructure available, with large storage facilities (in Germany, 20 billion m^3, corresponding to one-fifth of the annual consumption) and a pipeline network. However, on the systems level, it does not appear to be advantageous to use methane to power cars, considering that there are other options with probably better overall performance.

3.5.1.5 Hydrogen

Hydrogen in connection with proton exchange membrane fuel cells presently appears to be the only option that could provide similar driving ranges as cars powered by the internal combustion engine. Hydrogen has the highest gravimetric storage capacity (33.3 kWh kg^{-1}) of all chemical compounds, but the volumetric capacity is very low due to hydrogen being a gas (0.003 kWh L^{-1} at 1 bar). Storage of hydrogen onboard cars is one of the most important problems. High-pressure storage (700 bar) is currently state of the art, and despite intensive research efforts, no other option is forthcoming. Storage in liquefied form, adsorptive storage at liquid nitrogen temperature, storage in the form of hydrides, and storage in liquid chemical compounds (from which hydrogen is released by reforming) have been explored, but no viable solution superior to high-pressure storage is in sight. One advantage of hydrogen as a fuel is the fact that no CO_2 is emitted at the point of use. In addition, the efficiency of a fuel cell is appreciably higher (around 50% on the systems level) than that of an internal combustion engine, which leads to energy savings. However, as with hydrogen storage on a grid scale, one has to solve the question of hydrogen generation, which is associated with high energy losses. Whether hydrogen eventually will become an important fuel for mobile applications depends on advances in all key components of this technology. Possibly, hybrid vehicles using batteries for short driving ranges and a hydrogen-powered fuel cell for long-distance traffic will be the successful solution for transportation beyond the internal combustion engine.

3.5.2 Traction Batteries

Electric cars powered by batteries are an interesting option to provide mobility. Important advantages are the high efficiency and the fact that at the point of use this technology is CO_2 neutral. Depending on charge and discharge currents, efficiencies on the order of 80% can be achieved. The energy consumption of electric vehicles is an estimated 0.15 kWh km^{-1}, while internal combustion engines need about four times as much energy. However, the efficiency of electricity generation from primary energy has to be taken into account as well, and thus the difference is often not quite as high as it appears.

The biggest disadvantage of batteries for mobile applications is the low energy density of electrochemical energy storage. The most advanced battery types for mobile applications are lithium-ion batteries, which are used in essentially all laptops and cell phones. Maximum energy densities for this type of battery currently are around 250 Wh kg^{-1} and close to 700 Wh L^{-1}. The prospects for a substantial increase in energy density are not too bright for this type of battery since a high fraction of the weight is contributed by the positive electrode material. Although different types of electrode materials are being used, they all rely on transition metal oxides or phosphates or related compounds, in which the Li ions are intercalated. Thus, material with a formula mass of around 100 g mol^{-1} is used to store the Li ion with an atomic mass of about 7 g mol^{-1}, but to a first approximation only the latter carries the energy. Unless the "dead weight" of the matrix can be substantially reduced, the energy density of traction batteries will not improve significantly. In order to make progress, completely new battery concepts are needed.

One interesting aspect of large-scale deployment of traction batteries shall be mentioned: they could also be used as a distributed storage system for grid-scale electricity storage. Cars are parked most of the time, and if connected to the grid during this time, they could be used to balance fluctuating power. Whether this would be done unidirectionally (i.e., charging would only occur at high electrical energy supply) or bidirectionally (i.e., batteries would also be discharged during times of high electricity demand) depends on the cycle stability of traction batteries. If this is low, bidirectional use would be too expensive because it would lead to short battery life. Assuming a size of 30 kWh for traction batteries (between that of hybrid cars like the Opel Ampera and that of an all-electric car like the Tesla) and a fleet of 42 million cars in Germany, there would be an overall storage capacity of 1.26 TWh. This is very substantial compared to the 40 GWh storage capacity of pumped hydro in Germany, even if this capacity would not be fully utilized since the main purpose of cars is, after all, to provide mobility.

3.6 Systems Considerations

Future energy systems will be more dependent on energy storage than our current system. However, different elements cannot be considered in isolation; an overall systems view is necessary to develop an optimized solution. Furthermore, it is not only technical optimization that defines the choice of an energy system. Societal preferences, economic and political factors, and even chance can influence the development of an infrastructure. However, there are some basic points that should be taken into account on the systems level.

Future energy systems will depend much more on electricity than our current system. Conversions from electrical energy to other forms of energy and back are always associated with losses, and thus such conversion should be avoided – unless abundance of available energy is so high that losses do not matter. Grid extension helps to this end, and it also helps to couple the cheapest and most efficient form of transmission grid-scale storage, pumped hydro, into a continental electricity system. Mobility, especially passenger transport, should be tightly integrated into the electricity system, and battery-powered electromobility appears to be the most efficient option – although we will probably not be able to achieve the driving ranges we are used to nowadays. Thus, some behavioral changes, such as increased use of train transport, may be necessary. If excess electricity should be stored for long time periods, centralized hydrogen production and storage in caverns is probably the best solution. This hydrogen could be used to power fuel cell–driven cars or hybrid cars for long-distance car travel, or it could be directly reconverted to electricity, albeit at low cycle efficiency. For seasonal storage, methane also is an attractive option. However, methane should be produced from biomass via fermentation rather than from electrically generated hydrogen – hydrogen should be stored directly, not further converted. Biomass is also a suitable starting point for the production of hydrocarbons, which for quite some time will be indispensable as energy carriers for some kinds of transportation, such as aviation and heavy-duty trucks.

Finally, the heat market will be largely decoupled from the rest of the energy system. Heating energy demand will decrease due to better insulation standards, and part of the heating and cooling energy requirements will be met by heat storage systems. The rest of the heating and cooling power will be generated by electricity, and in combination with heat and cold storage, this will also have an equilibrating effect on the electricity system.

While progress in science and technology can always lead to unforeseen step changes, the planning of a new energy infrastructure has to rely on technologies at hand or at least in sight. From what we have and see, it seems clear that energy storage will be an important element in any future energy system.

Robert Schlögl

4 Chemical Energy Storage and Conversion: A Perspective

Abstract: About a decade after the energy transformation (Energiewende) started in Germany, it is now clear that the vision about a national all-electric energy system based upon solar and bio-based energy is not realistic. Such a hypothetical system cannot replace the function of a fossil-based system augmented with local renewable energy resources (RES) stemming from solar, wind, biomass, and hydroelectricity. Over the last few years, the vision grew globally that a combination of local RES and a global trade system for RES via hydrogen and its derivatives will be the functional solution for defossilizing the energy supply systems. The core element for this vision is the interconversion between free electrons and electrons bound in chemical bonds. This is impossible without catalysis becoming the central family of technologies for creating renewable molecular energy carriers. They combine high storage capacities with compatibility to the fossil liquid and gaseous energy carriers for which we have global technologies already in place. The interconversion of free electrons into bound electrons allows the transport of RES and solves the large-scale storage issue introduced by the volatility of local RES alone. Energy storage and chemical energy conversion become two fundamental capabilities of future energy systems resting both on the command and mega-scale realization of catalytic processes.

4.1 Chemical Energy Conversion

The energy system is a fundamental characteristic of a human society. All its individual and collective actions require the availability of free energy in various forms (mechanical, chemical, electrical). The sum of all man-made inorganic energy interconversion processes reaches at the end of the first decade in the twenty-first century the same dimension [1] as the biological processes needed to generate the food for mankind. Both process groups amount to about 15% of all biological solar energy conversion on the planet Earth. It is obvious that we are dealing with enormous dimensions that are difficult to comprehend, as they are so much larger than our individual frame of dimensions.

The role of chemistry is manifold in the field of energy. Traditional challenges in catalysis science deal with the increase in efficiency [2] or in the use of CO_2 as

Robert Schlögl, Fritz Haber Institute of the Fritz Haber Institute of the Max Planck Society, Faradayweg 4-6, 14195 Berlin, Germany Max Planck Institute for Chemical Energy Conversion, Stiftstr. 34-36, 45470 Mülheim an der Ruhr, Germany, e-mail: robert.schloegl@cec.mpg.de

https://doi.org/10.1515/9783110608458-004

feedstock in industry [3, 4]. These activities enable the future availability of materials from the chemical industry eventually without a fossil feedstock. It is needless to say that confining the use of fossil feedstock to chemical conversion would give us centuries of access to chemical products without having to invest enormous amounts of renewable energy resources (RES) for CO_2 reduction. CO_2 as the carbon source at large for chemical production will be unrealistic for a long time until we have sufficient access to RES. In Germany the industry producing basic chemicals emits 23 Mt CO_2 (in 2017). This can be used as a feedstock after hydrogenation with green hydrogen. For its conversion into synthetic naphtha 185 TWh of EE will be necessary. This equals to 35% of the total German electricity production or the total RES generation in Germany in 2018. It remains a critical task of chemical catalysis science to minimize the CO_2 emission from chemical industry to which the transformation of process heat generation from gas to hydrogen or electricity can make a substantial difference.

All these efforts deal with about 5% of the energy system. It is thus obvious that the highest priority lies in defossilizing the large sections of the energy system. Here chemical energy conversion has two main challenges. The first one is to provide green fuels for applications that are hard to electrify and as drop-in solution to continue using the vehicles, ships, and planes existing today. E-mobility, the present hyped solution for mobility, is a systemically desirable solution due to the unprecedented efficiency of an all-electric powertrain. The conversion of most forms of mobility on the whole globe will, however, be such a great challenge that it is not realistic to expect or to desire a complete exit from liquid fuels. In a complete systemically view they will have their role to play at least in high-energy applications like planes, ships, and certain heavy-duty vehicles and in remote areas of the world. E-fuels or syn fuels represent, thus, a large-scale challenge for chemical synthesis having to replace a good fraction of today´s petrochemistry at least in their upstream activities. The challenge is well recognized [5] including a critical view [6] on efficiency and prerequisites for true CO_2 reduction for synfuel synthesis.

The other challenge is to provide hydrogen and allow it to be stored and transported across the globe in the dimension of the global energy system. Chemical energy conversion comprises, thus, few chemical processes with "simple" molecular transformations, such as

$$2\,H_2O = O_2 + 2\,H_2$$

$$N_2 + 3\,H_2 \leftrightarrow 2\,NH_3$$

$$CO_2 + H_2 \leftrightarrow CO + H_2O$$

$$CO_2 + 3\,H_2 = CH_3OH + H_2O$$

$$CO + 3\,H_2 = CH_4 + H_2O$$

$$n\,CO + (2n+1)\ H_2 = C_nH_{2n+1} + n\,H_2O$$

$$C_2H_2 + H_2 = C_2H_4$$

Chemical science and technology can perform all these reactions today and this even in the multi-Mt a^{-1} scale. The challenge here is the increase in dimension that will be needed when we replace oil and gas by such synthetic energy carriers. Here chemical energy conversion science is required that delivers a firm and comprehensive understanding of the underlying physicochemical processes. Then a collaboration of theoretical, synthetic, and engineering sciences is needed to develop optimal processes in multiple realizations that are tailored to specific application scenarios differing in size (central-decentral) and in different regions of the world. In this challenge the distinction between "applied science" and fundamental knowledge disappears and we need to respect the statement of Max Planck:

Knowledge must precede application

In addition, chemical energy conversion or storage needs an aspect of material creation in order to generate novel concepts and ideas for performing its tasks. This refers to catalytic materials as well as to other energy carriers, be they solid or fluid. This part of chemical energy science is well underway with the large number of suggestions from nanoscience [7, 8], thermal engineering, and general material science [9–11]. It is not the purpose of this work to review or categorize these widespread applications. It should be stated, however, that it would be highly desirable if such work would consider a path forward to scaling their results. This can either be in scaling the production of functional materials [12–14] or in extracting functional concepts from such work and bringing it to bear as innovation ideas in areas where already scaled materials and processes are applied.

4.2 Quantitative Aspects of Energy Systems

The concept that oil and gas may disappear from the world energy supply is no common vision to all involved. The world energy council predicts that energy efficiency will increase as well as penetration of RES into the world energy system. The fractions are, however, way less than discussed in Europe or Germany. In Figure 4.1, the world energy primary consumption is shown for 2015 (mind the enormous dimensions) and in three different scenarios for 2050.

In the MJ and US scenarios an increase in electrification of the energy system is the main driver for CO_2 reduction. In all scenarios the contribution of chemical energy conversion is considered but with relatively small contributions in comparison to fossil resources. These scenarios can be considered as rather conservative extrapolating a realistic view on global interests. The German or European view on the

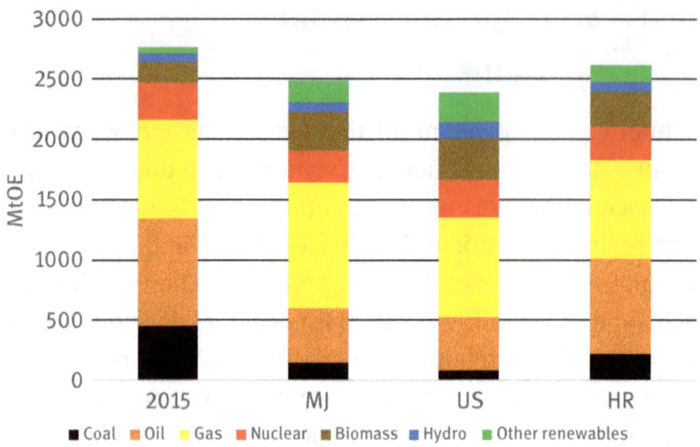

Figure 4.1: The primary energy consumption of the world in 2015 and for three scenarios in 2050: MJ for "Modern Jazz" representing RES penetration and energy saving at the present pace, US for "Unfinished Symphony" representing an aggressive approach to implement RES and HR for "Hard Rock" representing a conservative business as usual approach. "Other renewables" stands for wind, solar, and geothermal RES.

penetration of RES deviates from the Asian view. The US position was retarding in the past but begins to accelerate enormously since the election 2020 and may drive the phase out of fossil fuels faster than anticipated here. A strong argument for acceleration also comes from the cost of solar/wind energy converters. The following few numbers from the Fraunhofer PV report 2020 may indicate this: in 2019 142 GWp were newly installed. The energy payback time fell to 1 year. The module price fell to 0.2 Wp and the PV electricity price in Germany (!) fell to about 50€ per MWh.

The rather conservative proposed evolution (Figure 4.1) of the global energy system is in sharp contradiction to the concepts from climate research. A picture is drawn in [15], in which the penetration of the renewable grows exponentially with time as illustrated in Figure 4.2.

From the point of climate protection indicated in Figure 4.2, we need to stop growing the CO_2 emission immediately if we want to limit the climate change consequences to a "manageable extent." The German Energiewende has a milestone in 2021 with the end of nuclear power plants. Many technologies for energy conversion from primary electricity are in their infancy. Artificial photosynthesis [17] is yet a projection. No pressure for change comes from coal and gas in the next decades as large resources and still large reserves exist. A critical driver for a change in energy systems may come from oil. At the time of preparation of this text we are about at "peak oil" [18], at the maximum of oil production that will from now on decline as no more new reserves are being found and the easily accessible reservoirs are becoming depleted. This trend is seen in the data of Figure 4.3.

The decline in production is smoother than expected as the contribution of difficult-to-produce reservoirs is increasing with the inevitably increasing price. Even

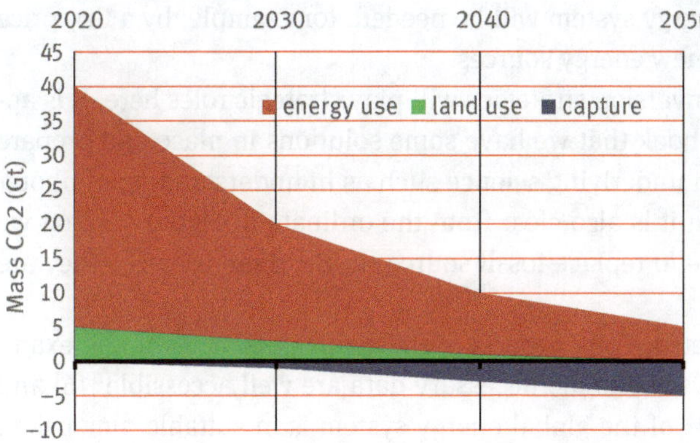

Figure 4.2: Required transformation for the world energy system expressed by the emission reduction of C = 2 from energy conversion and from land use. These changes are required to me. In the following we discuss a few aspects of these dimensions using the example of the German energy system as the necessary data are well accessible [16] and its absolute size of about 1% of the global energy system is in suitable dimension. These changes are required to meet the Paris agreement according to [15].

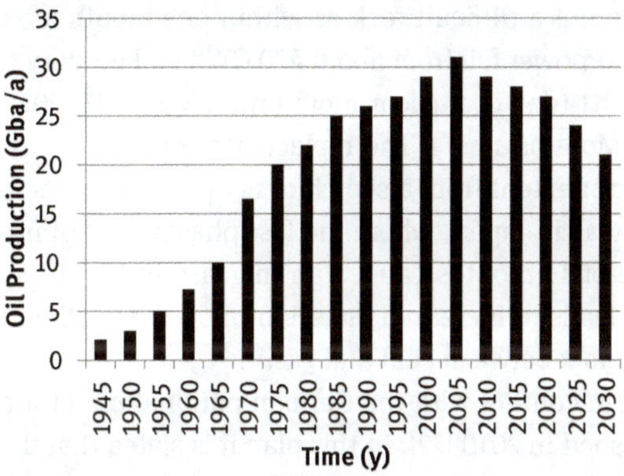

Figure 4.3: Peak oil. The graphics shows on a crude timescale the temporal evolution of oil production. The low resolution was chosen to identify the main trend irrespective of multiple events in shorter timescales such as the oil crisis. 1 Gba is 1.56×10^{11} l crude oil. Source: data from www.peakoil.com.

with large uncertainties about the future [18], it is clear that the enormous demand of the past 50 years led to a rapid exhaustion of the easily available oil resources. It may thus be expected that for oil-critical applications, such as mobility or chemical industry, the request for renewable sources of solar fuels may come faster than expected from the modification of the electricity system. Without knowing the fastest driver for changes in the energy system it is likely that in the coming decade severe pressures on the existing energy system will be needed, for example, by a CO_2 price to increase the pressure for new energy sources.

The chemical energy conversion strategies will play strategic roles here. It is apparent from the text of this book that we have some solutions in place and prepare for several breakthroughs in underlying science such as in understanding of photosynthesis or in catalysis. But it is also clear from the ordinate in Figure 4.3 that we are far apart from being able to replace fossil sources at the dimensions as they are consumed at present.

In the following we discuss a few aspects of these dimensions using the example of the German energy system as the necessary data are well accessible [16] and its absolute size of about 1% of the global energy system is in suitable dimension. We first describe the situation in the decade 2010 until 2020 and then turn the Wende 2.0 [19] presenting a more systemic approach with the National German Hydrogen Strategy [20].

The German "Energiewende" in the aftermath of the Fukushima events is a highly popularized impulse [21] on the future of an energy supply system. It comprises the termination of nuclear power generation within 10 years leading to the necessity to replace about 25% of the German power consumption of about 600 TWh by alternative sources. This by itself was not a difficult task as within one month after the decision the contribution of nuclear power fell from about 420 GWh on February 1, 2011 to below 100 GWh on May 23, 2011 (day of nuclear moratorium March 15, 2011) (source: BDEW, "Auswirkungen des Moratoriums"). The replacement came from extended use of fossil fuels and to a minor extent from a switch of net power exports to power imports much in the same way as in Japan, where the fast phasing out of nuclear power within one service cycle of 13 months caused an increase in fossil fuel consumption. In Germany this trend leads to increased emission of 60–100 million t CO_2 per annum depending on the fuel mix between coal and gas.

The Energiewende is thus partly in conflict with the German national plan for a change of the entire energy system issued in 2010 [22]. In this plan it is stated that the total German CO_2 emission should fall between 2010 and 2035 from about 1,000 million ton to 265 million tons. Electrical power generation, the subject of the Energiewende, has a 50% share in the total emission. This should be achieved by a massive implementation of renewable energy and a smaller contribution form increased energy efficiency through various measures. A key additional role in emissions plays house heating that should become unnecessary by 2050 according to the planning following improved building insulation. In terms of relative contributions to primary energy carriers, the situation was expect to develop as shown in Figure 4.4.

It is obvious from Figure 4.4 that achieving the ambitious goal of the National Energy Concept will require a deep-reaching change of the whole energy system and not only the replacement of nuclear energy in a given system. This qualifies the Energiewende indeed as an initial step rather than the great change in the energy system. The values for the year 2022 indicate that the political concept assumes a

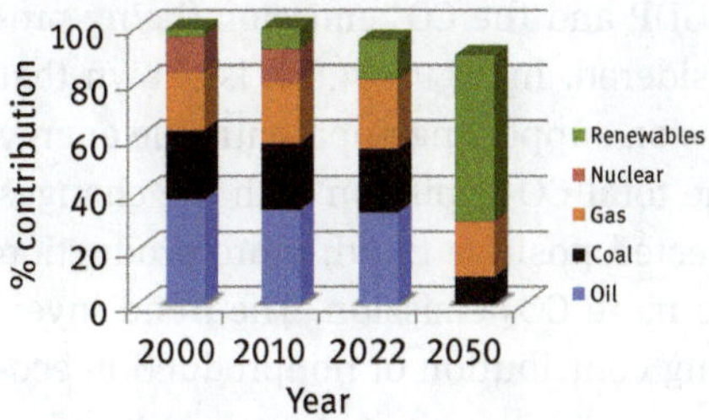

Figure 4.4: Relative total primary energy input into the German energy system. The values in 2022 relate to the completion of the "Energiewende," the data for 2050 for the completion of the National Energy Concept. Source: BMWI [16].

combination of three factors for replacing the nuclear energy: increased energy efficiency in parallel with more renewable energy generation and some additional power generation through gas combustion. In the following 35 years the contribution of renewable energy is expected to rapidly increase and to replace all oil and coal.

In order to estimate the dimension of changes necessary it is instructive to study the distribution of absolute primary energy consumption in Germany as indicated in Figure 4.5.

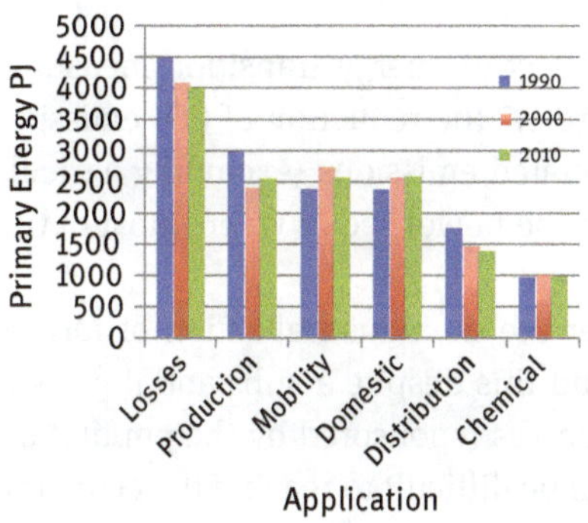

Figure 4.5: Distribution of primary energy streams into key application areas for Germany. Source: BMWI database 2011 [16].

The energy conversion losses are by far the largest sink of primary energy followed by three almost equal areas of industrial production, mobility, and transport and domestic energy consumption. Distribution of goods and services and the use of fossil energy carriers as chemical feedstock are slightly minor applications. The temporal evolution shows some changes following the industrial change with the unification in 1989 and a change in lifestyle following this event. We see very little energy saving or effects of energy efficiency increase that may well be hidden by rebound effects of increased mobility and more spacious housing. The apparent contradiction between assumed increased energy efficiency also "felt" by increasing changes towards a more energy-efficient lifestyle and the data in Figure 4.3 becomes clearer when the correlation between the German GDP and the CO_2 emission (being proportional to the fossil energy input) is considered. In Figure 4.6 it is shown that the total GDP inversely scales with CO_2 emission supporting apparently the energy saving trend. If, however, one correlates the total CO_2 emission with the changes in the production sector one finds the expected positive trend: more production means more primary energy use and hence more CO_2 emission. The trend inversion in Figure 4.6 is a clear sign of the strong contribution of nonproduction economic sectors to the GDP.

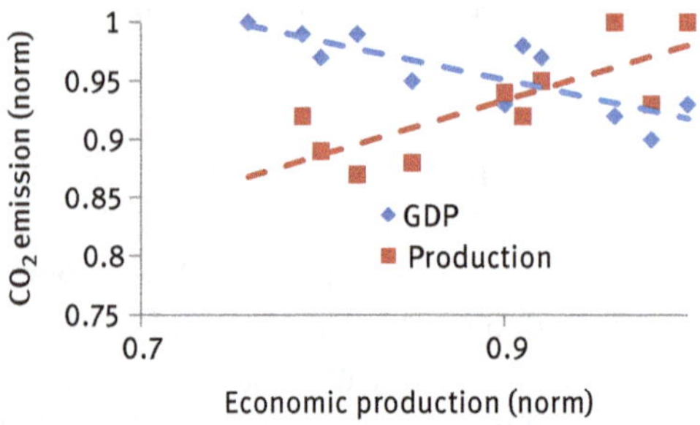

Figure 4.6: Relation between CO_2 emission in Germany and its economic activity. The normalized values are given for the whole GDP and for the production sector only. Source: BMWI database 2011 [16].

At this point it is instructive to check on the effect of energy transition measures with respect to climate change control. In Figure 4.7 the reduction of CO_2 emission in Germany over time is shown for all energy-related emissions (excluding agriculture and land use change contributions) and for the power sector where most of the national RES were penetrating.

Inspection of Figure 4.7 indicates that after the de-industrialization of former east Germany a steady reduction takes place and this despite a substantial growth of the national GDP. Despite this remarkable success (sponsored by the population with about 500 M€ per week) climate targets will be difficult to reach if the economy is intended to grow and no additional measures are introduced in the energy transformation. For the present context the dimensions (ordinate in Figure 4.7) need to be taken into account. The total chemical industry being a reference for chemical processes in energy conversion is a small factor (ca. 5%) on that scale. Suggestions for the use of CO_2 in any chemical production need to be considered with great care [23] when their intention is a contribution to climate change protection.

In the year 2021, the view about the German energy system is changing drastically without a clear tendency. It is now understood that the energy transition must be seen as coupled effort without shifting energy requests from one sector into another one. Most popular still is the request for massive electrification through installation of solar/wind converters in Germany. A typical value is the doubling of electricity generation from ca. 600 TWh (2020) to over 1,000 TWh (2050) with a massive increase

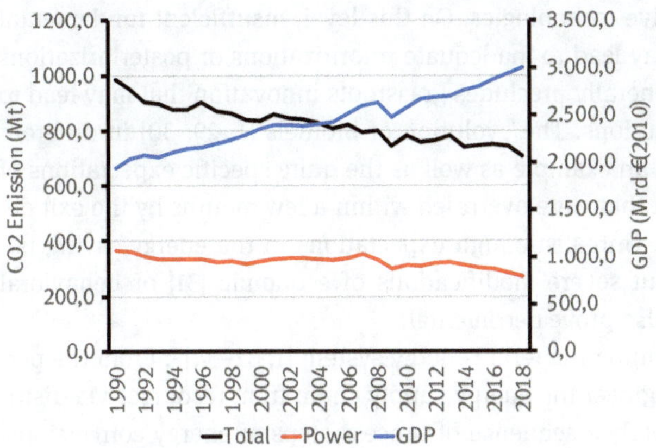

Figure 4.7: Energy-related CO_2 emissions from Germany. A model-free fit indicates that an average of about 10 Mt a^{-1} is saved with all energy transition measures currently in place. In red is shown the contribution from the power sector. In blue is shown the GDP related to a value of the Euro in 2010 (source: BMWI [16]).

in the current decade. In addition, hydrogen in free and derivatized forms comes into play with a fraction of 200 TWh to 800 TWh in 2050 with a strong growth after 2030. These number require restructuring of the mobility fleet into electrical power trains and import of materials like steel, cement, ammonia, and glass from global regions with high energy densities from solar sources. The uncertainty is very large in these scenarios as indicated by the example that the German chemical industry would need for a complete defossilization (in process heat and feedstock) ca. 700 TWh green electricity [24] in 2050.

Whatever the reality may look like in the decade from 2020, from perspective it is clear that massive amounts of molecular energy carriers will be needed. This is impossible without a highly efficient chemical energy conversion technology for which much science still has to be discovered.

4.3 Chemical Energy Conversion in a Wider Context

These coarse considerations indicate that energy systems [25–27] are interconnected in complex manners with multiple nonscientific aspects of utilization, generation, and distribution. Simple extrapolations of “trends” into evolutionary targets for individual technologies contributing to an energy system are dangerous and misleading. It is the role of politics to define global evolutionary targets for an energy system such as to be compatible with climate protection targets [28] or to refrain from nuclear energy options. Great care should be exerted when these macroscopic targets are underlined with microscopic targets delineating development avenues for individual and innovative technologies. On this level, insufficient fundamental or technical information may lead to inadequate prioritizations or posteriorizations of technical options and generally precludes grassroots innovation that may lead to completely unexpected solutions. The evolution of biofuels [1, 29, 30] in different parts of the world are such an example as well as the quite specific expectations of the German energy concept that were overruled within a few months by the exit decision from nuclear power. Unrealistic high expectations in the energy-saving potentials of a society without severe modifications of economic [31] or behavioral boundary conditions may also prove detrimental.

At this point we try defining the term "energy system [1, 32–34]". From the perspective of a scientist or engineer the energy supply chain from resources via distribution to application is merely a sequence of process steps of energy conversions. The free energy stored in an energy carrier is converted to other energy carriers or to mechanical/electrical energy. The challenge hereby is to use a minimum number of steps to minimize the losses of free energy to the universal heat sink and to likewise minimize such losses within every step. We accept that it is impossible to convert energy without losses and that the usefulness of thermal energy is related to its difference in temperature with respect to the environment of the process.

Within this realm a very large number of processes were suggested with few of them found their way to relevant technology. Nontechnological influences served as selection pressures and have thus to be considered as elements of the energy system in a more complete sense. The nonscientific elements of an energy system are depicted here with the relevant scientific/technical elements in Figure 4.8.

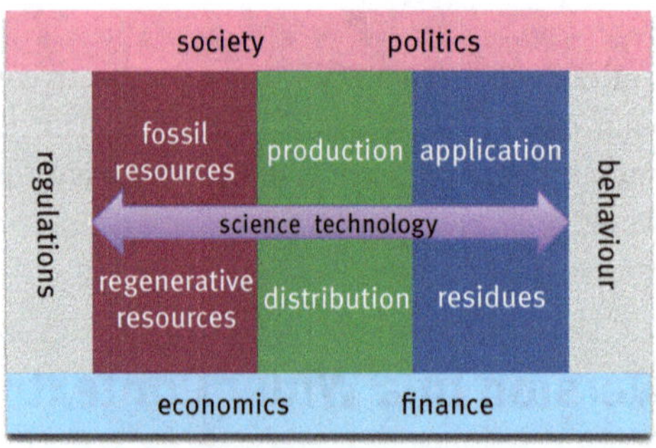

Figure 4.8: Scientific and nonscientific elements of the energy system. Multiple interfaces and control loops exist between the key elements. Science and technology are the enabling elements for all energy processes and additionally serve the important purpose of informing decision makers about necessary regulatory and behavioral boundary conditions.

It becomes evident that any meaningful treatment of science and technology for the energy system must be in contact with the nonscientific aspects. Science and technology further must transport their insights to those responsible for decision-making. This is clearly true for the aspects of implementing ready technologies. It is required in addition already at the earliest stages of fundamental scientific knowledge in order to account for the request of modern societies to be "knowledge societies" in which all members participate in detailed decisions on the functioning of the society. As energy supply in all its form is the cross-sectional enabling basis of all individual and collective activities of men it may be concluded that the evolution of the energy system is one of the central activities of organizing a society. The present book takes this into account by introducing selected aspects of nontechnical aspects of energy conversion in Chapter 1.

Keeping this broad view it is the intention of the present book to cover a tiny fraction of the scientific ground necessary for converting the present fossil-based energy system into one where regenerative primary energy is the main source of free energy. The following text thus concentrates not on recommendations of how to perform the Energiewende. It rather focuses on the contribution that chemistry can make to the evolution of the energy system. This contribution is twofold acting either on energy efficiency strategies of existing processes or providing solutions for the energy storage challenge being an essential ingredient into regenerative energy systems.

4.4 The Power of Nontechnical Influences on Energy Transitions

The political pressure for accelerating the climate protection efforts according to the Paris agreement has now led to the formulation of the "Green Deal" within the EU. This concept being put into binding national legislation supposes "climate neutrality" in 2050 as target. In Figure 4.9, the cumulative CO_2 emissions from EU 28 countries (EU Commission, DG Energy, Unit A4 ENERGY STATISTICS Energy datasheets: EU28 countries) are shown together with a linear extrapolation predicting a reduction of about 0.8% per year, the aspiration of a former 80% reduction target and the new target of climate neutrality in 2050. It is very evident that after the still ongoing debate about targets now a discussion about a plan for realization must follow.

The target of climate neutrality acknowledges that there will be greenhouse gas emissions from human activities that cannot be stopped (such as from agriculture and unavoidable technical sources). To compensate for these emissions' equivalent countermeasures of "sub-zero emission" technologies must be taken and verified. In this context chemistry comes into play by transforming CO_2 from direct air capture [35, 36]. Care should be taken, however, about the dimensions. A 10% contribution of CO_2 capture and storage or utilization would require about 500 Mt a^{-1} consumption. The logic of these requests is that the global ecosystems cannot tolerate any anthropogenic emissions. This is stated without proof and ignoring the fact that the ongoing climate change leads to emissions from increased wildfires and thawing of permafrost and other processes that are outside of human control. As important it would be to

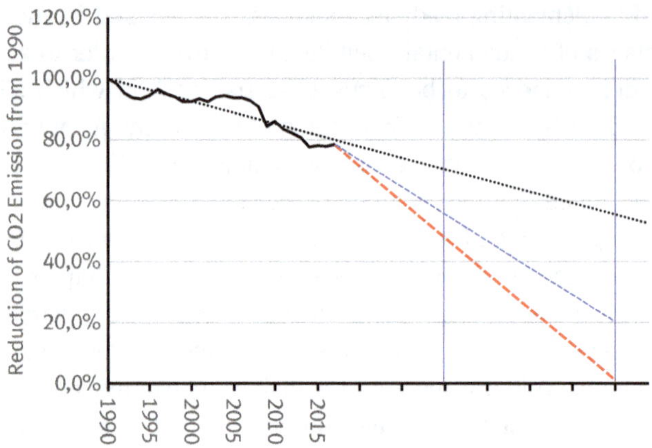

Figure 4.9: Relative CO_2 emission in Europe from fossil combustion. The reference value for 1990 is 5,772 Mt. A linear extrapolation model (black) is compared to the targets for an 80% reduction in 2050 (blue) and for climate neutrality (red).

maximize the speed of defossilizing the energy system as questionable is the request of climate neutrality. The fact that the atmospheric CO_2 concentration rises with a rate of above 2.3 ppm a^{-1} and involves more than 30 Gt a^{-1} CO_2 indicates again that chemical measures can only contribute to a small extent to the enormous challenge if they focus on CO_2 capture and use. The focus should rather be placed upon enabling the storage and transportation of RES. This may involve the creation of circular economies [37, 38] of energy carriers including carbon-containing species such as methanol or methane and hence the generation, capture, and reuse of CO_2 with this circle.

The societal pressure against fossil fuels leads to hectic activities without recognizing the dimensions and resulting timescales of many stakeholders, to unrealistic projections about measures and cost and all in all to a rather chaotic action in the energy transition. Rather, a well-planned and if possible coordinated action between countries creating first a global RES market and then quickly move out from fossil sources in large-scale applications would help the climate protection better. For chemistry, it is imperative to build the foundations of such a global RES market and form a broad and solid foundation of the required upcoming technologies.

4.5 The Role of Catalytic Chemistry in the Energy Challenge

This capacity of chemistry arises from the functions of chemistry being the science of inter-converting the structures of molecules and materials. All these inter-conversions are governed by the laws of thermodynamics requesting that all such processes are associated with irreversible losses of free (valuable) energy to the heat bath of the universe (not useable). Minimization of these losses through optimization of the pathway of transformation and through the optimization of functional properties of materials is one of the grand challenges of chemistry. A platform technology and science within chemistry to achieve this is the understanding and mastering of catalysis. Both our technological world of fuels and materials and nature with all its complex processes of life use this universal concept for controlling energy profiles and pathways of chemical transformations. In the following text we concentrate on catalytic processes and their science and leave the equally important aspects of material science aside. The reason for this limitation is the enormous diversity of application areas for energy-related functional and structural materials. It is felt that the aspects of catalysis in energy science are not fully recognized. In particular, the critical need for chemical energy conversion cannot be overestimated as the only measure to counteract the volatility of renewable primary electricity on a global and annual scale.

Catalysis [39] is the science of controlling the course of chemical reactions through managing the structure of transitions and intermediates on the pathway from educt to product systems. It uses functional materials called catalysts that bind in highly specific forms to reacting species and so control their reacting structures. These catalysts can be of the same phase as the reactants (homogeneous) or of a different phase (heterogeneous). Biological catalysis is a special form of homogeneous catalysis operating in the very restricted parameter space of life and using highly complex catalysts called enzymes.

All forms of catalysis contribute to the operation of energy systems. All fossil fuels were generated by biological catalysis as biomass. After its death this biomass was converted by a sequence of biological (homogeneous) and geochemical (heterogeneous) reaction sequences into the present energy carriers. It is noted here that the formation of oil seemed to have occurred only once in geological times as it is apparently not a continuous process. If we have consumed all oil [18] it will not be renewed by known to occur reactions. The utilization of fossil carriers in boilers (coal) or refineries (oil, gas) requires again heterogeneous catalytic processes. The conversion of chemical feedstock into the many thousands of materials we are using in modern life is also achieved by a combination of heterogeneous and homogeneous catalytic processes. Catalytic processes during combustion or fouling also support the final liberation of CO_2 from these molecules and materials at the end of their lifespan. Many of them occur unintentional during combustion through impurities or material–reactor interactions. In short, our existing energy system as well as much of our future energy system will require catalysis both for operating the relevant processes and for producing the required materials for energy conversion devices. Figure 4.10 indicates the universal role of chemistry [40] in the energy challenge both for efficiency optimization and for novel processes.

It is evident that chemistry stands at the center stage when it comes to optimization of the use of fossil energy as well as when it comes to solving the energy storage

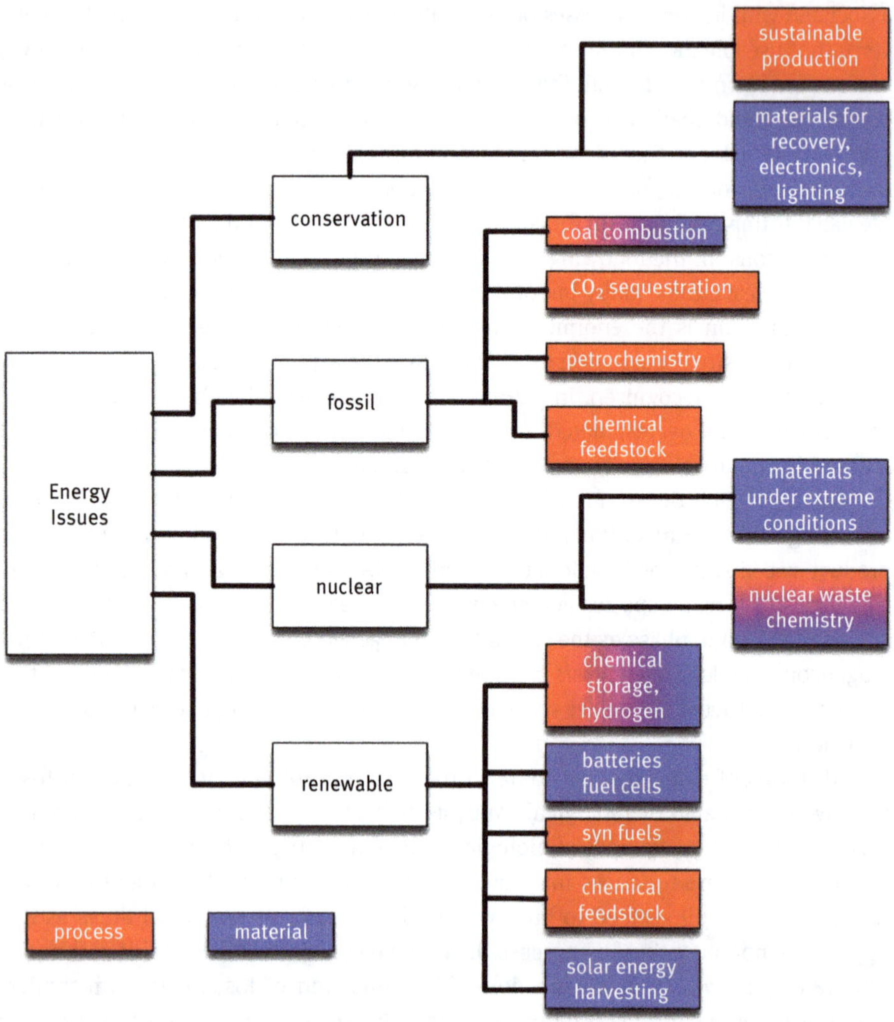

Figure 4.10: Chemistry is a platform science in the energy challenge. All major areas of energy conversion and use require materials or processes provided by chemistry.

issue at large scales. It is one purpose of the following chapters of the book to provide insight into the methodologies and concepts of how chemistry can perform these tasks. If it would become apparent that chemistry is indeed a central science for addressing the energy challenge then the book is a success. It seems still unclear to many stakeholders, how many chemical issues are hidden behind energy conversion and storage.

For the sake of completeness it should be stated that increasing energy efficiency is to significant extent not to be left to chemists and engineers increasing the energy efficiency of chemical processes; this reservoir of saving potential is being used already [41] since a long time and all easy steps have been taken in the interest of optimizing the production in chemical industry. The much larger portion of energy-saving strategies has to come from behavioral changes in energy utilization. Besides the easy steps such as house insulation also more difficult steps such as optimizing the energy efficiency of individual mobility are relevant. Strategies of goods logistics and production are optimized for economic and taxation processes but not for energy efficiency which is counteracted by multiple transportation subsidies. These examples indicate that multiple considerations stressed in Figure 4.8 outside of chemistry or science greatly affect our ability to design the changeover from fossil sources.

4.6 Chemical Reactions and Catalysis

Physical chemistry is the science that describes the course of a chemical reaction. We have a solid foundation in the theoretical description of chemical reactions. Reactions occur under the rule of thermodynamics that defines equilibrium states where in a closed system (no changes in the number of molecules) no net reaction occurs. In an open system with continuous changes of the number of molecules (flow) this state equals to a maximum conversion level that cannot be exceeded at given conditions. An example of such a system is the synthesis of ammonia [42] that is at technical reaction conditions heavily limited by equilibrium conversion.

The theory of chemical kinetics describes reactions in a multi-dimensional landscape of hills and valleys given by energy as function of the atom coordinates of the reacting system. A simplified energy profile occurs when the energy change is mapped against the reaction coordinate describing in a single number the changes of atom coordinates of the reacting system. For a very simple reaction the formation or cleavage of a single bond between two entities (atoms, molecular fragments) the resulting profile may look as shown in Figure 4.11.

The process in Figure 4.11 is called an elementary step as it refers to the change of one single chemical bond. The energy level of the nonreacted components (educts) is defined as relative zero point of the energy scale (dashed line in Figure 4.11). We see that approximation of the reactants first requires energy (the activation energy E_a) as the electron clouds are repelling the bodies. At a critical distance called the bond distance the electronic systems interact quantum chemically and form a chemical bond whereby the bond energy is liberated. If we try to compress the system further the total energy is rapidly increasing as the repulsion of the core electron systems of the atoms begins to dominate. The quantity ΔH is the enthalpy change of the process that is related through the second law of thermodynamics to the free energy change of the reaction.

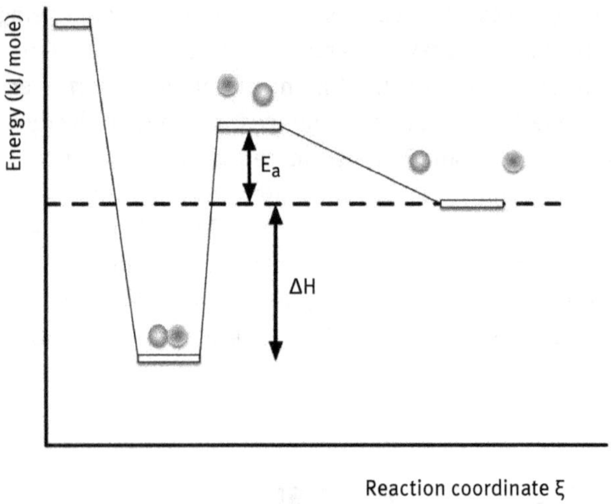

Figure 4.11: Reaction profile of a single elementary step being the formation of a diatomic molecule. The thin lines between the energy levels are guides to the eye and do not indicate the actual course of the changes of atomic coordinates.

The activation energy can be interpreted as resistance of the system to undergo a reaction, the enthalpy change is related to the extent to which the reaction can occur. In the present context it is important to note that chemical reactions that generate free energy from reacting energy carriers with oxygen (combustion) occur with large free enthalpy changes and thus always run to completion. This is unfortunately not the case for energy storing reactions in which CO_2 is reacted with a primary energy carrier such as hydrogen to form an energy storage molecule. These reactions that require the addition of energy both in form of heat flux from the environment and in form of chemical energy of the primary energy storage molecule are strongly limited by equilibrium conversions and require special measures such as removal of the product from the reaction system in order to run to significant conversions. This is illustrated in Figure 4.12 for the relevant example of hydrogenation of CO_2 to methanol [43]:

$$CO_2 + 3H_2 \rightarrow CH_3OH + H_2O$$

This reaction can also be considered as to occur from a first step of CO_2 reduction to CO followed by the methanol formation from CO being the more facile reaction:

$$CO_2 + H_2 \rightarrow CO + H_2O$$

$$CO + 2H_2 \rightarrow CH_3OH$$

Figure 4.12 indicates the maximal methanol yield as a function of temperature for various mixtures of CO_2/CO at 50 bar pressure. We see that indeed severe limitations exist for CO_2 hydrogenation unless extremely efficient catalysts are found that can operate at low temperatures. Such systems are presently unknown [44–46] as we operate at typically 523 K.

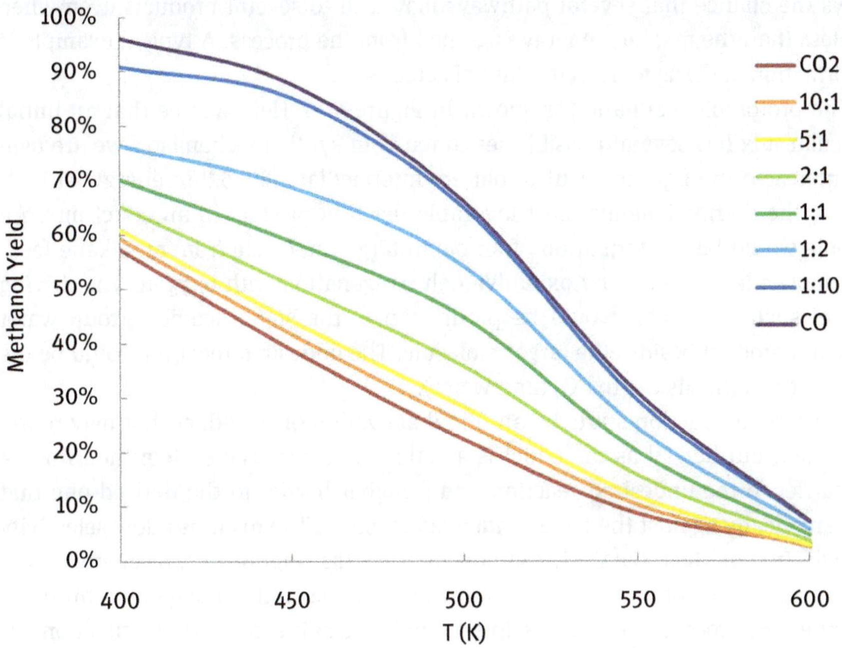

Figure 4.12: Equilibrium calculation for the hydrogenation of CO_2 to methanol at 50 bar pressure. The parameter is the fraction of CO in CO_2 facilitating the hydrogenation.

Without going into the details of thermodynamics here it should be noted that for chemical energy storage reactions the enthalpy change is not the relevant quantity as we are interested in the amount of work that can be done with a given amount of chemical energy storage material. This useful work is defined by the free energy and realistically by the exergy [1] of a process that relates the free energy to the actual process conditions. The free energy is independent from these conditions and thus the relevant quantity for scientific discussions, the energy is clearly related to specific realizations of a process.

Catalysis is changing the course of a reaction by modifying the path of the system through the energy hypersurface (the energy profile as function of all relevant atomic coordinates). This is achieved by creating activated complexes between the reactants and a functional material called catalyst. The elementary step reactions change under the influence of a catalyst but not the overall energetic difference between educts and products. This is equivalent to stating that a catalyst changes the velocity of a chemical reaction but not the maximal extent to which it can occur.

The statement is easy to understand when the reaction has only one product (or set of products). Typical examples are combustion reactions [47, 48] where always CO_2 and water are the final products. As molecules with many atoms (such as hydrocarbons) cannot combust in a single elementary process (not even methane) there is always the chance that several pathways may lead to several products even when then less than the maximal energy is gained from the process. A typical example is the formation of CO as toxic byproduct of combustion.

The prototypical situation is shown in Figure 4.13. Here we see that an initial molecular mix has several possibilities to react. In synthetic chemistry we are usually interested in target molecules that are intermediate [49–52] in energy with respect to the thermodynamic most favorable reaction product of an educt mix. An example would be the formation of an olefin target molecule from an alkane feedstock molecule accessible by oxidative dehydrogenation with oxygen. The driving force of such a reaction would be production of the stable leaving group water as second product besides the target molecule. The undesired reaction would be the combustion of the alkane to CO_2 (and water).

In practice, reaction starts by an initial activation of the educt that may represent a high hurdle [53] as the educt is a stable molecule. The system then finds a low barrier to the undesired reaction and a higher barrier to the desired one that loses kinetically against the faster undesired reaction. The result is a low selectivity to the desired product. Catalysis can now interfere (see Figure 4.13) by changing the reaction coordinate of the activated intermediate [54] such that the easy pathways to the undesired process is less accessible for only a small extra cost in activation energy. Then the process will slow down in velocity but gain substantially in selectivity.

The "secret" of the action of the catalyst is that by forming an activated complex with the activated reactant it changes the molecular structure of the reactant such that the undesired easy reaction path becomes unattractive. It is the specificity of the chemical bonding between substrate and catalyst that exerts the beneficial effect. In many catalytic processes an additional effect greatly changes the reaction rate: If more than one reactant is necessary then not only atoms but also electrons and atomic fragments need to be exchanged. A catalyst can provide the separation of these elementary steps in space and time. It can first activate one reactant and store the activated species at a reaction site for later use with the other reagent and it can lend and receive back redox equivalents (electrons) to participating species [55]. In this case the catalyst changes profoundly its own structure and it must be designed such that it can regain its initial structure when the catalytic turnover is completed.

A catalyst is thus a functional material that undergoes cyclic changes in its structure upon interaction with reactants. The design of catalysts thus has two targets:

It must ensure the specific modification of the reactant such that the desired reaction occurs with maximal selectivity and rate.

It must be flexible enough that the structure of the catalyst can undergo reversible changes ensuring that it is not consumed in the desired reaction. This stability

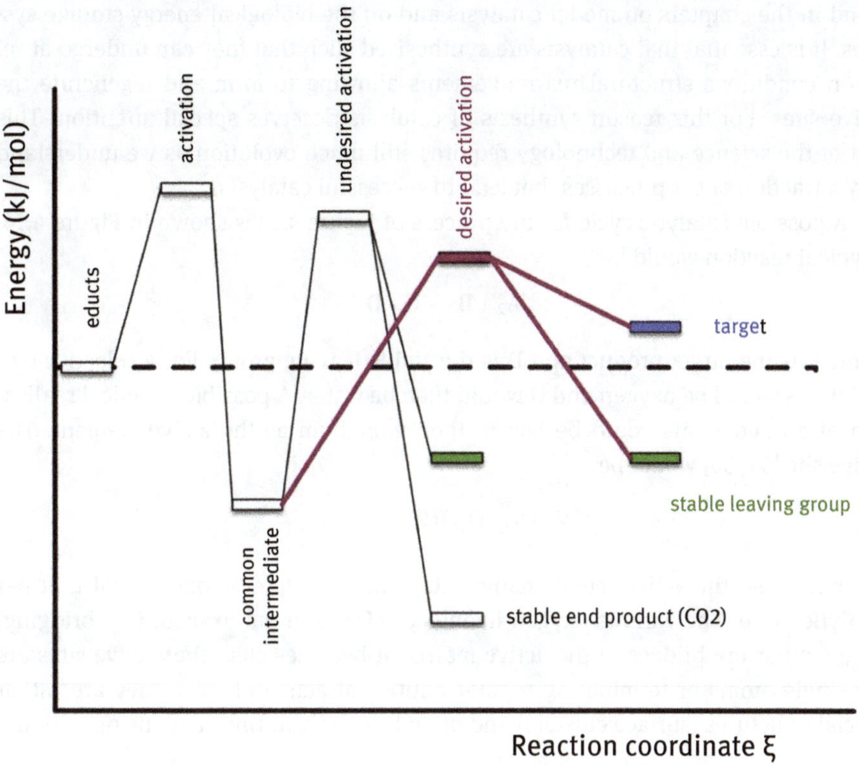

Figure 4.13: Energy profile of reaction with two product pathways and a common intermediate as primary activation product. Common to both reactions is further the stable leaving group (green). A desired catalytic pathway (red) to the target product (blue) circumvents the undesired unselective path to the most stable reaction product.

criterion is additionally made more difficult as also the reaction products must not strongly bind to the catalyst and thus poison it by irreversible interactions.

A central concept for the structure of catalysts is that of an active site. This active site is an ensemble of atoms guaranteeing the desired geometry of the reactant–catalyst bonding by a highly specific local arrangement of the constituting atoms. In order to ensure the regeneration of the site at the end of the catalytic cycle it is desirable that the active sites are embedded in a stabilizing matrix and that the embedment allows for geometric flexibility in order to minimize the activation energy for the rearrangement of the atoms in the active site.

This is either realized in molecular species where pocket structures of ligands around an ensemble of active atoms enable both specificity and flexibility. In solids, special "high energy sites" are terminating a matrix of an active phase providing by its rigid chemical and geometric structure directing interactions for the reactant binding (such as in pores or at steps of a solid). Examples of both strategies can be found in the chapters on model catalysts and on the biological energy storage systems. It is essential that catalysts are synthesized such that they can undergo at reaction conditions structural rearrangements allowing to form and regenerate the active sites. For this reason synthesis of catalysts deserves special attention. This part of the science and technology requires still much evolution as we understand only a fraction of the processes that lead to successful catalysts.

A possible catalytic cycle for the process of Figure 4.13 is shown in Figure 4.14. A typical reaction would be

$$A_2 + B \rightarrow C + D$$

where C is the target product and D is the stable leaving group. For a selective oxidation, A_2 would be oxygen and D would then be water. A possible chemical realization of an active site could be based upon vanadium as the active element. The active site [54, 56] would be

$$\left[V_2(O_t)_2(O_{br})\left(O_{sup}\right)_2\right]_2[O_2]$$

assuming that the active metal changes its oxidation state by one formal unit per catalytic cycle. The various oxygen ligands are O_t: terminal oxygen, O_{br}: bridging, O_{sup}: supporting bridges to the active matrix. It becomes clear that active sites are not single atoms or terminating regular entities of active phases. They are rather special structures (surface clusters) and often form only during catalytic operation.

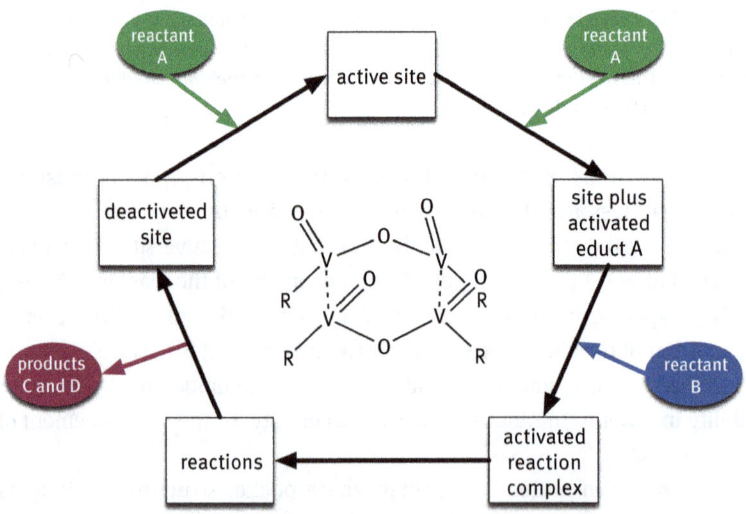

Figure 4.14: The catalytic cycle illustrates the steps of structural modification of a hypothetical active site represented in the centre.

A schematic representation for the energy-relevant process of methane combustion [57, 58] is given in Figure 4.15. Here a projection of the energy barriers on a selection of reaction coordinates is shown. The map can be read like a topographic map. The circular structures represent maxima (peaks) in the energy landscape whereas the arrows indicate pathways through saddle points (valleys) of the energy structure. We see immediately that two alternative reaction pathways are possible with one leading through decomposition of formaldehyde to the CO product. The other channel leads via complete dehydrogenation to deep combustion. The dashed lines indicate minor reaction paths via less deep saddle points. It becomes apparent how the facile combustion reaction is a complex process as soon as elementary steps are considered. It is also clear those catalysts such as Pd or Pt that strongly favor the dehydrogenation path [59, 60] will reduce the selectivity to CO. It is further illustrated by the inset in Figure 4.10 that these catalysts carry the risk for deactivation when the intermediate C atoms polymerize into soot along reaction coordinate 2 which eventually covers the active surface and kills the catalyst.

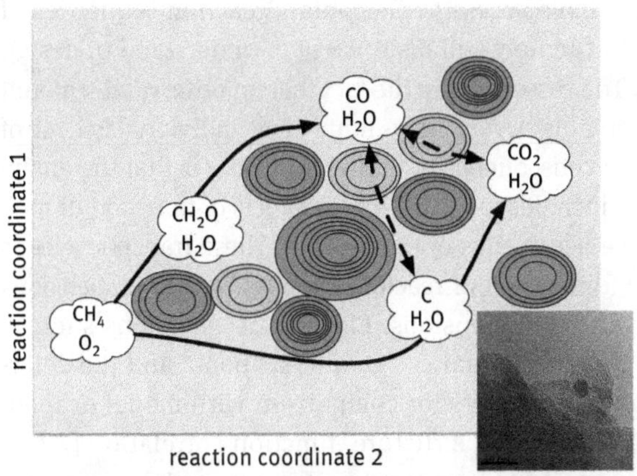

Figure 4.15: Schematic representation of a possible energy surface for methane combustion. The graphical impression is a projection of energy peaks onto a plane of reaction coordinates. The reactant systems (clouds) are not given with stoichiometric accuracy. There are many more intermediates and reaction pathways in the real gas combustion process. The inset shows a metal catalyst particle (arrow) that is fully covered by carbon stemming from the dissociation of a hydrocarbon species.

The understanding of catalysis at a level of complexity necessary to analyze the course of a chemical reaction is only achievable through the progress in studying model systems both with experimental [39, 61] and theoretical [62, 63] methods. The model approach was developed with structurally well-defined single crystal surfaces [64] allowing a comprehensive treatment of elementary step reactions. Most of these reactions were, however, adsorption reactions and only very limited conversion reactions such as dissociation reactions were possible. Introducing structural steps as high-energy sites improved the situation considerably as now reactions such as CO oxidation and ammonia synthesis became amenable to rigorous descriptions. More complex reactions are only recently under study by model systems [65, 66] since a combination of chemical complex oxide support systems with single crystalline quality and their decoration with nanoparticles from metals or oxides became experimentally accessible. In combination with an increasing understanding of the structure of free nanoparticles [67, 68] of active phases it is now possible to accurately describe complex models [69] and study elementary step reactions of kinetically demanding reactions such as alkane activation and C1 chemistry being of relevance in chemical energy conversion processes. Using the well-understood CO oxidation reaction [70] as proxy and probe reaction it becomes increasingly possible to extrapolate structural features from model systems to complex high performance catalysts where the direct structural analysis with the tools of surface science are not amenable for physical methodical limitations and theoretical models are still of insufficient complexity.

With this knowledge we now have access to elementary reaction sequences of individual molecules. This is unfortunately still insufficient to understand or design performance catalytic processes. The reason lies in the fact that an observable amount of a chemical (energy carrier) contains a very large number of individuals: 1 ml of water contains 10^{22} molecules. One consequence of this large scaling is that the analysis of observable catalytic reaction includes to a large extent effects of transport of molecules and of heat to and from the active sites of the catalyst. These transport effects scale themselves in macroscopic (dimension of reactor) and mesoscopic (dimensions of boundary layers and pore systems) dimensions. Elementary questions such as the comparison of catalysts for their performance "good" vs. "poor" and the determination of kinetic constants and parameters for comparison with model or theoretical predictions and the determination of structure–function correlation [71] by using in situ analytical techniques all are hampered by the unavoidable incorporation of transport effects. This multi-scale nature of catalysis [72, 73] is reflected in the reaction chain indicated in Figure 4.16.

The form of the activated complexes [74, 75] is substantially different in homogeneous and heterogeneous reactions. In homogeneous reaction the catalyst forms a molecular complex between the substrate that is recognized by its molecular shape and the single active site offering a binding site in the form of the replica of the substrate molecule. This is extremely specific in biocatalysts called enzymes and still highly specific in molecular catalysts where a limited set of ligand molecules defines the electronic and geometric constraints of substrate binding. In heterogeneous systems this process is less specific. Reactants are first chemisorbed on a broad range of surface sites. This interaction is weak with respect to the strength of chemical bonds to be activated. Only in cases of simple molecules such as H_2 and N_2 and at low coverage the approximations are valid that all adsorption sites are equal on a surface and

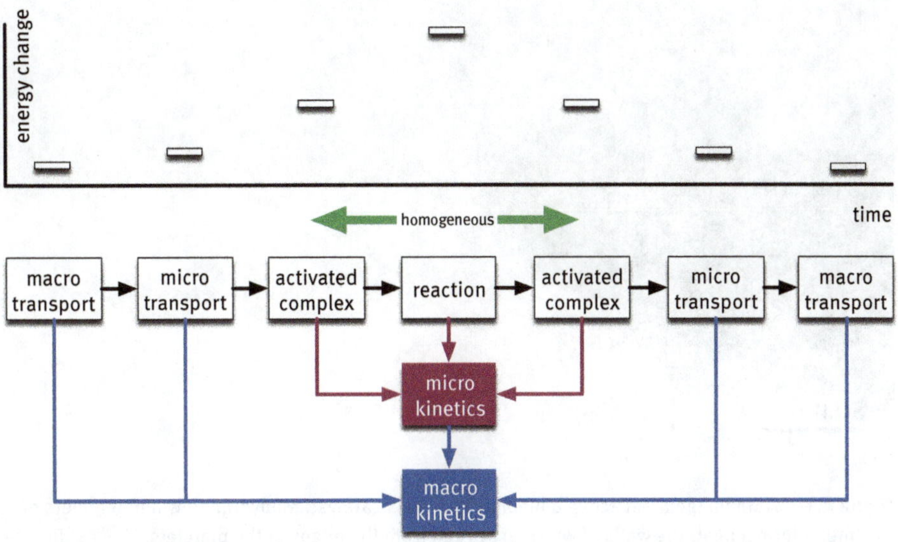

Figure 4.16: Steps of heterogeneous reactions. The individual processes comprising sequences of elementary step reactions are linked to a process sequence. The microscopic part is described by microkinetics, and the observable macroscopic performance by macrokinetics. A typical relative dimension of energy changes associated with the individual steps is indicated. In homogeneous reactions the transport parts are often ignored.

are not affected by neighboring adsorbed species. The total surface may thus be considered as "landing sites" for reactant molecules. They have to diffuse to special sites where reactions of breaking and making chemical bonds can occur. These high-energy sites [76, 77] are the few active sites discriminating any surface that may adsorb reactant molecules from catalysts that can convert more than once their adsorbates into products. The essentially flat surface location of active sites greatly reduces the specificity that can be executed by the surrounding atoms and chemical selectivity of reaction must be achieved only be exact matching the electronic structure [78] of the reaction site to the molecular electronic structure of the adsorbate. In reality it is a design target of performance catalysts to lift the flat nature of a reactive surface and create specific "roughness" on its surface allowing nesting of active sites or exposing sites at edges and corners of the material that are highly undercoordinated with neighbors in comparison to those sitting on terraces of a flat surface. Nanostructuring and utilization of multi-element compounds are strategies towards roughening active surfaces.

An example is shown in Figure 4.17. The mesoscopic structure of a bowl or nest is highly suited to accept and release molecular species. The stacking of thin platelets creates rough edges and corners with high-energy sites. The arrangement results form a special high temperature pretreatment of oxidic precursors giving rise

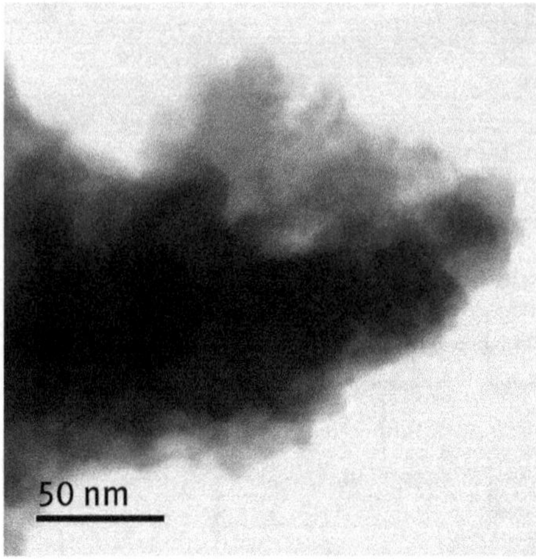

Figure 4.17: STEM image of an active ammonia synthesis catalyst. Many thin (few nm) platelets of iron metal form a nest, the walls of which are rough from the edges of the platelets. Tiny oxidic nanoparticles (round shadows) of "structural promoters" inhibit the collapse of the platelet structure. When inspecting the image consider that TEM delivers a 2-D projection of a 3-D object.

to the intimate mixing of the oxidic components [42] that would otherwise not interact so intimately as shown in Figure 4.12.

The timing of processes shown in Figures 4.13–4.16 is critical in heterogeneous reactions as a means to induce specificity. Besides transport that is controlled by the flow conditions and the hydrodynamic properties of the catalyst in a reactor (particle sizes, tube diameter, macro-meso-micropores) the kinetics of adsorption and diffusion are important factors. These depend on the nature of the chemical bond between catalyst and reactants. This leads to an extreme sensitivity of the catalytic performance to the details of the electronic structure of the solid controlled by many types of deviation [79] from a regular perfect translational order. The consequence of these defects on reactivity is called "structure sensitivity" [80, 81]. A complete description of a catalytic cycle can thus only be achieved by understanding both [82] the microkinetic and macrokinetic aspects [83] of the reaction.

The relative changes of energy with the individual steps in Figure 4.16 should not lead to the false assumption that steps with small energy changes are less relevant than the chemical reaction bringing about the usually largest energy change. If the molecules cannot access the active sites, they cannot perform their role. If the reaction energy cannot be distributed to the environment, side reactions can occur and poison the active site or damage can occur for the whole catalyst through the formation of hot spots.

Macrokinetics is the description and analysis of the performance of the functional unit catalyst plus reagents plus reactor. It leads to formal activation barriers called "apparent activation parameter" representing the superposition of several elementary barriers with transport barriers. It further delivers formal reaction orders and rates as function of the process conditions. These data can be modeled with formal mechanisms of varying complexity. In any case, these data can well describe the system performance but cannot be used to deduce the reaction mechanism.

This task is left to the microkinetic modeling [84] trying to observe a catalytic process unfolded from transport limitations. Here sequences of elementary steps are composed in reaction mechanisms with chemically meaningful process steps. These usually complex systems of equations with many unknown parameters are then approximated to performance observations. It is obvious that multiple observations obtained by profile reactor studies and/or by kinetic isotope exchange reactions will greatly enhance the significance of a microkinetic model adaptation. This approach is presently the state of the art in understanding the relevant energy storage reactions [85].

The reader is referred to a comprehensive [83] quantitative description of the intricate phenomenology of heterogeneous catalysis. The present simplified presentation may give an impression to what is needed for mastering the "design" of a catalyst or even making statements about the performance and thus "cost" of an energy conversion process. It is important to understand that many aspects of performing the abovementioned simple transformations that are so crucial for chemical energy conversion are not yet understood [86] or mastered to an extant allowing to state that here we deal with "mature" technologies that only need to be applied.

4.7 The Design of Catalysts and Processes

The present introduction and the respective chapters of the book may leave the impression that catalysis is a highly complex science and little can be done to systematically identify catalysts. This is a well-supported view as almost all presently industrially used catalysts were found and optimized by rather empirical methods. The advent of high-throughput experimentation [87] in combination with numerical methods for experimental design has greatly enhanced the empirical testing capacity. It has not led to higher productivities in discovering new catalysts as compared to traditional methods normalized to the same level or resource allocation.

It may be asked what makes us hope that the enormous challenges of chemical energy conversion being an order of magnitude larger in quantity than our present chemical industry may be based upon effective catalytic processes? The challenge is even larger as the sheer dimension of reactions excludes practically the application of our most powerful catalytic systems namely noble metal systems. Energy catalysis must after all be sustainable and scalable [88]. It must not occur that energy-related processes require more functional materials than globally available (economic and societal problems) nor must it be that energy-related reactions create waste and non-useable byproducts. We are thus limited in process design to systems that only leave uncritical wastes such as oxygen, nitrogen, and water, we must close all material loops of species with critical biological effect such as carbon, nitrogen oxides, and ammonia and we should render our process materials fit for recycling mineral resources such as metals. These boundary conditions limit the parameter space for designing catalysts.

Energy catalysis calls for a true effort in connecting the subdisciplines of catalysis such that cycles of system optimization are followed and not only fragments of such cycles without connecting them. One frequent fragment is the combination of (advanced) nanotechnological synthesis of nonscalable materials combined with performance testing. This combination led to a plethora of suggestions for novel catalysts in the scientific literature. Under the keyword "nanocatalysis" one finds in 2021, 882 papers with 204,000 citations. The combined search for "nanoparticle" and "catalysis" yields 18,123 papers with over 155,000 citations between 2016 and 2020. It is obvious that a very large science base exists for finding novel solutions. The concept of nanocatalysis [89] is fundamentally not new [90] but the variability and accessibility of chemical tools for nanostructuring have increased enormously. The challenge lies in scaling these findings to the "energy dimension."

A complementary theoretical approach [86] uses theory to preselect possible successful systems from its chemical composition. The concept uses [91] scaling relations between the surface electronic structure of materials and their adsorption properties [92] derived from quantum chemical calculations and relates these properties to the chemical kinetics of the reaction of interest using the Broenstedt–Evans–Polany relation [93]. An in-depth theoretical model of the reaction of interest is set up delivering critical parameters of adsorption and reaction. These critical data are then calculated for a large number of elements and surface geometries ignoring details of the chemical reaction and of the catalysts such as detailed structure, chemical dynamics, and deactivation. The resulting simplification in treatment allows for studying "simple reactions" as ammonia synthesis [94] as well as complex reaction networks [95] such as CO hydrogenation. As it turns out that many reactions are controlled by a small number of critical properties [96] of reactants and surfaces, a database can be built [97] allowing correlating hypothetical performances to fundamental properties of elements and compounds. This scaling can also include nonchemical variables such as the price of the constituting elements and one can predict the catalytic performance of a reaction against the price of the catalyst used [98]. From such predictions one may conclude that it is not impossible to find chemical solutions for energy conversion reactions that fulfill all boundary conditions of scalability and sustainability.

One reason why the problem is not solved practically (top performance catalysts found) lies in the complex challenge of making and stabilizing the required functional materials [99–101]. Here again homogenous and heterogeneous catalysis are widely different in their respective maturity. Related chapters of this book give insight into the state of affairs as well as into critical research desiderates [102] that need to be addressed to advance in this most important subfield [103–105] of chemical energy conversion.

At the time where this text is written a new strategy is upcoming that generates directions of catalysis development. "Digital catalysis" [106] uses methods of artificial intelligence and machine learning to extract from a comprehensive set of experimental observations on fully characterized existing catalysts complex structure–function relations. These "descriptors" [107–109] relate observable properties "actuators" between the dynamical working structure and the resulting performance. Reliable structure–function relations are complex, as catalysis is a dynamical multi-scale phenomenon [73] that is only poorly described with linear relations. The new high-quality descriptors allow extraction of critical properties of a catalyst for a given reaction. In this way the evolution is directed to optimize certain properties. The method cannot predict per se materials or chemical compositions. The method can, however, be used to search material databases that may contain only theoretical properties (such as the NOMAD database) and suggest materials with the required property profile generated from the structure–function analysis. The process is still in its infancy as many technical and conceptual hurdles of data availability and data quality are still to be overcome. The fact that this approach is neither simplifying the problem nor ignoring the intricacy of the catalyst material science brings this approach closer to the real problem than any development strategies suggested before. It must be seen that this approach is only possible as the earlier strategies provide insights and tools of theory and experimental science that were unavailable before. In this way the new approach unites the earlier concepts and makes use of the consolidated knowledge about catalysis.

The progress that catalysis science has made over the past 5 decades despite these shortcomings allows foreseeing that it will be possible to systematically design [74, 110, 111] catalysts for energy applications that are not built only upon the past empirical practice of "intuitive knowledge." The new systems will be built upon design criteria arising from the set of boundary conditions for energy processes. They will further use the insight into structure–function relations that are now accessible both from functional models and from high performance but unsustainable existing catalytic systems (such as systems based upon noble metals). The recently developed suite of in situ functional analysis [112] of working catalysts and the progress in atomically solving structures [113–115] of enormous complexity [116] are the tools that were not available in earlier phases of catalysts evolution. Also the conceptual understanding of catalysts as dynamical systems [117–119] with few differentiations between homogeneous and heterogeneous systems [73] is a new asset to the catalysis scientist. The strong pressure from chemical energy conversion upon the catalysis science and technology [120] will help advancing the academically available strategies and analytical technologies into a broad-based catalyst development effort conducted in collaboration between industry and academia. The history of chemical industry has seen such periods of highly fruitful and easy technology transferring periods before and after the wars in the twentieth century. It may well be that we are at the onset of another such phase of revolutionary development. It is understood that evolutionary strategies and incremental developments will not be sufficient to respond to the challenge of chemical energy conversion and storage.

4.8 The Biological Origin of Our Present Energy System

Our present energy system is largely built upon fossil energy carriers dominated globally by oil. These energy carriers are hydrocarbon molecules with a largely varying hydrogen content between C:H = 4 for methane and C:H = 0.1 for coal variants. These hydrocarbons stem from biomass that was converted in several steps from oxygen-rich structures typically $C_6H_{12}O_6$ to polymers (humic acid) and then by deoxygenation to hydrocarbons. In cellulosic biomass the lignin content is already much closer to the condensed structure of coal leading to the fact that lignin from ancient trees was a predominant precursor to coal.

All fossil materials were initially generated through photosynthesis [121] leading to glucose as a primary source of biomass. The photosynthesis process using water as reductant and thus liberating di-oxygen as waste is the largest chemical process on our planet creating all life, all food, and all energy carriers from sunlight. Its homogenous catalytic reactions are described in detail in a chapter of the book as they are of prototypical character for energy storage processes. It is noted that this process operates isothermally and is highly endothermic:

$$6CO_2 + 6H_2O \rightarrow C_6H_{12}O_6 + 6O_2$$

$$\Delta G' + 2,792\ \mathrm{kJ\ mol^{-1}} = \Delta H\ (2,813\ \mathrm{kJ\ mol^{-1}}) - T\Delta S\ (293\mathrm{K} \times 182\mathrm{J\ mol^{-1}})$$

80 g of sugar (glucose) is well sufficient to nourish one person for one day. It is evident that nature uses a dense form of energy storage. The free energy ΔG` is related to biological conditions of 293 K and pH 7 in water. The two molecules water and CO_2 are thus the key elementary ingredients into the energy system of life that requires several other critical inorganic ingredients in trace amounts to build the

molecular systems catalyzing the conversion of sunlight into the storage molecules. A systemic relation between relevant processes is given in Figure 4.18.

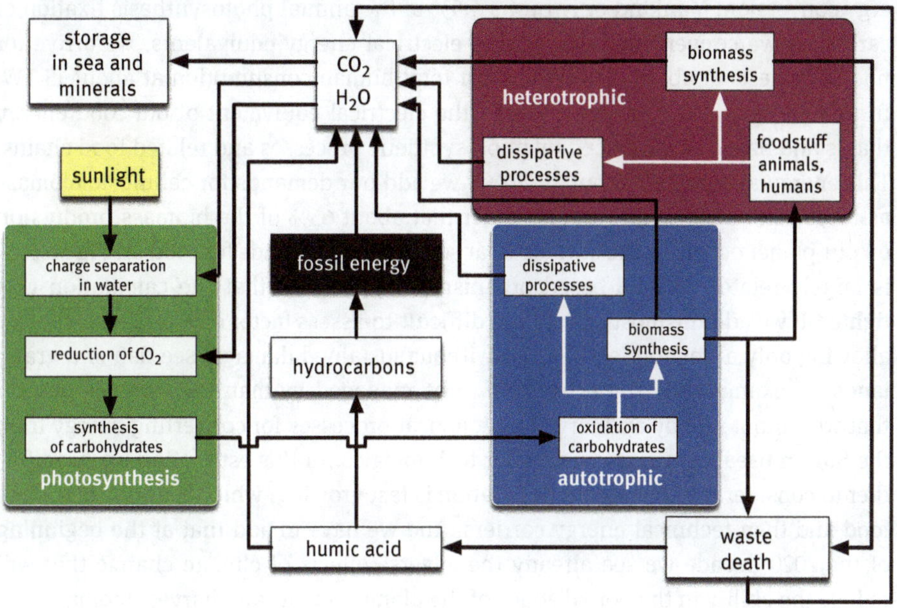

Figure 4.18: A crude systemic picture of the biogeochemical processes from which fossil fuels derive as nonoxidized storage system of nature.

The dimensions of the material and energy flow are enormous on dimensions of our planet. The table taken from the IPCC report gives some crude estimates.

Geosphere compartment	Stored carbon in Gt	Exchanged carbon in Gt a^{-1}
Deep sea	13,800	
Surface sea	1,020	
Sea atmosphere		92
Fossil energy carriers	4,000	
Combustion to atmosphere		5.5
Atmosphere	750	
Vegetation	610	
Photosynthesis		121
Soil	1,580	

These data underline the enormous dimension of photosynthesis and the biosphere for energy conversion. From these figures it could appear that there is no need for any artificial chemical energy conversion as there is abundant natural energy conversion. Mankind only burns 4.5% of the annual photosynthesis fixation of carbon. If we convert this into annual electrical energy equivalents, we arrive for photosynthesis at about 250 TW and for total human consumption at about 15 TW. In this argument we first have to add the electrical equivalent of our food energy that is and has to be produced by photosynthetic processes and related food chains. This energy amounts to about 25 TW. If we add our demands for cellulosic biomass as feedstock (wood, pulp) and consider that about 65% of the biomass production on our planet occurs in the surface-near sea but our demands for food and feedstock is largely related to land-based organisms then we see that the calculation gets tighter. If we add the most critical but difficult-to-assess factor of ecosystem stability allowing only a fraction (about 30% without sustained damage, see fish and grasslands for example) of any ecosystem to be managed by man then we see quickly that we cannot rely by and large on biological processes for converting energy from the Sun in useable energy carriers for technologies. In this estimation we have further to consider that the global population is fast growing, which requires first more food and then technical energy carriers. And we have to add that at the beginning of the 2020 decade we see already the adverse effects of climate change that will reduce the ability in the boreal zones of the planet to grow and harvest biomass.

Another issue is the required energy density. One metric ton coal as 1 t SKE contains 7,000,000 kcal. One ton of dry glucose contains 3,720,000 kcal. Biomass does not occur as dry glucose but rather as diluted and/or polymeric forms of glucose. Several steps of energy-intensive concentration and refining are required until we obtain glucose for technical applications. The relevant chapter in the book discusses potentials and difficulties of this aspect being often overlooked. That is also true for the many concepts of using algae or other aquatic plants requiring harvesting drying and refining before it can be used as energy carrier material.

The apparent initial success of biomass-derived energy carriers used in biofuels [1, 30] and fuel additives (E10) stems from a nonsustainable form of biomass harvesting. These materials are generated from the fruit of plants. These fruits are developed to carry a high-energy biomass such as starch or oil in order to support the reproduction of the plant. In fact, only a small fraction of the total biomass of the plant is concentrated in the fruits. In addition, the extensive biocatalytic processes of biomass formation for these fruits (Figure 4.18) is highly energy-intensive for the plant: a large fraction of the effectively harvested solar energy is thus "wasted" in the biological processes generating the fruits for generating biofuels (of the first generation). This form of energy farming is suitable as niche solution in areas where an excessive abundance of arable land is available in relation to the needs of the regional population. As a global solution it is inappropriate also due to many nonscientific consequences of energy farming (see also Figure 4.8).

Large-scale application of energy farming further rises the question about sustainability of land use, water use, and availability of mineral fertilizer components such as phosphate and soluble nitrogen. It is foreseeable that competitive situations of these limiting resources on the planet may occur against food production. Even if we solve the land use issue by appropriate land management (not yet practiced) the other resources may prevent the use of energy farming as a main source of renewable energy generation.

Whatever biomass may be spared from either keeping the ecosystems stable or from feeding the global population, it may be used as feedstock for chemical processing rather than as fuel for combustion processes. In this way the complex structures nature has generated during biomass formation (Figure 4.18) can be partly preserved and may be brought to use in material applications such as polymer formation or direct synthesis of intermediates with complex molecular structures. In order to follow this highly constructive use of biomass it is essential to understand the underlying chemistry of biomass transformation [122–124] being largely different from our well-studied hydrocarbon transformation. In the latter case we need to enhance the functionality of hydrocarbon structures (gas and oil) whereas in biomass transformations we need to de-functionalize over-functionalized molecules leading upon activation to poly-condensation processes with unwanted poly-oxo structures (such as humic acid).

An in-depth analysis of the potential of biomass to sustainably cure the CO_2 emission problem arising from burning the fossil resources revealed [125] that this is no permanent measure as compared to reducing the irreversible emission of CO_2 from fossil sources. "Although terrestrial ecosystems can be managed to reduce carbon emissions and increase carbon sink size significantly, such increased carbon uptake can offset fossil fuel emissions only temporarily – on a time scale from decades to a century. Terrestrial carbon sinks are thus best viewed as buying valuable time to address the most significant anthropogenic perturbation of the carbon cycle – fossil fuel emissions [126]."

4.9 Chemical Energy Storage: One Long-Term Solution

It is now obvious that we have to refrain from irreversible CO_2 emission as fast as we possibly can (Figure 4.2) [1, 127, 128]. This is a long-term trend and not a shorthand political wave. As the dimension of this challenge is so large, we are advised best if we use all options that fulfill the conditions of scalability and sustainability in order not to create a new energy system with a similar flaw in it as our present fossil system. In this sense chemical energy storage is one important option.

On these grounds it is foreseeable that solar energy with enormous but rather dilute energy flux onto our planet will be a central source of future energy. Besides the dilute form (200 W m^{-2}) the volatility of its flux at ground is a major obstacle. It is impossible to sustain uninterrupted energy flux at demand if we use solar energy directly converted into primary electricity. This approach is despite its volatility efficient to supply free electrons as energy carriers, as we need about 50% of our fossil primary energy carriers for generating electricity with an efficiency of ca. 40%. If electrical energy comes as RES without a CO_2 bill and without cost for fossil fuel it will greatly relieve the energy system.

Collecting RES is also quite efficient: In Germany 1 ha land with solar panels generates about 1,000,000 kWh a^{-1} electricity. Would the same land be planted with corn, then the energy equivalent from biomass to biogas conversion plus conversion into electricity at demand would generate 15,000 kWh a^{-1}. Would this field host a wind turbine in addition then an additional solar power of 2,085,000 kWh could be added. If we assume that such a "combi-energy-field" could generate 3 GWh a^{-1} ha^{-1} then in Germany an area of about 2,000 km^2 could generate as sum over the annual demand all electrical power needed in Germany. This is possible with existing technology. The investment would be substantial [129, 130] with about 1,500€ kW^{-1} for solar and about half as much for wind and biogas. As solar energy is free, low operation cost would partly compensate for this investment that is crudely estimated twice as high as the investment in conventional fossil power stations with a mix of coal and gas.

If we extrapolate the availability of solar energy to global dimensions then we obtain [131] very large numbers. Mankind consumes about 18,5 TWy a^{-1}. Only 5% are RES and most of it nonsustainable biomass. The following table lists the estimated available fossil energy resources and contrasts it to the ***annual*** RES supply.

Resource	Reserve (TWy)	Annual availability (TWy a^{-1})
Natural gas	220	
Oil	335	
Coal	830	
Uranium	185	
Solar irradiation on all land		23,000
PV electricity		4,600
Wind		100
Biomass		5
Hydroelectricity		4

The data reveal that from a supply perspective there is no pressure in the coming 3 decades to change from fossil to RES and this is reflected in the scenarios from

Figure 4.1. On the other hand, the enormous excess of RES as an annual free gift from the sun calls by itself from tapping into this reserve and harvest the double advantage of "infinite and free supply." The core problem that needs to be solved by chemistry at large is the wrong distribution in space and time of RES on our planet.

Even when solar power and wind power are combined and exchanged over the area of central Europe this would not be sufficient to guarantee the uninterrupted supply. Solar power has in Europe a pronounced annual fluctuation [132] and wind is quite unpredictable onshore and still not always available off shore. This behavior is measured in the so called "capacity factor" that gives the fraction of a year during which an energy harvesting or conversion system delivers work, measured in full load hours. In Germany this capacity factor is about 0.17 combined for sun and wind, whereas in top quality sites on the planet values above 0.6 can be reached [133]. The cost of RES is proportional to the capacity factor with a strong nonlinear relation for small factors due to the adverse effect of long payback times for the investments.

With this background it is highly obvious that conversion of RES into molecular storage systems and there into hydrogen as platform molecule is an indispensable task for chemical science and technology in the energy transition. The complaint about efficiency losses through conversion is correct. There is, however, no alternative to it if we strive for full and stable defossilization. The traditional counterargument of using carbon capture and storage [35, 134, 135] as apparently fossil technology that greens these energy carriers is in systemic view an oversimplification as it combines the use of two exhaustible resources, namely fossil energy carriers (see above) and storage sites of reliable stability (empty sites from natural gas production).

Volatility has several scales in time. These trends are indicated with data from the German electricity production in Figures 4.19. The total contribution of regenerative power to the German energy system was about 45% in 2020. Comparing to an average total demand of about 70 GW, we see in the episode from spring 2021 that on certain days the contribution of regenerative power is already above 50%. There are also periods with little sunshine and no wind where with high rates of change regenerative sources contribute to the power supply. In many instances, it is fortunate to combine PV with wind power to equalize the volatility. From the color shades in Figure 4.19 one can estimate the regional differentiation of RES in the four supply zones established in Germany. The equalization of power throughout the country ("Kupferplatte") is still far from perfect after 10 years of energy transition due to a slow addition of power line infrastructure.

The issue of volatility has also a dimension on much longer timescales than days. This is shown in Figure 4.19 top, displaying the energy mix for the whole year 2020. The CORONA effect in spring can be seen from the supply with base load coal power displaying nicely the weekday–weekend pattern in the total load. In the top green/orange part the contribution of RES and their volatility can be seen as well as the slow change between winter (wind, green) and summer (solar, yellow). The data clearly reveal that future energy systems relying to a large fraction on regenerative

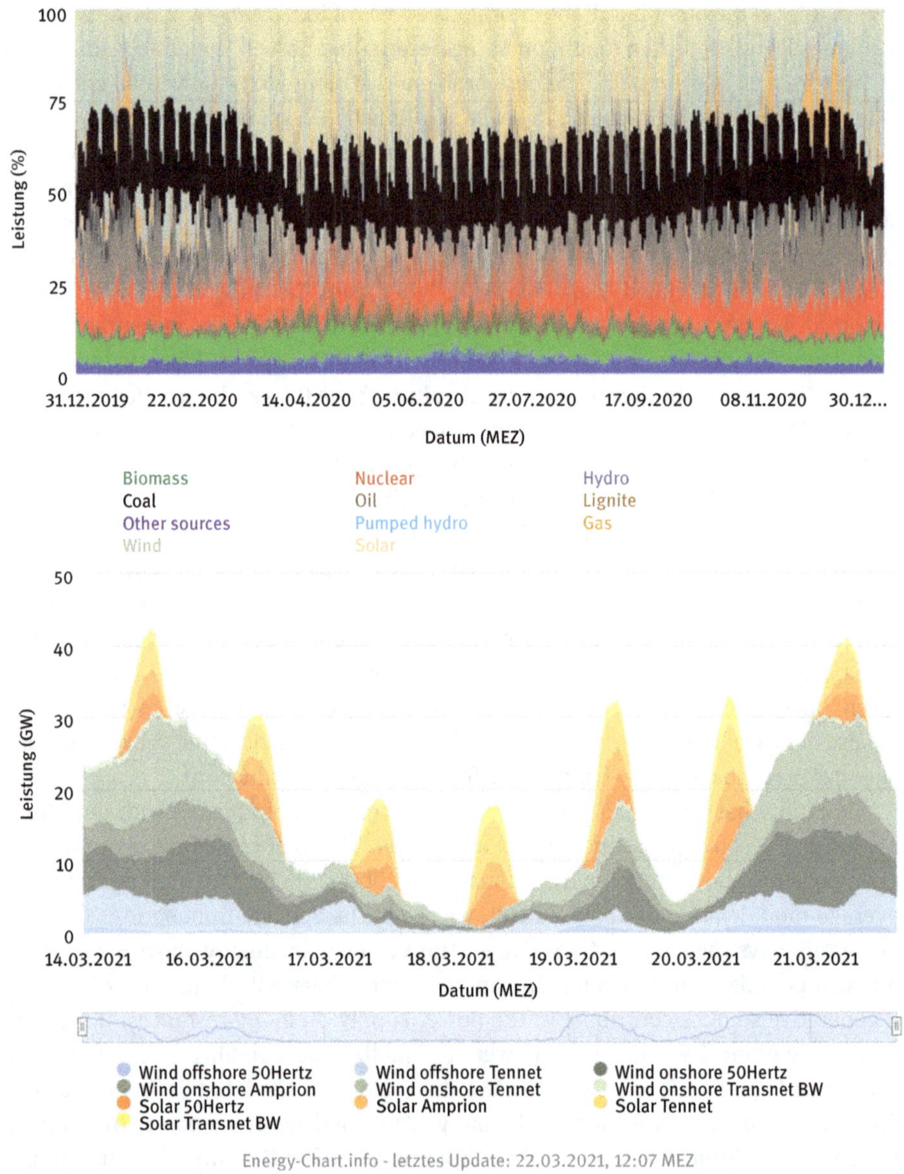

Figure 4.19: Energy mix in the German power grind. Top: the whole year 2020, bottom: the week 11 in 2021 (names in the legend are transmission operators). Source: Fraunhofer ISE Energy Charts (2021).

primary energy will need substantial capacities for energy storage. These storage systems need different response times from minutes to months and are thus combinations of physical (short-term) and chemical (long-term) storage solutions. It is also clear that an intelligent management structure of the energy system is needed deciding which storage option is used at what demand/supply ratio. The sheer amount of energy that needs to be stored certainly creates strong boundary conditions on economics, efficiency, and scalability of the solution.

The substantial contribution of volatile power sources will change the operation scheme of the whole energy system as the traditional distribution in continuous base load and discontinuous peak load does not apply. All elements of the energy system will need faster response times creating not only issues for the economic operation of large-scale installations and transmission capacities but also creating new demands on structural and functional materials having to cope with frequent changes of operational loads. For chemical energy conversion the new challenge arises that catalytic processes may have to be operated under dynamical conditions [136] much as we know it from the car exhaust catalyst. These reactions lead to thermodynamically stable end products whereas in energy storage catalysis the target molecules are thermodynamically unstable with respect to the educts. It is a largely unexplored field both conceptually and phenomenologically how the energy storage reactions respond to this challenge.

Chemical energy storage and transport will benefit from the availability of large-scale conversion of electrical energy [41, 137] in artificial solar fuels that can counteract the volatility of primary electricity in central and distributed installations. This has, however, a price tag [31, 33, 129, 138] arising from complexity and new optimization criteria. In Figure 4.20, a simplified scheme of the present energy system in Germany is shown to illustrate the existing key components.

By enlarge, we generate our free energy from combustion of fossil fuels with largely irreversible CO_2 emission. The existing grid manages fossil electricity with nuclear electricity in a common system and guarantees the uninterrupted supply according to demand. Some mechanical storage in pump hydroelectric stations, export/import across national borders, and increasing use of CHP systems help stabilizing the power distribution and efficiency. The contribution from renewable sources is substantial in this hybrid power system as exemplified in Figure 4.19.

This system is optimized with respect to provide electricity according to traditional large-scale consumers in locations of power stations and transmission lines. It is not suitable for a national distributed power generation from renewable sources and from the exchange of large amounts of renewable energy to locations where massive nuclear power installations exist. It is further optimized to use economical primary energy carriers such as coal and lignite for base load throughout the year and to operate gas power stations just for meeting fluctuating demands. The overall optimization criteria were price of electricity and a maximum availability of the grid service.

The criteria and targets for the design and optimization of infrastructure [26, 31] and services require modifications when the abandonment of nuclear power and the minimization of CO_2 emission are additional new boundary conditions. According

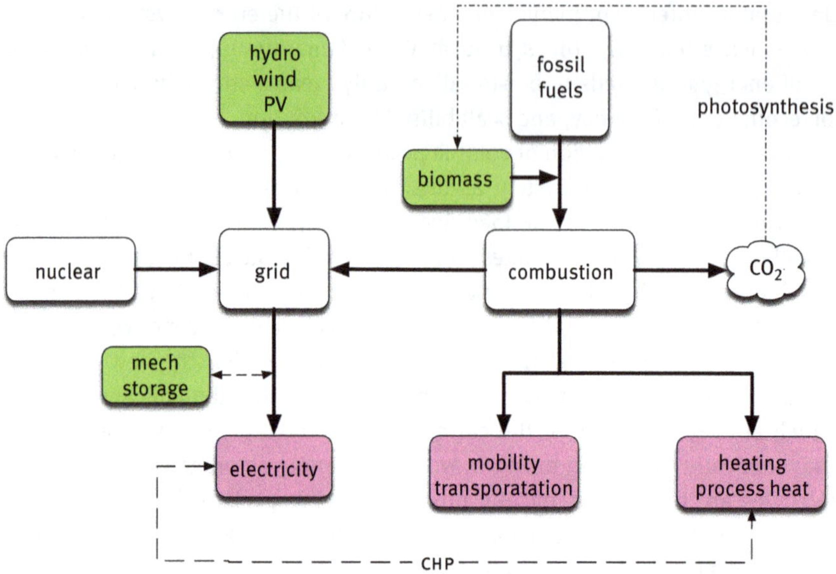

Figure 4.20: Schematic representation of the German (European) energy system.

to interconnections illustrated in Figure 4.8 it is first essential that all stakeholders agree on these new conditions. If we retain the condition of maximum availability of service then it becomes quite clear that new architectures are needed. This in turn will cause massive investments and incentives for industry to install new technologies. Some of them do exist and need rollout such as primary electricity generation, mechanical energy storage, and a smart grid, some others need to be created. This is a long process and will also need financial investments as well novel industrial structures. The total cost of such a system transition is hard to estimate as against the new cost savings have to accounted for arising from not purchasing primary energy carriers, from saving of cost for socialized risks of resource availability and cost for mitigation of climate change and cost for maintenance of the existing infrastructure. Such assessments are clearly not the task of scientific discourse but firm, reliable, and scalable scientific results will be needed in these assessments [6] in order to come up with reliable estimates of requirements and cost.

The prioritization of research in terms of cost expectations of a target technology is quite a limitation that needs to be conducted with great care. The technological revolutions of the past illustrate impressively (imagine the cost for cellular telephones and data transmission or the cost of the internet and their business cases) the unpredictability of scientific progress. Energy science should thus concentrate on creating options for solutions including all the data required for reliable life cycle analysis and cost estimates and then inject this information in societal processes taking decisions. As such societal processes control science funding, a nonnegligible feedback exists between the two spheres, which requires science to be aware of the interfaces shown in Figure 4.8. A particular difficulty with this feedback loop is the inherent volatility of societal processes being much faster that the timescales needed for fundamental and applied science breakthroughs. The history of biofuels with its premature decisions and resulting detrimental consequences for science and society is a vivid exemplification of this extra hurdle in energy science.

In any case a working hypothesis for energy systems should exist into which new scientific inventions and achievements in fundamental understanding can be brought in. In Figure 4.21 one example of such a hypothesis is sketched. Several conceptually scalable pathways from solar irradiation to electricity are indicated. Auxiliary options for eliminating the volatility through residual fossil fuels, through nuclear fusion or through biomass combustion are integrated.

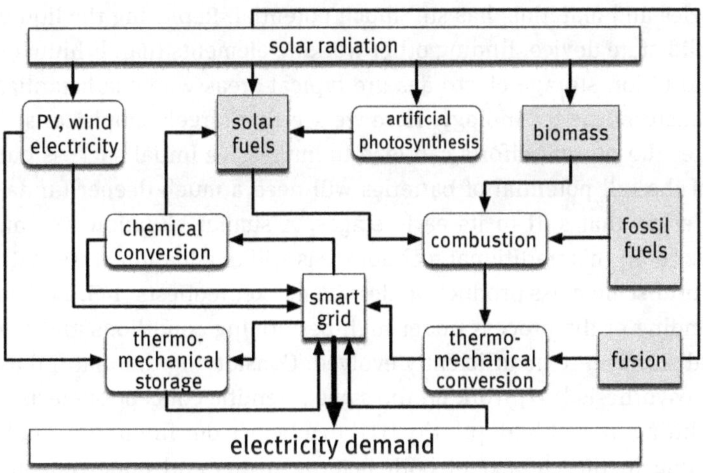

Figure 4.21: A possible structure of the electricity part of a future energy system.

Multiple feedback loops and the "intelligent" transmission and distribution system managing the generation and storage of electricity characterize the system. It includes the present fossil fuels to allow for gradual adaptation. It assumes the existence of several physical and chemical storage systems accommodating the varying requests for storage times and storage amounts. Both physical storage (existing) and chemical storage (in development) are assumed to be operative. The system also assumes feedback between energy demand and generation allowing noncritical or price-conscious applications to be managed in their time structure. This has the consequence that also large-scale electricity users may restructure their processes to accommodate temporal fluctuations. A core consequence will be that the traditional structure in base and peak load disappears as one can see already in the present German power system exemplified in Figure 4.19.

The process "chemical conversion" in Figure 4.21 is of central importance in the overall scheme. It ensures the storage of electrical energy in chemical bonds. For this task principally the operation of accumulators "batteries" and the formation of molecular energy carrier are solutions. The concept of a redox-flow battery [139] is a hybrid between the two concepts. Battery science [140, 141] is a crucial component although it is evident that grid amounts of electricity will not be stored in batteries. Their fast temporal response, their all-electric control process, and their ability to operate easily in distributed or mobile systems are however attractive properties. Their potential in e-mobility applications [142] is a further attraction. This application is not large with respect to the total energy system but it covers an important functional aspect, namely the satisfaction of mobility demands in urban areas where a good fraction of the total global population lives. Although the fundamentals of battery applications are already well-understood [143] their implementation in sustainable technologies and materials has still much potential. Replacing the liquid electrolyte by a solid-state device, finding other working elements than lithium or developing optimal Li ion storage electrodes are typical areas where substantial innovation can transform the technology. Here we see that largely empirical science that dominates the present efforts can lead to impressive initial success but the exploitation of the full potential of batteries will need a much deeper fundamental grounded effort that still in its early stages. A similar situation we can state in electrolysis [144]: this traditional technology is still in a manufacture state rather than in a world-scale mass production despite market requests [145, 146] as both the understanding of the process under high-performing conditions and the respective material science are insufficiently evolved. Considering that natural as well artificial photosynthesis [147] work on the same scientific concept of electrochemical water splitting, it is of high priority [111] to advance our full understanding of water splitting in the context of both heterogeneous and homogeneous electrocatalytic processes.

For chemistry the central target besides enabling energy-saving strategies is to generate options and technologies for chemical energy conversion of electricity, for converting the resulting primary energy carrier hydrogen into solar fuels and for back-converting these fuels in energy efficient ways into electricity. In addition, the enormous task to develop artificial photosynthesis needs still massive fundamental research efforts. The possible improvement of exploitation of natural photosynthetic energy conversion strategies both at the level of modifying living structures and at the processing of the resulting biomass is another area of fundamental as well as applied research for chemistry. It is obvious that all disciplines of chemistry are involved in this challenge. The timelines for solving these challenges is decades rather than years. This does, however, not imply that in selected areas where we can build on longstanding research efforts rapid breakthroughs and emerging technologies may occur within the first decade of concentrated and interdisciplinary research that is about to come into operation [148] at the time when this account is being prepared.

4.10 Concluding Remarks

In summary, if we envisage a sustainable generic future energy system and take into consideration the idea that we need trajectories from fossil to renewable energy then we arrive at the following generic structure of such a system indicated in Figure 4.22. The purpose is here to highlight the multiple roles of chemistry, catalysis or as interdisciplinary effort of energy science (green functions in Figure 4.22).

The diagram highlights the dual role of hydrogen as molecular platform and of free electrons from RES. They are transmitted and controlled by respective infrastructures (yellow in Figure 4.22). Electrolysis and power stations (turbines and motors in large and small units) are the conversion devices. The key application areas of energy (gray in Figure 4.22) are fed with energy by the respective most efficient forms. A carbon cycle (blue lines in Figure 4.22) interconnects unavoidable sources and effective users. The carbon leakage through mobile applications needs compensation by capture from air by physical adsorption or through biomass. Fossil energy resources will prevail for quite some time and disappear only gradually (see Figure 4.1). The possible role of nuclear fusion is seen as heat source for process heat as well as controllable electricity source. As its future is highly uncertain, we must plan without it but we can well integrate it into an energy system that must be essentially fee of fossil fuel much earlier before fusion can be of relevance for energy systems.

In conclusion, the present chapter gives a crude overview of what it takes from the side of chemical science to engage into the energy challenge. Future energy systems will be based substantially on renewable solar primary energy but cannot be operated without a suite of technologies of chemical energy conversion dealing with storage and interconversion of energy carriers. The strategic and central science in this endeavor is catalysis [86, 149] much in the same way as it now enables the operation of chemical and petrochemical industries. The insight and progress in understanding this physicochemical phenomenon allows predicting that the challenge will be met at a level higher than that enabling the evolution of the present energy industry. Despite promising activities a still enormous scientific, technological, and economic effort is needed [148] to initiate the turn of the energy system [150] apart from the dominance of fossil energy carriers. The technologies of physical energy conversion by PV and wind and other innovative concepts are far ahead of the chemical technologies. The combined efforts of science can be made faster and better targeted if the multiple interfaces between science and nonscientific influences are recognized and dealt with by the individual person carrying responsibility in the energy science challenge. May this book help to create awareness for the interdisciplinary needed to create new and sustainable energy systems.

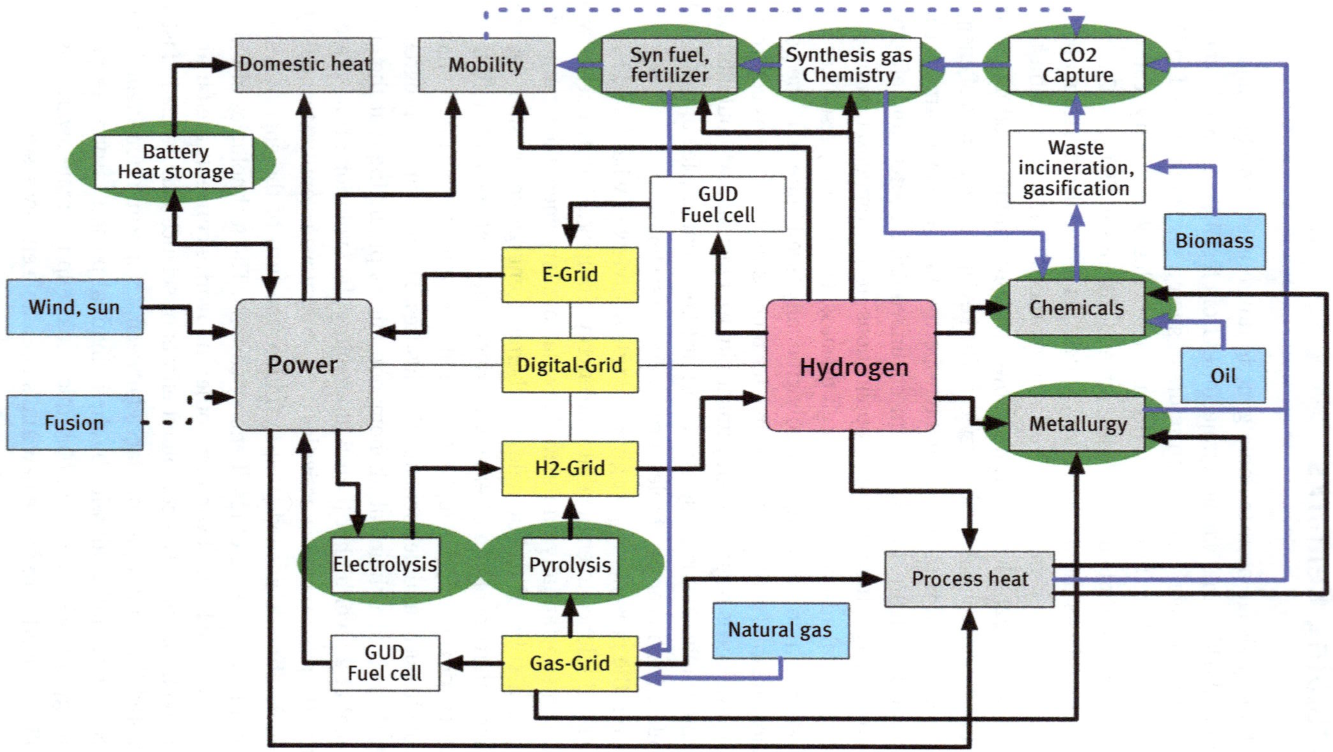

Figure 4.22: A generic energy system based upon RES in the form of free electrons and in "green" molecules. The key user groups are indicated in light grey. The green ovals mark elements of the system that depend on chemical processes and their suitable scalable technologies.

References

[1] Liao W, Heijungs R, Huppes G. Natural resource demand of global biofuels in the Anthropocene: A review. Renew Sustain Energy Rev. 2012;16(1):996–1003.

[2] Centi G, Misono M. New possibilities and opportunities for basic and applied research on selective oxidation by solid catalysts: an overview. Catal Today. 1998;41(4):287–96.

[3] Perathoner S, et al. Looking at the future of chemical production through the european roadmap on science and technology of catalysis the EU Effort for a long-term vision. Chemcatchem. 2017;9(6):904–09.

[4] Klankermayer J, et al. Selective catalytic synthesis using the combination of carbon dioxide and hydrogen: catalytic chess at the interface of energy and chemistry. Angew Chem Int Ed. 2016;55(26):7296–343.

[5] Deutz S, et al. Cleaner production of cleaner fuels: wind-to-wheel – environmental assessment of CO2-based oxymethylene ether as a drop-in fuel. Energy Environ Sci. 2018;11(2):331–43.

[6] Artz J, et al. Sustainable conversion of carbon dioxide: an integrated review of catalysis and life cycle assessment. Chem Rev. 2018;118(2):434–504.

[7] Simpson CD, et al. From graphite molecules to columnar superstructures – an exercise in nanoscience. J Mater Chem. 2004;14:494–504.

[8] Bell AT. The Impact of Nanoscience on Heterogeneous Catalysis. Vol. 2003, 1688–1691.

[9] Buller S, Strunk J. Nanostructure in energy conversion. J Energy Chem. 2016;25(2):171–90.

[10] Faber MS, Jin S. Earth-abundant inorganic electrocatalysts and their nanostructures for energy conversion applications. Energy Environ Sci. 2014;7(11):3519–42.

[11] Somorjai GA, Frei H, Park JY. Advancing the frontiers in nanocatalysis, biointerfaces, and renewable energy conversion by innovations of surface techniques. J Am Chem Soc. 2009;131(46):16589–605.

[12] Yang F, Antonietti M. Artificial humic acids: sustainable materials against climate change. Adv Sci. 2020;7(5.

[13] Fellinger T-P, et al. Mesoporous nitrogen-doped carbon for the electrocatalytic synthesis of hydrogen peroxide. J Am Chem Soc. 2012;134(9):4072–75.

[14] Titirici MM, Antonietti M, Baccile N. Hydrothermal carbon from biomass: a comparison of the local structure from poly- to monosaccharides and pentoses/hexoses. Green Chem. 2008;10(11):1204–12.

[15] Rockstrom J, et al. Climate Policy a roadmap for rapid decarbonization. Science. 2017;355 (6331):1269–71.

[16] BMWI, http://www.bmwi.de/Navigation/Technologie-und-Energie/Energiepolitik/energie daten.html.

[17] Beller JN, Beller M. Spiers memorial lecture. Artificial photosynthesis: an introduction. Faraday Discuss. 2019;215:9–14.

[18] Aleklett K, et al. The Peak of the Oil Age – Analyzing the world oil production Reference Scenario in World Energy Outlook 2008. Energy Policy. 2010;38(3):1398–414.

[19] Schlögl R. The Revolution Continues: Energiewende 2.0. Angew Chem Int Ed. 2015;54: 4436–39.

[20] Bundesregierung, NWS, BMWI, Editor. 2020.

[21] www.leopoldina.org/fileadmin/user_upload/Politik/Empfehlungen/Nationale_Empfehlun gen/Ad-hoc-Stellungnahme_Energie_Juni_2011.pdf

[22] Bundesregierung. Energiekonzept. 2010.

[23] Haas T, et al. Technical photosynthesis involving CO2 electrolysis and fermentation. Nat Catal. 2018;1(1):32–39.

[24] DECHEMA. Roadmap Chemie 2050. 2019.
[25] Deshmukh MK, Deshmukh SS. Modeling of hybrid renewable energy systems. Renew Sustain Energy Rev. 2008;12(1):235–49.
[26] Ghanadan R, Koomey JG. Using energy scenarios to explore alternative energy pathways in California. Energy Policy. 2005;33(9):1117–42.
[27] Doroodian K, Boyd R. The linkage between oil price shocks and economic growth with inflation in the presence of technological advances: a CGE model. Energy Policy. 2003;31(10): 989–1006.
[28] Heimann M, Reichstein M. Terrestrial ecosystem carbon dynamics and climate feedbacks. Nature. 2008;451(7176):289–92.
[29] Clark JH, Deswarte FEI, Farmer TJ. The integration of green chemistry into future biorefineries. Biofuel Bioprod Biorefin Biofpr. 2009;3(1):72–90.
[30] Demirbas A. Progress and recent trends in biofuels. Prog Energy Combust Sci. 2007;33(1):1–18.
[31] Campbell H, et al. Efficient energy utilization and environmental issues applied to power planning. Energy Policy. 2011;39(6):3630–37.
[32] Kemp R. Technology and the transition to environmental sustainability – the problem of technological regime shifts. Futures. 1994;26(10):1023–46.
[33] Lewis NS. Toward cost-effective solar energy use. Science. 2007;315(5813):798–801.
[34] Riahi K, Gruebler A, Nakicenovic N. Scenarios of long-term socio-economic and environmental development under climate stabilization. Technol Forecast Soc Change. 2007;74(7):887–935.
[35] Nocito F, Dibenedetto A. Atmospheric CO2 mitigation technologies: carbon capture utilization and storage. Curr Opin Green Sustainable Chem. 2020;21:34–43.
[36] Mazzotti M, et al. Direct air capture of CO2 with chemicals: optimization of a two-loop hydroxide carbonate system using a countercurrent air-liquid contactor. Clim Change. 2013;118(1):119–35.
[37] Grim RG, et al. Transforming the carbon economy: challenges and opportunities in the convergence of low-cost electricity and reductive CO2 utilization. Energy Environ Sci. 2020;13(2):472–94.
[38] Koytsoumpa EI, Bergins C, Kakaras E. The CO2 economy: Review of CO2 capture and reuse technologies. J Supercrit Fluids. 2018;132:3–16.
[39] Ertl G, Freund HJ. Catalysis and Surface Science. Phys Today. 1999;52(1):32–38.
[40] Schlögl R. The role of chemistry in the energy challenge. ChemSusChem. 2010;3(2):209–22.
[41] Centi G, Perathoner S. Opportunities and prospects in the chemical recycling of carbon dioxide to fuels. Catal Today. 2009;148(3–4):191–205.
[42] Schlögl R. Ammonia Synthesis. In: Handbook of Heterogeneous Catalysis. Ertl G, et al. editors. Wiley VCH Verlag: Weinheim; 2008, 2501–75.
[43] Olah GA. Beyond oil and gas: The methanol economy. Angew Chem Int Ed. 2005;44(18): 2636–39.
[44] Raudaskoski R, et al. Catalytic activation of CO2: Use of secondary CO2 for the production of synthesis gas and for methanol synthesis over copper-based zirconia-containing catalysts. Catal Today. 2009;144(3–4):318–23.
[45] Tang QL, Hong QJ, Liu ZP. CO2 fixation into methanol at Cu/ZrO2 interface from first principles kinetic Monte Carlo. J Catal. 2009;263(1):114–22.
[46] Liu Y, et al. Efficient conversion of carbon dioxide to methanol using copper catalyst by a new low-temperature hydrogenation process. Chem Lett. 2007;36:1182–83.
[47] Baldi M, et al. Catalytic combustion of C3 hydrocarbons and oxygenates over Mn3O4. Appl Catal B. 1998;16(1):43–51.

[48] Favre A, et al. Catalytic combustion of methane over barium hexaferrites. Catal Letters. 1998;49(3,4):207–11.
[49] Zavyalova U, et al. Morphology and microstructure of Li/MgO catalysts for the oxidative coupling of methane. Chemcatchem. 2011;3(6):949–59.
[50] Su DS, et al. Nanocarbons in selective oxidative dehydrogenation reaction. Catal Today. 2005;102:110–14.
[51] Frank B, et al. Oxygen Insertion Catalysis by sp(2) Carbon. Angew Chem Int Ed. 2011;50(43): 10226–30.
[52] Sanfiz AC, et al. Dynamics of the MoVTeNb Oxide M1 Phase in Propane Oxidation. J Phys Chem. 2010;114(4):1912–21.
[53] Rozanska X, Sauer J. Oxidative dehydrogenation of hydrocarbons by V3O7+ compared to other vanadium oxide species. J Phys Chem A. 2009;113(43):11586–94.
[54] Schlögl R. Active sites for propane oxidation: some generic considerations. Top Catal. 2011;54(10–12):627–38.
[55] Herrmann J-M. The electronic factor and related redox processes in oxidation catalysis. Catal Today. 2006;112(1–4):73–77.
[56] Gruene P, et al. Role of dispersion of vanadia on SBA-15 in the oxidative dehydrogenation of propane. Catal Today. 2010;157(1–4):137–42.
[57] Geske M, et al. In-situ investigation of gas phase radical chemistry in the catalytic partial oxidation of methane on Pt. Catal Today. 2009;142:61–69.
[58] Guelder OL, et al. Unified behaviour of maximum soot yields of methane, ethane and propane laminar diffusion flames at high pressures. Combust Flame. 2011;158(10):2037–44.
[59] Hickman DA, Schmidt LD. Production of syngas by direct catalytic-oxidation of methane. Science. 1993;259(5093):343–46.
[60] Gelin P, Primet M. Complete oxidation of methane at low temperature over noble metal based catalysts: a review. Appl Catal B Environ. 2002;39(1):1–37.
[61] Ertl G, Jennings JR. Elementary Steps in Ammonia Synthesis: The Surface Science Approach. In: Catalytic Ammonia Synthesis: Fundamentals and Practice, Fundamental and Applied Catalysis. Plenum Press; 1991. 1.
[62] Reuter K, Scheffler M. First-principles kinetic Monte Carlo simulations for heterogeneous catalysis: Application to the CO oxidation at RuO2(110). Phys Rev B. 2006;73(4).
[63] Stampfl C, et al. Catalysis and corrosion: the theoretical surface-science context. Surf Sci. 2002;500(1–3):368–94.
[64] Somorjai GA, Park JY. Concepts, instruments, and model systems that enabled the rapid evolution of surface science. Surf Sci. 2009;603(10–12):1293–300.
[65] Baron M, et al. Interaction of Gold with Cerium Oxide Supports: CeO2(111) Thin Films vs CeOx Nanoparticles. J Phys Chem C. 2009;113(15):6042–49.
[66] Freund HJ. Adsorption of gases on complex: Solid surfaces. Angew Chem Int Ed Engl. 1997;36(5):452–75.
[67] Williams WD, et al. Metallic corner atoms in gold clusters supported on rutile are the dominant active site during water-gas shift catalysis. J Am Chem Soc. 2010;132(40): 14018–20.
[68] Dobler J, Pritzsche M, Sauer J. Oxidation of methanol to formaldehyde on supported vanadium oxide catalysts compared to gas phase molecules. J Am Chem Soc. 2005;127(31): 10861–68.
[69] Sauer J, Freund HJ. Models in catalysis. Catal Letters. 2015;145(1):109–25.
[70] Freund H-J, et al. CO oxidation as a prototypical reaction for heterogeneous processes. Angew Chem Int Ed. 2011;50(43):10064–94.

[71] Schlögl R. In situ Characterisation of Practical Heterogeneous Catalysis. In: Basic Principles in Applied Catalysis. Baerns M, editor. Springer Verlag; Berlin Heidelberg New York: 2004, 321–60.
[72] Prasad V, et al. High throughput multiscale modeling for design of experiments, catalysts, and reactors: Application to hydrogen production from ammonia. Chem Eng Sci. 2010;65(1): 240–46.
[73] Schlögl R. Heterogeneous Catalysis. Angew Chem Int Ed. 2015;54(11):3465–520.
[74] Thomas J, Raja R. The advantages and future potential of single-site heterogeneous catalysts. Top Catal. 2006;40(1):3–17.
[75] Thomas JM. Turning-points in catalysis. Angew Chem Int Ed. 1994;33(9):913–37.
[76] Zambelli T, et al. Identification of the "active sites" of a surface-catalyzed reaction. Science. 1996;273(5282):1688–90.
[77] Homann K, Kuhlenbeck H, Freund HJ. N 2 Adsorption and Discussion on Thin Iron Films on W(110). Surf Sci. 1995;327:216–24.
[78] Dahl S, et al. Electronic factors in catalysis: the volcano curve and the effect of promotion in catalytic ammonia synthesis. Appl Catal A-Genl. 2001;222(1–2):19–29.
[79] Cuenya BR, Behafarid F. Nanocatalysis: size- and shape-dependent chemisorption and catalytic reactivity. Surf Sci Rep. 2015;70(2):135–87.
[80] Teschner D, et al. Alkyne hydrogenation over Pd catalysts: A new paradigm. J Catal. 2006;242 (1):26–37.
[81] Goodman DW. Catalysis – from single-crystals to the real-world. Surf Sci. 1994;299 (1–3):837–48.
[82] Cambell CT, et al. Micro- and Macro-Kinetics: Their Relationship in Heterogeneous Catalysis. Top Catal. 1994: J. C. Baltzer AG. 353–66.
[83] Chorkendorff IN, J.W. Concepts of Modern Catalysis and Kinetics. Weinheim: Wiley-VCH; 2007.
[84] Aparicio LM, et al. Ammonia synthesis kinetics: surface chemistry, rate expressions, and kinetic analysis. Front Catal Ammonia Synth Beyond. 1994, J. C. Baltzer AG, 233.
[85] Hellman A, et al. Predicting catalysis: Understanding ammonia synthesis from first-principles calculations. J Phys Chem. 2006;110(36):17719–35.
[86] Seh ZW, et al. Combining theory and experiment in electrocatalysis: Insights into materials design. Science. 2017;355(6321):146-+.
[87] Schlögl R. Combinatorial chemistry in heterogeneous catalysis: a new scientific approach or "the kingïs new clothes"? Angew Chem Int Ed. 1998;37(17):2333–36.
[88] Schlögl R. The role of chemistry in the energy challenge. Chemsuschem. 2010;3(2):209–22.
[89] Bell AT. The impact of nanoscience on heterogeneous catalysis. Science. 2003;299(5613): 1688–91.
[90] Schlögl R, Abd Hamid SB. Nanocatalysis: The recap of mature science or something really new? Angew Chem Int Ed. 2004;43(13):1628–37.
[91] Greeley J, et al. Computational high-throughput screening of electrocatalytic materials for hydrogen evolution. Nat Mater. 2006;5(11):909–13.
[92] Xin HL, et al. Effects of d-band shape on the surface reactivity of transition-metal alloys. Phys Rev B. 2014;89(11).
[93] Nørskov JK, et al. Towards the computational design of solid catalysts. Nat Chem. 2009;1(1): 37–46.
[94] Vojvodic A, et al. Exploring the limits: A low-pressure, low-temperature Haber-Bosch process. Chem Phys Lett. 2014;598:108–12.
[95] Medford AJ, et al. Activity and selectivity trends in synthesis gas conversion to higher alcohols. Top Catal. 2014;57(1–4):135–42.

[96] Nørskov JK, et al. Universality in heterogeneous catalysis. J Catal. 2002;209(2):275–78.
[97] Lausche AC, et al. Application of a new informatics tool in heterogeneous catalysis: Analysis of methanol dehydrogenation on transition metal catalysts for the production of anhydrous formaldehyde. J Catal. 2012;291:133–37.
[98] Studt F, et al. Identification of non-precious metal alloy catalysts for selective hydrogenation of acetylene. Science. 2008;320(5881):1320–22.
[99] Behrens M, et al. Knowledge-based development of a nitrate-free synthesis route for Cu/ZnO methanol synthesis catalysts via formate precursors (vol 47, pg 1701, 2011). Chem Commun. 2011;47(48):12885–12885.
[100] Hutchings GJ. Heterogeneous catalysts-discovery and design. J Mater Chem. 2009;19(9): 1222–35.
[101] Antipov EV, Abakumov AM, Istomin SY. Target-Aimed Synthesis of Anion-Deficient Perovskites. 2008. 8543–52.
[102] Trunschke A, et al. Towards experimental handbooks in catalysis. Top Catal. 2020;63 (19–20):1683–99.
[103] Behrens M, Schlögl R. How to prepare a good Cu/ZnO catalyst or the role of solid state chemistry for the synthesis of nanostructured catalysts. Z Anorg Allg Chem. 2013;639(15): 2683–95.
[104] Tompos A, et al. Combinatorial optimization and synthesis of multiple promoted MoVNbTe catalysts for oxidation of propane to acrylic acid. Catal Today. 2019, in Press.
[105] Schlögl R, Abd Hamid SB. Nanocatalysis: The recap of mature science or something really new? Nanocatalyse: Alter Wein aus neuen Schl„uchen oder etwas wirklich Neues? Angew Chem Int Ed / (Angew Chem). 2004;43(116, 13):1628/(1656)–1637/(1667).
[106] Jones G. Industrial computational catalysis and its relation to the digital revolution. Nat Catal. 2018;1(5):311–13.
[107] Yarulina I, et al. Structure-performance descriptors and the role of Lewis acidity in the methanol-to-propylene process. Nat Chem. 2018;10(8):804–12.
[108] Calle-Vallejo F, et al. Why is bulk thermochemistry a good descriptor for the electrocatalytic activity of transition metal oxides? ACS Catal. 2015;5(2):869–73.
[109] Grabow LC, et al. Descriptor-Based Analysis Applied to HCN Synthesis from NH3 and CH4. Angew Chem Int Ed. 2011;50(20):4601–05.
[110] Rinaldi R, Schuth F. Design of solid catalysts for the conversion of biomass. Energy Environ Sci. 2009;2(6):610–26.
[111] Lubitz W, Reijerse EJ, Messinger J. Solar water-splitting into H-2 and O-2: design principles of photosystem II and hydrogenases. Energy Environ Sci. 2008;1(1):15–31.
[112] Knop-Gericke A, et al. X-Ray photoelectron spectroscopy for investigation of heterogeneous catalytic processes. Adv Catal. 52 2009;52:213–72.
[113] Sasaki T, et al. Performance of low-voltage STEM/TEM with delta corrector and cold field emission gun. J Electron Microsc. 2010;59:S7–S13.
[114] Pyrz WD, et al. The effect of Nb or Ta substitution into the M1 phase of the MoV(Nb,Ta)TeO selective oxidation catalyst. Catal Today. 2009;142(3–4):320–28.
[115] Su DS, et al. Surface chemistry of Ag particles: Identification of oxide species by aberration-corrected TEM and by DFT calculations. Angew Chem Int Ed. 2008;47(27):5005–08.
[116] Sun YN, et al. CO adsorption and dissociation on iron oxide supported Pt particles. Surf Sci. 2009;603(20):3099–103.
[117] Mette K, et al. Redox dynamics of Ni catalysts in CO_2 reforming of methane. Catal Today. 2015;242(Part A):1010–110.
[118] Cao J, et al. In situ observation of oscillatory redox dynamics of copper. Nat Commun. 2020;11(1):3554.

[119] Plodinec M, et al. Insights into chemical dynamics and their impact on the reactivity of Pt Nanoparticles during CO oxidation by operando TEM. ACS Catal. 2020;10(5):3183–93.
[120] Schlögl R. Catalysis 4.0. ChemCatChem. 2017;9(4):533–41.
[121] Barber J. Photosynthetic energy conversion: natural and artificial. Chem Soc Rev. 2009;38(1): 185–96.
[122] Lange JP. Lignocellulose conversion: an introduction to chemistry, process and economics. Biofuel Bioprod Biorefin Biofpr. 2007;1(1):39–48.
[123] Rinaldi R, et al. Which controls the depolymerization of cellulose in ionic liquids: the solid acid catalyst or cellulose? Chemsuschem. 3(2), 266–76.
[124] Chheda J,N, Huber GW, Dumesic JA. Liquid-phase catalytic processing of biomass-derived oxygenated hydrocarbons to fuels and chemicals. Angew Chem Int Ed. 2007;46(38):7164–83.
[125] Rosillo-Calle F. A review of biomass energy-shortcomings and concerns. J Chem Technol Biotechnol. 2016;91(7):1933–45.
[126] Steffen W, et al. The terrestrial carbon cycle: Implications for the Kyoto Protocol. Science. 1998;280(5368):1393–94.
[127] Archer D, Brovkin V. The millennial atmospheric lifetime of anthropogenic CO2. Clim Change. 2008;90(3):283–97.
[128] Solomon SD, et al. IPCC 2007 Climate Change The Physical Science Base. Cambridge University Press; 2007. 996.
[129] Knittel CR, Roberts MR. An empirical examination of restructured electricity prices. Energy Econ. 2005;27(5):791–817.
[130] De Vries BJM, Van Vuuren DP, Hoogwijk MM. Renewable energy sources: Their global potential for the first-half of the twenty-first century at a global level: An integrated approach. Energy Policy. 2007;35(4):2590–610.
[131] Agency IE. A-Fundamental-Look-at-Supply-Side-Energy-Reserves-for-the-Planet. 2015.
[132] Jensen SH, Larsen PH, Mogensen M. Hydrogen and synthetic fuel production from renewable energy sources. Int J Hydrogen Energy. 2007;32(15):3253–57.
[133] Philibert C. Renewable Energy for Industry. Paris: International Energy Agency; 2017. 65.
[134] Haszeldine RS. Carbon capture and storage: how green can black be? Science. 2009;325 (5948):1647–52.
[135] Bruhn T, Naims H, Olfe-Krautlein B. Separating the debate on CO2 utilisation from carbon capture and storage. Environ Sci Policy. 2016;60:38–43.
[136] Zurbel A, et al. Methanol synthesis by CO2 hydrogenation overCu/ZnO/Al2O3 catalysts under fluctuating conditions. Chem Ing Tech. 2018;90(5):721–24.
[137] Nath K, Das D. Production and storage of hydrogen: Present scenario and future perspective. J Sci Ind Res. 2007;66:701–09.
[138] Clarke RE, et al. Direct coupling of an electrolyser to a solar PV system for generating hydrogen. Int J Hydrogen Energy. 2009;34(6):2531–42.
[139] De Leon CP, et al. Redox flow cells for energy conversion. J Power Sources. 2006;160(1): 716–32.
[140] Chu S, Cui Y, Liu N. The path towards sustainable energy. Nat Mater. 2017;16(1):16–22.
[141] Tarascon JM, Armand M. Issues and challenges facing rechargeable lithium batteries. Nature. 2001;414(6861):359–67.
[142] Altenburg T, Schamp EW, Chaudhary A. The emergence of electromobility: Comparing technological pathways in France, Germany, China and India. Sci Public Policy. 2016;43(4): 464–75.
[143] Jamnik J, Maier J. Nanocrystallinity effects in lithium battery materials – Aspects of nano-ionics. Part IV. Phys Chem Chem Phys. 2003;5(23):5215–20.

[144] Rossmeisl J, et al. Electrolysis of water on oxide surfaces. J Electroanal Chem. 2007;607 (1–2):83–89.
[145] Ayers K. The potential of proton exchange membrane-based electrolysis technology. Curr Opin Electrochem. 2019;18:9–15.
[146] Lorenzi G, et al. Exergo-economic analysis of a direct biogas upgrading process to synthetic natural gas via integrated high-temperature electrolysis and methanation. Energy. 2017;141: 1524–37.
[147] Rossmeisl J, et al. Comparing electrochemical and biological water splitting. J Phys Chem. 2007;111(51):18821–23.
[148] Nørskov J, et al. Research needs report - Energy-X 2020.
[149] Schlögl R. Put the sun in the tank: future developments in sustainable energy systems. Angew Chem Int Ed. 2019;58(1):343–48.
[150] Deerberg G, Oles M, Schlögl R. Carbon2Chem (R) – A Successful cross-industrial network for sustainable climate protection and preservation of competitiveness. Chem Ing Tech. 2020;92(10):1379–1379.

2 Electrochemical

Karl Doblhofer

5 Electrochemical Concepts: A Practical Guide

5.1 Introduction

Generations of students have been introduced to electrochemistry with the glass apparatus shown schematically in Figure 5.1.

Two platinum electrodes operate in a U-shaped electrochemical cell containing an acidic electrolyte. If a voltage of approximately 1.8 V is applied between the two electrodes, current flows that produces both oxygen and hydrogen according to the following reactions:

$$2H_2O \rightarrow O_2 + 4H^+ + 4e^- \tag{5.1}$$

$$2H^+ + 2e^- \rightarrow H_2 \tag{5.2}$$

The overall process is the electrochemical "decomposition" of water:

$$2H_2O \rightarrow O_2 + 2H_2, \quad \Delta G^0 = 474.4\ \text{kJmol}^{-1} \tag{5.3}$$

In eq. (5.3), the standard Gibbs free energy, ΔG^0, is given. $\Delta G^0 > 0$ means that the reaction is endothermic. The value of ΔG^0 defines the thermodynamic (minimum) electric energy required to decompose 2 mol liquid water according to the chemical equation (5.3) under standard conditions. An electrochemical process in which an endothermic reaction is forced to proceed by the imposition of an external voltage is termed *electrolysis*, and the cell is said to operate as an electrolytic cell. The electric energy W (in joules, J) consumed in an electrolytic process is defined by the product of applied voltage U (in volts, V) and the electric charge Q (in coulombs, C) that has passed through the cell, as shown in eq. (5.4). The electric charge is $Q = I\ t$ (I is the electric current in amperes, A, and t is the time in seconds, s).

$$W = Q\ U \tag{5.4}$$

The electric work W required for an electrochemical process in a practical electrolytic cell will be larger than ΔG^0. For example, the indicated value of ΔG^0 in eq. (5.3) corresponds to a thermodynamic standard potential difference of $E^0 = 1.23$ V, while the voltage typically applied to the cell of Figure 5.1 is approximately $U = 1.8$ V.

Acknowledgments: Valuable comments on this work by Prof. Waldfried Plieth, Technical University Dresden, are gratefully acknowledged.

Karl Doblhofer, Fritz-Haber-Institut der, Max-Planck-Gesellschaft, Department of Physical Chemistry, Faradayweg 4-6, 14195 Berlin, Germany, e-mail: acsek@fhi-berlin.mpg.de

https://doi.org/10.1515/9783110608458-005

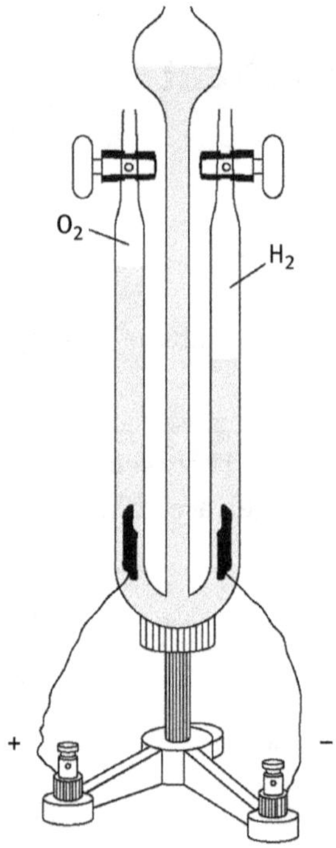

Figure 5.1: Apparatus for the electrochemical "decomposition" of water [1].

The overvoltage ($U - E^0$) is the driving force for the reaction to proceed at a desirable rate. However, it also causes an energy loss, which leads to heating of the cell.

The electrochemical *water decomposition* reaction can be reversed according to

$$O_2 + 2H_2 \rightarrow 2H_2O, \quad \Delta G^0 = -474.4\ \text{kJmol}^{-1} \tag{5.3a}$$

This is achieved electrochemically by supplying hydrogen and oxygen to platinum electrodes, such as the ones in Figure 5.1. A voltage is established between the two electrodes that may drive electric current across an external load. Such an electrochemical cell constitutes a *galvanic* cell. Fuel cells and batteries are examples of galvanic cells of technical interest. The term *fuel cell* has been established, because the reactants (fuels) are supplied continuously to the electrodes from outside the cell. The term *storage battery* or *battery* is used if the reactants and products are confined to the cell. The electric work done by galvanic cells is again defined by eq. (5.4).

One may consider the energy required for the water decomposition reaction to be stored as chemical energy in the products oxygen and hydrogen. The stored energy may be recovered in a fuel cell via eq. (5.3a). Figure 5.2 illustrates this concept of electrochemical energy storage [2]. A great number of electrochemical systems can be used for energy conversion and storage according to this concept [3, 4].

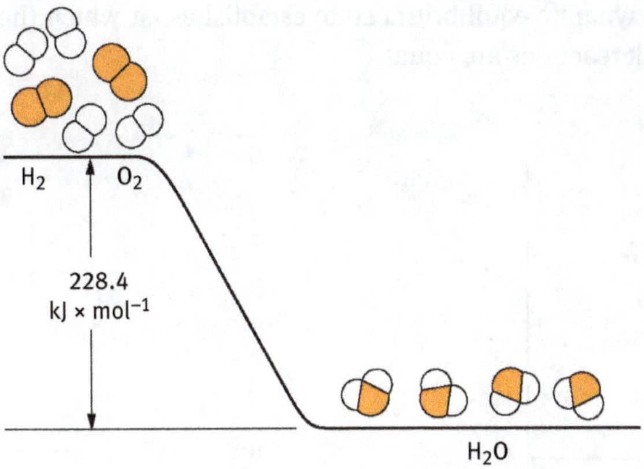

Figure 5.2: The concept of (electro-)chemical energy storage [2]. The indicated energy value corresponds to ΔG^0 for the decomposition of 1 mol gaseous H_2O according to eq. (5.3).

5.2 Electrodes in Electrolytes

The following types of electrodes are considered in the following discussions of electrochemical systems:

a. Gas electrodes
Examples are the oxygen electrode, eq. (5.1), or the hydrogen electrode (i.e., the H^+/H_2 reaction), eq. (5.2). Electron transfer proceeds regularly with the reacting species adsorbed at an electrode surface. In many cases, electrocatalysis by the electrode metal plays an important role. Associated chemical reactions, such as protonation and dissociation, render the reaction mechanisms complex. This is true in particular for the oxygen electrode.

b. Metal/metal ion electrodes
A metal (Me) acts as an electrode in the electrolyte containing its ions. The metal is said to be electroactive according to $Me \leftrightarrow Me^{n+} + n\,e^-$. An example is the Ag/Ag^+ electrode represented in Figure 5.3 (i.e., a silver metal electrode immersed in an electrolyte containing silver ions). The Zn/Zn^{++} electrode is another metal/metal ion electrode. It is widely used in batteries.

c. Redox electrodes
An inert conductor, such as platinum, gold, or graphite, is immersed in an electrolyte containing a dissolved redox couple (e.g., Fe^{3+}/Fe^{2+}). Electron transfer proceeds between the electrode and the redox species:

$$Fe^{3+} + e^- \leftrightarrow Fe^{2+} \tag{5.5}$$

At the equilibrium potential, a dynamic equilibrium state establishes, at which the rates of the forward and the back reactions are equal.

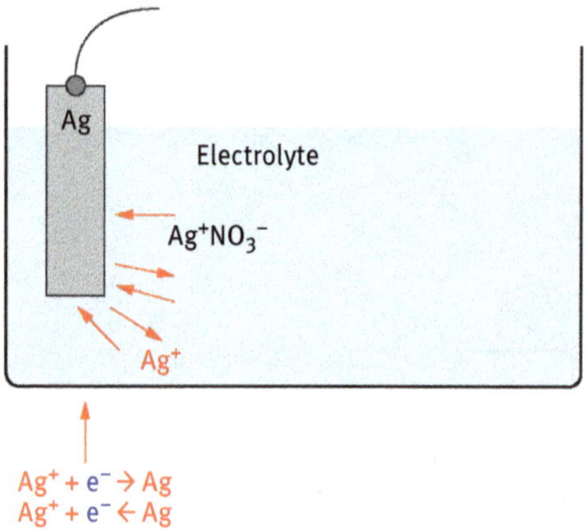

Figure 5.3: The Ag/Ag^+ electrode: an example of a metal/metal ion electrode. The dynamic equilibrium state is shown (i.e., both the forward and back reaction proceed at the same rate).

d. Concentration cells
Electrochemical cells may consist of two electrodes of the same type, but with different activities of the electroactive species in the electrolyte. Such systems will be discussed subsequently in the context of the Nernst equation.

e. Ideal polarizable electrodes
If an inert conductor operates as an electrode in an inert electrolyte (e.g., KCl), there exists a potential region in which no interfacial charge transfer takes place. Still, the variation of the applied electrode potential is associated with the flow of a capacitive (*nonfaradaic*) current. It is caused by charging and discharging of the electric double layer (i.e., the accumulation of electronic excess charge in the metal surface and ionic excess charge of the opposite sign at the solution side of the interface). The double-layer capacitance is on the order of 10–100 μF cm^{-2}. High-surface-area electrodes (e.g., carbon) can store substantial quantities of electric charge, and thus of electric energy.

5.3 Energetics of Electrode Reactions

Electron transfer reactions, such as the one in Figure 5.3, $Ag^+ + e^- \leftrightarrow Ag$, are associated with a substantial energy change. It is important to thoroughly understand this energy change, in particular in the context of energy conversion. With the help of a thermodynamic cycle (Born–Haber cycle), it is possible to obtain values of the Gibbs free energy change, ΔG^0, associated with the electrochemical transformation of 1 gram-equivalent (g-equivalent) of an electroactive species. For convenience, in the following analysis all reactants and products are present in their *standard state* (i.e., as pure solids, or at unit activity if dissolved). The superscript 0 in the symbol of the quantity concerned indicates the standard state.

The analysis of the Ag/Ag^+ electrode of Figure 5.3 is presented in Figure 5.4. The electrode is at *open circuit* (i.e., no external potential is imposed on the electrode). At the metal/electrolyte interface, the reaction $Ag^+ + e^- \leftrightarrow Ag$ proceeds both in the forward and backward directions. The following thermodynamic cycle leads to a value of ΔG^0 for the reaction $Ag \rightarrow Ag^+ + e^-$. It is conducted using energy data from various sources, in particular from Goodisman [5]. The first step is the sublimation of 1 mol metallic silver. The silver atoms are then ionized, for which the ionization potential (IP) must be supplied. The electrons are transferred back into the metal phase with the liberation of the free energy corresponding to the work function, Φ. At last, the standard free energy of solvation of Ag^+ is determined from the corresponding enthalpy (ΔH^0) and entropy (ΔS^0) values. Note that

$$\Delta G = \Delta H - T\Delta S \tag{5.6}$$

The data given in Figure 5.4 may be summarized as follows in Table 5.1.

The previous analysis clearly shows that a substantial amount of energy has to be supplied to the metallic silver in order to make it dissolve as Ag^+. On the other hand, Ag^+ from the electrolyte will spontaneously react at the silver surface to form metallic Ag: $\Delta G^0(Ag^+ \rightarrow Ag) \approx -95\ kJ\ mol^{-1}$. Indeed, this is why silver is considered a noble metal.

The establishment of a stable equilibrium potential between the metal electrode and the electrolyte can be straightforwardly explained as follows. As soon as the neutral silver electrode gets in contact with the electrolyte, the reaction $Ag^+ + e^- \rightarrow Ag$ will proceed, while the rate of the back reaction is negligibly small. The excess positive charge injected into the silver electrode will render the potential of the electrode (relative to the electrolyte solution) more and more positive. Consequently, the rate of the reaction $Ag^+ + e^- \rightarrow Ag$ decreases while the back reaction is accelerated. Eventually, the dynamic equilibrium state is attained at which both the silver deposition and dissolution reactions proceed at the same rate, as illustrated in Figure 5.3. This state corresponds to a stable positive potential drop across the metal/electrolyte interface. The experimental investigation of such interfacial potential drops requires methods as used in *double-layer* studies. While this is an important and interesting theme, it is not the subject of this work.

Rather, electrochemical cells containing a second electrode will be considered in the following. These are the electrochemical systems of practical interest, as shown in Figure 5.1. Still, one should keep in mind that the behavior of a two-electrode cell is the combination of the two individual electrodes. And indeed, experimental methods have been developed, in particular potentiostatic methods, by which the effects of the second electrode are virtually eliminated so that one electrode of interest may be studied [6].

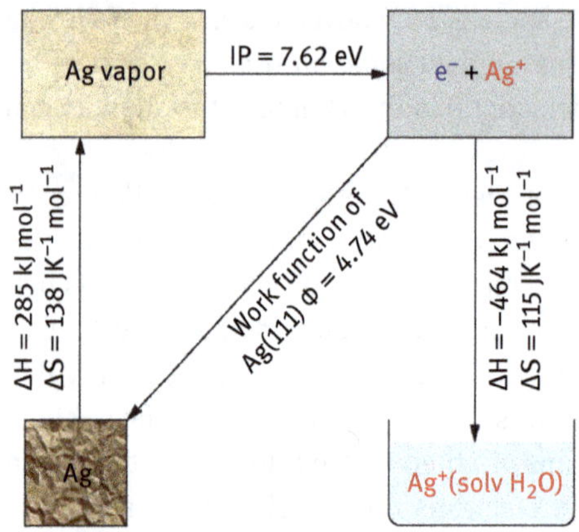

Figure 5.4: Born–Haber cycle for determining the standard free energy change associated with the reaction $Ag \rightarrow Ag^+$(solvated). The data are shown as they are normally reported in the literature: 1 eV/atom = 96.5 kJ mol^{-1}. For more details, see the text.

Table 5.1: Summary of the data from Figure 5.4.

Δ*G*(sublim)	**=**	**+247 kJ mol^{-1}**
IP	=	+735 kJ mol^{-1}
Φ	=	−457 kJ mol^{-1}
ΔG^0(solv Ag^+)	≈	−430 kJ mol^{-1}
$\Delta G^0(Ag \rightarrow Ag^+_{solv})$	≈	+95 kJ mol^{-1}
	≈	+1 eV atom^{-1}

5.4 The Electrochemical Cell

5.4.1 The Concept

Electrochemical cells of practical interest contain two electrodes, defining two *half cells*. Consider, for example, the Ag/Ag^+ and H_2/H^+ electrodes shown schematically in the cell of Figure 5.5. At the interfaces between the electrodes and the electrolyte, the following electrochemical reactions proceed:

$$1/2H_2 \rightarrow H^+ + e^- \quad (5.2a)$$

$$Ag^+ + e^- \rightarrow Ag \quad (5.7)$$

Driven by the energy produced in these reactions, electrons move spontaneously through the external electric circuit containing the *load* (i.e., a device powered by electric current). According to the definition, the system in the configuration of Figure 5.5 operates as a galvanic cell. The Ag electrode is termed the *cathode* because from the outside circuit the electrons enter into it, causing *cathodic reduction* of an electroactive species (Ag^+, in the electrolyte). On the other hand, from the H_2/H^+ electrode the electrons exit into the external circuit. Thus, it constitutes the *anode*, at which *anodic oxidation* of H_2 takes place.

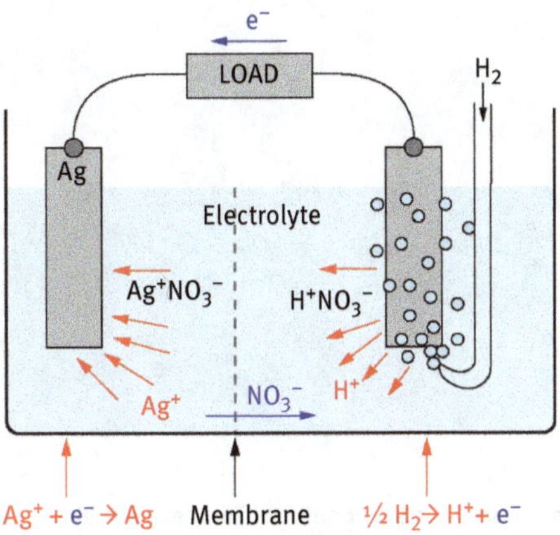

Figure 5.5: Electrochemical cell consisting of an Ag^+/Ag and an H_2/H^+ electrode. Electronic current (e^-) flows through the external circuit containing the "load." Inside the cell, the electric charge is shown to be transported by nitrate ions ("NO_3^-").

There is an important feature common to all electrochemical cells. Provisions have to be made to prevent the electroactive components from reaching the opposite electrodes. Typically, this is achieved by a membrane between the anode and the cathode compartments. The membrane must be permeable to ions (NO_3^- in Figure 5.5). Interestingly, the water decomposition cell of Figure 5.1 operates without a membrane. This has the consequence that some dissolved hydrogen and oxygen will be transported by diffusion and convection to the respective opposite electrodes. There, both reactions (5.1) and (5.2) will proceed in the back direction. In the device of Figure 5.1, this is tolerable because this cell is designed for demonstration; efficiency is not an issue.

The two blue arrows, marked as "NO_3^-" (nitrate ions) and as "e^-" (electrons), point to the continuous flow of negative electric charge across the entire electric circuit, consisting both of the cell and the external load. Ions are the charge carriers in the electrolyte, while electrons transport the charge in the metal and the external load. The transition from electronic to ionic charge transport occurs at the electrode/electrolyte interface upon electron transfer between the electrode and an electron acceptor or donor in the electrolyte.

In order to characterize the equilibrium state of the electrochemical cells, the electronic conductor connecting the electrodes of Figure 5.5 is removed. Now, both electrodes may establish their individual charge-transfer equilibrium state. A stable equilibrium potential difference, E, is established between the two electrodes.

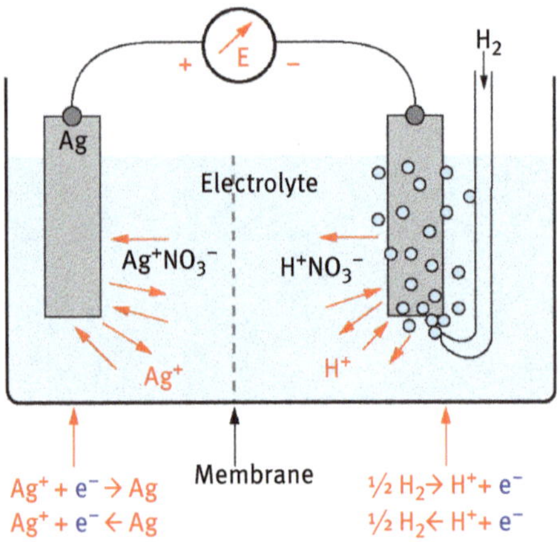

Figure 5.6: The electrochemical cell of Figure 5.5 under equilibrium condition. The device that measures the equilibrium cell voltage E has such large an internal resistance that the current flow across the cell is practically zero.

If both electrode processes operate under standard conditions, this voltage is E^0, the equilibrium standard electrode potential difference. Values of E and E^0 may be conveniently measured with electrometers of so large an internal resistance that the current flow is nearly zero. Figure 5.6 illustrates the measurement and the equilibrium state. The value of E^0 is a most significant quantity characterizing the thermodynamics of an electrochemical cell. Various important features of E and E^0 will be addressed in the following chapters.

Consider replacing the load of the cell of Figure 5.5 by an electric power supply. If a voltage of $U = E$ (or $U = E^0$) is applied to the electrodes, the cell equilibrium voltage will be exactly compensated (i.e., one observes zero current flow), and the equilibrium state will establish, as illustrated in Figure 5.6. If this power supply is set to apply a voltage of $U > E$ (or $U > E^0$) to the cell, the cell operation changes in a fundamental way. The direction of the current flow is reversed. Both electrode reactions are reversed (i.e., the H_2/H^+ electrode is now the cathode at which H_2 is produced). At the Ag/Ag^+ electrode, silver is oxidized to Ag^+; it is now the anode. At $U > E$, the cell operates as an electrolytic cell. If this system were used as a storage battery, this operation at $U > E$ would be termed *charging the battery*, while under the conditions of Figure 5.5 the battery would be discharged at $U < E$.

5.4.2 Chemical and Electric Energy

Michael Faraday (1791–1867) was the first to realize that electrochemical processes are stoichiometrically associated with straightforward chemical reactions. Consider the cell of Figure 5.5. The two electrode reactions of the cell have been represented previously in eqs. (5.2a) and (5.7). Thus, the spontaneous overall chemical reaction is

$$Ag^+\,(\mathrm{soln}) + 1/2H_2 \rightarrow Ag + H^+\,(\mathrm{soln}) \tag{5.8}$$

The cell compartment containing the electron donor (in the system of Figure 5.5: H_2) is the one in which electron transfer proceeds from the solution into the electrode. The electrons may pass through the external load to the cathode in the second half cell, with electron transfer to the acceptor (Ag^+ in the electrolyte). Any electric work generated in a galvanic cell is energy produced in the corresponding chemical reaction. Consider a chemical reaction in the general form

$$aA\ +\ bB \rightarrow cC\ +\ dD \tag{5.9}$$

It is conventional to write the cell reaction so that it proceeds spontaneously from left to right. The equilibrium constant K for reaction (5.9) may be formulated as follows:

$$K = e^{-\frac{\Delta G^\circ(\mathrm{reaction})\cdot}{RT}} = \left(\frac{a_C^c a_D^d}{a_A^c a_B^b}\right)_{\mathrm{equil}} \tag{5.10}$$

where R is the gas constant, and instances of a are the equilibrium activities of the compounds defined by the subscripts. As reaction (5.9) is considered to proceed spontaneously to form the products $C + D$, at equilibrium the numerator of the activity quotient of eq. (5.10) will be larger than the denominator – that is, the chemical equilibrium constant K will be $K > 1$, and the standard reaction free energy, ΔG^0(reaction), will have a negative sign, as indicated in the context of eqs. (5.3) and (5.3a). Values of ΔG^0(reaction) for 1 g-equivalent can be conveniently determined as follows:

$$\Delta G^0(\mathrm{reaction}) = \sum \Delta G^0 f(\mathrm{products}) - \sum \Delta G^0 f(\mathrm{reactants}) \tag{5.11}$$

where $\Delta G^0{}_f$ represents the standard free energies of formation. According to eq. (5.6), $\Delta G^0{}_f$ values may be determined from standard enthalpies and entropies of formation. These data are conveniently available in data collections such as the *Handbook of Chemistry and Physics* (CRC Press). The data for the cell of Figure 5.5 are listed in Table 5.2.

In the last column of Table 5.2 ("Δ"), the standard reaction enthalpies and entropies (products – reactants) are presented. Using the definition of the Gibbs free energy (5.6), one obtains for this reaction at $T = 298$ K:

$$\Delta G^0(\mathrm{reaction\ equation\ (5.8)}) = -77.2\ \mathrm{kJmol}^{-1} \tag{5.12}$$

Table 5.2: Values of the standard enthalpies and entropies of formation for the products and reactants of the chemical reaction proceeding in the electrochemical cell of Figure 5.5 (see eq. (5.8)). Note that the listed values for H^+_{soln} are zero because by convention the hydrogen electrode under standard condition is the reference system.

Chem. reaction	Ag^+_{soln}	½ H_2	→	Ag	H^+_{soln}	Δ
ΔH^0_f kJ mol^{-1}	105.8	0		0	0	−105.8
S^0_f J mol^{-1} K^{-1}	73.3	½ (130.6)		42.7	0	−95.9

5.4.3 The Maximum Electric Energy Produced and the Equilibrium Cell Voltage

As indicated previously, the electric charge passing through the cell and the external circuit is strictly proportional to the masses of reactants that have reacted at the electrodes. For example, if the Ag^+/Ag reaction as represented in eq. (5.7) proceeds such that 1 g-equivalent (107.9 g Ag) is deposited, 1 mol of electrons is required, which is the charge of an electron (1.6021×10^{-19} C) multiplied with the Avogadro constant, that is, $N = 6.022 \times 10^{23}$ mol^{-1},

$$\text{the charge of 1 mol electrons is } 96,487 \text{ C} = 1 \text{ Faraday} = 1 \text{ F} \tag{5.13}$$

Considering eq. (5.4), it is clear that one has to operate a galvanic cell at the maximum possible cell voltage in order to maximize the electric energy yield. As shown previously (Figure 5.6), the maximum value of U is the equilibrium electrode potential difference E or E^0. Thus, one may formulate the fundamental relationship between chemical and electric energy:

$$\Delta G\,(\text{reaction}) = -nFE \tag{5.14}$$

$$\Delta G^0\,(\text{reaction}) = -nFE^0 \tag{5.15}$$

Consider again our model system with both the Ag/Ag^+ and H_2/H^+ electrodes at equilibrium. Under standard conditions, the Gibbs reaction free energy was shown above to be $\Delta G^0 = -77.2$ kJ (g-equivalent)$^{-1}$. Thus, according to eq. (5.15), $E^0 = -\Delta G^0/nF = 0.80$ V. And indeed, this standard equilibrium cell voltage may be conveniently measured with the cell in the configuration of Figure 5.6.

One may short circuit galvanic cells (e.g., the one in Figure 5.5) by connecting the electrodes with an electronic conductor of zero resistance. This forces the potential difference between the electrodes to $U = 0$ V. In this situation, the electrode reactions will proceed at maximum rate (i.e., at maximum galvanic current flow). This current will be limited only by the internal resistance of the cell – that is, by the interfacial charge-transfer resistances and by the ohmic resistance of the electrolyte. Note that under this condition, no electric energy is harvested, as $W = QU = 0$. The energy dissipates as in the regular chemical reaction. The cell heats up.

Clearly, in galvanic cells of practical interest (i.e., in batteries and fuel cells), the voltage of operation U has to be a compromise. It must be smaller than E for obtaining an adequate current flow. On the other hand, it has to be close to E in order to recover as much as possible electric work, W.

Analogous considerations apply if the cell is used as an electrolytic cell. In this case, the minimum energy is required at a very small overvoltage ($U - E \approx 0$, but positive). For practical systems, the overvoltage required depends on the kinetics of the process. Consider again the water decomposition reaction of Figure 5.1. While E^0 for this reaction is $E^0 = 1.23$ V, a cell voltage of $U \approx 1.8$ V is typically applied. This means that instead of the 474 KJ (g-equivalent)$^{-1}$ thermodynamically required, actually 695 KJ mol^{-1} is used for the production of 1 g-equivalent $O_2 + 2\ H_2$ according to eq. (5.3). Clearly, this is a considerable energy loss.

5.5 Concentration Dependence of *E*: The Nernst Equation

5.5.1 The Nernst Equation

The Gibbs free energy of a dissolved species varies with the activity, a, and in the case of a gas, with the partial pressure, p. Consider the equation for the reaction isotherm (16) for the general chemical reaction equation (5.9):

$$\Delta G(\text{reaction}) - \Delta G^\circ(\text{reaction}) = RT \ln\left(\frac{a_C^c a_D^d}{a_A^c a_B^b}\right) \tag{5.16}$$

Equation (5.16) may be formulated for the equilibrium cell voltage, E:

$$E - E^0 = -\frac{RT}{nF} \ln\left(\frac{a_C^c a_D^d}{a_A^c a_B^b}\right) \tag{5.17}$$

Equation (5.17) is valid if interfacial charge-transfer equilibrium between the electrodes and both the reactants (A,B) and products (C,D) has been established. For illustration, consider two relevant special conditions: If the underlying chemical reaction is at equilibrium, as characterized by the activity quotient of eq. (5.10), the driving force for the chemical, and thus for the electrochemical reaction, is zero – that is, ΔG (reaction) $= -nFE = 0$. On the other hand, if standard conditions prevail, then ΔG (reaction) $= \Delta G^0$(reaction) (i.e., $E = E^0$). Equation (5.17) is valid for any combination of two electrodes making up a complete cell.

As indicated previously, it is desirable to consider the individual electrode reactions independently. One might suppose that this could be achieved by characterizing the individual electrodes as described in Section 3.1.3. However, for reasons of sound thermodynamics, another method has been established. It was decided to relate all electrode reactions to one common reference electrode. Electrochemists have chosen the H^+/H_2 reaction under standard conditions ($a_{H+} = 1$ M; $p_{H2} = 1$ bar) as such a general reference electrode. It is termed the *normal hydrogen electrode* or the *standard hydrogen electrode* (SHE). Thus, whenever E and E^0 values are presented for individual electrode reactions (half cells), it is understood that these values pertain to a complete cell in which the SHE constitutes the second electrode.

To illustrate this, consider again the model cell of Figure 5.6. According to eq. (5.17), the equilibrium cell potential, E, may be formulated as follows:

$$E = E^0 = -\frac{RT}{nF}\ln\frac{a_{Ag}a_{H^+}}{a_{H^+}\sqrt{pH_2}} \tag{5.18a}$$

$$= E^0 = -\frac{RT}{nF}\ln\frac{a_{Ag}}{a_{Ag^+}} - \frac{RT}{nF}\ln\frac{a_{H^+}}{\sqrt{pH_2}} \tag{5.18b}$$

In eq. (5.18b), the activity quotient is separated into the terms relating to the silver electrode and the hydrogen electrode. We assume that both electrodes (Ag^+/Ag and H^+/H_2) operate under the standard condition (i.e., the H^+/H_2 electrode of our cell happens to constitute the SHE). This means that the equilibrium voltage of the cell of Figure 5.6 is identical with the half-cell equilibrium potential $E^0(Ag^+/Ag) = 0.80$ V. Furthermore, we note that the activity of the element silver is per definition unity. As the stoichiometric number of electrons transferred is one, the Nernst equation for the Ag^+/Ag electrode can be formulated in the following convenient and standard way:

$$E = E^0 + \frac{RT}{F}\ln a_{Ag^+} \tag{5.19}$$

As shown in eq. (5.19), it is conventional for Nernst equations to represent the sum of E^0 and the logarithmic term, rather than the difference. Consequently, in the activity quotient, the activity of the oxidized form of the electroactive species has to be written as the numerator. Standard potentials of individual electrode reactions are conveniently available in textbooks of physical chemistry and electrochemistry, and in relevant handbooks. A small selection is presented in Table 5.3.

The standard equilibrium cell voltage resulting from a combination of any two electrodes is the difference between the two standard potentials, $E^0(2) - E^0(1)$. For instance, the standard cell equilibrium voltage of the combination F_2/F^- with the Li^+/Li electrode would be 5,911 V. Correspondingly, the standard free energy change of the underlying chemical reaction, $1/2\,F_2 + Li \rightarrow F^- + Li^+$, is $\Delta G^0 = -570$ kJ (g-equivalent)$^{-1}$.

If the electrode reactants and products are not in the standard state, the equilibrium cell voltage will be the difference between the E values (i.e., the corresponding activity terms have to be included). Consider again the cell consisting of the half reactions Ag^+/Ag and H^+/H_2. The equilibrium potential difference between the two electrodes, $E = E(2) - E(1)$, is represented in eq. (5.20). Considering that $a_{Ag} = 1$ and $E^0(H^+/H_2) = 0$ V, eq. (5.20) is identical to eq. (5.18):

$$E = E^0(Ag^+/Ag) + \frac{RT}{F}\ln a_{Ag^+} - \left(E^0(H^+/H_2) + \frac{RT}{F}\ln\frac{a_{H^+}}{\sqrt{pH_2}}\right) \tag{5.20}$$

Table 5.3: Selected standard electrode potentials at 25 °C.

Electrode reaction	**E^0 in volt**
$Li^+ + e^- \leftrightarrow Li$	−3.045
$Na^+ + e^- \leftrightarrow Na$	−2.714
$Al^{3+} + 3\,e^- \leftrightarrow Al$	−1.662
$Zn^{++} + 2\,e^- \leftrightarrow Zn$	−0.763
$Pb^{++} + 2\,e^- \leftrightarrow Pb$	−0.126
$2\,H + + 2\,e^- \leftrightarrow H_2$	**0.000**
$Cu^{++} + 2\,e^- \leftrightarrow Cu$	+0.337
$I_2 + 2\,e^- \leftrightarrow 2\,I^-$	+0.535
$Fe^{3+} + e^- \leftrightarrow Fe^{++}$	+0.770
$Ag^+ + e^- \leftrightarrow Ag$	+0.799
$Cl_2 + 2\,e^- \leftrightarrow 2\,Cl^-$	+1.360
$F_2 + 2\,e^- \leftrightarrow 2\,F^-$	+2.866

5.5.2 Concentration Cells

Electrochemical cells may consist of two electrodes of the same type, but with different concentrations of the electroactive species in the electrolyte. Such cells are known as concentration cells. For example, two platinum electrodes operate in two H^+/H_2 solutions of different activity, separated by a membrane. The equilibrium cell voltage is defined by eq. (5.21a). As the standard potential is the same for both electrode reactions, the measurable cell voltage will depend only on the activity ratios, eq. (5.21b). If in this system both electrolytes were in equilibrium with the same H_2 pressure, the measured E would respond linearly to the pH difference between the two electrolytes, eq. (5.21c) (i.e., a pH electrode):

$$E = \left[E^0(H^+/H_2) + \frac{RT}{F}\ln\frac{a_{H^+}(2)}{\sqrt{pH_2(2)}} - E^0(H^+/H_2) + \frac{RT}{F}\ln\frac{a_{H^+}}{\sqrt{pH_2}}\right] \tag{5.21a}$$

$$= \frac{RT}{F} \ln \frac{a_{H^+}(2)\sqrt{pH_2(1)}}{\sqrt{pH_2(2)}a_{H^+}(1)} \tag{5.21b}$$

$$= \frac{RT}{F}[pH(1) - pH(2)] \tag{5.21c}$$

The technical process of copper refining is an important example of an electrolytic concentration cell. Two copper electrodes operate in the same Cu^{++}-containing electrolyte, one as anode, the other as cathode. Clearly, at equilibrium, $E = 0$. The current flow at the anode will dissolve Cu as Cu^{2+}, raising the concentration there. On the other hand, at the cathode, Cu^{2+} deposits as Cu. Consequently, near the cathode, the Cu^{2+} concentration decreases. Thus, a cell voltage will establish that opposes the applied voltage and leads to a loss of energy. Note that the aim of the process is the purification of the metal. Some of the contaminants present in the anode copper will remain in the electrolyte; others form a solid deposit in the cell, so that the copper deposited on the cathode (electrolytic copper) contains fewer impurities than the original metal.

5.6 The Temperature Dependence of the Equilibrium Cell Voltage, *E*

Consider eq. (5.6), that is, the definition of ΔG. On this basis, one may formulate the equilibrium potential of an electrochemical cell in the following way:

$$E = -\Delta H(\text{reaction})/nF + T\Delta H(\text{reaction})/nF \tag{5.22}$$

If one makes the reasonable assumption that ΔH(reaction) and ΔS(reaction) do not significantly depend on the temperature, eq. (5.22) may be used conveniently to calculate the temperature dependence of E. Consider, for example, Table 5.2, in which the values of ΔH^0(reaction) and ΔS^0(reaction) for the cell consisting of the Ag^+/Ag and H_2/H^+ electrodes were determined. For room temperature (i.e., 298 K), a value of $E^0 = 0.80$ V was found. Raising the temperature by 100 K (i.e., to $T = 398$ K), a new value of $E^0 = 0.70$ V results. Note the significant effect of temperature on the equilibrium potential.

Differentiating eq. (5.22), one obtains

$$\partial E/\partial T = \Delta S(\text{reaction})/nF \tag{5.23}$$

Clearly, the temperature dependence of the equilibrium cell voltage is associated with the entropy change of the underlying chemical reaction. In the galvanic cell considered previously (Ag^+/Ag and H_2/H^+ electrodes), hydrogen is transformed into dissolved H^+. This means a reactant in the gas phase becomes a product in the condensed phase. This is a typical process associated with a negative change of reaction entropy, and thus with a negative temperature coefficient of the equilibrium potential.

Numerous electrode reactions require substantial overvoltages to proceed at acceptable rates. A method for enhancing the reaction rate at a given overvoltage is operation at higher temperature. One has to consider that this might be a futile effort if the cell voltage (i.e., the equilibrium state of the system) changes unfavorably.

5.7 Conclusion

It was the purpose of this chapter to present fundamental thermodynamic concepts for evaluating electrochemical processes for energy conversion and storage. To illustrate the merits of these concepts, two highly exothermic chemical reactions are presented in Table 5.4, with the corresponding thermodynamic data.

Table 5.4: Selected thermodynamic data of chemical processes relevant for electrochemical energy conversion.

	ΔH^0(reaction), kJ/mol	ΔS^0(reaction), J/mol K	ΔG^0(reaction), kJ/mol at 298 K	n
$2H_2 + O_2 \rightarrow 2H_2O_{(liq)}$	−572	−326.3	−474.4 (83% of ΔHO)	4
$C + O_2 \rightarrow CO_{2(gas)}$	−393	+2.9	−394	4

Consider first the reaction:

$$O_2 + 2H_2 \rightarrow 2H_2O \tag{5.3a}$$

It constitutes the reaction underlying the most widely used fuel-cell system. Note first that this reaction is associated with a relatively large negative entropy change. According to eq. (5.6), the free energy that might be converted into electric work at room temperature, ΔG^0(reaction), is only 83% of the reaction enthalpy, ΔH^0(reaction). This means that the gain in efficiency resulting from energy conversion by the electrochemical process relative to the heat engine is reduced substantially. Dividing the reported value of ΔG^0(reaction) by −4 F, the standard equilibrium cell voltage $E^0 = 1.23$ V is obtained. Unfortunately, practical fuel cells operate at voltages on the order of 0.85 V, which reduces the efficiency further by approximately one-third. Operation of the fuel cell at elevated temperature will indeed accelerate the sluggish cathodic O_2 reduction reaction. However, this effect will be partially compensated by the negative temperature coefficient of the equilibrium cell voltage, caused by the negative ΔS^0(reaction) term. Thus, the evaluation of the oxygen/hydrogen fuel cell on the basis of such fundamental concepts makes this system appear far from ideal.

By contrast, consider the reaction of carbon with oxygen:

$$C + O_2 \rightarrow CO_2(\text{gas}) \tag{5.24}$$

In this case, the entropy change is positive – that is, the ΔG^0(reaction) exceeds the value of the reaction enthalpy. Thus, conducting this reaction in an efficient fuel cell would mean an enormous gain in energy conversion efficiency in comparison with burning the coal in a heat engine. Clearly, it would be most desirable to develop an efficient process in which carbon acts as an electron donor at an electrode. This fact was pointed out by Wilhelm Ostwald as early as 1894 [7]. Unfortunately, since the days of Ostwald, no significant progress has been made in developing such systems. It would certainly be a rewarding subject of future research.

References

[1] Gobrecht H. Bergmann Schäfer Lehrbuch der Experimentalphysik. Vol. II, Elektrizität und Magnetismus. Berlin, Germany: Walter de Gruyter & Co; 1971.

[2] Dickerson RE, Geis I. Chemie. Weinheim, Germany: VCH; 1990.

[3] Vielstich W, Lamm A, Gasteiger HA. editors. Handbook of Fuel Cells. Vol. I, Fundamentals and Survey of Systems. Chichester, UK: John Wiley; 2003.

[4] Bard AJ, Stratmann M, editors. Encyclopedia of Electrochemistry. Vol. I, Thermodynamics and Electrified interfaces. Vol. VII, Inorganic Electrochemistry. Weinheim, Germany: Wiley-VCH; 2006.

[5] Goodisman J. Electrochemistry, Theoretical Fundamentals. New York: John Wiley & Sons; 1987.

[6] Bard AJ, Faulkner LR. Electrochemical Methods: Fundamentals and Application. New York: Wiley; 2001.

[7] Ostwald W. Die wissenschaftliche Elektrochemie der Gegenwart und die technische der Zukunft. Zeitschrift Elektrotechnik Elektrochemie. 1894;4:122–25.

Jan Rossmeisl

6 Water-Splitting Conceptual Approach

6.1 Foreword to 2020 Edition

The concepts and considerations described a decade ago that still hold the fundamentals of water splitting are exactly the same and so are the limitations. However, since then the field has gotten more attention and is often referred to as power2X. In 2019, I was, together with many of the authors of this book, part of an European initiative "Energy-X" [1] and the outcome of a two-day workshop was a report on the research needs within catalysis. I was specifically looking into the research needs on water splitting. It is clear from that report and others that water splitting will become the bottleneck for scaling power2X to a global scale.

There are different electrolyzer technologies and they all have different pros and cons and let me, as an example, focus on polymer electrolyte membrane (PEM) acid electrolysis. On the cathode, the hydrogen side, Pt with a low loading can be applied and with an annual production of Pt of a few hundred tons that is not a problem to scale. However, on the anode, the oxygen side, only Ir oxide is a viable catalyst material having an acceptable stability combined with an acceptable activity. By comparing the annual production of Ir, which is just a few tons and the required loading, it is obvious that Ir is not scalable. Our global power demand is around 20 TW, any scalable energy technology have to have an impact on that scale. Assuming that 1 ton of Ir should correspond to 0.1 TW power stored in H_2, it can be calculated that present day technology is a factor of ~50 off [2]. Today PEM acid electrolysis is simply not a scalable technology. For other electrolysis technologies, the challenges are different, but none of them is satisfying and there is an urgent need for breakthroughs. The question is what kind of breakthroughs can we hope for?

For energy conversion reactions, we are used to fundamental efficiency limitations such as the Carnot efficiency for heat engines, Betz limit for wind turbines, and the Shockley–Queisser limit for solar cells. This chapter illustrates a fundamental limit for low-temperature water splitting which represents an upper bond to the efficiency. Identifying the limitation will not solve the challenge, but it might help us understand the boundary conditions for the technologies.

Jan Rossmeisl, University of Copenhagen, Department of Chemistry Universitetsparken 5, 2100 København Ø, Denmark, e-mail: jan.rossmeisl@chem.ku.dk

https://doi.org/10.1515/9783110608458-006

6.2 Introduction

The elements and concepts of electrocatalysis were already in place more than 50 years ago. However, the tools to investigate electrochemical process have developed dramatically since. Recently, ab initio (mostly density functional theory (DFT)) simulations on systems of a size relevant for electrochemisty are emerging. DFT provides a method to calculate interaction energies, reaction barriers, and reaction mechanisms from first principles. Therefore, many of the fundamental concepts of electrochemistry are now being revisited and broad into a quantitative form based on electronic structure simulations. The new microscopic insight can hopefully help in solving some of the technological challenges in electrocatalysis and especially related to electrochemical energy conversion. One of the main challenges is the energy efficiency of the water-splitting reaction [3]. See also Chapter 7 by P. Strasser of this book.

The ab initio simulations should be used there they really can make a difference. In this context, it is important to realize that trends and differences are much better described in DFT than absolute numbers. Even semi-quantitatively trends can be obtained. In spite of the importance of having an accurate description of the real electrochemical environment for obtaining absolute values, it seems that many trends and relative features can be obtained within a somewhat simpler framework [4].

In this chapter I will focus on trends in electrocatalysis of the water-splitting reaction or the oxygen evolution reaction (OER), which is the reaction at the anode side in an electrolysis cell. Furthermore, I will present a simple framework for addressing OER applying DFT simulations. For further reading there are previous book chapters where the approach has been reviewed [5–7].

6.3 Fundamentals

The overall splitting of water into hydrogen and oxygen reads:

$$2H_2O(1) + \text{energy} \rightarrow O_2(g) + 2H_2(g) \tag{6.1}$$

This simple reaction is maybe the most studied, and it has been known for a very long time [8]. The amount of energy that has to enter in reaction (6.1) should at least cancel difference in enthalpy between water and hydrogen and oxygen which is 5.92 eV at standard conditions. As there are three diatomic molecules on the product side and only two liquid water molecules on the initial side, the entropy is higher at the product side. This means that some of the energy can be provided as heat without violating the second law of thermodynamic. The total work that has to be provided is 4.92 eV. This is in most cases the relevant energy to look at if the reaction takes place at constant temperature and pressure. However, it can be a difficult engineering issue to keep the temperature constant inside the electrolysis cell as the reaction is running if only the needed work is provided.

In the electrochemical cell the redox reaction is separated into two half-cell reactions, namely the reduction at the cathode and oxidation at the anode. The sum of the two reactions yields reaction (6.1):

$$4H^+ + 4e^-(\text{cathode}) \rightarrow 2H_2(g) \tag{6.2}$$

$$2H_2O(l) \rightarrow O_2(g) + 4H^+ + 4e^-(\text{anode}) \tag{6.3}$$

By applying a sufficiently high potential difference between the cathode and the anode the chemical potential of the electrons can be changed so that both half-cell reactions become possible. At the anode electrode the Fermi level has to be low enough that it can accept electrons from the water molecules, on the cathode electrode the Fermi level has to be high so that the cathode can donate electrons to protons forming hydrogen. In the overall reaction four electrons are involved per oxygen molecule, that means that the work that has to be done on each electron is $\Delta G^{\ominus}/4 = \underline{E}^{\ominus} = 1.23$ eV, which means that the potential difference of at least 1.23 V has to applied. However, due to losses and sluggish electrocatalysis a substantial additional potential must in reality be applied to accelerate the reaction.

This extra potential provides a driving force for the water-splitting reaction and is called the overpotential, η. The energy efficiency of the electrolysis can be defined as $(E^{\ominus} + \eta)/E^{\ominus}$. It is therefore important to design the electrolysis cell so that the overpotential is as small as possible. At the same time the current should be as high as possible, as the current defines how large and thereby how expensive the electrolysis cell has to be in order to produce hydrogen at a given rate.

6.4 Standard (Reversible) Hydrogen Electrode

To model the water-splitting reaction and any electrochemical reaction, it is necessary to include the potential or bias. As long as only reaction energies are considered it is possible to avoid explicit modeling of the electrochemical interface. Within this approach, barriers for charge transfer reactions cannot be treated. In this section, I will introduce the reference for the potential.

The work function of the metal electrode does not influence the bias. Imagine two electrodes in contact with each other and with the electrolyte. As all three elements are conducting they will have a common Fermi level. At both sides, we let the equilibrium of the reversible hydrogen electrode evolve:

$$2H^+ + 2e^- \leftrightarrow H_2(g) \tag{6.4}$$

If the pH of the electrolyte and the pressure of hydrogen is constant the chemical potential of protons and hydrogen molecules are also constant. This means that the chemical potential (the Fermi level) of the electrons is constant, which defines a reference for the potential. When a bias is applied the potentials are hereafter measured relative to this reversible hydrogen electrode reference, which evidentially is independent on the electrode material.

The work function, however, influences the amount of transferred charge when the Fermi levels are aligned. This in turn determines the surface charge and the field over the narrow electrochemical interface. In general it has to be tested to what degree the surface chemistry is dependent on the field. For the intermediates involved in water electrolysis the adsorption energies are only very weakly influenced by the field and normally this effect can be neglected [9].

The reversible hydrogen electrode reference is useful for calculating the free energy change of a reaction involving proton and electrons [10]. Instead of simulating protons in solution we can reference the energy to gas phase hydrogen molecule. The adsorption of hydrogen can of cause also be referenced to gas phase hydrogen, which means that the free energy difference between protons in solution and adsorbed hydrogen can be calculated.

6.5 The Cathode Half-Cell Reaction

At the cathode where the hydrogen is produced one can think of three different reaction steps – Volmer, Tafel, and Herovsky:

$$H^+ + e^- + {}^* \rightarrow H^* \text{ Volmer} \tag{6.5}$$

$$H^+ + e^- + H^* + {}^* \rightarrow 2H^* \rightarrow H_2 + 2^* \text{ Tafel} \tag{6.6}$$

$$H^+ + e^- + H^* \rightarrow H_2 + {}^* \text{ Herovsky} \tag{6.7}$$

This results in two different possible reaction paths, Volmer–Tafel or Volmer–Herovsky. No-matter the reaction mechanism the intermediate is always H^*. This makes hydrogen evolution a very simple case for rational catalyst material design. If H binds too weakly to the surface, H^+ cannot adsorb from the dissolved phase and if it binds too strongly, it will have difficulty leaving the surface for the gas phase. One would expect the optimal rate when hydrogen at the surface is as stable as gas-phase hydrogen – which by definition has the same free energy as solvated protons and electrons at zero potential relative to the reversible hydrogen electrode [11].

Hydrogen evolution has been studied intensively using DFT [12–14] and new catalyst materials have been suggested [1, 15]. However, the hydrogen evolution catalysis is very efficient on many noble metal surfaces and in most cases it is not the surface catalysis that limits the cathode reaction.

From a catalysis point of view the hydrogen evolution reaction can in principle be catalyzed perfectly, and in reality it is seen that there is almost no overpotential on the cathode side. The main challenge in terms of catalysis is related to the more complex OER at the anode electrode [16, 17].

6.6 The Anode Half-Cell Reaction

The anode is the main focus of this chapter (see Figure 6.1). The total half-cell reaction involves four charge transfer steps.

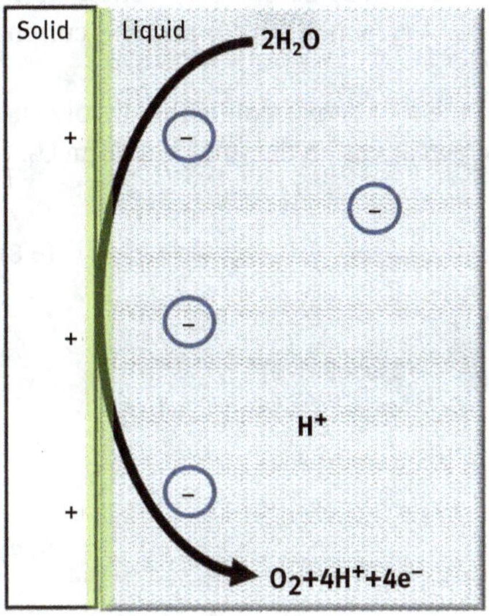

Figure 6.1: Schematic picture of the anode interface between electrode and electrolyte at which the OER takes place.

Let us call the different intermediate states for S_0–S_4 where S_0 corresponds to water and S_4 to O_2. Between each state an electron is removed and from S_0–S_4 also four protons are removed.

For a molecular catalysts the oxidation is driven by an oxidizing agent, which removes the electrons, when four electrons are removed the molecular catalyst should be so high in energy that it can react with water and form oxygen. The oxidizing agent needs to be able to accept electrons for all the intermediate states otherwise the oxidation will stop. The biggest difference in energy between two states will dictate how strong an oxidizing agent is needed and thereby the overpotential. In the perfect case the oxidizing agent only needs to be a little stronger than oxygen. Protons and electrons very often leave the catalyst together, but they do not need to. For the molecular catalyst the number of electrons at a state is important. It is possible to image a reaction there only a proton is removed keeping the number of electrons constant, as the electrons has to be transferred to the oxidizing agent, which is a reaction with a finite delay.

For an electrode surface it is not meaningful just to remove electrons, as for the electrode surface the number of electrons is not constant. Only the chemical potential for the electrons is constant, as the electrode is an electronic conductor the electrons can flow to and from the surface without any barrier or delay. In order to get the four different oxidation states a proton together with an electron is removed in each reaction step and thereby the oxidation state of the oxygen atoms on the surface is changed. Otherwise the picture is similar to that of the molecular catalyst. If the chemical potential is not sufficiently low to directly accept the electrons from each of every state then reaction will have to overcome a barrier and the current will be small.

On the electrode surface (see Figure 6.1) the four steps could either involve an HOO* intermediate or combination of two oxygen atoms on the surface to form O_2.

Reaction mechanism 1

$$\begin{aligned}&H_2O + {}^* \rightarrow HO^* + H^+ + e^- \\ &HO^* \rightarrow O^* + H^+ + e^- \\ &O^* + H_2O \rightarrow HOO^* + H^+ + e^- \text{ (associative)} \\ &HOO^* \rightarrow O_2 + {}^* + H^+ + e^-\end{aligned} \tag{6.8}$$

Reaction mechanism 2

$$\begin{aligned}&H_2O + {}^* \rightarrow HO^* + H^+ + e^- \\ &HO^* \rightarrow O^* + H^+ + e^- \\ &O^* + {}^* + H_2O \rightarrow HO^* + O^* + H^+ + e^- \text{ (dissociative)} \\ &HO^* + O^* \rightarrow 2O^* + H^+ + e^- \rightarrow O_2 + 2^*\end{aligned} \tag{6.9}$$

The reaction mechanism the catalyst will follow may depend on the electrode material. The difference is that in reaction mechanism (6.2), the last reaction step does not include an electron and proton. It is clear that reaction mechanism (6.2) can only be relevant if the barrier for recombining the two oxygen atoms is small. Later I will argue why including mechanism (6.2) in the analysis does not change the conclusion.

6.6.1 Free Energy Diagram

Focusing on reaction mechanism (6.1), I plot the free energy of the each intermediate S_0–S_4. The change in free energy for the four oxidation step is ΔG_{1-4}. Each reaction step involves one electron, which means that the change in free energy for one reaction step depends on the potential: $\Delta G_1 = S_1 - S_0 - eU$, $\Delta G_2 = S_2 - S_1 - eU$, $\Delta G_3 =$

$S_3 - S_2 - eU$, $\Delta G_1 = S_4 - S_3 - eU$. See Figure 6.2 where the free energy diagram for RuO_2 is plotted for $U = 0$ versus RHE, 1.23 V, E°, versus RHE and 1.60 V, ΔG^{OER} which is the smallest potential at which all reaction steps are downhill in free energy. The rate of the reaction will not increase if the potential is increased beyond this potential.

It is the state from which it is most difficult to remove a proton and electron that determines the smallest potential needed to obtain a high current. This is often referred to as the potential determining step. The size of the potential determining step can be estimated by just the biggest change in free energy in the free energy diagram [18]:

$$\Delta G^{OER} = \mathrm{Max}[\Delta G_1, \Delta G_2, \Delta G_3, \Delta G_4] \tag{6.10}$$

For RuO_2, ΔG_3 is the potential determining step. The overpotential calculated by this analysis is thus the additional potential needed compared to the equilibrium potential:

$$h^{OER} = (\Delta G^{OER}/e) - 1.23\mathrm{V} \tag{6.11}$$

This is a lower limit for the size of the actual overpotential as there could be additional barriers for the reaction steps, but these are not included in the analysis. The assumption is that trends are captured by the lower limit.

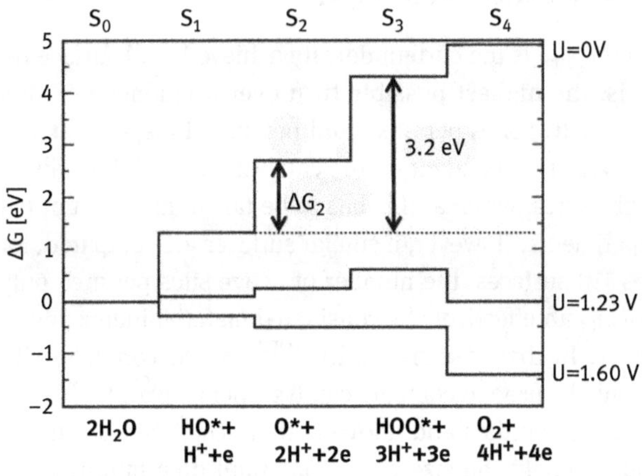

Figure 6.2: The free energy diagram of oxygen evolution on 110 facet of rutile RuO_2 at three different potentials, $U = 0$ V versus the RHE, the equilibrium potential, E°, $U = 1.23$ V and the smallest potential where all reaction steps are downhill, ΔG^{OER}, $U = 1.60$ V. The 3.2 eV difference between OH and HOO is shown by the arrow to the right, the smaller arrow to the left indicates the universal descriptor for OER.

6.6.2 Tafel Equation and ΔG^{OER}

The size of the potential determining step obtained from the free energy diagram above is a lower limit estimate of the size of the limiting potential. This potential is easy to calculate looking at the free energy diagram, but it is by no means easy to measure in experiments. However, it is linked to the experimental observables.

A good efficient catalyst, molecular as well as an electrode surface, needs both a low overpotential and a high current which means a low barrier for the rate-determining step. The current as a function of the potential can be expressed by the Tafel equation [19]:

$$j_a = j_0 \exp(\alpha\eta e/kT) \tag{6.12}$$

where j_a is the kinetic anode current density, j_0 the exchange current, η is the overpotential, and α is the transfer coefficient. Here I only consider $\alpha = 1$, but the model could be made for other values of α. The overpotential is written as $\eta \equiv U_0 - U$ in order to rewrite the Tafel equation:

$$j_a = j_0 \exp[\alpha(eU_0 - eU)/kT] = j_{limit} \exp[\alpha(eU^{OER} - eU)/kT] \tag{6.13}$$

where U^{OER} is the lowest potential at which j_a is maximum, so it is a parameter that only depends on the catalyst material. j_{limit} can be written as:

$$j_{limt} = j_0 \exp[\alpha(eU_0 - eU^{OER})/kT] \tag{6.14}$$

The physical meaning of the term j_{limit} is the current density achieved if all surface reactions are exothermic, that is, the highest possible turn-over-frequency per site. j_{limit} is dependent on number of active sites per area, and potential independent surface reactions. This means that j_{limit} is dependent on the catalyst material. If catalysts have a different number of active sites per area, this has to be taken into account before comparing theory and experiments. However, if similar surfaces are compared, for example, a set of rutile oxides 110 surfaces, the number of active sites per area only vary with a few presents and j_{limit} can effectively be considered material-independent, unlike exchange current density. In that case trends in U^{OER} should correlate with trends in activity, j_a. U^{OER} can only be used to compare catalysts per active site.

The assumption made in many studies and above is that eU^{OER} can be substituted with its own lower bond namely the size of the potential determining step, ΔG^{OER}, without changing trends. This corresponds to assuming that additional barriers are material independent or at least scales with ΔG^{OER}.

There is a difference between the concept of rate determining step and that of the potential determining step. It is clear that the exponential contains in eq. (6.13) the potential determining step whereas j_{limit} could contain the rate determining step. The potential determining step is also rate determining at low overpotentials, but as the driving force of the reaction is increased via the potential the size of this step is decreased until the point where the reaction rate is no longer increased by increasing the overpotential, at this point the rate is determined by the rate determining step. In experiments this is not very often seen, since the rate at some point is given by other issues than catalysis such as transport.

To summarize: Optimizing electrocatalytic activity is a matter of achieving the highest possible current at the lowest possible overpotential. That means that there are two things to consider the highest possible rate at the lowest possible potential. The highest possible current is often determined by number of sites per area, transport of reactants to and product from the surface, which is related to electrode structure, particle size, and many other things, but less related to the catalyst material per say.

Differences in electrocatalytic performance per active site between different materials are therefore mostly related to differences in overpotential. The step along the reaction path that gives rise to the highest overpotential is the potential determining step.

In this analysis, finding a good electrocatalyst becomes a matter of making the largest step ΔG_{1-4} as small as possible. If it was possible to tune all the level S_{1-3} independent on each other, then ΔG^{OER} could be only a bit larger than the equilibrium potential. However, the energy levels are not independent on each other.

6.6.3 Scaling Relations

The binding energies of different intermediates binding to a surface via the same kind of atom, for example, oxygen tends to scale linear with each other as the electrode material is varied (see Figure 6.3). The slopes of these scaling relations are determined by the number of bonds to the surface or more precisely the maximum number of hydrogen the molecule can covalently bind to. This is not always the same has number of bonds to the surface. Carbon is sp^3 hybridized and can coordinate to four hydrogen atoms, however, at a close packed surface one of the orbitals is pointing away from the surface and is not directly participating in the bond to the surface [20]. However, if a hydrogen atom is added to this C forming CH at the surface there are still three bonds the surface, but they are a bit weaker than for carbon alone. This is related to the coupling strength, which in the end ensures that the scaling for carbon species on the surface change with ¼ per hydrogen. This also ensures that molecules scale in the same manner independent on coordination of the metal atom it is binding to.

The slope is thus determined just by counting numbers of bonds to the surface. The intercept of the scaling relation is affected by the binding site geometry. Imagine a scaling relation between an intermediate that bind via two bonds to the surface as a function of the intermediate with one bonds to the surface, for example, $\Delta G^{O^*}(\Delta G^{HO^*})$. The slope will be two because HO* binds through one bond and O* via two bonds. Oxygen has a preference for hollow site binding, whereas HO* binds ontop or bridge. If we plot the most stable for each we get one intercept of the scaling relation. Now we change the surface and look at defects, an add atom. Both O* and HO* binds stronger to the add atom, but oxygen has to pay the prize of moving to the ontop binding to gain the increased interaction with the under-coordinated site. HO*, on the other hand, already prefers ontop binding why it gains the whole interacting. Please note that the slope of the scaling is conserved, which means that only the intercept changes. The scaling relation is therefore moved a bit toward strong HO* binding at a given oxygen binding strength.

That means that only two things can change the scaling the number of bonds and the binding site. The number of bonds is related to the intermediates, the binding site can be changed by defects or by isolating metals atoms like in oxides where only ontop binding is possible [21]. However, comparing two intermediates (e.g., HO* and HOO*) where the number of bonds to the surface is the same and they prefer the same binding geometry there is nothing what can be varied. The binding of both HO* and HOO* can change, but the scaling relation, $\Delta G^{HOO*}(\Delta G^{HO*})$, stays the same. This is a universal scaling relation. As HO* and HOO* binds via one bond and prefer the same binding site this means that $\Delta G^{HOO*}(\Delta G^{HO*}) = \Delta G^{HO*} + 3.2$ eV on all metal and metal oxide surfaces.

Over the last years this scaling relation has been confirmed in simulations, regardless of the parameters of calculations [22]. In Figure 6.3, the universal scaling is shown; points represents calculated data collected in the literature. The level of theory and the calculations parameters vary, however, the scaling seems robust. This is further supported by a study on the systematic error related to DFT for similar gas-phase reactions and adsorption on a few metals, arriving at the same relationship between HO- and HOO- species [23].

ΔG^{O*} and ΔG^{HOO*} as functions of ΔG^{HO*} are sketched in Figure 6.4. As example is the ΔG^{HO*} value for RuO_2 indicated, each crossing between the vertical line and the lines representing the free energy of H_2O, G^{HO*}, ΔG^{O*}, ΔG^{HOO*}, and O_2 are the sizes of the reaction steps for RuO_2 for $U = 0$ V and it is the same information as Figure 6.2.

6.6.4 Universal Scaling and Trends in Activity

It is clear that the universal scaling relations are in the end what limit the efficiency. Interestingly, the constant difference between the adsorption energies of HO* and HOO* of 3.2 eV [24] regardless of the binding energy of O* defines a lower limit for the OER overpotential on any surface. Since two proton and electron transfer step [25] separate the two intermediates, the perfect separation in terms of energy should be $2e \times 1.23\ V = 2.46$ eV. The difference in the energetic of these two steps between an actual catalysts and an ideal one (3.2–2.46 eV)/2e gives a minimum overpotential of 0.4–0.3 eV. Even if we could find a material where the O* level is placed optimally between those of HO* and HOO*. The thermochemically ideal catalyst is

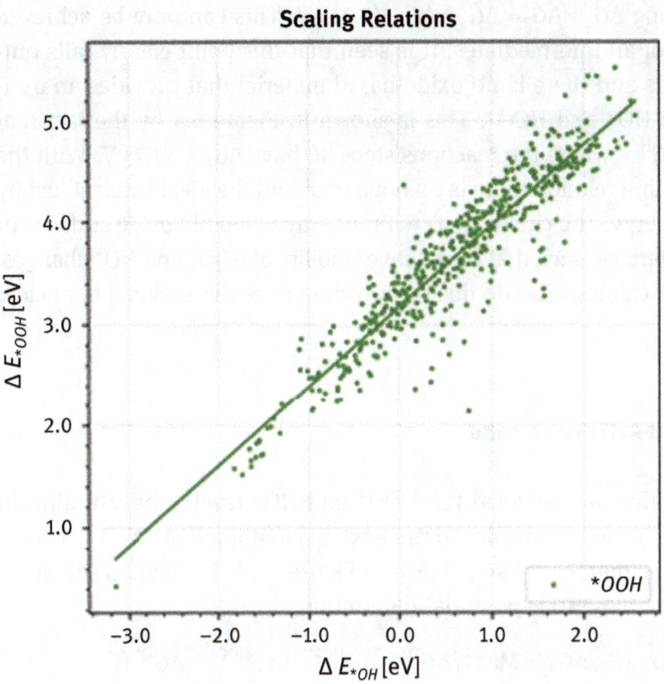

Figure 6.3: The scaling relation between HO* and HOO* from a collection of calculated data from the literature. The line represents the scaling of $\Delta E^{HOO^*} = \Delta E^{HO^*} + 3.2$ eV. The figure is from [22].

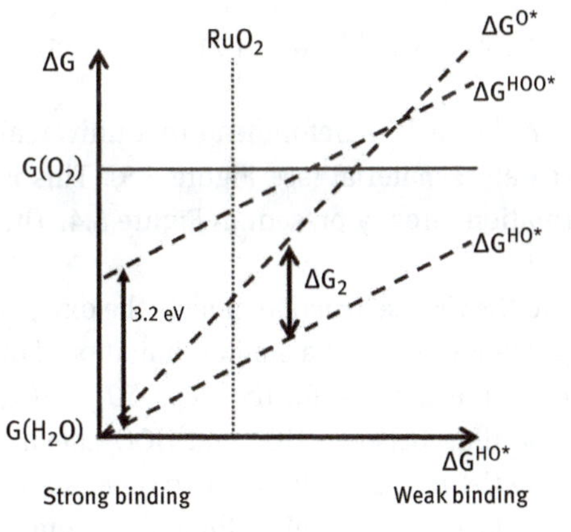

Figure 6.4: The scaling relations ΔG^{O^*}, ΔG^{HOO^*} as functions of ΔG^{HO^*} are represented with the dashes lines. The constant difference between ΔG^{HOO^*} and ΔG^{HO^*} of 3.2 eV is seen. RuO_2 is indicated by the dotted line. The sizes of the reaction steps for RuO_2 are the four sections of the dotted line, it is the same information as Figure 6.1 for $U = 0$ V. Moving the dotted line along the x-axis will give the reaction step sizes for different materials with different binding of ΔG^{HO^*}.

characterized by having $\Delta G_1 = \Delta G_2 = \Delta G_3 = \Delta G_4 = 1.23$ eV. This can only be achieved at a specific binding of all intermediates. It is seen that this point clearly falls outside the general trends and there is no oxide-based material that provides an optimum binding of both HO* and HOO*. This is shown in Figure 6.4 by the fact that there is no value of ΔG^{HO^*} where the reactions steps all have $\Delta G_{1-4} = 1.23$ V. With the universality of the scaling relation there is no hope to obtain the ideal catalyst just by turning the binding energy. The challenge is to find a way to modify oxide surfaces or the electrochemical interface, such that the relative stability of HOO* and HO* changes. Maybe some molecular catalysts can do this by providing three-dimensional bi-nuclear reactive sites.

6.6.4.1 Descriptor and Activity Volcano

Given the constant difference between the HOO* and HO* levels, the variation in the overpotential, η^{OER} from one surface to the next is determined by the O* adsorption energy. This means that, either step 2 (ΔG_2 in Figure 6.4 is the biggest) or step 3 (ΔG_3 in Figure 6.4 is the biggest) is potential-determining:

$$\begin{aligned}\Delta G^{OER} &= \mathrm{Max}[\Delta G_2, \Delta G_3] = \mathrm{Max}\left[\left(\Delta G^{O^*} - \Delta G^{HO^*}\right), \left(\Delta G^{HOO^*} - \Delta G^{O^*}\right)\right] \\ &\approx \mathrm{Max}\left[\left(\Delta G^{O^*} - \Delta G^{HO^*}\right), 3.2\mathrm{eV} - \left(\Delta G^{O^*} - \Delta G^{HO^*}\right)\right] \end{aligned} \tag{6.15}$$

The difference, $\Delta G_2 = (\Delta G_{O^*} - \Delta G_{HO^*})$, is therefore a unique descriptor for the OER activity, and the theoretical overpotential at standard conditions is:

$$\eta^{OER} = \left\{\mathrm{Max}\left[\left(\Delta G^{O^*} - \Delta G^{HO^*}\right), 3.2\mathrm{eV} - \left(\Delta G^{O^*} - \Delta G^{HO^*}\right)\right]/e\right\} - 1.23\mathrm{V} \tag{6.16}$$

Plotting η^{OER} as a function of $\Delta G_2 = \Delta G^{O^*} - \Delta G^{HO^*}$ will therefore lead to a universal volcano relationship independent on the catalyst material (see Figure 6.5). This is just a convenient way of plotting the information already present in Figure 6.4. The optimum will be at $\Delta G_2 = 1.6$ eV.

The exact same analysis can be applied to the reverse reaction namely the oxygen reduction reaction (ORR) where the scaling relation imposes a similar limitation. For the ORR the potential determining step will be the smallest for the steps $\Delta G_1 - \Delta G_4$, which most often is either ΔG_1 or ΔG_4. The scaling between HO* and HOO* means that $\Delta G_2 + \Delta G_3 = 3.2$ eV, which means that only the rest is left for the two other steps, 1.7 eV = $\Delta G_1 + \Delta G_4$. The two steps can at best be 1.7 eV/2 = 0.85 eV without compromising each other. This leads to an ORR volcano that looks like the OER volcano in Figure 6.5, but with an optimum at $\Delta G_1 = 0.85$ eV. Combining the OER and ORR volcano will show important fingerprints of catalysts limited by the scaling relation: there will be an overpotential for both the optimal ORR and the optimal OER catalyst of ~0. 3–0.4 V and the two optimal catalysts will not be the same. The two volcanos are

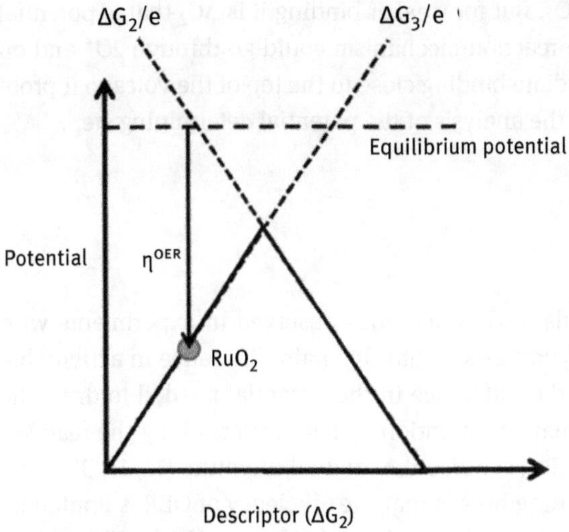

Figure 6.5: The volcano for OER. The potential and overpotential for RuO_2 is indicated by the spot and the arrow.

shifted so that the ORR has its optimum for catalysts binding stronger than the optimal catalyst for OER. This means that a good ORR catalyst is a bad OER catalyst and vice versa [26].

When comparing to experiments one has to keep in mind that no measured quantity directly corresponds to η^{OER}, however, the assumptions were that differences in measured catalytic activity should scale with the size of the potential determining step and that barriers should not significantly change that. It is challenging to confirm the scaling directly by experiments as the HOO* is unstable, however, systematic experiments shows the consequences of the scaling relation. A landmark paper in Science 2011 showed the measured trends in OER and ORR activity on a range of perovskites [27]. From that paper it is clear that the volcanos for OER and ORR in experiments ensamples the theory as shown in Figure 6.5 very closely. Even the best catalysts have an overpotential and the best OER materials are not good catalysts for ORR. Furthermore, it was also observed that the good ORR catalysts has a stronger binding than the good OER catalysts. These effects are fingerprints of the limitations imposed by the scaling relation from Figure 6.3. Furthermore, well-controlled experiments on Ir [28] and Ru [29] di-oxide 110 surfaces uses the measured ΔG_2 descriptor for investigating in activity trends as the pH is varied. In conclusion; there is no direct experimental measurements of the scaling relation, but the observed limitations and trends in activity correspond closely to those imposed by the scaling relation supporting the theory.

Very often the reaction path 1 applied in the analysis is considered another assumption. However it is not. Reaction mechanism 2 can only be relevant if $2O^* \rightarrow O_2$ ($2\Delta G^{O^*} > \Delta G(O_2)$, see Figure 6.3) is downhill in free energy, and this is the case for weaker bindings than that on RuO_2. But for weaker binding it is ΔG_2 that is potential determining. This means that the reaction mechanism could go through 2O* and on some of the surfaces with intermediate binding close to the top of the volcano it probably will, but this will not change the analysis of the potential determining step.

6.7 Conclusion

It is possible to capture the trends in OER activities observed in experiments with the simple analysis above. The hypothesis is that the main difference in activity between two catalysts is related to the difference in the potential needed to drive the most difficult reaction step. The main assumption is that barriers along the reaction scales with the reaction free energies, see also the original literature [13, 17, 30].

The somewhat depressing finding here is that the efficiency of OER is limited by a fundamental scaling relation between the binding of HO* and HOO*. The way to conceive this finding is that it is a limitation for energy efficiency in electrolysis in the same manner as the Betz's limit is for wind turbines and the Shockley–Queisser limit is for solar cells. The view on the water-splitting reaction presented here does not look at all the losses, but rather it simplify the problem to reveal the fundamental limitations. Just as for Betz's limit and Shockley–Queisser's limit, this limitation can be circumvented by changing the axioms. If we can find a system that does not follow the scaling relation or facilitate and other reaction pathways it may be possible to go beyond the limitations imposed by the scaling relations. The challenge of making electrolysis more energy efficient can thus be reformulated: it is a matter for changing the axioms away from surfaces and reaction mechanisms, which are limited by the scaling relations. Another route is to accept the limitations and look for scalable and cheaper catalyst materials.

References

[1] https://www.energy-x.eu/
[2] Bernt M, Siebel A, Gasteiger HA. J Electrochem Soc. 2018;165:F305–F314.
[3] Whitesides GM, Crabtree GW. Science. 2007;315:796.
[4] Nørskov JK, Bligaard T, Rossmeisl J, Christensen CH. Nat Chem. 2009;1. DOI: 10.1038/NCHEM.121.
[5] Rossmeisl, JG, Karlberg GS. Fuel Cell Catalysis: a Surface Science Approach, Chapter 3. Koper M, ed. Wiley-VCH; 2009.
[6] Koper MTM, Heering HA. Fuel Cell Science. Wieckowski A, Nørskov JK. ed. Chapter 2. Wiley-VCH; 2010.
[7] Bagger A, Castelli IE, Hansen MH, Rossmeisl J. Handbook of Materials Modeling: Applications: Current and Emerging Materials. Springer International Publishing; 2020. 1473–503.

[8] Dau H, Limberg C, Reier T, Risch M, Roggan S, Strasser P. Chemcatchem. 2010;2:724.

[9] Rossmeisl J, Nørskov JK, Taylor CD, Janik MJ, Neurock M. J Phys Chem B. 2006;110:21833–39.

[10] Nørskov JK, Rossmeisl J, Logadottir A, Lindqvist L, Kitchin JR, Bligaard T, Jónsson H. J Phys Chem B. 2004;108:17886.

[11] Parsons R. The rate of electrolytic hydrogen evolution and the heat of adsorption of hydrogen. Trans Faraday Soc. 1958;54:1053–63.

[12] Nørskov JK, Bligaard T, Logadottir A, Kitchin JR, Chen JG, Pandelov S, Stimming U. J Electrochem Soc. 2005;152:J23.

[13] Skulason E, Karlberg GS, Rossmeisl J, Bligaard T, Greeley J, Jonsson H, Nørskov JK. PCCP. 2007;9:3441–3250.

[14] Skulason E, Tripkovic V, Bjorketun ME, Gudmundsdottir S, Karlberg G, Rossmeisl J, Bligaard T, Jonsson H, Nørskov JK. J Phys Chem C. 2010;114:18182–97.

[15] Björketun ME, Bondarenko AS, Abrams BL, Chorkendorff I, Rossmeisl J. PCCP. 2010. DOI: 10.1039/C003826C.

[16] Marshall A, Tsypkin M, Børresen B, Hagen G, Tunold R. Mater Chem Phys. 2005;94:226–32.

[17] Trasatti S. Electrochim Acta. 1984;29:1503.

[18] Rossmeisl J, Qu Z-W, Zhu H, Kroes G-J, Nørskov JK. J Electroanalytical Chem. 2007;607:83–89.

[19] Rossmeisl J, Karlberg GS, Jaramillo T, Nørskov Faraday JK. Discuss. 2008;140:337.

[20] Abild-Pedersen F, Greeley J, Studt F, Rossmeisl J, Munter TR, Moses PG, Skulason E, Bligaard T, Nørskov JK. Phys Rev Lett. 2007;99:016105.

[21] Fernandez E, Moses PG, Toftelund A, Hansen HA, Martinez JI, Abild-Pedersen F, Kleis J, Hinnemann B, Rossmeisl J, Bligaard T, Nørskov JK. Angew Chem Int Ed. DOI: 10.1002/anie.200705739.

[22] Divanis S, Kutlusoy T, Boye IMI, Man IC, Rossmeisl J. Chem Sci. 2020;11(11):2943–295.

[23] Christensen R, Hansen HA, Dickens CF, Nørskov JK, Vegge T. J Phys Chem C. 2016;120(43): 24910–16.

[24] Man IC, Su H-Y, Calle-Vallejo F, Hansen HA, Martínez JI, Inoglu NG, Kitchin J, Jaramillo TF, Nørskov JK, Rossmeisl J. CHEMCATCHEM. 2011;3:1159–65.

[25] Koper MTM. Article in Press. J Electroanal Chem. 2010. DOI: 10.1016/j.physletb.2003.10.071.

[26] Busch M, Halck NB, Kramm UI, Siahrostami S, Krtil P, Rossmeisl J. Nano Energy. 2016;29: 126–35.

[27] Suntivich J, May KJ, Gasteiger HA, Goodenough JB, Shao-Horn Y. Science. 2011;334(6061): 1383–85.

[28] Kuo DY, Kawasaki JK, Nelson JN, Kloppenburg J, Hautier G, Shen KM. J Am Chem Soc. 2017;139 (9):3473–79.

[29] Kuo DY, Paik H, Kloppenburg J, Faeth B, Shen KM, Schlom DG, Hautier G. J Am Chem Soc. 2018;140(50):17597–605.

[30] Rossmeisl J, Logadottir A, Nørskov JK. Chem Phys. 2005;319:178–84.

Peter Strasser

7 Fuel Cells

The first half of the nineteenth century witnessed the advent of important electrochemical energy storage and energy conversion devices. In 1800, Volta developed his famous "pile" that enabled, for the first time in history, continuous electrical currents. Around the same time, Nicolson, Carlisle, and Ritter [1] used the Volta pile to split water into its constituents, hydrogen, and oxygen. The elusive origin of the continuous electric currents generated by the Volta pile created a long-lasting controversy among leading scientists ("Volta controversy") about whether electricity is caused by metal contact action or by a chemical process. Scientists explored a large number of variations of the original voltaic cell ("voltaic battery") involving many combinations of different electrode materials and electrolytes. These efforts resulted in the invention of the Leclanche element and the lead-acid battery, which are the basis of two important types of energy storage batteries of our times.

In 1838, between court appearances, a 27-year-old lawyer and later judge, Sir William Robert Grove, experimented with platinum wires, mineral-acid electrolytes, and voltaic piles. In a postscript of a letter from December 1838 [2], in which Grove reported additional experiments on the Voltaic Series, he describes "an important illustration of the combination of gases by platinum." Two electrochemically cleaned platinum strips were brought in contact with a common mineral-acid electrolyte. As soon as the upper half of one Pt strip was exposed to hydrogen gas, while the other was exposed to oxygen, a galvanometer needle indicated the flow of electricity. Grove called this galvanic cell device a "gas voltaic battery" [3]; today, we refer to it as a hydrogen/oxygen fuel cell [1, 4, 5]. A few years later, Grove proved that his gas voltaic battery, where hydrogen and oxygen combined to form water, can be used to drive the opposite reaction, the splitting of water into oxygen and hydrogen (Figure 7.1). At around the same time, Christian Friedrich Schoenbein recognized that it was a chemical action between the gases and the platinum wires rather than the metallic contacts that caused the combination of hydrogen and oxygen [6, 7].

During the 1890s and the following decades, galvanic cells received full recognition as highly efficient energy conversion devices. In a 1894 keynote address to chemists, physicists, and chemical engineers, the German physical chemist Wilhelm

Acknowledgments: The author thanks Professor Ifan Stephens, Professor Ib Chorkendorff, Dr. Shirlaine Koh, Dr. Ratndeep Srivastava, Dr. Prasanna Mani, Dr. Nastaran Ranjbar, Dr. Lin Gan, Dr. Camillo Spoeri, and Dr. Koteswara Vuyurru for help with the graphical material. The author is deeply indebted to Professor Anders Nilsson and Robert Schlögl for their past encouragement and support.

Peter Strasser, Technical University Berlin, The Electrochemical Energy, Catalysis, and Materials Science Laboratory, Department of Chemistry, Chemical Engineering Division,Straße des 17. Juni 124, 10623 Berlin, Germany, e-mail: pstrasser@tu-berlin.de

https://doi.org/10.1515/9783110608458-007

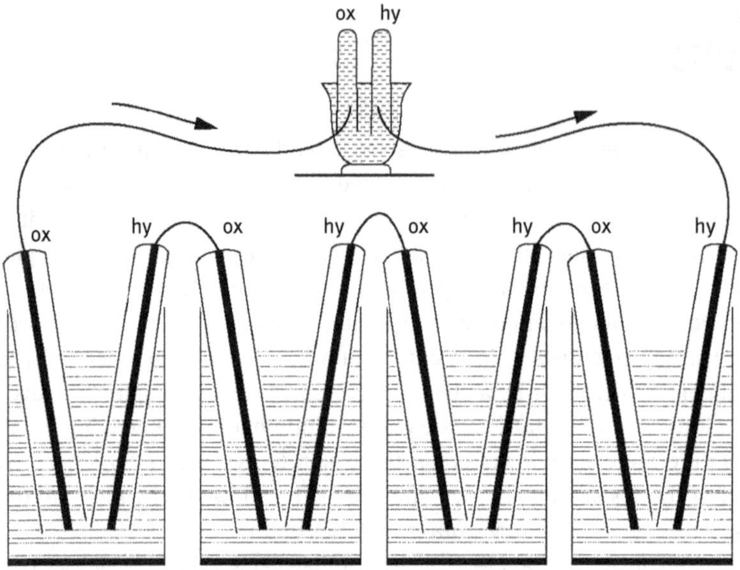

Figure 7.1: Scheme of a series of gas voltaic batteries (lower portion) consisting of four individual galvanic elements involving pairs of platinum wires (black lines), partially exposed to hydrogen and oxygen gas (upper portion of wire) and partially exposed to a liquid mineral acid (lower portion of wire). The cell voltage is sufficient to electrochemically split water into hydrogen and oxygen in a separate set up (upper portion). Arrows are to indicate the flow of electricity.

Ostwald characterized galvanic cells as a technical revolution far superior to the steam engine [8]. In particular, scientists and engineers focused on the realization of coal–air fuel cells, which remained elusive despite decades of intensive research work. During the 1930s and 1940s, research work shifted from coal to hydrogen as fuel, combined with alkaline liquid electrolytes. KOH electrolytes offered the advantage of reduced corrosion and the possibility of employing less expensive non noble metal electrocatalysts, such as Ni. In the early 1960s, finally, fuel cell technology reached a turning point when the National Aeronautics and Space Administration decided to develop alkaline fuel cells to power auxiliary units of the Apollo space modules. In parallel, work on novel solid polymer membrane fuel cells intensified, which replaced the liquid electrolyte. Early ion-exchange membranes were based on polystyrene and carried acid sulfonic groups and thus represented a highly acidic proton exchange ionomer. Now nonnoble metals had to be replaced by acid-stable noble metals, mostly platinum. Subsequent milestones in polymer electrolyte membrane fuel cell (PEMFC) technology included the development of chemically resistant perfluorinated proton exchange membranes, such as Nafion® [9], and the use of Nafion-containing fuel cell electrodes incorporating high-surface area-supported platinum nanoparticles rather than low-surface-area platinum black [10]. Since the early 1990s, driven by public and automotive industry funding, PEMFCs have been receiving increased attention as an alternative power source for vehicles.

7.1 What Is a Fuel Cell?

A fuel cell is a two-electrode galvanic cell device that converts chemically stored energy directly into electrical energy, that is, in a continuous flow of electrons as well as some dissipative heat (Figure 7.2). Depending on the chemical nature of the fuel and oxidant on which the fuel cell is operated, different chemical products can be generated. A hydrogen–oxygen fuel cell shown in Figure 7.2 generates only water as the by-product in the overall chemical reaction:

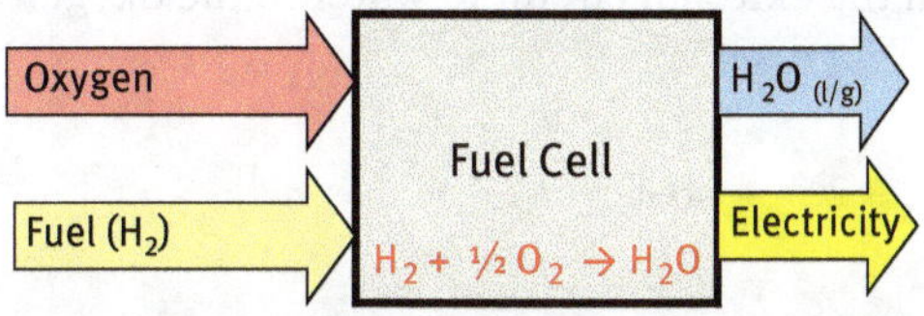

Figure 7.2: Principle of a fuel cell as an electrochemical energy conversion device. Inside a fuel cell, fuel, for example, hydrogen, and an oxidant, typically oxygen, combine electrochemically to form products, for example, water, and electricity and some excess heat (not shown). Figure adapted from ref [4].

$$H_2 + \frac{1}{2}O_2 \rightarrow H_2O, \qquad \Delta G^\circ = -236 \text{ kJ mol}^{-1}$$

In accordance with thermodynamic laws, only the Gibbs free energy, ΔG°, of the overall fuel cell reaction can be converted into the equivalent electric cell potential, ΔE°; these two quantities are linked via

$$\Delta G^\circ = -nF\Delta E^\circ$$

where F is the Faraday constant and n denotes the number of electrons exchanged in the overall chemical process.

The thermodynamic efficiency of a fuel cell is defined as the ratio between ΔG° and the enthalpy of reaction, ΔH°, $\eta = \Delta G^\circ/\Delta H^\circ$, and is not, unlike thermal external or internal combustion engines, limited by the ideal Carnot cycle.

7.2 Components of a Fuel Cell

Figure 7.3 schematically depicts the basic structure of an electrochemical fuel cell device. Generally, in electrochemical cells the overall chemical redox reaction proceeds via two coupled, yet spatially separated half-cell redox reactions at two separate electrodes.

Fuel, hydrogen gas (red), comes in contact with a catalytically active electrode (the anode), on the surface of which the hydrogen molecule splits into protons and electrons in the hydrogen oxidation reaction (HOR) according to

$$H_2 \longrightarrow 2H^+ + 2e^-$$

The protons travel across the ion-conducting (liquid) electrolyte to the opposite electrode (the cathode), where they recombine with the oxidant, here oxygen (blue), and the electrons that traveled through the external circuit to water in the oxygen-reduction reaction (ORR) according to

$$\frac{1}{2}O_2 + 2H^+ + 2e^- \longrightarrow H_2O$$

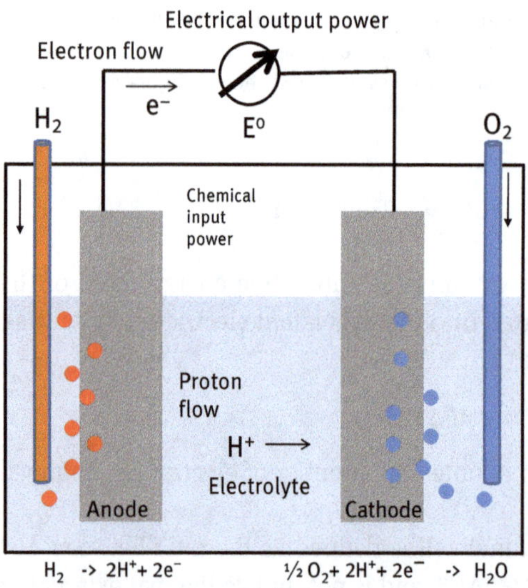

Figure 7.3: Principle of a hydrogen and oxygen fuel cell galvanic element. Hydrogen and oxygen are fed to the anode and cathode electrocatalyst, respectively, giving rise to a cell potential, E^o. Upon closing the external circuit, protons and electrons are continuously created at the anode. Protons migrate through the electrolyte to the cathode, while electrons flow through the external wiring to the cathode where they recombine with oxygen and protons to water. The flow of electrons generates an external electric current, *i*.

Table 7.1 shows an overview of the most common types of today's fuel cells. Fuel cells are typically categorized by the type of ion conductor or electrolyte employed.

An important class of fuel cells is based on proton-conducting (acidic) electrolytes, either in the form of a solid membrane (PEMFCs) or a liquid acid, possibly absorbed inside a polymer matrix (phosphoric acid fuel cells). The PEMFC has been the subject of tremendous research since the early 1990s, with impressive progress achieved in the areas of stability of the ion-conducting polymer and in the area of design and fundamental understanding of the Pt and Pt-alloy electrocatalysis of oxygen reduction, hydrogen oxidation, and the oxidation of small organic molecules [11–13].

A second class of fuel cells employs hydroxide-conducting (alkaline) electrolytes, again either in form of a solid membrane (anion exchange membrane (AEM) fuel

Table 7.1: Overview of the principal fuel cell types and their characteristics.

Fuel cell type	Electrolyte	Charge carrier	Operating temperature	Fuel	Electric efficiency (system)	Power range/ application
Alkaline fuel cell	KOH	OH^-	60–120 °C	Pure H_2	35–55%	<5 kW, niche markets (military, space)
Anion exchange membrane fuel cell	Solid polymer	OH^-	50–100 °C	Pure H_2, liquid small organics (limited CO_2 tolerance)	<35%	Portable, under research
Proton exchange membrane fuel cell[a]	Solid polymer (Nafion)	H^+	50–100 °C	Pure H_2 (tolerates CO_2), liquid small organics	35–45%	Automotive, stationary (5–250 kW), portable, under research
Phosphoric acid fuel cell	Phosphoric acid	H^+	~220 °C	Pure H_2 (tolerates CO2, approx. 1% CO)	40%	Stationary (200 kW)
Molten carbonate fuel cell	Lithium and potassium carbonate	CO_3^{2-}	~650 °C	H_2, CO, CH_4, other hydocarbons (tolerates CO_2)	>50%	Stationary (200 kW–MW)
Solid oxide fuel cell	Solid oxide electrolyte (yttria, zirconia)	O^{2-}	~1,000 °C	H_2, CO, CH_4, other hydocarbons (tolerates CO_2)	>50%	2 kW–MW range

a. Also known as a polymer electrolyte membrane fuel cell.
Table adapted with permission from [1].

cells) or a liquid electrolyte (alkaline fuel cells). While the modern era of fuel cells began with the latter type, the former type is under intense research today because a stable, highly conducting alkaline membrane with good CO_2 tolerance has remained elusive to date.

Finally, there are two fuel cell types operated at such high temperatures that organic polymers or liquid electrolytes could not withstand them. Their electrolytes are either CO_3^{2-}-conducting molten carbonates or O^{2-}-ion-conducting oxidic solids. Depending on the electrolyte and temperature, different fuels and catalysts can or must be employed. While hydrogen or a liquid organic compound are the fuels of choice for the low-temperature fuel cells, high-temperature fuel cells can be operated on a wider variety of fuels, including small hydrocarbons or CO-contaminated feeds, because the high temperatures allow for additional conversion chemistry of the hydrocarbon fuels and mitigate catalyst poisoning. Acidic electrolytes generally require more noble-metal electrocatalysts, while alkaline fuel cells tolerate less expensive catalysts. High-temperature fuel cells often incorporate oxidic electrocatalytic materials.

Table 7.1 also indicates suitable power ranges of the various fuel cell types and some of their typical applications. PEMFCs are clearly the most versatile class with strong focus on portable and automotive applications. High-temperature fuel cells are more often employed for stationary power generation.

Figure 7.4 provides electron microscopic cross-sectional images of an individual membrane-electrode assembly (MEA) of a PEMFC (Figure 7.4A). The proton-conducting membrane is at the center of the MEA with typical thicknesses of 25–100 µm. On either side of the membrane are the electrocatalyst layers with thicknesses of only approximately 5–10 µm. Next on either side of the catalyst layers are the gas diffusion layers, typically porous carbon fiber materials that help distribute the fuel and oxidants uniformly across the catalyst layer, help remove reaction products, and, as electronic conductors, are part of the external circuit for the electron flow. Figure 7.4B reveals the detailed structure of the cathode electrocatalyst layer showing nanoscale (2–10 nm) Pt particles that are supported on high-surface-area carbon black supports. Figure 7.4C shows the atomic fine structure of an individual Pt nanoparticle. A regular arrangement of individual Pt atoms is evident. Pt surface atoms (black arrow) act as active sites for the ORR at the cathode of hydrogen/oxygen fuel cells.

The detailed chemical structure of the most common proton-conducting polymer, Nafion®, is shown in Figure 7.5A. Stabilized and tethered using backbone chains of perfluorinated polyethylene (Teflon), sulfonic groups are dangling inside the ionomer and help transport protons across the membrane. As illustrated schematically in Figure 7.5B, the microscopic structure of a hydrated Nafion membrane reveals hydrophilic water channels in between self-organized hydrophobic Teflon domains. The sulfonic groups are reaching into the water channels and mediate the diffusion and migration transport of hydrated hydronium ions (H_3O^+). Ionomer membranes

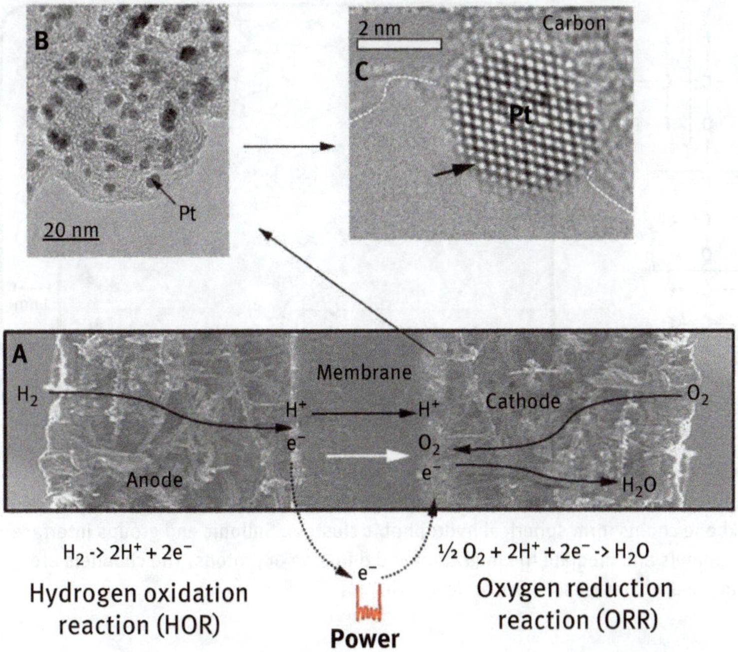

Figure 7.4: (A) Scanning electron microscopic cross-sectional view of an individual MEA of a low-temperature PEMFC. The Nafion membrane is sandwiched between the electrode layers (left: anode; right: cathode) followed by the two layers of porous carbon (gas diffusion layers). Hydrogen diffuses from left across the gas diffusion layer to the anode where the electrocatalytic oxidation of molecular hydrogen occurs. Protons migrate across the Nafion membrane, and electrons travel through the external circuit (dotted arrows). Oxygen enters the MEA from the right. (B) Enlargement of the cathode layer reveals the platinum nanoparticle electrocatalyst supported on a porous carbon support material. (C) Further enlargement of the cathode electrocatalyst shows an individual Pt nanoparticle consisting of well-ordered Pt atoms at the edge of the carbon support. Black arrow indicates a Pt surface atom where the electrocatalysis occurs.

are characterized by an equivalent weight parameter, which is defined as the average molecular weight of the (dry) polymer per sulfonic group. Typical values range from 800 to 1,200. The larger the equivalent weight, the less soluble the polymer is in aqueous solutions, but the lower its conductivity tends to be.

The MEA (Figure 7.4) is the heart of every individual PEMFC. Operated on hydrogen and oxygen, the maximum individual cell voltage is approximately 1.23 V. In order to obtain higher cell voltages, individual MEAs are put in series in so-called PEMFC stacks as shown in Figure 7.6. In between each individual MEA, there are electrically conductive flow plates that provide flow paths for the fuel and oxidant.

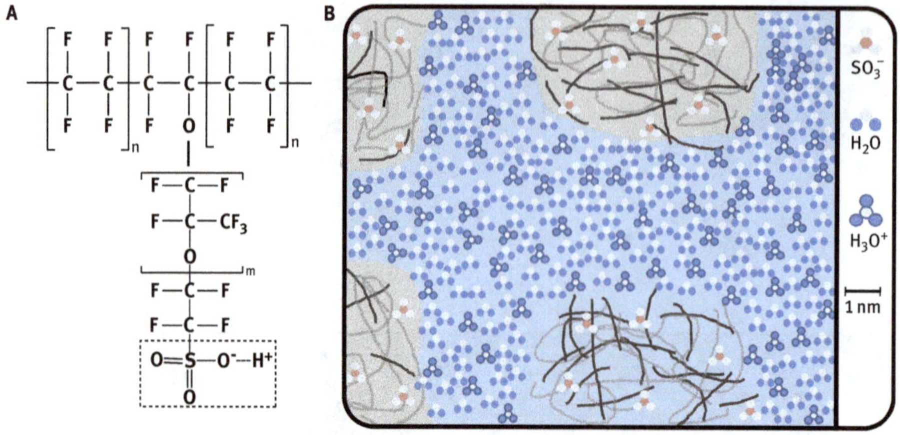

Figure 7.5: (A) Chemical structure of sulfonated perfluorinated polyethylene (Nafion®). (B) Schematic illustration of the microscopic structure of hydrated Nafion membrane: perfluorinated polyethylene backbone chains form spherical hydrophobic clusters. Sulfonic end groups interface with water-filled channels and mediate the migration and diffusion of protons. The channels are filled with water and hydronium ions. Figure adapted from [4].

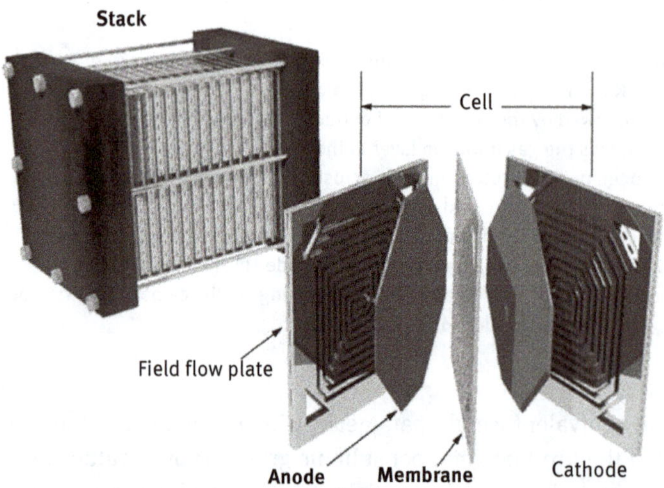

Figure 7.6: Illustration of the layered component structure of an individual technological PEMFC. Serial combination of individual PEMFCs results in fuel cell stacks. Figure reproduced from [1].

7.3 Performance Characteristics of a Fuel Cell

According to eq. (7.1) the theoretical, thermodynamic reversible cell voltage of a hydrogen/oxygen fuel cell at zero current (open circuit potential conditions) and standard conditions is approximately $E^o = 1.23$ V. Upon current flow (i.e., actual power generation), a number of different voltage losses occur in fuel cells lowering the observable cell voltage, V. At some current density, the cell voltage of every fuel cell becomes zero and so does its power generation. The cell voltage losses, also referred to as overvoltages or overpotentials, where the individual anode or cathode potential is concerned, can be classified into (i) activation polarization losses, η_{act}; (ii) ohmic losses, η_{ohmic}; and (iii) concentration polarization losses, η_{conc}. Activation overpotentials are due to the sluggish chemical kinetics of fuel and/or oxidant on the surface of the fuel cell catalysts. Ohmic overpotentials are ohmic voltage drops across the membrane and the external circuit caused by their combined ionic and electronic resistance, R_{ohmic}. Concentration overpotentials become noticeable only at higher current densities where large amounts of fuel and oxidants are converted at the catalysts. Under these conditions, mass transport of fuel and oxidant can become rate limiting and result in a loss of cell voltage at a given current density. Figure 7.7 illustrates how the total overvoltage of a fuel cell (i.e., the difference between the theoretical cell voltage, E^o, and the real cell voltage, V) splits into the three overvoltage contributions. Mathematically, the experimentally observed cell voltage, V, can be expressed and modeled according to the relation

$$V = E^0 - \eta_{act,\,cathode,\,ORR} - \eta_{act,\,anode,\,HOR} - iR_{ohmic} - \eta_{conc,\,anode} - \eta_{conc,\,cathode} \tag{7.1}$$

Figure 7.7 demonstrates that the activation overvoltages are by far the largest cause of loss of voltage and hence efficiency of a fuel cell. This is why over the past decade a great deal of research has focused on the discovery and the fundamental understanding of improved fuel cell electrocatalysts. The anode and the cathode reaction are not contributing equally to the total activation overvoltage. This is illustrated in Figure 7.8, which plots the experimental current density versus the electrode potential near their standard electrode potential for the hydrogen electrode (H_2/H^+, standard equilibrium potential 0 V) and the oxygen electrode (O_2/H_2O, standard equilibrium potential 1.23 V). These current potential characteristics of redox systems can be well approximated by exponential expressions first derived by Butler and Volmer, assuming the Gibbs free energy of activation of the forward and reverse reaction being a linear function of the electrode potential [14]. Figure 7.8 schematically illustrates how activation overpotentials of the individual electrode reaction depend on the shape of their Butler–Volmer-type j-E relation. For the HOR, a given positive, anodic current density, j_{cell}, causes a relatively small overpotential, η_{HOR}, while for the much less reversible ORR, the corresponding negative reductive cell current results in a much larger ORR overpotential, η_{ORR}. In fact, η_{ORR} typically exceeds

η_{HOR} by at least two orders of magnitude. This is because there is no material for which there is a measurable ORR current at the equilibrium potential of 1.23 V. This makes the *j-E* curve of ORR very broad and flat around its standard potential. Again, consistent with Figure 7.7, the difference between the two standard redox potentials of 1.23 V (equivalent to the standard cell voltage) and the two activation overpotentials combined generates the experimental cell voltage, *V*. Figure 7.8 suggests that developing an improved ORR electrocatalyst offers much larger gains in cell voltages and hence efficiencies, which is why much effort in the area of fuel cell catalysis is currently directed toward the identification of more active electrocatalyst materials for the ORR.

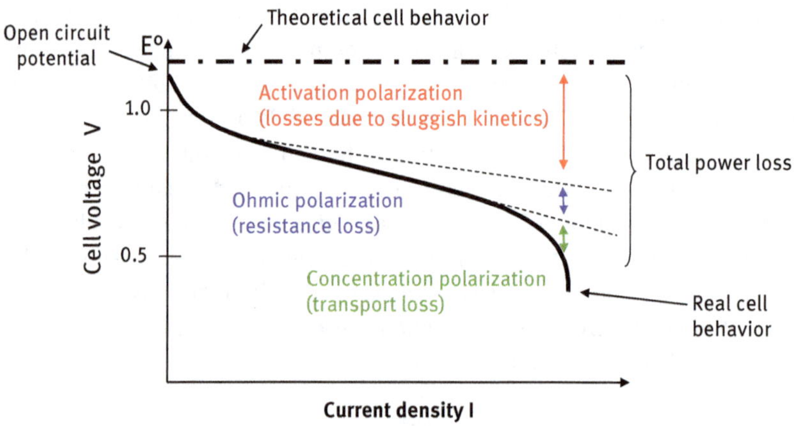

Figure 7.7: Theoretical (dashed dotted) and real (solid) cell voltage (*V*) – current density (*I*) performance characteristics of a fuel cell. Overpotentials are responsible for the difference between theoretical and real performance and cause efficiency losses. They split into (i) activation polarization overpotentials at anode and cathode due to slow chemical kinetics, (ii) ohmic polarization overpotential due to ohmic voltage losses along the circuit, and (iii) concentration polarization overpotentials due to mass-transport limitations. The activation overpotentials of the cathode are typically the largest contribution to the total overvoltage.

The rational design of improved electrocatalyst materials requires first and foremost a fundamental understanding of the origin of the kinetic overpotential. Over the past decade, new theoretical-computational approaches, most prominently density functional theory (DFT), combined with in situ surface-sensitive experimental methods, involving photons and electrons, have been developed and successfully utilized to study and understand the electrocatalytic interface at the atomic scale.

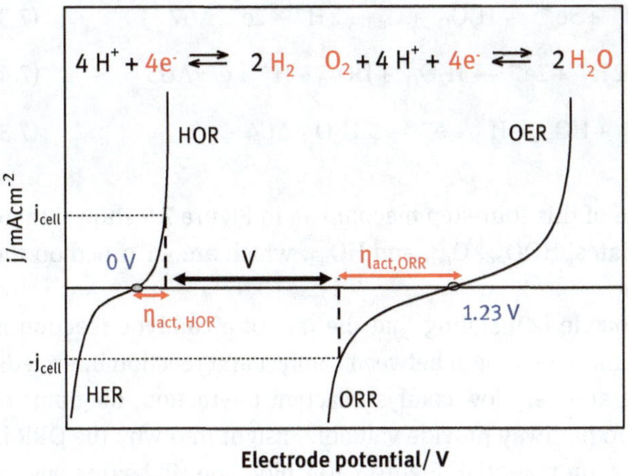

Figure 7.8: Schematic illustration of the origin of activation overpotentials in a hydrogen–oxygen fuel cell. The solid curves represent exponential analytic current densities versus electrode potential of the hydrogen electrode (standard potential 0 V) and the oxygen electrode (standard potential 1.23 V). Relevant for a PEMFC fuel cell are the HOR (anode) and the ORR (cathode) branches. To satisfy a cell current (j_{cell}), the anode potential moves more positive by $\eta_{act,HOR}$, while the cathode potential moves more negative by $\eta_{act,ORR}$. As a result of this, the observed cell potential is V, which is smaller than 1.23 V. The shape of the individual characteristics is such that the cathode overpotentials are larger than those at the anode.

7.4 The Electrocatalysis of Oxygen Reduction at Fuel Cell Cathodes

7.4.1 Understanding the Electrode Potential Dependence of the ORR

The identification of the mechanistic origins of the sluggishness and irreversibility of the oxygen redox electrochemistry involving the oxidative evolution of molecular oxygen from water and the reduction of molecular oxygen to water (ORR) has been a major challenge for many decades [15–17].

Recently, a simple theoretical DFT-based electrochemical framework was proposed [18] to account for the severe overpotential of the ORR. This framework incorporates a mechanistic hypothesis for the ORR consisting of four elementary steps and involving a number of oxygenated surface intermediates that are widely believed to contribute to the ORR chemistry [18, 19]:

$$O_2(g) + 4\,H^+ + 4e^- \rightarrow HOO_{ad} + 3\,H^+ + 3e^- \quad \Delta G1 \tag{7.2}$$

$$HOO_{ad} + 3H^+ + 3e^- \rightarrow H_2O_{(1)} + O_{ad} + 2H^+ + 2e^- \; \Delta G2 \tag{7.3}$$

$$H_2O_{(1)} + O_{ad} + 2H^+ + 2e^- \rightarrow H_2O_{(1)} + HO_{ad} + H^+ + e^- \; \Delta G3 \tag{7.4}$$

$$H_2O_{(1)} + HO_{ad} + H^+ + e^- \rightarrow 2H_2O_{(1)} \Delta G4 \tag{7.5}$$

As the schematic illustration of this four-step mechanism in Figure 7.9 shows, there are three reaction intermediates, HOO_{ad}, O_{ad}, and HO_{ad}, which are adsorbed on the surface of the electrocatalyst.

Based on Sabatier's principle [20] stating that the rate of a catalytic reaction is maximized at intermediate binding strength between catalyst and reaction intermediates and drops at very large and very low catalyst-reactant interaction, diagrams of the Gibbs free energy reaction pathway provide valuable insight into why the ORR is not proceeding at its equilibrium potential. Figure 7.10A plots the Gibbs free energy of the four elementary reaction steps (7.2) to (7.5) over the reaction coordinate for a Pt-containing electrocatalyst. At the equilibrium electrode potential of +1.23 V (red pathway), the Gibbs energies of the chemical and electrical species are such that reactions (7.2), (7.4), and (7.5) are endergonic (positive driving force, $\Delta G_i > 0$). This keeps practical reaction rates negligibly low. Only reaction (7.3) ($\Delta G_2 < 0$) could proceed spontaneously.

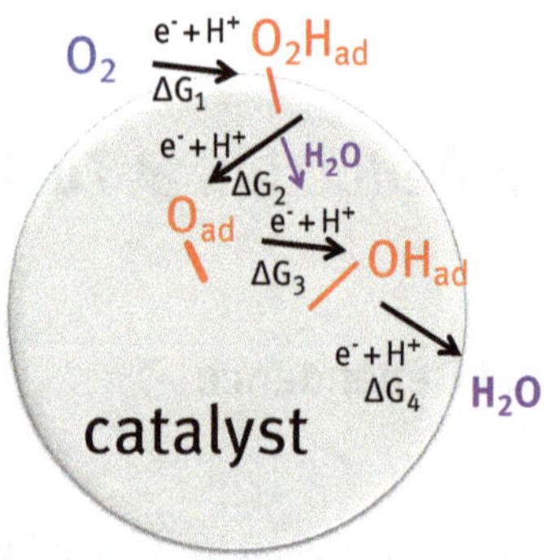

Figure 7.9: A simple four-step reaction mechanism of the ORR involving three adsorbed reactive intermediates, HOO_{ad}, O_{ad}, and HO_{ad}.

To model the effect of the electrode potential, the electrochemical framework links changes in the Gibbs free energy of the electrons, ΔG, to variations in the electrode potential, ΔE, through a simple linear relation, $\Delta G = -nF\Delta E$, where n is the number of electrons on the left side of reactions (7.2) through (7.5). Thus a decrease of the applied electrode potential lifts all the Gibbs energy levels in Figure 7.10A, however, at different rates. As a result of this, ΔG_1, ΔG_3, and ΔG_4 gradually decrease until, at +0.81 V, all steps have become downhill in energy and hence can proceed with measurable rates. Lowering the electrode potential further does not result in enhanced rates. In the context of this simple electrochemical framework, the last step to become downhill is referred to as the "potential-determining step" [18, 19].

The dependence of the Gibbs free energy pathway on electrode potential (Figure 7.10A) manifests itself directly in the experimental current potential characteristic illustrated in Figure 7.10B. At 1.23 V, no ORR current is measureable, while with decreasing electrode potentials the ORR current increases exponentially until at +0.81 V, processes other than surface kinetics (e.g., mass transport) begin to limit the overall reaction rate. Figure 7.10B represents a typical performance characteristic of a Pt or Pt-alloy electrocatalyst for the ORR.

7.4.2 Understanding and Predicting Trends in ORR Activity on Transition-Metal Catalysts

Rational design of catalysts requires a fundamental understanding of which molecular characteristics of the catalyst (so-called descriptors) control the macroscopic reaction rate. In principle, all surface catalytic phenomena are electronic in nature and hence could be understood in terms of an atomic-scale electronic description of catalyst and reacting molecules; however, despite much-improved reliability of energetic modeling by modern DFT, detailed atomic-scale electronic descriptors are often unknown, which is why other descriptors, such as the chemisorption energies of individual intermediates or the geometry of the catalyst surface, are used instead. Plotting catalytic reactivity over an atomic-scale descriptor of the interaction of catalyst and reactants/intermediates typically yields a volcano-type trend relation, in which very large and very small values of the descriptors result in low catalytic activity, yet intermediate descriptor values yield maximum activity [15, 18, 21].

Using reactions (7.2) through (7.5), an oxygen chemisorption descriptor-based trend (volcano) analysis of the ORR activity on various transition metals has been developed [15, 18, 22]. The ORR activity analysis only involves a single molecular descriptor, the oxygen chemisorption energy, ΔE_O, because the chemisorption energies of HOO_{ad}, O_{ad}, and HO_{ad} are linearly dependent on each other. This relation is referred to as a "scaling relationship" [19], and it is this scaling relation between the oxygenates that precludes the possibility of independently tuning the Gibbs energies (chemisorption strengths) of HOO_{ad}, O_{ad}, and HO_{ad} in order to arrive at an all-downhill energy pathway at +1.23 V (Figure 7.10A), which would represent the perfect overvoltage free ORR catalyst.

Figure 7.11 shows the predicted volcano-shaped ORR activity trend for various transition metals as a function of the chemisorption energy of O_{ad}. Following Sabatier's principle, early transition metals (W, Mo, Fe, Co, etc.) adsorb and bind oxygen too strongly and exhibit very low ORR activity because the desorption of HO_{ad} becomes rate limiting. Late transition metals such as Au or Ag, on the other hand, adsorb molecular oxygen very weakly. They show low ORR activity because the activation of molecular oxygen to HOO_{ad} (reaction 2) is slow and rate limiting. The optimal oxygen-binding energy is predicted to be slightly weaker than that of pure Pt (red

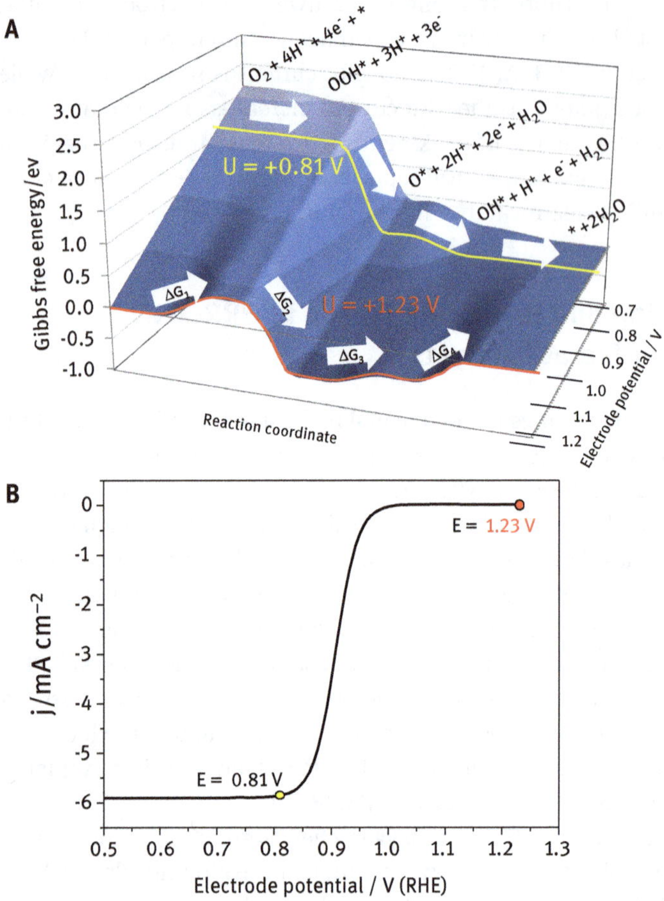

Figure 7.10: (A) The electrode potential dependence of the Gibbs free energy reaction pathway of the ORR. While the overall reaction has elementary steps that are energetically uphill at +1.23 V (red pathway), all elementary steps become downhill at +0.81 V (yellow pathway) (i.e., at an overpotential of approximately −0.42 V. At this point, the reaction is not limited by kinetics anymore. (B) The experimentally observed current-potential (*j–E*) relation of the ORR is consistent with the computational conclusions from (A): between +1.23 V and +0.81 V the *j–E* curve shows an exponential behavior, while at electrode potentials below +0.81 V, the ORR reaction rate becomes oxygen mass-transport limited, which is reflected by a flat (*j-E*) profile. Figure adapted with permission from [19].

vertical dotted line through top of volcano); however, there is no metal in the periodic table that exhibits that optimal binding energy.

How can we modify or even deliberately tune surface chemisorption energies of catalytic surfaces? A popular strategy is the mixing of two dissimilar metals in the surface of the catalyst. The distinct electronic structures of the two metals interact and, depending on their relative atomic ratio, can generate a surface with a new electronic structure, which is different from either one of the components. This alloying

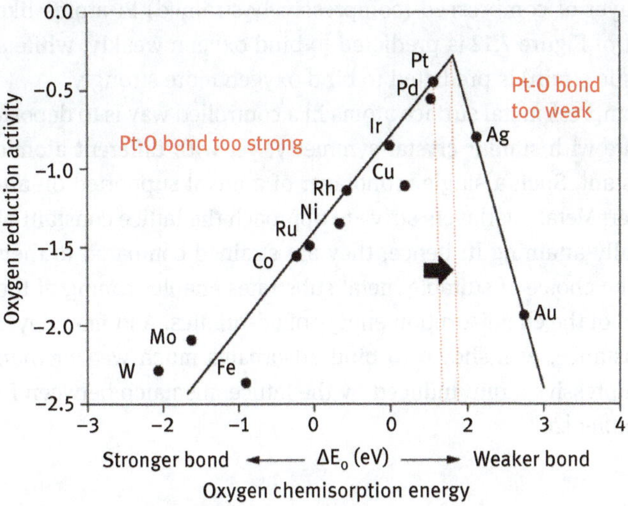

Figure 7.11: Computational volcano curve of the ORR activity of well-defined metal surfaces versus their oxygen chemisorption energy, ΔE_O. Metals on the left bind oxygen too strongly, resulting in low ORR activity; metals on the right bind too weakly, also resulting in low ORR rates. Pt is the most active monometallic ORR catalyst. The top of the volcano curve represents an unknown optimal catalyst and can be achieved by lowering the O chemisorption energy of Pt somewhat (arrow and red dotted lines). Figure adapted from [15].

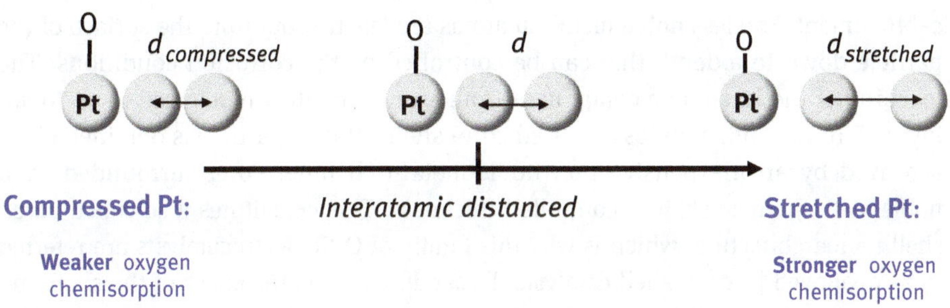

Figure 7.12: The effect of surface lattice strain on adsorbate chemisorption energies. Lattice strain modulates chemisorption and can be used to tune the reactivity of electrocatalysts. Compressed Pt surface layers (right portion) bind adsorbates more weakly; stretched Pt layers bind adsorbates more strongly.

strategy works well, though the interaction is not simple. For instance, mixing Au and Co in various ratios would not necessarily generate a surface with intermediate oxygen chemisorption properties closer to the top of the volcano.

A more controlled strategy to deliberately tune chemisorption energies is displayed in Figure 7.12. Theory predicts that the interatomic distance of surface atoms (surface lattice constant) affects the chemisorption energies in a nearly linear fashion [23]. A surface layer of compressed (compressively strained) Pt atoms like the one shown to the left of Figure 7.12 is predicted to bind oxygen weakly, while a stretched Pt surface (tensile strain) is predicted to bind oxygen more strongly.

A way to stretch or compress metal surface atoms in a controlled way is to deposit them on top of a substrate with similar crystal symmetry, yet with different atomic diameter and lattice constant. Such a single monolayer of a metal supported on another is called an overlayer. Metal overlayers strive to approach the lattice constant of their substrate without fully attaining it; hence, they are strained compared to their own bulk state [24, 25]. The choice of suitable metal substrates enables tuning of the strain in the overlayer and of the chemisorption energy of adsorbates. A Pt monolayer on a Cu substrate, for instance, was shown to bind adsorbates much weaker than bulk platinum due to compressive strain induced by the lattice mismatch between Pt and Cu, with Cu being smaller [26].

7.4.3 Nanostructured Pt Core–Shell and Shape-Controlled Electrocatalysts for the ORR

There are a number of different ways to prepare metal overlayers in practical nanoscale fuel cell electrocatalysts. One approach is schematically shown in Figure 7.13A, where a spherical alloy nanoparticle consisting of a noble metal (gray), platinum, and a nonnoble metal (red), such as copper, is subject to a corrosive electrochemical treatment. The less noble metal Cu atoms are leached out from the surface of the particle down to a depth that can be controlled by the corrosion conditions. The leaching of the less noble component from a uniform alloy is referred to as "dealloying." In the end, a nanostructured core–shell catalyst particle is obtained characterized by an inexpensive non noble-metal-rich inner core, surrounded by a noble-metal particle shell of controlled thickness. This constitutes a so-called core–shell nanoarchitecture, which is why this family of ORR electrocatalysts are referred to as dealloyed Pt core–shell catalysts. Depending on the thickness of the shell and the atomic diameter of the nonnoble element, the noble-metal surface atoms are kept in a state of compressive or tensile strain, depending on whether the nonnoble-metal atom is smaller or larger than the noble one, respectively.

Figure 7.14 offers atomistic insights in the structure and morphology of spherical dealloyed Pt–Ni nanoparticles obtained from acid leaching of $PtNi_3$ nanoparticles. Shown are scanning transmission electron microscopy (STEM) images of dealloyed Pt–Ni nanoparticles of 7, 9, 11, and 20 nm size. The images were taken in the Z-contrast mode, which makes heavier elements, such as Pt, appear brighter than lighter ones. Also, thicker portions of the particles appear brighter. Above the STEM images, elemental Energy Dispersive X-ray line scans of characteristic X-ray fluorescence of Pt (red) and Ni (green) are depicted. It is evident that only the two smallest dealloyed Pt–Ni nanoparticles (7 and 9 nm) feature the characteristic compact core–shell

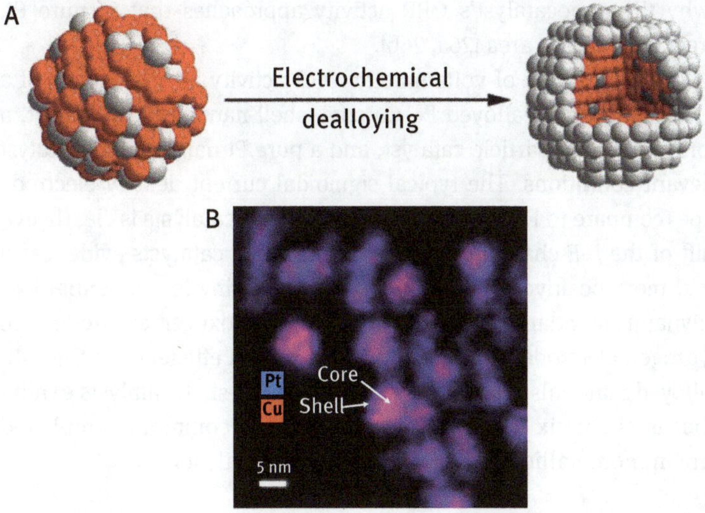

Figure 7.13: (A) Pt core–shell nanoparticle ORR electrocatalysts are prepared by (electro)chemical selective removal of a less noble metal, M, from a Pt-M nanoparticle alloy. The near-surface region is Pt enriched, while the particle core remains bimetallic. (B) Transmission electron microscopic elemental map of dealloyed Pt–Cu core–shell nanoparticle ORR electrocatalysts.

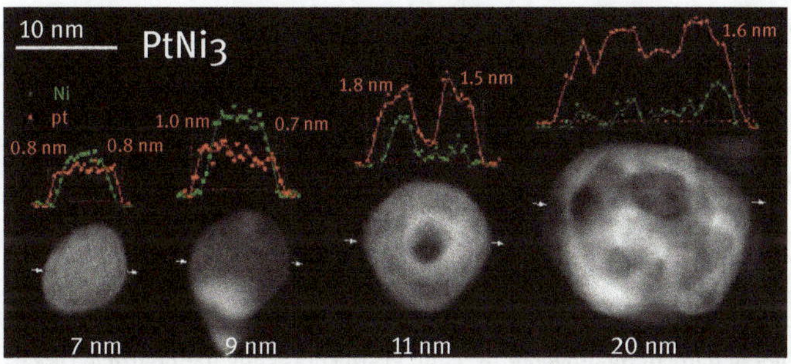

Figure 7.14: Scanning transmission electron microscopic (STEM) images of 4 dealloyed Pt–Ni nanoparticle electrocatalysts for the oxygen reduction reaction (ORR) at hydrogen fuel cell cathodes. The nanocatalyst differs in particle diameter ranging from 7 to 20 nm. Above each micrograph an elemental, cross-sectional energy-dispersive X-ray (EDX) analysis scan across each particle (see white arrows) is shown. The red and green EDX map denotes the local intensity of Pt and Ni, respectively. The red numbers indicate the thickness of the Pt shell derived from the elemental mappings.

structure with a ca. 1 nm thick pure Pt shell surrounding a Ni-rich Pt–Ni alloy core. The larger 11 and 20 nm sized dealloyed Pt–Ni nanoparticles are characterized by a severe Ni depletion and an increasingly porous morphology. The exposure of particle core regions to the acidic electrolyte is the origin of the enhanced Ni depletion. The reduced Ni content has detrimental effects on the compressive strain of the surface Pt atoms, which is why the nanocatalyst's ORR activity approaches that of pure Pt, while offering a larger real surface area [26a, 26b].

Figure 7.15 displays the results of voltammetric ORR activity measurements of a carbon-supported (/C) spherical dealloyed Pt–Cu core–shell nanoparticle catalyst, a dealloyed Pt–Ni core–shell nanoparticle catalyst, and a pure Pt nanoparticle catalyst under fuel cell relevant conditions. The typical sigmoidal current density electrode potential (*j*–*E*) shape (compare to Figure 7.10B) of the Pt-based catalysts is clearly evident. The large shift of the *j*–*E* characteristic of the core–shell catalysts evidences a catalytic reactivity at more positive electrode potentials, implying less potential loss versus the thermodynamic standard potential of 1.23 V for the oxygen electrode with reference to the hydrogen electrode; this also implies improved efficiency of the ORR process on the dealloyed materials. More specifically, the core–shell catalysts exhibit an ORR activity that is about six to eight times higher than commonly employed state-of-the-art pure monometallic Pt nanoparticle electrocatalysts.

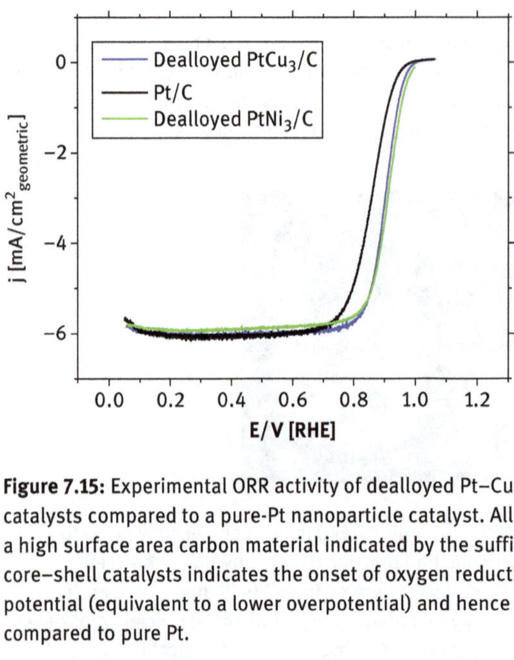

Figure 7.15: Experimental ORR activity of dealloyed Pt–Cu and Pt–Ni core–shell nanoparticle ORR catalysts compared to a pure-Pt nanoparticle catalyst. All three catalyst particles are supported on a high surface area carbon material indicated by the suffix /C. The shift of the *j*–*E* curve of the core–shell catalysts indicates the onset of oxygen reduction catalysis at a more anodic electrode potential (equivalent to a lower overpotential) and hence represents improved ORR reactivity compared to pure Pt.

Subsequent deployment of the new catalyst in the cathode layer of small-area MEAs first, then large-area MEAs, and finally fuel cell stacks represents the typical series of performance tests to check the practical viability of novel ORR electrocatalyst materials. Figure 7.16A shows the experimental cell voltage–current density characteristics (compare to Figure 7.7) of three dealloyed Pt–M (M = Cu, Co, Ni) nanoparticle ORR cathode electrocatalysts compared to a state-of-the-art pure-Pt catalyst. At current densities above 0.25 A cm^{-2}, the Co- and Ni-containing cathode catalysts perform comparably to the pure-Pt standard catalyst, even though the amount of noble metal inside the catalysts is lower than that of the pure-Pt catalyst by a factor of two to three. The dealloyed Pt-Cu catalyst is even superior to Pt at reduced metal loading.

To inspect and compare the activation overvoltage of the three catalysts in more detail, so-called Tafel plots are used, which plot the cell voltage as a function of the logarithm of the current density. Figure 7.15B shows the Tafel plots derived from Figure 7.16A. At a cell voltage of 0.9 V, where the overall reaction rate is limited by the chemical surface catalysis, the dealloyed core–shell catalysts perform three to five times more actively than pure Pt, confirming the ORR performance results obtained in liquid electrolytes presented in Figure 7.15.

More recent ORR fuel cell cathode electrocatalyst concepts include the class of shape-controlled octahedral Pt–Ni nanoparticles [29]. The underlying idea of this catalyst family rooted in studies of the structure sensitivity of well-defined PtNi(111) bimetallic model electrode surfaces, which were reported to exhibit unprecedented ORR catalytic reactivity. Given the face center cubic crystal lattice of PtNi alloys, the octahedral shape of PtNi bimetallic nanoparticles was to ensure the exposure of (111) facets only. Figure 7.17 displays TEM images of experimentally prepared octahedral PtNi nanoparticles. Their controlled morphology is apparent. Such a synthetic control is made possible using organic capping ligands that direct the energy and rate of formation of competing surface facets. As the EDX mapping in Figure 7.17B evidences, the elemental composition of the particle surface is nonuniform across the particle surface. Due to their formation mechanism [30] the vertices and edges are Pt-rich, while the (111) facets are Ni-rich. Thermal annealing treatments can trigger atomic surface diffusion and help make the nanoparticles compositionally more uniform. However, thermal treatments can be detrimental to the defined morphologies [31].

Figure 7.18 demonstrates the impressive experimental ORR mass activities of shape-controlled PtNi-based nanoparticles in direct comparison with the spherical dealloyed PtNi catalysts. Controlled catalyst morphologies caused ORR mass activity improvements of ca. 5× over spherical dealloyed PtNi catalysts, and of ca. 40× over state-of-the-art pure Pt nanoparticles (solid line). The realization of such high ORR mass activities in a realistic cathode catalyst layer of a PEM fuel cell, however, has remained elusive, which is one of the key challenges of current fuel cell catalyst research activities [35].

7.4.4 Noble-Metal-Free ORR PEMFC Electrocatalysts

The reduction of molecular oxygen is an important chemical step in biochemical aerobic respiration. Over billions of years, nature has designed a protein-embedded catalyst involving a porphyrin nitrogen-coordinated Fe center (the heme group) for the adsorption and reduction of oxygen molecules under ambient conditions. Unlike the “human-made” Pt-containing ORR catalysts, the biological C–N–Fe catalyst

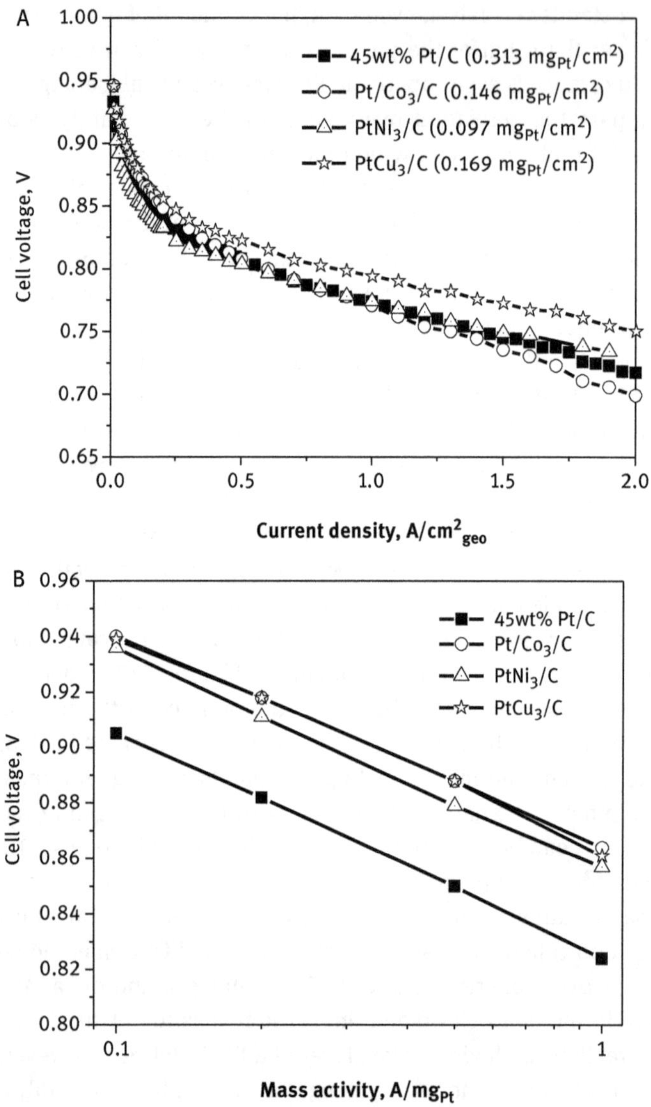

Figure 7.16: (A) Cell voltage–current density (*E-j*) profiles of a 10 cm^2 single PEMFC incorporating dealloyed Pt–M (M = Cu, Co, Ni) ORR catalysts at the cathode. The active catalysts were prepared by electrochemical dealloying of PtM_3 alloys. The alloy performance is shown in comparison to a reference PEMFC with a pure-Pt nanoparticle catalyst at the cathode. Dealloyed Pt–Cu nanoparticles outperform Pt and the other alloys over a wide current density range. (B) (*E*–log *j*) Tafel plots of the PEMFC profiles of (A). Pt-based mass activity (A/mg_{Pt}) is chosen as activity measure. Dealloyed Pt-Cu and dealloyed Pt-Co outperform pure Pt by a factor of three to four times at 900 mV cell voltage. Conditions: $T_{cell} = T_{anode} = T_{cathode}$ = 80 °C; H_2/O_2, fuel/oxygen excess = 2/10; total pressure = 150 kPa; MEA active area = 10 cm^2; membrane = NR212. Figure taken from [27].

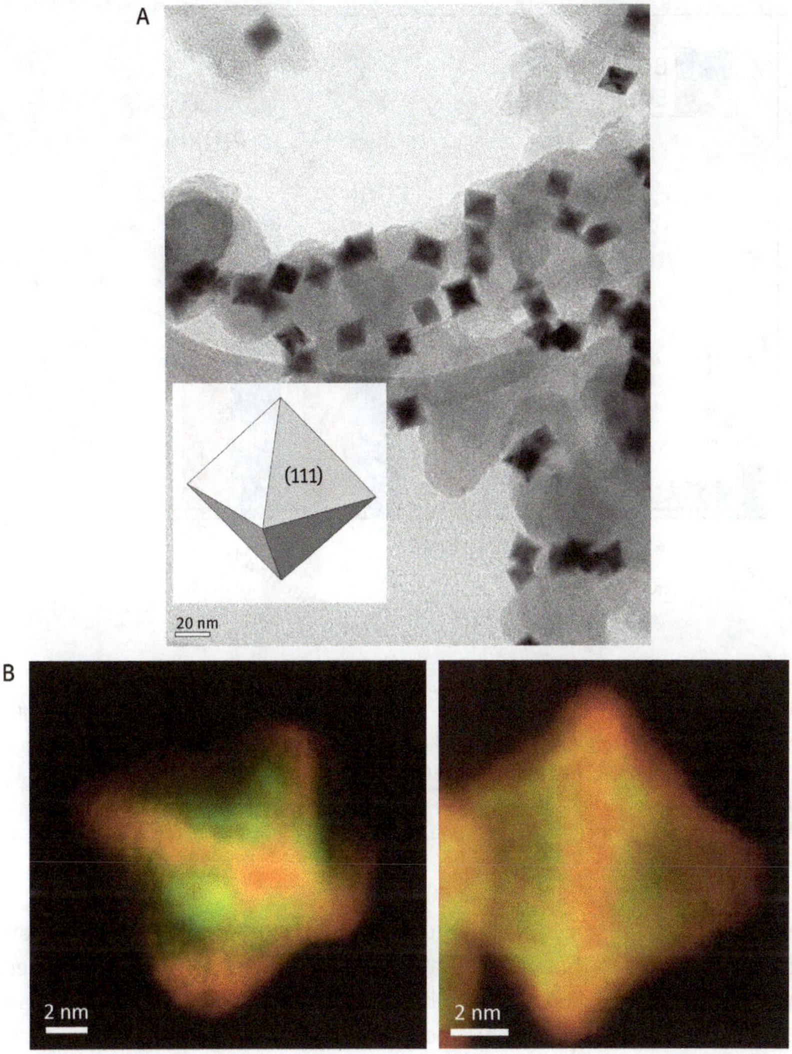

Figure 7.17: (A) Transmission electron microscopy (TEM) image of shape-controlled octahedral Pt–Ni ORR electrocatalyst nanoparticles that are supported on a high surface area conductive graphitic carbon support (light gray larger particle background). Inset: an illustration of an ideal octahedrally shaped particle. If the material has a face centered cubic atomic structure the eight particle facets display (111) geometry. (B) Elemental energy-dispersive X-ray (EDX) Analysis mapping of a single shape-controlled octahedral Pt–Ni fuel cell catalyst nanoparticle. Pt atoms are shown in red, while Ni atoms are shown in green. An aniosotropic distribution of the elements is evident.

can fulfill its function without the use of expensive, rare noble metals such as platinum. Inspired by this system, carbon-nitrogen-iron porphyrin compounds and their derivatives have long been investigated as ORR electrocatalysts. Over the past

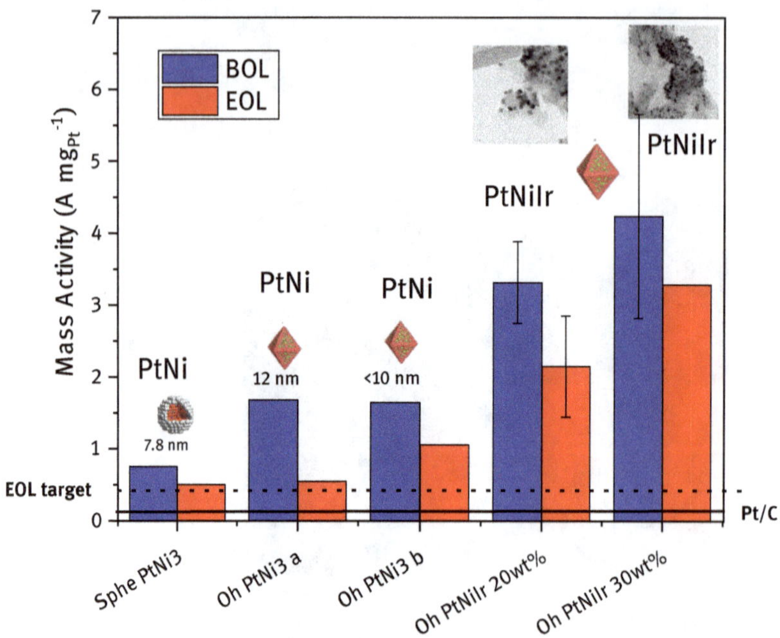

Figure 7.18: Catalytic ORR Pt mass activities (A/mg$_{Pt}$) of spherical dealloyed Pt–Ni nanocatalysts (Sphe PtNi3), compared to a number of shape-controlled octahedral PtNi nanoelectrocatalysts (Oh PtNi3 a 12 nm, Oh PtNi3 *b* < 10 nm, Oh PtNiIr 20 wt%, Oh PtNiIr 30 wt%). The blue and red bars denote the beginning of life (BOL) and end of life (EOL) mass activities, respectively. The mass activity of pure Pt nanoparticles is indicated by the solid line. The technological EOL target is shown by the dashed line. The mass activities were obtained in liquid electrolyte 3 electrode tests at 0.9 V_{RHE} electrode potential.

decade, much progress has been made in the development of highly active Fe-containing active centers, even though the exact nature of the active site is still subject to controversy [28]. Figure 7.18A shows voltammetric current density–electrode potential (*j*–*E*) profiles during the ORR of a noble-metal-free C–N–Fe at various stages of its preparation and activation. Curves 1 and 2 indicate that a porous carbon black material used as the supporting structure of the active sites has no ORR activity neither before nor after thermal annealing. Addition of a nitrogen-source to the porous carbon, here polyaniline (PANI) (Curve 3), followed by a thermal annealing step to 900 °C (Curve 4) increases the ORR activity stepwise. The addition of an Fe precursor prior to the thermal annealing step (Curve 5) suddenly increases the ORR activity significantly. The ORR activity of the resulting powder can be boosted further by a second thermal annealing treatment (Curve 6). Most likely, metallic Fe particles formed during the thermal treatment in the reductive carbon environment are removed during the acid leaching leaving behind catalytically active ORR sites. The electorcatalytic activity of curve 6 comes very close to that of a pure Pt electrocatalyst (curve Pt). Thus, a

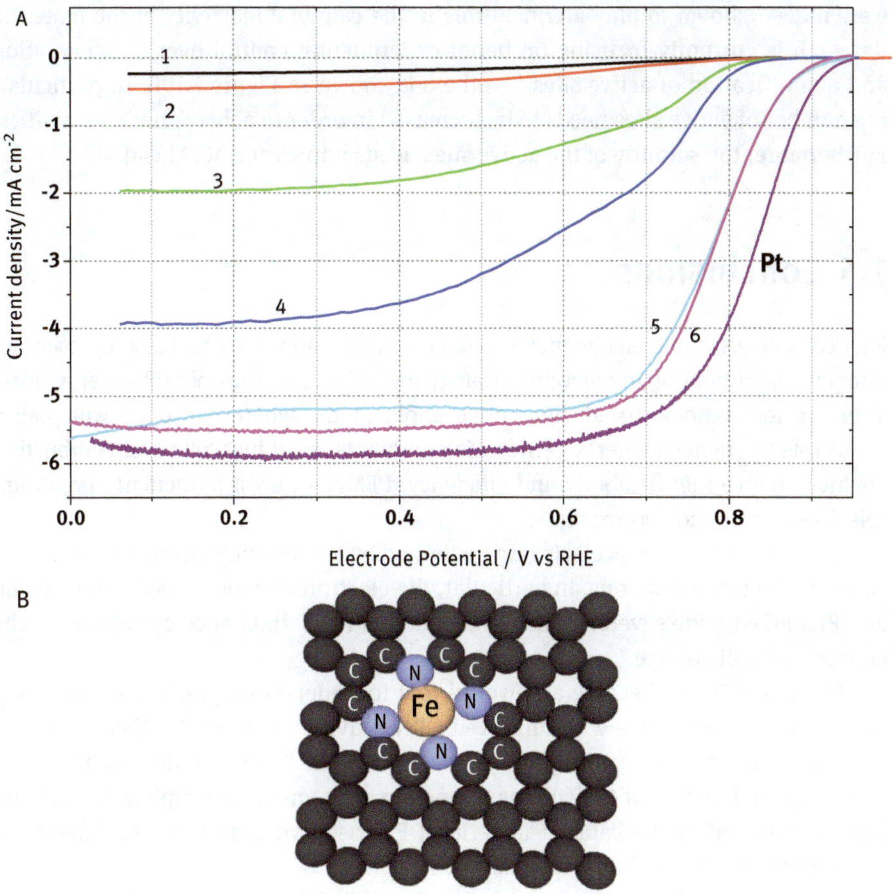

Figure 7.19: (A) Current density–potential plots of the ORR activity of noble-metal-free carbon–nitrogen–Fe electrocatalyst at various stages of their preparation. ORR activity of 1: A commercial carbon black support (Ketjen black 600), 2: The carbon support after thermal annealing to 900° C in nitrogen, 3: polyaniline (PANI) supported on the carbon black (PANI/Ketjen), 4: PANI/Ketjen after thermal annealing to 900° C, 5: PANI/Ketjen doped with Fe (PANI/Ketjen/Fe) after thermal annealing 900° C and single acid leaching, 6: PANI/Ketjen/Fe after a second thermal annealing, Pt: Pt nanoparticles (supported on carbon, 20% by weight) reference ORR catalyst. (B) Schematic illustration of the hypothesized active center of the C–N–Fe ORR electrocatalysts [28].

highly active biomimetic analogue of the biological heme-based ORR catalyst has been realized.

Recent investigations of the nature of the active sites in the biomimetic C–N–Fe ORR catalyst have highlighted the structural motive shown in Figure 7.19B [32]. A central Fe ion is fourfold coordinated in between the terminal pyridinic nitrogens at opposite edges of individual graphene layers. The space between graphene edges represents micropores inside the material, the presence and distribution of which were indeed shown to play a critical role in the catalytic reactivity of the material. Research is currently focusing on better experimental control over the generation and quantification of active sites like the one shown in Figure 7.19B. In particular, the number of active sites needs to be increased in order to achieve higher activities. Furthermore, the stability of the active sites needs improvement [33, 34].

7.5 Conclusions

Fuel cells are versatile electrochemical energy conversion devices. They represent an integral component of an emerging strategy toward a future sustainable energy infrastructure for stationary and automotive applications, where hydrogen will play a major role as chemical energy carrier. Major advantages of fuel cells over combustion engines involve fuel flexibility and efficiency. PEMFCs play a particularly important role due to their wide power range.

The efficiency of fuel cells is largely limited by the kinetic barriers of the surface catalytic electrode reactions. In particular, the electroreduction of molecular oxygen at a PEMFC cathode severely limits high reaction rates and hence currents near the equilibrium cell voltage.

Modern DFT has become a powerful tool to understand, predict, and discover electrochemical catalysts with improved ORR activity and stability. Computational free energy reaction diagrams provide insight into the “potential-determining” elementary reaction step of the ORR as a function of atomistic descriptors (surface-related properties) of the catalyst material. DFT-based volcano relations have been established pointing to improved catalyst systems.

From combined theoretical and experimental insights, nanostructured Pt “core–shell” electrocatalyst architectures as well as nanostructured shape-controlled PtNi architectures have recently emerged as promising, cost-effective cathode fuel cell catalysts. In the former, Pt-enriched multilayer surface shells surround Pt-poor cores that modify the reactivity of the surface Pt layer, while in the latter, preferential exposure of very active facets enables unprecedented ORR reactivity.

Beyond “human-made” Pt- or noble-metal-based ORR electrocatalysts, there exist very active biomimetic carbon–nitrogen–iron ORR electrocatalysts that show great potential for use in PEMFC cathodes, even rivaling the catalytic activity of pure Pt.

References

[1] Hoogers G. Fuel Cell Technology Handbook. Boca Raton, FL: CRC Press; 2003.

[2] Grove WR On voltaic series and the combination of gases by platinum. Philos Mag J Sci Ser 3. 1839;14:127–30.

[3] Grove WR On a gaseous voltaic battery. Philos Mag J Sci Ser 3. 1842;21:417–20.

[4] O'Hayre R, Cha S-W, Colella W, Prinz FB. Fuel Cell Fundamentals. New York: Wiley, 2006.

[5] Vielstich W, Lamm A, Gasteiger H. Handbook of Fuel Cells – Fundamentals, Technology, and Applications. Chichester, UK: Wiley, 2003.

[6] Schoenbein CF On the voltaic polarization of certain solid and fluid substances. Philos Mag J Sci. 1839;14:43–45.

[7] Schoenbein CF On the theory of the gaseous voltaic battery. Philos Mag J Sci Ser 3. 1843;22:165–66.

[8] Ostwald W Die Wissenschaftliche Elektrochemie der Gegenwart und die Technische der Zukunft. Z Elektrotech Elektrochem. 1894;4:122–25.

[9] Grot W Perfluorierte Ionenaustauscher-Membrane von hoher chemischer und thermischer Stabilität. Chem Ing Tech. 1972;44:167–69.

[10] Raistrick ID. Modified gas diffusion electrode for proton exchange membrane fuel cells. In: Van Zee JW, White RE, Kinoshita K, Barney HS editors, Proceedings of the Symposium on Diaphragms, Separators, and Ion-Exchange Membranes. Pennington, NJ: The Electrochemical Society, vol. PV–86, 1986:172.

[11] Lipkowski J, Ross PN. Electrocatalysis. New York: Wiley-VCH; 1998.

[12] Koper M. Fuel Cell Catalysis – A Surface Science Approach. Hoboken: NJ: Wiley; 2009.

[13] Vielstich W, Gasteiger HA, Yokokawa H. Handbook of Fuel Cells: Advances in Electrocatalysis, Materials, Diagnostics and Durability. Chichester, UK: John Wiley & Sons, vols 5–6, 2009.

[14] Bard AJ, Faulkner LR. Electrochemical Methods. New York: Wiley; 1980.

[15] Norskov JK, Rossmeisl J, Logadottir A et al.. Origin of the overpotential for oxygen reduction at a fuel-cell cathode. J Phys Chem B. 2004;108:17886–92.

[16] Hoare JP. The Electrochemistry of Oxygen. New York: Wiley; 1968.

[17] Kinoshita K. Electrochemical Oxygen Technology. New York: Wiley, 1992;448.

[18] Rossmeisl J, Karlberg GS, Jaramillo TF, Nørskov JK Steady state oxygen reduction and cyclic voltammetry. Faraday Discuss. 2008;140:337–46.

[19] Stephens IEL, Bondarenko AS, Grønbjerg U, Rossmeisl J, Chorkendorff I Understanding the electrocatalysis of oxygen reduction on platinum and its alloys. Energy Environ Sci. 2012;5:6744–62.

[20] Sabatier P Hydrogénations et deshydrogénations par catalyse. Ber Dtsch Chem Ges. 1911;44:1984–2001.

[21] Man IC, Su H-Y, Calle-Vallejo F et al.. Universality in oxygen evolution electrocatalysis on oxide surfaces. Chem Cat Chem. 2011;3:1159–65.

[22] Greeley J, Stephens IEL, Bondarenko AS et al.. Alloys of platinum and early transition metals as oxygen reduction electrocatalysts. Nat Chem. 2009;1:552–56.

[23] Mavrikakis M, Hammer B, Norskov JK Effect of strain on the reactivity of metal surfaces. Phys Rev Lett. 1998;81:2819–22.

[24] Kibler LA, El-Aziz AM, Hoyer R, Kolb DM Tuning reaction rates by lateral strain in a palladium monolayer. Angew Chem Int Ed. 2005;44:2080–84.

[25] Zhang JL, Vukmirovic MB, Xu Y, Mavrikakis M, Adzic RR Controlling the catalytic activity of platinum-monolayer electrocatalysts for oxygen reduction with different substrates. Angew Chem Int Ed. 2005;44:2132–35.

[26] Strasser P, Koh S, Anniyev T et al.. Lattice-strain control of the activity in dealloyed core – shell fuel cell catalysts. Nat Chem. 2010;2:454–60.

[26a] Gan L, Heggen M, O'Malley R, Theobald B, Strasser P. Understanding and Controlling Nanoporosity Formation for Improving the Stability of Bimetallic Fuel Cell Catalysts. Nano Lett. 2013;13:1131–38. DOI: 10.1021/nl304488q.

[26b] Gan L, Heggen M, Rudi S, Strasser P. Core-Shell Compositional Fine Structures of Dealloyed PtxNi1-x Nanoparticles and Their Impact on Oxygen Reduction Catalysis. Nano Lett. 2012;12:5423–30.

[27] Mani P, Srivastava R, Strasser P Dealloyed binary PtM_3 (M = Cu, Co, Ni) and ternary $PtNi_3M$ (M = Cu, Co, Fe, Cr) electrocatalysts for the oxygen reduction reaction: performance in polymer electrolyte membrane fuel cells. J Power Sources. 2011;196:666–73.

[28] Lefèvre M, Proietti E, Jaouen F, Dodelet J-P Iron-based catalysts with improved oxygen reduction activity in polymer electrolyte fuel cells. Science. 2009;324:71–74.

[29] Cui C, Gan L, Heggen M, Rudi S, Strasser P. Compositional segregation in shaped Pt alloy nanoparticles and their structural behaviour during electrocatalysis. Nat Mater. 2013;12 (8):765–71.

[30] Gan L, Cui C, Heggen M, Dionigi F, Rudi S, Strasser P Element-specific anisotropic growth of shaped platinum alloy nanocrystals. Science. 2014;346:1502–06.

[31] Dionigi F, Weber CC, Primbs M, Gocyla M, Bonastre AM, Spori C, Schmies H, Hornberger E, Kuhl S, Drnec J, Heggen M, Sharman J, Dunin-Borkowski RE, Strasser P. Controlling Near-Surface Ni Composition in Octahedral PtNi(Mo) Nanoparticles by Mo Doping for a Highly Active Oxygen Reduction Reaction Catalyst.. Nano Lett. 2019;19(10):6876–85.

[32] Chung HT, Cullen DA, Higgins D, Sneed BT, Holby EF, More KL, Zelenay P. Direct atomic-level insight into the active sites of a high-performance PGM-free ORR catalyst. Science. 2017;357 (6350):479–84.

[33] Sahraie NR, Kramm UI, Steinberg J, Zhang Y, Thomas A, Reier T, Paraknowitsch J-P, Strasser P Quantifying the density and utilization of active sites in non-precious metal oxygen electroreduction catalysts. Nat Commun. 2015;6:8618.

[34] Primbs M, Sun Y, Roy A, Malko D, Mehmood A, Sougrati M-T, Blanchard P-Y, Granozzi G, Kosmala T, Daniel G, Atanassov P, Sharman J, Durante C, Kucernak A, Jones D, Jaouen F, Strasser P. Establishing reactivity descriptors for platinum group metal (PGM)-free Fe–N–C catalysts for PEM fuel cells. Energ Environ Sci. 2020;13(8):2480–500.

[35] Sun Y, Polani S, Luo F, Ott S, Strasser P, and Dionigi F, “Advancements in Cathode Catalyst and Cathode Layer Design for Proton Exchange Membrane Fuel Cells”, *Nat Commun* **12**, 5984 (2021).

Nicholas Cox, James A. Birrell, Wolfgang Lubitz

8 Molecular Concepts of Water Splitting and Hydrogen Production: Nature's Approach

8.1 Introduction

Photosynthesis is the most important chemical process on our planet [1, 2]. The invention of water splitting by early photosynthetic organisms led to an oxygen-rich atmosphere and the protecting ozone layer in the stratosphere, both of which were important for the development of all higher life on Earth. Photosynthesis stores the sun's energy in the form of energy-rich organic compounds (carbohydrates) via CO_2 reduction. It is thus the only basic source of food, biomass (hay, wood, cotton, etc.), and fossil fuels such as coal, oil, and natural gas. The fast exploitation of these fossil fuels over the last one to two centuries, which are not renewable on a human time scale, has led to inevitable shortages. Furthermore, the burning of carbon-rich fuels has increased the concentration of greenhouse gases such as CO_2 in the atmosphere, leading to climate change with many adverse effects for our planet and human society.

Human population and economic growth, in particular in fast developing countries like India and China, will lead to further increases in energy demand. The economic, social, and political consequences of a shortage of fossil fuels are already felt today. It is, therefore, imperative to identify alternative sustainable energy sources. An obvious choice is to exploit solar energy, which has an enormous potential as a clean, abundant, and economical energy source. The problem is developing technologies that allow this energy source to be efficiently captured and stored in a useful form, such as heat, electricity, and in chemical fuels. Whereas the first two forms are already quite well developed, the use of solar energy to generate a practical chemical fuel still represents a major challenge [3].

Acknowledgments: Rita Groever and Birgit Deckers are acknowledged for their help with the text and artwork. The authors thank Holger Dau (FU Berlin) for helpful discussions on the OEC models. The authors' own work, cited in the manuscript, was financially supported by the Max Planck Society.

Nicholas Cox, Research School of Chemistry, The Australian National University, Canberra, Australia
James A. Birrell, Wolfgang Lubitz, Max Planck Institute for Chemical Energy Conversion, Mülheim an der Ruhr, Germany

https://doi.org/10.1515/9783110608458-008

Chemical bonds are the best way to store energy from excess electrical energy produced by renewable sources, for example wind power or photovoltaic devices. Thus, the efficient production of a clean, storable "solar fuel" would represent an important breakthrough in scientific research. Such a fuel must be made from abundant, inexpensive, and non-toxic materials such as water (H_2O), which can be split into molecular oxygen (O_2) and hydrogen (H_2). Hydrogen is considered the ideal primary fuel of the future [4, 5], since its combustion (e.g., in fuel cells) generates only water as a waste product. Furthermore, it can be converted to many other important energy-rich small molecules, for example, ammonia (NH_3), methane (CH_4), and methanol (CH_3OH) – for further storage and transport. These "fuels" can also have other uses in the industrial and agricultural sectors such as NH_3 for fertilizers (see Figure 8.1). All these processes, however, require appropriate catalysts.

A promising starting point for the development of a synthetic light-driven water-splitting catalyst is to use nature's approach as inspiration. A simple schematic of the multiple steps seen in oxygenic photosynthesis, the process via which water is split into O_2 and protons (H^+), and hydrogen production, the process via which protons are reduced to H_2, is shown in Figure 8.1. Oxygenic photosynthesis primarily occurs in a single biological super-complex, photosystem II (PSII), which is found in green plants, algae, and cyanobacteria. Many green algae and bacteria also contain an enzyme that allows hydrogen evolution, the hydrogenase. The efficient coupling of these two biological processes in vivo is currently being investigated in many laboratories for the development of biohydrogen technologies [6, 7]:

$$2\,H_2O \xrightarrow{\text{photosynthesis}} O_2\uparrow\ + 4\,H^+ + 4\,e^- \xrightarrow{\text{hydrogenase}} 2\,H_2 \tag{8.1}$$

The implementation of "artificial photosynthesis" [8–10] is a major challenge as scientists have great difficulties to design and synthesize devices that are able to mimic the natural process in a cost-efficient way, that is, with low energy waste (overpotential), high rates (turnover frequencies, TOF), and long lifetimes (turnover numbers, TON) of the catalyst. One problem is the little understood coupling of the light-induced one-photon/one-electron charge separation to the multi-electron catalytic processes leading to water oxidation and fuel production. In addition, nature uses a *smart matrix* – the protein – to optimize the enzymatic process. This matrix also provides protection and allows repair in vivo, that is, the lifetimes of in vitro isolated native enzymes assembled as artificial systems are typically very short. In this respect, chemically/biochemically synthesized systems are currently inferior to the natural ones. The key lesson to be kept in mind and learned from nature is that catalysts are not static – on the contrary, they are dynamically involved in the reaction process.

In the last few decades high-resolution X-ray crystallographic structures of PS II [11, 12] and various hydrogenases [13–15] have been obtained and additional spectroscopic and electrochemical experiments have very significantly increased our knowledge about the structure and function of the native enzymes. Slowly, a mechanistic understanding of these systems is emerging in all its details. It is clear that resolving the basic principles of water oxidation and hydrogen conversion in nature is of major importance in scientific research and for our society. This would probably allow us to design better photosynthetic organisms capable of (bio)hydrogen production and at the same time advance the field of *artificial photosynthesis*, aiming at synthesizing new catalysts for large-scale water splitting and hydrogen production and for energy storage in chemical bonds in general, processes that are of key importance for a sustainable future "hydrogen economy."

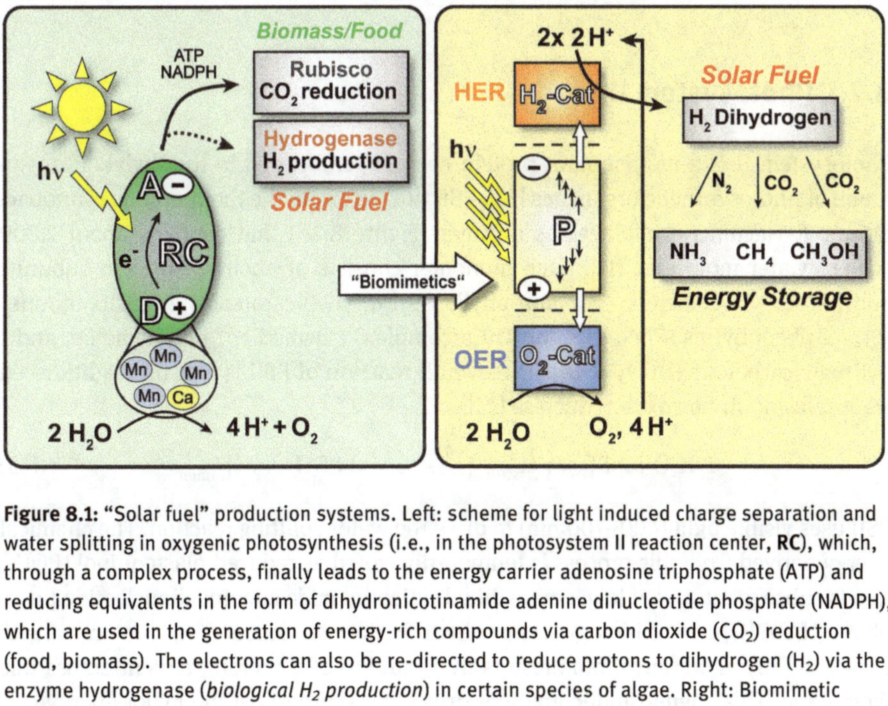

Figure 8.1: "Solar fuel" production systems. Left: scheme for light induced charge separation and water splitting in oxygenic photosynthesis (i.e., in the photosystem II reaction center, **RC**), which, through a complex process, finally leads to the energy carrier adenosine triphosphate (ATP) and reducing equivalents in the form of dihydronicotinamide adenine dinucleotide phosphate (NADPH), which are used in the generation of energy-rich compounds via carbon dioxide (CO_2) reduction (food, biomass). The electrons can also be re-directed to reduce protons to dihydrogen (H_2) via the enzyme hydrogenase (*biological H_2 production*) in certain species of algae. Right: Biomimetic system showing multiple light-induced charge separation in a photo(electro)catalyst **P** and transport of redox equivalents to a light-driven water-splitting catalyst (oxygen evolution reaction, OER), and a light-driven H_2-producing catalyst (H_2 evolution reaction, HER). The primary "solar fuel" produced is H_2, in which the Sun's energy is stored. H_2 can be directly used, for example, in combustion engines or in fuel cells to generate electricity, or can be converted for storage, transport or other uses. The examples given involve ammonia production from nitrogen and reactions with CO_2 to yield methane or methanol. These reactions also require efficient catalysts.

8.2 Water Oxidation

Oxygenic photosynthesis uses a single enzyme for water splitting, the water oxidizing complex or oxygen-evolving complex (OEC) embedded in PSII. It appears in many different organisms such as cyanobacteria, algae, and higher plants. Thus far only very minor differences between photosynthetic organisms have been found at the level of the OEC, the site in PSII where water splitting occurs. This feature is unique within biology and highlights the high level of complexity that is required to perform this strongly oxidizing process. Several important features of the water-splitting reaction in PSII have already been unraveled. These will be briefly outlined below with reference to their significance for the development of efficient artificial catalysts for the water-splitting half-reaction:

$$2\ H_2O \longrightarrow O_2 + 4\ H^+ + 4\ e^- \tag{8.2}$$

8.2.1 Photosystem II

Photosystem II is a multisubunit protein complex embedded in the thylakoid membrane of photosynthetic organisms [22]. PSII purified from the thermophilic cyanobacterium *T. vulcanus* crystallizes as a dimer (Figure 8.2A) that contains about 2,800 solvent water molecules [11]. Each monomer consists of about 20 protein subunits and harbors 77 cofactors: 35 chlorophylls (Chl *a*), 11 β-carotenes, 2 plastoquinones (PQ), 2 pheophytins (Pheo *a*), 1 Mn_4O_5Ca complex, 2 heme Fe, 1 non-heme Fe, and 1 hydrogencarbonate (HCO_3^-/CO_3). The overall reaction of PSII is that of a light-driven water:plastoquinone oxidoreductase [23]:

$$2\ H_2O + 2\ PQ + 4\ H^+_{stroma} \overset{4\,h\nu}{\rightarrow} O_2 + 2\ PQH_2 + 4\ H^+_{lumen} \tag{8.3}$$

PSII uses visible light (400–700 nm) to drive the water-splitting reaction. The chemical energy derived from this process is temporarily stored as reduced plastoquinol (PQH_2) that is subsequently used to generate dihydronicotinamide adenine dinucleotide phosphate (NADPH), one of biology's major hydride donors. Concomitant proton transfer across the thylakoid membrane results in a buildup of a proton gradient across this membrane. This chemiosmotic gradient drives the synthesis of the major energy carrier of biology, adenosine triphosphate (ATP). Both NADPH and ATP are required for carbon fixation; thus, PSII, together with photosystem I (PSI) (which is not dealt with here), is primarily responsible for the majority of biological energy storage that is derived from sunlight.

The reactions that occur in PSII can be divided into four processes:

i) *Light harvesting and energy transfer* by the chlorophyll and carotenoid molecules of the antenna complexes to the reaction center (RC) of PS II;

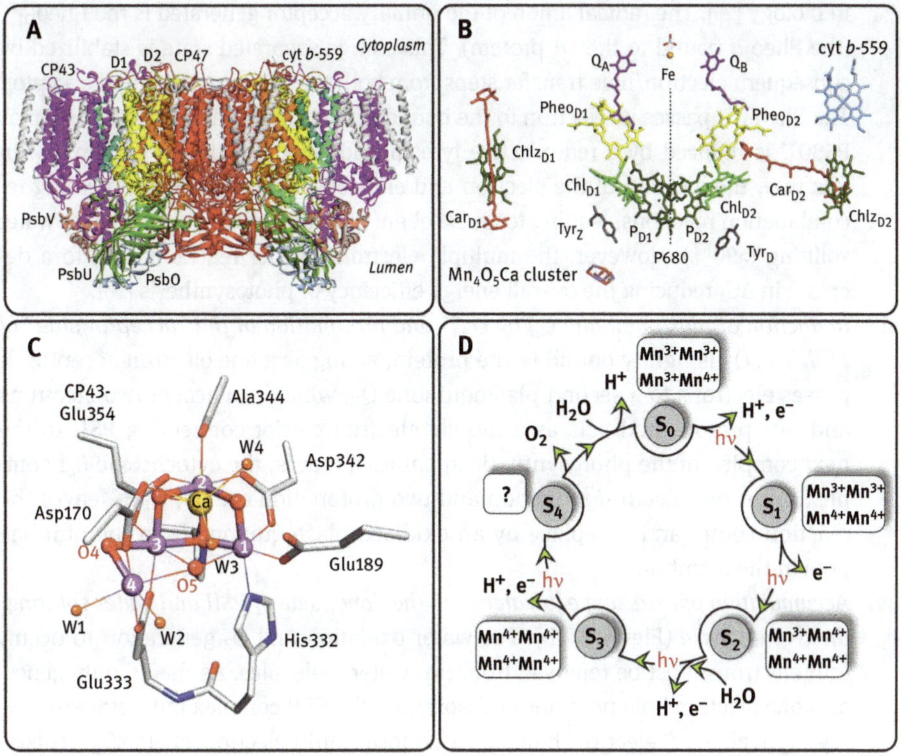

Figure 8.2: **(A)** X-ray crystallographic structure of the PSII dimer from *Thermosynechococcus (T.) vulcanus* [11] at 1.9 Å resolution showing the major protein subunits in color and **(B)** the arrangement of the cofactors in the central D1/D2 proteins involved in charge separation, electron transport and water oxidation (for details see text). **(C)** Structure of the Mn_4O_5Ca cluster with amino acid ligands and the four bound water molecules W1 to W4 [11]. The four Mn ions (purple) are numbered 1–4; the Ca (yellow) and the bridging oxygens (red) are indicated. **(D)** The catalytic cycle (S-state cycle) of water oxidation proposed by Kok et al. [16]. The different states of the Mn_4O_5Ca cluster that are attained after single oxidation events driven by short light flashes (hν) [17] are termed S_i states, where i = 0, . . ., 4 indicates the number of oxidizing equivalents stored in the cluster. The oxidation states of the Mn ions are given for each S-state (high valance case) [18–20]. The water binding events are indicated and the release of protons is also shown [19, 21].

ii) *Excitation and primary charge separation* of the RC, a multi-pigment assembly of four Chl *a* and two Pheo *a* molecules (Figure 8.2B), which are bound to the D1 and D2 protein scaffold. Light excitation initiates a single electron transfer within the multi-pigment assembly resulting in a charge separated (radical pair) state [24]. The radical cation (primary donor) generated is thought to predominantly reside on a single *Chl a* pigment (P_{D1}), which is commonly referred to as the pigment radical $P680^{\bullet+}$ according to the position of its optical absorption maximum. It has an estimated oxidizing potential of +1.2 to +1.3 V, one of the highest known in biology [25]. The radical anion of the primary acceptor generated is the $Pheo_{D1}^{\bullet-}$ (the Pheo *a* bound to the D1 protein). This charge separated state is stabilized by subsequent electron/hole transfer steps from both the primary donor and acceptor. The $Pheo_{D1}^{\bullet-}$ passes its electron to the bound plastoquinone molecule Q_A, whereas $P680^{\bullet+}$ is reduced by a redox-active tyrosine side chain of the D1 protein, Y_Z. In this way, the location of the electron and electron hole increases suppressing recombination reactions, leading to an excellent quantum efficiency of PSII for water splitting (>90%). However, the multiple electron transfer reactions lead to a decrease in ΔG, reducing the overall energy efficiency of photosynthesis [26].

iii) *Reduction of plastoquinone Q_B by $Q_A^{\bullet-}$ and protonation at the "acceptor side" of PSII.* The Q_A is tightly bound to the protein, acting as a one electron acceptor. It passes electrons to a second plastoquinone Q_B, which can accept two electrons and two protons and acts as a mobile electron carrier connecting PSII to the next complex of the photosynthetic apparatus, that is, the cytochrome b_6f complex. After two electron reduction and two protonation events, Q_BH_2 leaves the reaction center and is replace by an oxidized plastoquinone from the quinone pool in the membrane.

iv) *Accumulation of oxidizing equivalents on the donor side of PSII and water splitting: the S-state cycle* (Figure 8.2D). For water oxidation and O_2 generation to occur, four electrons must be removed from two water molecules. As the RC only generates one electron hole per photon absorption, the PSII complex must store oxidizing equivalents ("electron holes") to perform multi-electron chemistry. This is achieved at the site of the tetranuclear manganese complex. To retain charge neutrality, protons are released during the accumulation of the oxidizing equivalents (see Figure 8.2D). The oxidized tyrosine radical Z, $Y_Z^{\bullet}$ (D_1–Y_{161}), acts as a one electron hole carrier, connecting the primary donor $P680^{\bullet+}$ to the manganese cluster. The OEC consists of an inorganic Mn_4O_5Ca cluster embedded in and interacting with its surrounding protein matrix (Figure 8.2C). Y_Z is considered part of the functionally important protein matrix, which is involved in proton transfer reactions – coupling substrate conversion with proton release [21, 27]. The water-splitting reaction thus proceeds via a series of distinct redox intermediates steps, termed the S_i states, where the index indicates the number of stored oxidizing equivalents (i = 0–4) in the Mn_4O_5Ca cluster [16], see Figure 8.2D. The S_4 state, which has not yet been spectroscopically identified, rapidly decays to the S_0 state upon the release of molecular triplet oxygen and rebinding of one substrate water molecule. It remains unclear when the second water binds; however, it is postulated that this substrate is only loosely associated with the OEC until late in the reaction cycle, that is, S_3 [28], see Section 8.2.3 The dark stable or resting state of the enzyme is S_1. Higher S-states, S_2–S_3, decay back to this state on a timescale of seconds at room temperature. The S_0 state is of sufficient reduction potential that it can be oxidized via a second redox-active tyrosine residue Y_D (D2–161) [29], which does not participate further in the S-state cycle.

It is important to note that the timescales of reaction center photochemistry and water-splitting chemistry differ by more than nine orders of magnitude. The generation of the oxidant for the water-splitting reaction ($P680^{\bullet+}$) occurs on a time scale of about 3 ps [30]. In contrast, water oxidation chemistry is slow, requiring substrate binding, proton translocation, and, ultimately, O–O bond formation. Early S-state transitions occur with a halftime of ≈100 µs with the $S_3 \rightarrow S_4 \rightarrow S_0$ taking as long as 1–2 ms [21]. Thus, the secondary and tertiary electron transfer events that occur on both the acceptor and donor sides of PSII (see ii) are critical for water splitting to occur, as they stabilize an oxidative state of the complex that has a lifetime long enough for water-splitting chemistry to occur.

8.2.2 Structure of the Resting States of the Oxygen-Evolving Complex (OEC): The S_0, S_1, S_2 States

The structure of PSII was determined at a resolution of 1.9 Å by Umena et al. [11] using conventional X-ray diffraction (XRD) and later refined by X-ray free electron laser (XFEL) technologies (see [12] for a recent review). At this resolution, the position of all four Mn ions and Ca can be identified and the network of bridging oxygens connecting the metal ions is resolved. The structure of the inorganic core resembles a "distorted chair" where the base is formed by a µ-oxo-bridged cuboidal Mn_3O_4Ca unit (Figure 8.2C); three Mn and one Ca ion form the vertices of the cube. The fourth "outer" Mn, Mn_4, is attached to this core structure via a µ-oxo-bridged ligation (O4) to one of the oxygen corners of the cuboidal unit. A second linkage between the "outer" Mn and the cube is provided by a 5th oxygen O5, which in the original crystal structure was proposed to be a hydroxo bridge to the central Mn_3. More recent XFEL structures suggest it is a bridging µ-oxo between Mn_4 and Mn_3 [31, 32]. In addition to the µ-oxo-bridged network, the Mn_4O_5Ca is held together by six carboxylate ligands, four of which (Asp342, Asp170, Glu333, CP47-Glu354) form µ-carboxylato bridges between Mn sites. The cluster has only one nitrogen ligand, from the His332 (Figure 8.2C). Water-splitting activity is conferred to the OEC by the Ca^{2+} and Cl^- ions. Removal of either the Ca^{2+} or Cl^- ions inhibits water splitting by blocking the $S_2 \rightarrow S_3$ transition [33, 34], however, surrogates are capable of restoring water splitting albeit at slower turnover rates: Sr^{2+} can replace Ca^{2+}, Br^- and I^- can replace Cl^-. The replacement of both Ca^{2+} and Cl^- by surrogate ions, leads to an additive effect on the turnover frequency, slowing the $S_3 \rightarrow S_4 \rightarrow S_0$ by as much as 30-fold [35]. Both chloride and calcium have been proposed to influence proton transfer away from the OEC. Ca^{2+} may also play a role for substrate water binding.

The S-state of the cluster as measured by XRD is expected to represent the dark stable S_1 state; however, this assignment is at present unclear. Mn-Mn, Mn-Ca and Mn-O/N distances seen in the crystal structure are ≈0.1 Å longer as compared to those determined from experimental extended X-ray absorption fine structure (EXAFS) [36].

Similarly, the central O5 atom has unusually long bonds (2.5–2.6 Å) to three Mn ions and to the Ca ion, which are atypical for $Mn^{III,IV}$-O ligands/bridges, regardless of their protonation state (Figure 8.3, top). This suggests that the cluster underwent some degree of radiation-induced modifications, that is, photo-reduction of the Mn to lower oxidation states [37], during data collection and thus may not represent a physiological S-state. To solve this problem Shen and coworkers repeated this study using XFEL sources [38, 39], which, by design, should be free of radiation induced artefacts. The XRD and XFEL structures were, however, quite similar suggesting precise structural constraints cannot yet be fully resolved by this approach.

Regardless, the general pattern of three short and one long Mn-Mn distances and four Mn-Ca distances, as observed in EXAFS measurements, is preserved [36] and, thus, in this way, the crystal structure is similar to earlier literature models, including those derived from (polarized) EXAFS measurements and the computational models of Kusunoki [40], Siegbahn [41] and Ames et al. [42]. In these models, O5 is assigned to a μ-oxo bridge. This is found to be both energetically favorable and engendered Mn–Mn distances and magnetic properties that are consistent with EXAFS and electron paramagnetic resonance (EPR)/electron-nuclear double resonance (ENDOR) data.

Spectroscopic measurements of the S-states complement the structural data collected. These provide details of the electronic structure, that is, the net oxidation states of the S-states, the order in which the four Mn are oxidized and the level to which the Mn ions interact (via spin exchange, J), which has consequences for the redox properties of the cluster, the binding of the substrates and O_2 and H^+ release. As an example, magnetic spectroscopies such as EPR and related double resonance methods such as ^{55}Mn-ENDOR/ELDOR (electron-electron double resonance)-detected NMR (nuclear magnetic resonance) (EDNMR) were able to resolve the net oxidation states of the S_0 and S_2 states to be: Mn_4(III,III,III,IV) and Mn_4(III,IV,IV,IV), respectively [18, 43]. No Mn(II) has been detected in any of the S states. EPR techniques also allow the oxidation states and ligand fields of specific Mn ions to be deduced. These assignments compare favorably with complimentary spectroscopic measurements such as X-ray absorption near-edge spectroscopy (XANES) [44, 45], K_β-fluorescence spectroscopy [45] and resonant inelastic X-ray scattering (RIXS) spectroscopy [46].

Spectroscopic measurements also show that the structure of the cofactor becomes dynamic/fluxional in the S_2 state. EPR evidence for this is the observation of two interconvertible ground state signals of differing total spin, namely S_2^A (S = 1/2) and S_2^B (S = 5/2), see Figure 8.3. The observed spin state is a consequence of how the Mn ions interact through next-neighbor (spin exchange) interaction, with the two different spin states correlating with two distinct connectivities/topologies of the cofactor. The A type topology – which gives rise to the well know S_2 multiline signal – resembles the crystallographic model placing O5 as a bridge between Mn_4 and Mn_3. The B type topology – which gives rise to the g = 4.1 signal – instead places O5 inside the cubane unit as a ligand to Mn_3 and Mn_1. In the two almost isoenergetic structures the Mn(III) in the cluster is changing position, as indicated in Figure 8.3. It is postulated that the flexible coordination of O5 is important for cofactor activation, see below.

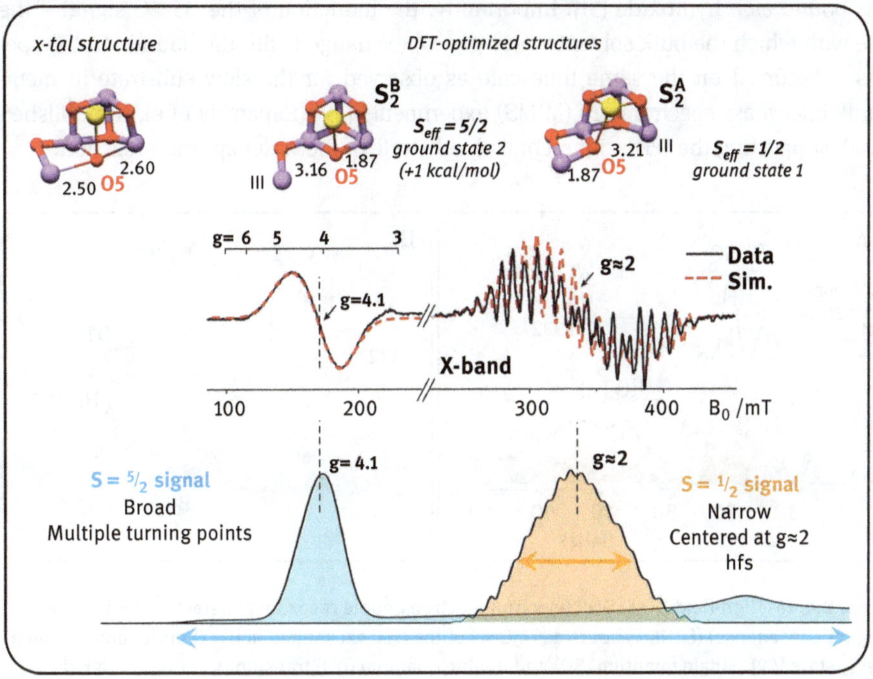

Figure 8.3: Top: Core structure of the Mn_4O_5Ca cluster (OEC) [11] and density functional theory (DFT) optimized structures of the OEC for the S_2 state [47]. These structures are more than 100 kcal/mol lower in energy than the crystal structure (left). The distances between oxygen O5 and Mn4 and Mn1 are given for the three structures. The two ground states S_2^A and S_2^B have almost equal energy; note that the position of the Mn(III) ion in the two states is different; spin coupling leads to different total (effective) spin ground states (S = 1/2 and S = 5/2) and to different EPR signatures at X-band (at g = 2, multiline signal with ^{55}Mn hyperfine structure; at g = 4.1 broad unstructured signal). Spectral simulations are shown in red and at the bottom a simulation of the EPR absorption for the two states is shown. Adapted with permission from [48]. Copyright 2016 American Chemical Society.

The identification of O5 as a ligand derived from the first substrate water has also been done by EPR techniques. EPR spectroscopy is a powerful tool for determining the identity ($H_2O/OH^-/O^{2-}$) and location of water-derived species bound in the vicinity of the Mn_4O_5Ca cluster. This can be achieved via the detection of the coupling of magnetic nuclei, for example, ^{17}O ($I = 5/2$), to the paramagnetic states (e.g., S_2) of the Mn_4O_5Ca cluster. Using high-field ^{17}O-EDNMR [49] three classes of nuclei were assigned: i) a μ-oxo bridge; ii) terminal Mn_4-OH/OH_2; and iii) Ca/2nd shell-$^-OH/H_2O$

ligand(s), by comparison to ^{17}O labeled model complexes. The exchangeable μ-oxo bridge was assigned to O5, which links the outer Mn_4 to the Mn_3O_3Ca open-cuboidal unit (Figure 8.4). A similar μ-oxo bridge signal was observed in Q-band ^{17}O-ENDOR experiments on S_2 [50]. ^{17}O-EDNMR also detected the O5 in the lowest S-state (S_0), where it is bound as a hydroxide [51]. Importantly, the induction of the O5 ^{17}O signal – the rate with which the bulk solvent water ($H_2{}^{17}O$) exchanged with the bound Mn-^{16}O species – occurred on the same timescale as observed for the slow substrate in membrane-inlet mass spectrometry (MIMS) experiments [49] (Rapatskiy et al. unpublished data), supporting the O5 assignment as a substrate of the water-splitting reaction.

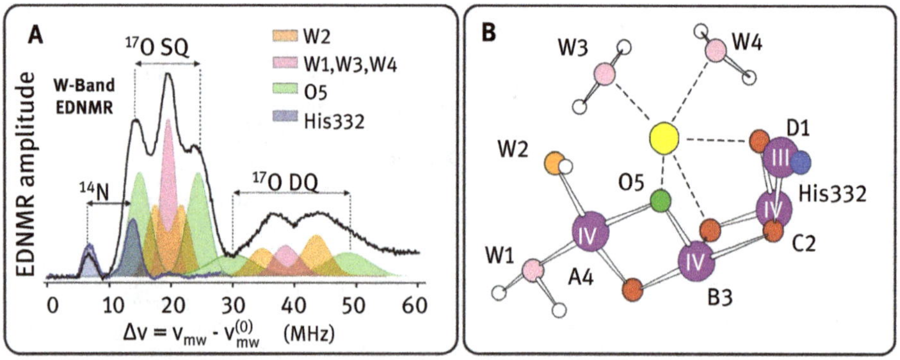

Figure 8.4: (A) High-field ^{17}O ELDOR (electron-electron double resonance)-detected NMR (nuclear magnetic resonance) (EDNMR) spectra (94 GHz) of the OEC exchanged in $H_2{}^{17}O$ buffer and poised in the S_2 state [49]. Single quantum (SQ) and double quantum (DQ) transitions of the ^{17}O signals are visible stemming from three types of oxygen ligands (see legend). The largest coupling is assigned to the μ-oxo bridge O5 (green). Superimposed, an EDNMR spectrum is also shown for an $H_2{}^{16}O$ exchanged sample showing only ^{14}N signals (from His332) and no ^{17}O. **(B)** Structure of the Mn_4O_5Ca cluster with μ-oxo bridges and bound water molecules W1 to W4. Color code: Mn – purple, Ca – yellow, N – blue, H – white, O (non-exchangeable) – red, and O (exchangeable) – pink/orange/green. Adapted with permission from [49]. Copyright 2012 American Chemical Society.

8.2.3 Structure of the Activated OEC: The S_3 and [S_4] States

It has long been thought that the Mn_4O_5Ca cluster undergoes structural changes during the catalytic cycle [44, 52, 53], particularly during the $S_2 \rightarrow S_3$ transition. Thus, the structure of the Mn_4O_5Ca cluster in the lower oxidation states, S_0-S_2, is expected to be different from that in the S_3 to S_4 states in which the O–O bond is formed. Recent spectroscopic and crystallographic results [31, 32, 54] now definitively demonstrate this notion is correct, showing that the stoichiometry of the cofactor changes upon formation of the S_3 state owing to the insertion of a water molecule into the cubane unit.

High-field EPR measurements constrained both the coordination number and oxidation states of the four Mn ions in the S_3 state. These measurements are best performed at higher magnetic fields owing to the cofactor now having an integer spin ground state (S = 3). The narrow EPR spectrum observed and ^{55}Mn-EDNMR data are consistent with the cluster containing only octahedral Mn(IV) ions, requiring the binding of a water ligand to the open coordination site of the Mn_1 (see Figure 8.5) [54]. Subsequent XFEL studies have confirmed this result, resolving additional electron density near the Mn_1 ion following progression to the S_3 state assigned to a new oxygen ligand (O6) [31, 32, 38]. Initial XFEL data supported O6 already forming a bond to the existing O5 (bridging peroxo ligand) [38]. This has been revised to instead assign O6 as a terminal oxyl radical ligand or more likely hydroxo ligand, more consistent with high-field EPR measurements (Figure 8.5, right) [31].

While there is now a consensus that a water binding event occurs during the S_2 to S_3 transition there is still an active debate as to how this insertion takes place. Crystallography studies support the direct binding of O6 to Mn_1 via the Ca^{2+} water channel [32]. This contrasts with spectroscopic data which instead resolves a number of structurally distinct intermediates. EPR measurements performed in the presence of substrate analogues (methanol) or on chemically modified samples (Sr^{2+} substituted) identify a precursor S_3 state (labeled S_3', Figure 8.5) which is assigned to

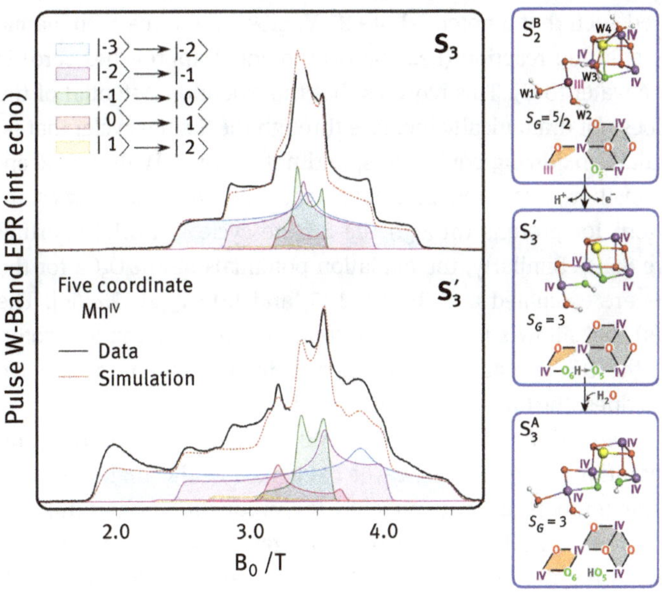

Figure 8.5: High-field (94 GHz) electron paramagnetic resonance (EPR) of two forms of the S_3 state [54, 55]. The narrower signal (top) represents the final S_3 state with all Mn(IV) ions 6-coordinated, whereas the wider signal (bottom) results from the precursor S_3' state with Mn_4(IV) only 5-fold coordinated. The spectral analysis is given and simulations are shown. On the right the respective structures and exchange-coupling schemes are depicted. Adapted with permission from [55].

a cluster containing only Mn(IV) ions but retaining one five coordinate Mn ion [55]. These data are consistent with a more complicated insertion mechanism proposed from theoretical calculations in which O5 migrates to Mn_1 followed by water binding at Mn_4, suggesting it is the high spin S_2 state, S_2^B, that is the catalytically active form (Figure 8.5, right).

8.2.4 Function of the OEC

The PSII/OEC simultaneously performs multiple chemical functions throughout the S-state cycle. These can be grouped into three classes:

(i) *Oxidation equivalent accumulation/redox-leveling*. The oxidant for the water-splitting reaction $P680^{\bullet+}$ has an oxidizing potential of approx. + 1.25 V [25]. Whilst among the highest known for biological systems, this potential is insufficient to directly oxidize water in bulk solution; the first step of the water oxidation reaction requires the transformation of water to the hydroxyl radical, which occurs at a potential above +2 V. This, however, is the most energy expensive way to split water. By using two concerted two-electron steps that involve a transient peroxide intermediate, or a concerted four-electron mechanism, where electron transfer is coupled to proton release, the reaction barrier to O–O bond formation is significantly lowered such that a potential of +1.2 V, close to the thermodynamic limit for the water-splitting reaction (i.e., the overpotential approaches zero) is sufficient to oxidize water to O_2. This requires that the oxidation potential of the Mn_4O_5Ca cluster does not dramatically increase through the S-state cycle; that is, the cluster accumulates oxidizing equivalents, and not charges. With the exception of the $S_1 \rightarrow S_2$ transition, this is indeed what occurs. The experimentally established release pattern for protons through the S-state cycle is 1:0:1:2 (starting from S_0, see Figure 8.2D). Similarly, the oxidation potentials of Mn_4O_5Ca for the individual S_i-states are estimated at 0.85, 1.1, 1.15, and 1.0 eV [21, 26, 56]. The exact mechanism(s) that allow substrate deprotonation and subsequent proton translocation from the OEC are currently a major topic of research and involve multiple protein residues, the Ca^{2+}/Cl^- cofactors, and so on.

(ii) *Rapid reassembly/repair of the OEC from Mn^{2+}*. Excitation of chlorophyll pigments (triplet generation) in the presence of oxygen, and the high-oxidative power of $P680^{\bullet+}$, inevitably leads to damage via the production of harmful side products in the reactions of PSII such as: singlet oxygen, superoxide radicals or peroxides. This limits the lifetime of the PSII-OEC to less than 30 min in full sunlight [57]. To combat this, photosynthetic organisms have devised an efficient repair cycle that selectively replaces the damaged protein unit, D1 protein, to which the main redox cofactors are bound. As the OEC is also bound to this subunit, it must also be replaced. Unlike many other cofactors, the OEC is synthesized in situ from Mn^{2+} and Ca^{2+} in solution via a light-driven process known as photo-activation [58], which requires no chaperone or maturation factors. This allows the PSII to rapidly recover water-splitting capability over the course of the assembly–degradation–repair cycle. It should be noted though that over the course of 30 min, the OEC has a maximum TON of close to 10^6 (TOF ≈500 s^{-1}). This is significantly higher than that achieved by most artificial first row transition metal molecular water-splitting catalysts and is only surpassed by a few precious metal catalysts [59].

(iii) *The selective binding of substrate water and O–O bond formation.* The OEC represents a genuine molecular catalyst with only two sites for substrate binding. This directs selective single product formation, namely O_2, as opposed to other dangerous reactive oxygen species ($H_2O_2/HO_2^-/O_2^{\bullet-}$), which presumably explains its large TON. While the sites of substrate binding remain under debate, the kinetics of substrate binding are well understood as determined from time-resolved MIMS studies, which employ a rapid jump in $H_2{}^{18}O$ concentration within specific flash sequences [28]. These experiments have identified a possible role for the Ca^{2+} in binding one substrate water as substrate binding is influenced by biochemical Ca^{2+}/Sr^{2+} exchange. Similarly, they also establish that the two substrate molecules bind at separate sites, as their exchange rates vary independently with S_i state. The slower exchanging substrate water (W_S) can be detected in all S_i states suggesting it binds earlier in the S-state cycle (S_0). This has been corroborated by $H_2{}^{17}O$ exchange and ^{17}O EDNMR experiments performed on S_2 and S_0 [49, 51]. The faster exchanging substrate (W_f) has only been resolved in the S_2 and S_3 states. EPR [54] and Fourier transform infrared (FTIR) [60] measurements support these findings, suggesting that one water molecule, possibly W_f, binds during the $S_2{\rightarrow}S_3$ transition, as described in Section 2.3.

In Figure 8.6 a possible reaction cycle for water oxidation in PSII is shown that details the stepwise oxidation of the Mn ions in the cluster, the substrate water binding, proton release and finally the O–O bond formation. The open cubane (A) and closed cubane (B) type structures and their significance are given. The first substrate water binds as hydroxide in S_0 and is further deprotonated in S_1. Another water enters in the $S_2{\rightarrow}S_3$ transition, as suggested by EPR and XFEL experiments [31, 32, 38, 54], which could serve as the second substrate. Further oxidation and deprotonation leads to the transient S_4 state. For the O–O bond formation and O_2 release different models exist, see below.

8.2.5 Suggested Mechanisms of O–O Bond Formation

There are several proposed mechanisms for O–O bond formation by the OEC; for reviews see [12, 61–64]. Essentially a proposed mechanism has to address two questions:

(i) What are the two substrate sites of the OEC?

(ii) What is the chemical mechanism by which these adjacent substrate oxygens couple?

Experimental evidence already described above limits the scope of question (i). One substrate site, presumably the slow substrate W_s bound in the early S–states (e.g., S_0), is very likely the μ_2-oxo bridge O5. The question is then what is the second substrate. If it represents a terminal ligand of Mn_4 (W2) or the nearby water of the Ca^{2+} (W3), O–O bond formation probably proceeds via a nucleophilic attack. Alternatively, if the second substrate is the new water ligand bound to the cofactor during the S_2 to S_3 transition (O6), O–O bond formation probably proceeds via a radical coupling mechanism. Both are described briefly below:

Nucleophilic attack in the S_4 state. A terminal oxygen, W2 or the μ_2-oxo bridge O5, is progressively deprotonated during the S-state cycle such that in S_4 it is a strong electrophile. In the simplest (terminal ligand) case this species can be considered a Mn(V)=O, however, it may equally be described as a Mn(IV)≡O^+ or Mn(IV)–$O^\bullet$ species. During the $S_3(Y_Z^\bullet)\rightarrow S_4\rightarrow S_0$ transition the electrophile is attacked by a nucleophilic water molecule. This may be a water/hydroxo bound to either the Ca^{2+} (W3) or the Mn_4 (W2) (Figure 8.2C). This type of mechanism requires a Mn-centered oxidation during the $S_2\rightarrow S_3$ transition and Mn-centered or a ligand-centered (substrate) oxidation during the $S_3\rightarrow S_4$ transition. The nucleophilic attack mechanism has previously been observed in Mn model systems that perform O–O bond formation. However, these systems display turnover rates orders of magnitude slower than that of the OEC [65].

Oxo/oxyl radical coupling in the S_4 state. A number of radical coupling mechanisms have been detailed in the literature: for reviews see refs. [12, 61, 63]. Here we limit our discussion to the most detailed and rigorous proposal at present, the mechanism originally proposed by Siegbahn [41]. In this catalytic cycle, shown in Figure 8.6 (bottom), the slowly exchanging substrate is considered to be the O5 μ-oxo bridge between Mn_4 and Mn_3. The fast substrate water binds then at the open coordination site on the Mn_1 (O6) as water/hydroxo in S_3, forming, in concert with deprotonation, an oxyl radical in [S_4] [41]. This readily reacts with the adjacent O5 to form a peroxo species between Mn_3 and Mn_1. Cleavage of the two Mn-O bonds finally leads to release of $^3O_2$. Concomitant with binding of a new water molecule and a deprotonation step the S_0 state is then reformed. The total spin states of the cluster and the spin alignment during the reaction sequence are given (Figure 8.6, bottom); for a detailed description see [41, 63]. In contrast to the nucleophilic attack, the radical coupling mechanism has no precedence in manganese chemistry. It is, however, the energetically favorable pathway for efficient O–O bond formation in non-earth-abundant (precious) metal catalysts such as the Ru blue dimer (see ref [66]).

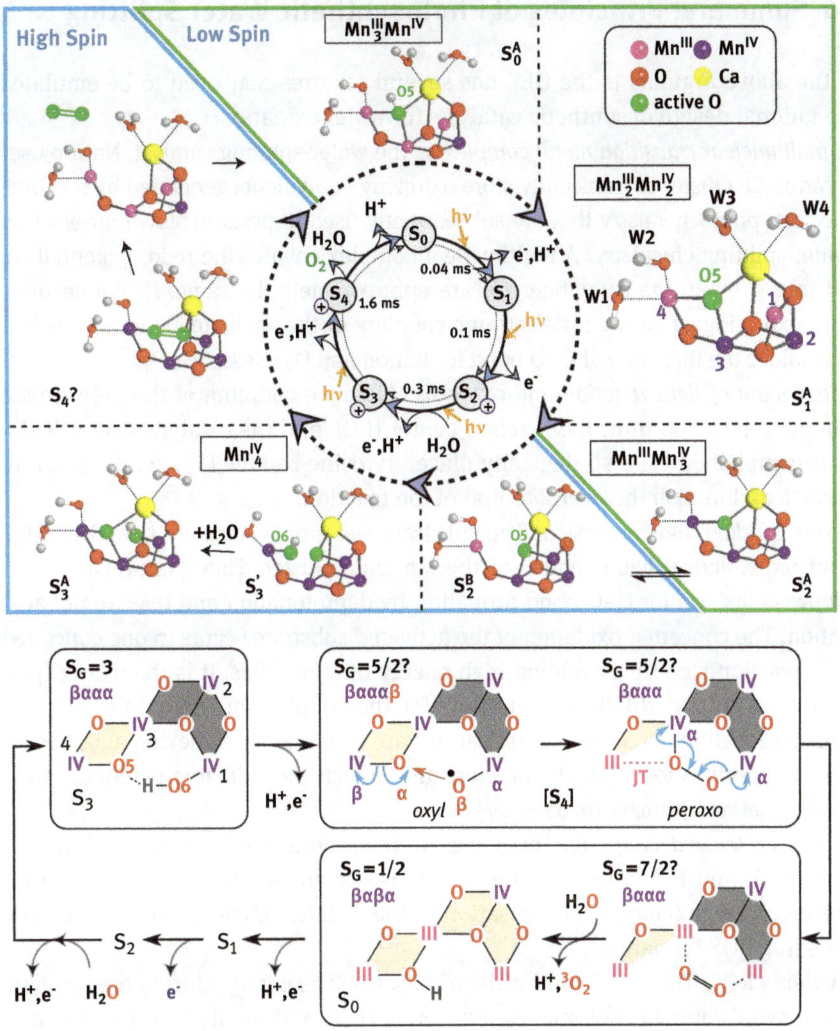

Figure 8.6: Top: Water oxidation cycle in PSII developed from experimental and computational work [48, 54]. In the center the Kok cycle is depicted [16] showing the five S-states S_0 to S_4 (cf. Figure 8.2D), kinetic data for each state transition are given. For every S-state the proposed structure of the Mn-cluster is shown, including the Mn oxidation states and the bound substrate oxygens O5 and O6 (in green). S_2 exists in two conformations S_2^A and S_2^B; two such states are also possible for S_3 (here the intermediary S_3' state is shown). At the S_2 state the cycle is switching from an effective low spin to a high spin configuration – and back to low spin in the S_4 to S_0 transition [48]. This switching may be necessary for the release of $^3O_2$ as the final product. **Bottom:** Total spin ground states S_G and spin alignment in the S_3, S_4 and S_0 states leading to O–O bond formation and O_2 release at low energy costs according to Siegbahn [41, 67]. Oxidation and spin states of the individual Mn ions are given (in purple), the spin of the oxyl radical in [S4]-oxyl is shown in red and the Jahn-Teller axis of the Mn(III) is indicated. For details of the mechanism, see text and [63]. Adapted with permission from [48]. Copyright 2016 American Chemical Society.

8.2.6 Summary: Principles of Photosynthetic Water Splitting

From the above arguments the OEC has several features that need to be emulated for the rational design of synthetic catalysts for water oxidation:

1. *A multinuclear transition metal complex* as the water-splitting catalyst. Nature uses a Mn_4O_5Ca cluster to transiently store oxidizing equivalents generated by fast one-electron photochemistry that are subsequently used to perform slow four-electron water-splitting chemistry. A multinuclear complex ensures the redox potentials of all four steps in the catalytic cycle are approximately the same. Its connectivity ($\mu_{2,3}$-oxo bridges) allows for electronic coupling of the active metal ions, which is crucial for the final step of O–O bond formation and O_2 release.
2. *Interfacing of light reaction and catalysis:* Effective coupling of the light-driven charge separation unit, the reaction center (RC), and the catalytic center (OEC) is accomplished via a strategically placed tyrosine residue (Y_Z) acting as an interface to diminish the overpotential of the reaction.
3. *Water binding and conversion:* The substrate water molecules are bound to adjacent redox-active metal centers of the Mn_4O_5Ca cluster. They are prepared in a stepwise fashion for O–O bond formation by deprotonation and (electronic) activation. The concerted oxidation of the activated substrate occurs in one concerted 4 e^- reaction step thus avoiding high energy intermediates. It is the matrix (protein) and the Ca^{2+}/Cl^- ions that allow for the coupling of proton and electron transfer reactions to occur. These features are essential for achieving about equal redox potentials for all oxidation steps that match the oxidizing potential of the light-generated primary oxidant.
4. *Product release decoupling:* The release of the products of the water-splitting reaction (O_2 and H^+) from the catalytic site is decoupled, which is important for the mechanism (charge compensation in the S states; O_2 release in a final step, avoiding ROS formation).
5. *Avoiding ROS:* The details of the electronic structure change during the catalytic cycle avoid formation of reactive oxygen species and finally lead to release of triplet dioxygen $^3O_2$.
6. *Structural dynamics:* The Mn_4O_5Ca cluster undergoes several changes of its geometric and electronic structure during the catalytic cycle, which are very significant for the mechanism of water oxidation, its regulation and oxygen release.
7. *Dedicated water/product channels:* Efficient access of the substrate water and binding to the active metals, including efficient water channels, is required.
8. *Self-healing of catalyst:* The Mn_4O_5Ca cluster is readily assembled from abundant Mn^{2+}/Ca^{2+} in solution without chaperones or maturation factors. As the PSII protein itself must be constantly repaired due to photodamage, rapid recovery of the water-splitting capability, that is, the Mn_4O_5Ca cluster, is of critical importance.

8.2.7 Synthetic Water-Splitting Catalysts

Based on the structure of the OEC (Mn_4O_xCa), and the principles of water splitting in natural photosynthesis, many researchers have tried to design and synthesize water oxidation catalysts that fulfil the above criteria. These could be used in water electrolyzers (alkaline or proton-exchange membrane (PEM) cells) or in photoelectrochemical (PEC) cells replacing precious metals commonly used in such systems. This has not been an easy task but progress has been made in recent years that is briefly described here. This field is authoritatively reviewed in [59, 68–72]. Synthetic water-splitting catalysts can broadly be divided into *molecular systems* and *metal oxides*, which may function in solution (homogeneous) or immobilized on a surface, for example, attached to suitable electrodes.

Molecular systems: The first well-characterized molecular water oxidizing catalyst was the "blue dimer" described by T. Meyer 40 years ago [73], a mono μ-oxo bridged dinuclear Ru^{III} complex with four bipyridyl and two water ligands. The initially low activity (TOF 4.2×10^{-3} s^{-1}) and stability (TON 4–5) of this catalyst under operation was greatly improved over the following decades by variation of the nuclearity of the metal complex and the structure and type of ligands. This led to an increase of the TOF by about seven orders of magnitude, even surpassing that of the natural OEC in PSII (≈500 s^{-1}) [59], a spectacular success of catalyst design and synthesis. Furthermore, a model for the water oxidation mechanism was elucidated [74, 75] for these Ru catalysts. In parallel Ir^{III} complexes were developed as water oxidizing catalysts [76]. Ir and Ru are precious metals and, therefore, show limited applicability for water oxidation systems on a (very) large scale. They have, however, the advantage of being kinetically stable in aqueous solution even at high potentials, where first row transition metals (Mn, Fe, Co, Ni and Cu) often undergo substitutions and/or decompose – depending on the type of co-ligands and solution pH. At low pH, protonation and loss of ligands are frequently observed; at high pH, most complexes decompose and form hydroxides and/or oxides [59, 69]. Since some of the formed oxide and hydroxide films on surfaces/electrodes were found to be very active in water oxidation under certain conditions (*vide infra*) a careful analysis of the process is necessary to determine the true catalytic species to avoid misinterpretations [59]. At present, it is generally accepted that most molecular first-row transition metal systems (in particular Mn-complexes) decompose rapidly and that the resulting amorphous oxides that form represent the real catalyst [69]. Nevertheless, there are exceptions and several molecular systems are still discussed as water oxidizing catalysts in the literature, see [72] for leading references. Among them are a number of interesting mono- and also multinuclear complexes of Ni, Fe, Co, and in particular Cu [77–86]. Many of these systems have been stabilized and functionalized on electrodes or surfaces or via encapsulation, for example, as polyoxometallates (POMs) [72].

Metal oxides: Increasingly researchers have focused in recent years on first row metal oxides that, in the form of nanosized oxo-clusters bear some similarities to the Mn_4O_xCa complex in PSII [87]. In addition to the criteria listed above (1–8) for nature's water-splitting catalyst, for a metal oxide to be considered as a water oxidizing catalyst it needs to have:

9. *A high amorphous surface area*, increasing the number of active centers present for catalysis; and
10. *Stability over a large pH range.* This is crucial for certain applications of the water oxidizing catalyst and is not obtained for all metal oxides – but is met for MnO_x.

Their inclusion in an electrochemical device (like an artificial leaf) also adds further constraints. These include: (i) choice of a (semi)conductive material as electrode (e.g., various forms of carbon); (ii) light transparent electrodes for photoelectrochemical (PEC) water oxidation cells (e.g., indium(III) tin(IV) oxides (ITO) and related materials); (iii) interfacing of catalysts to the electrode (coating/immobilization procedure); (iv) choice of the right electrolyte and pH; and (v) protection of the catalysts (e.g., by matrices and optimized working conditions). Details can be found for example in ref. [69].

Which metal oxides are most suitable and why? Manganese, nickel and cobalt oxides and related composite materials show superior water-splitting activity as compared to those of other first row transition metals [68–72]. In this brief review we can only present very few examples:

Manganese catalysts It has been known for some time [87] that the Mn_4O_xCa cluster has close analogs in nature. The bioinorganic core of the OEC can be considered a small fragment cut out of the extended structure of a MnCa-oxide mineral such as marokite ($CaMn_2O_4$) [88]. Such minerals together with various other Ca^{2+}-doped Mn-oxides have come in the focus of researchers looking for synthetic water oxidizing catalysts. These minerals (birnessite, marokite, etc.) indeed exhibit some water oxidation activity [88, 89].

Following the original work of Morita et al. in 1977 [90] on MnO_2 and Harriman et al. in 1988 [91] using Mn_2O_3 as water oxidizing materials, many more Mn oxides and related doped materials have been investigated (see table 2 in [69]) as water-splitting catalysts. A very recent review summarizes this extensive work and showed that even under highly optimized conditions using the best materials – MnO, α-Mn_2O_3, Mn_3O_4, γ-MnO_2 (ramsdellite), δ-MnO_2 (birnessite) – a lower activity is often obtained than for other metal oxides, for example, from Co and Ni. However, the accessible pH window of these materials is very large, ranging from pH 1 to 14, so that the MnO_x catalysts can more universally be employed. It was noticed that doping, especially with Ca^{2+} increases the activity [88, 92]. It was also recognized that the presence of Mn^{III} sites is important for activity [93, 94]; in the catalytic cycle Mn^{III}/Mn^{IV} is playing the major role. An interesting finding reported by Melder et al. [69] is that probably most of the active manganese starting materials (pre-catalysts) are converted on the electrode under applied potential [95–97] into the same catalytically active phase, which is an amorphous MnO_x with a birnessite-type layered structure.

This indicates formation of the catalyst by self-assembly and shows that probably a rather specific MnO_x structure is necessary for good catalytic activity, all properties that are also found for the OEC in PSII.

Cobalt catalysts The Pourbaix diagram predicts that cobalt oxides are only stable at positive potentials in alkaline solution [68]. However, it was found by Kanan and Nocera that CoPi (Co^{II} phosphate), formed by electrodeposition on ITO electrodes, is stable indefinitely at pH 7 and showed water oxidation activity at low overpotentials [98]. It is particularly robust and can operate in brine and river water, suggesting that it is not inhibited by naturally occurring concentrations of other ions [99]. Structural data [100, 101] show that the cobalt has near-octahedral coordination with only oxygen ligands and have been interpreted in terms of molecular-sized units (Co_4O_X), as opposed to an extended CoO_2 structure. Thus, while it is strictly an inhomogeneous catalyst, many of the structural and functional features it contains are reminiscent of the OEC. By spectroscopy it was shown that Co^{IV} is most probably the active species [102]; a model for O–O bond formation was presented [103]. The amorphous CoP_i film self-assembles and has the capability for self-healing upon damage [104]. The catalysts have been extensively studied and improved by Nocera and others and are key to the "artificial leaf" concept [105]; they can be used in "artificial photosynthesis" [10].

Frei and coworkers [106] have developed a complementary inhomogeneous catalyst, a multi cobalt system, using nanoclusters (spinel particles) of Co_3O_4 in mesoporous silica. These systems display very high TOF in the range ≈1,000 to 3,500 s^{-1} (number of O_2 per second per nanocluster), operating at mild pH (5.8) and temperature with an overpotential of 350 mV and good stability, all comparable to PS II. Hill and coworkers [107] have created a homogeneous Co-catalyst that self-assembles in water from inorganic salts (Co, W, and P). It employs a polytungstate (a polyoxometallate, POM) ligand, which is oxidatively resistant, thus enhancing catalyst durability. As the complex self-assembles in solution, like the Nocera catalyst, it can be considered self-repairing. These systems display TOF of >5 s^{-1} and TON ≈75, operating at mild pH and temperature with an overpotential of 350 mV.

Nickel catalysts Ni oxides have also been identified as water oxidizing catalysts [108]. Interestingly, small amounts of iron improve the activity of these NiO_x catalysts dramatically [109, 110]. Boettcher et al. performed a careful study on this effect [111] and mechanistic studies identified the active Ni oxidation states as Ni^{III} and Ni^{IV} [112]. The enormous potential of $FeNiO_x$ (deposited on a Ni foam) was demonstrated by Lu and coworkers [113]. The catalyst requires an overpotential of only 240 mV to deliver a current density of 500 mA/cm^2 (in 10 M KOH). The structure was determined to be $Ni_3Fe(OH)_9$, with the active species having Ni^{III}/Ni^{IV}. The long term stability of the catalyst is very good. A disadvantage is that it only works in alkaline solution. As one of the most efficient current first row transition metal catalysts for oxygen evolution it is used in alkaline water electrolyzers as an anode material on an industrial scale.

Other metal oxides Iron oxides FeO_x are appealing systems due to the large abundance and low price of this metal. The redox transitions are, however, not well suited for the water oxidation reaction [69]; it is likely that higher oxidation states (e.g., Fe(IV)=O) are involved in cases where activity has been reported, for example, in hematite [72]. A large number of other metal oxides from V, Cr, Zn, Cu, and so on have also been tested for water oxidation capability [68]. They suffer from similar problems. Mixed metal and composite systems could, however, change the picture. In this context the use of polyoxometallates (POMs) based on first-row transition metals are interesting that contain several metal ions and are stable even at high oxidation states, see for example the work of Hill [107]. Further, special layered double hydroxide structures were synthesized showing high water oxidation activity [68].

Why has nature chosen Mn? At first sight it is not clear why nature has chosen Mn over other earth abundant metals to perform the water oxidation reaction. Several Co and Ni oxides, for example, also show excellent properties for water splitting and O_2 release, better than that of the corresponding Mn oxide. Nature most likely uses a manganese cofactor as this metal was highly abundant in sea water at prebiotic times; in contrast the bioavailability of Ni and Co, was several orders of magnitude smaller (Mn^{2+} *versus* Ni^{2+} and Co^{2+}) [114]. In addition, manganese readily forms a range of semi-stable oxidation states allowing stabilization of the negatively charged and polarizing (hydr)oxide ligands, but which, nonetheless, act as strong oxidants. Thus, MnO_x-based catalysts are the best compromise that could be taken by nature with respect to coordination and redox chemistry, self-assembly, catalyst performance near neutral pH and, in particular, metal ion availability.

8.3 Hydrogen Production and Conversion

Nature has developed its own set of catalysts to utilize hydrogen – or to produce hydrogen from protons – the hydrogenases [7, 14, 115–121]. The appearance of this class of enzymes dates back to the early stages of life on the Earth, when the atmosphere was highly reducing, and may have been hydrogen-rich [122, 123]. In the absence of oxygen, some primitive organisms may have relied on protons as a terminal electron acceptor, while others may have utilized hydrogen as an energy source.

Nature's concept of hydrogen conversion – or the reverse process of hydrogen generation – at these highly optimized catalytic centers is based on a *heterolytic* splitting mechanism:

$$H_2 \rightleftharpoons H^- + H^+ \rightleftharpoons 2\,H^+ + 2\,e^- \tag{8.4}$$

The acidity of H_2, which is extremely low, is dramatically increased by performing this process at a metal site. Many of the currently used catalysts for anthropogenic utilization of hydrogen involve precious metals such as platinum, while nature's catalysts are based on earth-abundant (and thus cheap) first row transition metals (Ni, Fe). An important feature of the native hydrogenase enzymes is the very high turnover frequency (TOF), which often reaches numbers close to $10^4\ s^{-1}$ for H_2 production [124] and $10^5\ s^{-1}$ for H_2 oxidation [125]. Moreover, hydrogenases from (hyper)thermophilic organisms [126–130] show good stability at elevated temperatures (up to ≈100 °C). The pertinent problem of the high oxygen-sensitivity of many hydrogenases is solved in species (e.g., Knallgas bacteria) with specifically modified cofactors that show a greatly enhanced tolerance versus the deleterious effects of molecular oxygen [131–135]. All these features demonstrate how important it is to study and understand nature's concept of hydrogenase structure and function, when designing biomimetic (or bioinspired) systems for efficient hydrogen generation or utilization. This section will give insight into this class of enzymes.

8.3.1 Classification of Hydrogenases

Hydrogenases are found in many microorganisms, in particular in archaea and bacteria, but also in some eukaryotes [119]. Three phylogenetically distinct classes are known today; these are the [NiFe], the [FeFe], and the [Fe] hydrogenases [119]. The first two classes contain binuclear metal cores [136] with an unusual ligand sphere as catalytic centers (Figure 8.7, top), whereas the third class harbors a mononuclear iron next to a special organic cofactor [15, 137, 138]. In the latter the reaction mechanism is different; that is, the molecular hydrogen reacts only in the presence of an additional substrate and indirectly with the iron metal center. The [Fe] hydrogenases will therefore not be further discussed here. The nitrogenase also produces significant amounts of hydrogen [139] but, as this is not its primary function, it will also not be discussed.

The hydrogenases have also been classified according to their location in the cell (periplasm/cytosol, membrane-bound/soluble), their redox partners or their function (catalytic or regulatory/sensory hydrogenases). For the present consideration the classification according to the catalytic metal center is most appropriate. Quite a few bacteria contain more than one type of hydrogenase. The family of sulfate reducing bacteria (*Desulfovibrio)* have for example [FeFe] as well as [NiFe] hydrogenases [119], and the Knallgas bacterium *Ralstonia eutropha* contains four types of [NiFe] hydrogenases (membrane-bound, soluble, regulatory, and actinobacterial-type) [133, 140]. It is assumed that they are involved in different metabolic functions in different cell compartments. The genetic regulation and interplay between these hydrogenases is not yet completely understood.

8.3.2 Structure of [NiFe] and [FeFe] Hydrogenases

At present, there are nearly 100 structures available for [NiFe] hydrogenases (for a few examples see [13, 135, 141–143]), 22 for the subclass of [NiFeSe] hydrogenases (for a few examples see [144–146]), and 27 for the [FeFe] hydrogenases (for a few examples see [147–153]), with many more expected in the near future. In all catalytic hydrogenases the active site is deeply buried within the protein molecule (Figure 8.7). The hydrogenases possess three types of channels or pathways – for gas-access, proton transfer, and electron transfer – that allow substrate and educts to move between the active site and the molecular surface [116]. The protons produced by the heterolytic cleavage of molecular hydrogen at the active site are transferred to the surface of the protein via several hydrogen-bonding networks and the electrons are transported via a chain of iron-sulfur clusters (Figure 8.7). A common feature of all hydrogenases is that they contain at least one Fe atom in the active site that carries very unusual non-protein ligands, that is, carbon monoxide (CO) and cyanide (CN^-), never before found in biological systems.

[NiFe] hydrogenases: In Figure 8.7A, the structure of the [NiFe] hydrogenase from *Desulfovibrio vulgaris* Myazaki F is shown [154]. The active site is found in the large subunit (≈65 kDa) of the enzyme, whereas the smaller subunit (≈38 kDa) harbors one [3Fe–4S] cluster flanked by two [4Fe–4S] centers. The distance between the FeS-clusters is in the range of 9–12 Å, which is consistent with biological electron transfer [155]. The Ni atom in the active site is bound by four cysteine residues of the protein, two of which form a bridge to the Fe atom which, in turn, is coordinated by two CN^- and one CO ligand, originally identified by FTIR spectroscopy [156]. In the most oxidized state of the enzyme (as isolated) an additional bridging ligand "X" between Fe and Ni is modeled in the crystal structure [157], while in the active reduced states this electron density is absent [158, 159] or much decreased [154]. Spectroscopic evidence suggested that this is an OH^- species in the "ready" state [160] but the presence of sulfur may also be possible under certain conditions [157, 161, 162]. The much lower electron density in the reduced active states is due to absence of this bridging ligand or replacement with a bridging hydride [154, 163, 164]. The Ni atom has a distorted square pyramidal coordination with a *vacant site* believed to be the primary binding position of the H_2 substrate. The hydrophobic gas channel leads directly from the surface of the protein to the open coordination site of the Ni atom [165]. It is assumed that this gas channel is also involved in the reactions with CO and O_2, which inhibit the enzyme. Indeed, comparative crystallography showed that CO- and O_2-inhibited [NiFe] hydrogenases show additional di-atomic electron density at the open fifth coordination site at Ni [166]. In the oxidized "unready" state of the enzyme, an additional elongated electron density was observed at the bridging ligand "X," suggesting it to be a hydroperoxide OOH^- [167, 168]. It is suggested that the peroxo bridging ligand causes the enzyme to be "locked" in the unready state, which can be reactivated only very slowly. However, recent data have suggested this state may have an oxidized cysteine residue in addition to an OH^- as in the "ready" state [169, 170]. Another metal center has been assigned to a magnesium ion located at the C-terminus [171] and is probably involved in proton transfer between the [NiFe] active site and the molecular surface. Recent, ultra-high-resolution structures have revealed not only the presence of a bridging hydride between the Ni and Fe ions in the active reduced state but also the initial proton binding site on a coordinating cysteine during heterolytic splitting of hydrogen [154].

The subclass of [NiFeSe] hydrogenases contains a selenocysteine replacing one of the terminal cysteines coordinating the Ni, notably the ligand that becomes oxidized in the "unready" state and that also is thought to accept a proton during H_2 splitting. The ratio of H_2 production to H_2 oxidation activity of this enzyme is higher than that for the other [NiFe] hydrogenases, which makes this enzyme particularly interesting for biohydrogen technologies. Furthermore, the [NiFeSe] hydrogenases exhibit a higher oxygen tolerance and can be more easily activated than the [NiFe] hydrogenases. The overall structure of this [NiFeSe] hydrogenase [146] is very similar to that of the [NiFe] hydrogenases (Figure 8.7), except for the presence of three [4Fe–4S] clusters in the electron transfer chain and an iron ion replacing the magnesium.

In oxygen-tolerant hydrogenases (e.g., from *Aquifex aeolicus* and *Ralstonia eutropha*) [132, 172] a novel type of proximal [4Fe] cluster has been found that is ligated by 6 cysteines and has a [4Fe–3S] core (see Figure 8.11). It has special redox properties and is essential for the defense mechanism against molecular oxygen [132, 173]. In some [NiFe] hydrogenases the proximal [4Fe–4S] cluster has one cysteine substituted for an aspartic acid, which may also confer some oxygen tolerance [135].

[FeFe] hydrogenases: Crystal structures of [FeFe] hydrogenases have been obtained from the bacteria *Clostridium (C.) pasteurianum* (*Cp*I or *Cp*HydA1) (e.g., [147, 148, 150, 151, 174–178], *D. desulfuricans* (*Dd*H or *Dd*HydAB) [152, 153, 179], and the alga *Chlamydomonas (C.) reinhardtii* (*Cr*HydA1) [149, 180, 181]. These three enzymes are not only the best structurally characterized [FeFe] hydrogenases but also those for which the majority of spectroscopic and functional information is available. Figure 8.7B shows the structure of the [FeFe] hydrogenase isolated from *C. pasteurianum* in its oxidized form. The active site, the "hydrogen-converting cluster" (H-cluster) – a unique [6Fe–6S] cluster – is located in the center of the protein. Three additional [4Fe–4S] clusters and one [2Fe–2S] cluster (F-clusters) form the electron pathway to the external redox partner. Generally, [FeFe] hydrogenases show a highly conserved active site but differ remarkably in the number of additional iron–sulfur centers. In *Dd*HydAB two [4Fe–4S] clusters serve this task [152] while *Cr*HydA1 lacks the accessory FeS clusters altogether [149]. In *Cr*HydA1, the H-cluster interacts directly with the ferredoxin redox partner. Hydrophobic gas channels have also been modeled [182], which extend from the protein surface to the most remote Fe atom (Fe_d) of the active site (see Figure 8.7B). Fe_d is, thus, believed to be the site where H_2 initially binds (or H_2 is released) and also where the inhibitors CO [150] and O_2 bind [183–185]. Finally, a

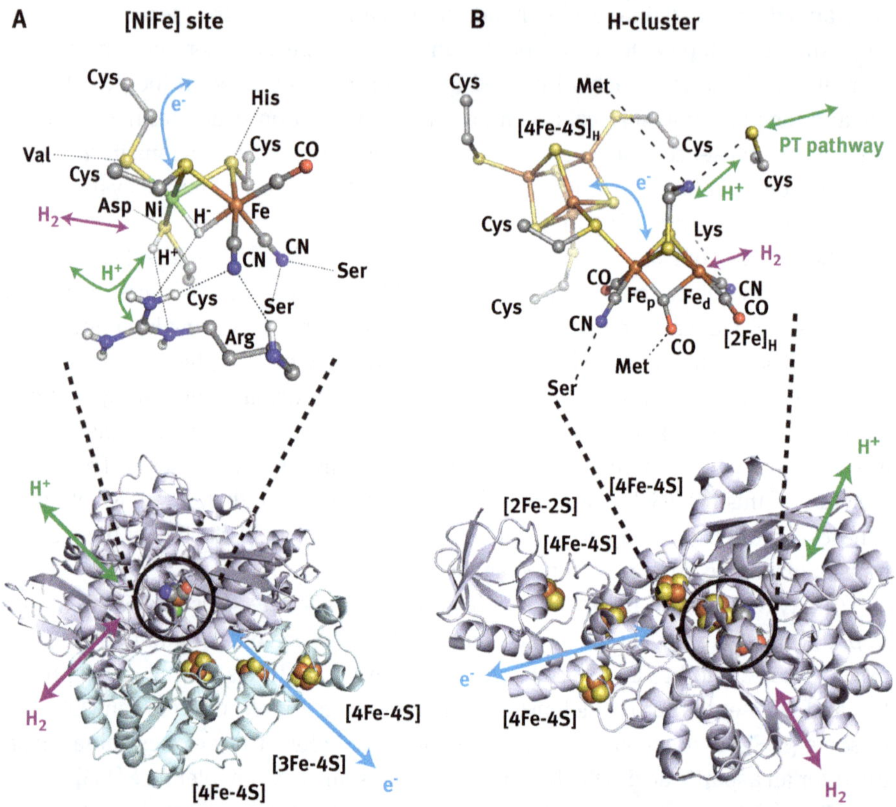

Figure 8.7: (A) Bottom: Ultra-high-resolution three-dimensional crystal structure of the [NiFe] hydrogenase from *Desulfovibrio vulgaris* Miyazaki F in the Ni-R state (PDB 4U9H) [154]. The small and large subunits are shown in cyan and purple, respectively. The [NiFe] active site and the three accessory FeS-clusters required for electron transport are depicted as spheres, and the protein backbone is shown as ribbons. The putative pathways for proton and electron transfer are indicated by green and blue arrows, respectively. The putative H_2 diffusion pathways are shown in magenta arrows. Top: Close-up of the [NiFe] active site and its surrounding amino acid residues, shown as balls and sticks. The possible hydrogen bonds to nearby amino acids are depicted with black dotted lines. The putative pathways for proton and electron transfer are indicated by green and blue arrows, respectively. The third bridging ligand between Ni and Fe has been assigned to a hydride [154, 163]. **(B)** Bottom: Structure of the [FeFe] hydrogenase from *Clostridium pasteurianum* (PDB 4XDC) [148]. The active site H-cluster and the four accessory FeS-clusters required for electron transport are depicted as spheres. The putative pathways for proton and electron transfer are indicated by green and blue arrows, respectively. The putative H_2 diffusion pathways are shown in magenta arrows. Top: Ball and stick representation of the H-cluster and the surrounding amino acid residues. The possible hydrogen bonds to nearby amino acids are depicted with black dashed lines. The putative pathways for proton and electron transfer are indicated by green and blue arrows, respectively.

chain of amino acid residues is identified, which appear to act as a proton transfer pathway [147].

The active site of [FeFe] hydrogenase, the H-cluster, harbors a classical cubane [4Fe–4S] subcluster connected to the "catalytic" $[2Fe]_H$ subcluster through a cysteine bridge. This subcluster consists of two Fe atoms (one octahedrally coordinated, the other square pyramidal with an open coordination site) and five diatomic non-protein ligands, that is, CO, CN^-. The CO and CN^- ligands in the crystal structure have been assigned making use of FTIR spectroscopy as well as taking into account the possibility for hydrogen bonding of the CN^- groups to nearby amino acid residues [152]. The CO ligands are located in a hydrophobic pocket, whereas specific amino acid residues provide H-bonds to the CN^- ligands (Figure 8.7B). It is noteworthy that the only covalent attachment of the $[2Fe]_H$ subcluster to the protein is through the cysteine bridging the subclusters. Furthermore, a dithiolate bridge exists between the two iron atoms. The dithiolate ligand was originally assigned to a propanedithiolate (PDT) [152], but was later revised to a aza-propanedithiolate (ADT). Magnetic resonance techniques [186] have confirmed that the apical atom of the dithiolate bridge is a nitrogen, and semisynthetic production of [FeFe] hydrogenases by artificial maturation of the H-cluster with synthetic diiron precursor complexes demonstrated that only ADT gives a functional enzyme [187].

8.3.3 Intermediate States and Reaction Mechanisms

The redox chemistry of hydrogenases shows several intermediate states that are involved in the enzyme's activation/inactivation, inhibition, light-sensitivity, and in the actual catalytic cycle [116, 188, 189]. These can often be trapped or stabilized for sufficiently long times to allow their spectroscopic characterization. If these states are paramagnetic, EPR techniques can be employed yielding detailed information about the electronic and also the geometrical structure [116, 117]. Both EPR-active and -inactive states are amenable to infrared spectroscopy, in which the CO and CN^- ligands at the Fe serve as structural probes [116, 156, 189, 190]. The combination with electrochemical techniques allows the pH-dependent determination of the redox transitions [191–193]. More recently, time-resolved IR (TRIR) has been used to study how the catalytic states evolve on sub-turnover timescales [194, 195]. Together with quantum chemical calculations these and other techniques, like nuclear magnetic resonance (NMR) spectroscopy, Mössbauer and the related technique, nuclear resonance vibrational spectroscopy (NRVS), or X-ray absorption spectroscopy (XAS), are helpful in elucidating the details of the reaction mechanism of hydrogenases. Reviews detailing the importance of spectroscopy [116, 117, 188, 189] and concomitant electrochemical methods like protein film electrochemistry (PFE) [116, 120, 196–200] have been published in recent years (for examples see Figure 8.8).

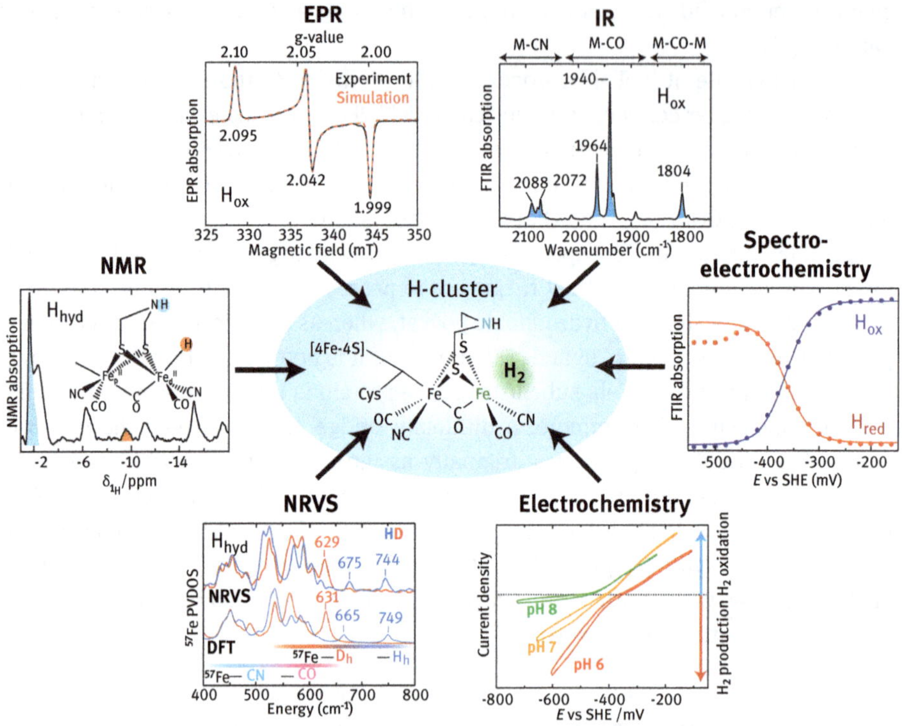

Figure 8.8: Physical techniques employed in hydrogenase research. The structure of the H-cluster of [FeFe] hydrogenase is shown in the center: (i) the EPR spectrum of the paramagnetic ($S = 1/2$) H_{ox} state (yielding the principal g values) in *Cr*HydA1 [116, 117]; (ii) the IR spectrum of the H_{ox} state (probing vibrational bands of the CO and CN ligands of the H-cluster) in *Cr*HydA1 [116]; (iii) IR-monitored electrochemical titration curves for the H_{ox} to H_{red} transition in *Cr*HydA1 with the PDT ligand in the [2Fe] subcluster to yield the redox potential (Adapted with permission from [201]. Copyright 2020 American Chemical Society); (iv) protein film electrochemistry (PFE): cyclic voltammetry (CV) curves at 3 pH values showing reversible switching (no overpotential) between H_2 oxidation and H_2 production with *Cr*HydA1 (Adapted with permission from [181]. Copyright 2018 American Chemical Society); (v) nuclear resonance vibrational spectroscopy (NRVS) on the H_{hyd} state in *Cr*HydA1 resolving vibrations of the hydride attached to Fe_d and comparison with DFT calculated spectra (Adapted with permission from [202]. Copyright 2017 American Chemical Society) and (vi) paramagnetic ^1H NMR spectrum of H_{hyd} in *Cr*HydA1 in liquid solution detecting resonances from the Fe-bound hydride and the proton on the N-bridgehead (both H/D exchangeable) (Adapted with permission from [203]. Copyright 2018 American Chemical Society). Note that spectroscopy delivers additional and complementary information to X-ray crystallography.

[NiFe] hydrogenase: In the aerobically "as isolated" form, the [NiFe] hydrogenases are inactive and have to be activated with H_2 or other reductants (Figure 8.9). The activation process has been studied in detail using EPR and other spectroscopies [116, 117, 204]. Ni-A is the "unready" state of the enzyme and takes up to an hour to reactivate. In contrast, Ni-B can be reactivated within a few minutes. It should be noted that in oxygen-tolerant species (e.g., *R. eutropha, A. aeolicus*) the Ni-A state is not observed. It is generally considered as the oxygen-inhibited state; however, one study demonstrated formation of this state in the absence of oxygen [205].

The current model of the catalytic cycle of [NiFe] hydrogenase encompasses 4 well characterized states: Ni–SI$_a$, Ni-L, Ni–C, and Ni–R, which are interconverted by one electron/one proton equilibria (Figure 8.9) [207–211]. In the catalytic hydrogen converting process the approaching H_2 is attached to the open coordination site of Ni (II) in the Ni-SI$_a$ state. Most studies in the literature suggest the Ni(II) states are low spin (diamagnetic) [154, 163, 212–215]. This is important as the spin state plays a conclusive role in many of the hydrogenase reactions [216]. Density functional theory

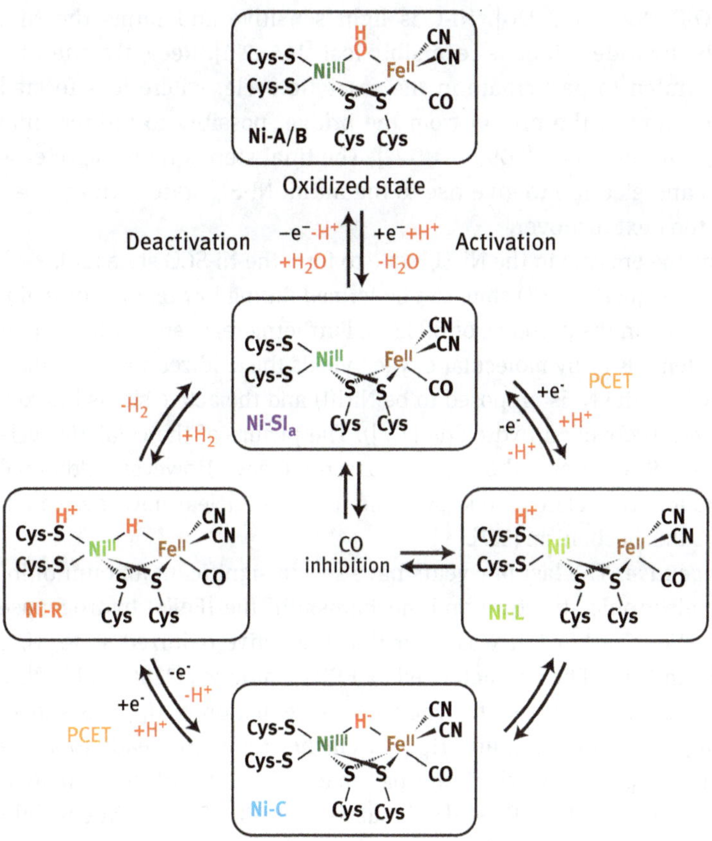

Figure 8.9: Proposed reaction scheme for a catalytic [NiFe] hydrogenase detailing the catalytic cycle and the activation pathway starting from the oxidized state(s) Ni$_{A/B}$ [13, 116, 121, 206]. The formal oxidation states of the Ni and the Fe are given for all intermediates, along with an indication of the type of bridging ligand, which changes identity in the cycle (see text for details). PCET: proton coupled electron transfer.

(DFT) calculations of the electronic structure of the Ni-Fe cluster have been published giving information about possible redox states of the individual Ni and Fe atoms [217, 218]. It is shown that Fe remains Fe(II) throughout all redox accessible states, which is caused by the strong field CO and CN^- ligands. The H-H bond is polarized followed by base-assisted heterolytic cleavage of the H_2 molecule leading to the Ni-R state. This state is EPR-silent and thought to be Ni(II), with a hydride bridging Ni and Fe, identified by X-ray crystallography [154] as well as NRVS [163]. One of the candidates for acting as a base is a terminal cysteine at the Ni. Alternatively, a water molecule bound to the iron and a nearby arginine residue have been proposed [206, 217]. Subsequent deprotonation and electron transfer to the proximal [4Fe–4S] cluster then leads to the Ni-C state, which has also been shown to carry the hydride bridge between Ni(III) and Fe(II) using advanced EPR techniques (HYSCORE, ENDOR) [164, 219, 220]. Ni-C is light sensitive and forms the Ni-L state, in which the hydride bridge is reversibly lost [164, 221]. Recently, the Ni-L state was demonstrated to participate in the catalytic cycle, where it is formed from Ni-C by relocation of the proton from the bridge, possibly to the terminal cysteine, creating the Ni-L state [209, 210, 222]. The final step is then the release of another proton and electron to give rise to the initial Ni-SI_a state with an open bridge, ready for the next turnover.

CO gas inhibits the enzyme in the Ni-SI_a state to form the Ni-SCO state [223, 224]. Interestingly, a paramagnetic Ni-CO state can be formed during low temperature illumination of Ni-C to Ni-L in the presence of CO [225]. Furthermore, overoxidation either by high applied potentials or by molecular oxygen yields the oxidized inactive states Ni-A and Ni-B in which the Ni is proposed to be Ni(III) and the active site is blocked by a bridging ligand (probably a hydroxide [226]). The picture of the catalytic cycle presented in Figure 8.9 represents the current consensus view. However, additional forms of the Ni-SIa and Ni-R states are known and it is still unclear how these additional states fit into the mechanism [227].

[FeFe] hydrogenase: The last few years have seen a significant amount of information on the electronic structure and mechanism of the [FeFe] hydrogenase become available. Very early on it was clear that the active oxidized state, H_{ox}, was paramagnetic and could be characterized by EPR techniques [228]. CO binding to H_{ox} also yields a paramagnetic state. One electron reduction of H_{ox} was shown to give the diamagnetic states H_{red} and $H_{red}H^+$. Further reduction leads to states called $H_{sred}H^+$ and H_{hyd}, both of which are paramagnetic, but with the unpaired electron highly localized on the [4Fe–4S] subcluster [229–231]. Thus, while insight into the electronic structure of the $[2Fe]_H$ subcluster from EPR can only be obtained for the H_{ox} and H_{ox}-CO states, FTIR, Mössbauer, NMR, NRVS and XAS spectroscopy have given crucial information on the other intermediate states of the H-cluster (see Figure 8.8 for examples) [202, 203, 232–236]. An important aspect is that proton coupled electron transfer (PCET) plays a crucial role in all these processes [127, 181, 191, 192].

An important breakthrough in the understanding of the active site has been the determination of nitrogen as the bridgehead atom in the dithiolate moiety of the $[2Fe]_H$ subcluster, initially by detection of ^{14}N using pulse EPR (HYSCORE) methods [186]. This nitrogen is placed in perfect position to act as a pendant base for the heterolytic splitting of molecular hydrogen or for the reverse reaction, that is, formation of hydrogen at the H-cluster (Figure 8.7). The crucial catalytic function of the nitrogen in the bridgehead position was demonstrated through semi-synthetic production of the [FeFe] hydrogenase [187, 237]. In this approach, the hydrogenase is produced heterologously in *Escherichia (E.) coli*, as an incomplete enzyme (known as the apo-hydrogenase), lacking the [2Fe] subcluster [238], followed by reconstitution of the [2Fe] subcluster with synthetic diiron precursor complexes, a process known as artificial maturation [187]. Normally, the [2Fe] subcluster is synthesized by a pair of radical SAM family enzymes called HydE and HydG [239], and a third protein HydF is both the scaffold upon which the cofactor is built [240] and the chaperone that inserts the cofactor into the apo-hydrogenase [241]. However, *E. coli* does not produce an [FeFe] hydrogenase itself, and so this bacterium lacks the [2Fe] subcluster biosynthetic and insertion machinery. It was discovered that reconstitution of an active H-cluster could be achieved using chemically synthesized [2Fe] precursor compounds. These were first loaded onto the scaffold protein HydF and then the HydF-[2Fe] complex was mixed with the apo-hydrogenase to give active hydrogenase, indistinguishable from the native enzyme [187]. It was subsequently demonstrated that the apo-hydrogenase could be reconstituted even without HydF [237, 242]. The only [2Fe] complex to give significant activity was that containing an ADT bridging ligand. Reconstitution with complexes in which this ligand was substituted with PDT or 2-oxapropane-1,3-dithiolate (ODT) could be achieved but yielded enzymes with much lower activity. This was also true for most of the $[2Fe]_H$ cluster models that were synthesized and introduced in the apo-enzyme [175, 242–244]. These results demonstrate the high degree of optimization of these very active enzymes in nature.

In Figure 8.10, the various states of the catalytic cycle, which have so far been identified are listed together with the proposed oxidation states of the iron atoms in the bi-nuclear subcluster. The active oxidized state (H_{ox}) has a characteristic rhombic EPR spectrum (Figure 8.8), indicating that the binuclear subcluster is in a paramagnetic $[Fe_p(II)Fe_d(I)]$ configuration, while the [4Fe–4S] subcluster is in its oxidized (2+) state. One-electron reduction produces the active reduced state (H_{red}), which is EPR silent, but can be identified in FTIR and Mössbauer spectroscopy. There are two possible electronic configurations for the H_{red} state: one in which the diiron subcluster is reduced to a diamagnetic Fe(I)Fe(I) state, and one in which the [4Fe–4S] cluster is reduced to its paramagnetic 1+ state with antiferromagnetic coupling between the two subclusters producing a diamagnetic ground state overall. Based on pH-dependent FTIR studies [191, 192] and resonance Raman studies [245] it is believed that the H_{red} state is actually a mixture of both of these configurations. The interconversion between these two states is pH dependent and suggests that protonation of the $[2Fe]_H$ subcluster allows electron transfer from the [4Fe–4S] to the $[2Fe]_H$ subcluster [191, 192]. Hence, these states are referred to as the H_{red} and $H_{red}H^+$ states. It should be noted that alternative explanations for the two forms of the H_{red} state have been proposed, as well as an alternative nomenclature [246]. In the F-cluster containing [FeFe] hydrogenases, the H_{red} state (with the reduced [4Fe–4S] subcluster) is destabilized by reduction of the proximal F-cluster, leading to enhanced electron transfer to $[2Fe]_H$ [192, 201]. This explains why the H_{red} state was not observed in earlier studies of the [FeFe] hydrogenase from *D. desulfuricans*, *Dd*HydAB [247].

As indicated in Figure 8.10, the H_{ox}, H_{red} and $H_{red}H^+$ states are considered to take part in the catalytic cycle [194]. For *Cr*HydA1, an additional reduced state $H_{sred}H^+$ ("super reduced"), has been identified using FTIR monitored spectroelectrochemistry and further studied by EPR spectroscopy [229, 248]. In this state the cubane subcluster is reduced to $[4Fe–4S]^+$, and the diiron subcluster is in the Fe(I)Fe(I) configuration. This state was also identified in *Dd*HydAB but was found to occur only at very negative potential and to be unstable [247]. The low redox potential of the $H_{sred}H^+$ state in *Dd*HydAB is also a consequence of the F-clusters [192]. Both $H_{red}H^+$ and $H_{sred}H^+$ states show very unusual properties of their bridging CO ligands. For many years these states were thought to completely lack a bridging CO, leading to suggestions that these states could contain alternative bridging ligands such as hydrides [236]. However, at cryogenic temperatures the bridging CO ligand can be observed [249–251]. The $H_{sred}H^+$ state can rearrange to give a state with a terminal hydride bound to Fe_d (H_{hyd}) [176, 230, 231, 251]. This key state is thought to have an over-oxidized [2Fe] subcluster and a reduced [4Fe–4S] cluster. The presence of a terminal hydride on Fe_d was originally inferred from an H/D isotope effect on the IR band of the bridging CO in this state [231]. More recently, the terminal hydride has been directly observed using NRVS [202, 235] and NMR spectroscopy [203]. The final stages of the cycle leading to formation of H_2 have not been studied in detail. However, based on the events taking place in the one-electron reduced H_{red} state, it is hypothesized that protonation of $[2Fe]_H$ triggers electron transfer from $[4Fe–4S]_H$ to $[2Fe]_H$, to yield a $H_{hyd}H^+$ state. This would then be followed by H_2 bond formation to give a $H_{ox}(H_2)$ state, and H_2 release reforming H_{ox} as indicated in Figure 8.10. Finding evidence for a H_2 bound intermediate is a hot subject of hydrogenase research at the moment.

Enzymatic activity can be effectively inhibited by CO, which binds to the open coordination site on Fe_d [150]. The (H_{ox}-CO) state has the same electronic configuration as H_{ox} and shows a characteristic axial EPR spectrum. The inhibiting CO ligand is photolabile and can be photo-dissociated (<40 K) leading to formation of the H_{ox} state [150, 254, 255]. Reduction of the H_{ox}-CO state leads first to the formation of the H_{red}-CO state with a reduced [4Fe–4S] subcluster. Further reduction of the H_{red}-CO state either releases the CO, forming the $H_{sred}H^+$ state [253], or releases the [2Fe] subcluster giving the apo-form of the hydrogenase [256, 257]. There is evidence from DFT calculations, that a ligand rearrangement at Fe_d occurs in the H_{ox}-CO state leading

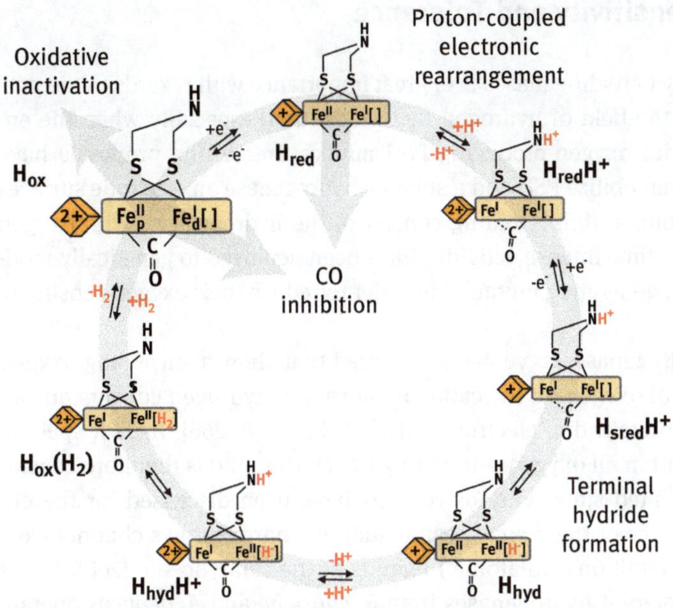

Figure 8.10: Simple model for the catalytic cycle of [FeFe] hydrogenase according to Refs. [194, 201, 250]. The pathways for oxidative inactivation (H_{inact} and H_{trans} states) [153, 252] and CO inhibition (H_{ox}-CO and H_{red}-CO states) [253] are not shown. The $[2Fe]_H$ and the attached $[4Fe–4S]_H$ subclusters are shown schematically (in light and dark orange, respectively). Furthermore, the proposed oxidation states of the Fe ions in $[2Fe]_H$ are given, as well as the putative protonation of the bridgehead nitrogen; the open coordination site at Fe_d is indicated by a bracket ([ ]); in some states a hydrogen species is bound (H^- = hydride, H_2 = dihydrogen). The oxidation state of the $[4Fe–4S]_H$ is also shown (1 +/2 +). In the $H_{red}H^+$ and $H_{sred}H^+$ states the bridging CO ligand is shown as moving to a semi-bridging position. Note that, at the time of writing, the catalytic cycle of [FeFe] hydrogenase is still under debate and the authors' favored model is presented. For further details see text and refs [189, 246, 251].

to an apical CN arrangement [258]. However, recent data from pulsed TRIPLE ENDOR spectroscopy are consistent with the apical CO arrangement [259]. The H-cluster is irreversibly destroyed by molecular oxygen (see below) where the O_2 is believed to also first attack the open coordination site at Fe_d [180, 183, 185]. Lastly, the inactive states H_{inact} and H_{trans} in the [FeFe] hydrogenase from *D. desulfuricans* (*Dd*HydAB) contain an over-oxidized Fe(II)Fe(II) [2Fe] subcluster, with an additional SH ligand on Fe_d [153, 252]. H_{inact} has an oxidized [4Fe–4S] subcluster while H_{trans} (a transition state between H_{inact} and H_{ox}) has a reduced [4Fe–4S] subcluster. Recently, a similar H_{inact} state in the [FeFe] hydrogenase from *Clostridium beijerinckii* was shown to have a cysteine ligand bound to Fe_d [260, 261]. The H_{inact} state is particularly interesting as it is protected from oxygen.

8.3.4 Oxygen Sensitivity and Tolerance

The oxygen sensitivity of hydrogenases is of great importance with regards to the possible applications in the field of hydrogen fuel production, especially when the enzyme is combined with oxygen-producing PSII mimics. Interesting proposals have been put forward to immobilize PSII and a suitable hydrogenase on electrode surfaces separated by a membrane, thus avoiding contact of the hydrogenases with oxygen [262, 263]. At the same time intense activities have been deployed to genetically modify hydrogenases (i.e., generating mutants) in order to reduce their oxygen sensitivity [264, 265].

A few [NiFe] hydrogenases have been identified that show a surprising oxygen tolerance. The effect of oxygen on the catalytic process of hydrogen conversion can best be followed by protein film electrochemistry (PFE) [198, 266]. In EPR spectra, the Ni-A state is absent in all oxygen-tolerant hydrogenases, and is therefore considered the oxygen-inhibited state. Various reasons have been discussed for the enhanced oxygen-tolerance of these enzymes, including a narrower gas channel [267] and the existence of additional (diatomic) ligands at the catalytic site [268]. It has been found that O_2-tolerant hydrogenases from *R. eutropha* and *A. aeolicus* operate at significantly higher redox potentials [129, 269]. Redox titrations of the FeS clusters [172] revealed an additional high potential iron–sulfur species that explained some peculiarities of the EPR spectra and could be assigned to the proximal FeS cluster.

X-ray crystallographic structures of several O_2-tolerant hydrogenases were reported [131, 134]. The basic structure of both enzymes is very similar to those of O_2-sensitive hydrogenases, except for the proximal FeS-cluster. This cluster is modified; it is not a classical [4Fe–4S] cluster ligated by four cysteines but rather a novel [4Fe–3S] cluster harboring six cysteines (Figure 8.11) that is able to undergo two redox transitions in a narrow potential range [172]. This unique property, never found before in any natural FeS cluster, enables the hydrogenase to work in the presence of molecular oxygen. The explanation is that O_2-tolerant [NiFe] hydrogenases can also act as "oxygenases," that is, reduce the potentially harmful oxygen to water; one molecule is released and one retained in the active site as OH^- forming the bridging ligand in Ni-B (see Figure 8.11) [266, 270]. The process depends on the potential availability of reducing equivalents (electrons) and can lead – when electrons are not abundant – to peroxide, thus creating a powerful and potentially damaging oxidant to attack the Ni-Fe center. This is supposed to generate the Ni-A state, possibly with a bridging OOH^- bridging ligand or with oxidized cysteines (S=O, S–OH groups) [198]. In support of this mechanism, mutagenesis studies have shown that replacement of the supernumerary cysteines next to the proximal [4Fe] cluster leads to an increased oxygen-sensitivity [132, 173]. Furthermore, water channels have been found in the crystallographic structure [131] that could serve for water transport. A possible mechanistic scenario of such an enzyme working both as *hydrogenase* and under O_2

attack also as *oxygenase* is depicted in Figure 8.11. Although a final interpretation leading to a detailed mechanism is still under discussion [133, 172] it is clear that a modification of the proximal cluster in the electron transport chain is likely to be a major factor leading to oxygen tolerance of these [NiFe] hydrogenases. However, there are examples of O_2-tolerant [NiFe] hydrogenases lacking the specialized proximal cluster [135, 140, 271].

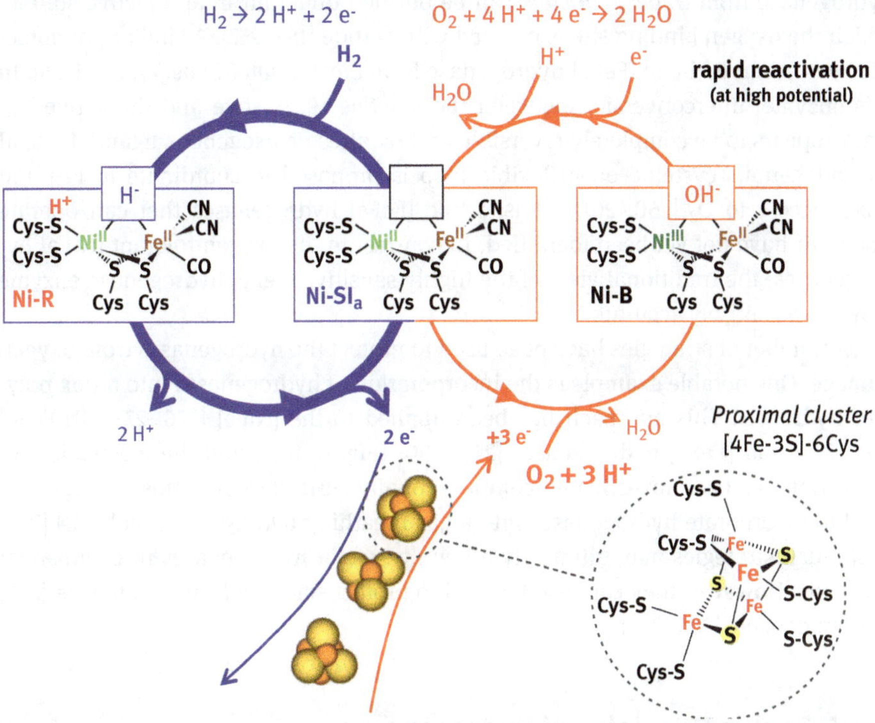

Figure 8.11: Simplified scheme for O_2-tolerant [NiFe] hydrogenases that show both hydrogenase (blue circle) and oxygenase activity (red circle) [116, 132, 266]. In the first case, the Ni-SI$_a$ state (center) reacts with H_2 and forms the Ni-R state (left) that returns (via Ni-C and Ni-L, not shown) back to Ni-SI$_a$, thereby 2 H^+ and 2 e^- are released. Upon O_2 attack of Ni-SI$_a$ one electron from Ni(II) and three from the FeS clusters (two from the proximal cluster!) are delivered quickly, together with 3 H^+, to fully reduce oxygen to water avoiding ROS. OH^- remains in the active side bridging Ni and Fe in the Ni-B state (right). A key role is played here by the novel proximal [4Fe–3S]–6Cys cluster shown below [131, 132, 134, 172, 270]. Efficient electron delivery from the FeS clusters also assures rapid reactivation of the enzyme from the oxidized state.

The oxygen sensitivity of [FeFe] hydrogenases is even more severe since the H-cluster is irreversibly destroyed upon reaction with oxygen [183]. The Fe centers in the subcluster are probably oxidized to Fe(III) and loose their CO ligands. Nevertheless, some [FeFe] hydrogenases seem to show a remarkable oxygen tolerance

[272, 273]. This behavior might be correlated with the properties of the hydrophobic gas channel. It is assumed that H_2 can diffuse much easier through the protein and reach the active site than O_2 and CO. Therefore, much effort has been put into the modification of the amino acids in the gas channel in particular of the hydrogenase from *C. reinhardtii* [274]. But this is certainly not the only effect responsible for the oxygen tolerance of these hydrogenases; also the geometric and electronic structures of the active site are thought to play a significant role. Interestingly, the [FeFe] hydrogenase from *D. desulfuricans* can be purified under air in an inactive state in which the oxygen binding site is blocked with sulfide [153, 252]. A similar phenomenon was observed in an [FeFe] hydrogenase from *Clostridium* (*C.*) *beijerincki* [275]. In this enzyme, interconversion between the inactive H_{inact} state and the active H_{ox} state appears to be completely reversible and requires no exogenous ligand. Instead an endogenous cysteine on a flexible loop is proposed to coordinate to Fe_d and block access to O_2 [260, 261]. Thus, while [FeFe] hydrogenases that can operate under air have not yet been identified, in contrast to the oxygen-tolerant [NiFe] hydrogenases, the traditional view of the highly sensitive [FeFe] hydrogenase enzyme may not be entirely accurate.

A number of strategies have been used to protect the hydrogenases from oxygen damage. One notable example is the incorporation of hydrogenases into redox polymer hydrogels. This approach has been applied to the [NiFe] [276, 277], [NiFeSe] [277, 278] and [FeFe] hydrogenases [279, 280] (Figure 8.12) and the mechanism of protection has been investigated computationally [281]. Recently, this strategy was used to incorporate hydrogenases into a gas-breathing hydrogen/air fuel cell) [282, 283]. Such strategies may ultimately allow the use of hydrogenases in commercial fuel cells. However, they can also be used to protect small molecular catalysts from oxygen damage [284, 285].

8.3.5 Design Principles of Hydrogenases

For a better understanding of the design principles of the native enzymes a comparison of the two major classes of hydrogenases is useful.

The two groups of hydrogenases have similar active site structures and mechanisms, which provide a great example of convergent evolution. However, they are totally unrelated to each other in terms of protein sequence/structure, maturation mechanisms and distribution ([NiFe] hydrogenases are found in archaea and bacteria, [FeFe] hydrogenases are found in anaerobic bacteria and some eukaryotes) [118]. The [FeFe] hydrogenases are, in general, most active in H_2 production while [NiFe] hydrogenases are more tuned to H_2 oxidation. Both types are, however, bidirectional. Most [FeFe] hydrogenases are extremely oxygen sensitive and will be inhibited irreversibly under O_2. [NiFe] hydrogenases are, in general, more oxygen-tolerant and some enzymes (e.g., from Knallgas bacteria) can both oxidize and

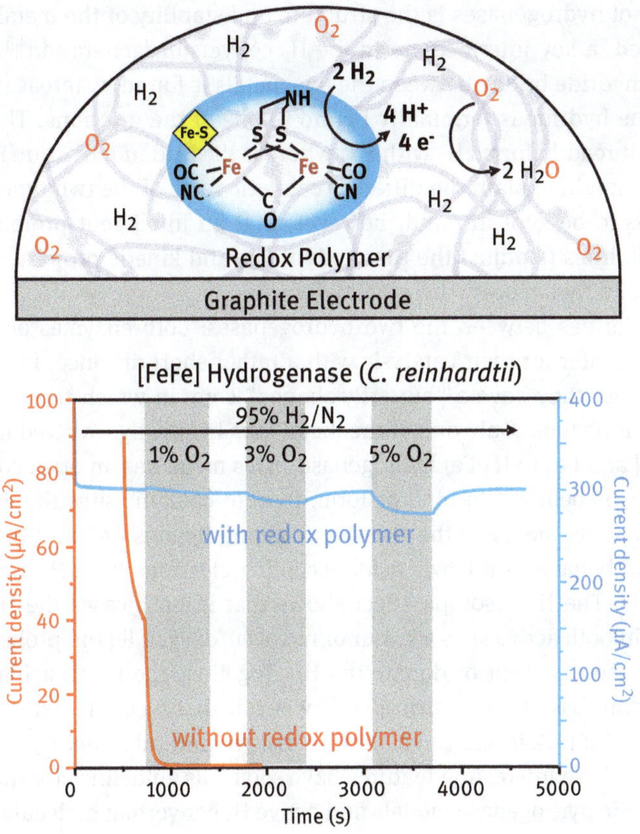

Figure 8.12: Top: Proposed model for protection of an [FeFe] hydrogenase on a graphite electrode against molecular oxygen by embedding it in a tailor-made viologen-modified redox polymer (hydrogel). The active hydrogenase oxidizes H_2 in the middle of the polymer. The electrons are transferred to the viologen moieties of the polymer. Some of these electrons flow toward the electrode surface generating positive current, while some of the electrons flow toward the surface of the polymer where they can react with O_2 generating H_2O as the product. (for details see [276, 279–281]). **Bottom:** Chronoamperometry for evaluation of the oxygen tolerance. The red curve is showing the experiment without the redox polymer in direct electron transfer (DET), the light blue curve is with the redox polymer. In the latter case the catalytic current (proportional to the hydrogenase activity) is only weakly affected by adding increasing amounts of O_2 to the electrochemical cell, whereas it very quickly drops to zero without protection. The decay in the current in the absence of O_2 is predominantly due to film loss from the electrode surface. Adapted from [279] with permission from *John Wiley and sons.*

evolve H_2 in the presence of O_2. Next to the different structures of the active sites of these two classes of hydrogenases, a difference lies also in the FeS centers of the electron transport chain. It is remarkable that the dinuclear iron center of the [FeFe] hydrogenase is linked to one [4Fe–4S] cluster, which acts as a redox ligand and directly affects the electronic structure of the $[2Fe]_H$ site. Another difference between the two types of hydrogenases is the structure and stability of the metal-hydride complex formed, a key intermediate in the H_2 conversion/generation. In [NiFe] hydrogenases a hydride bridge between the two metals is found, whereas in [FeFe] hydrogenases the hydride is terminally bound to one of the iron ions. The latter is more reactive, it readily forms H_2 with H^+, whereas the bridging hydride is much more stable. This might explain the different catalytic rates of the two types of hydrogenases. It has to be kept in mind, however, that an intelligent protein matrix has many possibilities to adjust the thermodynamic and kinetic properties of these intermediates and thus modify their reactivity.

There are also similarities between the two hydrogenases: both enzymes employ a bimetallic active center for their catalysis with a rather short distance, indicating a metal-metal bond, they have a "butterfly-shaped" core, in which the two metals are bridged by SR-ligands. Only one of the metal ions is directly involved in binding H_2 (Ni in [NiFe] and Fe_d in [FeFe] hydrogenase). This metal has an open coordination site where H_2 is believed to bind or form. In both catalytic sites the Fe atoms are kept in low valence states by the strongly donating ligands CN^- and CO. This allows the metal to behave as a Lewis acid, accepting electrons from H_2 and forming a metal-hydride. The H/D isotope effect shows that in both cases the H_2 splitting is heterolytic. In both active sites a sulfur or nitrogen/oxygen ligand probably acts as a pendant base to accept or donate the H^+. Together, the Lewis acidic metal site and the pendant base form a frustrated Lewis pair that facilitate heterolytic splitting of H_2 [116, 286]. Both enzymes are O_2 sensitive and inhibited by CO (exceptions are known). These interesting features have served as guidelines for the construction of biomimetic hydrogenase models and active H_2 converting molecular catalysts, as discussed below.

8.3.6 Molecular Catalysts for H_2 Conversion and Production

The extensive structural and functional work on native hydrogenases over the last decades has stimulated the synthesis of a variety of metal-based model systems that might be used as electrocatalysts for hydrogen production or in fuel cells for hydrogen conversion. Considering the high price and low abundance of platinum, which is commonly used for such conversions, there is an urgent need for using sustainable metals such as iron and nickel as employed by nature in the hydrogenases. Important criteria for such catalysts are (i) robustness, that is, a long-term stability, (ii) activity (high turnover frequency) and (iii) efficiency (a low overpotential). However, these criteria are not easy to meet in model systems [244, 287–294].

The work on biomimetic models for [NiFe] and [FeFe] hydrogenase has been described in several review articles [244, 288] [287, 291, 295–306]. In this work many of the structural features important for proper function found in the native systems have been successfully incorporated, for example the bimetallic Ni–Fe or Fe–Fe core with rather short metal-metal distances and an open coordination site at one metal, the sulfur-rich environment (terminal and bridging thiolate ligands), CO/CN ligation of the iron(s), the incorporation of a base for acceptance of the proton, and more recently of hydrides. Still existing problems of many model systems are the O_2 sensitivity, the very high overpotentials, and lack of activity (low turnover rates).

Inspired by [NiFe] hydrogenase Rauchfuss and coworkers [307] described a complex, which features a Ni(diphosphine) group linked to a $Fe(CO)_3$ unit by two bridging thiolate ligands. In this complex, a bridging hydride as a key functional intermediate could be established that reacts with acids and evolves dihydrogen. However, this complex and almost all subsequent NiM complexes do not appear to show activity at the Ni site but at the other metal (M) site. Recently, Brazzolotto et al. published a [NiFe] bipyridine–bisthiolate complex with a TOF of 225 s^{-1} that shows Ni-centered activity [308]. However, DFT calculations from Tang and Hall suggest the formation of a semi-bridging hydride that is still Fe-centered [309]. Thus, generating effective catalysts that are close structural and functional mimics of the [NiFe] hydrogenase active site remains a challenge.

Significant progress has also been made with [FeFe] hydrogenase models, where in particular the issues of Fe valence states, pendant base and hydride chemistry have been addressed [244, 287, 298]. It was also recognized early on that the attached [4Fe–4S] cluster in the native enzyme is important. Thus, in 2005 Pickett and coworkers described a novel tetra-iron thiolate carbonyl assembly with a "butterfly" configuration that evolves H_2 electocatalytically [310]. A complete system requires a combination of all three major functional components of the active site that couples acid–base to redox reactions: (i) a reactive diiron center with open coordination site, (ii) an amine base (azadithiolate) as proton relay, and (iii) a redox-active ligand modeling the [4Fe–4S] center of the native H-cluster. In 2012, Camara and Rauchfuss developed a diiron aminopropane dithiolate complex, with a ferrocene moiety for electron transfer, which shows catalytic activation of hydrogen [311]. More recently, Becker et al. reported an O_2-tolerant synthetic mimic containing a redox-active phosphole ligand, whose electron-rich redox states are stabilized through ligand protonation. In dilute acid, the catalyst showed an impressive TOF of 70,000 s^{-1} [312]. Thus, synthetic mimics of [FeFe] hydrogenase are showing promise as molecular catalysts and highlight the valuable contribution from studies of the natural enzyme.

Inspiration from the native hydrogenases suggests that catalytic activity might require a bimetallic site. However, a closer look shows – at least for the [FeFe]-hydrogenase – that proton or hydrogen substrate binding and also the hydride-proton reaction exclusively occurs at the iron distal to the [4Fe–4S]-cluster, suggesting that mononuclear iron complexes might also be viable catalysts. Consequently, Ott and coworkers have synthesized and characterized some stable pentacoordinated Fe(II) complexes with five ligands that nicely mimic the native ones and exhibit an open coordination site [313–315]. This approach avoids the formation of less reactive bridging hydrides that are found in many of the dinuclear complexes [298]. Catalytic H_2 formation from weak acids at low overpotentials with promising TOF and catalyst stability could be demonstrated [313].

A similar approach was used earlier by DuBois and coworkers who synthesized a number of mononuclear Ni complexes with cyclic diphosphine ligands and pendant amine functionalities. This approach combines in a beautiful way structural and functional features of both the [FeFe] and [NiFe] hydrogenases. Figure 8.13 shows a comparison of the active sites of these two enzymes with the DuBois Ni-catalyst in the proposed modes of H_2 binding, with H· · ·H (polarized). In the DuBois catalyst Ni, acting as a Lewis acid, is carrying the hydride and the proton is attached to the pendant amine base that is positioned perfectly forming the frustrated Lewis pair required for efficient H_2 formation. It is remarkable that for the first time a catalyst of this type showed a very high TOF ($\approx 10^5$ s^{-1}) for electrocatalytic H_2 production in the presence of water [316]. Such catalysts are often working in both directions. The rather high overpotential of these complexes has recently been improved. Artero et al. succeeded in grafting a catalyst of this type on a nanostructured electrode (multiwalled carbon nanotubes) and obtained hydrogen evolution from acids exhibiting very low overpotentials and high stability – comparable to the native enzyme [317]. More recently, the same group has been developing the approach even further with impressive results [289, 318, 319]. An alternative approach to build an extended proton-relay has also allowed these catalysts to function efficiently in water at low overpotential [304, 305, 320–323].

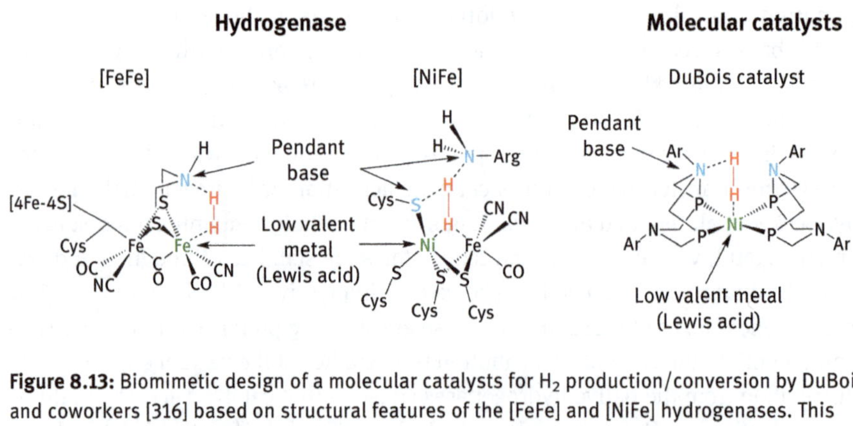

Figure 8.13: Biomimetic design of a molecular catalysts for H_2 production/conversion by DuBois and coworkers [316] based on structural features of the [FeFe] and [NiFe] hydrogenases. This catalyst is based on the concept of a frustrated Lewis acid-base pair and uses a low valent metal (Ni) (Lewis acid) for binding a hydride and exactly positioned (flexible) pendant ligands with basic functionalities (nitrogens) acting as proton acceptors. For details of this concept see [304, 305, 316, 320]. DuBois and coworkers reported a turnover frequency of this type of catalyst of over 100,000 s^{-1} for H_2 production [316]. Subsequently, a number of studies have improved the properties of this type of catalyst by incorporation of amino acids in the second coordination sphere [304, 305].

One remaining problem is that synthesizing molecular catalysts is chemically demanding and expensive. A perfect solution would thus be a system that self-assembles in solution – preferentially from inexpensive materials – and thus might even have the ability of self-repair in case of damage. In this respect systems like the one described by Xu et al. [324] are promising, who described a simple Ni(II)-thiolate complex that self-assembles in aqueous solution and, with a sacrificial reagent, produces H_2 under visible light when sensitized with a dye. This presents a promising approach toward the development of simple, low-cost, and benign solar hydrogen-producing materials.

In this chapter only a few models closely resembling the native hydrogenases have been mentioned. From this work it is clear that nature has designed enzymes with active sites that are robust, highly active, and use abundant metal ions. The hydrogenase field is a beautiful example of nature's chemistry, which provides inspiration for the synthesis of biomimetic compounds. There exist many more metal catalysts for H_2 oxidation and production using other metals, ligands and structures; see, for example, references [288, 290, 325–330] for selected examples and reviews. Future work will show which systems are superior as catalysts for H_2 conversion and production.

8.4 Conclusions

During recent years our understanding of the water-oxidizing complex in photosynthesis and the hydrogenase enzymes has made considerable progress. This is due to intensive efforts of basic research in these important scientific fields. The availability and combination of structural result from reliable high-resolution X-ray crystallography with structural and functional multifrequency spectroscopy, electrochemistry, and kinetic measurements proved to be extremely valuable. Furthermore, the enormous progress in theoretical chemistry now allows us to calculate structures and properties of reaction intermediates and reaction mechanisms even of larger (biological) systems [41, 67, 331–334].

Based on studies of the natural systems, much has also been learned concerning the design principles required for biomimetic catalysts for water splitting and hydrogen evolution/conversion. In summary, these include the use of abundant and inexpensive metals, the effective activation and protection of the catalytic sites in functional environments, repair/replacement of active components in case of damage, and the optimization of reaction rates. Biomimetic chemistry aims to mimic all these features; many labs are working toward this goal by developing new approaches in the design and synthesis of such systems, encompassing not only the catalytic center but also smart matrices and the assembly via self-organization [98, 324, 335].

More stable catalysts that do not require self repair may be obtained by fully artificial (inorganic) catalytic systems that are totally different from the biological ones and only apply some basic principles learned from nature. Furthermore, alternative metals apart from Mn/Ca, Fe, and Ni could be used for catalysis (e.g., Co), as well as new ligand spheres and other matrices. For light harvesting, charge separation/stabilization, and the effective coupling of the oxidizing/reducing equivalents to the redox catalysts, different ways have been proposed; for example covalently linked molecular donor-acceptor systems, photovoltaic devices, semiconductor-based systems, and photoactive metal complexes (for a review, see [288, 330, 336]).

The aim of all these approaches is to develop catalytic systems that split water with sunlight into hydrogen and oxygen, while displaying a high efficiency and long-term stability. Such a system –biological, biomimetic, or bioinspired – has the potential to be used on a large scale to produce "solar fuels," for example, hydrogen or secondary products thereof.

References

[1] Wydrzynski T, Satoh K, editors. Photosystem II. The light-driven water: plastoquinone oxidoreductase. As a series in: Advances in photosynthesis and respiration. Govindjee. editor, Vol. 22. Dordrecht, The Netherlands: Springer; 2005.

[2] Blankenship RE. Molecular mechanisms of photosynthesis, 2nd ed. Chichester, UK: Wiley-Blackwell; 2014.

[3] Dau H, Fujita E, Sun L. Artificial photosynthesis: Beyond mimicking nature. ChemSusChem. 2017;10(22):4228–35. DOI: 10.1002/cssc.201702106.

[4] Staffell I, Scamman D, Velazquez Abad A, et al. The role of hydrogen and fuel cells in the global energy system. Energ Environ Sci. 2019;12(2):463–91. DOI: 10.1039/C8EE01157E.

[5] Nazir H, Louis C, Jose S, et al. Is the H_2 economy realizable in the foreseeable future? Part I: H_2 production methods. Int J Hydrog Energy. 2020;45(27):13777–88. DOI: 10.1016/j.ijhydene.2020.03.092.

[6] Stripp ST, Happe T. How algae produce hydrogen – news from the photosynthetic hydrogenase. Dalton Trans. 2009;(45):9960–69. DOI: 10.1039/b916246a.

[7] Cammack R, Frey M, Robson R, editors. Hydrogen as a fuel: Learning from nature. London, UK: Taylor & Francis; 2001.

[8] Collings AF, Critchley C, editors. Artificial photosynthesis: from basic biology to industrial application. Weinheim, Germany: Wiley-VCH Verlag GmbH; 2005.

[9] acatech – National Academy of Science and Engineering, German National Academy of Sciences Leopoldina, Union of the German Academies of Sciences and Humanities (Eds.) (2018): Artificial photosynthesis. Munich, 74 pages.

[10] Dogutan DK, Nocera DG. Artificial photosynthesis at efficiencies greatly exceeding that of natural photosynthesis. Acc Chem Res. 2019;52(11):3143–48. DOI: 10.1021/acs.accounts.9b00380.

[11] Umena Y, Kawakami K, Shen JR, et al. Crystal structure of oxygen-evolving photosystem II at a resolution of 1.9 Å. Nature. 2011;473(7345):55–60. DOI: 10.1038/nature09913.

[12] Cox N, Pantazis DA, Lubitz W. Current understanding of the mechanism of water oxidation in photosystem ii and its relation to XFEL data. Ann Rev Biochem. 2020;89(1):795–820. DOI: 10.1146/annurev-biochem-011520-104801.

[13] Ogata H, Lubitz W, Higuchi Y. Structure and function of [NiFe] hydrogenases. J Biochem. 2016;160(5):251–58. DOI: 10.1093/jb/mvw048.

[14] Peters JW, Schut GJ, Boyd ES, et al. [FeFe]- and [NiFe]-hydrogenase diversity, mechanism, and maturation. Biochim Biophys Acta Mol Cell Res. 2015;1853(6):1350–69. DOI: 10.1016/j.bbamcr.2014.11.021.

[15] Shima S, Pilak O, Vogt S, et al. The crystal structure of [Fe]-hydrogenase reveals the geometry of the active site. Science. 2008;321(5888):572–75. DOI: 10.1126/science.1158978.

[16] Kok B, Forbush B, McGloin M. Cooperation of charges in photosynthetic O_2 evolution. Photochem Photobiol. 1970;11(6):457–76. DOI: 10.1111/j.1751-1097.1970.tb06017.x.

[17] Joliot P, Barbieri G, Chabaud R. Un nouveau modele des centres photochimiques du systeme II*. Photochem Photobiol. 1969;10(5):309–29. DOI: 10.1111/j.1751-1097.1969.tb05696.x.

[18] Kulik LV, Epel B, Lubitz W, et al. Electronic structure of the Mn_4O_xCa cluster in the S_0 and S_2 states of the oxygen-evolving complex of Photosystem II based on pulse ^{55}Mn-ENDOR and EPR spectroscopy. J Am Chem Soc. 2007;129(44):13421–35. DOI: 10.1021/ja071487f.

[19] Renger G, editor. Primary processes of photosynthesis, part 2. Principles and apparatus. Cambridge, UK: RSC Publishing; 2008.

[20] Krewald V, Retegan M, Cox N, et al. Metal oxidation states in biological water splitting. Chem Sci. 2015;6(3):1676–95. DOI: 10.1039/C4SC03720K.

[21] Dau H, Haumann M. The manganese complex of photosystem II in its reaction cycle-basic framework and possible realization at the atomic level. Coord Chem Rev. 2008;252(3–4):273–95. DOI: 10.1016/j.ccr.2007.09.001.

[22] Kargul J, Barber J. Structure and function of photosynthetic reaction centres. In: Wydrzynski TJ, Hillier W, editors. Molecular Solar Fuels. Cambridge, UK: RSC Publishing; 2012. 107–42.

[23] Satoh K, Wydrzynski TJ, Govindjee. Introduction to photosystem II. In: Wydrzynski TJ, Satoh K, Freeman JA, editors. Photosystem II: The light-driven water: plastoquinoneoxidoreductase. Dordrecht: Springer Netherlands; 2005. 11–22.

[24] Holzwarth AR, Müller MG, Reus M, et al. Kinetics and mechanism of electron transfer in intact Photosystem II and in the isolated reaction center: Pheophytin is the primary electron acceptor. Proc Natl Acad Sci USA. 2006;103(18):6895–900. DOI: 10.1073/pnas.0505371103.

[25] Rappaport F, Guergova-Kuras M, Nixon PJ, et al. Kinetics and pathways of charge recombination in Photosystem II. Biochemistry. 2002;41(26):8518–27. DOI: 10.1021/bi025725p.

[26] Messinger J, Renger G. Photosynthetic water splitting. In: Renger G, editor. Primary processes of photosynthesis, part 2: principles and apparatus. Comprehensive series in photochemical and photobiological sciences. 9. Cambridge, UK: The Royal Society of Chemistry; 2008. 291–349.

[27] Tommos C, Babcock GT. Oxygen production in nature: A light-driven metalloradical enzyme process. Acc Chem Res. 1998;31(1):18–25. DOI: 10.1021/ar9600188.

[28] Hillier W, Wydrzynski TJ. ^{18}O-water exchange in photosystem II: Substrate binding and intermediates of the water splitting cycle. Coord Chem Rev. 2008;252(3–4):306–17. DOI: 10.1016/j.ccr.2007.09.004.

[29] Styring S, Rutherford AW. In the oxygen-evolving complex of photosystem II the S_0 state is oxidized to the S_1 state by D^+ (signal II_{slow}). Biochemistry. 1987;26(9):2401–05. DOI: 10.1021/bi00383a001.

[30] Renger G, Holzwarth AR. Primary electron transfer. In: Wydrzynski TJ, Satoh K, editors. Photosystem II: The Light-Driven Water: Plastoquinoneoxidoreductase. Advances in Photosynthesis and Respiration, Vol. 22. Dordrecht, The Netherlands: Springer; 2005. 139–75.

[31] Suga M, Akita F, Yamashita K, et al. An oxyl/oxo mechanism for oxygen-oxygen coupling in PSII revealed by an x-ray free-electron laser. Science. 2019;366(6463):334–38. DOI: 10.1126/science.aax6998.

[32] Kern J, Chatterjee R, Young ID, et al. Structures of the intermediates of Kok's photosynthetic water oxidation clock. Nature. 2018;563(7731):421–25. DOI: 10.1038/s41586-018-0681-2.

[33] Boussac A, Rutherford AW. Nature of the inhibition of the oxygen-evolving enzyme of photosystem II induced by sodium chloride washing and reversed by the addition of Ca^{2+} or Sr^{2+} Biochemistry. 1988;27(9):3476–83. DOI: 10.1021/bi00409a052.

[34] Boussac A, Setif P, Rutherford AW. Inhibition of tyrosine Z photooxidation after formation of the S_3-state in Ca-depleted and Cl^-depleted photosystem-II. Biochemistry. 1992;31(4):1224–34. DOI: 10.1021/bi00119a036.

[35] Ishida N, Sugiura M, Rappaport F, et al. Biosynthetic exchange of bromide for chloride and strontium for calcium in the Photosystem II oxygen-evolving enzymes. J Biol Chem. 2008;283(19):13330–40. DOI: 10.1074/jbc.M710583200.

[36] Yano J, Kern J, Sauer K, et al. Where water is oxidized to dioxygen: Structure of the photosynthetic Mn_4Ca cluster. Science. 2006;314(5800):821–25. DOI: 10.1126/science.1128186.

[37] Yano J, Kern J, Irrgang KD, et al. X-ray damage to the Mn_4Ca complex in single crystals of Photosystem II: A case study for metalloprotein crystallography. Proc Natl Acad Sci USA. 2005;102(34):12047–52. DOI: 10.1073/pnas.0505207102.

[38] Suga M, Akita F, Sugahara M, et al. Light-induced structural changes and the site of O=O bond formation in PSII caught by XFEL. Nature. 2017;543(7643):131–35. DOI: 10.1038/nature21400.

[39] Suga M, Akita F, Hirata K, et al. Native structure of photosystem II at 1.95 Å resolution viewed by femtosecond X-ray pulses. Nature. 2015;517(7532):99–103. DOI: 10.1038/nature13991.

[40] Kusunoki M. Mono-manganese mechanism of the photosytem II water splitting reaction by a unique Mn_4Ca cluster. Biochim Biophys Acta Bioenerg. 2007;1767(6):484–92. DOI: 10.1016/j.bbabio.2007.03.012.

[41] Siegbahn PEM. Structures and energetics for O_2 formation in photosystem II. Acc Chem Res. 2009;42(12):1871–80. DOI: 10.1021/ar900117k.

[42] Ames W, Pantazis DA, Krewald V, et al. Theoretical evaluation of structural models of the S_2 state in the oxygen evolving complex of Photosystem II: Protonation states and magnetic interactions. J Am Chem Soc. 2011;133(49):19743–57. DOI: 10.1021/ja2041805.

[43] Peloquin JM, Campbell KA, Randall DW, et al. ^{55}Mn ENDOR of the S_2-state multiline EPR signal of Photosystem II: Implications on the structure of the tetranuclear Mn cluster. J Am Chem Soc. 2000;122(44):10926–42. DOI: 10.1021/ja002104f.

[44] Haumann M, Müller C, Liebisch P, et al. Structural and oxidation state changes of the photosystem II manganese complex in four transitions of the water oxidation cycle ($S_0 \rightarrow S_1$, $S_1 \rightarrow S_2$, $S_2 \rightarrow S_3$, and $S_{3,4} \rightarrow S_0$) characterized by X-ray absorption spectroscopy at 20 K and room temperature. Biochemistry. 2005;44(6):1894–908. DOI: 10.1021/bi048697e.

[45] Messinger J, Robblee JH, Bergmann U, et al. Absence of Mn-centered oxidation in the $S_2 \rightarrow S_3$ transition: Implications for the mechanism of photosynthetic water oxidation. J Am Chem Soc. 2001;123(32):7804–20. DOI: 10.1021/ja004307+.

[46] Glatzel P, Bergmann U, Yano J, et al. The electronic structure of Mn in oxides, coordination complexes, and the oxygen-evolving complex of Photosystem II studied by resonant inelastic X-ray scattering. J Am Chem Soc. 2004;126(32):9946–59. DOI: 10.1021/ja038579z.

[47] Pantazis DA, Ames W, Cox N, et al. Two interconvertible structures that explain the spectroscopic properties of the oxygen-evolving complex of photosystem II in the S_2 state. Angew Chem Int Ed. 2012;51(39):9935–40. DOI: 10.1002/anie.201204705.

[48] Krewald V, Retegan M, Neese F, et al. Spin state as a marker for the structural evolution of nature's water-splitting catalyst. Inorg Chem. 2016;55(2):488–501. DOI: 10.1021/acs.inorgchem.5b02578.

[49] Rapatskiy L, Cox N, Savitsky A, et al. Detection of the water-binding sites of the oxygen-evolving complex of Photosystem II using W-band ^{17}O electron-electron double resonance-detected NMR spectroscopy. J Am Chem Soc. 2012;134(40):16619–34. DOI: 10.1021/ja3053267.

[50] McConnell IL, Grigoryants VM, Scholes CP, et al. EPR-ENDOR characterization of (^{17}O, ^{1}H, ^{2}H) water in manganese catalase and its relevance to the oxygen-evolving complex of Photosystem II. J Am Chem Soc. 2012;134(3):1504–12. DOI: 10.1021/ja203465y.

[51] Lohmiller T, Krewald V, Sedoud A, et al. The first state in the catalytic cycle of the water-oxidizing enzyme: Identification of a water-derived µ-hydroxo bridge. J Am Chem Soc. 2017;139(41):14412–24. DOI: 10.1021/jacs.7b05263.

[52] Robblee JH, Messinger J, Cinco RM, et al. The Mn cluster in the S_0 state of the oxygen-evolving complex of Photosystem II studied by EXAFS spectroscopy: Are there three di-µ-oxo-bridged Mn_2 moieties in the tetranuclear Mn complex?. J Am Chem Soc. 2002;124(25):7459–71. DOI: 10.1021/ja011621a.

[53] Liang WC, Roelofs TA, Cinco RM, et al. Structural change of the Mn cluster during the $S_2 \rightarrow S_3$ state transition of the oxygen-evolving complex of Photosystem II. Does it reflect the onset of water/substrate oxidation? Determination by Mn X-ray absorption spectroscopy. J Am Chem Soc. 2000;122(14):3399–412. DOI: 10.1021/ja992501u.

[54] Cox N, Retegan M, Neese F, et al. Photosynthesis. Electronic structure of the oxygen-evolving complex in photosystem II prior to O-O bond formation. Science. 2014;345(6198):804–08. DOI: 10.1126/science.1254910.

[55] Chrysina M, Heyno E, Kutin Y, et al. Five-coordinate Mn^{IV} intermediate in the activation of nature's water splitting cofactor. Proc Natl Acad Sci USA. 2019;116(34):16841–46. DOI: 10.1073/pnas.1817526116.

[56] Rappaport F, Diner BA. Primary photochemistry and energetics leading to the oxidation of the Mn_4Ca cluster and to the evolution of molecular oxygen in Photosystem II. Coord Chem Rev. 2008;252(3–4):259–72. DOI: 10.1016/j.ccr.2007.07.016.

[57] Vass I. Molecular mechanisms of photodamage in the photosystem II complex. Biochim Biophys Acta. 2012;1817(1):209–17. DOI: 10.1016/j.bbabio.2011.04.014.

[58] Dasgupta J, Ananyev GM, Dismukes GC. Photoassembly of the water-oxidizing complex in photosystem II. Coord Chem Rev. 2008;252(3–4):347–60. DOI: 10.1016/j.ccr.2007.08.022.

[59] Matheu R, Garrido-Barros P, Gil-Sepulcre M, et al. The development of molecular water oxidation catalysts. Nat Rev Chem. 2019;3(5):331–41. DOI: 10.1038/s41570-019-0096-0.

[60] Noguchi T. FTIR detection of water reactions in the oxygen-evolving centre of photosystem II. Phil Trans R Soc B. 2008;363(1494):1189–95. DOI: 10.1098/rstb.2007.2214.

[61] Pantazis DA. Missing pieces in the puzzle of biological water oxidation. ACS Catal. 2018;8(10):9477–507. DOI: 10.1021/acscatal.8b01928.

[62] Vinyard DJ, Brudvig GW. Progress toward a molecular mechanism of water oxidation in photosystem II. Ann Rev Phys Chem. 2017;68(1):101–16. DOI: 10.1146/annurev-physchem-052516-044820.

[63] Lubitz W, Chrysina M, Cox N. Water oxidation in photosystem II. Photosynth Res. 2019;142(1):105–25. DOI: 10.1007/s11120-019-00648-3.

[64] Junge W. Oxygenic photosynthesis: History, status and perspective. Q Rev Biophys. 2019;52: E1. DOI: 10.1017/s0033583518000112.
[65] Gao Y, Åkermark T, Liu J, et al. Nucleophilic attack of hydroxide on a Mn^V oxo complex: A model of the O-O bond formation in the oxygen evolving complex of Photosystem II. J Am Chem Soc. 2009;131(25):8726–27. DOI: 10.1021/ja901139r.
[66] Liu F, Concepcion JJ, Jurss JW, et al. Mechanisms of water oxidation from the blue dimer to Photosystem II. Inorg Chem. 2008;47(6):1727–52. DOI: 10.1021/ic701249s.
[67] Blomberg MRA, Borowski T, Himo F, et al. Quantum chemical studies of mechanisms for metalloenzymes. Chem Rev. 2014;114(7):3601–58. DOI: 10.1021/cr400388t.
[68] Hunter BM, Gray HB, Müller AM. Earth-abundant heterogeneous water oxidation catalysts. Chem Rev. 2016;116(22):14120–36. DOI: 10.1021/acs.chemrev.6b00398.
[69] Melder J, Bogdanoff P, Zaharieva I, et al. Water-oxidation electrocatalysis by manganese oxides: Syntheses, electrode preparations, electrolytes and two fundamental questions. Z Phys Chem. 2020;234(5):925–78. DOI: 10.1515/zpch-2019-1491.
[70] Najafpour MM, Renger G, Hołyńska M, et al. Manganese compounds as water-oxidizing catalysts: From the natural water-oxidizing complex to nanosized manganese oxide structures. Chem Rev. 2016;116(5):2886–936. DOI: 10.1021/acs.chemrev.5b00340.
[71] Najafpour MM, Zaharieva I, Zand Z, et al. Water-oxidizing complex in Photosystem II: Its structure and relation to manganese-oxide based catalysts. Coord Chem Rev. 2020;409:213183. DOI: 10.1016/j.ccr.2020.213183.
[72] Li J, Triana CA, Wan W, et al. Molecular and heterogeneous water oxidation catalysts: Recent progress and joint perspectives. Chem Soc Rev. 2021;50(4):2444–85. DOI: 10.1039/d0cs00978d.
[73] Gersten SW, Samuels GJ, Meyer TJ. Catalytic oxidation of water by an oxo-bridged ruthenium dimer. J Am Chem Soc. 1982;104(14):4029–30. DOI: 10.1021/ja00378a053.
[74] Concepcion JJ, Jurss JW, Templeton JL, et al. One site is enough. Catalytic water oxidation by $[Ru(tpy)(bpm)(OH_2)]^{2+}$ and $[Ru(tpy)(bpz)(OH_2)]^{2+}$. J Am Chem Soc. 2008;130(49):16462–63. DOI: 10.1021/ja8059649.
[75] Duan L, Bozoglian F, Mandal S, et al. A molecular ruthenium catalyst with water-oxidation activity comparable to that of photosystem II. Nat Chem. 2012;4(5):418–23. DOI: 10.1038/nchem.1301.
[76] Hull JF, Balcells D, Blakemore JD, et al. Highly active and robust Cp* iridium complexes for catalytic water oxidation. J Am Chem Soc. 2009;131(25):8730–31. DOI: 10.1021/ja901270f.
[77] Garrido-Barros P, Funes-Ardoiz I, Drouet S, et al. Redox non-innocent ligand controls water oxidation overpotential in a new family of mononuclear Cu-based efficient catalysts. J Am Chem Soc. 2015;137(21):6758–61. DOI: 10.1021/jacs.5b03977.
[78] Barnett SM, Goldberg KI, Mayer JM. A soluble copper–bipyridine water-oxidation electrocatalyst. Nat Chem. 2012;4(6):498–502. DOI: 10.1038/nchem.1350.
[79] Gerlach DL, Bhagan S, Cruce AA, et al. Studies of the pathways open to copper water oxidation catalysts containing proximal hydroxy groups during basic electrocatalysis. Inorg Chem. 2014;53(24):12689–98. DOI: 10.1021/ic501018a.
[80] Zhang T, Wang C, Liu S, et al. A biomimetic copper water oxidation catalyst with low overpotential. J Am Chem Soc. 2014;136(1):273–81. DOI: 10.1021/ja409267p.
[81] Fisher KJ, Materna KL, Mercado BQ, et al. Electrocatalytic water oxidation by a copper(II) complex of an oxidation-resistant ligand. ACS Catal. 2017;7(5):3384–87. DOI: 10.1021/acscatal.7b00494.
[82] Wasylenko DJ, Ganesamoorthy C, Borau-Garcia J, et al. Electrochemical evidence for catalytic water oxidation mediated by a high-valent cobalt complex. Chem Commun. 2011;47(14):4249–51. DOI: 10.1039/C0CC05522K.

[83] Rigsby ML, Mandal S, Nam W, et al. Cobalt analogs of Ru-based water oxidation catalysts: Overcoming thermodynamic instability and kinetic lability to achieve electrocatalytic O_2 evolution. Chem Sci. 2012;3(10):3058–62. DOI: 10.1039/C2SC20755A.

[84] Gimbert-Suriñach C, Moonshiram D, Francàs L, et al. Structural and spectroscopic characterization of reaction intermediates involved in a dinuclear Co–Hbpp water oxidation catalyst. J Am Chem Soc. 2016;138(47):15291–94. DOI: 10.1021/jacs.6b08532.

[85] Han Y, Wu Y, Lai W, et al. Electrocatalytic water oxidation by a water-soluble nickel porphyrin complex at neutral pH with low overpotential. Inorg Chem. 2015;54(11):5604–13. DOI: 10.1021/acs.inorgchem.5b00924.

[86] Funes-Ardoiz I, Garrido-Barros P, Llobet A, et al. Single electron transfer steps in water oxidation catalysis. Redefining the mechanistic scenario. ACS Catal. 2017;7(3):1712–19. DOI: 10.1021/acscatal.6b03253.

[87] Sauer K, Yachandra VK. A possible evolutionary origin for the Mn_4 cluster of the photosynthetic water oxidation complex from natural MnO_2 precipitates in the early ocean. Proc Natl Acad Sci USA. 2002;99(13):8631–36. DOI: 10.1073/pnas.132266199.

[88] Najafpour MM, Ehrenberg T, Wiechen M, et al. Calcium manganese(III) oxides ($CaMn_2O_4 \cdot x$ H_2O) as biomimetic oxygen-evolving catalysts. Angew Chem Int Ed. 2010;49(12):2233–37. DOI: 10.1002/anie.200906745.

[89] Frey CE, Kurz P. Water oxidation catalysis by synthetic manganese oxides with different structural motifs: A comparative study. Chem Eur J. 2015;21(42):14958–68. DOI: 10.1002/chem.201501367.

[90] Morita M, Iwakura C, Tamura H. The anodic characteristics of manganese dioxide electrodes prepared by thermal decomposition of manganese nitrate. Electrochim Acta. 1977;22(4):325–28. DOI: 10.1016/0013-4686(77)85081-0.

[91] Harriman A, Pickering IJ, Thomas JM, et al. Metal oxides as heterogeneous catalysts for oxygen evolution under photochemical conditions. J Chem Soc Faraday Trans. 1988;84 (8):2795–806. DOI: 10.1039/F19888402795.

[92] Najafpour MM, Nayeri S, Pashaei B. Nano-size amorphous calcium–manganese oxide as an efficient and biomimetic water oxidizing catalyst for artificial photosynthesis: Back to manganese. Dalton Trans. 2011;40(37):9374–78. DOI: 10.1039/C1DT11048A.

[93] Zaharieva I, Chernev P, Risch M, et al. Electrosynthesis, functional, and structural characterization of a water-oxidizing manganese oxide. Energ Environ Sci. 2012;5(5):7081–89. DOI: 10.1039/C2EE21191B.

[94] Morgan Chan Z, Kitchaev DA, Nelson Weker J, et al. Electrochemical trapping of metastable Mn^{3+} ions for activation of MnO_2 oxygen evolution catalysts. Proc Natl Acad Sci USA. 2018;115 (23):E5261–E8. DOI: 10.1073/pnas.1722235115.

[95] Pickrahn KL, Gorlin Y, Seitz LC, et al. Applications of ALD MnO to electrochemical water splitting. Phys Chem Chem Phys. 2015;17(21):14003–11. DOI: 10.1039/C5CP00843C.

[96] Tesch MF, Bonke SA, Jones TE, et al. Evolution of oxygen–metal electron transfer and metal electronic states during manganese oxide catalyzed water oxidation revealed with in situ soft x-ray spectroscopy. Angew Chem Int Ed. 2019;58(11):3426–32. DOI: 10.1002/anie.201810825.

[97] Huynh M, Shi C, Billinge SJL, et al. Nature of activated manganese oxide for oxygen evolution. J Am Chem Soc. 2015;137(47):14887–904. DOI: 10.1021/jacs.5b06382.

[98] Kanan MW, Nocera DG. *In situ* formation of an oxygen-evolving catalyst in neutral water containing phosphate and Co^{2+}. Science. 2008:3211072-5. DOI: 10.1126/science.1162018.

[99] Esswein AJ, Surendranath Y, Reece SY, et al. Highly active cobalt phosphate and borate based oxygen evolving catalysts operating in neutral and natural waters. Energ Environ Sci. 2011;4(2):499–504. DOI: 10.1039/c0ee00518e.

[100] Risch M, Khare V, Zaharieva I, et al. Cobalt-oxo core of a water-oxidizing catalyst film. J Am Chem Soc. 2009;131(20):6936–37. DOI: 10.1021/ja902121f.
[101] Kanan MW, Yano J, Surendranath Y, et al. Structure and valency of a cobalt-phosphate water oxidation catalyst determined by *in situ* X-ray spectroscopy. J Am Chem Soc. 2010;132 (39):13692–701. DOI: 10.1021/ja1023767.
[102] McAlpin JG, Surendranath Y, Dincă M, et al. EPR evidence for Co(IV) species produced during water oxidation at neutral pH. J Am Chem Soc. 2010;132(20):6882–83. DOI: 10.1021/ ja1013344.
[103] Risch M, Klingan K, Ringleb F, et al. Water oxidation by electrodeposited cobalt oxides – role of anions and redox-inert cations in structure and function of the amorphous catalyst. ChemSusChem. 2012;5(3):542–49. DOI: 10.1002/cssc.201100574.
[104] Costentin C, Nocera DG. Self-healing catalysis in water. Proc Natl Acad Sci USA. 2017;114 (51):13380–84. DOI: 10.1073/pnas.1711836114.
[105] Nocera DG. The artificial leaf. Acc Chem Res. 2012;45(5):767–76. DOI: 10.1021/ar2003013.
[106] Jiao F, Frei H. Nanostructured cobalt oxide clusters in mesoporous silica as efficient oxygen-evolving catalysts. Angew Chem Int Ed. 2009;48(10):1841–44. DOI: 10.1002/anie.200805534.
[107] Yin Q, Tan JM, Besson C, et al. A fast soluble carbon-free molecular water oxidation catalyst based on abundant metals. Science. 2010;328(5976):342–45. DOI: 10.1126/science.1185372.
[108] Oliva P, Leonardi J, Laurent JF, et al. Review of the structure and the electrochemistry of nickel hydroxides and oxy-hydroxides. J Power Sources. 1982;8(2):229–55. DOI: 10.1016/0378-7753 (82)80057-8.
[109] Młynarek G, Paszkiewicz M, Radniecka A. The effect of ferric ions on the behaviour of a nickelous hydroxide electrode. J Appl Electrochem. 1984;14(2):145–49. DOI: 10.1007/ BF00618733.
[110] Corrigan DA. The catalysis of the oxygen evolution reaction by iron impurities in thin film nickel oxide electrodes. J Electrochem Soc. 1987;134(2):377–84. DOI: 10.1149/1.2100463.
[111] Trotochaud L, Young SL, Ranney JK, et al. Nickel–iron oxyhydroxide oxygen-evolution electrocatalysts: The role of intentional and incidental iron incorporation. J Am Chem Soc. 2014;136(18):6744–53. DOI: 10.1021/ja502379c.
[112] Lyons MEG, Brandon MP. The oxygen evolution reaction on passive oxide covered transition metal electrodes in aqueous alkaline solution. Part 1-nickel. Int J Electrochem Sci. 2008;3 (12):1386–424. http://hdl.handle.net/2262/33829.
[113] Lu X, Zhao C. Electrodeposition of hierarchically structured three-dimensional nickel–iron electrodes for efficient oxygen evolution at high current densities. Nat Commun. 2015;6 (1):6616. DOI: 10.1038/ncomms7616.
[114] Williams R. The biodistribution of metal ions. In: Kraatz H-B, Metzler-Nolte N, editors. Concepts and models in bioinorganic chemistry, 1st ed. Weinheim, Germany: Wiley-VCH; 2006.
[115] Lubitz W, Ogata H, Reijerse E, et al. Structure and function of hydrogenase enzymes. In: Wydrzynski TJ, Hillier W, editors. Molecular Solar Fuels. Cambridge, UK: RSC Publishing; 2012. 288–325.
[116] Lubitz W, Ogata H, Rüdiger O, et al. Hydrogenases. Chem Rev. 2014;114(8):4081–148. DOI: 10.1021/cr4005814.
[117] Lubitz W, Reijerse E, Van Gastel M. [NiFe] and [FeFe] hydrogenases studied by advanced magnetic resonance techniques. Chem Rev. 2007;107(10):4331–65. DOI: 10.1021/cr050186q.
[118] Vignais PM, Billoud B, Meyer J. Classification and phylogeny of hydrogenases. FEMS Microbiol Rev. 2001:25455–501. DOI: 10.1111/j.1574-6976.2001.tb00587.x.
[119] Vignais PM, Billoud B. Occurrence, classification, and biological function of hydrogenases: An overview. Chem Rev. 2007;107(10):4206–72. DOI: 10.1021/cr050196r.

[120] Armstrong FA, Evans RM, Hexter SV, et al. Guiding principles of hydrogenase catalysis instigated and clarified by protein film electrochemistry. Acc Chem Res. 2016;49(5):884–92. DOI: 10.1021/acs.accounts.6b00027.

[121] Ash PA, Kendall-Price SET, Vincent KA. Unifying activity, structure, and spectroscopy of [NiFe] hydrogenases: Combining techniques to clarify mechanistic understanding. Acc Chem Res. 2019;52(11):3120–31. DOI: 10.1021/acs.accounts.9b00293.

[122] Payne RC, Brownlee D, Kasting JF. Oxidized micrometeorites suggest either high pCO_2 or low pN_2 during the Neoarchean. Proc Natl Acad Sci USA. 2020;117(3):1360–66. DOI: 10.1073/pnas.1910698117.

[123] Zahnle K, Schaefer L, Fegley B. Earth's earliest atmospheres. Cold Spring Harb Perspect Biol. 2010;2(10):a004895–a. DOI: 10.1101/cshperspect.a004895.

[124] Glick BR, Martin WG, Martin SM. Purification and properties of the periplasmic hydrogenase from *Desulfovibrio desulfuricans*. Can J Microbiol. 1980;26(10):1214–23. DOI: 10.1139/m80-203.

[125] Hatchikian EC, Forget N, Fernandez VM, et al. Further characterization of the [Fe]-hydrogenase from *Desulfovibrio desulfuricans* ATCC 7757. Eur J Biochem. 1992;209(1):357–65. DOI: 10.1111/j.1432-1033.1992.tb17297.x.

[126] Eberly JO, Ely RL. Thermotolerant hydrogenases: Biological diversity, properties, and biotechnological applications. Crit Rev Microbiol. 2008;34(3–4):117–30. DOI: 10.1080/10408410802240893.

[127] Chongdar N, Birrell JA, Pawlak K, et al. Unique spectroscopic properties of the H-cluster in a putative sensory [FeFe] hydrogenase. J Am Chem Soc. 2018;140(3):1057–68. DOI: 10.1021/jacs.7b11287.

[128] Chongdar N, Pawlak K, Rüdiger O, et al. Spectroscopic and biochemical insight into an electron-bifurcating [FeFe] hydrogenase. J Biol Inorg Chem. 2020;25(1):135–49. DOI: 10.1007/s00775-019-01747-1.

[129] Pandelia M-E, Fourmond V, Tron-Infossi P, et al. Membrane-bound hydrogenase I from the hyperthermophilic bacterium *Aquifex aeolicus*: Enzyme activation, redox intermediates and oxygen tolerance. J Am Chem Soc. 2010;132(20):6991–7004. DOI: 10.1021/ja910838d.

[130] Fritsch J, Lenz O, Friedrich B. Structure, function and biosynthesis of O_2-tolerant hydrogenases. Nat Rev Microbiol. 2013;11(2):106–14. DOI: 10.1038/nrmicro2940.

[131] Fritsch F, Scheerer P, Frielingsdorf S, et al. The crystal structure of an oxygen-tolerant hydrogenase uncovers a novel iron-sulphur centre. Nature. 2011;479(7372):249–52. DOI: 10.1038/nature10505.

[132] Goris T, Wait AF, Saggu M, et al. A unique iron-sulfur cluster is crucial for oxygen tolerance of a [NiFe]-hydrogenase. Nat Chem Biol. 2011;7(5):310–18. DOI: 10.1038/nchembio.555.

[133] Lenz O, Ludwig M, Schubert T, et al. H_2 conversion in the presence of O_2 as performed by the membrane-bound [NiFe]-hydrogenase of *Ralstonia eutropha*. ChemPhysChem. 2010;11(6):1107–19. DOI: 10.1002/cphc.200901002.

[134] Shomura Y, Yoon KS, Nishihara H, et al. Structural basis for [4Fe-3S] cluster in the oxygen-tolerant membrane-bound [NiFe] hydrogenase. Nature. 2011;479(7372):253–57. DOI: 10.1038/nature10504.

[135] Schäfer C, Bommer M, Hennig SE, et al. Structure of an actinobacterial-type [NiFe]-hydrogenase reveals insight into O_2-tolerant H_2 oxidation. Structure. 2016;24(2):285–92. DOI: 10.1016/j.str.2015.11.010.

[136] Fontecilla-Camps JC, Volbeda A, Cavazza C, et al. Structure/function relationships of [NiFe]- and [FeFe]-hydrogenases. Chem Rev. 2007;107(10):4273–303. DOI: 10.1021/cr050195z.

[137] Huang G, Wagner T, Wodrich MD, et al. The atomic-resolution crystal structure of activated [Fe]-hydrogenase. Nat Catal. 2019;2(6):537–43. DOI: 10.1038/s41929-019-0289-4.

[138] Shima S, Thauer RK. A third type of hydrogenase catalyzing H_2 activation. Chem Rec. 2007;7 (1):37–46. DOI: 10.1002/tcr.20111 [doi].

[139] Seefeldt LC, Yang Z-Y, Lukoyanov DA, et al. Reduction of substrates by nitrogenases. Chem Rev. 2020;120(12):5082–106. DOI: 10.1021/acs.chemrev.9b00556.

[140] Schäfer C, Friedrich B, Lenz O. Novel, Oxygen-Insensitive Group 5 [NiFe]-Hydrogenase in *Ralstonia eutropha*. Appl Environ Microbiol. 2013;79(17):5137–45. DOI: 10.1128/AEM.01576-13.

[141] Beaton SE, Evans RM, Finney AJ, et al. The structure of hydrogenase-2 from *Escherichia coli*: Implications for H_2-driven proton pumping. Biochem J. 2018;475(7):1353–70. DOI: 10.1042/BCJ20180053.

[142] Volbeda A, Mouesca JM, Darnault C, et al. X-ray structural, functional and computational studies of the O_2-sensitive *E. coli* hydrogenase-1 C19G variant reveal an unusual [4Fe-4S] cluster. Chem Commun. 2018;54(52):7175–78. DOI: 10.1039/c8cc02896f.

[143] Shomura Y, Taketa M, Nakashima H, et al. Structural basis of the redox switches in the NAD^+-reducing soluble [NiFe]-hydrogenase. Science. 2017;357(6354):928–32. DOI: 10.1126/science.aan4497.

[144] Volbeda A, Amara P, Iannello M, et al. Structural foundations for the O_2 resistance of *Desulfomicrobium baculatum* [NiFeSe]-hydrogenase. Chem Commun. 2013;49(63):7061–63. DOI: 10.1039/c3cc43619e.

[145] Marques MC, Coelho R, De Lacey AL, et al. The three-dimensional structure of [NiFeSe] hydrogenase from *Desulfovibrio vulgaris* Hildenborough: A hydrogenase without a bridging ligand in the active site in its oxidised, "as-isolated" state. J Mol Biol. 2010;396(4):893–907. DOI: 10.1016/j.jmb.2009.12.013.

[146] Garcin E, Vernede X, Hatchikian EC, et al. The crystal structure of a reduced [NiFeSe] hydrogenase provides an image of the activated catalytic center. Structure. 1999;7(5):557–66. DOI: 10.1016/s0969-2126(99)80072-0.

[147] Duan J, Senger M, Esselborn J, et al. Crystallographic and spectroscopic assignment of the proton transfer pathway in [FeFe]-hydrogenases. Nat Commun. 2018;9(1):4726. DOI: 10.1038/s41467-018-07140-x.

[148] Esselborn J, Muraki N, Klein K, et al. A structural view of synthetic cofactor integration into [FeFe]-hydrogenases. Chem Sci. 2016;7(2):959–68. DOI: 10.1039/c5sc03397g.

[149] Mulder DW, Boyd ES, Sarma R, et al. Stepwise [FeFe]-hydrogenase H-cluster assembly revealed in the structure of HydA(DeltaEFG). Nature. 2010;465(7295):248–51. DOI: 10.1038/nature08993.

[150] Lemon BJ, Peters JW. Binding of exogenously added carbon monoxide at the active site of the iron-only hydrogenase (CpI) from *Clostridium pasteurianum*. Biochemistry. 1999;38 (40):12969–73. DOI: 10.1021/bi9913193.

[151] Peters JW, Lanzilotta WN, Lemon BJ, et al. X-ray crystal structure of the Fe-only hydrogenase (CpI) from *Clostridium pasteurianum* to 1.8 angstrom resolution. Science. 1998;282 (5395):1853–58. DOI: 10.1126/science.282.5395.1853.

[152] Nicolet Y, Piras C, Legrand P, et al. *Desulfovibrio desulfuricans* iron hydrogenase: The structure shows unusual coordination to an active site Fe binuclear center. Structure. 1999;7 (1):13–23. DOI: 10.1016/s0969-2126(99)80005-7.

[153] Rodríguez-Maciá P, Galle LM, Bjornsson R, et al. Caught in the H_{inact}: Crystal structure and spectroscopy reveal a sulfur bound to the active site of an O_2-stable state of [FeFe] hydrogenase. Angew Chem Int Ed. 2020;59(38):16786–94. DOI: 10.1002/anie.202005208.

[154] Ogata H, Nishikawa K, Lubitz W. Hydrogens detected by subatomic resolution protein crystallography in a [NiFe] hydrogenase. Nature. 2015;520(7548):571–74. DOI: 10.1038/nature14110.

[155] Page CC, Moser CC, Chen X, et al. Natural engineering principles of electron tunnelling in biological oxidation-reduction. Nature. 1999;402(6757):47–52. DOI: 10.1038/46972.

[156] Happe RP, Roseboom W, Pierik AJ, et al. Biological activation of hydrogen. Nature. 1997;385 (6612):126. DOI: 10.1038/385126a0.

[157] Higuchi Y, Yagi T, Yasuoka N. Unusual ligand structure in Ni-Fe active center and an additional Mg site in hydrogenase revealed by high resolution X-ray structure analysis. Structure. 1997;5(12):1671–80. DOI: 10.1016/s0969-2126(97)00313-4.

[158] Ilina Y, Lorent C, Katz S, et al. X-ray crystallography and vibrational spectroscopy reveal the key determinants of biocatalytic dihydrogen cycling by [NiFe] hydrogenases. Angew Chem Int Ed. 2019;58(51):18710–14. DOI: 10.1002/anie.201908258.

[159] Higuchi Y, Ogata H, Miki K, et al. Removal of the bridging ligand atom at the Ni–Fe active site of [NiFe] hydrogenase upon reduction with H_2, as revealed by X-ray structure analysis at 1.4 Å resolution. Structure. 1999;7(5):549–56. DOI: 10.1016/S0969-2126(99)80071-9.

[160] Carepo M, Tierney DL, Brondino CD, et al. ^{17}O ENDOR detection of a solvent-derived Ni-(OH_x)-Fe bridge that is lost upon activation of the hydrogenase from *Desulfovibrio gigas*. J Am Chem Soc. 2002:124281–86. DOI: 10.1021/ja010204v.

[161] Vincent KA, Belsey NA, Lubitz W, et al. Rapid and reversible reactions of [NiFe]-hydrogenases with sulfide. J Am Chem Soc. 2006;128(23):7448–49. DOI: 10.1021/ja061732f.

[162] Higuchi Y, Yagi T. Liberation of hydrogen sulfide during the catalytic action of *Desulfovibrio* hydrogenase under the atmosphere of hydrogen. Biochem Biophys Res Commun. 1999;255 (2):295–99. DOI: 10.1006/bbrc.1999.0210.

[163] Ogata H, Kramer T, Wang H, et al. Hydride bridge in [NiFe]-hydrogenase observed by nuclear resonance vibrational spectroscopy. Nat Commun. 2015:67890. DOI: 10.1038/ncomms8890.

[164] Brecht M, Van Gastel M, Buhrke T, et al. Direct detection of a hydrogen ligand in the [NiFe] center of the regulatory H_2-sensing hydrogenase from *Ralstonia eutropha* in its reduced state by HYSCORE and ENDOR spectroscopy. J Am Chem Soc. 2003;125(43):13075–83. DOI: 10.1021/ja036624x.

[165] Volbeda A, Fontecilla-Camps JC. The active site and catalytic mechanism of NiFe hydrogenases. Dalton Trans. 2003;(21):4030–38. DOI: 10.1039/B304316A.

[166] Ogata H, Mizogushi Y, Mizuno N, et al. Structural studies of the carbon monoxide complex of [NiFe] hydrogenase from *Desulfovibrio vulgaris* Miyazaki F: Suggestion for the initial activation site for dihydrogen. J Am Chem Soc. 2002:12411628–35. DOI: 10.1021/ja012645k.

[167] Ogata H, Hirota S, Nakahara A, et al. Activation process of [NiFe] hydrogenase elucidated by high-resolution X-ray analyses: Conversion of the ready to the unready state. Structure. 2005;13(11):1635–42. DOI: 10.1016/j.str.2005.07.018.

[168] Volbeda A, Martin L, Cavazza C, et al. Structural difference between the ready and unready oxidized states of [NiFe] hydrogenases. J Biol Inorg Chem. 2005:10239–49. DOI: 10.1007/s00775-005-0632-x.

[169] Volbeda A, Martin L, Barbier E, et al. Crystallographic studies of [NiFe]-hydrogenase mutants: Towards consensus structures for the elusive unready oxidized states. J Biol Inorg Chem. 2015;20(1):11–22. DOI: 10.1007/s00775-014-1203-9.

[170] Nishikawa K, Mochida S, Hiromoto T, et al. Ni-elimination from the active site of the standard [NiFe] hydrogenase upon oxidation by O_2. J Inorg Biochem. 2017:177435–37. DOI: 10.1016/j.jinorgbio.2017.09.011.

[171] Ogata H, Lubitz W, Higuchi Y. [NiFe] hydrogenases: Structural and spectroscopic studies of the reaction mechanism. Dalton Trans. 2009;37:7577–87. DOI: 10.1039/B903840J.

[172] Pandelia ME, Nitschke W, Infossi P, et al. Characterization of a unique [FeS] cluster in the electron transfer chain of the oxygen tolerant [NiFe] hydrogenase from *Aquifex aeolicus*. Proc Natl Acad Sci USA. 2011;108(15):6097–102. DOI: 10.1073/pnas.1100610108.

[173] Lukey MJ, Roessler MM, Parkin A, et al. Oxygen-tolerant [NiFe]-hydrogenases: The individual and collective importance of supernumerary cysteines at the proximal Fe-S cluster. J Am Chem Soc. 2011;133(42):16881–92. DOI: 10.1021/ja205393w.

[174] Pandey AS, Harris TV, Giles LJ, et al. Dithiomethylether as a ligand in the hydrogenase H-cluster. J Am Chem Soc. 2008;130(13):4533–40. DOI: 10.1021/ja711187e.

[175] Kertess L, Wittkamp F, Sommer C, et al. Chalcogenide substitution in the [2Fe] cluster of [FeFe]-hydrogenases conserves high enzymatic activity. Dalton Trans. 2017;46(48):16947–58. DOI: 10.1039/c7dt03785f.

[176] Winkler M, Senger M, Duan J, et al. Accumulating the hydride state in the catalytic cycle of [FeFe]-hydrogenases. Nat Commun. 2017:816115. DOI: 10.1038/ncomms16115.

[177] Duan J, Mebs S, Laun K, et al. Geometry of the catalytic active site in [FeFe]-hydrogenase is determined by hydrogen bonding and proton transfer. ACS Catal. 2019;9(10):9140–49. DOI: 10.1021/acscatal.9b02203.

[178] Artz JH, Zadvornyy OA, Mulder DW, et al. Tuning catalytic bias of hydrogen gas producing hydrogenases. J Am Chem Soc. 2020;142(3):1227–35. DOI: 10.1021/jacs.9b08756.

[179] Nicolet Y, De Lacey AL, Vernède X, et al. Crystallographic and FTIR spectroscopic evidence of changes in Fe coordination upon reduction of the active site of the Fe-only hydrogenase from *Desulfovibrio desulfuricans*. J Am Chem Soc. 2001;123(8):1596–601. DOI: 10.1021/ja0020963.

[180] Swanson KD, Ratzloff MW, Mulder DW, et al. [FeFe]-hydrogenase oxygen inactivation is initiated at the H cluster 2Fe subcluster. J Am Chem Soc. 2015;137(5):1809–16. DOI: 10.1021/ja510169s.

[181] Rodríguez-Maciá P, Kertess L, Burnik J, et al. His-ligation to the [4Fe-4S] subcluster tunes the catalytic bias of [FeFe] hydrogenase. J Am Chem Soc. 2019;141(1):472–81. DOI: 10.1021/jacs.8b11149.

[182] Nicolet Y, Cavazza C, Fontecilla-Camps JC. Fe-only hydrogenases: Structure, function and evolution. J Inorg Biochem. 2002;91(1):1–8. DOI: 10.1016/s0162-0134(02)00392-6.

[183] Stripp ST, Goldet G, Brandmayr C, et al. How oxygen attacks [FeFe] hydrogenases from photosynthetic organisms. Proc Natl Acad Sci USA. 2009;106(41):17331–36. DOI: 10.1073/pnas.0905343106.

[184] Mebs S, Kositzki R, Duan J, et al. Hydrogen and oxygen trapping at the H-cluster of [FeFe]-hydrogenase revealed by site-selective spectroscopy and QM/MM calculations. Biochim Biophys Acta Bioenerg. 2018;1859(1):28–41. DOI: 10.1016/j.bbabio.2017.09.003.

[185] Esselborn J, Kertess L, Apfel U-P, et al. Loss of specific active-site iron atoms in oxygen-exposed [FeFe]-hydrogenase determined by detailed X-ray structure analyses. J Am Chem Soc. 2019;141(44):17721–28. DOI: 10.1021/jacs.9b07808.

[186] Silakov A, Wenk B, Reijerse E, et al. ^{14}N HYSCORE investigation of the H-cluster of [FeFe] hydrogenase: Evidence for a nitrogen in the dithiol bridge. Phys Chem Chem Phys. 2009;11(31):6592–99. DOI: 10.1039/b905841a.

[187] Berggren G, Adamska A, Lambertz C, et al. Biomimetic assembly and activation of [FeFe]-hydrogenases. Nature. 2013;499(7456):66–69. DOI: 10.1038/nature12239.

[188] Pandelia M-E, Ogata H, Lubitz W. Intermediates in the catalytic cycle of [NiFe] hydrogenase: Functional spectroscopy of the active site. ChemPhysChem. 2010;11(6):1127–40. DOI: 10.1002/cphc.200900950.

[189] Land H, Senger M, Berggren G, et al. Current state of [FeFe]-hydrogenase research: Biodiversity and spectroscopic investigations. ACS Catal. 2020;10(13):7069–86. DOI: 10.1021/acscatal.0c01614.

[190] Bagley KA, Duin EC, Roseboom W, et al. Infrared-detectable groups sense changes in charge density on the nickel center in hydrogenase from *Chromatium vinosum*. Biochemistry. 1995;34(16):5527–35. DOI: 10.1021/bi00016a026.

[191] Sommer C, Adamska-Venkatesh A, Pawlak K, et al. Proton coupled electronic rearrangement within the H-cluster as an essential step in the catalytic cycle of [FeFe] hydrogenases. J Am Chem Soc. 2017;139(4):1440–43. DOI: 10.1021/jacs.6b12636.

[192] Rodríguez-Maciá P, Pawlak K, Rüdiger O, et al. Intercluster redox coupling influences protonation at the H-cluster in [FeFe] hydrogenases. J Am Chem Soc. 2017;139(42):15122–34. DOI: 10.1021/jacs.7b08193.

[193] Senger M, Laun K, Wittkamp F, et al. Proton-coupled reduction of the catalytic [4Fe-4S] cluster in [FeFe]-hydrogenases. Angew Chem Int Ed. 2017;56(52):16503–06. DOI: 10.1002/anie.201709910.

[194] Sanchez MLK, Sommer C, Reijerse E, et al. Investigating the kinetic competency of *Cr*HydA1 [FeFe] hydrogenase intermediate states via time-resolved infrared spectroscopy. J Am Chem Soc. 2019;141(40):16064–70. DOI: 10.1021/jacs.9b08348.

[195] Greene BL, Vansuch GE, Chica BC, et al. Applications of photogating and time resolved spectroscopy to mechanistic studies of hydrogenases. Acc Chem Res. 2017;50(11):2718–26. DOI: 10.1021/acs.accounts.7b00356.

[196] Flanagan LA, Parkin A. Electrochemical insights into the mechanism of NiFe membrane-bound hydrogenases. Biochem Soc Trans. 2016;44(1):315–28. DOI: 10.1042/BST20150201.

[197] Sensi M, Del Barrio M, Baffert C, et al. New perspectives in hydrogenase direct electrochemistry. Curr Opin Electrochem. 2017;5(1):135–45. DOI: 10.1016/j.coelec.2017.08.005.

[198] Vincent KA, Parkin A, Armstrong FA. Investigating and exploiting the electrocatalytic properties of hydrogenases. Chem Rev. 2007;107(10):4366–413. DOI: 10.1021/cr050191u.

[199] Del Barrio M, Sensi M, Orain C, et al. Electrochemical investigations of hydrogenases and other enzymes that produce and use solar fuels. Acc Chem Res. 2018;51(3):769–77. DOI: 10.1021/acs.accounts.7b00622.

[200] McPherson IJ, Vincent KA. Electrocatalysis by hydrogenases: Lessons for building bio-inspired device. J Braz Chem Soc. 2014:25427–41. DOI: 10.5935/0103-5053.20140042.

[201] Rodríguez-Maciá P, Breuer N, DeBeer S, et al. Insight into the redox behavior of the [4Fe–4S] subcluster in [FeFe] hydrogenases. ACS Catal. 2020;10(21):13084–95. DOI: 10.1021/acscatal.0c02771.

[202] Pelmenschikov V, Birrell JA, Pham CC, et al. Reaction coordinate leading to H_2 production in [FeFe]-hydrogenase identified by nuclear resonance vibrational spectroscopy and density functional theory. J Am Chem Soc. 2017;139(46):16894–902. DOI: 10.1021/jacs.7b09751.

[203] Rumpel S, Sommer C, Reijerse E, et al. Direct detection of the terminal hydride intermediate in [FeFe] hydrogenase by NMR spectroscopy. J Am Chem Soc. 2018;140(11):3863–66. DOI: 10.1021/jacs.8b00459.

[204] De Lacey AL, Fernandez VM, Rousset M, et al. Activation and inactivation of hydrogenase function and the catalytic cycle: Spectroelectrochemical studies. Chem Rev. 2007;107(10):4304–30. DOI: 10.1021/cr0501947.

[205] Hamdan AL, Burlat B, Gutierrez-Sanz O, et al. O_2-independent formation of the inactive states of NiFe hydrogenase. Nat Chem Biol. 2013;9(1):15–17. DOI: 10.1038/nchembio.1110.

[206] Evans RM, Brooke EJ, Wehlin SA, et al. Mechanism of hydrogen activation by [NiFe] hydrogenases. Nat Chem Biol. 2016;12(1):46–50. DOI: 10.1038/nchembio.1976.

[207] George SJ, Kurkin S, Thorneley RNF, et al. Reactions of H_2, CO, and O_2 with active [NiFe]-hydrogenase from *Allochromatium vinosum*. A stopped-flow infrared study. Biochemistry. 2004:436808–19. DOI: 10.1021/bi049853k.

[208] Kurkin S, George SJ, Thorneley RNF, et al. Hydrogen-induced activation of the [NiFe]-hydrogenase from *Allochromatium vinosum* as studied by stopped-flow infrared spectroscopy. Biochemistry. 2004:436820–31. DOI: 10.1021/bi049854c.

[209] Hidalgo R, Ash PA, Healy AJ, et al. Infrared spectroscopy during electrocatalytic turnover reveals the Ni-L active site state during H_2 oxidation by a NiFe hydrogenase. Angew Chem Int Ed. 2015;54(24):7110–13. DOI: 10.1002/anie.201502338.

[210] Murphy BJ, Hidalgo R, Roessler MM, et al. Discovery of dark pH-dependent H^+ migration in a [NiFe]-hydrogenase and its mechanistic relevance: Mobilizing the hydrido ligand of the Ni-C intermediate. J Am Chem Soc. 2015;137(26):8484–89. DOI: 10.1021/jacs.5b03182.

[211] Fontecilla-Camps JC, Amara P, Cavazza C, et al. Structure-function relationships of anaerobic gas-processing metalloenzymes. Nature. 2009;460(7257):814–22. DOI: 10.1038/nature08299.

[212] Caserta G, Lorent C, Ciaccafava A, et al. The large subunit of the regulatory [NiFe]-hydrogenase from Ralstonia eutropha – a minimal hydrogenase? Chem Sci. 2020;11 (21):5453–65. DOI: 10.1039/D0SC01369B.

[213] Breglia R, Greco C, Fantucci P, et al. Theoretical investigation of aerobic and anaerobic oxidative inactivation of the [NiFe]-hydrogenase active site. Phys Chem Chem Phys. 2018;20 (3):1693–706. DOI: 10.1039/C7CP06228A.

[214] Wang H, Patil DS, Gu W, et al. L-edge X-ray absorption spectroscopy of some Ni enzymes: Probe of Ni electronic structure. J Electron Spectrosc Relat Phenom. 2001;114:116855–63. DOI: 10.1021/ic7019878.

[215] Delcey MG, Pierloot K, Phung QM, et al. Accurate calculations of geometries and singlet–triplet energy differences for active-site models of [NiFe] hydrogenase. Phys Chem Chem Phys. 2014;16(17):7927–38. DOI: 10.1039/C4CP00253A.

[216] Pardo A, De Lacey AL, Fernandez VM, et al. Density functional study of the catalytic cycle of [NiFe] hydrogenases and the involvement of high-spin nickel(II). J Biol Inorg Chem. 2006;11 (3):286–306. DOI: 10.1007/s00775-005-0076-3.

[217] Stein M, Lubitz W. Quantum chemical calculations of [NiFe] hydrogenase. Curr Opin Chem Biol. 2002;6(2):243–49. DOI: 10.1016/S1367-5931(02)00313-7.

[218] Stein M, Lubitz W. Relativistic DFT calculation of the reaction cycle intermediates of [NiFe] hydrogenase: A contribution to understanding the enzymatic mechanism. J Inorg Biochem. 2004;98(5):862–77. DOI: 10.1016/j.jinorgbio.2004.03.002.

[219] Foerster S, Gastel M, Brecht M, et al. An orientation-selected ENDOR and HYSCORE study of the Ni-C active state of *Desulfovibrio vulgaris* Miyazaki F hydrogenase. J Biol Inorg Chem. 2005;10(1):51–62. DOI: 10.1007/s00775-004-0613-5.

[220] Foerster S, Stein M, Brecht M, et al. Single crystal EPR studies of the reduced active site of [NiFe] hydrogenase from *Desulfovibrio vulgaris* Miyazaki F. J Am Chem Soc. 2003;125(1):83–93. DOI: 10.1021/ja027522u.

[221] Medina M, Williams R, Cammack R, et al. Studies of light-induced nickel EPR signals in *Desulfovibrio gigas* hydrogenase. J Chem Soc Faraday Trans. 1994;90(19):2921–24. DOI: 10.1074/jbc.271.39.23799.

[222] De Lacey AL, Fernandez VM, Rousset M. Native and mutant nickel-iron hydrogenases: Unravelling structure and function. Coord Chem Rev. 2005:2491596–608. DOI: 10.1016/j.ccr.2005.03.009.

[223] Bagley KA., Van Garderen CJ, Chen M, et al. Infrared studies on the interaction of carbon monoxide with divalent nickel in hydrogenase from *Chromatium vinosum*. Biochemistry. 1994;33(31):9229–36. DOI: 10.1021/bi00197a026.

[224] Pandelia M-E, Ogata H, Currell LJ, et al. Inhibition of the [NiFe] hydrogenase from *Desulfovibrio vulgaris* Miyazaki F by carbon monoxide: An FTIR and EPR spectroscopic study. Biochim Biophys Acta Bioenerg. 2010;1797(2):304–13. DOI: 10.1016/j.bbabio.2009.11.002.

[225] Van Der Zwaan JW, Albracht SPJ, Fontijn RD, et al. EPR evidence for direct interaction of carbon monoxide with nickel in hydrogenase from *Chromatium vinosum*. Biochim Biophys Acta Prot Struct Mol Enzym. 1986;872(3):208–15. DOI: 10.1016/0167-4838(86)90273-6.

[226] Barilone JL, Ogata H, Lubitz W, et al. Structural differences between the active sites of the Ni-A and Ni-B states of the [NiFe] hydrogenase: An approach by quantum chemistry and single crystal ENDOR spectroscopy. Phys Chem Chem Phys. 2015;17(24):16204–12. DOI: 10.1039/c5cp01322d.

[227] Tai H, Higuchi Y, Hirota S. Comprehensive reaction mechanisms at and near the Ni–Fe active sites of [NiFe] hydrogenases. Dalton Trans. 2018;47(13):4408–23. DOI: 10.1039/C7DT04910B.

[228] Adams MWW. The structure and mechanism of iron-hydrogenases. Biochim Biophys Acta Bioenerg. 1990;1020(2):115–45. DOI: 10.1016/0005-2728(90)90044-5.

[229] Adamska A, Silakov A, Lambertz C, et al. Identification and characterization of the "super-reduced" state of the H-cluster in [FeFe] hydrogenase: A new building block for the catalytic cycle?. Angew Chem Int Ed. 2012;51(46):11458–62. DOI: 10.1002/anie.201204800.

[230] Mulder DW, Guo Y, Ratzloff MW, et al. Identification of a catalytic iron-hydride at the H-cluster of [FeFe]-hydrogenase. J Am Chem Soc. 2017;139(1):83–86. DOI: 10.1021/jacs.6b11409.

[231] Mulder DW, Ratzloff MW, Bruschi M, et al. Investigations on the role of proton-coupled electron transfer in hydrogen activation by [FeFe]-hydrogenase. J Am Chem Soc. 2014;136 (43):15394–402. DOI: 10.1021/ja508629m.

[232] Münck E, Popescu CV. Mößbauer studies of exchange coupled cluster assemblies in biological systems. Hyperfine Interact. 2000;126(1–4):59–67. DOI: 10.1023/A: 1012632409563.

[233] Popescu CV, Münck E. Electronic structure of the H cluster in [Fe]-hydrogenases. J Am Chem Soc. 1999;121(34):7877–84. DOI: 10.1021/ja991243y.

[234] Rumpel S, Ravera E, Sommer C, et al. ^{1}H NMR spectroscopy of [FeFe] hydrogenase: Insight into the electronic structure of the active site. J Am Chem Soc. 2018;140(1):131–34. DOI: 10.1021/jacs.7b11196.

[235] Reijerse EJ, Pham CC, Pelmenschikov V, et al. Direct observation of an iron-bound terminal hydride in [FeFe]-hydrogenase by nuclear resonance vibrational spectroscopy. J Am Chem Soc. 2017;139(12):4306–09. DOI: 10.1021/jacs.7b00686.

[236] Chernev P, Lambertz C, Brunje A, et al. Hydride binding to the active site of [FeFe]-hydroge. nase. Inorg Chem. 2014;53(22):12164–77. DOI: 10.1021/ic502047q.

[237] Esselborn J, Lambertz C, Adamska-Venkatesh A, et al. Spontaneous activation of [FeFe]-hydrogenases by an inorganic [2Fe] active site mimic. Nat Chem Biol. 2013;9(10):607–09. DOI: 10.1038/nchembio.1311.

[238] Kuchenreuther JM, Grady-Smith CS, Bingham AS, et al. High-yield expression of heterologous [FeFe] hydrogenases in *Escherichia coli*. PloS one. 2010;5(11):e15491. DOI: 10.1371/journal.pone.0015491.

[239] Shepard EM, Mus F, Betz JN, et al. [FeFe]-hydrogenase maturation. Biochemistry. 2014;53 (25):4090–104. DOI: 10.1021/bi500210x.

[240] Byer AS, Shepard EM, Ratzloff MW, et al. H-cluster assembly intermediates built on HydF by the radical SAM enzymes HydE and HydG. J Biol Inorg Chem. 2019;24(6):783–92. DOI: 10.1007/s00775-019-01709-7.

[241] Caserta G, Pecqueur L, Adamska-Venkatesh A, et al. Structural and functional characterization of the hydrogenase-maturation HydF protein. Nat Chem Biol. 2017;13 (7):779–84. DOI: 10.1038/nchembio.2385.

[242] Siebel JF, Adamska-Venkatesh A, Weber K, et al. Hybrid [FeFe]-hydrogenases with modified active sites show remarkable residual enzymatic activity. Biochemistry. 2015;54(7):1474–83. DOI: 10.1021/bi501391d.

[243] Sommer C, Richers. CP, Lubitz W, et al. A [RuRu] analogue of an [FeFe]-hydrogenase traps the key hydride intermediate of the catalytic cycle. Angew Chem Int Ed. 2018;57(19):5429–32. DOI: 10.1002/anie.201801914.

[244] Kleinhaus JT, Wittkamp F, Yadav S, et al. [FeFe]-Hydrogenases: Maturation and reactivity of enzymatic systems and overview of biomimetic models. Chem Soc Rev. 2021;50(3):1668–784. DOI: 10.1039/D0CS01089H.

[245] Katz S, Noth J, Horch M, et al. Vibrational spectroscopy reveals the initial steps of biological hydrogen evolution. Chem Sci. 2016;7(11):6746–52. DOI: 10.1039/C6SC01098A.

[246] Haumann M, Stripp ST. The molecular proceedings of biological hydrogen turnover. Acc Chem Res. 2018;51(8):1755–63. DOI: 10.1021/acs.accounts.8b00109.

[247] Roseboom W, De Lacey AL, Fernandez VM, et al. The active site of the [FeFe]-hydrogenase from *Desulfovibrio desulfuricans*. II. Redox properties, light sensitivity and CO-ligand exchange as observed via infrared spectroscopy. J Biol Inorg Chem. 2006;11(1):102–18. DOI: 10.1007/s00775-005-0040-2.

[248] Silakov A, Kamp C, Reijerse E, et al. Spectroelectrochemical characterization of the active site of the [FeFe] hydrogenase HydA1 from *Chlamydomonas reinhardtii*. Biochemistry. 2009;48 (33):7780–86. DOI: 10.1021/bi9009105.

[249] Ratzloff MW, Artz JH, Mulder DW, et al. CO-bridged H-cluster intermediates in the catalytic mechanism of [FeFe]-hydrogenase *CaI*. J Am Chem Soc. 2018;140(24):7623–28. DOI: 10.1021/jacs.8b03072.

[250] Birrell JA, Pelmenschikov V, Mishra N, et al. Spectroscopic and computational evidence that [FeFe] hydrogenases operate exclusively with CO-bridged intermediates. J Am Chem Soc. 2020;142(1):222–32. DOI: 10.1021/jacs.9b09745.

[251] Lorent C, Katz S, Duan J, et al. Shedding light on proton and electron dynamics in [FeFe] hydrogenases. J Am Chem Soc. 2020;142(12):5493–97. DOI: 10.1021/jacs.9b13075.

[252] Rodríguez-Maciá P, Reijerse EJ, Van Gastel M, et al. Sulfide protects [FeFe] hydrogenases from O_2. J Am Chem Soc. 2018;140(30):9346–50. DOI: 10.1021/jacs.8b04339.

[253] Adamska-Venkatesh A, Krawietz D, Siebel J, et al. New redox states observed in [FeFe] hydrogenases reveal redox coupling within the H-cluster. J Am Chem Soc. 2014;136 (32):11339–46. DOI: 10.1021/ja503390c.

[254] Albracht SP, Roseboom W, Hatchikian EC. The active site of the [FeFe]-hydrogenase from *Desulfovibrio desulfuricans*. I. Light sensitivity and magnetic hyperfine interactions as observed by electron paramagnetic resonance. J Biol Inorg Chem. 2006;11(1):88–101. DOI: 10.1007/s00775-005-0039-8.

[255] Silakov A, Wenk B, Reijerse E, et al. Spin distribution of the H-cluster in the H_{ox}-CO state of the [FeFe] hydrogenase from *Desulfovibrio desulfuricans*: HYSCORE and ENDOR study of ^{14}N and ^{13}C nuclear interactions. J Biol Inorg Chem. 2009;14(2):301–13. DOI: 10.1007/s00775-008-0449-5.

[256] Rodríguez-Maciá P, Reijerse E, Lubitz W, et al. Spectroscopic evidence of reversible disassembly of the [FeFe] hydrogenase active site. J Phys Chem Lett. 2017;8(16):3834–39. DOI: 10.1021/acs.jpclett.7b01608.

[257] Baffert C, Bertini L, Lautier T, et al. CO disrupts the reduced H-cluster of FeFe hydrogenase. A combined DFT and protein film voltammetry study. J Am Chem Soc. 2011;133(7):2096–99. DOI: 10.1021/ja110627b.

[258] Senger M, Mebs S, Duan J, et al. Stepwise isotope editing of [FeFe]-hydrogenases exposes cofactor dynamics. Proc Natl Acad Sci USA. 2016;113(30):8454–59. DOI: 10.1073/pnas.1606178113.

[259] Reijerse E, Birrell JA, Lubitz W. Spin polarization reveals the coordination geometry of the [FeFe] hydrogenase active site in its CO-inhibited state. J Phys Chem Lett. 2020;11(12):4597–602. DOI: 10.1021/acs.jpclett.0c01352.

[260] Winkler M, Duan J, Rutz A, et al. A safety cap protects hydrogenase from oxygen attack. Nat Commun. 2021;12(1):756. DOI: 10.1038/s41467-020-20861-2.

[261] Corrigan PS, Tirsch JL, Silakov A. Investigation of the unusual ability of the [FeFe] hydrogenase from *Clostridium beijerinckii* to access an O_2-protected state. J Am Chem Soc. 2020;142(28):12409–19. DOI: 10.1021/jacs.0c04964.
[262] Esper B, Badura A, Rögner M. Photosynthesis as a power supply for (bio-)hydrogen production. Trends Plant Sci. 2006;11(11):543–49. DOI: 10.1016/j.tplants.2006.09.001.
[263] Sokol KP, Robinson WE, Warnan J, et al. Bias-free photoelectrochemical water splitting with photosystem II on a dye-sensitized photoanode wired to hydrogenase. Nat Energy. 2018;3(11):944–51. DOI: 10.1038/s41560-018-0232-y.
[264] Koo J, Shiigi S, Rohovie M, et al. Characterization of [FeFe] hydrogenase O_2 sensitivity using a new, physiological approach. J Biol Chem. 2016;291(41):21563–70. DOI: 10.1074/jbc.M116.737122.
[265] Stapleton JA, Swartz JR. Development of an *in vitro* compartmentalization screen for high-throughput directed evolution of [FeFe] hydrogenases. PloS one. 2010;5(12):e15275. DOI: 10.1371/journal.pone.0015275.
[266] Armstrong FA, Belsey NA, Cracknell JA, et al. Dynamic electrochemical investigations of hydrogen oxidation and production by enzymes and implications for future technology. Chem Soc Rev. 2009;38(1):36–51. DOI: 10.1039/b801144n.
[267] Buhrke T, Lenz O, Krauss N, et al. Oxygen tolerance of the H_2-sensing [NiFe] hydrogenase from *Ralstonia eutropha* H16 is based on limited access of oxygen to the active site. J Biol Chem. 2005;280(25):23791–96. DOI: 10.1074/jbc.M503260200.
[268] Van Der Linden E, Burgdorf T, De Lacey AL, et al. An improved purification procedure for the soluble [NiFe]-hydrogenase of *Ralstonia eutropha*: New insights into its (in)stability and spectroscopic properties. J Biol Inorg Chem. 2006;11(2):247–60. DOI: 10.1007/s00775-005-0075-4.
[269] Vincent KA, Cracknell JA, Lenz O, et al. Electrocatalytic hydrogen oxidation by an enzyme at high carbon monoxide or oxygen levels. Proc Natl Acad Sci USA. 2005;102(47):16951–54. DOI: 10.1073/pnas.0504499102.
[270] Wulff P, Day CC, Sargent F, et al. How oxygen reacts with oxygen-tolerant respiratory [NiFe]-hydrogenases. Proc Natl Acad Sci USA. 2014;111(18):6606–11. DOI: 10.1073/pnas.1322393111.
[271] Horch M, Lauterbach L, Mroginski MA, et al. Reversible active site sulfoxygenation can explain the oxygen tolerance of a NAD^+-reducing [NiFe] hydrogenase and its unusual infrared spectroscopic properties. J Am Chem Soc. 2015;137(7):2555–64. DOI: 10.1021/ja511154y.
[272] Baffert C, Demuez M, Cournac L, et al. Hydrogen-activating enzymes: Activity does not correlate with oxygen sensitivity. Angew Chem Int Ed. 2008;47(11):2052–54. DOI: 10.1002/anie.200704313.
[273] Tosatto SCE, Toppo S, Carbonera D, et al. Comparative analysis of [FeFe] hydrogenase from *Thermotogales* indicates the molecular basis of resistance to oxygen inactivation. Int J Hydrog Energy. 2008;33(2):570–78. DOI: 10.1016/j.ijhydene.2007.10.010.
[274] Cohen J, Kim K, King P, et al. Finding gas diffusion pathways in proteins: Application to O_2 and H_2 transport in *CpI* [FeFe]-hydrogenase and the role of packing defects. Structure. 2005;13(9):1321–29. DOI: 10.1016/j.str.2005.05.013.
[275] Morra S, Arizzi M, Valetti F, et al. Oxygen stability in the new [FeFe]-hydrogenase from *Clostridium beijerinckii* SM10 (*Cb*A5H). Biochemistry. 2016;55(42):5897–900. DOI: 10.1021/acs.biochem.6b00780.
[276] Plumeré N, Rüdiger O, Oughli AA, et al. A redox hydrogel protects hydrogenase from high-potential deactivation and oxygen damage. Nat Chem. 2014;6(9):822–27. DOI: 10.1038/nchem.2022.

[277] Ruff A, Szczesny J, Marković N, et al. A fully protected hydrogenase/polymer-based bioanode for high-performance hydrogen/glucose biofuel cells. Nat Commun. 2018;9(1):3675. DOI: 10.1038/s41467-018-06106-3.

[278] Ruff A, Szczesny J, Zacarias S, et al. Protection and reactivation of the [NiFeSe] hydrogenase from *Desulfovibrio vulgaris* Hildenborough under oxidative conditions. ACS Energy Lett. 2017;2(5):964–68. DOI: 10.1021/acsenergylett.7b00167.

[279] Oughli AA, Conzuelo F, Winkler M, et al. A redox hydrogel protects the O_2-sensitive [FeFe]-hydrogenase from *Chlamydomonas reinhardtii* from oxidative damage. Angew Chem Int Ed. 2015;54(42):12329–33. DOI: 10.1002/anie.201502776.

[280] Oughli AA, Hardt S, Rüdiger O, et al. Reactivation of sulfide-protected [FeFe] hydrogenase in a redox-active hydrogel. Chem Commun. 2020;56(69):9958–61. DOI: 10.1039/D0CC03155K.

[281] Fourmond V, Stapf S, Li H, et al. Mechanism of protection of catalysts supported in redox hydrogel films. J Am Chem Soc. 2015;137(16):5494–505. DOI: 10.1021/jacs.5b01194.

[282] Szczesny J, Birrell JA, Conzuelo F, et al. Redox-polymer-based high-current-density gas-diffusion H_2-oxidation bioanode using [FeFe] hydrogenase from *Desulfovibrio desulfuricans* in a membrane-free biofuel cell. Angew Chem Int Ed. 2020;59(38):16506–10. DOI: 10.1002/anie.202006824.

[283] Szczesny J, Marković N, Conzuelo F, et al. A gas breathing hydrogen/air biofuel cell comprising a redox polymer/hydrogenase-based bioanode. Nat Commun. 2018;9(1):4715. DOI: 10.1038/s41467-018-07137-6.

[284] Li H, Buesen D, Dementin S, et al. Complete protection of O_2-sensitive catalysts in thin films. J Am Chem Soc. 2019;141(42):16734–42. DOI: 10.1021/jacs.9b06790.

[285] Oughli AA, Ruff A, Boralugodage NP, et al. Dual properties of a hydrogen oxidation Ni-catalyst entrapped within a polymer promote self-defense against oxygen. Nat Commun. 2018;9(1):864. DOI: 10.1038/s41467-018-03011-7.

[286] Fontaine F-G, Stephan DW. On the concept of frustrated Lewis pairs. 2017;375(2101):20170004. DOI: 10.1098/rsta.2017.0004.

[287] Schilter D, Camara JM, Huynh MT, et al. Hydrogenase enzymes and their synthetic models: The role of metal hydrides. Chem Rev. 2016;116(15):8693–749. DOI: 10.1021/acs.chemrev.6b00180.

[288] Dalle KE, Warnan J, Leung JJ, et al. Electro- and solar-driven fuel synthesis with first row transition metal complexes. Chem Rev. 2019;119(4):2752–875. DOI: 10.1021/acs.chemrev.8b00392.

[289] Huan TN, Jane RT, Benayad A, et al. Bio-inspired noble metal-free nanomaterials approaching platinum performances for H_2 evolution and uptake. Energ Environ Sci. 2016;9(3):940–47. DOI: 10.1039/C5EE02739J.

[290] Queyriaux N, Jane RT, Massin J, et al. Recent developments in hydrogen evolving molecular cobalt(II)-polypyridyl catalysts. Coord Chem Rev. 2015;304:3053–19. DOI: 10.1016/j.ccr.2015.03.014.

[291] Simmons TR, Berggren G, Bacchi M, et al. Mimicking hydrogenases: From biomimetics to artificial enzymes. Coord Chem Rev. 2014;270:271127–50. DOI: 10.1016/j.ccr.2013.12.018.

[292] Bullock RM, Chen JG, Gagliardi L, et al. Using nature's blueprint to expand catalysis with earth-abundant metals. Science. 2020;369(6505):eabc3183. DOI: 10.1126/science.abc3183.

[293] Caserta G, Roy S, Atta M, et al. Artificial hydrogenases: Biohybrid and supramolecular systems for catalytic hydrogen production or uptake. Curr Opin Chem Biol. 2015:2536–47. DOI: 10.1016/j.cbpa.2014.12.018.

[294] Onoda A, Hayashi T. Artificial hydrogenase: Biomimetic approaches controlling active molecular catalysts. Curr Opin Chem Biol. 2015:25133–40. DOI: 10.1016/j.cbpa.2014.12.041.

[295] Canaguier S, Artero V, Fontecave M. Modelling NiFe hydrogenases: Nickel-based electrocatalysts for hydrogen production. Dalton Trans. 2008;(3):315–25. DOI: 10.1039/b713567j.

[296] Darensbourg MY, Lyon EJ, Smee JJ. The bio-organometallic chemistry of active site iron in hydrogenases. Coord Chem Rev. 2000:206533–61. DOI: 10.1016/S0010-8545(00)00268-X.

[297] Dawson J, Ghiotto F, McMaster J, et al. Metal complex of hydrogenase active sites. In: Wydrzynski TJ, Hillier W, editors. Molecular Solar Fuels. Cambridge, UK: RSC Publishing; 2012. 326–86.

[298] Gloaguen F, Rauchfuss TB. Small molecule mimics of hydrogenases: Hydrides and redox. Chem Soc Rev. 2009;38(1):100–08. DOI: 10.1039/b801796b.

[299] Artero V, Fontecave M. Some general principles for designing electrocatalysts with hydrogenase activity. Coord Chem Rev. 2005;249(15–16):1518–35. DOI: 10.1016/j.ccr.2005.01.014.

[300] Sun LC, Åkermark B, Ott S. Iron hydrogenase active site mimics in supramolecular systems aiming for light-driven hydrogen production. Coord Chem Rev. 2005;249(15–16):1653–63. DOI: 10.1016/j.ccr.2005.01.013.

[301] Capon JF, Gloaguen F, Schollhammer P, et al. Catalysis of the electrochemical H_2 evolution by di-iron sub-site models. Coord Chem Rev. 2005;249(15–16):1664–76. DOI: 10.1016/j.ccr.2004.11.018.

[302] Ohki Y, Tatsumi K. Thiolate-bridged iron–nickel models for the active site of [NiFe] hydrogenase. Eur J Bioinorg Chem. 2011;2011(7):973–85. DOI: 10.1002/ejic.201001087.

[303] Tard C, Pickett CJ. Structural and functional analogues of the active sites of the [Fe]-, [NiFe]-, and [FeFe]-hydrogenases. Chem Rev. 2009;109(6):2245–74. DOI: 10.1021/cr800542q.

[304] Dutta A, Appel AM, Shaw WJ. Designing electrochemically reversible H_2 oxidation and production catalysts. Nat Rev Chem. 2018;2(9):244–52. DOI: 10.1038/s41570-018-0032-8.

[305] Laureanti JA, O'Hagan M, Shaw WJ. Chicken fat for catalysis: A scaffold is as important for molecular complexes for energy transformations as it is for enzymes in catalytic function. Sustain Energy Fuels. 2019;3(12):3260–78. DOI: 10.1039/C9SE00229D.

[306] Esmieu C, Raleiras P, Berggren G. From protein engineering to artificial enzymes – biological and biomimetic approaches towards sustainable hydrogen production. Sustain Energy Fuels. 2018;2(4):724–50. DOI: 10.1039/c7se00582b.

[307] Barton BE, Whaley CM, Rauchfuss TB, et al. Nickel–iron dithiolato hydrides relevant to the [NiFe]-hydrogenase active site. J Am Chem Soc. 2009;131(20):6942–43. DOI: 10.1021/ja902570u.

[308] Brazzolotto D, Gennari M, Queyriaux N, et al. Nickel-centred proton reduction catalysis in a model of [NiFe] hydrogenase. Nat Chem. 2016;8(11):1054–60. DOI: 10.1038/nchem.2575.

[309] Tang H, Hall MB. Biomimetics of [NiFe]-hydrogenase: Nickel- or iron-centered proton reduction catalysis? J Am Chem Soc. 2017;139(49):18065–70. DOI: 10.1021/jacs.7b10425.

[310] Tard C, Liu X, Ibrahim SK, et al. Synthesis of the H-cluster framework of iron-only hydrogenase. Nature. 2005;433(7026):610–13. DOI: 10.1038/nature03298.

[311] Camara JM, Rauchfuss TB. Combining acid–base, redox and substrate binding functionalities to give a complete model for the [FeFe]-hydrogenase. Nat Chem. 2012;4(1):26–30. DOI: 10.1038/nchem.1180.

[312] Becker R, Amirjalayer S, Li P, et al. An iron-iron hydrogenase mimic with appended electron reservoir for efficient proton reduction in aqueous media. Sci Adv. 2016;2(1):e1501014. DOI: 10.1126/sciadv.1501014.

[313] Beyler M, Ezzaher S, Karnahl M, et al. Pentacoordinate iron complexes as functional models of the distal iron in [FeFe] hydrogenases. Chem Commun. 2011;47(42):11662–64. DOI: 10.1039/C1CC14449A.

[314] Kaur-Ghumaan S, Schwartz L, Lomoth R, et al. Catalytic hydrogen evolution from mononuclear iron(II) carbonyl complexes as minimal functional models of the [FeFe] hydrogenase active site. Angew Chem Int Ed. 2010;49(43):8033–36. DOI: 10.1002/anie.201002719.

[315] Orthaber A, Karnahl M, Tschierlei S, et al. Coordination and conformational isomers in mononuclear iron complexes with pertinence to the [FeFe] hydrogenase active site. Dalton Trans. 2014;43(11):4537–49. DOI: 10.1039/c3dt53268b.

[316] Helm ML, Stewart MP, Bullock RM, et al. A synthetic nickel electrocatalyst with a turnover frequency above 100,000 s^{-1} for H_2 production. Science. 2011;333(6044):863–66. DOI: 10.1126/science.1205864.

[317] Le Goff A, Artero V, Jousselme B, et al. From hydrogenases to noble metal–free catalytic nanomaterials for H_2 production and uptake. Science. 2009;326(5958):1384–87. DOI: 10.1126/science.1179773.

[318] Reuillard B, Blanco M, Calvillo L, et al. Noncovalent integration of a bioinspired Ni catalyst to graphene acid for reversible electrocatalytic hydrogen oxidation. ACS Appl Mat Interfaces. 2020;12(5):5805–11. DOI: 10.1021/acsami.9b18922.

[319] Tran PD, Le Goff A, Heidkamp J, et al. Noncovalent modification of carbon nanotubes with pyrene-functionalized nickel complexes: Carbon monoxide tolerant catalysts for hydrogen evolution and uptake. Angew Chem Int Ed. 2011;50(6):1371–74. DOI: 10.1002/anie.201005427.

[320] Dutta A, DuBois DL, Roberts JAS, et al. Amino acid modified Ni catalyst exhibits reversible H_2 oxidation/production over a broad pH range at elevated temperatures. Proc Natl Acad Sci USA. 2014;111(46):16286–91. DOI: 10.1073/pnas.1416381111.

[321] Dutta A, Lense S, Hou J, et al. Minimal proton channel enables H_2 oxidation and production with a water-soluble nickel-based catalyst. J Am Chem Soc. 2013;135(49):18490–96. DOI: 10.1021/ja407826d.

[322] Lense S, Dutta A, Roberts JAS, et al. A proton channel allows a hydrogen oxidation catalyst to operate at a moderate overpotential with water acting as a base. Chem Commun. 2014;50 (7):792–95. DOI: 10.1039/C3CC46829A.

[323] Rodriguez-Maciá P, Dutta A, Lubitz W, et al. Direct comparison of the performance of a bio-inspired synthetic nickel catalyst and a [NiFe]-hydrogenase, both covalently attached to electrodes. Angew Chem Int Ed. 2015;54(42):12303–07. DOI: 10.1002/anie.201502364.

[324] Zhang W, Hong J, Zheng J, et al. Nickel–thiolate complex catalyst assembled in one step in water for solar H_2 production. J Am Chem Soc. 2011;133(51):20680–83. DOI: 10.1021/ja208555h.

[325] Kubas GJ. Fundamentals of H_2 binding and reactivity on transition metals underlying hydrogenase function and H_2 production and storage. Chem Rev. 2007;107(10):4152–205. DOI: 10.1021/cr050197j.

[326] Sala X, Escriche L, Llobet A. Molecular Ru and Ir complexes capable of acting as water oxidation catalysts. In: Wydrzynski T, Hillier W, editors. Molecular Solar Fuels. Cambridge, UK: RSC Publishing; 2012. 273–87.

[327] Matsumoto T, Kim K, Ogo S. Molecular catalysis in a fuel cell. Angew Chem Int Ed. 2011;50 (47):11202–05. DOI: 10.1002/anie.201104498.

[328] Dempsey JL, Brunschwig BS, Winkler JR, et al. Hydrogen evolution catalyzed by cobaloximes. Acc Chem Res. 2009;42(12):1995–2004. DOI: 10.1021/ar900253e.

[329] Eckenhoff WT, Eisenberg R. Molecular systems for light driven hydrogen production. Dalton Trans. 2012;41(42):13004–21. DOI: 10.1039/C2DT30823A.

[330] Zhang B, Sun L. Artificial photosynthesis: Opportunities and challenges of molecular catalysts. Chem Soc Rev. 2019;48(7):2216–64. DOI: 10.1039/C8CS00897C.

[331] Neese F. Prediction of molecular properties and molecular spectroscopy with density functional theory: From fundamental theory to exchange-coupling. Coord Chem Rev. 2009;253(5):526–63. DOI: 10.1016/j.ccr.2008.05.014.

[332] Barrozo A, Orio M. Molecular electrocatalysts for the hydrogen evolution reaction: Input from quantum chemistry. ChemSusChem. 2019;12(22):4905–15. DOI: 10.1002/cssc.201901828.

[333] Vaissier V, Van Voorhis T. Quantum chemical approaches to [NiFe] hydrogenase. Essays Biochem. 2017;61(2):293–303. DOI: 10.1042/EBC20160079.

[334] Segatta F, Cupellini L, Garavelli M, et al. Quantum chemical modeling of the photoinduced activity of multichromophoric biosystems. Chem Rev. 2019;119(16):9361–80. DOI: 10.1021/acs.chemrev.9b00135.

[335] Lutterman DA, Surendranath Y, Nocera DG. A self-healing oxygen-evolving catalyst. J Am Chem Soc. 2009;131(11):3838–39. DOI: 10.1021/ja900023k.

[336] Navarro RM, Alvarez-Galván MC, Villoria De La Mano JA, et al. A framework for visible-light water splitting. Energ Environ Sci. 2010;3(12):1865–82. DOI: 10.1039/C001123A.

Dominik Samuelis, Joachim Maier

9 Batteries: Concepts and Systems

9.1 Introduction

Imagine for a moment our world without electricity, and the significance of electrical energy in daily life will be quickly understood. This not only applies to energy supply or conversion as such, but also indirectly to powering modern communication technologies and thereby the basis of the nervous system of our society. As far as its key role for information, communication, and mobility is concerned, electricity is comparable with the role that bioelectrochemistry plays for our individual activity [1].

In the context of chemical energy storage, the transformation of electrical energy into chemical energy must be a matter of highest priority, the more so since electricity is hard to store and conversion into chemical energy represents a highly effective storage mode. This is particularly attractive if, within the same device, this can be reversed. Exactly this is possible in a rechargeable battery (also termed *secondary battery* or *accumulator*) [2].

Looking at the basic mechanism of energy storage in batteries, a simple comparison to a pumped-storage hydroelectric power plant can be made (see Figure 9.1). In this kind of power plant, water is stored at two different levels of gravitational energy. In the discharged state, all water is stored in a downhill basin. Upon *charging*, water is pumped to an uphill reservoir. In terms of energy, electrical energy drives a pump motor that produces mechanical energy, which ultimately is converted to potential energy. The *discharge* process works vice versa: water from the uphill basin drives a turbine connected to a generator producing electrical energy.

At a first glance, energy storage in batteries works in a very similar way. Just as the hydropower plant offers two levels of gravitational potential for storage of water, a battery offers two levels of chemical potential for an electrochemically active species, such as Li in a lithium battery. The chemical potential can be understood as a measure of how much a certain increase of species is disliked in a given system and

Acknowledgments: The authors would like to thank Dr Venkat Srinivasan (Lawrence Berkeley National Laboratory), Dr E. Peter Roth (Sandia National Laboratory), Dr Roland Wengenmayr, and Changbao Zhu (MPI-FKF) for providing figures. Thanks also to Dr Jelena Popovic for a critical reading of the manuscript.

Dominik Samuelis, High Performance Computing Center Stuttgart (HLRS), Nobelstr. 19, 70569 Stuttgart, Germany, e-mail: dominik.samuelis@heraeus.com
Joachim Maier, Max Planck Institute for Solid State Research, Heisenbergstraße 1, 70569 Stuttgart, Germany, e-mail: s.weiglein@fkf.mpg.de

https://doi.org/10.1515/9783110608458-009

therefore states how much (free) energy is released if the species is removed from that system (see Section 9.4 for more details).

The uphill reservoir, which is the state of high (gravitational) potential, corresponds to an electrochemical reservoir (= electrode), where an electrochemically active species is stored with high chemical potential (see Figure 9.10). This electrode is called negative electrode in a battery context, and during discharge, the anodic reaction takes place here.[1]

The corresponding downhill reservoir corresponds to an electrode where the electrochemically active species can be stored with a low respective chemical potential.

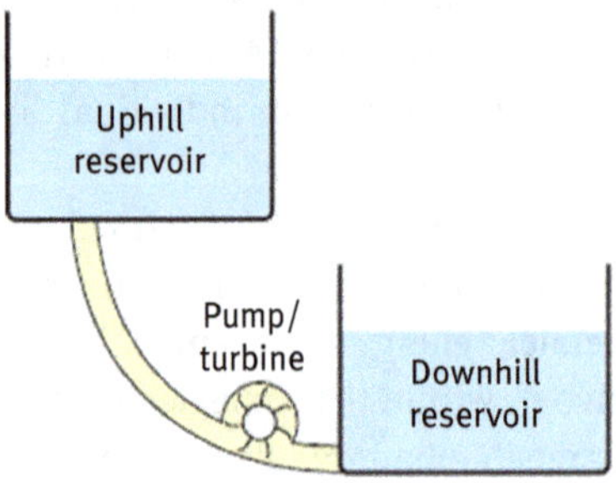

Figure 9.1: Pumped-storage hydroelectric plant.

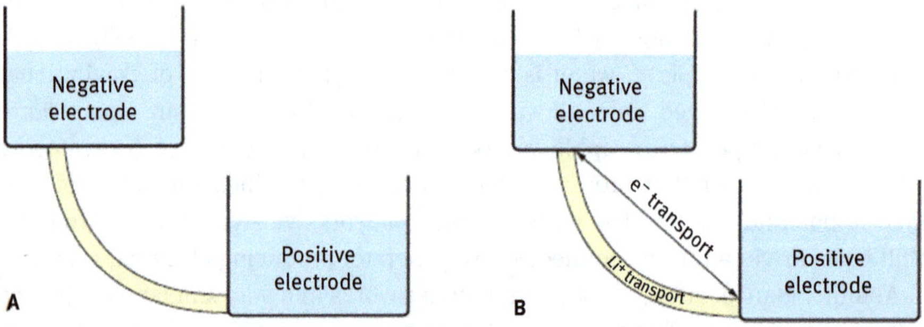

Figure 9.2: (A) Comparison battery – hydroelectric power plant; (B) an electrochemical storage system.

This is the positive electrode of a battery, with the cathodic reaction taking place during discharge [1].

In a hydroelectric power plant, the uphill and downhill reservoirs are connected via pipes and valves. Such a seemingly simple, direct connection does not apply for

1 Please note that the negative electrode strictly speaking acts as the anode only upon discharge, upon charge it acts as cathode. Still, in almost all of the battery literature, the terms "anode" and "negative electrode" are used synonymously, independent of the direction of current. Similarly, "cathode" and "positive electrode" are used synonymously as well.

an electrochemical energy storage device. If, in a charged battery, anode and cathode electrode came in direct contact, Li as the electrochemically active species would immediately flow from the negative to the positive electrode. The difference in chemical potential would be released in the form of heat. This heat could be used again to produce electricity, but only in a Carnot-limited process. The same would be true if the connecting phase (yellow in Figure 9.2A) was a material that is conductive for both Li^+ and e^-.

The secret of a battery is now to restrict the transport between positive and negative electrodes in a way that only the ionic species, but not electrons, can pass. For this, an electrolyte is put in place between the electrodes.

The electrons are then forced to flow through an external wire, where they directly can drive an external load (Figure 9.2B). In this way, the whole free-energy content of the electrochemical reaction theoretically can be converted to electrical energy.

Besides the efficiency, the enormous difference in energy density (per mass or volume) deserves to be emphasized. Assuming a height difference of 100 m, the gravimetric energy density is 981 N m kg^{-1}, or 0.27 W h kg^{-1} in units used in battery characterization. The enormously higher value of ~100 W h kg^{-1} for a typical electrochemical process expresses the high energy content of chemical bonds.

Since we will be concerned with rechargeable batteries in greater detail, we will briefly discuss – with an eye to Table 9.1 – electrochemical systems that fulfill the requirements of such a device only partly.

Table 9.1: Comparison of electrochemical cells used in energy conversion/energy storage applications.

Device	Conversion mode	Typical application
Electrolyzers	Electrical energy → chemical energy	Synthesis, storage
Primary batteries	Chemical energy → electrical energy	Consumer appliances, portable devices, medical applications (low cost, high reliability)
Fuel cells	Chemical energy → electrical energy (continuous supply of gaseous components)	Power plants, heat-power devices (efficient fuel conversion)
Primary metal-air	Chemical energy → electrical energy (chemical charging)	Fleet applications
Electrochemical capacitors	Chemical energy ⇄ electrical energy	High power density applications
Secondary batteries	Chemical energy ⇄ electrical energy	High energy density applications electromobility, portable devices, sustainable elements

Primary batteries are electrochemical elements that can only be discharged. After complete discharge, the device is flat and can only be chemically recycled. They are used whenever a certain advantage is so important that no compromise with regard to enabling rechargeability is tolerated. Examples are cost issues or reliability issues owing to low self-discharge and high stability. We only mention the Leclanché element (Li || MnO_2) or the classic heart-pacemaker battery [3] (Li || I_2 -adduct), the latter using the intrinsically formed LiI solid electrolyte, which self-heals on cracking. Related systems are fuel cells in which the active components are gases that can be continuously supplied. They are promising devices whenever chemical fuels (e.g. H_2) are to be efficiently "electrochemically burnt": The electrochemical cycle can be closed if the device is coupled with an electrolyzer that produces the active components out of the products. The classic Zn-air battery is an intermediate between fuel cell and primary battery in that the negative mass is solid zinc and the positive mass is gaseous oxygen. The product dissolves in the basic electrolyte (hydroxozincate in KOH). If such primary elements are used for a taxi fleet for instance, the chemical reforming into Zn and KOH is a sensible way of reloading [4].

In secondary batteries finally, on which the following consideration concentrates, the charging is done electrochemically by reversing the discharging process. This requirement is quite demanding in terms of kinetics and makes a compromise indispensable with respect to energy and power density. The requirement of reversibility renders well-defined solid matrix materials necessary as a consequence of which fuel cells typically outcompete batteries in terms of energy density. Nonetheless, the energy density of batteries is large compared with most other alternatives (let alone the hydropower plant) and is due to the high energy density of chemical bond breaking or forming in the solids. The resulting transport problems typically lead to comparatively low power densities. Electrical capacitors in which storage occurs at the electrolyte–electrode interface have quite opposite properties (fast charge–discharge, but low storage capacity), and combining both devices is advantageous in many examples. It is anticipated at this stage that modern nanotechnology makes both device concepts approach each other.

9.2 Secondary Battery Systems

Since 1803, when Ritter invented the "voltaic pile," rechargeable batteries have been known to exist [5]. The big breakthrough came in 1859, when Planté introduced the lead-acid battery [6]. Until today, this kind of secondary battery has been the most well-known electrochemical device. It is omnipresent in every car as a starter battery and also plays an important role in stationary energy storage (e.g. for uninterruptable power supplies).

Numerous other battery chemistries have evolved over time. The most prominent ones are assembled in Table 9.2. One possible categorization of battery technologies can be made according to the class of electrolyte they use. Here, we will distinguish between liquid aqueous, liquid nonaqueous, and solid electrolytes. To a certain degree, the phase state of the electrolyte determines the state of the electrodes. In general, it is advantageous to have a solid/liquid phase boundary between electrode and electrolyte because of much lower contact resistance in comparison to solid/solid contacts. Therefore, if the electrodes are solids, the electrolyte should be preferably liquid and vice versa.

As already mentioned, an important criterion to characterize the applicability of different battery technologies for certain applications is the energy density, which specifies how much electrical energy can be stored per mass or volume (gravimetric or volumetric storage capacity).

Another important criterion for the comparison of different storage technologies is the (gravimetric or volumetric) power density, which specifies how fast the electrode can be discharged or charged at a given mass or volume.

Within a given line of battery technology, energy and power density typically are inversely related. Increasing the energy density requires maximization of active electrode material mass per battery mass and thus requires minimization of the amount of additional components (such as current collector, conductive additives, or void space for optimizing electronic and ionic connectivity within an electrode material, see Section 9.5), most of which are beneficial for increased power densities.

This interdependence is captured in the so-called Ragone plot (Figure 9.3), where specific energy is plotted (usually double logarithmically) as a function of specific power.

Lead-acid, nickel-iron (Ni-Fe), nickel-cadmium (NiCd), and nickel-metal hydride (NiMH) batteries are the most important examples of batteries with aqueous electrolytes. In lead-acid batteries, the overall electrochemical reaction upon discharge consists of a comproportionation of Pb^0 and Pb^{4+} to Pb^{2+}. All nickel-containing battery reactions are based on the same cathodic reduction of Ni^{3+} to Ni^{2+}, but utilize different anodic reactions providing the electrons. Owing to toxicity and environmental concerns, the formerly widely used Cd^0/Cd^{2+} couple (NiCd cells) has been almost entirely replaced by H/H^+, with the hydrogen being stored in a special intermetallic compound (NiMH).

Nowadays, the most prominent battery technology is based on lithium storage and employs a nonaqueous liquid electrolyte. Before we focus on these systems, we will address selected rechargeable systems based on other elements.

Sodium-sulfur batteries use a solid Na^+ conductor ("Na-β-alumina") instead of a liquid electrolyte. This allows elevated temperatures (250 °C–300 °C) and liquid electrodes to be applied, namely molten sodium on the anode side, and molten sulfur on the cathode side (see Figure 9.4A). This not only ensures good contact but also allows one to accommodate the reaction products within the liquid cathode

Table 9.2: Compilation of the most important secondary battery technologies. Energy densities compiled from various manufacturers' data sheets and references [4, 7–9] (EC/DMC, ethylene carbonate/dimethyl carbonate; EMS (ethyl methyl sulfone)).

Battery technology	Electrolyte	Mobile species in electrolyte	Anode reaction during discharge	Cathode reaction during discharge	Standard cell potential/V	Gravimetric energy density/ Wh kg^{-1}	Notes
Lead acid	H_2SO_4 (aq)	SO_4^{2-}	$Pb + SO_4^{2-} \rightleftharpoons PbSO_4 + 2\,e^-$	$PbO_2 + SO_4^{2-} + 4\,H_3O^+ + 2e^- \rightleftharpoons PbSO_4 + 6\,H_2O$	2.1	30–40	
Li-ion *	$LiPF_6$ in EC/ DMC	Li^+	$LC_6 \rightleftharpoons C + Li^+ + e^-$	FePO4 + $Li^+ + e^- \rightleftharpoons$ LiFePO4	3.6	150–250	
				or: $CoO_2 + Li^+ + e^- \rightleftharpoons LiCoO_2$	4.2	150	
NiMH	KOH (aq)	OH^-	$2\,MH + 2\,OH^- \rightleftharpoons 2\,M + 2\,H_2O + 2e^-$	$2\,NiO(OH) + 2\,H_2O + 2\,e^- \rightleftharpoons 2\,Ni(OH)_2 + 2\,OH^-$	1.2	30–80	
NiCd	KOH (aq)	OH^-	$Cd + 2\,OH^- \rightleftharpoons Cd(OH)_2 + 2\,e^-$	$2\,NiO(OH) + 2\,H_2O + 2\,e^- \rightleftharpoons 2\,Ni(OH)_2 + 2\,OH^-$	1.2	40–60	
Ni–Fe	KOH (aq)	OH^-	$Fe + 2\,OH^- \rightleftharpoons Fe(OH)_2 + 2\,e^-$	$2\,NiO(OH) + 2\,H_2O + 2\,e^- \rightleftharpoons 2\,Ni(OH)_2 + 2\,OH^-$	1.2	50	
Na–sulfur	Na-β-alumina	Na^+	$2\,Na \rightleftharpoons 2\,Na^+ + 2\,e^-$	$2\,Na^+ + x\,S + 2\,e^- \rightleftharpoons NaS_x$	2.1	150	Operating temperature: ~300 °C
$Na/NiCl_2$ (ZEBRA)	Na-β-alumina, molten $NaAlCl_4$	Na^+	$2\,Na \rightleftharpoons 2\,Na^+ + 2\,e^-$	$2\,Na^+ + NiCl_2 + 2\,e^- \rightleftharpoons 2\,NaCl + Ni$	2.6	90	Operating temperature: ~250 °C
Zn–bromine flow battery	ZnBr (aq)	Zn^{2+}, Br^-	$Zn \rightleftharpoons Zn^{2+} + 2\,e^-$	Br_2 (liq) $+ 2\,e^- \rightleftharpoons 2\,Br^-$	1.8	75–85	

Vanadium redox flow	Nafion ®	H^+	$V^{2+} \rightleftharpoons V^{3+} + e^-$ (in H_2SO_4)	$VO_2^+ + 2\,H_3O^+ + e^- \rightleftharpoons VO^{2+} + 3\,H_2O$	1.2–1.6	25–35	
Zn-air	KOH (aq)	OH^-	$2\,Zn + 8\,OH^- \rightleftharpoons 2\,Zn(OH)4^{2-} + 4\,e^-$	$O_2 + 2\,H_2O + 4\,e^- \rightleftharpoons 4\,OH^-$	1.6	450	"Chemical" rechargeability by replacing the Zn anode.
Li-air	Solid Li^+ conductor	Li^+	$2\,Li \rightleftharpoons 2\,Li^+ + 2\,e^-$	$2\,Li^+ + 2\,e^- + O_2 \rightleftharpoons (g)\,Li_2O_2$	2.7	750–2000	
Li-sulfur	$LiPF_6$ in EMS	Li^+	$2\,Li \rightleftharpoons 2\,Li^+ + 2\,e^-$	$2\,Li^+ + x\,S + 2\,e^- \rightleftharpoons LiSx$	2.0	350	

*The wide variety of Li-ion battery chemistries is discussed in the following section "Lithium Batteries."
M = (La,Ce,Nd,Pr)(Ni,Co,Mn,Al)$_5$ alloy.

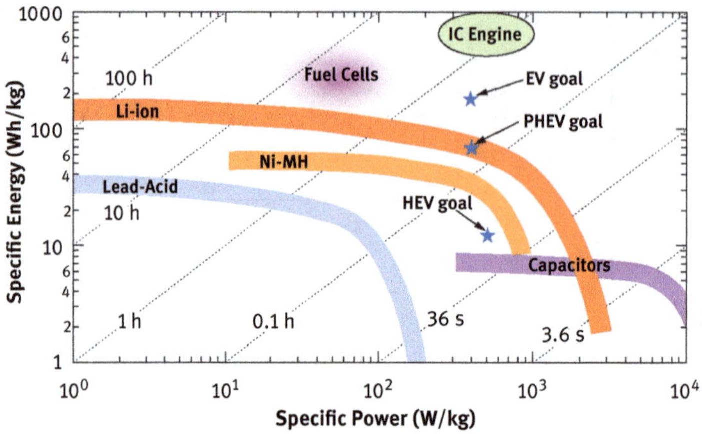

Figure 9.3: Ragone plot for different energy storage solutions. The stars show specific energy and power goals for different vehicle technologies (EV: electric vehicle; PHEV: plug-in hybrid EV; HEV: hybrid EV); compared to internal combustion (IC) engine. (V. Srinivasan, Berkeley Electrochemical Research Council). Used with author's permission.

phase. These systems show excellent energy and power densities along with excellent cyclabilities, having made them candidates for driving electric cars. Unfortunately, they suffer from potential cracking of the ceramic electrolytic membrane leading to catastrophic failure. The latter is greatly diminished in the sodium-nickel chloride (ZEBRA) cell (Figure 9.4B), which relies on the same negative electrode and electrolyte but is based on Na reacting with $NiCl_2$ to Ni and NaCl. On the cathode side, an additional $NiAlCl_4$ melt is introduced that not only accommodates the

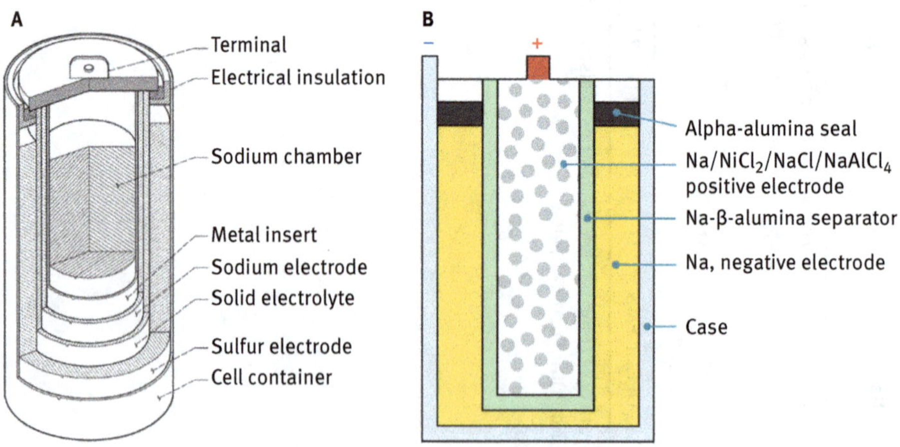

Figure 9.4: (A) Sodium sulfur cell. Source: NASA John Glenn Research Center, public domain. (B) Na/$NiCl_2$ cell. Source: [10].

NaCl but also, in the case of failure, converts any Na to NaCl and Al. The metallic phase formed prevents electrical interruption.

Zn-bromine flow and vanadium redox flow are special cases of secondary batteries. Here, liquid electrode materials are used on one (Zn-Br flow) or both sides (V redox flow) of the electrochemical cell. In contrast to regular batteries, which are typically completely closed systems, the liquid electrode materials in flow batteries are circulated and replenished from tanks (Figure 9.5). Therefore, the flow batteries possess large electrodes, the effective size of which is just limited by the volume of those tanks. This partly decouples energy and power capabilities of the batteries, allowing one to optimize both separately.

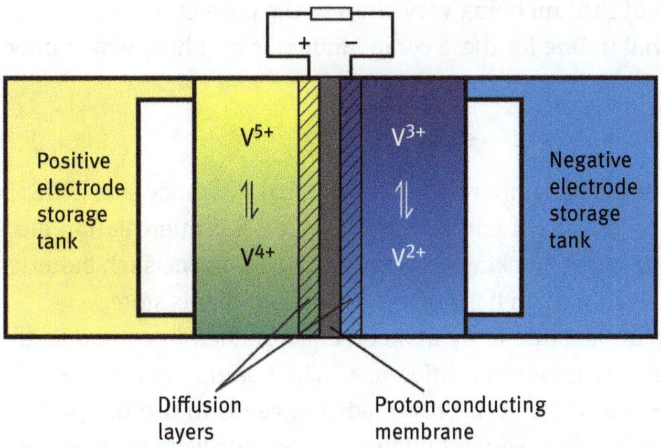

Figure 9.5: Vanadium redox flow battery.

These types of batteries fulfill the requirements for reversible storage clusters, which are needed for buffering fluctuations of solar or wind power plants. Here, cost, safety, and energy density, rather than power density, are the primary considerations.

9.3 Lithium Batteries

With regard to energy and power density, Li as the element to be stored is the element of choice because, according to its position in the periodic table, Li is light, small, and very electropositive. These are excellent prerequisites for low weight, high voltages, fast transport, and easy accommodation in solid phases.

Early lithium batteries were already being investigated in the 1970s, back then using TiS_2 as the positive electrode and lithium metal as the negative electrode material [11]. Here, Li is dissolved in the solid sulfide with Li^+occupying interstitial sites and e^- reducing Ti^{4+}. With an open circuit voltage (OCV) of only approximately

2.6 V, TiS_2 was soon to be replaced by the structurally similar CoO_2, which gave an OCV of 4.7 V versus Li metal [12].

At the same time, it was found that Li metal is an unsafe negative electrode material [11, 13]. Upon charging a battery, lithium is electrodeposited on the Li metal electrode. During this process, formation of dendrites was observed, and after multiple charge/discharge cycles, those dendrites penetrated the separator and led to an internal short circuit of the cell with intense heat formation and sometimes even combustion of the cell.

It was recognized that this dendrite problem could not be solved without completely changing the negative electrode. Instead of electrodepositing metal, the anode reaction of choice was found to be storage of lithium in compounds with chemical potentials of lithium being very close to the pure Li metal.

The first anode material in line for these compounds was graphite, which intercalates Li according to

$$6C + Li^{+} + e^{-} \rightleftharpoons LiC_6 \tag{9.1}$$

As the storage on the anode side is similar to the storage on the cathode side if TiS_2 is used, the lithium is rocked backward and forward between two intercalation electrodes. This gave rise to the name "rocking chair battery." Nowadays, such batteries are termed *Li-ion battery*, even though it is always $Li = Li^{+} + e^{-}$ that is stored.

The replacement of pure metallic Li by Li metal encapsulated in carbon as the anode led to a technologically important difference. While originally metallic lithium was the lithium source in the cell, it proved advantageous to use the positive electrode as the Li source. In other words, lithiated compounds were used as positive electrodes, and the cell was manufactured in the discharged state.

The first large-scale commercial lithium battery was built in the 1980s by Sony. It employed graphite as the negative electrode and $LiCoO_2$ as the positive electrode (see Figure 9.6). This is still the prevailing battery concept, especially in consumer electronics.

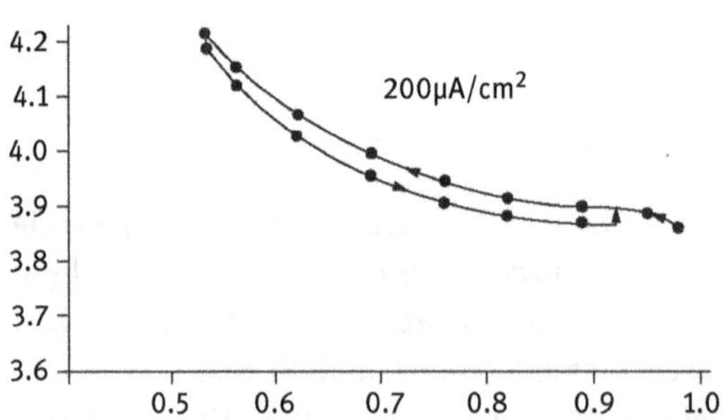

Figure 9.6: Discharge curve of Li_xCoO_2 positive electrode material, with Li metal as negative electrode. Reprinted from [12] with permission, copyright 1980 Elsevier.

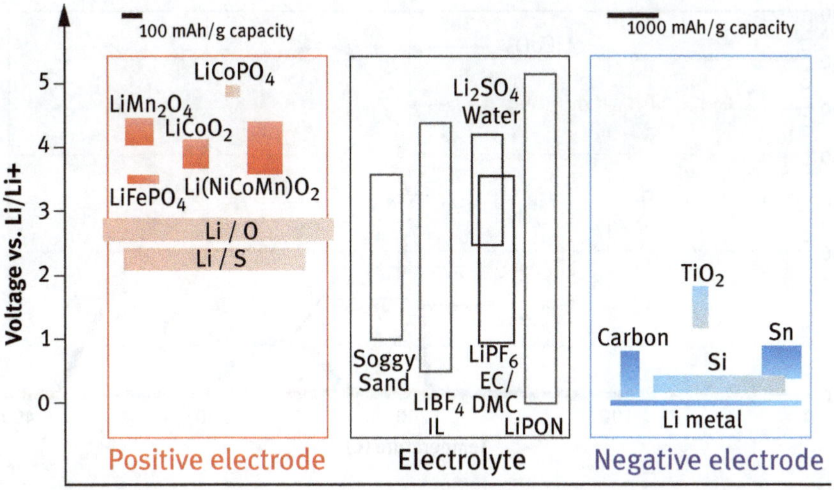

Figure 9.7: Selected cathode and anode materials and electrolytes used in today's lithium batteries (darker colors) and possibly useful future technologies (lighter colors). For electrode materials, the width of the boxes displays demonstrated capacities (note the different scales for positive (cathode) and negative (anode) electrode materials). The height of the boxes corresponds to the half-cell potential and its variation for different states of charge. References: [12–20].

Numerous other electrode materials for both negative and positive electrodes have been developed since then. A compilation of the most significant developments is shown in Figure 9.7.

Among the most significant novel developments for positive electrode materials are the layered oxide $LiNi_{1/3}Co_{1/3}Mn_{1/3}O_2$ (LiNCM) [20, 21] and the olivine-structured $LiFePO_4$ [14]. A main reason for moving away from $LiCoO_2$ (besides environmental concerns) is safety aspects. $LiCoO_2$ tends to release O_2 gas under overcharge conditions and – even worse – under high thermal load, possibly leading to exothermic reactions with the electrolyte inside the cell that increase the temperature even further (the so-called thermal runaway, see Figure 9.8). $LiFePO_4$ does not exhibit such behavior, and for LiNCM, the onset temperature of thermal runaway is increased by almost 70 K compared to $LiCoO_2$.

As shown in Figure 9.7, the capacities for most cathode materials do not differ drastically. From the 1980s when $LiCoO_2$ (125 mA h g^{-1}) was produced to recently, with the development of LiNCM (180 mA h g^{-1}), capacities were increased by less than 50%. From chemical and structural considerations, there is a clear reason not to expect drastic increases in electrode capacities within these classes of materials (and storage mechanisms). For example, the lithium insertion and extraction reaction in $LiFePO_4$ reads

$$LiFePO_4 \rightleftharpoons FePO_4 + Li^+ + e^- \tag{9.2}$$

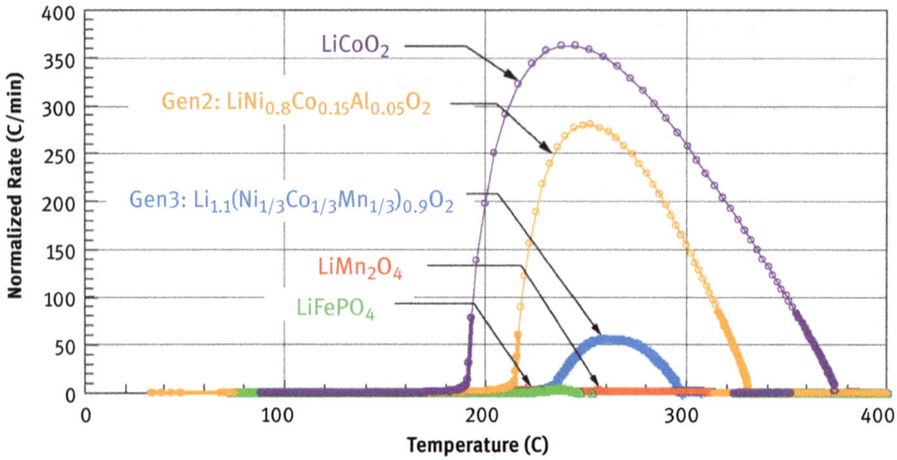

Figure 9.8: Thermal runaway behavior of the most common positive electrode materials upon heating at 5 K min^{-1}. Source: E. Peter Roth, Power Sources Technology Group, Sandia National Laboratories [22]. Used with author's permission.

During Li extraction, Fe^{2+} is oxidized to Fe^{3+}. Generally, the electrode material should have a transition metal as a constituent that could exist in at least two valence states. At the same time, the crystal structure of the electrode material must not change too drastically between the lithiated and delithiated state. This is trivially the case if no phase transition is involved as in $LiCoO_2$ [12].

Drastic increases in capacity could, for example, be achieved by using electrode materials, where the transition metal releases two electrons upon Li extraction. This was believed to be the case in the Li_2MnSiO_4 system [23], in which a change from Mn^{2+} to Mn^{4+} was postulated. Unfortunately, only one Li^+/e^- can really be extracted from this structure. At higher delithiation, the crystal structure of the electrode material is destroyed and does not regenerate upon Li reinsertion [24].

This is a general feature in positive electrode research. While many materials tolerate a one-electron transfer, a two-electron transfer usually leads to mechanical instability owing to volume variation or makes the material chemically unstable.

Therefore, another important degree of freedom in improving energy density comes to the fore, namely increasing the cell voltage. Using different redox couples in the transition metal cathode allows one to adjust the cell voltage. The Co^{3+}/Co^{4+} couple in $LiCoO_2$, for example, provides an operating potential of approximately 4.3 V versus Li/Li^+. Using a Co^{2+}/Co^{3+} couple in $LiCoPO_4$ changes the cell voltage to almost 5 V versus Li/Li^+. At the moment, however, there is a significant drawback with using such higher cell-potential cells. As shown in Figure 9.7, the cell potential of a possible $LiCoPO_4$ electrode is outside of the stability range of any of the given liquid electrolytes. Electrolytes based on ionic liquids [25] might be promising candidates for such a system, with additional benefits, such as higher safety due to not containing flammable organic solvents, but also drawbacks, mostly reduced conductivity (even if additional salts are used) due to higher viscosity of the medium.

Oxides such as TiO_2 are investigated as negative electrodes as well [26–32]. Advantages are environmental safety and easy nanostructuring; the disadvantage is the reduced cell potential as seen in Figure 9.7. High storage capacities are offered by Sn or Si [33, 34] because the formed Li alloys exhibit maximal storage stoichiometries of up to $Li_{4.4}Si$ (i.e. they accept more than 4 Li atoms per formula unit). Mastering the immense volume changes is the greatest challenge here.

As already mentioned, salt-containing liquid solvents are typically used as electrolytes. The most prominent example is $LiPF_6$ as a conductive salt, dissolved in a 1:1 mixture of ethylene carbonate (EC) and dimethyl carbonate (DMC) as 1 molar solution. It should be mentioned that this electrolyte is not thermodynamically stable in contact with lithium or, for example, LiC_6. Its success comes from the fact that it forms an extremely stable passivation layer on top of the electrode, the so-called solid–electrolyte interface (SEI) [35]. Key properties of such SEI layers are high Li^+ and very low e^- conductivity – that is, they act as additional electrolyte films, where the electrode potential drops to a level the liquid electrolyte can withstand [36].

As will be outlined later, high-performance electrodes use small particles, and it is hard to imagine that solid electrolytes can provide sufficiently good contact with respect to a nanopowder arrangement. Solid electrolytes [16] such as LiPON are, however, favorable for thin film cells for low-power applications or as additional separators, and they are stable even in contact with metallic Li [17].

An excellent compromise is provided by liquid–solid composites [37], in particular by the soggy sand electrolytes [19], which not only show partial shapability, but also even higher Li^+ conductivity. Ion-exchange polymers allow exchanging the conductive proton by Li^+ [38]. Both types of electrolytes also show very low counter-ion contributions to the conductivity, which diminishes the electrolyte polarization [39]. A large portion of counter-ion conductivity leads to a concentration gradient of the salt during current flow that might even lead to precipitation.

Typical commercial lithium battery cells nowadays have capacities >3 A h. At positive electrode capacities of ~150 mA h g^{-1}, and typical electrode loading on the order of 4×10^{-2} g cm^{-2} (double sided), electrode areas as high as 500 cm^2 are required. Figure 9.9 shows a cylindrical cell, where the electrodes are wound around a core. Similarly, in case of prismatic cells, electrodes are folded or stacked on top of each other. For both principles, the current collector foils are coated on both sides with electrode material, and winding or folding leads to an extended stack of single cells. Positive and negative electrodes are prevented from touching by a polymer separator, which is soaked with the liquid electrolyte. As mentioned in the introduction, a direct contact of positive and negative electrodes would lead to an uncontrolled release of the chemical energy stored in the battery in the form of heat. Preventing direct contact between positive and negative electrodes is therefore extremely important to cell safety, and multiple optimizations such as using polymer/ceramic composite materials as separators have evolved [40]. Alternatively, using a shapable electrolyte such as the above-mentioned soggy sand electrolytes, or polymer electrolytes [41], could make the separator dispensable.

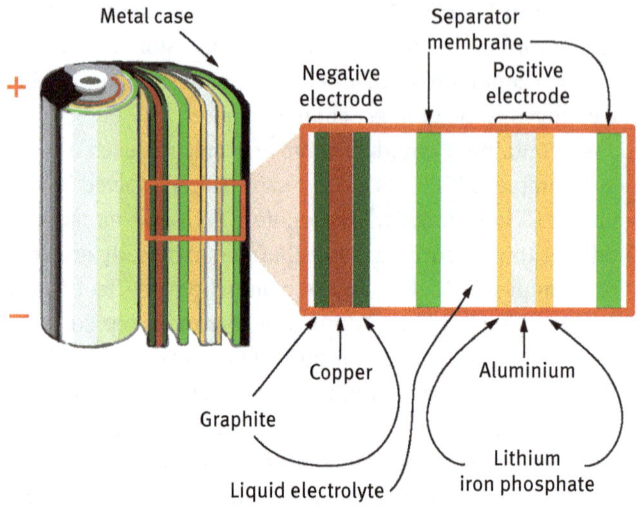

Figure 9.9: Building principle of a cylindrical lithium-ion battery.

In the next section, we will also address nonconventional storage modes (conversion reaction, interfacial storage), which can provide significant energy densities. Yet it remains to be seen whether or not they can be made reversible enough as to compete with single- and two-phase storage.

9.4 Thermodynamics of Electrochemical Energy Storage

The equilibrium situation in an electrochemical cell is obtained, if the electrical current is interrupted, if all local actions (e.g. transport in the electrode) have come to an end and no internal short circuits occur. Then, as mentioned (Figure 9.10), the cell voltage is determined by the difference in the lithium potential (chemical potential of lithium) between the left-hand side (*lhs*) and right-hand side (*rhs*) of the electrochemical cell (E = open cell voltage, F = Faraday constant):

$$EF = -(\mu_{\mathrm{Li}}[\mathrm{lhs}] - \mu_{\mathrm{Li}}[\mathrm{rhs}]) \tag{9.3}$$

This intelligible equation is obtained rigorously when splitting the chemical potential of lithium into the contributions of the lithium ions and the electrons ($\mu_{\mathrm{Li}} = \tilde{\mu}_{\mathrm{Li}^+} + \tilde{\mu}_{\mathrm{e}^-}$);

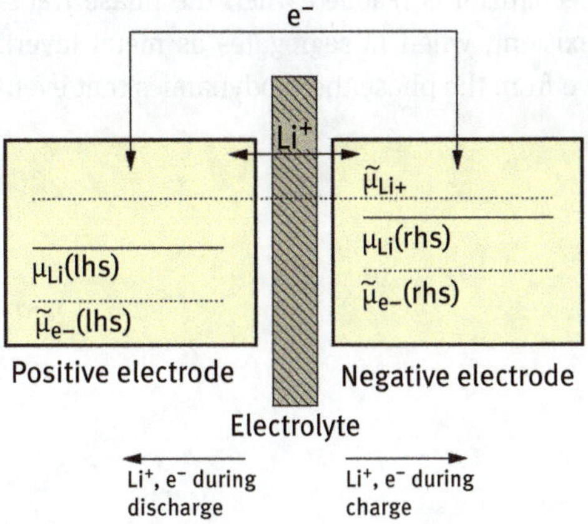

Figure 9.10: Chemical (μ)/electrochemical ($\tilde{\mu}$) potentials of Li, Li^+, and electrons.

the difference of the first vanishes due to the presence of the electrolyte, and the difference of the second determines the voltage measured.

The bracketed difference is the Gibbs energy of the reaction that would occur on direct contact. As the Gibbs energy differs from the reaction enthalpy by $T\Delta S$ (T: thermodynamic temperature, ΔS: reaction entropy), a positive ΔS would result in theoretical efficiencies greater than 100% (then the environment would cool). Usually, those efficiencies are approximately 100%, as ΔS is negligible. This high theoretical efficiency is another asset of electrochemical conversion devices.

The second quantity besides cell voltage that is important here is the equilibrium storage capacity. This quantity follows from the defect chemistry and phase thermodynamics.

Figure 9.11 shows the storage modes of relevance. The already mentioned intercalation mode belongs to the class of single-phase mechanism. This means homogeneous storage by a phase-absorbing Li (e.g. Li_xCoO_2). This is both a redox reaction (accumulation of e^- by introducing excess electrons and/or annihilating electron holes; in praxi it almost always coincides with reduction of transitions metal elements) as well as an acid–base reaction (in the sense of [42]; accommodation of Li^+ by occupying interstitial sites in the lattice and/or annihilating vacancies).

Let us for simplicity assume the storage to solely occur through

$$\text{Li} \rightleftharpoons \text{Li}^+\,(\text{interstitial site}) + \text{e}^-\,(\text{excess electron}) \tag{9.4}$$

If the two defects on the right-hand side are dilute, an ideal mass action law can be formulated. The defect concentrations can be connected with the nonstoichiometry ($\delta = i - \nu$; i: interstitial Lithium concentration; ν: Lithium vacancy concentration), which is proportional to the storage capacity, while the lithium activity (see the left side of eq. (9.4)) relates to the cell voltage. The maximum capacity (maximum nonstoichiometry) of the single-phase mechanism is reached when the phase transforms into a Li-richer phase (or if not existent, when Li segregates as metal layer). These concentrations of coexistence derive from the phase thermodynamics that eventually is also based on defect chemistry.

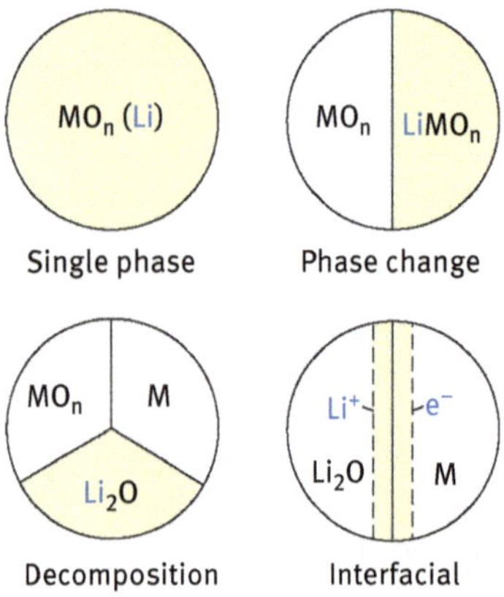

Figure 9.11: Relevant storage modes in lithium batteries. Reprinted from [73] with permission.

The end of solubility is not the end of storage as the transformation into neighboring phases can also be used as a storage mode. A prominent example is $FePO_4$/ $LiFePO_4$. If only Li is mobile, the voltage stays constant during the transformation, as μ_{Li} is invariant. It proves simpler in such cases to connect E with the phase reaction ($Li(s) + FePO_4 \rightleftharpoons LiFePO_4$). This is, however, identical with eq. (9.3) as the chemical potential of Li in the electrode can be written as the difference of the chemical potentials of the two phases. During the phase reaction, the individual nonstoichiometries of $FePO_4$ ($Li_\varepsilon FePO4$) and $LiFePO_4$ ($Li_{1-\delta}FePO_4$) do not change; rather it is the phase fraction that varies and determines the storage capacity. As the local nonstoichiometries define the transport coefficients, defect chemistry is also relevant for this storage mode, a point that is taken up in the Section 9.5.

If the Li-poor component has been fully transformed, then another single-phase mechanism sets in (dissolution of Li in the Li-richer phase) before a further phase reaction occurs. The abovementioned phase reaction is very simple as only $FePO_4$/$LiFePO_4$ interfaces are generated. The same occurs for RuO_2/$LiRuO_2$; further Li incorporation in the last case, however, leads to a true decomposition, namely, reduction to Ru and Li_2O. Here three phases are in equilibrium and a complex morphology evolves [43–46]. This conversion reaction to be used in a rechargeable mode is kinetically extremely demanding. Fortunately, these conversion reactions automatically lead to extreme nanostructuring. Nonetheless, RuO_2, Fe_2O_3, and FeF_3 are the only examples where a reasonable reversibility could be achieved. Performing further research is highly relevant as the storage capacity of this mechanism is huge. Having finally formed the metal, alloying is a further single-phase mechanism of relevance for negative electrodes (Sn, Si).

A last storage mode is of interest if nanocomposites are addressed – that is, excess storage at interfaces. A mechanism of general significance is the job-sharing mechanism that allows Li to be stored heterogeneously at a contact of two phases, α and β, while both individually would not store Li. Ru, for example, can store e^- but not Li^+, and Li_2O can store Li^+ but not e^-; still, the nanocomposite can heterogeneously store Li to quite an amount [47–49].

Again, defect chemistry can be exploited to give the voltage-capacity relation. Here, however, electroneutrality has to be replaced by Poisson's equation (for details, the reader is referred to the references [50–51]).

Naturally, also the very core of the interfaces might store Li at potentials still higher than metallic lithium. The same can happen at dislocation cores, pores, or frozen-in point defects (cf. Li-storage in carbon vacancies [52]).

The aptitude of a given storage mechanism is mainly determined by the cell voltage connected with it (positive electrodes are typically in the range of 2–5 V vs. Li/Li^+; negative electrodes in the range of 0–1.5 V), the respective storage capacity connected with it, and its reversibility.

9.5 Kinetics of Energy Storage

In contrast to Section 9.4, where equilibrium thermodynamics (i.e. a current-less state was discussed), this section deals with losses that occur while current is flowing. A nonzero current leads to losses (overpotentials) that can be translated into resistances. These kinetic contributions lead to $U < E$ (U: terminal voltage, I: current) for discharge ($I > 0$) and to $U > E$ for the charge process $(1 < 0)$.

$$U(I) = E - I \sum_j R_j(I) \tag{9.5}$$

In other words, under realistic conditions ($I = 0$), entropy is produced, with the positive entropy production being given by fluxes and forces related to the process j. Equation (9.5) assumes that all these processes are in series, which is mostly correct. The most obvious contributions are transport resistances, due to the finite conductivities of Li^+ and e^- in electrolyte and electrodes. For usual geometries, these resistances are constant to a good approximation, while for resistances stemming from impeded charge transfer and phase boundaries, the dependence on current can be severe.

Diffusion resistances can occur for Li in the electrode, but also for the salt in the electrolyte (if anion conductivity in the electrolyte is significant). Further effects are due to depletion of carriers at a phase boundary. In such cases, time dependencies of the electrical properties occur (in addition to Rs, effective capacitances Cs also appear). The same is true for impeded nucleation processes. Since any potential step of the electrochemical potential can be connected with current-dependent effective resistances and capacitances, the kinetic description is typically very specific and complex. As the storage processes in Li-based batteries are solid-state processes, the significance of solid-state diffusion should be emphasized. Indeed, the sheer chemical diffusion of lithium appears to be rate determining in many cases. As the diffusion relaxation time is proportional to (diffusion length)^2/diffusion coefficient, one can mitigate the diffusion problems by either modifying the materials parameters, or morphology and geometry. This brings us to the consideration of the optimization potential of materials research for batteries.

9.6 Materials Optimization: Adjusting Screws

There are multiple ways of optimizing thermodynamic and kinetic limitations in the battery materials, as described in Sections 9.4 and 9.5. A chemist would of course first attempt to generate entirely new compounds. However, due to the multitude of constraints, only very few electrode materials have proved successful over the years, as shown in Figure 9.7.

An often very helpful strategy therefore is to modify existing materials. Compared to new compounds, such attempts would primarily not address thermodynamic quantities, such as open cell voltage. Instead, kinetic parameters can be optimized, and quantities such as practical cell voltage, electrode capacity, and power density can be drastically improved.

All those quantities are highly dependent on the charge carrier concentrations and conductivities in the electrode materials. Increase of chemical complexity by doping, for example, gives means to change carrier concentrations and therefore materials properties over a wide range, as demonstrated in the properties of doped silicon.

Before doping can be applied as a tool for improving materials properties, the relations between carrier and dopant concentrations have to be understood. Systematic defect chemical investigations reveal the thermodynamic equilibria that describe those relations. Unlike materials in high-temperature chemistry, materials relevant for Li-based batteries largely contain frozen compounds (so Fe:P:O in $LiFePO_4$ does not vary during performance). Frozen native defects hence can be considered dopants (e.g. oxygen vacancies in TiO_2) [32]. Another characteristic low-temperature feature is the association of oppositely charged defects such as trapping of holes by lithium vacancies.

Figure 9.12 displays the results of such defect chemical investigations on $LiFePO_4$ and $FePO_4$, in the form of a so-called Kroger–Vink or Brouwer diagram [53–55]. Key information from this diagram is, for example, that acceptor doping can be used to increase the electronic conductivity of the material, at the expense of lithium conductivity, however, which in turn is decreased (cf. $V'_{Li} = Li^+$ vacancy and $h^\bullet$ = electron hole in $LiFePO_4$ or Li_i = excess Li^+ and e' = excess electron in $FePO_4$). This is quite a general result: For large Li excess as for large deficiency, the electronic and ionic defects of relevance are of opposite charge and hence affected in opposite ways by doping. As both carriers are needed for chemical Li diffusion, the effect of doping on chemical diffusion is thus limited.

Not only the increase of chemical, but also the increase of morphological complexity can lead to altered charge carrier concentrations in materials. This is because surface layers or interfaces to other phases in general generate charge redistribution, allowing for heterogeneously doping materials [56]. This of course works best in nanomaterials, where the interfacial density is high [51, 52]. Again, the effect of two oppositely charged carriers is opposite. Yet the potential of heterogeneous doping has been revealed in the generation of soggy sand electrolytes [19, 57]. In these semisolid–liquid materials, anions are adsorbed at the surface of oxide particles admixed to the salt-containing organic liquids. Hence, Li^+ conductivity is increased. The decreased counter-ion conductivity is very favorable here.

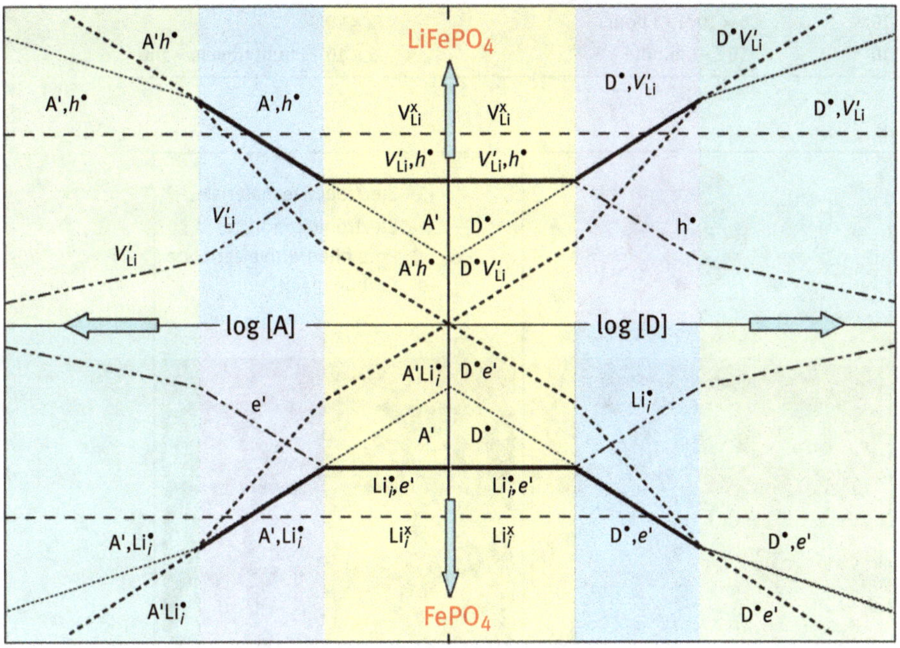

Figure 9.12: The variation of charge carrier ("defect") concentrations in $LiFePO_4$ and $FePO_4$ by acceptor [A] or donor [D] doping shows a wide range of modified properties for the materials. This is to illustrate the sheer complexity of the defect equilibria in battery materials. For details, the reader is referred to the references [53, 54].

In most cases, morphological optimizations are done not in order to vary materials properties, but rather to vary the electrodes' network topology. The effects achieved are quite often really drastic. As mentioned, solid-state chemical diffusion of lithium is a critical step for almost all solid battery electrodes. Hence, shortened transport length for solid-state diffusion is a key to better electrode kinetics and is particularly important for high-power applications. Note that downsizing electrodes from 1 mm to 10 nm reduces the diffusion time by 10 orders of magnitude (Figure 9.13A), but only if ions and electrons are brought quickly enough to the squillions of particles. This requires a very efficient ionic/electronic wiring. So nanostructuring combined with morphological preparation will be a key strategy for improving power density.

A

$\tilde{D}_{Li}$ / $cm^2\,s^{-1}$	t/s for 1 mm particles	t/s for 10 nm particles
10^{-20}	5×10^{17}: ~4x earth's age	5×10^{7}: ~duration of PhD work
10^{-15}	5×10^{12}: ~age of Homo sapiens species	5×10^{2}: ~10 minutes
10^{-10}	5×10^{7} : ~duration of PhD work	5×10^{-3}: ~in a trice
10^{-8}	5×10^{5} : ~1 week	5×10^{-5}
10^{-6}	5×10^{3} : ~1 hour	5×10^{-7}
10^{-4}	50 : ~1 minute	5×10^{-9}: light travels ~ 1m

B

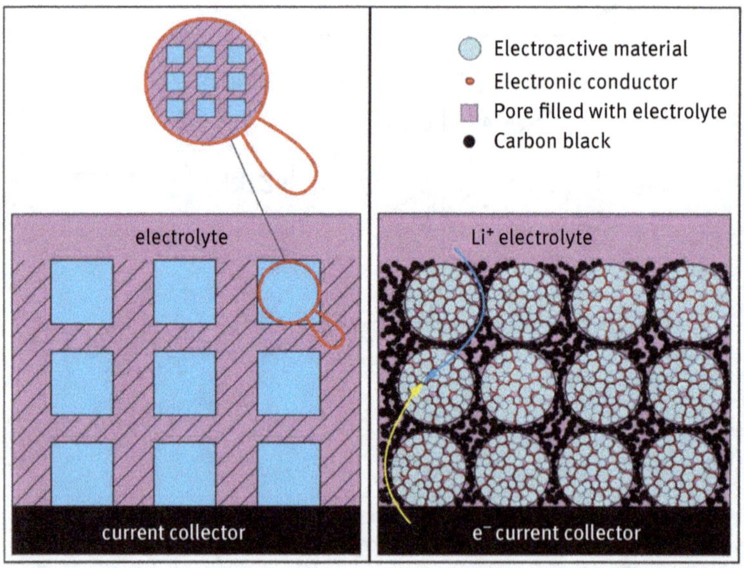

Figure 9.13: (A) Equilibrium times for diffusion on macroscopic (1 mm) and nanoscopic (10 nm) length scales. (B) Illustration of ionic and electronic wiring, with hierarchical porosity as Li^+ distribution network and a carbon second-phase e^- distribution network. Reprinted from [58] with permission, copyright 2007 John Wiley & Sons.

As far as theoretical cell voltage is concerned, the effect of nanostructuring on the theoretical voltage is quite small [59]. This may be different for amorphous materials as here the total long-range order is lost [47, 48]. (In a way, the thermodynamic effect of amorphous materials can be viewed as that of nanocrystallinity with atomistically small size [47].) Similarly, the storage capacity can be very different for small nanocrystals or amorphous materials. In addition, there is a general tendency of diminished if not vanished miscibility gaps [60–68].

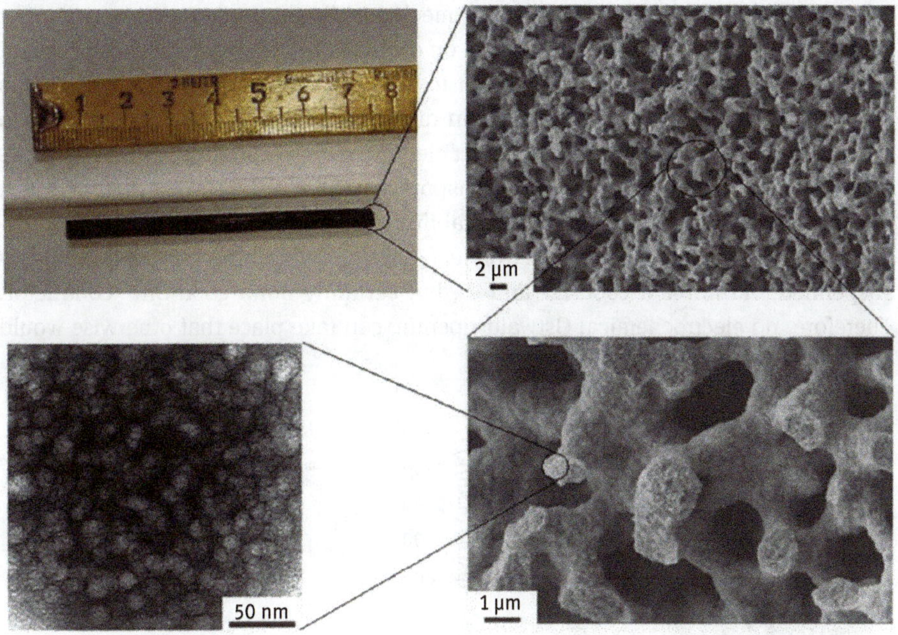

Figure 9.14: Carbon monolith with hierarchical porosity. Reprinted from [69] with permission, copyright 2007 Wiley-VCH.

Beyond nanosizing, the optimization of morphology is possible (e.g. by directed introduction of porosity into materials) [58, 69, 70]. In combination with a liquid electrolyte, which can penetrate the pores, distribution networks for Li^+ ions can be formed. With hierarchical pores, such distribution networks resemble the network of blood vessels in the human body, which by itself is a highly efficient distribution network (see Figure 9.14).

Especially for electrode materials with low electronic conductivity (such as $LiFePO_4$, TiO_2, and many others), not only lithium transport, but also electron transport between current collector and electrode particle is an issue. While the admixing of carbon is helpful on a micrometer scale, the electronic wiring on a nanometer scale requires special preparation procedures [31, 32, 71]. While the admixed electronic phases decrease the energy density, enhancing the electronic conductivity of the material (at the surface of in the whole material) is helpful here even if the chemical diffusion coefficient may only be slightly varied. An example is the elegant way of improving electronic conductivity by the introduction of a frozen-in oxygen nonstoichiometry into TiO_2 [32].

Figure 9.15 gives a particularly instructive example [33, 34]. It refers to Sn as anode material that even in the form of (commercial) nanopowders does not show a useful cyclability. This is in sharp contrast to the morphology shown in Figure 9.15. The morphology leads to optimization by improving not less than seven battery-relevant parameters: (1) Sn particles are mechanically decoupled and do not suffer from pulverization upon volume change on cycling. (2) Carbon provides an efficient way of electron transport along the fiber to the tin particle. (3) Li^+ transport occurs perpendicular to the surface; only a short distance in the Li-permeable carbon is to be traversed, leading to an almost perfect electrochemical coupling. (4) Tin particles are small enough to ensure quick Li transport therein. (5) The carbon fiber acts as a binder and guarantees morphological stability. (6) Carbon provides a stable passivation layer in contact with the electrolyte. (7) The growth of the tin nanoparticles is suspended; carbon is a good Li^+ and e^-, but certainly not a good Sn^{4+} conductor. Therefore, no electrochemical Ostwald ripening can take place that otherwise would cause the elemental particles to grow [72].

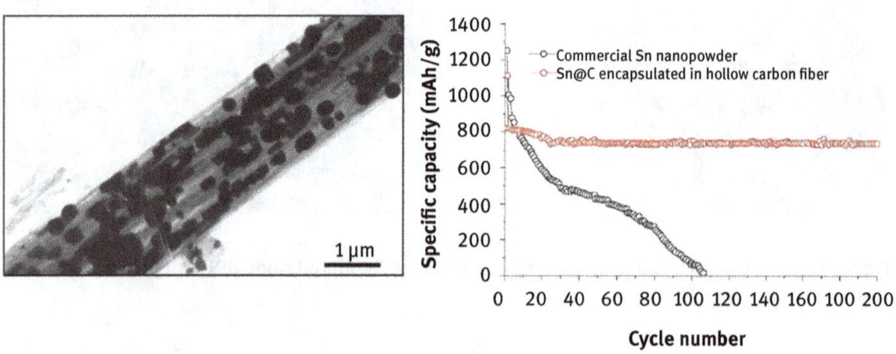

Figure 9.15: Tin nanoparticles in an electrospun carbon fiber for use as a negative electrode (lhs, reprinted with permission from [34], copyright 2009 John Wiley & Sons; rhs, reprinted with permission from [33], copyright 2009 American Chemical Society).

9.7 Outlook

The field of rechargeable batteries is – even though having experienced an enormous push – not at all a new technology. Owing to limits set by the periodic table, energy densities cannot be increased by orders of magnitude, unless one gives up the enormous reversibility achievable in modern batteries. Great improvements are, however, possible as regards kinetics and hence power density. The progress in this exciting field is to be achieved by the highly interdisciplinary research embracing chemistry, physics, and materials science, whereby solid-state electrochemistry will play the key role.

Dealing with the fundamentals – what this contribution does – has the advantage that its context is not getting outdated soon. This impression is corroborated after reading the text in the course of the newest edition. It appears that more novel applications focus on using solid electrolytes and Li-metal electrodes. Also the significance of Na-based batteries became more obvious. The reader is referred to Refs. [J. Janek and W. Zeier, Nat. Energy 1 (2016) 16141, T. Krauskopf, F.H. Richter, W.G. Zeier, and J Janek, Chem. Rev. 120 (2020) 7745], R. Usiskin, Y. Lu, J. Popovic, M. Law, P. Balaya, Y.-S. Hu, and J. Maier, Nature Review Materials 6 (2021) 1020. As far as conceptual points are concerned, job-sharing storage in artificial electrodes connected with generalization of bulk and boundary storage are to be mentioned [C.-C. Chen and J. Maier, Nat. Energy 3 (2018) 102, C.-C. Chen, L.-J. Fu, and J. Maier, Nature 436 (2016) 159] as well as the concept of wiring lengths which bring about the possibility to quantify the paramount influence of morphology [C. Zhu, R.-E. Usiskin, Y. Yu, and J. Maier, Science 358 (2017) eaao2808(1-8), R.E. Usiskin and J. Maier, J. Electrochem. Soc.167 (2020) 080505, R.E. Usiskin and J. Maier, Phys. Chem. Chem. Phys. 20 (2018) 16449].

References

[1] Gruss P, Schüth F. editors. Die Zukunft der Energie. München: C.H. Beck; 2008.

[2] Hamann CH, Hamnett A, Vielstich W. Electrochemistry. Weinheim: Wiley-VCH; 1998.

[3] Mallela VS, Ilankumaran V, Rao NS. Trends in cardiac pacemaker batteries. Ind Pacing Electrophysiol J. 2004;4(4):201–12.

[4] Winter M, Besenhard JO. Wiederaufladbare Batterien. Teil I: Akkumulatoren mit wäßriger Elektrolytlösung. Chemie in unserer Zeit. 1999;33(5):252–66.

[5] Ritter JW. Beiträge zur näheren Kenntniß des Galvanismus und der Resultate seiner Untersuchung. Jena: Frommann, 1800–05.

[6] Planté G. Storage of Electrical Energy. Birmingham. Parker & Hill; 1859.

[7] Winter M, Besenhard JO. Wiederaufladbare Batterien. Teil II: Akkumulatoren mit nichtwäßriger Elektrolytlösung. Chemie in unserer Zeit. 1999;33(6):320–32.

[8] Linden D, Reddy TB. editors. Handbook of Batteries. 3rd, New York: McGraw-Hill; 2002.

[9] Hamann CH, Vielstich W. Elektrochemie. 3rd. Weinheim: Wiley-VCH; 1998.

[10] Dvorak W. Sodium-nickel-chloride cell. In Wikimedia Commons CC-BY, 2011 Available from http://commons.wikimedia.org/wiki/File:Sodium-nickel-chloride_cell.svg.

[11] Whittingham MS. Lithium batteries and cathode materials. Chem Rev. 2004;104(10): 4271–301.

[12] Mizushima K, Jones PC, Wiseman PJ, Goodenough JB. Li_xCoO_2 ($0<x\leq1$): a new cathode material for batteries of high energy density. Mater Res Bull. 1980;15(6):783–89.

[13] Tarascon JM, Armand M. Issues and challenges facing rechargeable lithium batteries. Nature. 2001;414(6861):359–67.

[14] Padhi AK, Nanjundaswamy KS, Goodenough JB. Phospho-olivines as positive-electrode material for rechargable lithium batteries. J Electrochem Soc. 1997;144(4):1188–94.
[15] Wessels C, Ruff R, Huggins RA, Cui Y. Investigations of the electrochemical stability of aqueous electrolytes for lithium battery applications. Electrochem Solid-State Lett. 2010;13:A59–61.
[16] Knauth P. Inorganic solid Li ion conductors: an overview. Solid State Ionics. 2009;180; 14–16:911–16.
[17] West WC, Whitacre JF, Lim JR. Chemical stability enhancement of lithium conducting solid electrolyte plates using sputtered LiPON thin films. J Power Sources. 2004;126(1–2):134–38.
[18] Lee JM, Kim SH, Tak Y, Yoon YS. Study on the LLT solid electrolyte thin film with LiPON interlayer intervening between LLT and electrodes. J Power Sources. 2006;163(1):173–79.
[19] Bhattacharyya AJ, Maier J. Second phase effects on the conductivity of non-aqueous salt solutions: "soggy sand electrolytes.". Adv Mater. 2004;16:811–14.
[20] Thackeray MM, Kang S-H, Johnson CS, Vaughey JT, Benedek R, Hackney SA. Li_2 MnO_3-stabilized $LiMO_2$ (M = Mn, Ni, Co) electrodes for high energy lithium-ion batteries. J Mater Chem. 2007;17:3112–25.
[21] Kang K, Meng YS, Bréger J, Grey CP, Ceder G. Electrodes with high power and high capacity for rechargeable lithium batteries. Science. 2006;311(5763):977–80.
[22] Roth EP Comparative abuse response of Li-ion cells with $LiFePO_4$ and $LiMn_2O_4$ cathodes. Presented at: Fall 2008 ECS meeting; 2008; Honululu, HI. 2008.
[23] Dominko R, Bele M, Kokalj A, Gaberscek M, Jamnik J. Li_2MnSiO_4 as a potential Li-battery cathode material. J Power Sources. 2007;174(2):457–61.
[24] Nyten A, Kamali S, Haggstrom L, Gustafsson T, Thomas JO. The lithium extraction/insertion mechanism in Li_2FeSiO_4. J Mater Chem. 2006;16(23):2266–72.
[25] Armand M, Endres F, MacFarlane DR, Ohno H, Scrosati B. Ionic-liquid materials for the electrochemical challenges of the future. Nature mater. 2009;8(8):621–29.
[26] Huang SY, Kavan L, Exnar I, Gratzel M. Rocking chair lithium battery based on nanocrystalline TiO_2 (anatase). J Electrochem Soc. 1995;142(9):L142–4.
[27] Kavan L, Procházka J, Spitler TM et al. Li insertion into $Li_4Ti_5O_{12}$ (xpinel). J Electrochem Soc. 2003;150(7):A1000–7.
[28] Hu Y-S, Kienle L, Guo Y-G, Maier J. High lithium electroactivity of nanometer-sized rutile TiO_2. Adv Mater. 2006;18:1421–26.
[29] Armstrong AR, Armstrong G, Canales J, Carcía R, Bruce PG. Lithium-ion intercalation into TiO_2-B nanowires. Adv Mater. 2005;17:862–65.
[30] Wang J, Polleux J, Lim J, Dunn B. Pseudocapacitive contributions to electrochemical energy storage in TiO_2 (anatase) nanoparticles. J Phys Chem C. 2007;111:14925–31.
[31] Shin J-Y, Samuelis D, Maier J. Sustained lithium storage performance of hierarchical nanoporous anatase TiO_2 at high rates: emphasis on interfacial storage phenomena. Adv Funct Mater. 2011;21(18):3464–72.
[32] Shin J-Y, Joo J, Samuelis D, Maier J. Oxygen-deficient $TiO_{2-\delta}$ nanoparticles via hydrogen reduction for high rate capability lithium batteries. Chem Mater. 2012;24(3):543–1.
[33] Yu Y, Gu L, Wang C, Dhanabalan A, Van Aken PA, Maier J. Encapsulation of Sn@carbon nanoparticles in bamboo-like hollow carbon nanofibers as an anode material in lithium-based batteries. Angew Chem Int Ed. 2009;48:6485–89.
[34] Yu Y, Gu L, Zhu C, Van Aken PA, Maier J. Tin nanoparticles encapsulated in porous multichannel carbon microtubes: preparation by single-nozzle electrospinning and application as anode material for high-performance Li-based batteries. J Am Cheml Soc. 2009;131:15984–85.

[35] Peled E. The electrochemical behavior of alkali and alkaline earth metals in nonaqueous battery systems – the solid electrolyte interphase model. J Electrochem Soc. 1979;126: 2047–51.

[36] Balbuena PB, Wang Y. Lithium-ion batteries: solid-electrolyte interphase. 1st. London, UK: Imperial College Press; 2004.

[37] Armand MB, Bruce PG, Forsyth M, Scrosati B, Wieczorek W. Polymer electrolytes. In Bruce DW, O'Hare D, Walton RI editors, Energy Materials, Chichester, UK: John Wiley & Sons; 2011.

[38] Kreuer K-D, Wohlfarth A, De Araujo CC, Fuchs A, Maier J. Single alkaline-ion (Li+, Na+) conductors by ion exchange of proton-conducting ionomers and polyelectrolytes. Chem-PhysChem. 2011;12(14):2558–60.

[39] Maier J. Concentration polarization of salt-containing liquid electrolytes. Adv Funct Mater. 2011;21(8):1448–55.

[40] Arora P, Zhang Z. Battery separators. Chem Rev. 2004;104:4419–62.

[41] Armand M, Chabagno JM, Duclot M. Poly-ethers as solid electrolytes. In Vashishita P, Mundy JN, Shenoy GK editors, Fast ion transport in solids, Amsterdam: North-Holland Publishing, 1979.

[42] Maier J. Acid–base centers and acid–base scales in ionic solids. Chem Eur J. 2001;7(22): 4762–70.

[43] Badway F, Cosandey F, Pereira N, Amatucci GG. Carbon metal fluoride nanocomposites. J Electrochem Soc. 2003;150(10):A1318–27.

[44] Idota Y, Kubota T, Matsufuji A, Maekawa Y, Miyasaka T. Tin-based amorphous oxide: a high-capacity lithium-ion-storage material. Science. 1997;276(5317):1395–97.

[45] Poizot P, Laruelle S, Grugeon S, Dupont L, Tarascon JM. Nano-sized transition-metal oxides as negative-electrode materials for lithium-ion batteries. Nature. 2000;407(6803):496–99.

[46] Li H, Balaya P, Maier J. Li-storage via heterogeneous reaction in selected binary metal fluorides and oxides. J Electrochem Soc. 2004;151(11):A1878–85.

[47] Delmer O, Balaya P, Kienle L, Maier J. Enhanced potential of amorphous electrode materials: case study of RuO_2. Adv Mater. 2008;20(3):501–05.

[48] Delmer O, Maier J. On the chemical potential of a component in a metastable phase-application to Li-storage in the RuO_2-Li system. Phys chem cheml phyS. 2009;11(30):6424–29.

[49] Balaya P, Li H, Kienle L, Maier J. Fully reversible homogeneous and heterogeneous Li storage in RuO_2 with high capacity. Adv Funct Mater. 2003;13:621–25.

[50] Maier J. Mass storage in space charge regions of nano-sized systems (Nano-ionics. Part V). Faraday Discuss. 2007;134:51–66.

[51] Maier J. Nanoionics: ion transport and electrochemical storage in confined systems. Nature Mater. 2005;4(11):805–15.

[52] Kaskhedikar N, Maier J. Lithium storage in carbon nanostructures. Adv Mater. 2009;21; 25–26:2664–80.

[53] Amin R, Maier J, Balaya P, Chen DP, Lin CT. Ionic and electronic transport in single crystalline $LiFePO_4$ grown by optical floating zone technique. Solid State Ionics. 2008;179 (27–32):1683–87.

[54] Maier J, Amin R. Defect chemistry of $LiFePO_4$. J Electrochem Soc. 2008;155(4):A339–44.

[55] Sigle W, Amin R, Weichert K, Van Aken PA, Maier J. Delithiation study of $LiFePO_4$ crystals using electron energy-loss spectroscopy. Electrochem Solid-State Lett. 2009;12(8):A151–4.

[56] Maier J. Ionic conduction in space charge regions. Prog Solid State Chem. 1995;23(3):171–263.

[57] Balaya P, Bhattacharyya AJ, Jamnik J, Zhukovskii YF, Kotomin EA, Maier J. Nanoionics in the context of lithium batteries. J Power Sources. 2006;159(1):171–78.

[58] Guo Y-G, Hu YS, Sigle W, Maier J. Superior electrode performance of nanostructured mesoporous TiO_2 (anatase) through efficient hierarchical mixed conducting networks. Adv Mater. 2007;19(16):2087–91.

[59] Maier J. Thermodynamics of nanosystems with a special view to charge carriers. Adv Mater. 2009;21(25–26):2571–85.
[60] Wagemaker M, Singh DP, Borghols WJH et al. Dynamic solubility limits in nanosized olivine $LiFePO_4$. J Am Cheml Soc. 2011;133(26):10222–28.
[61] Gu L, Zhu C, Li H et al. Direct observation of lithium staging in partially delithiated $LiFePO_4$ at atomic resolution. J Am Cheml Soc. 2011;133(13):4661–63.
[62] Van Der Ven A, Garikipati K, Kim S, Wagemaker M. The role of coherency strains on phase stability in Li_xFePO4: needle crystallites minimize coherency strain and overpotential. J Electrochem Soc. 2009;156:A949–57.
[63] Tang M, Carter WC, Belak JF, Chiang Y-M. Modeling the competing phase transition pathways in nanoscale olivine electrodes. Electrochimica acta. 2010;56:969–76.
[64] Meethong N, Huang H, Speakman S, Carter WC, Chiang Y-M. Strain Accommodation during phase transformations in olivine-based cathodes as a materials selection criterion for high-power rechargeable batteries. Adv Funct Mater. 2007;17:1115–23.
[65] Dreyer W, Jamnik J, Guhlke C, Huth R, Moškon J, Gaberščеk M. The thermodynamic origin of hysteresis in insertion batteries. Nature Mater. 2010;9:448–53.
[66] Delmas C, Maccario M, Croguennec L, Le Cras F, Weill F. Lithium deintercalation in $LiFePO_4$ nanoparticles via a domino-cascade model. Nature Mater. 2008;7:665–71.
[67] Chen G, Song X, Richardson TJ. Metastable solid solution phases in the $LiFePO_4/FePO_4$ system. J Electrochem Soc. 2007;154:A627–32.
[68] Wagemaker M, Mulder FM. Van der Ven A. The Role of Surface and Interface Energy on Phase Stability of Nanosized Insertion Compounds. Adv Mater. 2009;21(25–26):2703–09.
[69] Hu YS, Adelhelm P, Smarsly B, Hore S, Antonietti M, Maier J. Synthesis of hierarchically porous carbon monoliths with highly ordered microstructure and their application in rechargeable lithium batteries with high-rate capability. Adv Funct Mater. 2007;17(12):1873–78.
[70] Hu YS, Guo YG, Sigle W, Hore S, Balaya P, Maier J. Electrochemical lithiation synthesis of nanoporous materials with superior catalytic and capacitive activity. Nature Mater. 2006;5(9): 713–17.
[71] Guo Y-G, Hu Y-S, Sigle W, Maier J. Superior electrode performance of nanostructured mesoporous TiO_2 (anatase) through efficient hierarchical mixed conducting networks. Adv Mater. 2007;19:2087–91.
[72] Schröder A, Fleig J, Gryaznov D, Maier J, Sitte W. Quantitative model of electrochemical Ostwald ripening and its application to the time-dependent electrode potential of nanocrystalline metals. J Phy Chem B. 2006;110(25):12274–80.
[73] J. Maier, Angew. Chem. Int. Ed. 52 (19) (2013) 4998.
[74] J. Janek and W. Zeier, Nat. Energy 1 (2016) 16141.
[75] T. Krauskopf, F.H. Richter, W.G. Zeier, and J Janek, Chem. Rev. 120 (2020) 7745.
[76] R. Usiskin, Y. Lu, J. Popovic, M. Law and P. Balaya, Y.-S. Hu, and J. Maier, Sodium-Based Batteries: Fundamentals, Status, and Perspectives, Nat. Rev. Mater. (2021).
[77] C.-C. Chen and J. Maier, Nat. Energy 3 (2018) 102.
[78] C.-C. Chen, L.-J. Fu, and J. Maier, Nature 436 (2016) 159.
[79] C. Zhu, R.-E. Usiskin, Y. Yu, and J. Maier, Science 358 (2017) eaao2808(1-8).
[80] R.E. Usiskin and J. Maier, J. Electrochem. Soc.167 (2020) 080505.
[81] R.E. Usiskin and J. Maier, Phys. Chem. Chem. Phys. 20 (2018) 16449.

3 Thermochemical

Sebastian Arndt, Reinhard Schomäcker

10 Chemical Kinetics: A Practical Guide

10.1 Theory

10.1.1 Introduction

Any chemical conversion requires time. The time-dependent development of a reaction, the reaction rate r, is expressed by its kinetics, which describes the correlation of r and the determining factors, such as concentration, pressure, temperature, diffusion, catalyst, mass and heat transfer, and so forth.

The reaction rate determines the application potential and/or the need of improvement to accelerate the conversion. For example, the production of biogas via fermentation of biomass and the onboard production of H_2 via steam reforming of CH_3OH require different times for a certain degree of conversion. However, for the biogas production the time is of less importance, as one can simply wait and/or increase the size of the fermentation vessel. This is obviously not possible for the onboard production of H_2 in a vehicle.

Therefore, it is important to determine a reaction rate. Moreover, it is important to know how to determine it because the reaction rate is influenced by several factors, hampering the comparison of reactions, reactors and catalysts. The next sections will shortly introduce the basic concepts for the description of the kinetics of a heterogeneously catalyzed reaction and their practical applications; followed by three real-research examples demonstrating the application and possible problems in detail.

10.1.2 Course of a Catalytic Reaction

Many chemical reactions are heterogeneously catalyzed, which has a strong impact on their kinetics. Therefore, the idealized course of a heterogeneously catalyzed reaction will be introduced here shortly, cf. Figure 10.1. In other, homogeneously catalyzed reactions for instance, certain points can be omitted simplifying the course.

Acknowledgment: This work is part of the Cluster of Excellence “Unifying Concepts in Catalysis” coordinated by the Technische Universität Berlin. Financial support by the Deutsche Forschungsgemeinschaft (DFG) within the framework of the German Initiative for Excellence is gratefully acknowledged.

Sebastian Arndt, Technische Universität Berlin, Department of Chemistry, Straße des 17. Juni 124, 10623 Berlin, Germany, e-mail: arndt@chem.tu-berlin.de
Reinhard Schomäcker, Technische Universität Berlin, Department of Chemistry, Straße des 17. Juni 124, 10623 Berlin, Germany, e-mail: schomaecker@tu-berlin.de

https://doi.org/10.1515/9783110608458-010

The heterogeneous catalyst particles in the reactor are surrounded by a boundary layer of gas or liquid, which can be considered as a static film around the particle. A reactant molecule has to diffuse through this boundary layer via film diffusion (1). As most catalysts have pores, the reactant molecule also has to diffuse through the pores in order to approach the active site, the pore diffusion process (2). Inside the pores, the reactant molecules adsorb at or near the active center and react (3, 4). The resulting product molecules desorb (5) and return back into the fluid phase via pore diffusion (6) and film diffusion (7). Further details on this can be found in general textbooks [1–3].

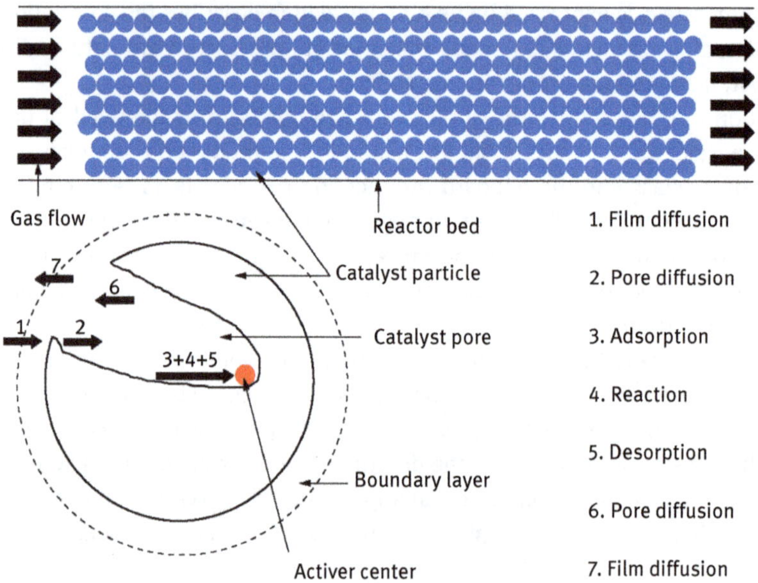

Figure 10.1: Delineation of the steps of a catalyzed chemical reaction. In an ideal case, the reactant molecule diffuses through the boundary layer and the pores. Near the active center, it adsorbs, reacts, and desorbs and is the being transported back via pore and film diffusion.

Depending on this, a concentration profile occurs as shown in Figure 10.2. The concentration in the gas phase is considered to be equal at every point. If the transport of the reactants through the boundary layer and the catalyst is much faster than the reaction no concentration profile occurs. However, if the reaction is faster than the transport of the molecules, a concentration profile forms, with a shape depending on the differences in velocity.

The rate of the product formation is determined by the slowest step of this sequence, regarded as the rate-determining step (RDS). In the development of a kinetic model most attention has to be focused on this step and its interaction with the other steps.

In a closed system and at a fixed temperature, the thermodynamic equilibrium constant of any reaction has a fixed value. A catalyst has an impact on the reaction

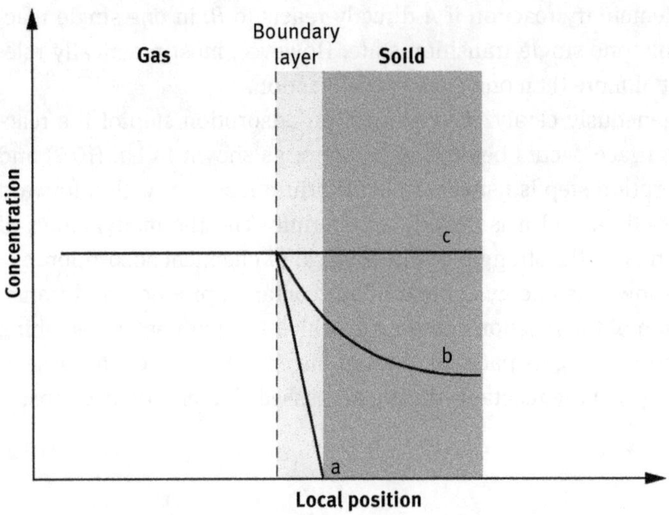

Figure 10.2: Concentration profile in the catalyst particle if the reaction is much faster (a) or of comparable rate (b) or much slower (c) than the transport of the reactants to the active site.

rate by lowering the activation energy, reducing the required time to achieve the thermodynamic equilibrium and it can have an impact on the reaction channel by favoring one of several possible transition states, but a catalyst does not influence the thermodynamic equilibrium itself. That means the maximal achievable yield cannot be higher than that predicted by thermodynamics. An important consequence of this limitation is that for a comparison of different catalysts the experimental conditions must not allow that the thermodynamic equilibrium is reached. Typical reaction conditions for this case would be: low reactant flow (setup then resembling a closed system) and/or a temperature allowing a very fast reaction. If the experimental conditions allow the reaction to approach the thermodynamic equilibrium, a comparison of different catalysts is usually impossible. For any kinetic studies the experimental conditions have to be chosen in a way that with all variations of parameters a kinetic control of the reaction is ensured.

10.1.3 Reaction Kinetics

10.1.3.1 Adsorption and Reaction

The most general and simple reaction is shown in eq. (10.1), with A being the reactant, B being the product and a and b their stoichiometric coefficients:

$$aA \rightarrow bB \tag{10.1}$$

This reaction is an elementary reaction if A directly reacts to B, in one single reaction step and with only one single transition state. However, most practically relevant reactions consist of more than one elementary reaction.

For most heterogeneously catalyzed reactions, an adsorption step of the reactant on the catalyst surface occurs before the reaction, as shown in eq. (10.2) and Figure 10.3. The adsorption step is usually an equilibrium reaction, with a forward reaction and back reaction, and it is usually exothermic. The thermodynamics of this equilibrium depends on the strength of the physical or chemical adsorption.

Figure 10.3 also shows the energy content of all reactants, products and transition states as a function of the reaction coordinate, which is a parameter describing the progress of a reaction along its pathway through its energy surface. This assembles all energetic changes of the reactants during all passed elementary reactions:

$$A \rightleftharpoons A_{ads.} \rightarrow B \tag{10.2}$$

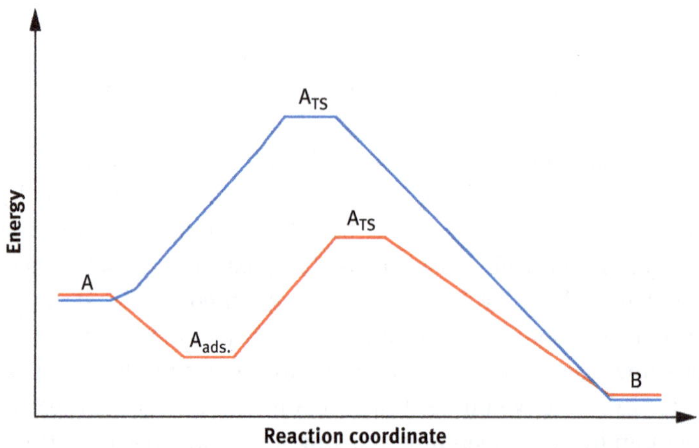

Figure 10.3: For the uncatalyzed reaction (blue), the activation energy required for the reaction is high. If catalysis takes place (red), a transition state in which the reactant has adsorbed on the catalyst surface exists. In such a case, the required activation energy for the conversion of A to B is considerably lower.

If a reaction has to be divided into more than one elementary reaction, it is called a reaction network. The complexity of such reaction networks can be very different, ranging from just two elementary reactions to a network consisting of parallel-, side-, subsequent-, and equilibrium reactions. Details about more complicated reactions, such as bimolecular reactions, reversible reaction steps and reactions with different kinds of adsorption (chemical, physical, dissociative, etc.), can be found in the typical literature [1–4].

10.1.3.2 Reaction Rate

The time dependent evolution of the concentrations of reactant c_A or product c_B, taking the stoichiometric coefficients into account, is a measure for the reaction rate or its velocity, cf. Equation (10.3). The convention requires a negative sign for reactants, indicating consumption, and a positive sign for the products, indicating formation:

$$r = -\frac{1}{a}\frac{dc_A}{dt} = \frac{1}{b}\frac{dc_B}{dt} \tag{10.3}$$

It is also possible to take additional factors into account to further characterize the reaction, for example, F being the reaction volume, the mass of the catalyst, the surface of the catalyst, or the interface of a two-phase system, and so forth, as shown in eq. (10.4):

$$r = -\frac{1}{aF}\frac{dc_A}{\times dt} = \frac{1}{bF}\frac{dc_B}{\times dt} \tag{10.4}$$

The differential eq. (10.3) is a general expression, as well as a measurement instruction. However, a mathematical model is required to solve it and to perform any kind of calculations, inter- and extrapolations. An analytical solution of eq. (10.3) is possible for elementary reactions, with a being the stoichiometric coefficient and the exponent of the concentration in the rate law. For a reaction network consisting of more than one elementary reaction, it is still possible to build a model with an appropriate number of differential equations for the key compounds. However, it might not be analytically but only numerically solvable. Details can be found in many text books on physical chemistry or catalysis [1, 3]. A common approach to express eq. (10.3) has the form of a polynomial with empirical exponent for the concentration, as shown in eq. (10.5):

$$r = k \times (c_A)^n \tag{10.5}$$

The rate constant k is independent of the concentration, but temperature dependent, the details are described in Section 10.1.3.3. n is the reaction order. For elementary reactions, it is the stoichiometric coefficient a. However, if the studied reaction is not an elementary reaction, the order of the reaction does not relate to the stoichiometric factors anymore. Moreover, the reaction order can then be a fractional number and it usually does not have a physical meaning. Nevertheless, the reaction order has a strong influence on the time dependent development of the concentration of the reactants, disregarding if it is an elementary reaction or not.

If a reaction network exists, for example, with a complex adsorption step, as described in Section 10.1.3.1, it is possible to define an overall reaction. However, in such a case, one also obtains an overall rate constant that is combined from the rate constant of the elementary reactions, the so called effective rate constant k_{eff}, cf. Section 10.1.3.3.

For technical and engineering purposes, a detailed reaction mechanism and a detailed model is usually not required. A formal kinetic model describing the time dependent evolution of the concentration with a mathematical function is fitted to the obtained experimental data. In such cases, the variables k and n do not have a real physical meaning based on the mechanism of the reaction.

10.1.3.3 Impact of the Temperature

The reaction rate r is the time dependent evolution as a function of the reactant concentration and the rate constant k. However, the dependency of the reaction rate on the temperature is also very important.

For an elementary reaction the temperature dependency of the rate constant is given by the Arrhenius equation, eq. (10.6), which accounts only for elementary reactions. It is important to note that this equation gives the dependency of the rate constant k on the temperature, not the dependency of r. The preexponential factor P also shows a dependency on the temperature, but its dependency is weak compared to the exponential dependency of k:

$$k = P \times e^{-\frac{Ea}{RT}} \tag{10.6}$$

P = pre-exponential factor
E_a = activation energy [J/mol]
R = universal gas constant, 8.314 [J/mol K]
T = absolute temperature [K].

For a reaction with a preceding adsorption step, two rate constants for the equilibrium reaction and one rate constant for the conversion to B exist, as shown in eq. (10.7):

$$A \rightleftharpoons A_{ads.} \xrightarrow{k_2} B \tag{10.7}$$

It is a reasonable assumption that the equilibrium reaction proceeds much faster than the conversion. The reaction rate of B is shown in eq. (10.8):

$$r = \frac{dc_B}{dt} = k_2 \times c_{Aads.} \tag{10.8}$$

The concentration of the adsorbed species can be obtained by compiling the formation term and the consumption term of the rate for the adsorbed species, as displayed in eqs. (10.9) and (10.10):

$$\text{Formation:} = \frac{dc_{A_{ads.}}}{dt} = k_1 \times cA \tag{10.9}$$

$$\text{Consumpation:} = \frac{dc_{A_{ads.}}}{dt} = k_{-1} \times c_{A_{ads.}} + k_2 \times c_{A_{ads.}} \tag{10.10}$$

In the steady state, these two equations are equal. The resulting equation can be converted, giving the concentration $c_{A_{ads.}}$

$$k_1 \times c_A = k_{-1} \times c_{A_{ads.}} \; k_2 \times c_{A_{ads.}} \tag{10.11}$$

$$c_{A_{ads.}} = \frac{k1}{k_{-1} + k_2} cA \tag{10.12}$$

The obtained expression for $c_{A_{ads.}}$ can now be used to obtain the reaction rate for *B*, as shown in eq. (10.13). If the rate constant of desorption (k_{-1}) is much higher than the rate of conversion to the product (k_2), $c_{A_{ads.}}$ can be expressed from the adsorption constant *K* by $c_{A_{ads.}} = K \times c_A$.

$$r = k_2 \times c_{A_{ads.}} = k_2 \times \left(\frac{\mathrm{k1}}{\mathrm{k}_{-1} + \mathrm{k}_2}\right) \mathrm{c_A} \tag{10.13}$$

$$\approx k_2 \times K \times c_A \tag{10.14}$$

$$= k_{\text{eff}} \times c_A \tag{10.15}$$

The product $k_2 \times K$ can now be comprised to a new rate constant k_{eff}, which is a rate constant for the overall reaction. In order to account for the temperature dependency of k_{eff}, the temperature dependencies for both, k_2 and *K*, have to be considered. While k_2 follows the temperature according to the Arrhenius equation, eq. (10.6), the temperature dependency of *K* is given by the van't Hoff Law, cf. Equation (10.16):

$$\frac{d(\ln K)}{dT} = \frac{\Delta H^{\ominus}}{RT^2} \tag{10.16}$$

A combination of both results in an overall temperature dependency expressed by an effective activation energy that is calculated from $E_{\text{eff}} = E - \Delta H_{\text{ads.}}$ Therefore, for k_{eff} we obtain:

$$k_{\text{eff}} = P_{\text{eff}} \times e^{-\frac{E_{\text{eff}}}{RT}} \tag{10.17}$$

If a complex reaction network of elementary reactions is present, the results of the calculation change; however, the principle ansatz of comprising the rate constants for several reactions remains the same. But in a complex reaction network one might not know all single rate constants. In such a case, it is useful to denominate them as apparent rate constant and apparent activation energy, as shown in eq. (10.18). However, the apparent activation energy might only be valid for a certain temperature range because of a change in the predominant step of the reaction, cf. Section 10.1.3.4:

$$k_{\text{app}} = P_{\text{app}} \times e^{-\frac{E_{\text{app}}}{RT}} \tag{10.18}$$

A linearization of the Arrhenius eqs. (10.6), (10.17), or (10.18) as shown in eq. (10.19) is called an Arrhenius plot. It is very useful for the discussion of several different aspects:

$$\ln(k_{\text{app}}) = \ln(P_{\text{app}}) - \frac{E_{\text{app}}}{R} \times \frac{1}{T} \tag{10.19}$$

The slope of eq. (10.19) is the activation energy or the effective or apparent activation energy. As shown in Figure 10.4, a comparison of different catalysts by comparing their activation energy is possible. A low slope, and therefore a small activation energy, is usually a desirable feature of a good catalyst. The uncatalyzed thermal reaction usually exhibits the highest activation energy.

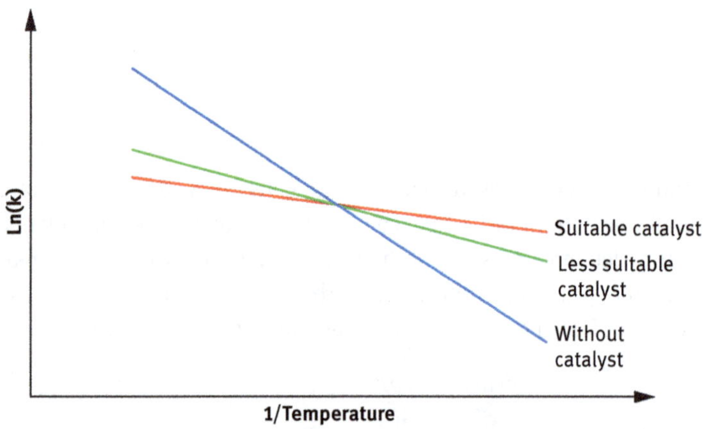

Figure 10.4: Delineation of activation energy for different catalysts. The uncatalyzed one has the highest slope and therefore the highest activation energy. The better catalyst has a smaller slope and therefore a smaller activation energy, as compared to the less suitable catalyst.

There are two options for the utilization of a catalyst providing a low activation energy for a reaction. It can be operated at the same temperature than other catalysts, achieving a higher productivity or the temperature of the reaction may be decreased.

However, care has to be taken in such cases because a different slope can also be an indication for different regimes of transport limitation, as shown in Figure 10.5. At very high temperatures (small values for $1/T$), the reaction is so fast that the mass transport into the particles is the limiting factor for the reaction. Because this process is nearly independent of the temperature, a temperature change does not affect the observed rate. At lower temperatures, the transport inside the pores becomes the limiting factor. The observed activation energy is approximately 1/2 of the real activation energy. However, at even lower temperatures (or higher $1/T$ values) the reaction rate is determined by the chemical reaction itself and not by any transport limitations. In this case, the slope is the highest and the determined activation energy is the "real one."

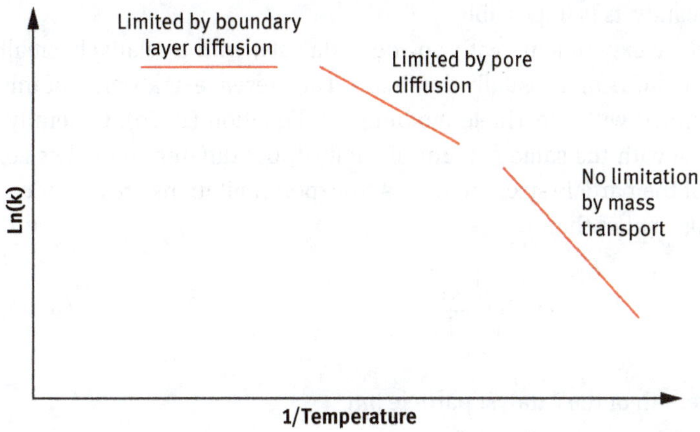

Figure 10.5: The activation energy for a given reaction in the case of mass transport limitation (boundary layer diffusion and pore diffusion, respectively) and in the case of the absence of mass transport limitation.

10.1.3.4 Limitations

The interaction of the kinetics of the chemical reaction and transport phenomena of the reactions are described as macrokinetics, in which transport phenomena, such as mass and heat transfer, adsorption, and desorption have a substantial impact on the time-dependent development of the reaction. The macro-kinetic is important for reactor engineering, construction, operation, and safety. However, for development and comparison of heterogeneous catalysts, the macrokinetic is not suitable, as the comparison of different catalytic materials is often hampered by the transport phenomena.

The intrinsic kinetics describes a reaction rate that is not influenced by such transport phenomena; therefore, it only depends on the factors concentration, pressure, temperature, and catalyst. For the comparison of the catalytic activity and the investigation of different catalysts, it is necessary to adjust the experimental conditions such that only the intrinsic kinetics is determined. If this is not the case, none of the obtained data are of use. The microkinetics is equivalent to the intrinsic-kinetic, with the difference that it consists of the elementary reactions.

10.1.3.4.1 Mass Transport

As described in Section 10.1.2, in most catalytic reactions, the reactant molecules diffuse through a boundary layer and through the pores to the active center, react, and diffuse back. If the velocity of any of these two diffusion processes is smaller than the conversion of the reactants at the active center, the overall reaction rate for the whole process is limited by the mass transport and not by the chemical reaction. If the reaction is influenced by mass transport effects, a comparison of the catalytic activity of different catalysts is impossible!

In many laboratory experiments with powder catalysts, with a relatively small particle size, the pore diffusion is usually dominant. The presence or absence of this effect can be determined with the Thiele modulus, cf. Equation (10.20). Generally, experiments are done with the same amount of catalyst, but different particle size. If r is independent of the particle size, then mass transport limitations are excluded, indicated by Φ being smaller than 1:

$$\Theta = l\sqrt{\frac{k \times c_s^{n-1}}{D_e}} \leq 1 \quad (10.20)$$

l = characteristic length of the catalyst particle $[m]$
k = intrinsic rate constant $[s^{-1}]$
c_s = surface concentration $\left[\frac{\mathrm{mol}}{\mathrm{L}}\right]$
n = order of the reaction [1]
D_e = effective diffusion coefficient $\left[\frac{\mathrm{m}^2}{\mathrm{s}}\right]$

10.1.3.4.2 Heat Transport

The heat transport is treated similar to the transport of mass. There are three ways for heat transport: thermal conduction, convection and radiation. If the heat transport is limited, hot spots can occur in the catalyst bed, causing deactivation of the catalyst. Moreover, if a hot-spot occurs, the temperature at which is the reaction occurs is unknown. However, in typical laboratory equipment (small reactor, small particle size, diluted catalyst, and limited conversion), this is usually not a problem.

10.1.3.4.3 Estimation Criteria

There are several criteria equations described in literature to check for the presence or absence of transport limitations. An overview of different criteria can be found here [[4–6]]. However, there are many more textbooks listing these criteria.

10.2 Practical Aspects

The practical aspects of heterogeneous catalysis and the application of theory will be introduced in the following chapter. However, it is important to note that in the performed experiments, the applied reactors and reaction conditions depend on the reaction, the catalyst, and the desired information.

10.2.1 Laboratory Reactors

There are four ideal reactors: the batch reactor (real counterpart: stirred tank reactor), semibatch reactor,[1] continuous stirred tank reactor (CSTR), and the plug flow tubular reactor (PFTR) (real counterpart: tube reactor). For production applications, there are also numerous other reactors [7–9]. An overview of typical and advanced laboratory reactors was given by Kapteijn and Moulijn [6].

The choice of the appropriate reactor applied for kinetic measurements is determined by the type of reaction (simple, parallel, or consecutive), the reaction heat and the phase state of the reaction mixture. In general, reactors with simple, almost ideal mixing behavior are preferred in order to obtain simple material balances.

10.2.2 Preliminary Tests

10.2.2.1 Catalyst Stability

There are various reasons for a catalyst to loose activity [10]. However, the determination of the various variables explained in Section 4.1.1, heavily depends on the catalyst stability. The kinetic parameters can only be determined for a stable catalytic performance otherwise only snapshots of the deactivation trajectory are measured.

Before determining the catalytic activity of any reaction or catalyst, it is very important to investigate the stability. A stability check should be done under conditions that are not too unrealistic for a practical application because a large dilution of the feed might suppress the deactivation. Once the stability or the stable time period has been determined, one can proceed to investigate the different kinetic parameters.

1 There is no agreement in the literature as to whether this is an ideal reactor.

10.2.2.2 Transport Limitation

For the estimation of the mass transport limitation, eq. (10.20) has an important drawback. In many cases neither the rate constant k nor the reaction order n is known. However, the Weisz-Prater criterion, cf. eq. (10.21), which is derived from the Thiele modulus [4, 8], can be calculated with experimentally easily accessible values, taking Φ 1 for any reaction without mass transfer limitations. However, it is not necessary to know all variable exactly, even for the Weisz-Prater criterion n can be unknown. Reasonable assumptions can be made, for example, $n = 1, 2, 3$, or 4 and l is the particle diameter instead of the characteristic length. For the gas phase, D_e can be calculated with statistical thermodynamics or estimated; common values are within the range of 10^{-5} to 10^{-7} m²/s. In the liquid phase, the estimation becomes more complicated. A common value of ϱ_c for solid catalysts is 1,300 kg/m³, but if the catalyst is diluted with an inert material, this value cannot be applied. A measurement would then be necessary. The Weisz-Prater criterion gives information whether transport limitation is absent, if the value is significantly below 1; however, it cannot be used to determine the degree of transport limitation:

$$\Psi = l^2 \times \frac{n+1}{2} \times \frac{r_e \varrho_c}{D_e c_{i,0}} \leq 1 \qquad (10.21)$$

l = characteristic length [m]
r_e = observed effective reaction rate $\left[\frac{\text{mol}}{\text{Kg} \times \text{s}}\right]$
ϱ_c = catalyst density $\left[\frac{\text{kg}}{\text{m}^3}\right]$
$c_{i,0}$ = concentration of the component i at the time $t = 0$ $\left[\frac{\text{mol}}{\text{m}^3}\right]$
D_e = effective diffusion coefficient $\left[\frac{\text{m}^2}{\text{s}}\right]$.

10.2.3 Comparative Studies

In a search for new catalysts, it is often necessary to compare the activity of different materials. For such a comparative study of different catalysts in the same experimental setup, it is necessary to obey certain rules.

1. The experimental conditions should be chosen such that the thermodynamic equilibrium is not approached.
2. Mass transfer limitations must be excluded.
3. Isothermal conditions should be ensured.
4. Proper flow-patterns should be ensured and the formation of strong concentration gradients should be avoided.
5. The contribution of the uncatalyzed reaction should be known and the application of inert reactors is advisable, although not always possible.
6. To avoid by-pass gas flow the reactor diameter should be at least 10 times larger than the particle diameter.

7. The length of the bed should be at least 5 to 10 times larger than the particle diameter. However, the catalyst bed should also not be too long, as one might face problems with pressure drop and the limited length of the isothermal zone of the furnaces, which adjusts the temperature of the reactors.
8. The weighed amount of catalysts and the gas flow should be adjusted to the essential features of the catalyst, assigned to the catalytic activity (e.g. surface area, active component, dopant, etc.).

Taking a close look at these rules, it becomes obvious that before the experiments one should consider in detail what one desires to measure, evaluate, and compare.

Furthermore, for comparative studies without determining kinetic parameters, it is necessary that one does not measure at 100% conversion for one of the reactants. At such high conversions, differences between the different catalytic materials cannot be observed under such conditions. For the determination of kinetic parameters, the conversion should be lower than 10%. However, for a comparison of the catalytic activity under a given set of experimental conditions, higher conversions are acceptable.

It is noteworthy that *catalytic activity* is not a general term with a fixed definition and no absolute quantity. In the literature this term can have several meanings. Usually, the activity of a catalyst is used to describe how much reactant is converted. Improving the activity means the catalyst converts more reactant, not considering if the desired product or an unwanted by-product is produced. Thus the term *catalytic performance* can be used, which is closely related to the yield, taking into account how much of the desired product was formed. But it also does not have an absolute quantity. A catalyst with a high conversion and a low selectivity and one with a low conversion and a high selectivity can give the same yield. The latter catalyst has the better catalytic performance because it produces fewer by-products.

For many kinetic studies the selectivities of different catalysts are compared at the same level of conversion. However, the main drawback is that sometimes very different reaction conditions (mass of catalyst, residence time, reactor temperature, feed gas composition) are required to achieve this. Another option is the comparison of conversion and selectivity under a given set of reaction conditions. The main drawback of this approach appears if catalysts show full or very different conversions, then the comparison of the selectivities is not appropriate. Despite this, it is still possible to conclude if a catalyst heavily favors one of several reaction pathways, for example, if partial or total oxidation is prevailing.

Therefore, to compare the catalytic activity, or better the catalytic performance, of a series of catalysts, several factors such as reaction conditions, conversion, and selectivity must be taken into account.

10.2.4 Development of Kinetic Models

10.2.4.1 Determination of Concentration

For a simple reaction, as shown in Equation (10.1), the reaction rate can actually be measured at any time during the reaction. However, one prefers to measure at small conversion to reduce the error of $\frac{dc_A}{dt}$ to a minimum and so that $dc_{A_{t=X}} \approx dc_{A_{t=0}}$ Many reactions consist of a reaction network, including parallel and subsequent reactions. For such cases, small conversions are necessary to avoid falsification of the desired reaction rate by undesired side and/or subsequent reactions.

The definition and calculation of the conversion and selectivity will be shown at the example of the oxidative coupling of methane, as this reaction will be discussed later in detail, cf. Section 10.3.1. The definition of the conversion X of CH_4 is given in eq. (10.24) and the selectivity for the desired products in eq. (10.25).[2] Generally, two different approaches exist for the calculation of conversion and selectivity. For this, the amount of reacted molecules can either be calculated by comparing the number of reactants after the reaction and before (making a measurement of the feed gas without reaction necessary) or by calculating the number of reacted molecules via the weighted addition of the detected products. The first method leads to large errors due to unavoidable fluctuations of the feed gas flow and composition. Therefore, the second method is preferred. However, for the second method, a mass balance should always be calculated and compared with a mass balance of the feed, making sure that it is closed, in order to detect deviations, which would lead the calculation of conversion and selectivity ad absurdum, because further reactions might be overlooked.

$$X_{CH4} = \frac{s_{i,0} - s_i}{s_{i,0}} \tag{10.22}$$

$$= \frac{(\sum(\text{products}) + s_i) - s_i}{\sum(\text{products}) + s_i} \tag{10.23}$$

$$S_{C_2H_6+C_2H_4} = \frac{s_{P_1} - s_{P_{1,0}}}{s_{i,0} - s_i} \times \text{Ratio of stoichiometric coefficients} \tag{10.24}$$

$$= \frac{s_{P_1} - s_{P_{1,0}}}{\sum(\text{products}) + s_i} \times \text{Ratio of stoichiometric coefficients} \tag{10.25}$$

The dilution of the feed can be useful to prevent hot-spots in the catalytic bed or to enable experiments at small conversions, which are necessary for the determination

2 The amount of substance is usually abbreviated with n; however, since in this manuscript n has been used for the reaction order, s has been chosen.

of kinetic parameters. However, for such cases it is important to check if the catalyst is stable with and without dilution of the feed.

However, measurements at small conversions are also limited by another factor, the analytic power. Regardless of the applied analytic methods (GC, MS, HPLC, etc.), there is a detection limit. For small conversions of reactants, the amount of products can be rather small, falling near the detection limit.

It is very useful to determine a mass balance for at least one element, comparing the amount of unconverted reactants and the amount of products. For a proper determination of kinetic parameters the mass balance should be closed at least up to 95%. Commonly carbon is used for the determination of the mass balance, as all carbon containing materials are rather easy to be quantified, compared to H_2 or H_2O, for example.

Moreover, it is useful to detect and quantify as many reaction products as possible and as accurate as possible. A gas chromatograph is usually to be preferred compared to a mass spectrometer and the applied detectors should be as sensitive as possible. For example, for the detection of hydrocarbons and CO_x, an FID in combination with a TCD + methanizer should be preferred to only a TCD, due to the higher sensitivity.

Furthermore, it is necessary to ensure that after sample taking, the reaction is being quenched immediately. This is usually not a big problem for reactions at higher temperatures, for example, the oxidative conversion of hydrocarbons. However, for reactions carried out in the liquid phase near room temperature or at only slightly elevated temperatures, this might very well be a problem.

10.2.4.2 Determination of Time

The exact determination of the kinetic parameters requires not only the exact determination of *dc* but also of the elapsed time *dt*. Taking samples from a batch reactor, the determination of *dt* is rather simple: it is the time span from the start of the reaction until the sampling, provided that the reaction is quenched in the sample.

If there is a volume flow into and out of the reactor, the residence time τ is taken into account, as defined in eq. (10.26). A very important remark is that the volume flow of a gas changes with the applied reactor temperature, having a great impact on the actual residence time:

$$\tau = \frac{V_R}{\dot{V}} \tag{10.26}$$

τ = residence time $\left[\frac{1}{\mathrm{s}}\right]$
V_R = volume of reactor $[\mathrm{m}^3]$
$\dot{V}$ = volume flow $\left[\frac{\mathrm{m}^3}{\mathrm{s}}\right]$

However, if the reactor is filled, for example, with a catalyst, the situation becomes more complicated. The V_R would be the empty volume of the reactor, which is then difficult to determine, for instance, using settled apparent densities. The residence time can also be experimentally determined, usually resulting in a residence time distribution; however, the experimental effort for such experiments is often large. Therefore, it is useful to apply a modified residence time, as shown in eq. (10.27), which defines the ratio of the mass of the catalyst and the gas flow, two easily measurable values:

$$\tau_{\text{modified}} = \frac{m_{\text{Catalyst}}}{\dot{V}} \tag{10.27}$$

τ_{modified} = modified residence time $\left[\frac{1}{\text{s} \times \text{kg}}\right]$
m = mass [kg]

It has to be stressed that the exact determination of dc and dt is necessary. If only one parameter is accurately measured and the other one not, the error of the calculated kinetic parameters will still be large.

10.2.4.3 Determination of Kinetic Parameters

10.2.4.3.1 Design of Experiments

To obtain experimental data suitable for the determination of kinetic parameters, two different operation modes of reactors should be considered, the differential and the integral mode. In the differential mode rates are calculated from small conversions (10%) within a fixed time span dt. For reactions with two reactants, it is suitable to use excess of one reactant to suppress undesired side reactions.

In the integral mode (e.g. in a batch reactor), the evolution of the concentration can be measured as a function of time. For a plug flow reactor, it is measured as a function of residence time, cf. Section 10.2.4.2.

10.2.4.3.2 Reaction Rate, Reaction Constant, and Reaction Order

For the simple elementary reaction, $A \rightarrow B$, the reaction rate r is correlated to the reaction conditions by the rate law composed of the rate constant k, the reactant concentration $c_{A,}$ and the order of the reaction a. This is analytically solvable for elementary reactions:

$$r = -\frac{1}{a} \times \frac{dc_A}{dt} = k \times (c_A)^a \tag{10.28}$$

In a power law approach for the description of a more complex reaction, a numerical fit of the data points is necessary and a is exchanged for n, the "apparent" order of the reaction. Moreover, if two and more reactants take part in the reaction, all of them will have their own, respective reaction order. A separate series of experiments is necessary for the determination of each reaction order.

10.2.4.3.3 Activation Energy

For the determination of the activation energy, it is necessary to determine the rate constant k for several different temperatures, as described previously. The linearization of eq. (10.6), for an elementary reaction, or eq. (10.18) for an overall reaction gives the activation energy or the apparent activation energy as the slope of an Arrhenius plot:

$$\ln(k) = -\frac{E_a}{R} \times \frac{1}{T} \tag{10.29}$$

$$\ln(k_{\text{eff}}) = -\frac{E_{\text{app}}}{R} \times \frac{1}{T} \tag{10.30}$$

10.2.4.3.4 Miscellaneous

The turnover frequency (TOF) and the turnover number (TON), as shown in eqs. (10.31) and (10.32), are frequently used parameters for the characterization of the catalyst performance. However, while the determination of the number of active centers is not too difficult for homogeneous catalysts, it can be a very intricate subject for heterogeneous catalysts. If it is useful to determine a TOF or TON for a catalyst, whose number of active center is unclear, is left to the readers opinion.

$$\text{TON} = \frac{\text{Number of Reacted Molecules}}{\text{Number of Active Centers}} \tag{10.31}$$

$$\text{TON} = \frac{\text{Number of Reacted Molecules}}{\text{Time} \times \text{Number of Active Centers}} \tag{10.32}$$

10.3 Examples

10.3.1 Oxidative Coupling of Methane

The oxidative coupling of methane (OCM) to ethane and ethylene is a very attractive reaction for the conversion of natural and biogas, see eq. (10.33). However, the reaction suffers from severe drawbacks, such as high reaction temperatures, lack of stable and selective catalysts, and an extremely complex reaction mechanism.

$$\begin{aligned} 2CH_4 + \frac{1}{2}O_2 &\rightarrow C_2H_6 + H_2O \\ 2CH_4 + O_2 &\rightarrow C_2H_4 + 2H_2O \end{aligned} \tag{10.33}$$

In a detailed academic study, the application potential of Li-doped MgO, prepared via different synthetic routes and with different loadings of Li, was investigated [11]. Catalysts were prepared via decomposition of single source precursors, wet impregnation, precipitation, and mixed milling. The materials were sieved to a particle size ≤ 200 µm to avoid mass transfer limitations. Certain preparations (e.g. precipitation) result in very fluffy materials; in such cases, the catalysts were pressed in advance of sieving.

In eq. (10.36), an estimation of the mass transfer with the Weisz-Prater criterion is given. By taking always reasonable estimations or overestimated values, one obtains a good conclusion if mass transfer is present or not. For the characteristic length, 200 µm as particle diameter is used. The reaction order usually has the value of 1 to 4; a value of 4 would therefore be a worst case scenario. The catalyst density can be measured, or the common estimation of 1.3 kg/m^3 can be used, which should not be too erroneous for Li-doped MgO. The observed reaction rate r_e is calculated from the concentration of CH_4 at the inlet of the reaction $c_{CH4,0}$ multiplied with the highest observed conversion of 25% (the highest initial value for all tested catalysts), divided by the inverse flow rate, corrected by the reactor temperature. The calculation of r_e is shown in equation (10.34):

$$
\begin{aligned}
r_e &= \frac{c_0 \times dX}{\text{Flowrate} \times \frac{\text{Reactor Temperature}}{298\text{K}}} \times \frac{1}{\text{Mass of Catalyst}} \\
&= \left(5.3\frac{\text{mol}}{\text{L}} \times 0.25\right) \times \left(\frac{1}{0.001\frac{\text{L}}{\text{s}} \times \frac{1023\text{K}}{298\text{K}}}\right) \times \left(\frac{1}{0.0001\text{kg}}\right) \\
&= 38.5 \times 10^{-3}\frac{\text{mol}}{\text{kg} \times \text{s}}
\end{aligned}
\tag{10.34}
$$

The concentration $c_{CH4,0}$ is obtained via the ideal gas law to $\frac{s}{V} = \frac{p}{RT}$ For the diffusion of gases, the two most important cases are molecular diffusion (diffusion in the gas phase) and Knudsen diffusion (diffusion through pores, while the number of collisions between the pore wall is larger than the number of collisions between the gas molecules). Since OCM catalysts have typically low surface areas and a low porosity, it is reasonable to assume that the importance of the Knudsen diffusion is low and the diffusion coefficient for the molecular diffusion is used. The molecular diffusion can easily be calculated with the help of the kinetic gas theory; however, typical values are within the range of 10^{-7}m^2/s to 10^{-5}m^2/s, with the latter value

3 Usually the labyrinth factor is abbreviated wit τ; however, this symbol was already used for the residence time.

being the "worst case scenario." For porous catalysts, the effective diffusion coefficient has to be considered, as shown in eq. (10.35):

$$D_e = D\frac{\in}{\omega} \tag{10.35}$$

ε = porosity
ω = labyrinth factor

Moreover, for high temperatures as applied in the OCM, the diffusion coefficients can be expected to be even higher.

$$\Psi = (0.0002\text{m})^2 \times \frac{4+1}{2} \times \frac{38.5\times10^{-3}\frac{\text{mol}}{\text{kgxs}}\times 1,300\frac{\text{kg}}{\text{m}^3}}{10^{-5}\frac{\text{m}^2}{\text{s}}\times 5.3\frac{\text{mol}}{\text{m}^3}} \leq 1 \tag{10.36}$$

$$= 0.01 \leq 1$$

l = 0.0002 m characteristic length
n = 4 reaction order
ϱ_c = $1,300\frac{\text{kg}}{\text{m}^3}$ catalyst density
$c_{CH4,0}$ = $5.3\frac{\text{mol}}{\text{m}^3}$ concentration of CH_4

D_e = $10^{-5}\frac{\text{m}^2}{\text{s}}$ effective diffusion coefficient
r_e = $38.5\times10^{-3}\frac{\text{mol}}{\text{kg}\times\text{s}}$ observed effective reaction rate

The obtained value is two orders of magnitude below 1; therefore, mass transport limitations can be considered to be absent. It is worth mentioning that the calculation for values such as ϱ_c is not absolutely correct; in particular, the estimation of $d\tau$ is not exact. But it is sufficient for an estimation, as done in with the Weisz Prater criterion.

For the catalyst testing, several important points were taken into consideration. The catalyst was diluted in quartz sand in order to avoid hot spots and to allow a proper plug-flow of the gas through the reactor bed. The reaction conditions (amount of catalyst, feed gas composition, temperature) were chosen, such that the conversion of CH_4 and O_2 is limited. The so called kinetic regime is valid. Usually, several preliminary experiments or a lot of experience is necessary to determine the appropriate reaction conditions, but this is hardly mentioned in the final publication of the results.

Once this was done, the stability of the catalysts was determined by time on stream experiments, as shown in Figure 10.6 for Li/MgO samples prepared via wet impregnation. The time on stream experiments revealed that the Li-doped MgO catalyst is instable, irrespective of the preparation procedure or the Li-loading. After 40 h time on stream, none of the catalysts had reached a stable state. Some catalysts seem to be stable after 20 h; however, this is only due to the scaling of the axis.

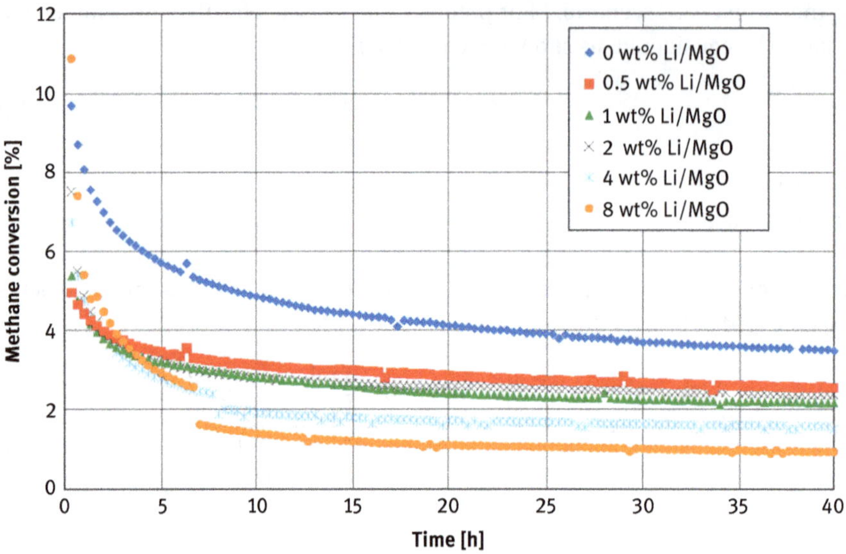

Figure 10.6: The CH_4 conversion versus time on stream for Li-doped MgO with different loadings of Li, prepared via wet impregnation. Reaction conditions: 750 °C; 100 mg catalyst; feed gas composition, CH_4:O_2:N_2 = 4:1:4. The strong deactivation is evident, apparent stability is only due to the scaling of the axis. A stable range has not been reached yet. Unchanged figure adapted from [11] Reprinted from Topics in Catalysis Vol. 54, S. Arndt, U. Simon, S. Heitz, A. Berthold, B. Beck, O. Görke, J. D. Epping, T. Otremba, Y. Aksu, E. Irran, G. Laugel, M. Driess, H. Schubert, R. Schomäcker, Li-doped MgO from different preparative routes for the oxidative coupling of methane, 1266–1285, Copyright (2011), with permission from Springer via the Copyright Clearance Center.

A gap in the trajectories of 4 wt% and 8 wt% Li/MgO appears at approximately 7 h. This can be explained by looking into the calculation of conversion and selectivity in detail. The definition of the conversion X of CH_4 is given in eq. (10.24), and the selectivity for the desired products in eq. (10.25).

There are two ways to obtain $s_{i,0}$ and s_i. In the first way, CH_4 is measured in the bypass mode before or during the reaction and a comparison with the amount of CH_4 after the gas passed the reactor is done. However, this procedure has drawbacks: the gas flow through the reactor is never absolutely stable due to uncertainties of the mass flow controllers; additionally, during such measurements one cannot measure the outlet gas of the reactor. Therefore, the obtained values have large errors, due to summation, especially at low conversions. In the second way, the sum of the remaining amount of CH_4 and all the products is used to obtain $s_{i,0,}$ and the sum of the reaction products is used to obtain $s_{i,0} - s_i$. Therefore, the calculation can result in much lower uncertainties. For the detection of the hydrocarbon species, a flame ionization detector (FID), and for CO and CO_2 a thermal conductivity detector (TCD), was used. When CO_x falls below the detection limit of the TCD, it drops out of the sum to calculate $s_{i,0}$ and $s_{i,0} - s_i$. In that moment, a part of the actual conversion is not detected/calculated anymore; therefore, the conversion suddenly drops as observed for 4 wt% and 8 wt% Li/MgO at 7 h time on stream. The FID is more sensitive than a TCD by approximately a factor of 1,000, which means that hydrocarbon products are still detected, while CO_x species are not, due to the small amounts. Therefore, the calculation of the selectivity is meaningless, as without the detection of the CO_x species, it is 100%. However, this is only a problem at very small conversions.

The duration of the time on stream experiments for the determination of the catalyst stability depends on the intended application. More than 12 h is often useful.

Due to the strong deactivation, it is close to impossible to determine kinetic parameters because all data measured are only snapshots on the deactivation trajectory. Moreover, since the active center of the Li-doped MgO catalyst is unknown [12], it is not reasonable to calculate values such as TON or TOF. That the apparent activation energy also depends on the state of deactivation is also shown in [11].

It is worth mentioning that the Li-doped MgO reacts with the quartz reactor, which is known in the literature [12]. The Li leaves the catalysts rather quickly as the volatile LiOH and it then reacts with the SiO_2 of the inert diluent or the reactor to form Li_2SiO_3. Therefore, the inertness of the reactor is another important subject, which can be, depending on the required reaction conditions and catalyst, a rather difficult task.

In the case of kinetic studies with Li/MgO the catalyst is only stable in the presence of ≥ 80% inert gas in the feed, only such conditions allow the evaluation of the kinetic parameters.

10.3.2 Decomposition of Ammonia

For the change of our energy supply towards renewable energy sources, H_2 is considered to be a potential future energy carrier. However, the storage and transport of H_2 is still an unsolved problem. A possible solution could be the chemical storage in chemical compounds, such as HCOOH, CH_3OH, or NH_3. However, for such applications, it is necessary that the chemicals can be easily decomposed for H_2 retrieval on demand.

The decomposition is a very important step because the decomposition has to be fast enough to ensure the appropriate energy supply in mobile devices, such as vehicles. Moreover, certain by-products can poison the fuel cell. Therefore, the decomposition of the possible hydrogen carrier NH_3 was investigated in detail [13], as it is expected that over ZrON as catalyst, no production of hydrazine occurs.

A reaction mechanism had previously been proposed by Soerijanto et al. [14, 15], as shown in Figure 10.7. Depending on the reaction mechanism, elementary reactions were assumed and the according reaction rates were established, shown in Table 10.1.

The reaction mechanism, and therefore the model, and depending on that the elementary reactions should be carefully chosen in order to obtain a suitable model.

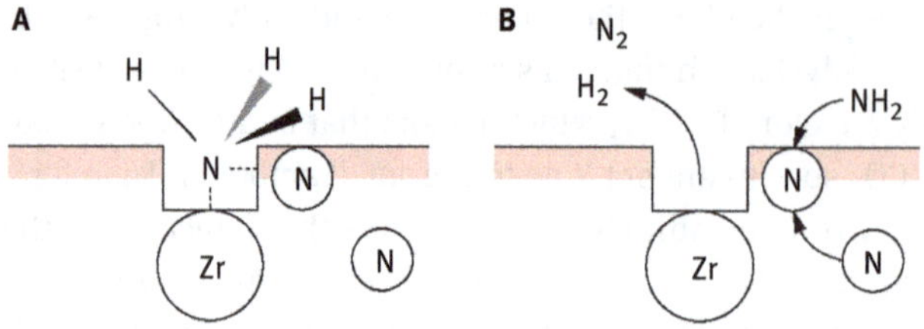

Figure 10.7: The proposed reaction mechanism of Soerijanto et al. assuming a Mars van-Krevelen–like reaction mechanism considering the structure of the catalyst, especially the oxygen vacancies. In (A), NH_3 adsorbs into a vacancy adjacent to a nitrogen atom of the bulk. After delivering of the hydrogen atoms, the remaining nitrogen reacts irreversibly with the neighboring atom to N_2 (B). In contrast to the surface reaction mechanism, the formed N_2 molecule consists of nitrogen from NH_3 and the catalyst. As a result of this reaction step, another vacancy is formed on the catalysts surface, which is filled with nitrogen of another NH_3 molecule, delivering more H_2. Unchanged figure adapted from [13]. Reprinted from Applied Catalysis A: General, Vol. 392, T. Otremba, N. Frenzel, M. Lerch, T. Ressler, and R. Schomäcker, Kinetic studies on ammonia decomposition over zirconium oxynitride, 103–110, Copyright (2011), with permission from Elsevier via the Copyright Clearance Center.

Reactions 1, 2, 4, and 5 are assumed to be fast and reversible reactions, therefore, an established equilibrium can be assumed, with the according equilibrium constants.

Reaction 3 was considered to be the RDS. Therefore, the reaction rates of 1, 2, 4, and 5 are zero, and the reaction rate of 3 represents the rate of the catalytic cycle. A rate for the overall reaction depending on the assumed mechanism can be calculated. To simplify the fitting, the reaction rate was further simplified by combining the first and the last reaction step and the according reaction rate, see eq. (10.37), or by taking the reaction steps 1, 2, 3 and a combination of 4 and 5 into account, see eq. (10.38), with $k_3{}^\star = k_3 \cdot [c_s]$.

$$R = \frac{K_{12} k_3^* K_{45} p(NH_3)^2}{K_{45} p(NH_3) \left(p(H_2)^{\frac{3}{2}} + K_{12} p(NH_3) \right) + p(H_2)^3} \tag{10.37}$$

$$R = \frac{K_1 K_2 k_3^* K_{45} p(NH_3)^2}{K_{45} p(NH_3) \left[K_1 (p(NH_3)(K_2 + p(H_2)^{\frac{3}{2}} + p(H_2)^{\frac{3}{2}} \right] + p(H_2)^3} \tag{10.38}$$

The development of the concentration as a function of the residence time was experimentally determined for several different reaction conditions, such as modified residence time, temperature, and feed gas composition. This is shown in Figure 10.8. Moreover, it is important to perform experiments adding product to the feed gas stream, to investigate if a product inhibition of the reaction can occur.

The fit of the experimental data to the developed model is done by varying the equilibrium constants and one rate constant, until the best possible fit is obtained by minimizing the least square method (the most commonly applied method). Such fitting can be done with a variety of simulation software tools.

Table 10.1: Illustration of the derivation of the elementary steps. Step 1: Adsorption of NH3 on catalyst surface; step 2: first release of H2; step 3: N2-formation of one N atom from catalyst and one from former NH3; steps 4 and 5: filling of one vacancy with another N atom from NH3. Unchanged table adapted from [13]. Reprinted from Applied Catalysis A: General, Vol. 392, T. Otremba, N. Frenzel, M. Lerch, T. Ressler, and R. Schomäcker, Kinetic studies on ammonia decomposition over zirconium oxynitride, 103–110, Copyright (2011), with permission from Elsevier via the Copyright Clearance Center.

Step	Elementary Reaction	Reaction Rate
1	$N'_0\ V_0^{00} + NH_3(g) \underset{k_{1,r}}{\overset{k_{1,f}}{\rightleftharpoons}} N'_0(NH_3)_0^{00}$	$r_1 = k_{1,f} \cdot \left[N'_0\ V_0^{00}\right] \cdot p(NH_3) - k_{1,r} \cdot \left[N'_0(NH_3)_0^{00}\right] = 0$
2	$N'_0(NH_3)_0^{00} \underset{k_{2,r}}{\overset{k_{2,f}}{\rightleftharpoons}} (N_0N_0)^0 + \frac{3}{2}H_2(g)$	$r_2 = k_{2,f} \cdot \left[N'_0(NH_3)_0^{00}\right] - k_{2,r} \cdot \left[(N_0N_0)^0 \cdot p(H_2)^{\frac{3}{2}}\right] = 0$
3	$(N_0N_0)^0 \xrightarrow{k_3} (V_0V_0)^0 + N_2(g)$	$r_3 = k_3 \cdot \left[(N_0N_0)^0\right] \neq 0$
4	$(V_0V_0)^0 + NH_3(g) \underset{k_{4,r}}{\overset{k_{4,f}}{\rightleftharpoons}} (V_0(NH_3)_0)^0_{k_{4,r}}$	$r_4 = k_{4,f}\left[(V_0V_0)^0\right] \cdot p(NH_3) - k_{4,r}\left[(V_0(NH_3)_0)^0\right] = 0$
5	$(V_0(NH_3)_0)^0 \underset{k_{5,r}}{\overset{k_{5,f}}{\rightleftharpoons}} N_0\ V_0^{00} + \frac{3}{2}H_2(g)_{k_{5,r}}$	$r_5 = k_{5,f}\left[(V_0(NH_3)_0)^0\right] - k_{5,r}\left[N'_0\ V_0^{00}\right] \cdot p(H_2)^{\frac{3}{2}} = 0$

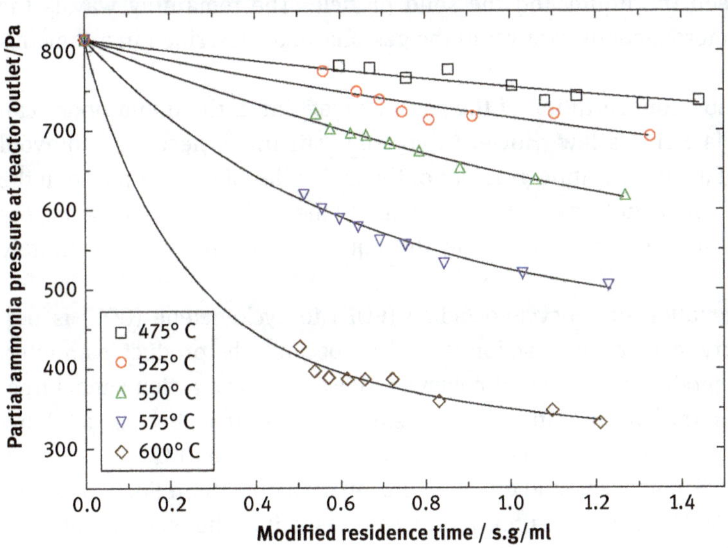

Figure 10.8: Residual NH_3 partial pressures after leaving reactor as a function of the modified residence time $(\tau \times p)/m_{cat}$. The comparison of the three curves shows the influence of H_2 on the catalyst. ($V = 50$ mL/min, $p = 1$ bar (a), $T = 600$ °C, $m_{cat} = 2.0$ g ZrON catalyst). Unchanged figure adapted from [13]. Reprinted from Applied Catalysis A: General, Vol. 392, T. Otremba, N. Frenzel, M. Lerch, T. Ressler, and R. Schomäcker, Kinetic studies on ammonia decomposition over zirconium oxynitride, 103–110, Copyright (2011), with permission from Elsevier via the Copyright Clearance Center.

If such a fit is impossible, the assumed reaction mechanism is incorrect and a different one has to be chosen or developed. However, in the present case, a suitable agreement was reported. Moreover, the enthalpies and their errors could be determined. Based on the reaction mechanism and the determined enthalpies, a reaction pathway was proposed, as shown in Figure 10.9. Its agreement with theoretical predictions supports the proposed mechanism.

10.3.3 Slurry Reaction

In the industry, many processes are performed as three-phase reactions, gas, liquid, and solid. Typically a slurry reactor is used for such purposes, as shown in Figure 10.10. In such a reactor, the catalyst is usually present in the form of a solid distributed in a liquid, which can be a solvent and reactant concomitant. The gas, as reactant, is also pressed into the reactor.

In such a three-phase system, a different spatial concentration profile occurs, as shown in Figure 10.11, as compared to a two-phase system, cf. Figure 10.2. The gaseous reactant molecules first have to cross the boundary layer between the gas and liquid phase. In the liquid phase, they have to diffuse through the boundary layer between the liquid and the solid particle. The remaining way is the same as for a heterogeneous catalyst in the gas phase, as described in a previous Section 10.1.2.

The saturation concentration of the gaseous reactant in the liquid phase can be calculated via Henry's law. However, vigorous stirring is necessary to avoid the formation of a concentration gradient of the molecules of the gas phase in the liquid phase. If this is not ensured, the obtained kinetic data are mostly useless because the saturation concentration in the liquid phase cannot be accurately determined.

The hydrogenation of 1,5-cyclo-octadien (COD) to cyclo-octene (COE) is performed in a slurry reactor. The reaction is relevant because the product is an intermediate for the production of special polymers. However, this reaction suffers from the drawback that the hydrogenation does not stop at cyclo-octene, because a full hydrogenation to cyclo-octane (COA) is possible, as shown in Figure 10.12.

Schmidt and Schomäcker studied the hydrogenation of COD on $Pd/\alpha\text{-}Al_2O_3$ in a slurry reactor [16]. H_2, as gaseous reactant, was pressed into the reactor, but mass transfer of the H_2 from the gas phase to the liquid phase is necessary. To avoid a concentration gradient, the reactor was vigorously stirred. Only when the reaction rate is independent of the stirring rate there is no mass transfer limitation and the solution is saturated with H_2.

Schmidt and coworkers tried to determine the activation energy for this reaction, by measuring the reaction rate for different temperatures. The expectation was an accelerated reaction rate at higher temperatures. However, for this system, the additional

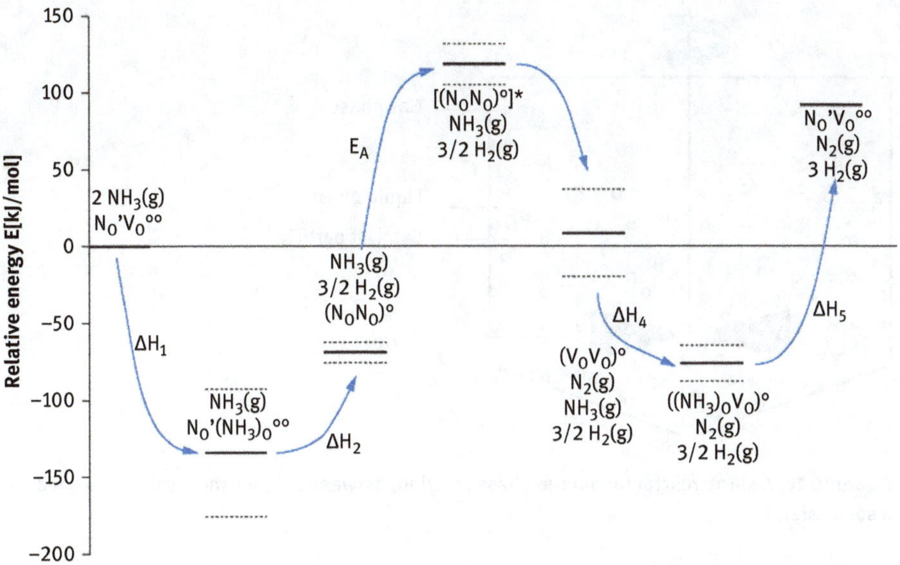

Figure 10.9: The energy level diagram for the assumed reaction mechanism illustrates the energy of the system as a function of the reaction coordinate. The dashed lines represent the error bars of the determined values. At the beginning, the energy of the system is zero, containing the active site and two NH_3 molecules. In the first step, NH_3 adsorbs at the active site and the energy ΔH_1 is freed. Next, the energy of ΔH_2 is needed to release the H_2 molecules. This energy level corresponds to $\Delta H_{1,2}$ because this is the sum of ΔH_1 and ΔH_2. After that, the activation energy E_a lifts the system into an excited stage. N_2 is released, and the systems energy drops. NH_3 adsorbs to the double vacancy, and the energy ΔH_4 is released. The last step is the release of H_2, which needs an energy of ΔH_1. Figure adapted from [13]. Reprinted from Applied Catalysis A: General, Vol. 392, T. Otremba, N. Frenzel, M. Lerch, T. Ressler, and R. Schomäcker, Kinetic studies on ammonia decomposition over zirconium oxynitride, 103–110, Copyright (2011), with permission from Elsevier via the Copyright Clearance Center.

factors (the mass transfer of H_2 into the liquid phase and the solubility of H_2 in the liquid phase) also need to be taken into account. They are also temperature dependent and they do not necessarily increase with increasing temperature.

It was found that the observed reaction rate did increase with increasing temperatures, see Figure 10.13, but by far not as much as expected considering the activation energy. This is a result of the reduced concentration of H_2 in the liquid phase, due to the increased temperatures.

For the kinetic investigation of every chemical reaction an individual procedure has to be developed, considering all special aspects of the reaction and the analytics of its components. No general procedure or standard equipment can be recommended. The three presented examples were selected to show the diversity of kinetic studies combined with aspects common to all catalytic reactions.

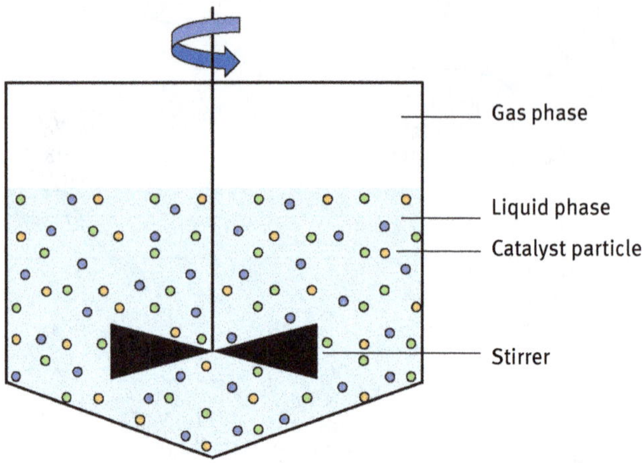

Figure 10.10: A slurry reactor for a three-phase reaction, between a liquid and a solid, catalyzed by a solid catalyst.

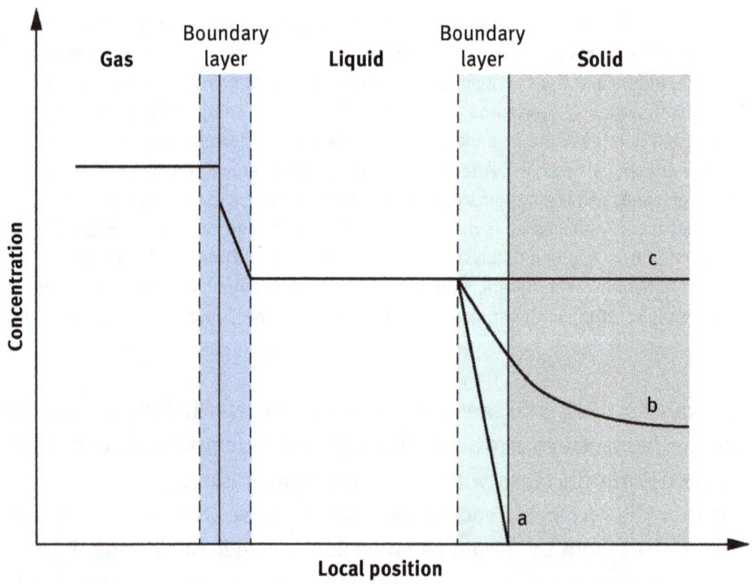

Figure 10.11: Concentration profile in a three-phase system. The absence of a concentration gradient in the liquid phase can only be assured by vigorous stirring. The concentration profile in the catalyst particle depends on whether the reaction is much faster (a) or of comparable rate (b) or much slower (c) than the transport of the reactants to the active site.

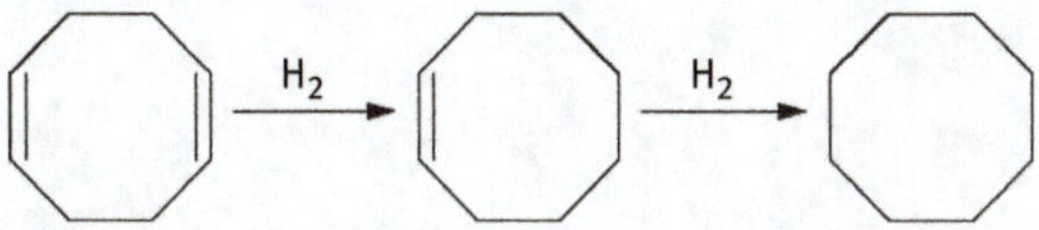

Figure 10.12: Reaction scheme of the hydrogenation of 1,5-cyclo-octadien.

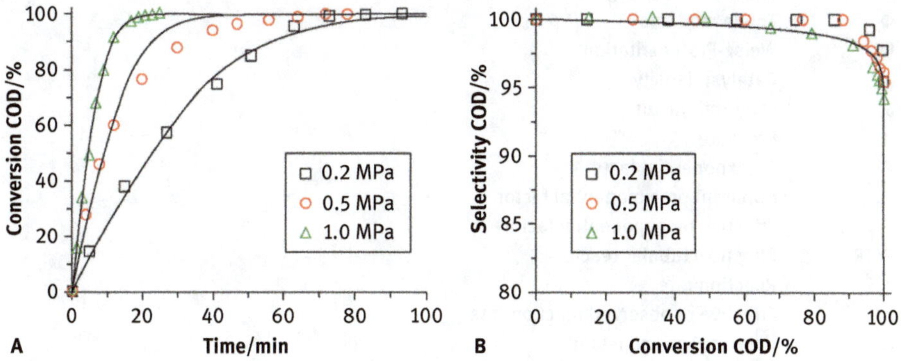

Figure 10.13: COD conversion (A) and COE selectivity (B) at different hydrogen pressures, 50 °C, and $c_{COD,0}$: 0.41 mol/L. Simulation (lines) and experimental data (symbols). Unchanged figure adapted from [16]. Reprinted from Industrial & Engineering Chemistry Research, Vol. 46, A. Schmidt and R. Schomäcker, Kinetics of 1,5-cyclooctadiene hydrogenation on Pd/α-Al_2O_3, 103–110, Copyright (2007), with permission from Elsevier via the Copyright Clearance Center.

Abbreviations

c	Concentration
$c_{i,0}$	Concentration of the component *i* at the time $t = 0$
c_s	Surface concentration
COA	Cylco-octane
COD	1,5-Cyclo-octadien
COE	Cyclo-octene
CSTR	Continuous stirred tank reactor
D	Diffusion coefficient in the gas phase
D_e	Effective or observed diffusion coefficient
ε	Porosity
E_a	Activation energy
E_{app}	Apparent activation energy
E_{eff}	Effective activation energy
FID	Flame ionization detector
GC	Gas chromatography
H	Enthalpy
HPLC	High-pressure liquid chromatography
K	Equilibrium constant

k	Rate constant
k_{app}	Apparent rate constant
k_{eff}	Effective rate constant
l	Characteristic length
MS	Mass spectrometry
m	Mass
n	Order of the reaction
OCM	Oxidation coupling of methane
Φ	Thiele modulus
Ψ	Weisz-Prater criterion
ϱ_c	Catalyst density
ω	Labyrinth factor
p	Pressure
P	Preexponential factor
P_{app}	Apparent preexponential factor
P_{eff}	Effective preexponential factor
PFTR	Plug flow tubular reactor
r	Reaction rate
r_e	Effective or observed reaction rate
R	Universal gas constant
RDS	Rate-determining step
S	Selectivity
s	Amount of substance
t	time
τ	Residence time
$\tau_{modified}$	Modified residence time
T	Temperature [in K or °C]
TCD	Thermal conductivity detector
TOF	Turnover frequency
TON	Turnover number
X	Conversion
$\dot{V}$	Volume flow
V_R	Volume of reactor

References

[1] Chorkendorff I, Niemantsverdriet JW. Concepts of Modern Catalysis and Kinetics. Weinheim: Wiley-VCH Verlag GmbH & Co. KGaA; 2007.

[2] Thomas JM, Thomas WJ. Principles and Practice of Heterogenous Catalysis. Weinheim: VCH; 1996.

[3] Van Santen RA, Niemantsverdriet JW. Chemical Kinetics and Catalysis. New York: Plenum Press; 1995.

[4] Baerns M, Behr A, Brehm A, et al. Technische Chemie. Weinheim: Wiley-VCH Verlag GmbH; 2006.

[5] Dittmeyer R, Emig G. Simultaneous heat and mass transfer and chemical reaction. In: Ertl G, Knözinger H, Weitkamp J eds, Handbook of Heterogeneous Catalysis, 2nd ed., Weinheim: Wiley-VCH Verlag GmbH & Co. KGaA; 2008:1727–84.

[6] Kapteijn F, Moulijn JA. Laboratory testing of solid catalysts. In: Ertl G, Knözinger H, Weitkamp J eds, Handbook of Heterogeneous Catalysis, 2nd edition, Weinheim: Wiley-VCH Verlag GmbH & Co. KGaA; 2008:2019–45.

[7] Hagen J. Industrial catalysis. Weinheim: Wiley-VCH Verlag GmbH & Co. KGaA; 2006.

[8] Ertl G, Knözinger H, Weitkamp J, editors. Handbook of heterogeneous catalysis. Weinheim: Wiley-VCH Verlag GmbH & Co. KGaA; 2008.

[9] Fogler HF. Elements of Chemical Reaction Engineering. Upper Saddle River: Prentice Hall International; 2004.

[10] Moulijn JA, Van Diepen AE, Kapteijn F. Deactivation and regeneration. In: Ertl G, Knözinger H, Weitkamp J editors. Handbook of heterogeneous catalysis, 2nd ed., Weinheim: Wiley-VCH Verlag GmbH & Co. KGaA; 2008:1–18.

[11] Arndt S, Simon U, Heitz S, et al. Li-doped MgO from different preparative routes for the oxidative coupling of methane. Top Catal. 2011;54(16–18):1266–85.

[12] Arndt S, Laugel G, Levchenko S. et al. A critical assessment of Li/MgO-based catalysts for the oxidative coupling of methane. Catal Rev Sci Eng. 2011;53(4):424–514.

[13] Otremba T, Frenzel N, Lerch M, Ressler T, Schomäcker R. Kinetic studies on ammonia decomposition over zirconium oxynitride. Appl Catal A Gen. 2011;392(1–2):103–10.

[14] Soerijanto H, Rödel C, Wild U, et al. The impact of nitrogen mobility on the activity of zirconium oxynitride catalysts for ammonia decomposition. J Catal. 2007;250(1):19–24.

[15] Soerijanto H. Catalytic on-board hydrogen production from methanol and ammonia for mobile application [PhD thesis]. Berlin: Technische Universität; 2007.

[16] Schmidt A, Schomäcker R. Kinetics of 1,5-cyclooctadiene hydrogenation on Pd/α-Al_2O_3. Ind Eng Chem Res. 2007;46(6):1677–81.

Annette Trunschke

11 Synthesis of Solid Catalysts

11.1 Introduction

11.1.1 Catalysis as an Enabling Technology in Energy Transformation and Storage

Nanostructured inorganic solids applied as heterogeneous catalysts have a major impact on the efficient production of chemical goods and the conversion and storage of energy [1–4]. By altering the mechanism of a chemical reaction, the catalyst in homogeneous or heterogeneous phase reduces the reaction temperature and directs the conversion of the reacting molecule selectively to the desired product rendering energy-intensive separation processes unnecessary, reducing emissions and avoiding the deposition of polluting residues. In this respect, catalysis is an integral part of energy-saving and environmentally friendly technologies. Important current applications of homogeneous and heterogeneous catalysis are energy-related technologies, such as the production of fuels, hydrogen, and synthesis gas, which are still mainly made from fossil raw materials [5–8]; and large-scale industrial processes for the synthesis of chemical consumer goods, like polymerization [9–12], selective oxidation [13], hydrogenation and dehydration [14–16], aromatic alkylation [17]; and the manufacture of methanol [18–20], ammonia [21–24], and many other commodities, intermediates, and increasingly also fine chemicals and pharmaceuticals [25]. In addition, the catalytic wastewater treatment and the purification of gases in stationary processes and transportation emerged as a very important field in heterogeneous catalysis [26].

The present and the future are waiting with new challenges for catalysis research. The classical areas of refining and petrochemistry will continue to face changes in the raw materials basis toward heavier fossil feedstocks [27], and the chemical utilization of light alkanes and C1 building blocks [28]. From a material's prospective, carbon continues to be an important component of consumables produced by

Acknowledgments: All coworkers of the Department of Inorganic Chemistry at the Fritz-Haber-Institut der Max-Planck-Gesellschaft and our collaboration partners (see http://www.fhi-berlin.mpg.de/acnew/welcome.epl) are greatly acknowledged for their contributions, which have been included in this chapter. In particular, I would like to thank Pierre Schwach, Thomas Lunkenbein, Wibke Frandsen, Eduard Fitz, Malte Behrens, Tom Cotter, Giulia Bellini, Hamideh Ahi, and Sabrina Jung for data shown in this chapter and Clara Patricia Marshall, Ezgi Erdem, and Daniel Delgado Munoz for discussion. I thank Robert Schlögl for discussions and his continuous support of our work.

Annette Trunschke, Fritz-Haber-Institut der Max-Planck-Gesellschaft, Department of Inorganic Chemistry, Faradayweg 4-6, 14195 Berlin, Germany, e-mail: trunschke@fhi-berlin.mpg.de

https://doi.org/10.1515/9783110608458-011

the chemical industry simply because of human nature and culture. But this is also why particular efforts must be undertaken to close the carbon cycle. In a first step, CO_2 formation must be avoided in the production of commodities. Released CO_2 must be captured to a great and reasonable extent and re-used applying catalysis technologies [29, 30]. Emissions of CO_2 are particularly high in fertilizer production, which is important for feeding mankind [31], and is typically made from nitric acid. Nitric acid is produced in three steps via steam reforming of methane to generate hydrogen, direct reduction of nitrogen in the Haber-Bosch process, and oxidation of ammonia to NO_2 in the Ostwald process. An important goal here should be to minimize the use of fossil raw materials as hydrogen suppliers and for energy production. Natural carbon sources (crude oil, natural gas, coal, biomass) should not be further burned, but should be used to produce carbon-containing functional materials for daily use and pharmaceuticals, possibly preserving the functional groups that nature already provides in biomass [32–36]. The energy sector should eventually be decarbonized or become essentially carbon neutral. Energy storage and conversion by thermal catalysis and electrocatalytic processes are important issues that need to be addressed to solve such challenging tasks.

The future energy system is expected to be based on hydrogen [37]. Hydrogen can be supplied in many ways using both unsustainable (gray H_2 sources: fossil, including coal and natural gas) and sustainable (blue H_2 sources: fossil with carbon capture and storage and/or use; green H_2 sources: from renewable resources) methods [38]. Hydrogen is obtained as a green energy carrier by water splitting when the necessary energy is supplied from renewable resources like sunlight or wind. Commercial solutions for the production of hydrogen by water electrolysis are already available [39]. The alkaline electrolysis (AEL) performed in KOH or NaOH using, for example, Ni-based electrodes is the most mature and robust technology [40]. Polymer electrolyte membrane (PEM) electrolysis in acidic medium is more efficient and flexible in terms of intermittent renewable power supply, produces hydrogen with high purity, but shows problems with respect to lifetime and is more expensive due to the sophisticated membrane and the use of noble metals in the electrodes [41]. Hydrogen itself stores the energy of wind and sun before it is converted back into electrical energy by combustion in thermal power plants using the existing infrastructure or in fuel cells when needed. The main types of fuel cells differ essentially in the electrolyte used and include proton exchange membrane fuel cells (PEMFC), alkaline fuel cells (AFC), phosphoric acid fuel cells (PAFC), molten carbonate fuel cells (MCFC), and solid oxide fuel cells (SOFC) [42]. In large scale and for transport of hydrogen over long distances, it seems more practicable to incorporate the hydrogen molecule into H-rich substances that are liquid under normal conditions, such as alcohols or ammonia [20, 43]. The existing petrochemical infrastructure can then be used for the transport of the energy carriers, which can be catalytically produced by hydrogenation of CO_2 and nitrogen, respectively, and decomposed again when hydrogen is needed or, in case of alcohols, used directly as fuel. Batteries represent a complementary, especially local energy storage technology, although their high costs prevent widespread use [44].

The difficulty in all this is that catalysis technologies for energy storage and conversion must work on a very large scale to meet the enormous needs of the energy sector. The same is true if solutions for the fixation of N_2 by direct oxidation in fertilizer production should be found [31]. The future usage of hydrogen as sustainable energy carrier and reactant in the production of chemicals and the decarbonization of our energy system depend, therefore, on the availability of efficient, robust and low-priced catalysts composed of abundant elements. The implementation of these constraints places high demands on material and catalyst synthesis.

11.1.2 Multidimensional Nature of Heterogeneous Catalysis

The synthesis of a heterogeneous catalyst significantly influences the catalytic performance, which is generally described by the rate of the catalytic reaction (activity), the proportion of the desired product in the product mixture if several products can be formed (selectivity), and the lifetime of the catalyst (stability) [45].

The catalyst performance is highly sensitive toward changes in the preparation procedure due to the very complex interplay between solid-state, surface, and gas- or liquid-phase chemistry during catalyst synthesis, pretreatment and under conditions of the catalytic reaction. Structure–function relationships are generally the focus of interest in thermal catalysis research [46]. In the increasingly studied field of photocatalysis for water purification or the synthesis of high-value-added products, the relationship between synthesis and derived parameters like chemical composition, structure, crystallinity, electronic properties, morphology and the performance of the material has not even been analyzed in detail. Therefore, upscaling of photocatalysts from the laboratory to the real world has largely failed so far [47]. Electrocatalysis involves catalytic processes occurring at the surface of an electrode. There is growing scientific interest in the direct electrochemical synthesis of important chemical intermediates, like hydrogen peroxide by electroreduction of oxygen, ammonia by electroreduction of nitrogen, or in the synthesis of value-added products such as carbon monoxide, hydrocarbons and oxygenates by direct electrocatalytic reduction of carbon dioxide. In this case, the chemical nature of the electrode material, crystallographic structure, band structure, morphology of the constituent particles, and surface defects significantly affect the efficiency of the catalytic process as well [48, 49].

Electrodes and solid catalysts applied in the synthesis of chemicals or in emission control are, generally, hierarchical systems comprising dimensions ranging from millimeter to nanometer scale, allowing for mass and heat transport within a reactor or an electrode, and molecular transport of reactants and products through a pore system of the catalyst particle, which is generally composed of the active phase and a binder. The molecules move on through the pore system of the active phase and are being subject to chemical reactions on the nanostructured, frequently multifunctional surface sites and ensembles (active sites) (Figure 11.1).

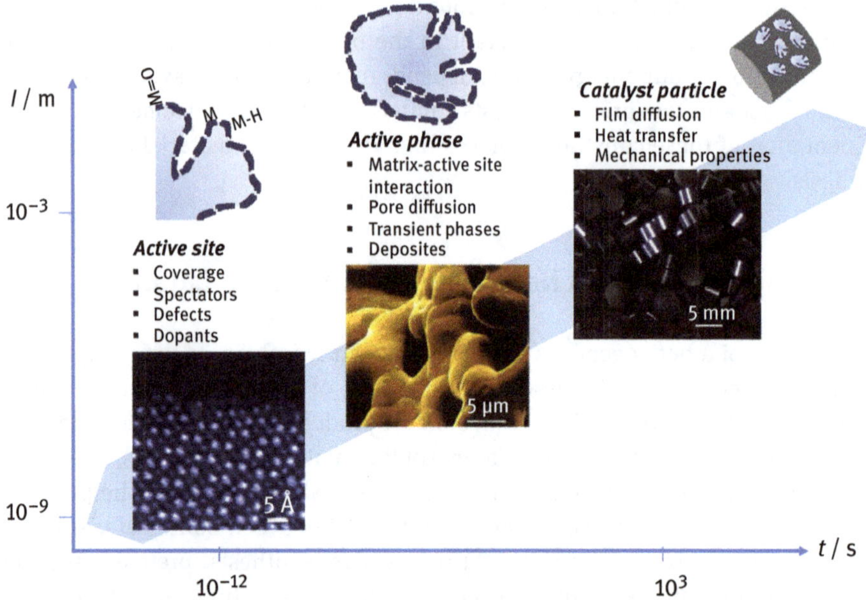

Figure 11.1: Schematic illustration of time and length scale in heterogeneous catalytic systems (illustrated by photographs, electron microscopy images and pictograms) that vary between mm-sized catalyst particles (active phase plus binder and additives), µm-sized regions with the active phase (bulk phase with surface layer), and nm-sized active sites (surface species in molecular dimensions), and between the minutes that the diffusion of a molecule from the gas phase through the pore system of the catalyst particle to the active site can take and picoseconds that elementary reactions last. Some important parameters that influence the rate of a reaction in different dimensions and at different time scales are listed above the pictures; adapted from A. Trunschke, et al., Towards Experimental Handbooks in Catalysis, Topics in Catalysis, 63 (2020) 1683–1699; Open access article, Springer [50].

According to the current understanding, the breaking and making of chemical bonds at atomic scale takes place at ***active sites*** [3, 51–55] that are subject to considerable dynamics either within one catalytic cycle [56, 57] or through the fluctuating emergence and disappearance of sites [58], which, due to their large number, on average provide stationary performance during the needed long-term stable runtime of the catalyst. In heterogeneous catalysis the molecule-sized active site is part of the ***terminating layer*** (dashed dark blue line in the pictograms in Figure 11.1) of the ***active phase*** (light blue area in the pictograms in Figure 11.1) of a dense or porous solid [3].

The terminating layer is the boundary layer between the volume of the active phase and the reacting medium.

The volume of the active phase is either a chemically different support (for supported catalysts) or a self-supporting crystalline or amorphous phase (for bulk catalysts). So, the volume of ***bulk catalysts*** contains the same elements, which are important constituents of the active site on the surface; that is, the active site is "self-supported." In the case of ***supported catalysts***, the surface layer is usually generated synthetically, for example, by impregnation, if the components of the active site are, for example, expensive or rare. Both the termination layer and the support strongly affect the function of the active site by electronic interactions or spatial constraints [59–61], which are adjustable through synthesis.

11.1.3 General Synthesis Strategies

Catalyst preparation yields always a catalyst precursor, whereas the active phase and the active sites are only formed in contact with the feed of the reacting molecules in time-scales ranging from seconds to weeks [3]. Therefore, the nature of the active site finally established at steady state depends not only on the structure of the catalyst precursor, but also on the operation conditions, for example, the reaction temperature and the chemical potential of the gas phase or the liquid phase, for example, determined by the pH value in an aqueous solution. The reacting molecules are activated on the catalyst surface by undergoing a chemical interaction implying that in the course of the catalytic cycle the molecular structure of the active surface species is subjected to electronic and structural modifications as well. In other words, heterogeneous catalysis is necessarily a highly dynamic phenomenon with respect to the structure of the active site and the solid-state catalyst structure that accommodates this site.

An example of catalyst dynamics under different conditions is shown in Figure 11.2. Polycrystalline mixed Mo-V oxide with an orthorhombic crystal structure (called M1 phase [62, 63]) is a very active and selective catalyst in the oxidative dehydrogenation of ethane to ethylene [64]. The annular dark-field scanning transmission electron microscopy images shown here uncover the defective surface of $(Mo,V)O_x$ viewed along the crystallographic *c*-axis. The bright spots correspond to the heavy Mo and V atoms in the complex crystal structure. With the help of the black hexagons, a tiling of the regular crystal structure was performed to simplify the identification of local deviations in the microstructure from the ideal crystal structure. The colored symbols illustrate various surface defects deviating from the ideal crystal structure as explained in the caption of Figure 11.2 and in the literature [65]. The pristine oxide as a product of hydrothermal synthesis (Figure 11.2a) has a lot of surface defects [65], which undergo modifications after thermal treatment in N_2 atmosphere at elevated temperature (Figure 11.2b). It is known that when heated under nitrogen, even the bulk crystal structure of (Mo,

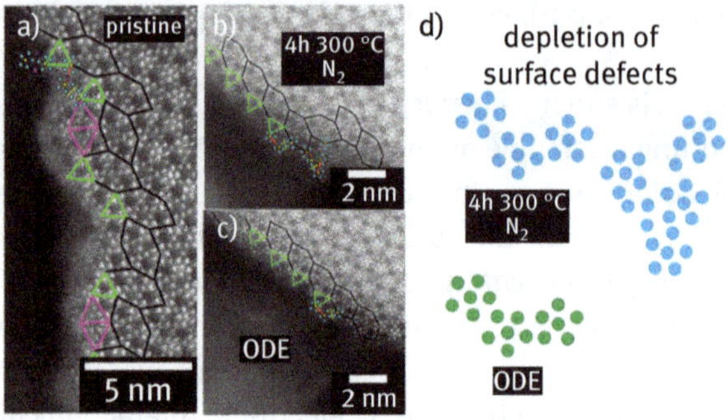

Figure 11.2: Dynamics of a mixed metal oxide in the oxidative dehydrogenation of ethane to ethylene: annular dark-field scanning transmission electron microscopy images of the defective surface of orthorhombic (Mo,V)O_x (M1 phase) viewed along the crystallographic *c*-axis: (a) pristine sample, (b) after thermal treatment in N_2 atmosphere at 300 °C for 4 h, (c) image of the identical location after exposure to conditions relevant for the oxidative dehydrogenation of ethane (ODE), and (d) schematic illustration of the transformation of surface defects during reaction; Black hexagons – tiling of the ideal M1 phase, green and pink triangular motifs, yellow line – mirrored motif, red arrow – translated motif, red and pink circle – shared and rotated motif, respectively. Reproduced from A. Trunschke, et al., Towards Experimental Handbooks in Catalysis, Topics in Catalysis, 63 (2020) 1683–1699; Open access article, Springer [50].

V)O_x changes in terms of metal site occupancy [66]. Depletion of defects to a large extent is observed after exposure of the catalyst to the reaction conditions of the oxidative dehydrogenation of ethane at the same temperature (Figure 2c). These changes are schematically illustrated in Figure 2d. Still, it is not clear, which surface motives are responsible for the superior catalytic performance, because the local surface structure of the primary (Mo,V)O_x particles is rather heterogeneous [67]. Here it is necessary to arrive at a more definite picture by means of further investigations. But it also becomes clear from the example shown in Figure 11.2 that a tailored catalyst synthesis requires both information about the nature of the active site and the knowledge of how the active site is formed, that is, how the kinetics of catalyst transformations under different conditions are correlated with the kinetics of chemical transformations in the environment of the catalyst and at the interphase between the solid and the reacting phases. The link is established by the local chemical potential and its variation during pretreatment procedures and under different reaction conditions.

In the unfavorable case, catalyst dynamics can also lead to more severe changes by transformation of the metastable active phase into a thermodynamically more stable phase [68], which might be one important reason for catalyst deactivation in all types of heterogeneous catalysis, in addition to poisoning by reactants, side products or impurities in the feed [69–72]. The responsible processes are particle reconstruction [73], agglomeration, Ostwald ripening or alloying/dealloying processes in multicomponent systems [74], but also dissolution, corrosion or particle detachment [71]. Stabilization strategies are generally aimed at embedding the active site or ensemble in a flexible matrix that supports the dynamics, which are necessary for the function of the catalyst. In addition, the matrix should prevent irreversible structural transformations. At the same time, however, these strategies affect catalyst stability and performance. The optimization of catalyst dynamics is particularly difficult, when the catalytic reactions are performed at high temperatures at which the atoms in the lattice of the solid feature already considerable mobility [75]. Under these conditions, contributions from homogeneous gas phase reactions have to be taken into account as well. Formation and depletion of involved radical species might be mediated by defective surface sites [76, 77], which further complicates the elucidation of structure–reactivity relationships.

In addition, catalyst synthesis must fine-tune material properties of the volume and the termination layer that are responsible for phenomena like solvation, heat transport, and mass transfer [78–81]. Parameters, for example, are:

- Concentration of hydrophobic or hydrophilic surface functional groups
- Heat capacity and thermal conductivity
- Particle size and surface area
- Porosity

Powders initially obtained in the synthesis of catalyst precursors are shaped into a structured body, such as a pellet or an extrudate, by using binders and additives. These substances and the shaping procedure can additionally modify the activity, the selectivity, and the stability of the catalyst. The shape of the catalyst body depends on the applied reactor technology and the reaction. The dimensions of the hierarchically structured catalyst span orders of magnitudes from the mm large pellets to the nm small active sites (Figure 11.1). This intricacy highlights that synthesis strategies must be designed to optimize the interaction of the catalyst components across all scales. Only then control is executed over the local chemical potential.

In summary, complexity and dynamics in heterogeneous catalysis require integrated approaches in catalyst development, that include not only the synthesis of the catalyst precursor, but also the intended reactor technology and the catalyst dynamics that exist under the given process conditions. It is particularly important to understand the chemical processes that lead to the formation of the active form of the catalyst and to incorporate this knowledge into the catalyst design right from the start.

In the following, general aspects of the synthesis of catalyst precursors (pre-catalysts) will be addressed and discussed covering the entire scale of the catalyst body involving the macro-, meso- and nanostructure. The technical unit operations of catalyst precursor synthesis are summarized in a simplifying manner in Figure 11.3. The first step involves the synthesis of the inorganic solid from solutions by, for instance, precipitation or sol–gel synthesis including filtration, washing, and drying procedures, or by solid-state synthesis techniques. After the manufacture of a shaped body (e.g., spray-dried spherical particles, split, and extrudates), in the second step,

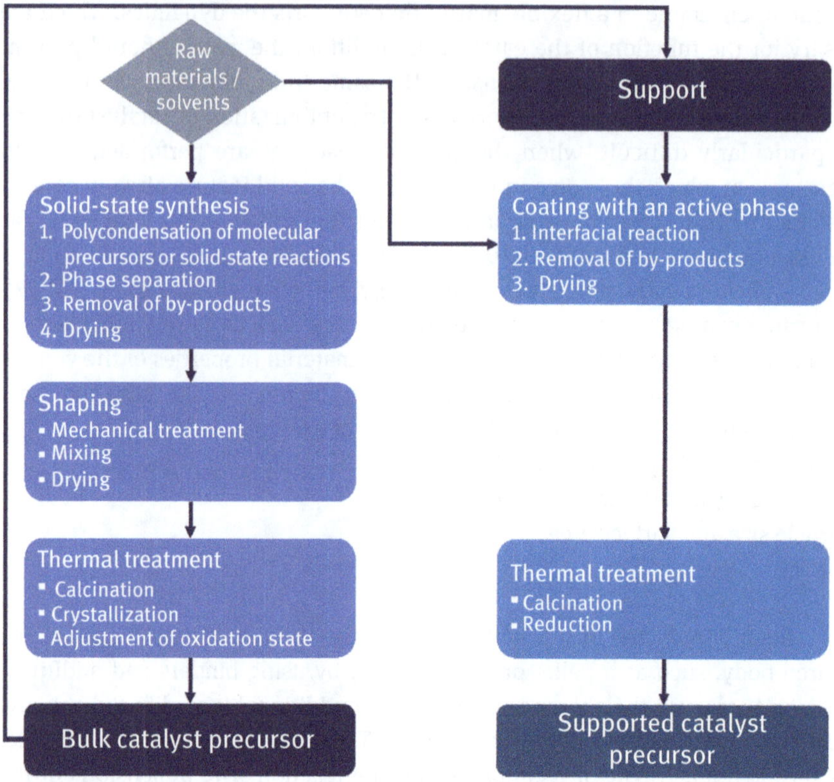

Figure 11.3: Simplified technical workflow of the preparation of catalyst precursors.

thermal treatment finally results in the catalyst precursor or in a carrier (support). In the synthesis of supported catalysts, the active phase is deposited by, for example, impregnation, atomic layer deposition or deposition–precipitation onto the support, again involving separation, purification and drying operations. The final step that yields a supported catalyst precursor is another thermal treatment. It should also be mentioned that in academic studies, the shaping step in the synthesis of bulk catalyst precursors or supports is often omitted. The solid-state formation is then immediately followed by the thermal treatment of the fine powder. For catalyst testing in the laboratory, the precursor is then pressed into split without binder, sieved into a sieve fraction (fifth-tenth part of the reactor diameter), filled into the test reactor, and another thermal treatment is performed just before starting the catalytic reaction. The catalyst formed then under catalysis conditions might show, however, very different catalytic performance compared to a catalyst that contains the same active phase, but was shaped before, e.g., impregnation and/or thermal treatment using additives and binders.

All unit operations in catalyst preparation have a strong impact on catalyst performance, sometimes described with the term “chemical memory.” Reproducible synthesis of catalyst precursors, therefore, requires strict process control by application of analytical tools. Understanding the synthesis and rational catalyst improvements are only possible by scientific analysis of all unit operations considering the catalyst as a hierarchical system and the catalyst design as a loop that comprises solid-state synthesis and activation under working conditions.

Within the scope of this short overview, a detailed description of the numerous synthetic approaches used in catalysis is impossible. Instead, some of the most important basic principles are outlined, referring to literature for elaboration of the concepts. Also, the comprehensive collection of original literature cannot be fully cited here. Reference is made primarily to some review articles that can serve the interested reader as a starting point for further literature studies.

11.2 Tailoring the Macro- and Meso-structure of Heterogeneous Catalysts

11.2.1 Surface Area and Pore Structure

The number of potential active sites on the surface of a heterogeneous catalyst strongly depends on its specific surface area, which can be controlled by synthetic means. The ***specific surface area*** (S [$m^2\ g^{-1}$]) is related to crystallite or particle size, particle morphology, surface texturing and porosity (Figure 11.4) [82–85]. The smaller the particle size of the primary catalyst particles, the larger the specific surface area. Surface defects also contribute to an increase in the surface area [86]. Porous materials generally exhibit much larger specific surface areas than dense solids [87]. Since the surface area is normalized to the mass of the catalyst, the number is affected by the chemical composition (atomic mass of components), and in case of the same composition, by the crystallographic density.

The porosity is referred to as the fraction of the total void volume with respect to the volume of the catalyst grain. The accessibility of the active sites requires pores within the solid that allow molecular transport. Such pores are formed by voids between agglomerated or aggregated primary catalyst particles [84] that represent the catalyst grains (Figure 11.1, active phase). Larger pores are determined by the packing of these catalyst grains, or occur in specifically synthesized hierarchical structures [91–92] and monoliths [93–97]. The active phase, as well as binder components, possess either irregular or regular pore patterns within amorphous particles (meso-structured materials) or a structural porosity of primary crystallites where the pore system is spanned by the atomic arrangements in the crystal structure of the solid, like in zeolites, metal-organic frameworks [87, 98], or crystalline transition

Factor	Low surface area		High surface area	
Size	MgO combustion synthesis $S = 12\ m^2g^{-1}$ $D_{XRD} = 85$ nm		$MnWO_4$ hydrothermal synthesis $S = 29\ m^2g^{-1}$ $D_{EM} = 24$ nm	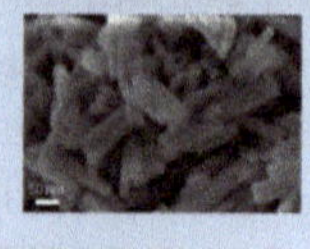
Surface roughness	MgO sol-gel synthesis $S = 39\ m^2g^{-1}$ $D_{XRD} = 11.4$ nm	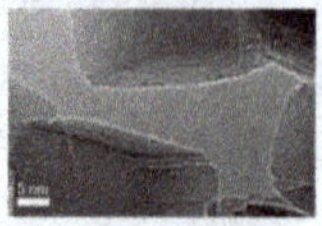	MgO microwave synthesis $S = 111\ m^2g^{-1}$ $D_{XRD} = 13.3$ nm	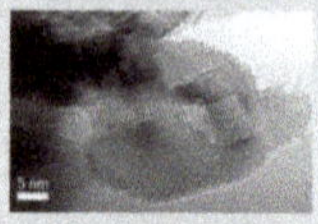
Porosity	Mo-based mixed oxide hydrothermal synthesis $S = 10\ m^2g^{-1}$	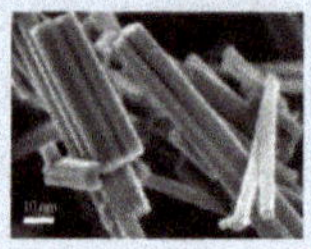	TiO_x supported on mesoporous silica SBA-15 sol-gel/grafting $S = 310\ m^2g^{-1}$	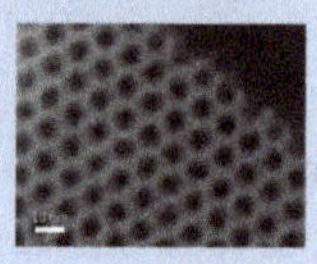

Figure 11.4: Factors that determine the specific surface area of a material illustrated by using the examples of MgO [86], nanostructured $MnWO_4$ [88], mixed Mo-based oxide (Mo-V-Nb-Te oxide) [89], and TiO_x supported on mesoporous silica [90]. *D* corresponds to the primary particle size in polycrystalline solids determined by X-ray diffraction (XRD) or electron microscopy (EM), respectively; for experimental details, please refer to the original publications; please note that when comparing the specific surface area of materials with different chemical composition, the atomic weight of the constituents also plays a role to the disadvantage of the material that consists of heavier elements.

metal oxides [62]. Frequently, catalysts possess a mixture of different pore classes (pore size distribution). The intra-particular pores involve pore diameters in sub-nanometer or nanometer scale. According to IUPAC, the pores are classified into macropores ($d > 50$ nm), mesopores (2 nm $\leq d \leq$ 50 nm), and micropores ($d < 2$ nm) [82, 85]. Pore size, pore size distribution, and shape of the pores constitute the ***texture*** of a catalyst. The ***total pore volume*** (V_P [$cm^3\ g^{-1}$]) is an important descriptive value of a catalyst.

Catalytic properties are controlled by texture and particle size due to mass transport issues. The concepts of "product shape selectivity" and "transition-state shape selectivity" come into play when the size of the pores is in the dimension of the size of reacting molecules [100]. This is, for example, the case in zeolites or inorganic-organic hybrid materials [101–102]. Microporous materials act as "molecular sieves" or prefer only one particular transition state over others and thus the selectivity of a reaction is affected. The stabilization of specific transition states can take place via van der Waals interactions within the micropores. Typical applications are the dewaxing of crude oil fractions by cracking and alkane isomerization or the isomerization of xylenes using the zeolite ZSM-5 [103].

For the transformation of more bulky molecules, larger pores are necessary to reduce diffusion limitations. Synthesis strategies for mesostructured materials, such as silica (e.g., MCM-41, MCM-48, and SBA-15), alumina, carbon and transition metal

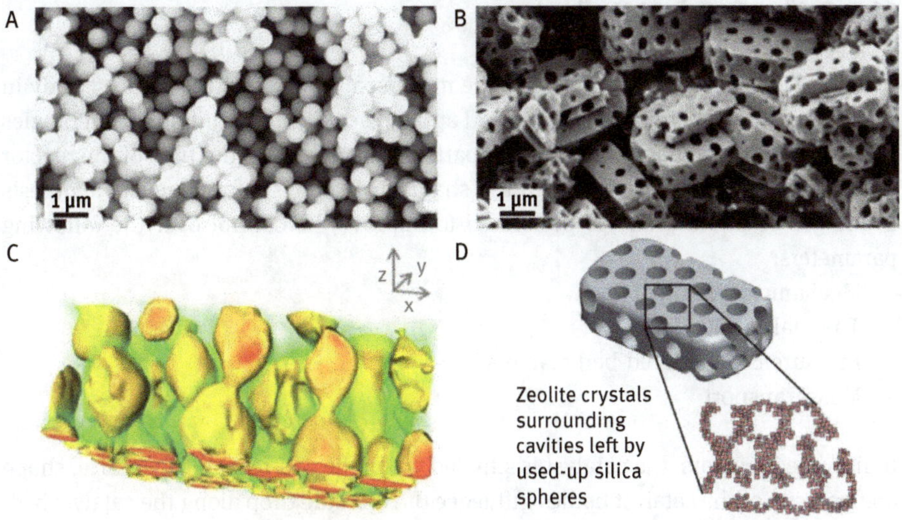

Figure 11.5: Synthesis of a hierarchically organized catalyst systems; mesoporous silica spheres used a template (A) to form microporous crystalline MFI-type zeolite particles with macropores (B); characterization of the zeolite crystals by STEM tomography – 3D visualization of inner pore space (yellow/red surface and volume rendering) and zeolite phase (green volume rendering); schematic representation of the hierarchical system (D). Adapted from A. G. Machoke, et al., Micro/Macroporous System: MFI-Type Zeolite Crystals with Embedded Macropores, Advanced Materials, 27 (2015) 1066; with permission from John Wiley and Sons [99].

oxides with irregular or highly ordered pore structures and pore dimensions in the mesopore-size range have been developed [104–107]. In contrast to zeolites, the pore walls of these mesoporous materials are amorphous and the catalytic properties are not satisfactory due to limited (hydro)thermal stability and acid/base strength [108]. In the context of texture engineering and supported by computational modelling [87], synthesis concepts leading to new zeolites with hierarchical pore architectures and extra-large pores have been developed [109–111]. Other approaches toward decreasing the diffusion path are to reduce the size of nanoporous crystalline particles [112], or to interconnect micro and mesopores by using macroscopic hard templates (inorganic materials with defined particle morphology, carbons, polymer spheres) in zeolite synthesis [99]. Figure 11.5 shows an example in which spherical mesoporous silica particles (Figure 11.5A) were used both as a silica source as well as a sacrificial template for macropore formation (Figure 11.5B and C) during crystallization of a zeolite [99]. The silica spheres are dissolved and used up for the formation of a MFI-type crystalline zeolite phase (Figure 11.5 D). The optimization of electronic and ionic mobility by synthesis of hierarchical structures is equally crucial to improve the properties of batteries and electrodes in fuel cells [113].

11.2.2 The Catalyst Body

Before its use in a catalytic reaction, the macroscopic shape of the catalyst body in the micro- to millimeter scale is adjusted applying catalyst formulation technologies [78, 114, 115]. The shape of the catalyst particles is adapted to the type of the reactor and the process conditions. Generally, shaping of catalysts for chemical synthesis in the gas or liquid phase is performed to improve the control over the following parameters:

- Mechanical stability
- Thermal conductivity
- Pressure-drop in fixed-bed reactors
- Mass transport

In fixed-bed reactors, the catalyst is situated within heated reactor tubes. Size, shape and porosity of the catalyst bodies influence the pressure drop along the catalyst bed, the efficiency of the radial heat transfer, internal and/or external mass transport limitations, and the packing density. All these factors have an impact on the kinetics of the chemical reaction and finally on the space–time–yield ($STY[g_{\mathrm{product}} \cdot g_{\mathrm{cat}}{}^{-1} \cdot \mathrm{h}^{-1}]$) of the product that can be achieved in the process. The spatial distribution of species and temperature gradients within tubular reactors can be, for example, measured in profile reactors (Figure 11.6) [116].

Sufficient mechanical and thermal stability is necessary to avoid attrition and crumbling while charging the reactor and crushing under the weight of the catalyst bed and due to thermal stress under operation [117]. For example, the selective oxidation of propene to acrolein and acrylic acid is performed in tubular fixed-bed reactors with inner diameter of 20–25 mm and a length of 5–7 m [118]. Tens of thousands of these reactors are bundled and placed within a molten salt bath as heat transfer medium. The reactor tubes are filled with the catalyst coated on ceramic spheres (Figure 11.7, schematic illustration of a tubular reactor on the left). The structure of such a catalyst is illustrated in Figure 11.7 showing scanning electron microscopy images of a mixed oxide catalyst which has been coated on steatite spheres with a diameter of approximately 4 mm (Figure 11.7, second from left). The primary, crystalline catalyst particles consist of rods, which are aggregated and/or agglomerated in spheres that have been formed as a result of the spray-drying process applied for drying after precursor precipitation (Figure 11.7, right). These spheres with a diameter of a few micrometers are embedded in a porous matrix of an inorganic binder, resulting in a thin layer of about 100 μm covering the steatite sphere (Figure 11.7, middle).

Common technologies to produce granular catalyst particles with diameters between 1 and 50 mm for fixed-bed and moving-bed applications are pelletizing, granulation, or extrusion [119]. Monoliths [94–97, 120], foams [121], and gauzes [122–123]

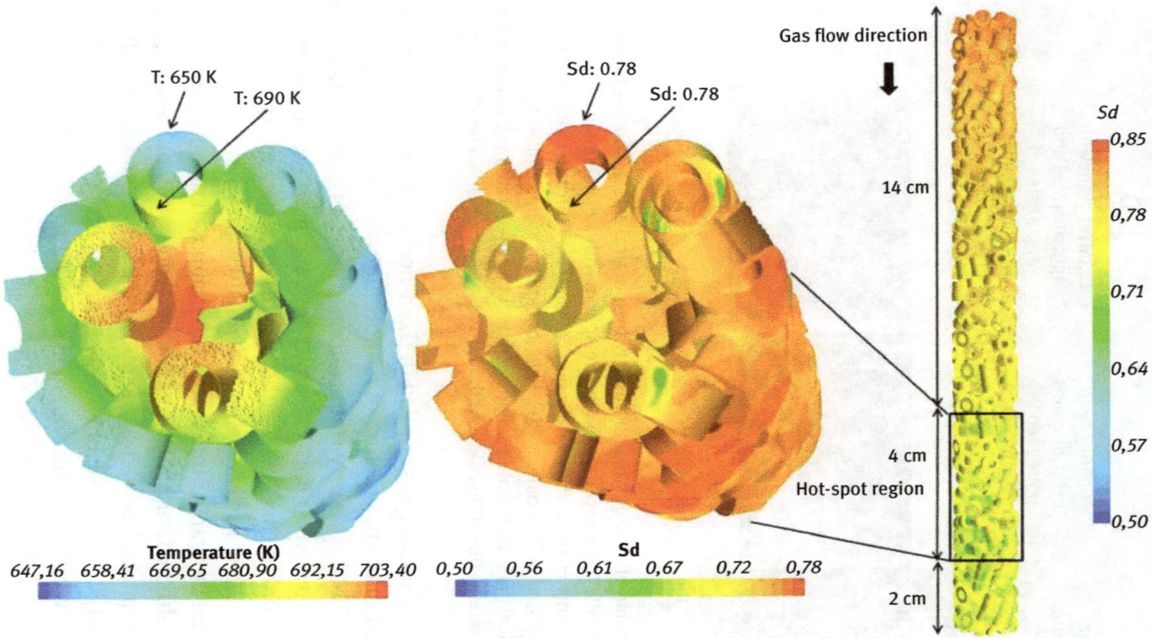

Figure 11.6: Three-dimensional plots of pellets of vanadyl pyrophosphate (VPP) stacked in a multitubular fixed-bed reactor (right) and applied in the oxidation of *n*-butane to the desired reaction product maleic anhydride and undesired oxidation products formed in consecutive and parallel reactions. The temperature profile of the pellets is compared with the computed differential selectivity *Sd* to maleic anhydride (for details see the original publication) at the surface of each pellet in the hot spot of the reactor formed between 14 and 18 cm inside the catalyst bed if the tube wall temperature is set to 370 °C. Reprinted from Y. Dong, et al., What happens in a catalytic fixed-bed reactor for *n*-butane oxidation to maleic anhydride? Insights from spatial profile measurements and particle resolved CFD simulations, Chemical Engineering Journal, 350 (2018) 799–811; reproduced with permission from Elsevier [116].

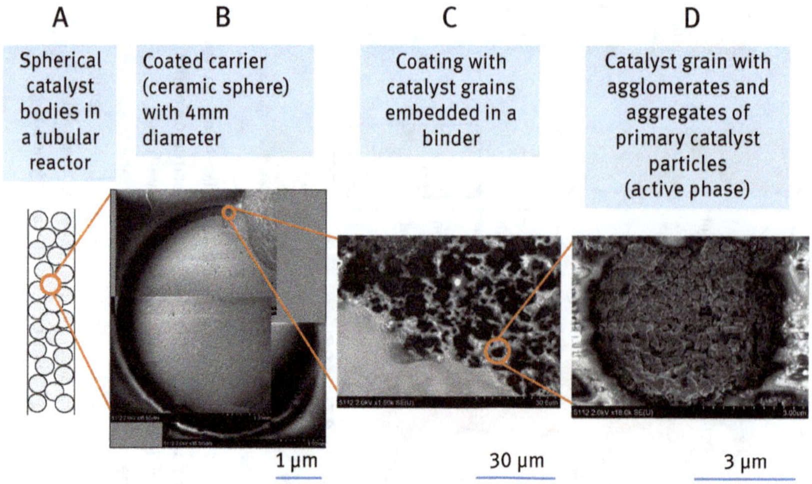

Figure 11.7: Schematic illustration of a tubular reactor filled with spherical catalyst bodies (A) and electron microscopy images of the cross section of a steatite sphere with a diameter of approximately 4 mm coated with a mixed metal oxide with increasing magnification from left to right. Composite image of one sphere (B), coated with a layer of catalyst grains embedded in a binder (C). An almost cylindrical grain that contains aggregated and agglomerated rod-like primary catalyst particles of approximately 200–300 nm in length and 150–180 nm in diameter (D).

are other examples of shaped catalysts applied in fixed-bed reactors. Recently, 3D printing has been applied in the production of monolithic catalysts [124–125].

Catalysts for fluidized bed or slurry applications consist of micro-spheroidal particles with diameters between 20 and 100 µm. These particles are subjected to strong mechanical stress within the fluidized bed. Attrition resistance is achieved either by embedding the catalyst particles homogeneously within a mechanical stable and porous matrix of an inorganic binder, which is composed of amorphous alumina, amorphous silica–alumina, or clays (e.g., kaolin, bentonite), or covering the active mass with a porous, protective layer of the binder (egg-yolk catalyst). Catalysts for fluid catalytic cracking (FCC) of petroleum fractions are generally composed of zeolite crystallites embedded within a mixture of amorphous aluminosilicates and clays [126]. Another prominent example is promoted bismuth molybdate for ammoxidation of propene [127], which is protected by a silica layer. For preparation of microspheroidal particles, the spray-drying technique is frequently applied. The technique is based on rapid evaporation of water from droplets, which are formed by feeding an aqueous sol, gel, or suspension through, for example, in the simplest construction, a nozzle together with a flow of hot air or any other suitable gas resulting in spherical catalyst particles of 10–100 µm diameter [128–129]. The spray of droplets can also be generated by discharging the liquid at high speed into hot air from a rotating wheel or disk. The shape of the particles and the particle size distribution depends on the operation conditions and the characteristics of the starting liquid feed.

The mechanical stress applied during shaping of solid catalysts can originate changes in the specific surface area and pore structure of the active phase. Stabilization techniques [130] or the direct synthesis of shaped catalyst bodies [130–131] are therefore beneficial, in particular, with regard to shaping strategies for meso- and microporous catalysts. The usage of formulation additives in catalyst shaping technologies has a strong feedback to the accessibility and nature of active sites [132]. Active sites might be modified or deactivated by the interaction with solvents, lubricants, binders, or carriers through leaching, coverage, or chemical reaction. In recent years, spatial-temporal spectroscopic and microscopic techniques have been used and even specifically developed to characterize inhomogeneities in the distribution and accessibility of active sites within catalyst particles [92, 133].

In addition to the requirements with respect to size, shape, and mechanical stability, the nature of the active phase has also to be adjusted when the same catalyst is applied in different reactor concepts, mainly due to differing process conditions. Vanadium phosphorous oxide (VPO) composed of the vanadyl pyrophosphate (VPP) phase $(VO)_2P_2O_7$ is an excellent catalyst for selective oxidation of *n*-butane to maleic anhydride [134–137]. This type of catalyst has been operated in, for example, fixed-bed reactors [116] and fluidized bed riser reactors [138]. In the different reactor types, different feedstock is applied, the feed being more rich in *n*-butane, that is, more reducible in the riser reactor technology, which requires different catalyst characteristics compared to the fixed-bed reactor [139].

The transport of electrons, protons, ions, and molecules are crucial design elements in the optimization of electrochemical reactions at electrodes [140–141]. The dispersion of metallic or oxidic catalyst nanoparticles on the surface of electrically conductive supports in alkaline or PEM electrolyzers, and the morphology and porosity of the active phase are determined by the synthesis method, the synthesis conditions and the additives used to fabricate the electrode, which also has an impact on mass transport phenomena and the mechanism of the detachment of gas bubbles from the surface. The mechanical stress due to blistering, and phenomena such as electronic stress from the applied current, increased current by blocked sites, the contact resistance between the catalyst and support, as well as the oxidation of the support, contribute to electrode degradation with impact on the long-term stability of the electrolyzer [39]. The optimization of the electrode structure and formulation significantly enhances the adhesion and corrosion resistance of the active mass. The same holds for the transport of charge carriers in batteries [142], and fuel cells. Molten carbonate fuel cells (MCFC) and solid oxide fuel cells (SOFC) operate at temperatures between 650 and 1000 °C. The high operation temperatures can cause chemical changes of the applied catalyst due to enhanced diffusion processes, which can lead to the segregation of phases. Due to the low operation temperature, the high power-density, and the rapid response to changing loads, the proton exchange membrane (PEM) fuel cell is preferred for its use in transportation and many other applications. In PEM fuel cells, the transport processes comprise the electron transport, the diffusion of protons from the membrane to the catalyst, and the diffusion of the reactant gases to the catalyst, which are again controlled by the catalyst structure, the structure of the gas diffusion layer, the design of the electrodes, and the overall fuel cell design [143].

11.3 From Molecular Species to the Condensed State – Synthesis of Active Phases

The entire tools of synthetic inorganic chemistry including, for example, high-temperature methods, precipitation, solvothermal synthesis, sol–gel chemistry, chemical vapor deposition, and soft matter techniques adapted in part from organometallic chemistry have been applied to synthesize the active phase of catalysts. Generally, the active phase is composed of a bulk phase and a thin surface layer with a thickness in the nanometer range (Figure 11.8, schematic drawing of the active phase on the right). The bulk phase is either a chemically different support (supported catalysts) or a phase that contains already the elements of the surface layer (bulk catalysts). The most frequently applied support materials in catalysis are alumina, silica, amorphous or nanoporous aluminosilicates, and carbon. For battery materials [144–145], and proton exchange membrane fuel cells (PEMFC) or direct methanol fuel cells (DMFC) [69, 140, 143, 146], nanostructured carbon supports are most commonly used, while ceramics or metals are applied, for example, in solid oxide fuel cells (SOFCs) [147–148]. In Figure 11.8, some frequently used methods for the synthesis of bulk phases are listed (see box "solid-state formation").

If required, highly dispersed metal or metal-X (X = for example, oxide, sulfide, carbide, nitride) species are deposited on the surface of the support (see box "Coating" in Figure 11.8), for example, by ion exchange, impregnation, grafting, atomic layer deposition (ALD), deposition–precipitation, or by spreading and wetting. The thin surface layer can also be formed from the bulk phase itself. The coating applied by synthetic methods is also modified under pretreatment and catalysis conditions (see box "Chemical potential during pretreatment and catalysis" in Figure 11.8). The chemical composition and the electronic structure of the surface generally change under the influence of the chemical potential of the environment (see Section 11.1.3). This means that the chemical and electronic properties of the surface layer of a working catalyst are only ever formed under reaction conditions. This takes place at the process conditions of the catalytic reaction through the interaction of the reaction medium (reacting molecules, products, solvents, additives, electrolytes) with the bulk and with the surface layer of the catalyst.

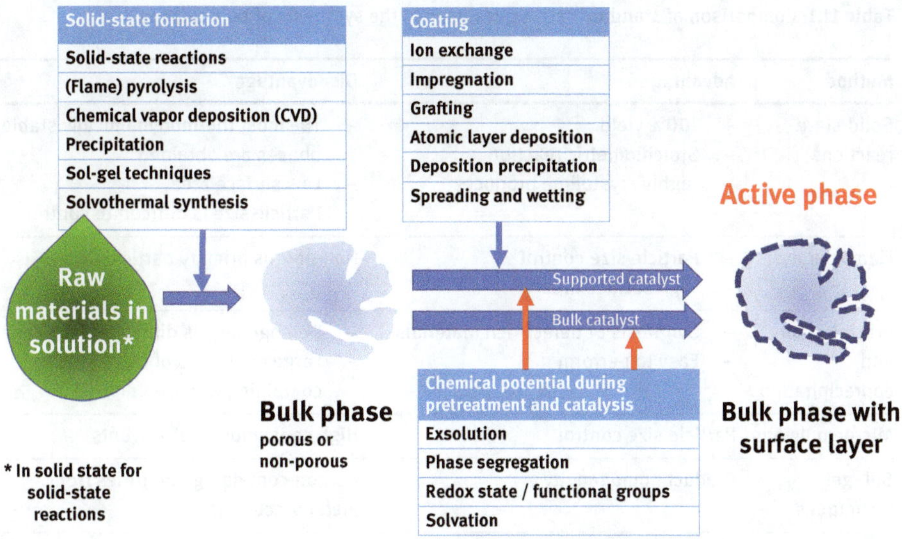

Figure 11.8: Schematic representation of the synthesis of the active phase (bulk phase with surface layer). In the boxes the synthesis methods most frequently used for bulk preparation and generation of the surface layer, respectively, are given. For supported catalysts, the bulk phase corresponds to the support and at least two steps including the coating step and further pretreatment step(s) are necessary to form the active phase. In case of bulk catalysts, the coating step is not necessary and the active phase is formed in pretreatment step(s) of the synthesized bulk phase, which contains already the active element(s).

11.3.1 Bulk Syntheses

Advantages and disadvantages of common bulk preparation techniques are summarized in Table 11.1. What are not mentioned in Table 11.1 are mechanochemical routes to bulk and supported catalysts. In mechanochemical reactions the energy input is supplied by mechanical energy. The solid reactants are mixed and treated mechanically, for example, by grinding or ball milling. The method requires little or no solvent and is in particular effective in introducing defects and strain. Even though the method has been comparatively little applied and studied, very interesting catalysts can still be synthesized in this way [149].

11.3.1.1 High-Temperature Techniques

High-temperature solid-state (ceramic) techniques are advantageous when multicomponent single-phase catalysts with atomically homogeneous distribution of the constituents are desired. In this method, solid reactants are intimately mixed in dry

Table 11.1: Comparison of standard techniques used in the synthesis of bulk phases.

Method	Advantage	Disadvantage
Solid-state reactions	– 100% yield – Stoichiometric reaction – Highly crystalline products	– The most thermodynamically stable phases are obtained – Low surface area – Particle size is difficult to control
Flame pyrolysis	– Particle size control – High surface area	Non-porous primary particles
Precipitation and coprecipitation	– Synthesis of defect-rich materials – Easy to perform	– Homogeneity is difficult to achieve – Large quantities of metal ion-containing wastewater
Microemulsions	Particle size control	High consumption of solvents
Sol–gel techniques	Product homogeneity	Carbon-containing impurities from metal precursors
Solvothermal synthesis	Kinetic control over the phases formed	Predictions are difficult

form and subjected to a temperature treatment. The chemical reaction in solid state requires the diffusion of ions, atoms or molecules within the bulk of the solid, on its surface or via the gas phase. Since the diffusion coefficient of, for example, ions in a solid is typically 10^{-13} cm^2 s^{-1} at 300 K, but increases exponentially with temperature, the reactions are performed at high temperatures, usually higher than two-thirds of the melting temperature of the reactant with the highest melting point in the mixture. One example is the synthesis of carbides from the elements, like vanadium carbide that is synthesized above 1,673 K. Consequently, thermodynamically stable and dense phases with very small specific surface areas and crystallographic rather perfect particles with low abundance of defects are obtained (Table 11.1). In the laboratory, the exothermic and stoichiometric reactions are performed in crucibles or ampoules in presence of a gas phase or under vacuum. The control of the reaction and product separation is difficult. The selectivity to a desired product can be optimized by systematic variation of the mixture of reactants, the reaction time, temperature and the heating and cooling rates. Since the reaction starts at the interface between particles, and defects have a strong influence on the diffusion, previous ball milling enhances the rate of reaction. Examples for catalysts synthesized by ceramic methods are nitrogen-conducting zirconium oxynitrides that have been investigated in the decomposition of ammonia [150] or non-platinum fuel cell catalysts [151]. Some heterogeneous catalysts are produced by metallurgic techniques, such as the fusion of oxidic and metallic components [152]. The most prominent example is the iron catalyst applied in the synthesis of ammonia from the elements [24].

Synthesis at high temperatures is beneficial for thermal stability and controlled crystallinity. On the other hand, high temperatures imply the formation of thermodynamically equilibrated compounds with small surface area and low abundance of defects. More recently, a number of high-temperature methods have become known that provide access to nanostructured, multicomponent single-phase catalysts. Some examples are listed below.

Temperature-programmed methods have been developed that allow the synthesis of high surface area carbides and nitrides [153]. The carbothermal reduction of oxide precursors, such as h-$Mo_{1-x}V_xO_3$, in flowing mixtures of hydrogen and hydrocarbons yields, for example, mesoporous carbides (Figure 11.9A) with exceptionally high surface areas of 80–110 $m^2\ g^{-1}$ in a synthesis that is scalable to the gram scale [154].

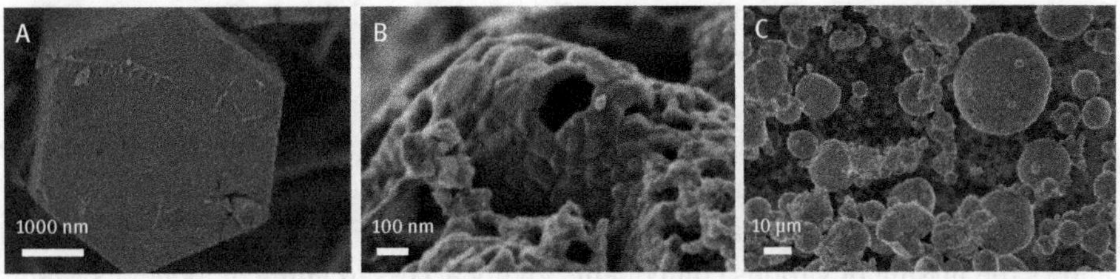

Figure 11.9: Scanning electron microscopy images of (A) β-$(Mo_{0.92}V_{0.08})_2C$ prepared by carbothermal reduction of a hexagonal Mo-V oxide (ID 6703), (B) of the foam-like aggregated and agglomerated particles of the perovskite $LaMnO_3$ prepared by solution-combustion synthesis and calcination in air at 800 °C (ID 30,649), and (C) fumed silica Aerosil® 300 (Evonik) coated with a K_2CO_3–V_2O_5 mixture (ID 22,233); the spherical Aerosil® 300 support particles are still clearly visible. The ID is a sample number that clearly assigns the batch of the synthesized material to the synthesis protocol in an internal sample database of the author's institution.

Solution combustion synthesis (SCS) is a simple method that yields high surface area materials in short reaction times. The procedure starts from an aqueous solution that contains the metal salt precursors in dissolved form. Metal nitrates and glycine as fuel are mixed in a stoichiometric ratio so that combustion after ignition of the resulting gel occurs without the participation of gas phase oxygen. Foams composed of crystalline primary particles are obtained (Figure 11.9B). The heat required for phase formation is supplied by the exothermic reaction, which is why thermal post-treatment can often be omitted. Despite the high reaction temperatures, particle growth is limited by the short reaction time, resulting in crystalline materials of high purity with considerable specific surface area. However, phase composition, morphology of the primary catalyst particles, and the cation distribution in the host lattice are difficult to control. Thus, important parameters that govern the functionality of the material, such as defect concentration, strain, and segregation behavior, are not easily adjustable by this synthesis [155].

Metal oxide nanoparticles synthesized by metalorganic chemical vapor synthesis (MOCVS) at high temperature under reduced pressure followed by quenching at low temperature are promising precursors for functional materials and model catalysts due to their highly defective, non-equilibrium structure [156]. This synthesis technique is especially suited for the preparation of doped or mixed metal oxides, because a particular homogeneous distribution of the components is achieved [157]. The disadvantage is that only small quantities can be produced.

Disordered multicomponent systems (high-entropy ceramics [158]) can be prepared continuously in flames or by applying fast pyrolysis techniques. Supports and catalysts prepared in flames combine thermal stability and defined crystallinity with purity, small particle size distribution, controlled aggregate size, well-defined surface chemistry and high specific surface area. Flame technologies, such as flame hydrolysis have been used in the large-scale manufacture of binary oxides, like fumed silica (Aerosil®, Evonik; Figure 9C), alumina, and titania. The standard TiO_2 photocatalyst P25 is manufactured using this method. In synthesis of fumed silica, $SiCl_4$, air and hydrogen are reacted in a continuously operated flame. The continuous operation is one advantage in view of large-scale production. The size of the non-porous primary nanoparticles can be controlled by the concentration and residence time of the reactants and the flame temperature. Flame spray pyrolysis (FSP) allows the mixing of various components giving access to multicomponent systems and nanocomposites [128, 159]. The deposition of layers on flexible and/or nonflexible substrates is also possible [160].

In the flameless nebulized spray pyrolysis (NSP) method [161], an aqueous solution, which is located in a glass chamber and contains a mixture of metal nitrates, is transformed into a mist using an ultrasonic generator. The mist is then transported by a carrier gas into a hot-wall reactor where the particles are formed at a slight negative pressure.

Chemical vapor deposition (CVD) is an established method for the synthesis of carbon nanotubes (CNTs) [162–163]. Supported transition metals that catalyze the growth of CNTs, such as iron, nickel, or cobalt, are situated in a tubular reactor and CNTs are grown at elevated temperature on the surface of the catalyst particles by decomposition of carbon-containing precursors. The catalyst particles have to be removed by chemical treatment/washing in order to obtain a metal-pure final product.

11.3.1.2 Precipitation

An important technique in the synthesis of support materials and catalysts is precipitation [164]. Binary oxides, like silica, and active aluminas are produced in large scale using this technique and applied as catalyst supports. Co-precipitation is performed in the synthesis of multicomponent materials [165]. Prominent examples of technical catalysts manufactured by co-precipitation are $Cu/ZnO/Al_2O_3$ catalysts for methanol synthesis [19] and vanadium pyrophosphate $(VO)_2P_2O_7$ catalysts for selective oxidation of *n*-butane to maleic anhydride [136].

Precipitation comprises phase separation of a solid from homogeneous solution induced either by physical or chemical means. In catalyst preparation, usually a precipitating agent is added that leads to high supersaturation by chemical reaction. Often, an amorphous solid is formed. Disadvantages associated with this technique are the necessary product separation and large volumes of salt-containing solutions that are generated (Table 11.1). Nevertheless, precipitation is the most important method used in synthesis of catalyst supports. Co-precipitation is applied, in particular, when a homogeneous distribution of different catalyst components in the catalyst precursor is required. In heterogeneous catalysis, a high catalytic activity per unit volume of catalyst is generally the target, which accounts for the stabilization of very small particles under reaction conditions. Stabilization against thermal sintering can be achieved by depositing the nanoparticles on the surface of a thermally stable, porous oxide as support. When loadings of the active mass higher than 10–20 wt% are required, deposition–precipitation is the method of choice [166]. Here, the precipitating agent is often generated in situ by hydrolysis of, for example, urea $(CO(NH_2)_2)$ initiated by increasing the temperature above 60 °C.

Despite its importance, fundamental understanding of precipitation processes remains fragmentary and the control of product properties is limited, which is due to restricted experimental access to the key steps of solid formation and to the multitude of influencing parameters [164]. Precipitation basically involves four steps, including:

1. Formation of a precursor, for example, the hydrolysis product of a metal ion in solution, which is able to condense and to form a solid phase
2. Creation of nuclei through condensation of precursor molecules
3. Growth of the nuclei through the addition of matter, until the primary particle stage is reached
4. Aging of the particles in suspension

In aqueous solution, the solvated aquo-, hydroxo-, or oxo-complexes (eq. (11.1)) of metal ions undergo hydroxylation yielding hydroxy precursors (eq. (11.2)) that are subjected to inorganic polymerization reactions. Depending on the chemical nature of the metal ion, either olation (eq. (11.3)) or oxolation (eq. (11.4)) occurs. Olation results in the formation of hydroxo-bridges between the central metal atoms involving polycations as intermediates (e.g., $[Al_{13}O_4(OH)_{24}(H_2O)_{12}]^{7+}$), while oxo-bridges are formed in oxolation via polyanions (e.g., $[Si_4O_{12}]^{4-}$):

$$\underset{\substack{\text{aquo}\\ \text{pH<7, z} \leq 4}}{[\mathrm{M(OH_2)}]^{z+}} \overset{\mathrm{H_2O}}{\rightleftharpoons} \underset{\substack{\text{hydroxo}\\ \text{pH>7, z} \leq 4}}{[\mathrm{M(OH)}]^{(z-1)+} + \mathrm{H}^{+}_{\mathrm{aq}}} \overset{\mathrm{H_2O}}{\rightleftharpoons} \underset{\substack{\text{oxo}\\ \text{z} > 4}}{[\mathrm{M{-}O}]^{(z-2)+} + 2\,\mathrm{H}^{+}_{\mathrm{aq}}} \tag{11.1}$$

$$[\mathrm{M\text{-}O}]^{-} + \mathrm{H_3O^+} \longrightarrow \underset{\text{precursor}}{\mathrm{M\text{-}OH} + \mathrm{H_2O}} \longleftarrow [\mathrm{M\text{-}OH_2}]^{+} + \mathrm{HO^-} \tag{11.2}$$

$$\mathrm{M\text{-}OH} + \mathrm{M\text{-}OH_2} \longrightarrow \mathrm{M\text{-}OH\text{-}M} + \mathrm{H_2O} \tag{11.3}$$

$$\mathrm{M\text{-}OH} + \mathrm{M\text{-}OH} \longrightarrow \mathrm{M{-}}\overset{\mathrm{H}}{\overset{|}{\mathrm{O}}}\mathrm{{-}M{-}OH} \longrightarrow \mathrm{M\text{-}O\text{-}M} + \mathrm{H_2O} \tag{11.4}$$

Hydrolysis equilibria and polycondensation chemistry of metals in aqueous solution are generally well known for single metal ions in diluted systems [167]. The complexity, however, increases substantially under conditions of catalyst precursor precipitation, which is usually performed at high concentration and often involving multicomponent mixtures. In situ spectroscopy and sampling techniques during precipitation titration can yield information about reaction intermediates contributing to rationalization of synthesis procedures that often have been optimized empirically [168–169]. The highly efficient coprecipitation of mixed hydroxycarbonate precursors of Cu/ZnO/Al_2O_3 catalysts for methanol synthesis from metal nitrate solutions is, for example, performed at temperatures between 60 and 70 °C, constant pH between 6 and 7, and approximately 1 M aqueous solution with respect to the total [Cu^{2+}, Zn^{2+}, Al^{3+}] concentration using 1.6 M aqueous sodium carbonate solution as precipitating agent. Investigation of the precipitation by titration under conditions, which are relevant for industrial catalyst manufacture, revealed that the essential parameter for successful preparation of homogeneous Cu–Zn precursors is the incorporation of Zn^{2+} into malachite $Cu_2(OH)_2CO_3$ under formation of zincian malachite. The step is difficult, because the solidification of the components occurs sequentially with increasing pH as shown in Figure 11.10 for Cu, Zn, and mixed Cu–Zn solutions (see also Section 11.5.5).

Copper is precipitated at pH = 3 (Figure 11.10a) and Zn precipitates near pH = 5 (Figure 11.10b) [169] The consecutive precipitation of Cu and Zn is clearly evident from Figure 11.10c. Increasing the temperature shifts the onset of Zn precipitation to lower pH, but favors on the other hand the formation of undesired CuO by oxolation. It has been shown that the nature of the counter ion changes the hydrolysis chemistry of the components by complexation, resulting in an increased Zn content in malachite when formate is used as anion [169]. The systematic study of the precipitation chemistry revealed that the optimized precipitation parameters found empirically [170], and applied in industry, represent a compromise that has been optimized with respect to performance of the final catalyst. Rationalization of precursor formation during precipitation of complex catalysts can provide new starting points for further optimization of catalyst synthesis leading to phase-pure and homogeneous

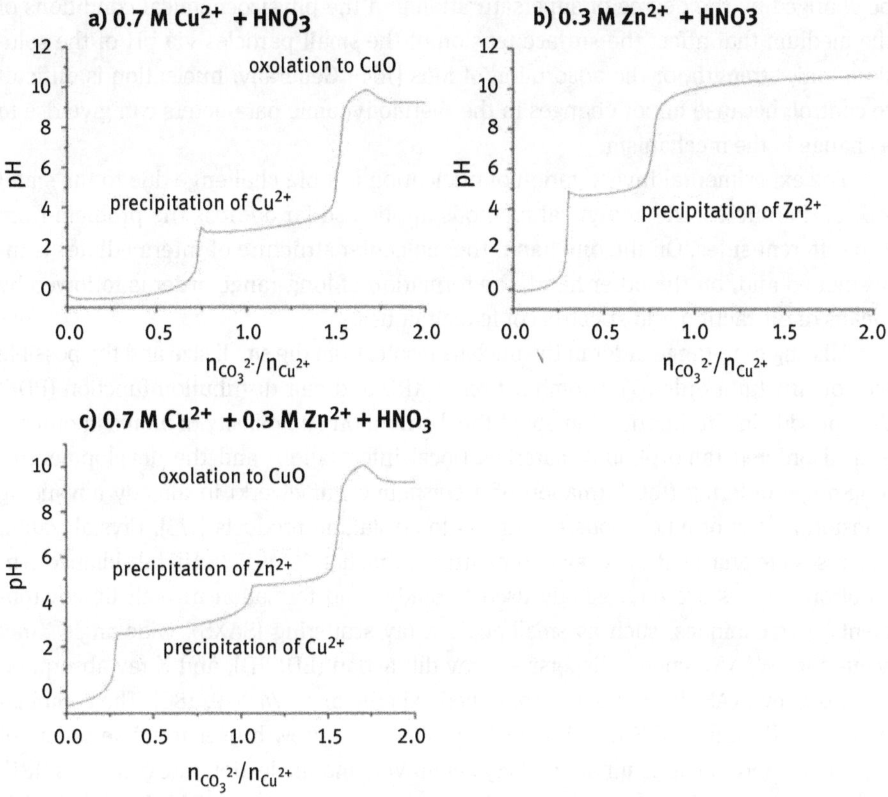

Figure 11.10: Precipitation titration of 1 M (a) Cu, (b) Zn, and (c) Cu–Zn nitrates in aqueous solutions at 65 °C using 1.6 M Na_2CO_3 as precipitation agent. Adapted from M. Behrens, et al., Understanding the complexity of a catalyst synthesis: co-precipitation of mixed Cu, Zn, Al hydroxycarbonate precursors for $Cu/ZnO/Al_2O_3$ catalysts investigated by titration experiments, Applied Catalysis A: General, 392 (2011) 93–102; reproduced with permission from Elsevier [169].

catalysts and must be performed from the ground up for every chemically different system that is to be prepared with this method.

The condensation of the molecular precursors by chemical reactions finally leads to the formation of a solid. The growth requires the formation of zero-charged precursor complexes (eq. (11.2)), because the condensation of charged polycations or polyanions is limited. The smallest solid-phase compound consisting of atoms, molecules or ions, which is formed during a precipitation, and which is capable of spontaneous growth, is called a nucleus [171]. Homogeneous nucleation happens only from clear, supersaturated solutions in absence of seeds or any other solids. The formation of the nucleus is the key step that determines the particle size distribution within the final product. Based on thermodynamic considerations, the radius of the nuclei can be changed by the degree of supersaturation and the physicochemical conditions of the medium that affect the surface tension of the small particles via pH of the solution, ionic strength, or the adsorption of ions [164]. Generally, nucleation is difficult to control, because minor changes in the thermodynamic parameters can give rise to a change in the mechanism.

The experimental investigation of nucleation is a big challenge due to the small size of the nuclei. The analytical methods applied so far address the problem from two different sides. On the one hand, the molecular structure of intermediates is investigated and, on the other hand, the formation of long-range order is followed by means of diffraction and spectroscopic techniques.

Missing long-range order in the nucleus results from the small size and the possible lack of structural order. The combination of XRD and pair distribution function (PDF) data provide in situ information about the distribution and density of atoms in ordered and disordered (amorphous) materials (local information) and the development of long-range ordering (the formation of a crystalline framework) in already advancing transformations of amorphous precursors to crystalline products [173]. Crystallization processes are analyzed by X-ray and neutron diffraction [174–178]. High-brilliance synchrotron sources are increasingly used to study solid formation in realistic environments by techniques, such as small-angle X-ray scattering (SAXS), wide-angle X-ray scattering (WAXS), energy-dispersive X-ray diffraction (EDXRD), and X-ray absorption spectroscopy (XAS) in single and combined experiments [174, 179, 180]. The combination of SAXS and WAXS enables tracking of particle growth over multiple orders of magnitude and the determination of crystal-growth and nucleation rate constants [181]. Light scattering techniques can be used to track the change in particle size distribution with time [182, 183]. Early stages of solid formation have been investigated by high-resolution cryogenic transmission electron microscopy (cryo-TEM) assuming that the observed species are quenched in the hydrated state as in solution [184]. Studies in liquid phase are difficult, because reactive radicals created by electron beam radiolysis during liquid phase transmission electron microscopy (LP-TEM) affect the crystallization and the conclusions might not be directly transferable to the nucleation outside the electron beam [185]. Figure 11.11 shows electron microscopy images of dried semicrystalline precipitation products formed at elevated temperature and pressure under hydrothermal conditions (see Section 11.3.1.4). The target structure is an orthorhombic phase called "M1" (diffraction patterns and schematic representation of the structure in Figure 11.11, top right). In the M1 structure, the metal atoms exist in octahedral and pentagonal–bipyramidal coordination with oxygen atoms. The M–O polygons are connected via edges and corners in the crystallographic *ab* plane, and the *ab* planes are stacked along the crystallographic *c* direction with a distance of 0.4 nm. The images of the precipitation products reveal that long-range ordering in the *ab* plane (images on the left) is still limited, but proceeds already along the *c* direction (sharp 001 peak in XRD and TEM images right, bottom). The formation of similar precipitation products for different metal salt concentrations in solution (250 mM for MoV-Org, 43.75 mM

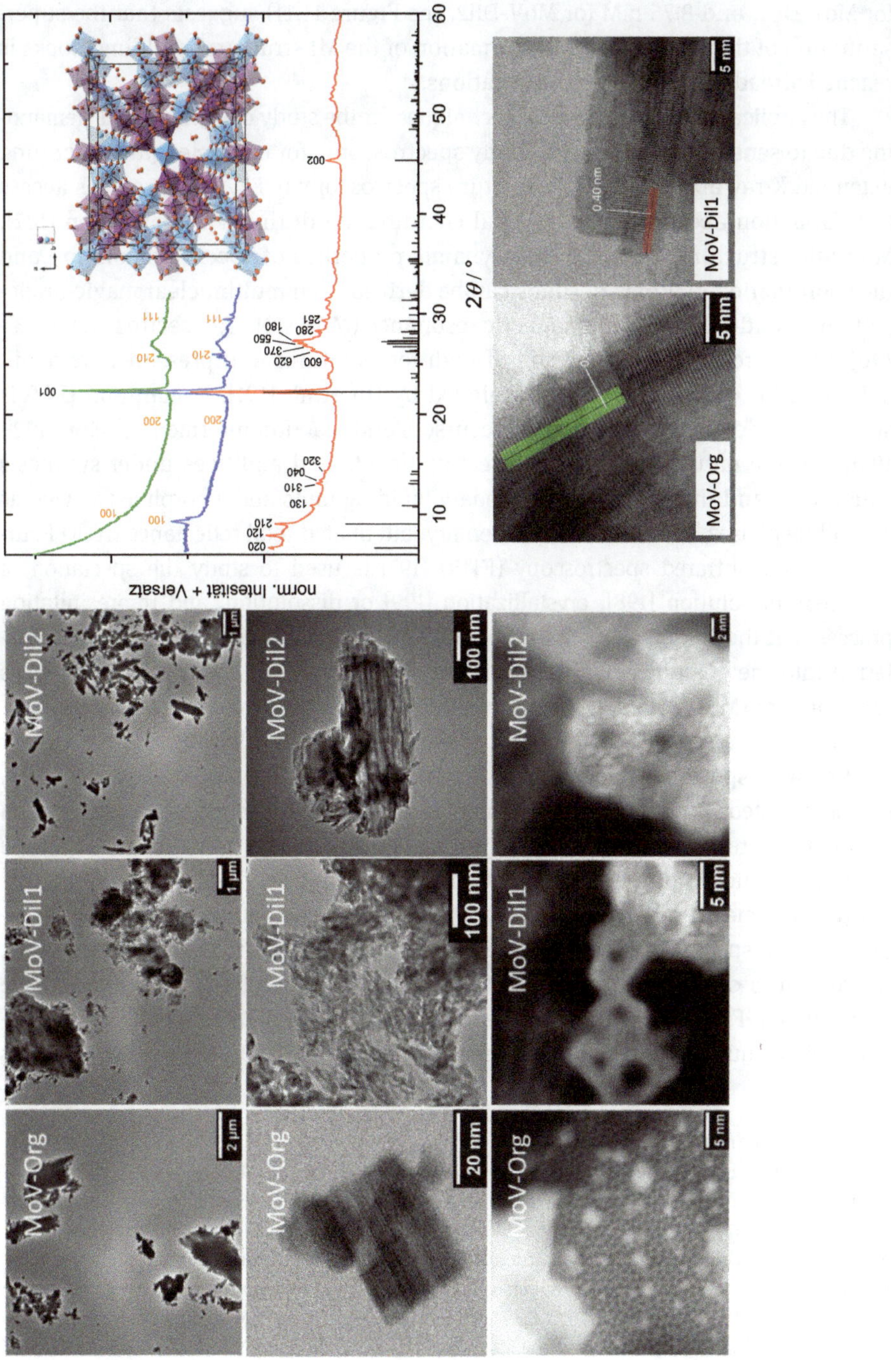

Figure 11.11: Electron microscopy images with different magnification and along different orientations of the semicrystalline hydrothermal products of Mo-V oxide synthesized at 175 °C with

for MoV-Dil1, and 8.75 mM for MoV-Dil2, see Figure 11.11) suggests that the supersaturation of the precursor for the formation of the M1 structural building blocks is reached already at very low concentrations.

The application of spectroscopic techniques for the study of nucleation is demanding due to sensitivity limits [186]. X-ray spectroscopy, for example, quick-scanning extended X-ray absorption fine-structure spectroscopy (QEXAFS), provides access to information about changes in local coordination during solid formation [187]. Molecular structure, chemical and dynamic properties of species in solution and the coordination of atoms in solids can be derived from multinuclear magic-angle-spinning solid-state nuclear magnetic resonance (MAS ssNMR) spectroscopy [188–190]. Sealed rotors for application of high temperatures and pressures are available, see, for example, the cell developed by Hu et al. [191]. Raman and UV/Vis spectroscopy can probe molecular precursors and reaction intermediates [66, 192–193], and the nature of structure directing agents and additives under synthesis conditions, and the structure of secondary building units and amorphous as well as crystalline phases [194–196]. Complementary, attenuated total reflectance (ATR) Fourier transform infrared spectroscopy (FTIR) [197] is used to study the speciation of oligomers in solution [198], crystallization [199] or dissolution, and re-precipitation processes at the solid–liquid interface [200]. Cluster ions from solution can be transferred into the gas phase with electrospray ionization (ESI) and analyzed by mass spectrometry (MS) [201–202]. The advantage of mass spectrometry over NMR or vibrational spectroscopy is that not only selected elements (with NMR active nuclei) or clusters with specific symmetry (with Raman or IR active vibrations), respectively, can be detected. However, ESI-MS can only be applied at low concentrations, which are often not representative of the conditions of catalyst synthesis. In addition, the clusters in aqueous phase are transferred to the gas phase. Combined with ion mobility spectrometry–mass spectrometry (IMS–MS), infrared multiple photon dissociation (IRMPD) spectroscopy, and infrared action spectroscopy in helium nanodroplets, the dynamics of oligomerization reactions in the absence of solvation effects have been studied (Figure 11.12) revealing transitions in the coordination of charged molecular molybdenum oxide units with increasing cluster size [203]. The method provides

Figure 11.11 (continued)
a Mo:V ratio of 4:1 in aqueous solution and a standard total metal concentration of 250 mM (200 mM Mo, 50 mM V) (MoV-Org), as well as diluted synthesis gels of 43.75 mM (MoV-Dil1), and 8.75 mM (MoV-Dil2). The X-ray diffraction (XRD) patterns of the products (top right) show broadened peaks of the target crystal structure "M1" (ICSD 55,097, schematic illustration of the M1 structure in the inset) [62]. For the products obtained under diluted conditions additional reflexes of a hexagonal $(Mo,V)O_3$-type phase (ICSD 75,417) are present. In the TEM images below the XRD patterns, the direction of the *c*-axis of the M1 crystal structure is depicted as a colored bar. Adapted from Jung, S., Speciation of molybdenum- and vanadium-based polyoxometalate species in aqueous medium and gas-phase and its consequences for M1 structured MoV oxide synthesis. Doctoral Thesis, Technische Universität Berlin, Berlin, 2018 [172].

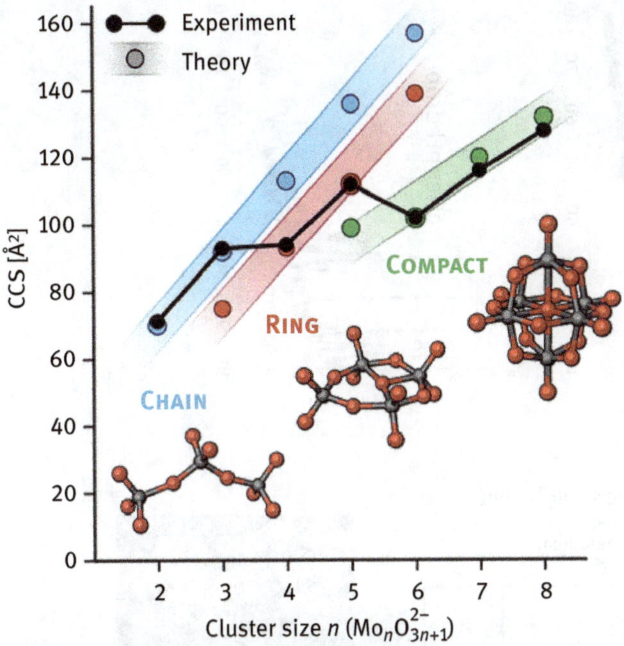

Figure 11.12: Experimental versus theoretical collision cross section (CCS) values measured by IMS-MS for doubly charged polyoxomolybdate clusters $Mo_nO_{3n+1}^{2-}$ (n = 2–8) and plotted as a function of the number of molybdenum atoms in the cluster n, with experimental values shown in black, and predicted CCS values for model structures with a chain, ring, and compact motif shown in blue, red, and green, respectively. Reprinted with permission from Marianski, M.; Seo, J.; Mucha, E.; Thomas, D. A.; Jung, S.; Schlögl, R.; Meijer, G.; Trunschke, A.; von Helden, G., Structural Characterization of Molybdenum Oxide Nanoclusters Using Ion Mobility Spectrometry–Mass Spectrometry and Infrared Action Spectroscopy, The Journal of Physical Chemistry C, 123 (2019) 7845–7853; Copyright 2019 American Chemical Society [203].

access to IR spectra of every individual species separated by IMS-MS and thus to the structure of individual clusters and their structural dynamics in absence of a solvent. Thus, the technique can provide important information for the modeling of oligomerization reactions.

This brief overview of various in situ methods applied to study nucleation and growth in catalyst synthesis shows that complementary tools are needed to approach the nature of the nucleus from both sides: The molecular structure in solution and the periodic or irregular structure in the solid.

Subsequent to nucleation, the particles grow by incorporation of precursor molecules, which might be the same complex, polynuclear species previously involved in the nucleation or simple monomers. Growth and nucleation can also occur simultaneously. The rate of nucleation and particle growth are strongly temperature- and concentration-dependent. The size distribution of the final particles is determined by the

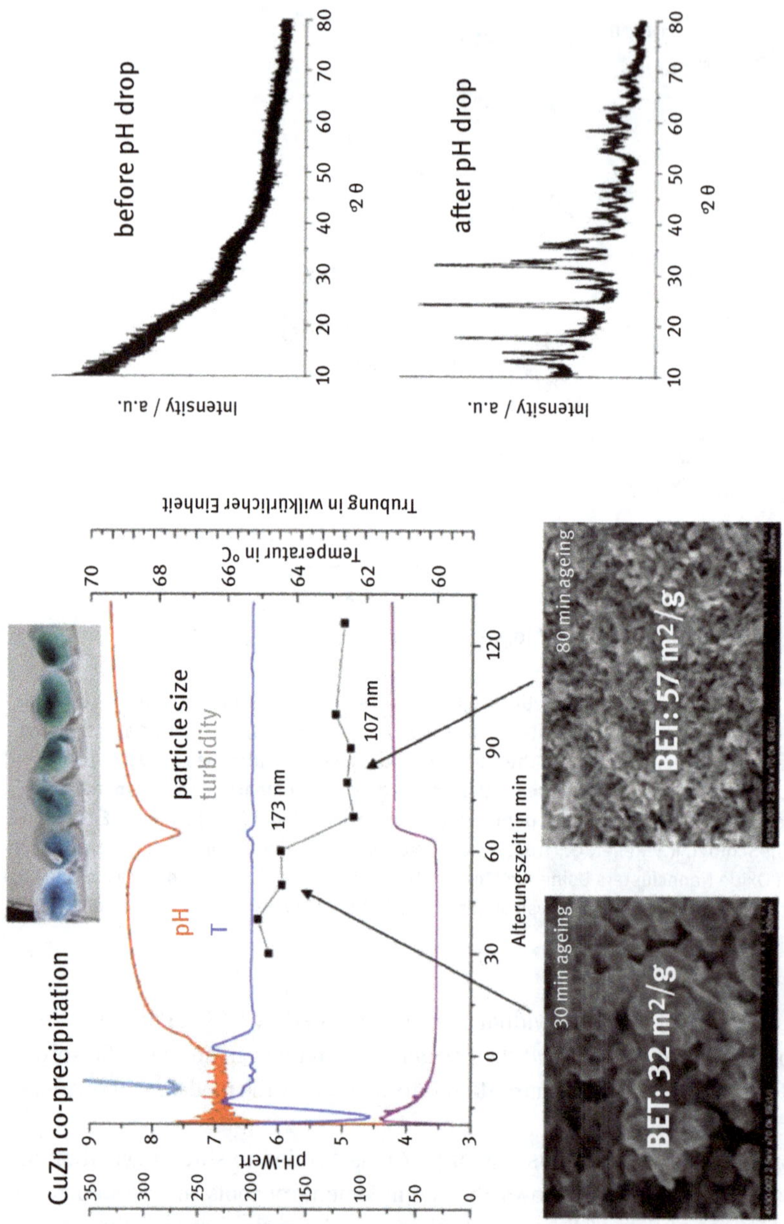

Figure 11.13: Aging of precipitated (Cu, Zn) hydroxycarbonates (Cu:Zn = 70:30); In an automated reactor, temperature (blue), pH (red) and turbidity (purple) were continuously measured. The color change of samples with time is shown in the photograph on the top; The measurement of the hydrodynamic particle diameter was performed after sampling (black squares). Recording of scanning electron microscopy (SEM) images (bottom), determination of the specific surface area according to the Brunauer–Emmett–Teller (BET) method (numbers in the SEM images), and the

relative rates of nucleation and growth, which depends on the growing mechanism. The kinetic description of the formation of complex catalyst precursors applying classical nucleation and growth models [204] often fails because simple assumptions, like a spherical shape of the nucleus, are not always applicable in case of complex systems.

In the precipitation of complex catalyst precursors, amorphous products are often formed in a kinetically controlled manner. Aging the precipitate in the mother liquor allows the system to reach increased thermodynamic stability. During aging various changes may take place, such as further particle growth by dissolution–precipitation or aggregation processes, crystallization or phase transformations and changes in the particle morphology. Figure 11.13 illustrates the modifications that happen when co-precipitated Cu–Zn hydroxy carbonates are allowed to age in the mother liquor at a temperature of 65 °C. The process was monitored in an automated laboratory reactor (LabMax, Mettler-Toledo) using probes for measuring temperature, pH, conductivity (not shown), and turbidity. Crystallinity and average particle size were analyzed by X-ray diffraction and dynamic light scattering investigating extracted samples. After about 60 min, the transformation of the initially amorphous precipitate into a phase mixture of malachite $(Cu,Zn)_2(OH)_2CO_3$ and aurichalcite $(Zn,Cu)_5(OH)_6(CO_3)_2$ occurs, which is indicated by a color change, a temporarily drop in the pH and an increase in turbidity of the mixture. The particle size of the rounded, aggregated particles of the fresh precipitate decreases and smaller, needle-like crystallites are formed, which is reflected in a higher specific surface area of the aged product.

Filtration is a necessary unit operation when a catalyst is prepared by precipitation. Careful washing has to be performed when the counter ions cannot be decomposed by the subsequent calcination. Therefore, carbonate or ammonium ions are preferred counter ions in catalyst synthesis. In the case of nitrates, the evolution of nitrous gases during calcination and the formation of nitrate containing wastewater have to be taken into consideration. In academic research, washing and filtration are frequently rather ill-defined steps, because they cannot be automated easily in laboratory scale. Reproducibility is reduced, for example, due to dissolution–re-precipitation processes during washing [206, 207].

As illustrated above, the thermodynamic process conditions and many other parameters, such as purity and properties of the raw materials, stirring speed and design of the stirrer, geometry of the precipitation vessel, mixing sequence and process design (batch or continuous, constant pH or decreasing/increasing pH) [208], or the addition of complexing agents or organic additives have a strong impact on the

Figure 11.13 (continued)
measurement of X-ray diffraction (XRD) patterns (right) were performed after sampling before and after the drop in the pH after 30 and 60 min. The peaks in the diffractogram of the crystalline material (bottom, right) indicate that the amorphous precipitate (XRD, top) was transformed to a mixture of the Cu,Zn hydroxycarbonates aurichalcite and zinician malachite during aging. Adapted from Master Thesis Fitz, E., Dynamische Lichtstreuung zur Partikelgrößenbestimmung von CuZnAl Hydroxykarbonaten. Beuth Hochschule für Technik Berlin, Berlin, 2008 [205].

quality of the final catalyst precursor prepared by precipitation. Uniform particle size distribution and phase purity of the product can be achieved by optimizing the process conditions of continuous precipitation processes as has been shown for the synthesis of layered double hydroxides [208].

In summary, reproducible catalyst synthesis by precipitation requires a careful mode of operation and online control of process parameters as shown in Figure 11.13. Desired properties, like narrow particle size distribution and homogeneity are currently mainly empirically achieved and the prediction by mathematic modeling becomes limited when the complexity of the system increases. Systematic experimental studies and the application of in situ spectroscopic techniques will contribute in the future to improve catalyst synthesis on a rational basis.

11.3.1.3 Sol–Gel Synthesis

Sol–gel techniques offer some advantages over precipitation with respect to the control of surface area and texture of the resulting material (Table 11.1) [209–214]. In sol–gel synthesis, no real phase separation occurs, differing here from precipitation or flocculation.

The first step of sol–gel processing consists in the preparation of a ***sol*** by primary condensation of molecular precursors into colloidal particles that comprise dimensions between 10 and 100 nanometers. In the case of silica-based sols, ring structures are present as in vitreous silica (Figure 11.14A). The ***sol*** is a stable, clear solution that contains these colloidal particles in form of invisible clusters (micelles). The micelles are charged and those repelling forces prevent coagulation. Hydrated metal ions or metal alkoxide molecules can serve as molecular precursors. Tetraethyl orthosilicate $Si(OEt)_4$ (TEOS) is a frequently used alkoxide precursor in the sol–gel synthesis of silica-containing materials [215]. Due to the low solubility of alkoxides in water, the synthesis is frequently performed in mixtures of organic solvents and water. As shown for silicon alkoxides in eqs. (11.5) and (11.6), the formation of the sol occurs by hydrolysis with the release of an alcohol (eq. (11.5)) and condensation with the release of water (eq. (11.6)):

$$\equiv Si-OR+H_2O \rightleftharpoons \equiv Si-OH+R-OH \tag{11.5}$$

$$\equiv Si-OH+HO-Si\equiv \rightleftharpoons \equiv Si-O-Si\equiv +H_2O \tag{11.6}$$

Because of the low polarity of the Si–O bond, the reaction requires the presence of an acid or a base as catalyst in both hydrolysis and (poly)condensation reactions. Silica sols in water are formed, for example, by acidification of the solution of sodium silicate with a mineral acid forming silicic acid as an intermediate. The acid-catalyzed hydrolysis of a silicon alkoxide, for example, is shown as follows [216]:

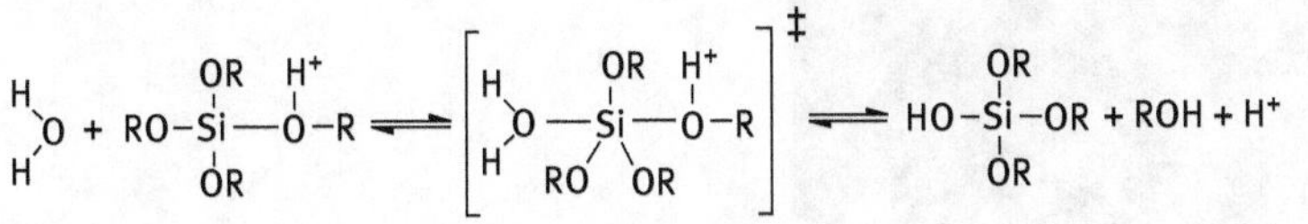

(11.7)

Non-aqueous preparation routes involve the condensation of metal halides, alkoxides or acetylacetonates using alcohols, ethers, ketones, or aldehydes as oxygen donors (Figure 11.14B) [217].

The condensation progresses and after some time, the sol converts into a ***gel*** ("hydrogel" or "liogel") by further three-dimensional cross-linking of the colloidal particles in the sol. The ***gel*** formed is a three-dimensional network of polymeric chains that occupies the entire preparation vessel and is characterized by elasticity when touched (see, e.g., the blue gel in Figure 11.14C). The gelation time in aqueous media depends on the chemistry of the involved elements, the micelle concentration, the pH, and the ionic strength. The process itself may occur within minutes or even days. The catalyst used is also very important. Acid–base catalysis allows the control over the relative rates of hydrolysis and condensation, and, hence, the texture of the resulting colloidal gel.

Within the polymeric network of the gel, solvent molecules and by-products of the condensation reactions are trapped. Before drying, the by-products are discharged by washing. Xerogels are obtained when the solvent molecules are removed by evaporation. If the solvent-free products are obtained after supercritical drying, these are called aerogels.

The versatile methods of sol–gel processing are widely applied today in the synthesis of binary oxides, multimetal mixed oxides and composites, in the preparation of coated monoliths, the synthesis of membranes and thin films [219]. Mesoporous materials are prepared in presence of surfactants, which control the polymerization of silica species so that regular meso-structures are formed (Figure 11.14D) [104, 220]. Catalytically active molecular structures and nanoparticles may be entrapped during synthesis [221]. One disadvantage of the method is the organic chemistry that is often involved and that may cause undesired contamination of the final catalyst with carbon. In particular with respect to the synthesis of transition metal oxides, residual carbon may interfere with the redox chemistry of the metals giving rise to unwanted reduction.

11.3.1.4 Solvothermal Synthesis

Hydrothermal or solvothermal routes are applied to synthesize metastable, more complex structures involving smaller enthalpy and entropy changes than adjustable under standard conditions. The term hydrothermal or solvothermal synthesis is applied for any heterogeneous reaction carried out in a closed system in the presence

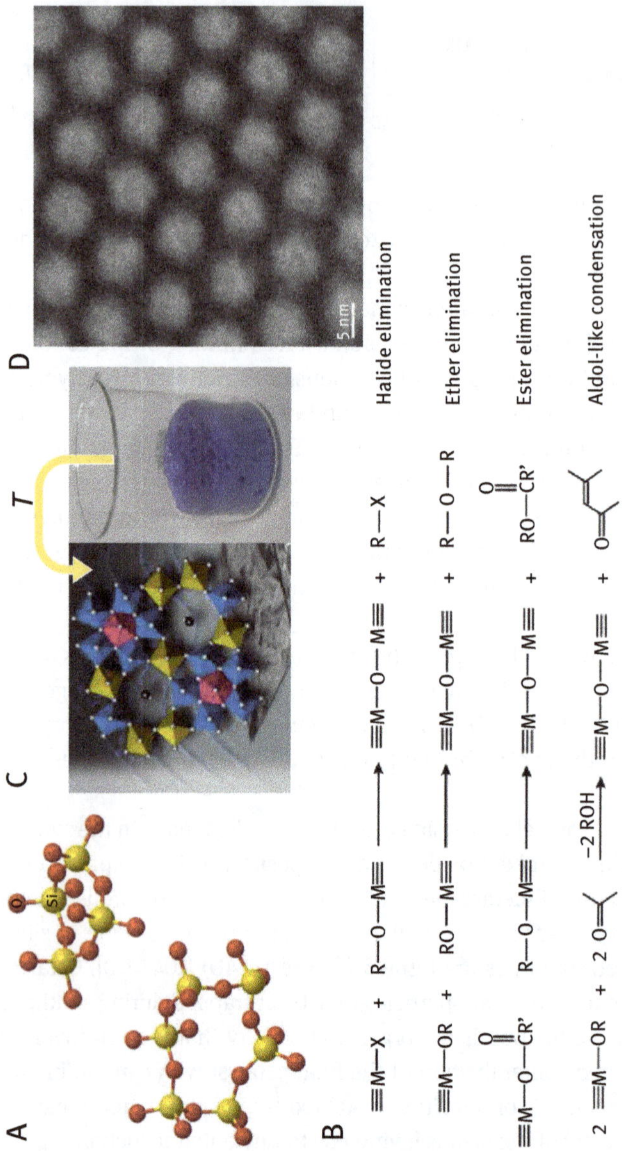

Figure 11.14: (A) Schematic representation of SiO_4 structural units in silica sols, which resemble the primary structural units of vitreous silica (yellow: silicon, red: oxygen). Adapted with permission from R. Ciriminna, et al., The Sol–Gel Route to Advanced Silica-Based Materials and Recent Applications, Chemical Reviews, 113 (2013) 6592–6620; Copyright 2013 American Chemical Society [214]. (B) Condensation steps leading to M–O–M bonds in nonaqueous sol–gel processes [217]. (C) A beaker with a blue hydrogel obtained by acrylamide gelation in presence of Mo, V, Te, and Nb ions; Acrylamide and *N,N'*-methylenebisacrylamide cross-linker were added to a clear

of water or any other solvent at temperatures above the boiling point of water or the solvent used, respectively. Often, no external pressure, but the pressure that develops in the system at the reaction temperature (autogenous pressure) is applied. Solvothermal synthesis is used to prepare a broad variety of material families including oxides, halogenides, chalcogenides, nitrides, carbides, phosphides, metallic, intermetallic, open framework, and hybrid materials [216, 222–229].

Under the particular physical and chemical conditions in an autoclave or bomb, the synthesis of new phases or the stabilization of new inorganic compounds with specific nanostructures, like rods, wires, or tubes, and particular properties, like a narrow particle size distribution and a homogeneous microstructure, is possible. The specific reaction pathways that prevail under solvothermal conditions are in part related to the strongly modified solvent properties [230, 231]. The critical point of water corresponds to a temperature of 374.12 °C and a pressure of 221.2 bar. Under sub-critical conditions the density of water changes dramatically with only slight changes in temperature and pressure. When the temperature rises, the hydrogen bonds are progressively broken, which affects the viscosity of water that decreases as the temperature increases. The decreased viscosity at higher temperatures causes an accelerated mobility of dissolved species. The dielectric constant of water decreases with increasing temperature and increases as pressure increases. All these changes affect the solvation properties of water. Under supercritical conditions, for instance, water molecules cannot screen the metal and counter ions sufficiently, and ion pairs are formed. The acid–base properties are altered as well. The ionic product of water strongly increases with temperature. Subcritical water is, therefore, a rich source of protons and hydroxyl ions. Due to the modified solvent properties and the increased temperature, the hydrolysis of metal ions is forced under hydrothermal conditions accelerating the kinetics of the heterogeneous reactions within an autoclave. Since the condensation mechanism of metal ions is linked to their acid–base characteristics, the changes in the acid–base properties of the reaction medium will have an impact on the polymerization and polycondensation reactions of the metals. Therefore, metal ions with unusual coordination environment can be stabilized under hydrothermal conditions. Furthermore, the redox-chemistry of the metals changes with increasing pressure and as a function of the pH [230]. In addition, hydrolysis of organic counter ions, like acetates, formats or oxalates involves the release of carbon monoxide, which exhibits increased solubility in water under hydrothermal conditions. Acting as a reducing agent, CO can lead to the formation of nanoparticles containing the metals in unusual oxidation

Figure 11.14 (continued) solution of Mo/Te/V/Nb oxalates/citrates. A mixed oxide with an orthorhombic crystal structure (structure schematically shown in the scheme) was formed after transformation of the 3D tangled polymer network into a solid carbon-based xerogel by microwave drying, and subsequent calcination [218]. (D) Electron microscopy image of the regular pore structure of mesoporous silica SBA-15.

states. On the other hand, oxidizing agents, like H_2O_2, $HClO_4$, HNO_3, Cl_2, Br_2, and O_2 can be added to adjust the redox potential. Reactants that do not dissolve easily can be forced to dissolve by addition of mineralizers, which are substances that support dissolution by formation of complexes. Alkali metal hydroxides, for instance, are used as mineralizers in the synthesis of silicate materials.

Specific structures, morphologies, and porosities can be achieved by using structure directing agents or templates. The template-assisted hydrothermal synthesis of crystalline, microporous aluminosilicates (zeolites) is the classical example [224, 225, 232]. Larger pores in the mesopore range are obtained applying sol–gel-related techniques under solvothermal or normal conditions, while templating relies on supramolecular arrays, in which micellar systems are formed using surfactants or block copolymers [91, 108, 213, 232–235].

Solvothermal treatment can also be used to improve the crystallinity, to transform amorphous material into crystalline material, to induce phase changes or spontaneous formation of porous oxides in absence of templates [236]. Microwave radiation can significantly reduce the time that is required to complete the hydrothermal reaction [237–241]. Combined sonochemical/hydrothermal treatments can accelerate crystallization, increase the content of thermodynamically stable phases, and initiate redox reactions [241–243].

The synthesis of metal oxide nanoparticles has also been studied under supercritical conditions [211, 222, 244–248]. In flow processes (Figure 11.15), unique morphologies and metastable structures can be formed, which are not accessible in batch processes [249–251]. Nanoparticles with high specific surface areas can be obtained.

Most of the investigations devoted to the synthesis of inorganic materials by solvothermal synthesis are restricted to the impact of the starting material or the reaction conditions on final product properties. In order to have the ability to control the physical and chemical properties of the synthesis products, the reaction paths and kinetics of the inorganic reactions must be understood. This requires the application, adaptation or development of advanced in situ analysis tools as outlined above in Section 11.3.1.2. The analysis of corresponding spectroscopic data is generally a challenging task in view of the complexity of many systems of practical relevance [195]. Especially with respect to the synthesis of new materials for electrochemical application, such investigations are seldom performed. However, a more detailed understanding of the reaction mechanisms is required for the rational development of next generation materials with advanced specifications. The schematic drawing in Figure 11.16A represents an analytical autoclave. The system was developed to monitor inorganic reactions under hydrothermal conditions using spectroscopic methods (Raman spectroscopy) and to record important process parameters (temperature, pressure, pH, force applied by the stirrer) [196]. The system is fully automated and pre-defined recipes can be processed. Raman spectra of the synthesis

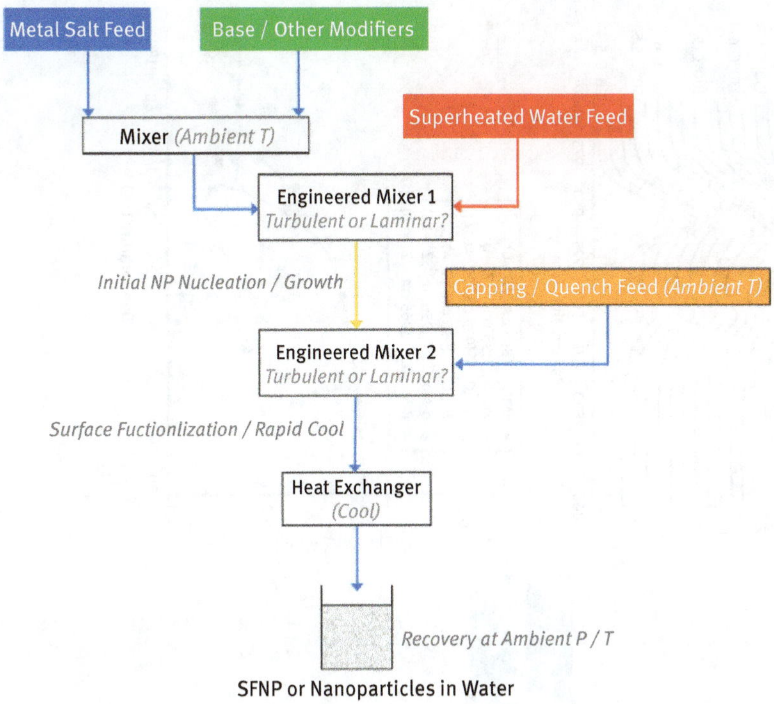

Figure 11.15: Simplified block diagram of a continuous hydrothermal flow synthesis (CHFS). Reprinted with permission from Darr, J. A.; Zhang, J.; Makwana, N. M.; Weng, X., Continuous Hydrothermal Synthesis of Inorganic Nanoparticles: Applications and Future Directions. *Chemical Reviews* **2017,** *117*, 11,125–11,238. Copyright 2017 American Chemical Society [249].

gel can be measured at elevated pressures and temperatures. Figure 11.16B and C exemplarily shows how the speciation of molybdenum oxide species analyzed by Raman spectroscopy changes with increasing temperature and pressure [196]. Depending on the speciation in solution, the formation of different Mo-based mixed oxides was observed [196]. The experiment shows that conclusions about the reaction pathways in hydrothermal syntheses are not possible just based on the knowledge that exists about the situation under normal conditions. In the analytical autoclave comparatively large batches of hydrothermal products can be synthesized. More importantly, under hydrothermal conditions, the synthesis pathway can be influenced and the reaction can be directed to alternative products by the addition of further chemicals using HPLC pumps [195]. This interactive synthesis method, which requires tracking of process data and spectroscopic information, opens up entirely new possibilities for controlled and knowledge-based preparation of catalyst precursors and improves the reproducibility of catalyst synthesis.

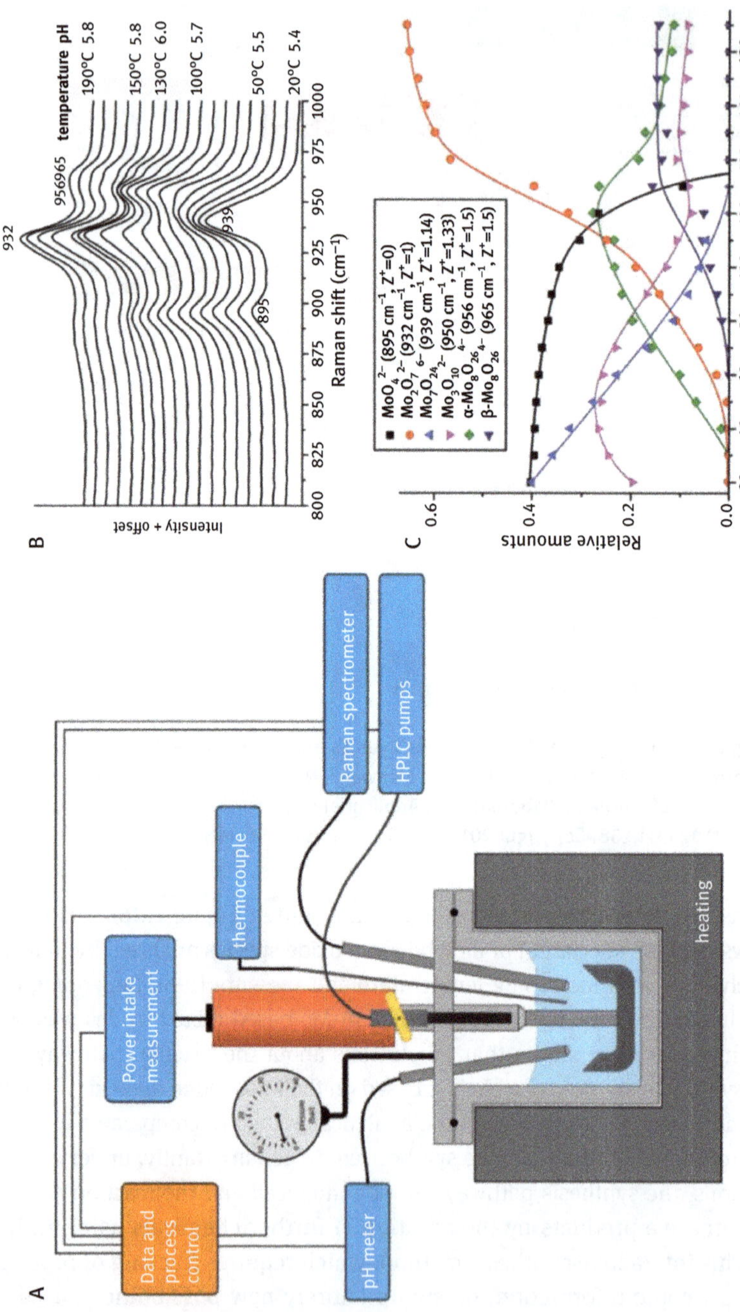

Figure 11.16: Automated analytical autoclave (A) designed together with Premex Reactor AG, Switzerland, for allowing simultaneous analysis of the reaction by Raman spectroscopy, pH measurements, and power intake of the stirrer at temperatures up to 210 °C; The reactor vessel with a

11.3.2 Surface Layers

Under working conditions, an active surface layer is formed by interaction of the reacting phase (reacting molecules, intermediates, products, solvents and additives) with the catalyst (Figure 11.8). In thermal catalysis this occurs under the chemical potential of the gas phase and frequently at elevated temperature, in electrocatalysis under the influence of the applied potential in the presence of the solvent or reactant, gases, and the electrolyte.

For bulk catalysts prepared by the methods described above, the interaction between catalyst and reaction medium occurs at the termination of the crystal or the amorphous solid. Supported catalysts contain a termination layer, which is chemically different from the support and which was added by synthetic means [252]. This layer is also subjected to modifications under working conditions of the catalyst. Supported catalysts are usually prepared, when the precursor that contains the catalytically active element, is costly or when just one monolayer of the active phase is desired. In Figure 11.8, some frequently used synthetic methods for the preparation of supported catalysts are listed in the box "coating." In the following, the most important techniques are described briefly. Subsequently, possible processes during thermal pretreatments and under reaction conditions are shortly addressed for both bulk and supported catalysts (Box "Chemical potential during pretreatment and catalysis" in Figure 11.8).

11.3.2.1 Synthetic Surface Modification

11.3.2.1.1 Ion-Exchange and Impregnation

When the surface of a support bears charged ions, ion exchange can be employed to coat the carrier with the active element. An equilibrium is established between an ion inherently attached to the solid support and an ion in a diluted solution followed by filtration and washing.

Zeolites or clay silicates (e.g., kaolinite and montmorillonite) are cation exchangers as a consequence of their structure containing protons to compensate the negatively

---

Figure 11.16 (continued)
total volume of 400 mL made of Hastelloy C-22 (2.4602) is equipped with a Raman probe (RAMANRXN1, immersion optic 1/4"OD (HC-276); Kaiser Optical Systems), a pH probe (ZrO_2 probe Model A2 and Ag/AgCl reference electrode, both with a ½" outer tubing made from C-276; Corr Instruments), a valve to add gas or to take samples, and two valves to add liquid reactants into the vessel at any given time by use of two HPLC pumps (Smartline pump 1050; Knauer); in situ Raman spectra recorded during heating of a 0.2 M ammonium heptamolybdate solution from 20 °C to 190 °C with the measured pH values (B) and species distribution plot derived from spectral deconvolution of the Raman spectra recorded simultaneously (C). The Z^+ value describes the number of protons consumed per molybdenum atom in the condensation reaction; adapted with permission from John Wiley and Sons [196].

charged framework. The protons can be exchanged by other positively charged ions, for example, alkali ions or organic cations. The same holds for layered double hydroxides as anion exchangers, which can be represented by the general formula $[M^{2+}{}_{1-m}M^{3+}{}_{m}(OH)_2][(A^{n-})_{m/n}\bullet xH_2O]$, where M are divalent (e.g., Mg^{2+}, Ni^{2+}, Cu^{2+}, Zn^{2+}) and trivalent (e.g., Al^{3+}, Fe^{3+}, Rh^{3+}) cations, respectively, and A stands for anion, such as OH^- or $CO_3{}^{2-}$, but also bigger anions, such as $[Mo_7O_{24}]^{6-}$. Such structural ion exchangers possess a constant number of exchangeable surface sites, irrespective of the pH of the aqueous solution in which the ion exchange is performed subject to stability of the phase. The ions can be exchanged in the entire structural pore system (bulk) of the crystalline materials, not only on the external surface.

But, also the surface of binary oxides, like silica, can be functionalized in order to achieve ion exchange capacity [131, 253]. In Figure 11.17, an example is shown. Grafting of 3-aminopropyl-trimethoxysilane on the surface Si-OH (silanol) groups gives surface amino groups that can be converted into anchored propylammonium chloride species by reaction with hydrochloric acid resulting in a surface that is capable to exchange the chloride anions [254, 255]. Chloride can be exchanged, for example, by vanadate or molybdate species or any other negatively charged ion in solution, resulting in a high dispersion of the adsorbed species on the surface of the support [255, 256]. Finally, the residual organic groups on the functionalized support must be removed by calcination in air at 550 °C (Figure 11.17A). The latter step is comparatively ill-defined. Depending on the loading, it leads to the support being covered by a sub-monolayer or a monolayer of anchored metal oxide species and residual hydroxyl groups, respectively. This is schematically illustrated for silica-supported molybdenum oxide species in Figure 11.17A at the bottom. The characterization of (sub-) monolayer MoO_x/SiO_2 catalyst precursors has shown that with increasing amount of surface molybdenum oxide species (increasing intensity of bands due to Mo=O and Mo–O–Si stretching vibrations in Figure 11.17B), the proportional concentration of H-bonded silanol groups increases (Figure 11.17C), but the total concentration of the remaining free silanol groups decreases (Figure 11.17D, blue data). The interaction of the surface molybdenum oxide species with residual OH groups results in the formation of new acid sites on the surface of the supported catalyst precursor both of Lewis and Brønsted type (Figure 11.17D, black and red data, respectively) [257].

On the surface of non-functionalized mineral oxides, electrostatic adsorption (equilibrium adsorption) rather than ion exchange is possible. Equilibrium adsorption is the method of choice when small amounts of expensive noble metals should be deposited onto the support, like Pt supported on alumina for catalysts applied in hydrocarbon conversions. In aqueous solution, the oxides are covered by hydroxyl groups, which undergo hydrolysis resulting in a pH-dependent adsorption capacity based on electrostatic interactions according to the following equation:

$$[M-O]^- + H_3O^+ \rightleftharpoons M-OH + H_2O \rightleftharpoons [M-OH_2]^+ + HO^- \tag{11.8}$$

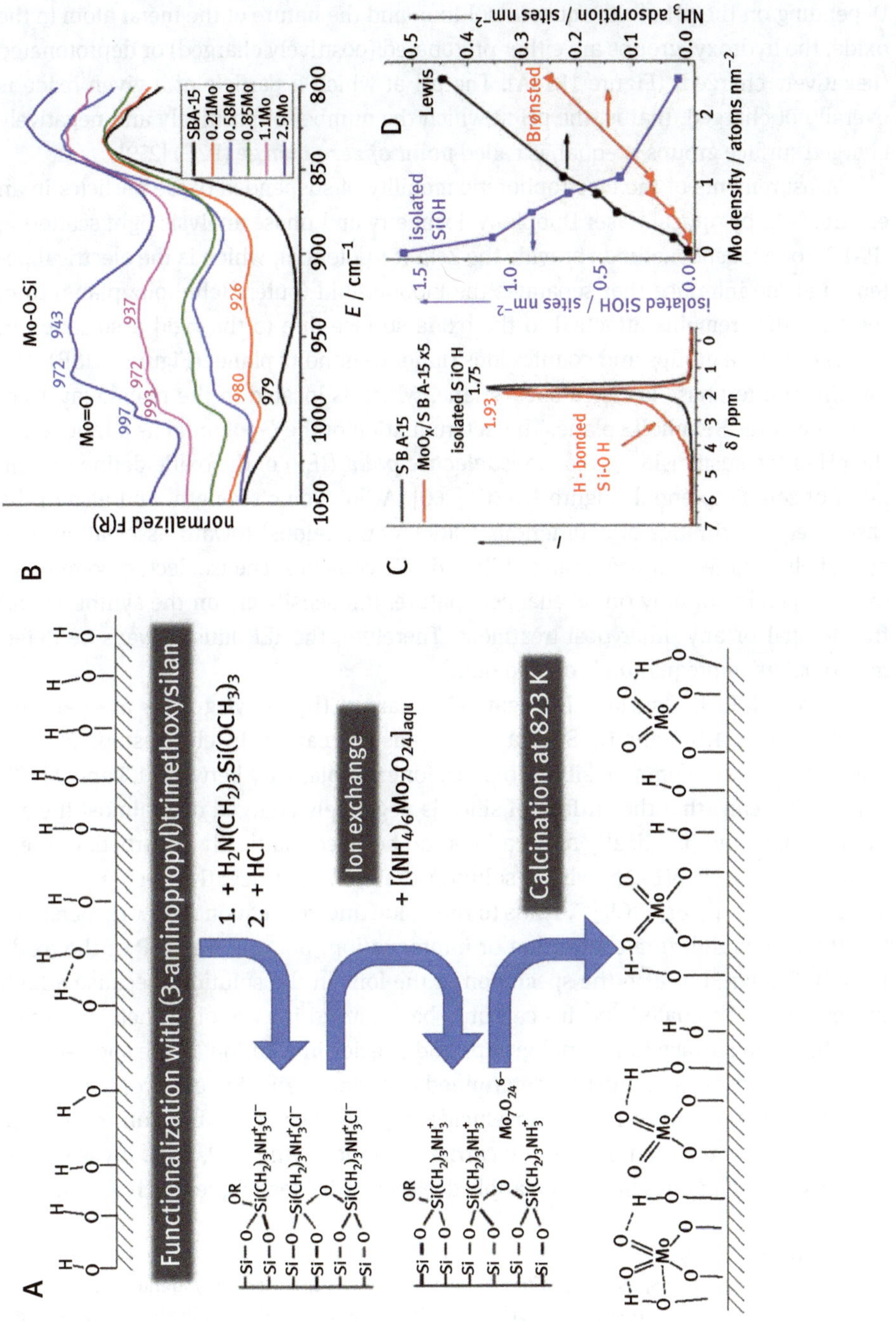

Figure 11.17: Ion exchange on the surface of functionalized silica. (A) Scheme adapted from Hess et al. and Amakawa et al., which shows the functionalization of the silanol groups (Si-OH) on the surface of silica, subsequent ion exchange, and the formation of dispersed di-oxo Mo(VI) surface

Depending on the pH of the aqueous solution and the nature of the metal atom in the oxide, the hydroxyl groups are either protonated (positively charged) or deprotonated (negatively charged) (Figure 11.18A). The pH at which a particle of a given oxide is overall not charged, that is, the pH at which the number of positively and negatively charged surface groups is equal, is called ***point of zero charge*** (PZC) [259].

Measurements of the electrophoretic mobility of suspended oxide particles in an electric field by optical (laser Doppler velocimetry and phase analysis light scattering (PALS)) or acoustic methods provide the Zeta (ζ)-potential, which is the electrical potential at the interface that separates the mobile fluid (outer Helmholtz plane) from the fluid that remains attached to the oxide surface due to the rigid association of charged surface groups and counter ions (inner Helmholtz plane) (Figure 11.18B). The imaginary interphase is called shear plane, which is located in the mobile layer beyond the outer Helmholtz plane. The determination of the ζ-potential as a function of the pH of the suspension gives the ***isoelectric point*** (IEP) of the oxide defined as the point of zero ζ-potential (Figure 11.18C) [260]. At low ionic strength, and also in the case of equal affinities of counterions (cations and anions) toward association with oppositely charged surface groups, IEP and PCZ coincide. The isoelectric point of an oxide depends not only on its chemical nature, but sensitively on the synthesis and the thermal or any other post-treatment. Therefore, the IEP must always be determined before impregnation is carried out.

At pH > IEP, the surface is negatively charged (hydroxyl groups are deprotonated) (Figure 11.18A and C). Such a surface adsorbs cations by electrostatic interaction. IEP values of common silicon oxides, for example, vary between 1.5 and 3 [259, 261]. That means that the surface of silica is negatively charged over almost the entire pH range and practically only cations can be electrostatically adsorbed. Impregnation of silica at pH > 1.5 with a solution that contains negatively charged ions, such as $[Mo_7O_{24}]^{6-}$ or $[PtCl_6]^{2-}$, leads to repulsion and consequently low dispersion.

Prior to equilibrium adsorption or impregnation, not only the IEP of the oxide must be known, but also the speciation of the ions in the solution, because a high dispersion of the coated species can only be achieved in case of attractive interactions between surface functional groups and species in solution. If the speciation is not known, adsorption must be determined experimentally. Spectroscopic methods have proven to be very helpful in elucidating the speciation in solution and the changes during interaction with the carrier or while drying [262, 263]. Invasive and noninvasive techniques have been applied. Noninvasive techniques such as magnetic

Figure 11.17 (continued)
species by calcination [255, 256]. (B) Infrared spectra of silica-supported molybdenum oxide catalyst precursors with increasing Mo surface density (Mo atoms/nm^2) as indicated in the legend and by the different colors. (C) ^{1}H MAS- NMR spectra of SBA-15 and MoO_x/SBA-15 (1.27 Mo_{atoms} nm^{-2}). The spectrum of MoO_x/SBA-15 is magnified by factor of 5. (D) Density of isolated silanol groups measured by FTIR spectroscopy and ammonia adsorption sites as estimated by FTIR spectroscopy of adsorbed ammonia; adapted with permission from John Wiley and Sons [258].

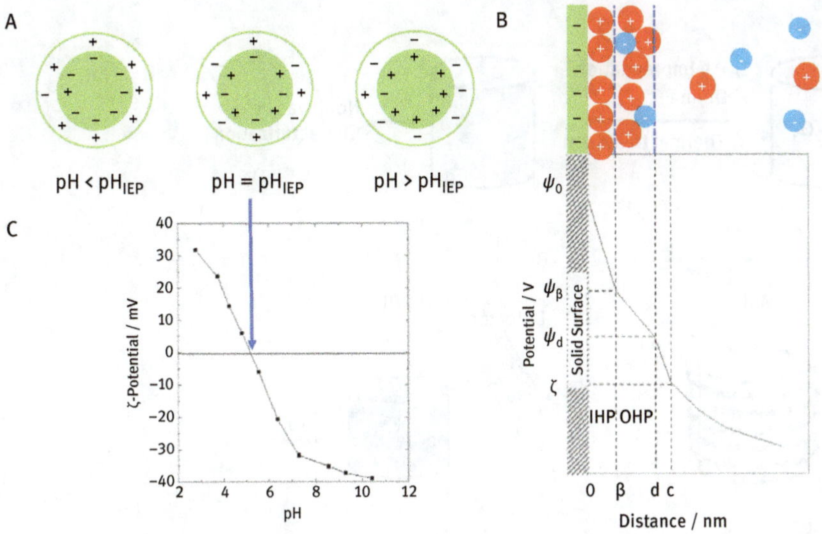

Figure 11.18: (A) Schematic representation of the charge of the surface M-OH groups (in the white ring) of an oxide particle (green) depending on the pH of the aqueous suspension. When oxide particles are suspended in aqueous solutions with $pH < pH_{IEP}$, the surface is positively charged (protonation of M-OH under formation of $[M\text{-}OH_2]^+$) and anions can be adsorbed; In case of $pH > pH_{IEP}$, the oxide surface is negatively charged (deprotonation of M-OH under formation of $[M\text{-}O]^-$) and cations can be adsorbed. (B) Model of the interfacial charged layer; In the inner Helmholtz plane (IHP), the charged surface groups are located (distance β from the surface). The outer Helmholtz plane (OHP) is the beginning of the diffuse layer (distance d from the surface) ($0 \le \beta \le d$) governed by electrostatic forces; within the diffuse layer, the potential at the c-plane is the electrokinetic or ζ -potential; adapted with permission from Springer. (C) The isoelectric point is determined by measuring the ζ-potential as a function of the pH of the suspension; adapted with permission from Springer [260].

resonance imaging (MRI), and 2D nuclear magnetic resonance (NMR) relaxation time measurements, and synchrotron-based methods, such as tomographic energy dispersive diffraction imaging (TEDDI) and X-ray microtomography, have the advantage that the sample is not manipulated and all stages of catalyst preparation can be investigated (Figure 11.19) [263].

If higher loadings are to be achieved, impregnation is the method of choice. In this technique, the pores of the support are filled with a solution that contains a precursor of the active species and the solvent is then removed by drying without previous washing. During impregnation, it is equally important to ensure electrostatic attraction between surface functional groups and charged ions in solution, as explained above for ion exchange and equilibrium adsorption. Two different procedures are frequently used. In the so-called "dry impregnation" or "incipient wetness" technique, the volume of the precursor solution corresponds to the volume of the pores. The solution fills the pores due to capillary forces. To reduce the mechanical stress, which the escaping gas

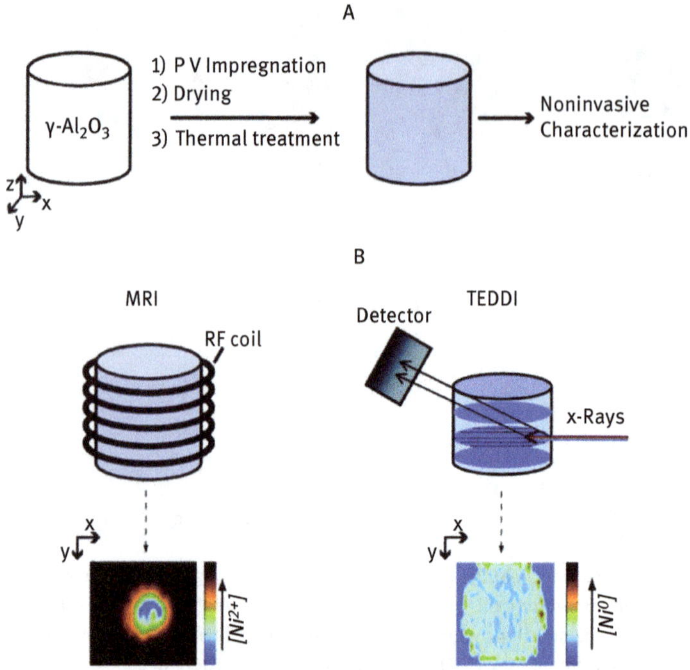

Figure 11.19: (A) Noninvasive characterization techniques during the impregnation, drying, or thermal treatment steps of a catalyst body. (B) The MRI image (left) represents an xy slice of a γ-Al_2O_3 extrudate 5 min after impregnation with a 0.5 M $[Ni(H_2O)_6]^{2+}$ solution; The outer red region indicates a high concentration of the complex, while the blue core indicates the presence of pure water. The TEDDI image (right) represents an xy slice of a Ni/γ-Al_2O_3 pellet after thermal treatment in N_2. A green external ring can be observed, which indicates an egg-shell distribution of the metallic Ni (200) reflection at 57.6 keV. Reprinted with permission from Espinosa-Alonso, L.; Beale, A. M.; Weckhuysen, B. M., Profiling Physicochemical Changes within Catalyst Bodies during Preparation: New Insights from Invasive and Noninvasive Microspectroscopic Studies. *Accounts of Chemical Research* **2010,** *43*, 1279–1288. Copyright 2010 American Chemical Society [263].

bubbles exert on the pore walls, incipient wetness is either performed in vacuum, or surfactants are added to the solution. In an alternative technique ("wet impregnation"), the pores are already filled with the solvent and the driving force for the diffusion of the precursor into the pores is the concentration gradient. When the interaction between the surface and the precursor is weak, such process might need hours to take place. The diffusion processes in catalyst bodies have been tracked by in situ microspectroscopic (UV/vis-NIR, IR, and Raman) techniques [263]. The strength of the interaction between the precursor and the support surface is affected by numerous parameters, including the pH of the solution, the concentration, and the temperature. All these factors determine the charging of the surface on the one hand and the speciation of the precursor on the other hand, which is, for example, particularly important for oxo-species of transition metals, like molybdates, chromates or vanadates, that can undergo pH and concentration dependent condensation reactions, but also, in general for all metal ions in aqueous solution, which can undergo hydrolysis. Other important factors that affect the precursor–support interaction are the nature of ligands or the presence of complexing agents. Competitive adsorption has to be taken into account when the impregnation of more than one metal is intended. Furthermore, the type of the counter ion has to be considered in particular in view of oxidizing (e.g., NO_3^-) or reducing (e.g., NH_4^+) properties during the subsequent calcination that may change the nature of the active species on the surface of the final catalyst.

The equilibrium adsorption or impregnation is always followed by a drying step, which strongly affects the properties of the catalyst precursor [252]. During removal of the solvent the concentration of the solution in the pores of the support increases. This can lead to changes in the speciation and eventually, if the supersaturation concentration is reached, to precipitation. The drying step can have an impact on the distribution of the active phase on the support at macroscopic and/or nanoscopic scales. Instead of classical drying in the oven, freeze-drying, for example, can provide a more homogenous distribution of the active component [264]. In the freeze-drying process, the solution enclosed in the pores of the support is frozen and the solvent is removed by sublimation at reduced temperature and pressure.

If an even higher loading is to be achieved, the surface layer can be applied by deposition–precipitation (see Section 11.3.1.2) [166]. The conditions of the precipitation must be adapted in such a way that the precursor at first migrates into the pores of the support and precipitate formation outside the pore system is avoided. Furthermore, an attractive interaction of the nuclei with the surface of the support is a prerequisite for a successful deposition–precipitation. Precipitation exclusively on the support is achieved by generating the precipitating agent in situ, for example, by decomposition of urea $CO(NH_2)_2$ above 60 °C.

A classical technique in particular in electrochemistry for the preparation of supported metals is the electrochemical metal deposition at electrode–electrolyte interfaces under the influence of an electric field [265]. Overpotential, metal–substrate interactions, and defects on the surface of the substrate determine the mechanism of nucleation and growth and the formation of specific nanostructures [265, 266]. Metal oxides, chalcogenides (sulfides and selenides), and phosphides can also be prepared by electrodeposition [267].

Well-defined particle size distribution of metal nanoparticles and a homogeneous distribution of the particles on the support can be obtained by application of colloidal nanoparticles. However, the surfactants used for stabilization of the colloidal particles in solution are generally of disadvantage for catalyst purity and functionality, which has led to the development of surfactant-free approaches to prepare supported colloidal metal nanoparticles. These systems can serve as models for heterogeneous catalysts [268, 269].

11.3.2.1.2 Grafting and Atomic Layer Deposition (ALD)

An improved control with respect to coverage of the surface with an active species is possible when the precursor is grafted by chemical reactions with surface functional groups, such as surface hydroxyl groups, forming a strong chemical bond [90, 270, 271]. Precursors can be, for example, alkoxides, such as titanium(IV) isopropoxide $Ti(OCH(CH_3)_2)_4$, dissolved in an organic solvent (grafting) or supplied via the gas phase (atomic layer deposition (ALD)).

The complexation of the surface species, that is, whether a molecular metal precursor reacts with one or more OH groups of the support, can be controlled via the density of the hydroxyl groups on the support. The average OH density can be adjusted in the synthesis of the carrier and during its thermal treatment, and can be estimated by using analytical methods, like thermal analysis, infrared spectroscopy or NMR spectroscopy [256]. The determination of the amount of residual functional groups on the surface by spectroscopic techniques applied after the grafting process allows the measurement of adsorption isotherms [272], and enables conclusions with respect to the surface structure of the grafted complex that can be deduced from the stoichiometry of the reaction [273, 274]. The molecular structure will, however, change during the subsequent calcination that is necessary to remove residual ligands.

In the alternative concept of surface organometallic chemistry (SOMC) [57, 275], a suspected intermediate of a catalytic reaction is prepared from the chemical interaction of an organometallic or coordination compound with the support material. The well-defined surface organometallic fragment (SOMFs) prepared in this way is believed to be the active site of the catalytic reaction. This may be close to reality only under mild reaction conditions.

Atomic layer deposition (ALD) is a ***self-limited*** surface modification at the solid–gas interphase [276]. A volatile metal precursor, such as trimethylaluminum (Figure 11.20), is admitted in pulses. In the first half-reaction, the precursor reacts with surface functional groups of the support, such as OH groups of silica. After completion of a monolayer, excess of unreacted precursor is removed. In the second half-reaction, residual ligands of the grafted metal precursors are eliminated and the initial surface functional groups are re-formed on top of the first monolayer. So, for example, the reaction of water with grafted trimethylaluminum (Figure 11.20) leads to cross-linking of the Al species and Al-OH formation by condensation reactions under formation of methane. In this way, the surface of a substrate can be modified layer by layer with excellent thickness control (Figure 11.20). More complex structures can also be prepared by using various precursors and growing up different layers. The reactions can also involve redox reactions [277]. The method is particularly useful to modify catalysts with the aim to stabilize the particle size of metal-nanoparticles, to introduce metal–promoter interactions or to block undesired surface sites [278].

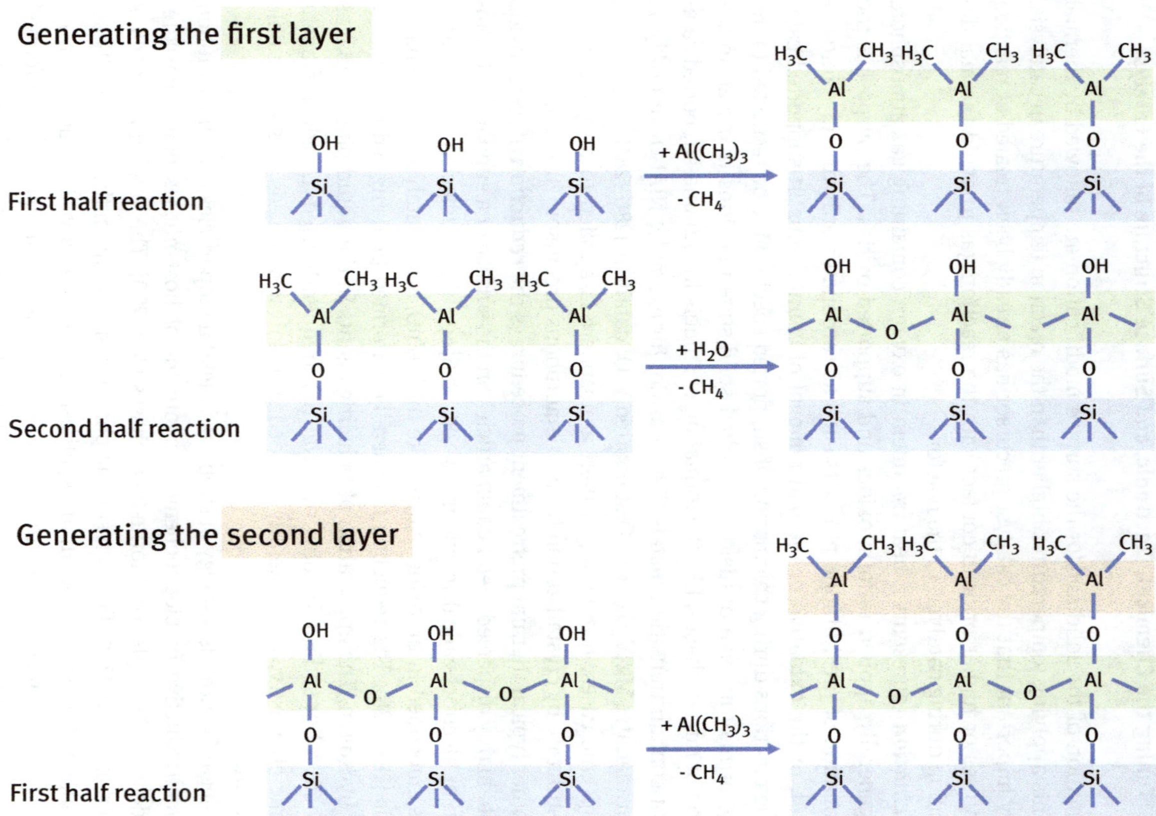

Figure 11.20: Layer-by-layer modification of surfaces by ALD. The example shows schematically the coating of silicon dioxide with aluminum oxide layers.

11.3.2.2 Tuning the Chemical, Electronic, and Surface Structure of the Catalyst

The formation of the solid and/or the surface modification are followed by thermal treatments applying temperatures higher than the reaction temperature in catalysis. It is very important that the catalyst precursor has already been treated at reaction temperatures or higher prior to contact with the reaction gas to avoid thermal decomposition during catalysis in the reactor.

Calcination represents a heat treatment in oxygen containing gas atmosphere that is generally applied to form oxides and supported oxides. The purpose is the transformation of precipitation products such as carbonates or hydroxides into oxides [279], or the oxidative or thermal removal of counter ions in supported catalysts. The reactions during calcination of supported species may be restricted to the interface or may involve diffusion-controlled solid-state reactions that lead to the formation of mixed phases, like the formation of spinels in alumina-supported catalysts that contain transition metals like Ni or Co, for example, $NiAl_2O_4$ or $CoAl_2O_4$, or orthovanadates $Mg_3V_2O_8$ in MgO-supported V_xO_y catalysts [280–281].

In specific cases, ***annealing*** in inert atmosphere is applied to form a desired crystal structure by crystallization from an amorphous precursor [282–284].

Another typical thermal pretreatment procedure is the ***reduction*** of metal oxide species to highly dispersed metal nanoparticles on the surface of a support for hydrogenation or electrochemical reactions [14, 285, 286], or highly defective metal nanoparticles embedded in an oxide matrix, like in $Cu/ZnO/Al_2O_3$ catalysts for methanol synthesis [287, 288]. The reduction is generally performed by using reducing gases, such as hydrogen. The application of reducing agents in the liquid phase, such as ethylene glycol or sodium borohydride ($NaBH_4$), frequently used as reducing agent in organic chemistry, are generally avoided in catalyst synthesis due to possible contaminations.

An elegant synthesis strategy of supported metal nanoparticles is their exsolution by thermal/reductive or electrochemical treatment of host phases like perovskite, spinel, delafossite, rutile, or fluorite-type phases [289, 290]. The general approach is illustrated in Figure 11.21. First, crystalline phases, which contain the catalytically active metal, such as Fe, Cu, or Ni, are synthesized. The reductive treatment then removes some of the metal atoms from the lattice and metal nanoparticles are formed on the surface (Figure 11.21A). The exsolution process is controlled by the solid-state properties of the host, such as phase composition, crystal orientation, vacancy concentration and lattice strain, as well as by the environmental conditions. The nanoparticles generally possess a narrow particle size distribution and are characterized by their ability to self-regenerate in redox processes, which is advantageous for the stability of the derived catalysts compared to nanoparticles deposited by impregnation, which frequently suffer from sintering (Figure 11.21B) [291].

The as-prepared catalyst precursor is usually not yet active for the desired catalytic reaction. The step in catalyst preparation that transforms the catalyst precursor

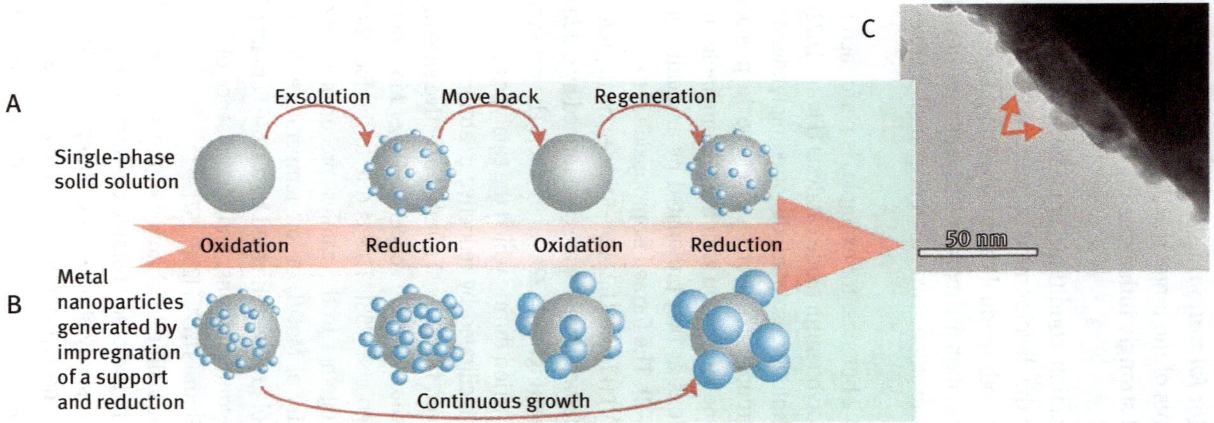

Figure 11.21: General approach for synthesis of supported metal nanoparticles by exsolution and schematic illustration of the stability of the system under consecutive oxidative and reductive environments (A) compared with conventional catalysts prepared by impregnation (B); adapted with permission from Zhang, J.; Gao, M.-R.; Luo, J.-L., In Situ Exsolved Metal Nanoparticles: A Smart Approach for Optimization of Catalysts. *Chemistry of Materials* **2020,** *32*, 5424–5441. Copyright 2020 American Chemical Society [289]. Transmission electron micrograph of $La_{0.6}Sr_{0.4}FeO_3$ that has been reduced at 600 °C (C). The surface of the perovskite is decorated with small (about 10 to 15 nm in diameter, marked by arrows) particles consisting of exsolved metallic iron; adapted with permission from Götsch, T.; Schlicker, L.; Bekheet, M. F.; Doran, A.; Grünbacher, M.; Praty, C.; Tada, M.; Matsui, H.; Ishiguro, N.; Gurlo, A.; Klötzer, B.; Penner, S., Structural investigations of La0.6Sr0.4FeO3–δ under reducing conditions: kinetic and thermodynamic limitations for phase transformations and iron exsolution phenomena. *RSC Advances* **2018,** *8*, 3120-3131 – Published by The Royal Society of Chemistry [292].

into an active material is called ***activation*** (or ***formation***) [293]. Generally the activation is performed in the feed of the reactants of the target reaction. The activation is a crucial process in catalyst synthesis and strongly influences the catalyst performance and stability [3].

The dynamics of the catalyst under reaction conditions must be investigated using *operando* methods in order to advance the development of improved catalysts in a knowledge-based manner. This applies to both thermal catalysis [294] and electrocatalysis [295]. The changes during catalyst formation are more or less pronounced and depend on both the catalyst precursor and the reaction. Two different examples are explained below for illustration.

The metathesis of propene to ethene and 2-butenes, for example, is performed at mild conditions (50 °C) on silica-supported molybdenum oxide catalysts (Figure 11.22). By using multiple in situ characterization techniques, isolated, distorted molybdenum di-oxo species in close vicinity to two silanol groups have been shown to be the precursor of the active site, which is only formed by the interaction with the reacting molecule propene [256–258, 296] The propene molecule is at first protonated by an acidic OH group on the catalyst surface (Figure 11.22A, 1–3). The formed isopropoxide is oxidized to acetone whereby Mo(VI) is reduced to Mo(IV) (Figure 11.22A, 4–5). Acetone migrates to a neighboring silanol group and the active carbene species are then formed by oxidative addition of propene to the free Mo(IV) sites (Figure 11.22A, 6–9) [296]. The catalyst formation process is reflected in an induction period during which the catalytic activity increases (Figure 11.22B). The process can be facilitated by reductive treatment of the catalyst precursor with methanol [258]. Moreover, the highest number of active sites on the surface is found only with an optimal medium surface concentration of Mo atoms (1.1 Mo atoms/nm^2), (Figure 11.22B, red data points) [258]. This is due to the fact that various favorable circumstances must come together in the catalyst precursor for the active catalyst to form, such as the reducibility of Mo(VI) at low temperature due to strain in Mo dioxo species, the acidity of silanol groups generated by a close contact of OH groups and Mo oxide species in strained configurations, and the presence of free silanol groups for adsorption of by-products (Figure 11.22C) [256, 257].

The dynamics of bulk catalysts are often visible in changes of the surface and bulk chemical and electronic structure even in the case of stable chemical composition and crystal structure. This has been intensively studied, for example, on Mo-V oxides with M1 structure (Figures 11.2, 11.11, and 11.14) in alkane activation [62, 64, 65, 67]. This type of catalyst is highly active and selective in the oxidative dehydrogenation of ethane to ethene at a reaction temperature of 300 °C [297], but shows also interesting properties in other oxidation reactions [298]. The Mo-V oxide is a p-type semiconductor. The average oxidation state of the redox active element vanadium in the catalyst bulk is approximately 4.3. The conductivity of Mo-V oxide is sensitively varying with the gas composition, both in terms of the composition of the reactants (oxidation of different alkanes with different chain lengths) and the

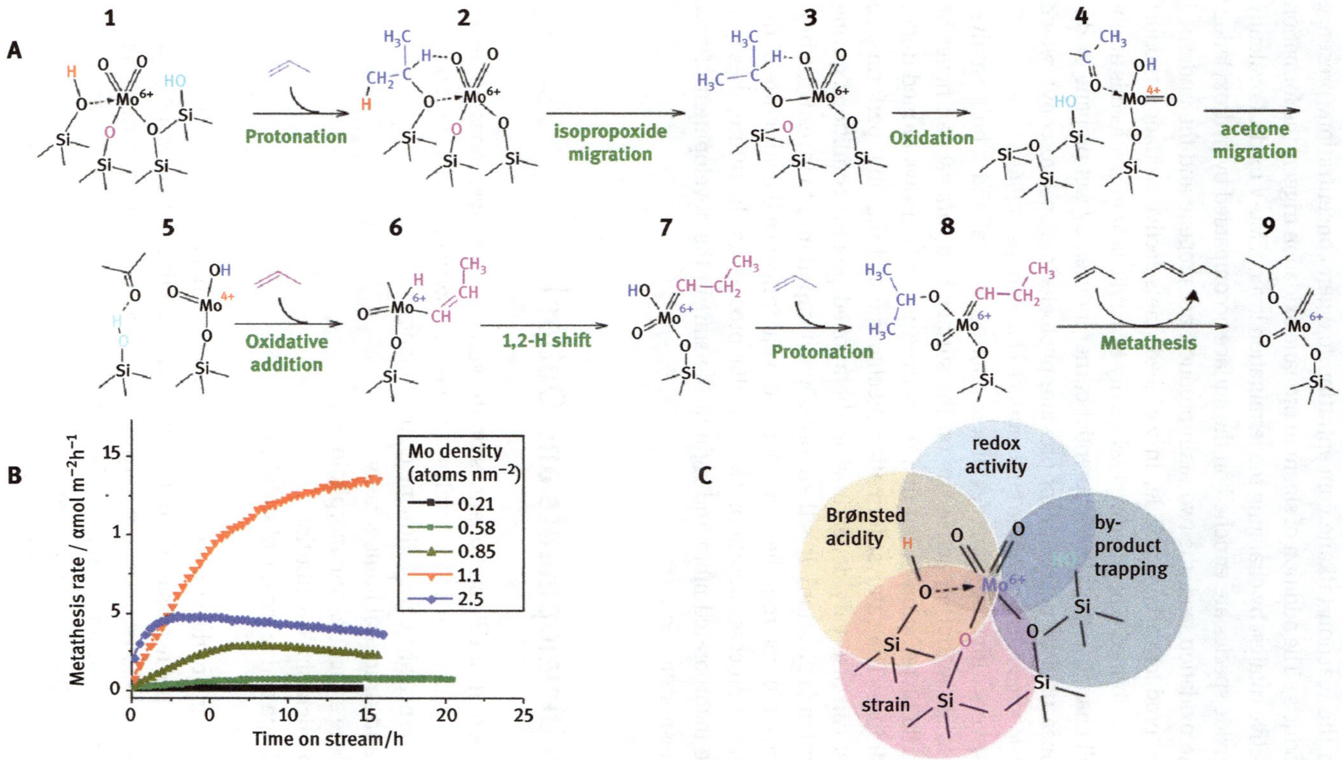

Figure 11.22: Schematic representation of the chemical reactions leading to the formation of an active catalyst starting from a MoO_x/SiO_2 catalyst precursor (Figure 11.17) under reaction conditions of propene metathesis (A), which is reflected in an induction period with increasing catalytic activity before the steady state is reached (B), and which requires the coincidence of several favorable structural properties in the catalyst precursor (C). (A) is reprinted with permission from John Wiley and Sons [258].

conversion in the reaction (alkane concentration, oxygen concentration, presence of reaction products). The addition of steam in propane oxidation triggers the formation of a more selective catalyst by changing the termination layer of Mo-V oxide. Covalently bonded V(V) oxide species are enriched on the surface accompanied by hydroxylation. In dry propane oxidation feed, the two main products are propene and the undesired total oxidation product carbon dioxide. In wet feed, the selective oxidation product acrylic acid is also formed. The improved selectivity to acrylic acid in wet feed is attributed to a small change in the surface potential barrier induced by wet propane oxidation feed (as measured by near-ambient pressure photoelectron spectroscopy), which can contribute to a modification of the bulk–surface charge transfer [64].

These two very different examples of a supported catalyst and a bulk catalyst are intended to show that an active site on the surface of a catalyst can never be created by synthetic techniques, but that the active species is always formed under reaction conditions. The catalyst formation mechanism and the final performance of the catalyst can also vary depending on the applied process conditions during catalyst formation (activation). It is therefore of great importance to investigate how the active catalyst forms from the precursor and what influence the structure of the precursor and the process conditions exert on this process. In the future, these aspects must be incorporated more and right at the start in the development of new catalyst synthesis strategies.

11.4 New Developments and Outlook

The wide range of desired functionalities on the surface of catalysts does not make it easy to derive general rules for catalyst synthesis. However, there is no question that good catalysts feature the following characteristics:

- Chemical and structural homogeneity
- Adapted pore structures and morphologies
- Uniform particle size distribution
- Suitable shape and mechanical stability

The nanostructuring of the surface is essential in many applications, but nanostructured surfaces are not necessarily required in every catalytic reaction. Many applications, especially in electrochemistry, rely on the use of expensive and rare metals, like Pt, Li, and rare-earth elements. In the synthesis of micro- and mesoporous materials, costly structure directing agents and additives are frequently applied. The reduction of catalyst mass and the prevention of waste formation, for instance, by recycling of synthesis additives and solvents, are therefore highly topical research issues. The practicability of future technologies based on catalysis will depend on the availability of efficient catalysts composed of abundant elements prepared by robust, preferentially aqueous-based synthesis methods, the reduction of environmental impacts arising from catalyst manufacture, and the possibility of reprocessing all metallic elements, also base metals, in spent catalysts.

Present synthesis concepts are generally aimed at creating the ***active site*** itself already in the synthesis of the catalyst material. As outlined in this chapter, this strategy faces two difficulties. First, the structure of the active site in the working heterogeneous catalyst is usually not known in detail. Second, the product of catalyst synthesis is generally a ***catalyst precursor***. The actual working catalyst is formed from the catalyst precursor only in contact with the reaction medium under reaction conditions (e.g., temperature, pressure, potential, chemical composition of the reaction medium, external impacts such as electric and magnetic fields).

The close interdependence of solid-state kinetics of the catalyst on the one side and reaction kinetics of the reacting molecules on the other side is responsible for the fact that structure–function relationships are never straightforward in catalysis. Future synthesis concepts thus require integrated approaches that include the reaction conditions and the type of the intended reactor technology from the very beginning. The chemistry and mechanism leading from the catalyst precursor to the active form of the catalyst need to be built into the design.

In the future, conventional synthesis planning needs to be reconsidered to take into account the improved understanding of the formation of active sites. This redesign should be compatible with the requirements from digital catalysis imposing novel concepts of documentation of unit operations (Figure 11.23) [50]. Synthesis strategies should be inspired by theory at the beginning of a new project [87, 299–301]. This also applies to insights from studies on well-defined model systems [302], or inverse catalyst models [303], as well as microfabricated 2D and 3D nano- and microstructures of various shapes, produced, for example, by lithographic techniques [304], (electro)chemical etching [305], inkjet printing [125], ALD [276], or CVD [306], or thin films and specific nanostructures fabricated for the application of alternative heating concepts (induction heating) in catalysis [307].

Since it is not possible to synthesize the active species itself, it is first necessary to understand how the active site is formed from the catalyst precursor. *Operando* and quasi in situ methods in microscopy, diffraction, chemical analysis, X-ray, and molecular spectroscopy have shown great potential and must be applied extensively to clarify the solid-state kinetics of working catalysts during transformation of the freshly prepared catalyst precursor into the active state (see also Section 11.3.1.2) [134, 294, 308, 309]. The knowledge gained in this way should then be incorporated into the development of synthetic toolboxes including both new and classical preparation techniques, which will allow the synthesis of the most suitable catalyst precursors.

The synthesis methods used differ depending on whether a self-supported active phase (bulk catalyst) or an active phase on the surface of a chemically different support (supported catalyst) is prepared. Impregnation, for example, is the most

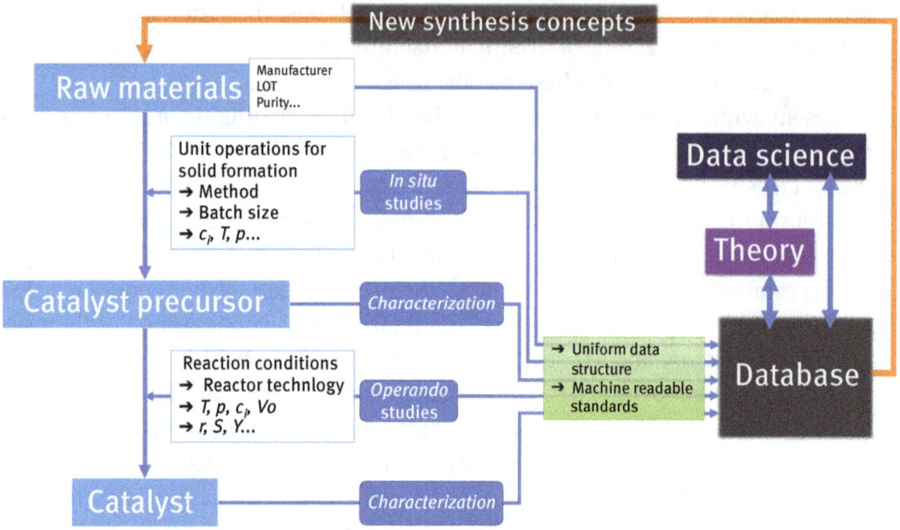

Figure 11.23: The workflow in digital catalyst synthesis.

frequently applied preparation technique of supported active phases. Here, it is well known that the dispersion of the active phase and hence the performance of the resulting catalyst are strongly affected by the impregnation parameters [81, 262]. Important factors are the nature of the support surface in contact with an aqueous solution of a specific pH and ionic strength, the speciation of the dissolved species in solution, the kinetics of their interaction with the support, and the diffusion within the pores of the support or pellet filled with the solvent. When all possible physical and chemical quantities are measured during unit operations like impregnation, they can be used to understand and control the synthesis. This will improve the reproducibility of catalyst synthesis and give new impetus to the further development of in situ techniques in inorganic preparative chemistry and material science. Such unit operations, which up to now have mainly been carried out manually in academic research, for example, filtration, washing, drying, or shaping, pose a particular challenge.

Catalyst synthesis must become reproducible and scalable. Small batch size and difficulties in upscaling of promising materials discovered in academic laboratories will prevent such catalyst precursors from ever being considered for any type of application. New materials are often prepared only on a milligram scale. However, catalysis studies and comprehensive characterization of the catalyst require gram-scale quantities also in academic investigations. Reproduction of literature syntheses often fails when only methods for synthesis of minute amounts are reported there, not to mention that upscaling is often not possible. The scientific literature is also frequently ambiguous and incomplete in particular in terms of the description of the catalyst synthesis. This hinders the validation of published results and the use of synthesis data by data science. The key factor impeding the digitalization of catalysis is the lack of publicly accessible reports and databases that contain standardized information including transparent recipes for catalyst synthesis in acceptable quantity. These considerations suggest some important measures that should be implemented:

- Synthesis of catalysts in larger batch sizes (minimum about 10 g) also in research projects
- Elaboration of generally accepted and strict guidelines for the publication of synthesis details in scientific literature and databases using a uniform data structure
- Development of hardware and software for recording all data and metadata generated during catalyst synthesis in machine-readable standards

This ultimately leads to an automated and digital catalyst synthesis (Figure 11.23). The data obtained from *operando* methods, kinetic studies, catalyst synthesis, and basic characterization can then be analyzed by using mathematical algorithms. To overcome the hurdles of complexity in catalysis, artificial intelligence can help to identify descriptors that reveal possibly intricate correlations between catalyst performance and relevant materials properties. The important information obtained then serves as target for catalyst synthesis.

References

[1] Centi G, Čejka J. Needs and gaps for catalysis in addressing transitions in chemistry and energy from a sustainability perspective. ChemSusChem. 2019;12:621–32.

[2] Nørskov JK, Weckhuysen B, Centi G, Chorkendorff I, Schlögl R, Marin G, Grimaud A, Rossmeisl J, Strasser P, Koper M, Roldan B, Behrens M, Bowker M, Hutchings G, Bordiga S, Lercher J, Nilsson A, Stephens I, Cantat T, Leitner W, DeBeer S, Groot HD, Jacobsen KW, Scheffler M, Moses PG, Schreiber M, Gazzani M, Mazzotti M, Iaquaniello G, Schouten KJ, Growitsch C, Schmidt C Research needs towards sustainable production of fuels and chemicals. https://www.energy-x.eu/wp-content/uploads/2019/10/Energy-X-Research-needs-report.pdf.

[3] Schlögl R. Heterogeneous catalysis. Angew Chem Int Ed. 2015;54:3465–520.

[4] Somorjai GA, Frei H, Park JY. Advancing the frontiers in nanocatalysis, biointerfaces, and renewable energy conversion by innovations of surface techniques. J Am Chem Soc. 2009;131:16589–605.

[5] Abbas HF, Wan Daud WMA. Hydrogen production by methane decomposition: A review. Int J Hydrogen Energy. 2010;35:1160–90.

[6] Pakhare D, Spivey J. A review of dry (CO2) reforming of methane over noble metal catalysts. Chem Soc Rev. 2014;43:7813–37.

[7] Vogt ETC, Whiting GT, Chowdhury AD, Weckhuysen BM. Zeolites and Zeotypes for Oil and Gas Conversion. In: Jentoft FC, Ed. Advances in Catalysis. Vol. 58. San Diego: Elsevier Academic Press Inc; 2015, 58, 143–314.

[8] Vermeiren W, Gilson JP. Impact of Zeolites on the Petroleum and Petrochemical Industry. Top Catal. 2009;52:1131–61.
[9] Britovsek GJP, Gibson VC, Wass DF. The Search for New-Generation Olefin Polymerization Catalysts: Life beyond Metallocenes. Angew Chem Int Ed. 1999;38:428–47.
[10] Resconi L, Cavallo L, Fait A, Piemontesi F. Selectivity in propene polymerization with metallocene catalysts. Chem Rev. 2000;100:1253–345.
[11] Grubbs RH. Olefin metathesis. Tetrahedron. 2004;60:7117–40.
[12] Franssen NMG, Reek JNH, De Bruin B. Synthesis of functional 'polyolefins': state of the art and remaining challenges. Chem Soc Rev. 2013;42:5809–32.
[13] Grant JT, Venegas JM, McDermott WP, Hermans I. Aerobic oxidations of light alkanes over solid metal oxide catalysts. Chem Rev. 2018;118:2769–815.
[14] Teschner D, Borsodi J, Wootsch A, Revay Z, Hävecker M, Knop-Gericke A, Jackson SD, Schlögl R. The roles of subsurface carbon and hydrogen in palladium-catalyzed alkyne hydrogenation. Science. 2008;320:86–89.
[15] Sattler JJHB, Ruiz-Martinez J, Santillan-Jimenez E, Weckhuysen BM. Catalytic dehydrogenation of light alkanes on metals and metal oxides. Chem Rev. 2014;114:10613–53.
[16] Carrero CA, Schlögl R, Wachs IE, Schomäcker R. Critical literature review of the kinetics for the oxidative dehydrogenation of propane over well-defined supported vanadium oxide catalysts. ACS Catal. 2014;4:3357–80.
[17] Perego C, Ingallina P. Recent advances in the industrial alkylation of aromatics: new catalysts and new processes. Catal Today. 2002;73:3–22.
[18] Bowker M. Methanol synthesis from CO2 hydrogenation. ChemCatChem. 2019;11:4238–46.
[19] Behrens M, Studt F, Kasatkin I, Kühl S, Hävecker M, Abild-Pedersen F, Zander S, Girgsdies F, Kurr P, Kniep B-L, Tovar M, Fischer RW, Nørskov JK, Schlögl R. The Active Site of Methanol Synthesis over Cu/ZnO/Al2O3 Industrial Catalysts. Science. 2012;336:893–97.
[20] Ruland H, Song H, Laudenschleger D, Stürmer S, Schmidt S, He J, Kähler K, Muhler M, Schlögl R. CO2 hydrogenation with Cu/ZnO/Al2O3: A benchmark study. ChemCatChem. 2020;12:3216–22.
[21] Ertl G. Reactions at surfaces: From atoms to complexity (Nobel Lecture). Angew Chem Int Ed. 2008;47:3524–35.
[22] Marakatti VS, Gaigneaux EM. Recent advances in heterogeneous catalysis for ammonia synthesis. ChemCatChem. 2020;12:5838–57.
[23] Schlögl R. Catalytic synthesis of ammonia – A "never-ending story"?. Angew Chem Int Ed. 2003;42:2004–08.
[24] Schlögl R. Ammonia Synthesis. In: Handbook of Heterogeneous Catalysis. Wiley-VCH Verlag GmbH & Co. KGaA; 2008.
[25] Climent MJ, Corma A, Iborra S. Heterogeneous catalysts for the one-pot synthesis of chemicals and fine chemicals. Chem Rev. 2010;111:1072–133.
[26] Granger P, Parvulescu VI. Catalytic NOx abatement systems for mobile sources: From three-way to lean burn after-treatment technologies. Chem Rev. 2011;111:3155–207.
[27] Speight JG. New approaches to hydroprocessing. Catal Today. 2004;98:55–60.
[28] Schwarz H. Chemistry with methane: concepts rather than recipes. Angew Chem Int Ed. 2011;50:10096–115.
[29] Schittkowski J, Ruland H, Laudenschleger D, Girod K, Kähler K, Kaluza S, Muhler M, Schlögl R. Methanol Synthesis from Steel Mill Exhaust Gases: Challenges for the Industrial Cu/ZnO/Al2O3 Catalyst. Chemie Ingenieur Technik. 2018;90:1419–29.
[30] Artz J, Müller TE, Thenert K, Kleinekorte J, Meys R, Sternberg A, Bardow A, Leitner W. Sustainable Conversion Of Carbon Dioxide: An Integrated Review Of Catalysis And Life Cycle Assessment. Chem Rev. 2018;118:434–504.

[31] Chen JG, Crooks RM, Seefeldt LC, Bren KL, Bullock RM, Darensbourg MY, Holland PL, Hoffman B, Janik MJ, Jones AK, Kanatzidis MG, King P, Lancaster KM, Lymar SV, Pfromm P, Schneider WF, Schrock RR. Beyond fossil fuel–driven nitrogen transformations. Science. 2018;360: eaar6611.

[32] Rinaldi R, Schüth F. Design of solid catalysts for the conversion of biomass. Energy Environ Sci. 2009;2:610–26.

[33] Lanzafame P, Centi G, Perathoner S. Catalysis for biomass and CO2 use through solar energy: opening new scenarios for a sustainable and low-carbon chemical production. Chem Soc Rev. 2014;43:7562–80.

[34] Corma A, Iborra S, Velty A. Chemical routes for the transformation of biomass into chemicals. Chem Rev. 2007;107:2411–502.

[35] Mika LT, Cséfalvay E, Németh Á. Catalytic conversion of carbohydrates to initial platform chemicals: chemistry and sustainability. Chem Rev. 2018;118:505–613.

[36] Sun Z, Fridrich B, De Santi A, Elangovan S, Barta K. Bright side of lignin depolymerization: toward new platform chemicals. Chem Rev. 2018;118:614–78.

[37] The European Green Deal. Communication from the Commission to the European Parliament, the European Council, the Council, The European Economic and Social Committee and the Committee of the Regions. Brussels: Eurepean Comission; 2019.

[38] Al-Qahtani A, Parkinson B, Hellgardt K, Shah N, Guillen-Gosalbez G. Uncovering the true cost of hydrogen production routes using life cycle monetisation. Appl Energy. 2021;281:115958.

[39] Spöri C, Kwan JTH, Bonakdarpour A, Wilkinson DP, Strasser P. The stability challenges of oxygen evolving catalysts: towards a common fundamental understanding and mitigation of catalyst degradation. Angew Chem Int Ed. 2017;56:5994–6021.

[40] Götz M, Lefebvre J, Mörs F, McDaniel Koch A, Graf F, Bajohr S, Reimert R, Kolb T. Renewable power-to-gas: A technological and economic review. Renewable Energy. 2016;85:1371–90.

[41] Carmo M, Fritz DL, Mergel J, Stolten D. A comprehensive review on PEM water electrolysis. Int J Hydrogen Energy. 2013;38:4901–34.

[42] Armaroli N, Balzani V. The hydrogen issue. ChemSusChem. 2011;4:21–36.

[43] Schüth F, Palkovits R, Schlögl R, Su DS. Ammonia as a possible element in an energy infrastructure: catalysts for ammonia decomposition. Energy Environ Sci. 2012;5:6278–89.

[44] Dunn B, Kamath H, Tarascon J-M. Electrical energy storage for the grid: a battery of choices. Science. 2011;334:928–35.

[45] Gallei EF, Hesse M, Schwab E. Development of Industrial Catalysts. In: Handbook of Heterogeneous Catalysis. Wiley-VCH Verlag GmbH & Co. KGaA; 2008.

[46] Nørskov JK, Bligaard T. The catalyst genome. Angew Chem Int Ed. 2013;52:776–77.

[47] Parrino F, Bellardita M, García-López EI, Marcì G, Loddo V, Palmisano L. Heterogeneous photocatalysis for selective formation of high-value-added molecules: some chemical and engineering aspects. ACS Catal. 2018;8:11191–225.

[48] Guerrini E, Trasatti S. Recent developments in understanding factors of electrocatalysis. Russ J Electrochem. 2006;42:1017–25.

[49] Seh ZW, Kibsgaard J, Dickens CF, Chorkendorff I, Nørskov JK, Jaramillo TF. Combining theory and experiment in electrocatalysis: Insights into materials design. Science. 2017;355:eaad4998.

[50] Trunschke A, Bellini G, Boniface M, Carey SJ, Dong J, Erdem E, Foppa L, Frandsen W, Geske M, Ghiringhelli LM, Girgsdies F, Hanna R, Hashagen M, Hävecker M, Huff G, Knop-Gericke A, Koch G, Kraus P, Kröhnert J, Kube P, Lohr S, Lunkenbein T, Masliuk L, Naumann d'Alnoncourt R, Omojola T, Pratsch C, Richter S, Rohner C, Rosowski F, Rüther F, Scheffler M, Schlögl R, Tarasov A, Teschner D, Timpe O, Trunschke P, Wang Y, Wrabetz S. Towards experimental handbooks in catalysis. Top Catal. 2020;63:1683–99.

[51] Taylor HS, A theory of the catalytic surface. Proceedings of the Royal Society of London. Series A 1925, 108, 105–11.
[52] Boudart M. Turnover rates in heterogeneous catalysis. Chem Rev. 1995;95:661–66.
[53] Somorjai GA, Park JY. Molecular factors of catalytic selectivity. Angew Chem Int Ed. 2008;47:9212–28.
[54] Zambelli T, Wintterlin J, Trost J, Ertl G. Identification of the “Active Sites” of a Surface-Catalyzed Reaction. Science. 1996;273:1688–90.
[55] Thomas JM. Design, synthesis, and in situ characterization of new solid catalysts. Angew Chem Int Ed. 1999;38:3588–628.
[56] Mitchell S, Vorobyeva E, Pérez-Ramírez J. The multifaceted reactivity of single-atom heterogeneous catalysts. Angew Chem Int Ed. 2018;57:15316–29.
[57] Samantaray MK, Pump E, Bendjeriou-Sedjerari A, D’Elia V, Pelletier JDA, Guidotti M, Psaro R, Basset J-M. Surface organometallic chemistry in heterogeneous catalysis. Chem Soc Rev. 2018;47:8403–37.
[58] Mars P, Van Krevelen DW. Oxidations carried out by means of vanadium oxide catalysts. Chem Eng Sci. 1954;3(Supplement 1):41–59.
[59] Tauster SJ, Fung SC, Garten RL. Strong metal-support interactions. Group 8 noble metals supported on titanium dioxide. J Am Chem Soc. 1978;100:170–75.
[60] Freund H-J, Pacchioni G. Oxide ultra-thin films on metals: new materials for the design of supported metal catalysts. Chem Soc Rev. 2008;37:2224–42.
[61] Fu Q, Yang F, Bao X. Interface-confined oxide nanostructures for catalytic oxidation reactions. Acc Chem Res. 2013;46:1692–701.
[62] Trunschke A, Noack J, Trojanov S, Girgsdies F, Lunkenbein T, Pfeifer V, Hävecker M, Kube P, Sprung C, Rosowski F, Schlögl R. The impact of the bulk structure on surface dynamics of complex Mo–V-based oxide catalysts. ACS Catal. 2017;7:3061–71.
[63] Sadakane M, Kodato K, Yasuda N, Ishikawa S, Ueda W. Thermal behavior, crystal structure, and solid-state transformation of orthorhombic Mo–V oxide under nitrogen flow or in air. ACS Omega. 2019;4:13165–71.
[64] Wernbacher AM, Kube P, Hävecker M, Schlögl R, Trunschke A. Electronic and dielectric properties of MoV-Oxide (M1 Phase) under alkane oxidation conditions. J Phys Chem C. 2019;123:13269–82.
[65] Masliuk L, Heggen M, Noack J, Girgsdies F, Trunschke A, Hermann KE, Willinger MG, Schlögl R, Lunkenbein T. Structural complexity in heterogeneous catalysis: cataloging local nanostructures. J Phys Chem C. 2017;121:24093–103.
[66] Sadakane M, Yamagata K, Kodato K, Endo K, Toriumi K, Ozawa Y, Ozeki T, Nagai T, Matsui Y, Sakaguchi N, Pyrz WD, Buttrey DJ, Blom DA, Vogt T, Ueda W. Synthesis of orthorhombic Mo-V-Sb oxide species by assembly of pentagonal Mo6O21 polyoxometalate building blocks. Angew Chem Int Ed. 2009;48:3782–86.
[67] Masliuk L, Schmidt F-P, Hetaba W, Plodinec M, Auffermann G, Hermann K, Teschner D, Girgsdies F, Trunschke A, Schlögl R, Lunkenbein T. Compositional Decoupling of Bulk and Surface in Open-Structured Complex Mixed Oxides. J Phys Chem C. 2020;124:23069–77.
[68] Qi W, Thermodynamics N. Acc Chem Res. 2016;49:1587–95.
[69] Shao M, Chang Q, Dodelet J-P, Chenitz R. Recent Advances in Electrocatalysts for Oxygen Reduction Reaction. Chem Rev. 2016;116:3594–657.
[70] Argyle M, Bartholomew C. Heterogeneous Catalyst Deactivation and Regeneration: A Review. Catalysts. 2015;5:145.
[71] Hodnik N, Dehm G, Mayrhofer KJJ. Importance and challenges of electrochemical in situ liquid cell electron microscopy for energy conversion research. Acc Chem Res. 2016;49:2015–22.
[72] Bartholomew CH. Mechanisms of catalyst deactivation. Appl Catal A Gen. 2001;212:17–60.

[73] Zhu B, Meng J, Yuan W, Zhang X, Yang H, Wang Y, Gao Y. Reshaping of metal nanoparticles under reaction conditions. Angew Chem Int Ed. 2020;59:2171–80.

[74] Huang X, Teschner D, Dimitrakopoulou M, Fedorov A, Frank B, Kraehnert R, Rosowski F, Kaiser H, Schunk S, Kuretschka C, Schlögl R, Willinger M, Trunschke A. Atomic-Scale observation of the metal-promoter interaction in Rh-Based syngas upgrading catalysts. Angew Chem Int Ed. 2019;58:8709–13.

[75] Werny MJ, Wang Y, Girgsdies F, Schlögl R, Trunschke A. Fluctuating Storage of the Active Phase in a Mn-Na2WO4/SiO2 Catalyst for the Oxidative Coupling of Methane. Angew Chem Int Ed. 2020;59:14921–26.

[76] Chiesa M, Giamello E, Che M, Characterization EPR. and Reactivity of Surface-Localized Inorganic Radicals and Radical Ions. Chem Rev. 2009;110:1320–47.

[77] Avdeev VI, Bedilo AF. Electronic Structure of Oxygen Radicals on the Surface of VOx/TiO2 Catalysts and Their Role in Oxygen Isotopic Exchange. J Phys Chem C. 2013;117:14701–09.

[78] Mitchell S, Michels N-L, Pérez-Ramírez J. From powder to technical body: the undervalued science of catalyst scale up. Chem Soc Rev. 2013;42:6094–112.

[79] Grunwaldt J-D, Wagner JB, Dunin-Borkowski RE. Imaging Catalysts at Work: A Hierarchical Approach from the Macro- to the Meso- and Nano-scale. ChemCatChem. 2013;5:62–80.

[80] Hartmann M, Machoke AG, Schwieger W. Catalytic test reactions for the evaluation of hierarchical zeolites. Chem Soc Rev. 2016;45:3313–30.

[81] Zandbergen MW, Beale AM, Weckhuysen IBM. On the Microdistributions of Cr-Ion Complexes within mm-Sized γ-Al2O3 Catalyst Bodies upon Impregnation as Studied by UV/Vis and Raman Microspectroscopy. ChemCatChem. 2012;4:217–27.

[82] Sing KSW. Reporting physisorption data for gas/solid systems with special reference to the determination of surface area and porosity. Pure Appl Chem. 1985;57:603–19.

[83] Rouquerol J, Baron G, Denoyel R, Giesche H, Groen J, Klobes P, Levitz P, Neimark AV, Rigby S, Skudas R, Sing K, Thommes M, Unger K. Liquid intrusion and alternative methods for the characterization of macroporous materials (IUPAC Technical Report). Pure Appl Chem. 2011;84:107–36.

[84] Rouquerol J, Avnir D, Fairbridge CW, Everett DH, Haynes JM, Pernicone N, Ramsay JDF, Sing KSW, Unger KK. Recommendations for the characterization of porous solids (Technical Report). Pure Appl Chem. 1994;66:1739–58.

[85] Thommes M, Kaneko K, Neimark AV, Olivier JP, Rodriguez-Reinoso F, Rouquerol J, Sing KSW. Physisorption of gases, with special reference to the evaluation of surface area and pore size distribution (IUPAC Technical Report). Pure Appl Chem. 2015;87:1051–69.

[86] Schwach P, Frandsen W, Willinger M-G, Schlögl R, Trunschke A. Structure sensitivity of the oxidative activation of methane over MgO model catalysts: I Kinetic study. J Catal. 2015;329:560–73.

[87] Slater AG, Cooper AI. Function-led design of new porous materials. Science. 2015;348:aaa8075.

[88] Li X, Lunkenbein T, Kröhnert J, Pfeifer V, Girgsdies F, Rosowski F, Schlögl R, Trunschke A. Hydrothermal synthesis of bi-functional nanostructured manganese tungstate catalysts for selective oxidation. Faraday Discuss. 2016;188:99–113.

[89] Kolen'ko YV, Zhang W, d'Alnoncourt RN, Girgsdies F, Hansen TW, Wolfram T, Schlögl R, Trunschke A. Synthesis of MoVTeNb Oxide Catalysts with Tunable Particle Dimensions. ChemCatChem. 2011;3:1597–606.

[90] Hamilton N, Wolfram T, Tzolova Müller G, Hävecker M, Kröhnert J, Carrero C, Schomäcker R, Trunschke A, Schlögl R. Topology of silica supported vanadium-titanium oxide catalysts for oxidative dehydrogenation of propane. Catal Sci Technol. 2012.

[91] Corma A. From microporous to mesoporous molecular sieve materials and their use in catalysis. Chem Rev. 1997;97:2373–420.

[92] Mitchell S, Michels N-L, Kunze K, Pérez-Ramírez J. Visualization of hierarchically structured zeolite bodies from macro to nano length scales. Nat Chem. 2012;4:825–31.
[93] Govender S, Friedrich HB. Monoliths: A Review of the Basics, Preparation Methods and Their Relevance to Oxidation. Catalysts. 2017;7:62.
[94] Avila P, Montes M, Miró EE. Monolithic reactors for environmental applications: A review on preparation technologies. Chem Eng J. 2005;109:11–36.
[95] Heck RM, Gulati S, Farrauto RJ. The application of monoliths for gas phase catalytic reactions. Chem Eng J. 2001;82:149–56.
[96] Yang J, Salman AD, Blanco-García P. A review of measurement techniques of mechanical properties of the catalyst layer in catalytic converters. Johnson Matthey Technol Rev. 2019;63:177–90.
[97] Nijhuis TA, Beers AEW, Vergunst T, Hoek I, Kapteijn F, Moulijn JA. Preparation of monolithic catalysts. Catal Rev. 2001;43:345–80.
[98] Cheetham AK, Férey G, Loiseau T. Open-Framework Inorganic Materials. Angew Chem Int Ed. 1999;38:3268–92.
[99] Machoke AG, Beltrán AM, Inayat A, Winter B, Weissenberger T, Kruse N, Güttel R, Spiecker E, Schwieger W. Micro/Macroporous System: MFI-Type Zeolite Crystals with Embedded Macropores. Adv Mater. 2015;27:1066–70.
[100] Degnan TF. The implications of the fundamentals of shape selectivity for the development of catalysts for the petroleum and petrochemical industries. J Catal. 2003;216:32–46.
[101] Diercks CS, Yaghi OM. The atom, the molecule, and the covalent organic framework. Science. 2017;355:eaal1585.
[102] Yang D, Gates BC. Catalysis by Metal Organic Frameworks: Perspective and Suggestions for Future Research. ACS Catal. 2019;9:1779–98.
[103] Martínez C, Corma A. Inorganic molecular sieves: Preparation, modification and industrial application in catalytic processes. Coord Chem Rev. 2011;255:1558–80.
[104] Zhao D, Feng J, Huo Q, Melosh N, Fredrickson GH, Chmelka BF, Stucky GD. Triblock Copolymer Syntheses of Mesoporous Silica with Periodic 50 to 300 Angstrom Pores. Science. 1998;279:548–52.
[105] Gu D, Schuth F. Synthesis of non-siliceous mesoporous oxides. Chem Soc Rev. 2014;43:313–44.
[106] Ciesla U, Schüth F. Ordered mesoporous materials. Microporous Mesoporous Mater. 1999;27:131–49.
[107] Möller K, Bein T. Talented Mesoporous Silica Nanoparticles. Chem. Mater. 2017;29:371–88.
[108] Perego C, Millini R. Porous materials in catalysis: challenges for mesoporous materials. Chem Soc Rev. 2013;42:3956–76.
[109] Li J, Corma A, Yu J. Synthesis of new zeolite structures. Chem Soc Rev. 2015.
[110] Verboekend D, Pérez-Ramírez J. Design of hierarchical zeolite catalysts by desilication. Catal Sci Technol. 2011;1:879–90.
[111] Chen L-H, Sun M-H, Wang Z, Yang W, Xie Z, Su B-L. Hierarchically Structured Zeolites: From Design to Application. Chem Rev. 2020;120:11194–294.
[112] Valtchev V, Tosheva L. Porous Nanosized Particles: Preparation, Properties, and Applications. Chem Rev. 2013;113:6734–60.
[113] Trogadas P, Ramani V, Strasser P, Fuller TF, Coppens M-O. Hierarchically Structured Nanomaterials for Electrochemical Energy Conversion. Angew Chem Int Ed. 2016;55:122–48.
[114] Baldovino-Medrano VG, Le MT, Van Driessche I, Bruneel E, Alcázar C, Colomer MT, Moreno R, Florencie A, Farin B, Gaigneaux EM. Role of shaping in the preparation of heterogeneous catalysts: Tableting and slip-casting of oxidation catalysts. Catal Today. 2015;246:81–91.
[115] Campanati M, Fornasari G, Vaccari A. Fundamentals in the preparation of heterogeneous catalysts. Catal Today. 2003;77:299–314.

[116] Dong Y, Geske M, Korup O, Ellenfeld N, Rosowski F, Dobner C, Horn R. What happens in a catalytic fixed-bed reactor for n-butane oxidation to maleic anhydride? Insights from spatial profile measurements and particle resolved CFD simulations. Chem Eng J. 2018;350: 799–811.

[117] Wu D, Zhou J, Li Y. Mechanical strength of solid catalysts: Recent developments and future prospects. AIChE J. 2007;53:2618–29.

[118] Grasselli RK, Burrington JD. Oxidation of Low-Molecular-Weight Hydrocarbons. In: Ertl G, Knözinger H, Schüth F, Weitkamp J, editors. Handbook of Heterogeneous Catalysis. 2008, 3479–89.

[119] Schüth F, Hesse M. Catalyst Forming. In: Handbook of Heterogeneous Catalysis. Wiley-VCH Verlag GmbH & Co. KGaA; 2008.

[120] Meille V. Review on methods to deposit catalysts on structured surfaces. Appl Catal A Gen. 2006;315:1–17.

[121] Twigg MV, Richardson JT. Fundamentals and Applications of Structured Ceramic Foam Catalysts. Ind Eng Chem Res. 2007;46:4166–77.

[122] Sadykov VA, Isupova LA, Zolotarskii IA, Bobrova LN, Noskov AS, Parmon VN, Brushtein EA, Telyatnikova TV, Chernyshev VI, Lunin VV. Oxide catalysts for ammonia oxidation in nitric acid production: properties and perspectives. Appl Catal A Gen. 2000;204:59–87.

[123] Goetsch DA, Schmidt LD. Microsecond catalytic partial oxidation of alkanes. Science. 1996;271:1560–62.

[124] Bogdan E, Michorczyk P. 3D printing in heterogeneous catalysis – the state of the art. Materials. 2020;13:4534.

[125] Maleki H, Bertola V. Recent advances and prospects of inkjet printing in heterogeneous catalysis. Catal Sci Technol. 2020;10:3140–59.

[126] Vogt ETC, Weckhuysen BM. Fluid catalytic cracking: recent developments on the grand old lady of zeolite catalysis. Chem Soc Rev. 2015;44:7342–70.

[127] Grasselli R, Tenhover M. Ammoxidation. In: Ertl G, Knözinger H, Schüth F, Weitkamp J, editors. Handbook of Heterogeneous Catalysis. 2008, 3489–517.

[128] Debecker DP, Le Bras S, Boissière C, Chaumonnot A, Sanchez C. Aerosol processing: a wind of innovation in the field of advanced heterogeneous catalysts. Chem Soc Rev. 2018;47: 4112–55.

[129] Shoinkhorova T, Dikhtiarenko A, Ramirez A, Dutta Chowdhury A, Caglayan M, Vittenet J, Bendjeriou-Sedjerari A, Ali OS, Morales-Osorio I, Xu W, Gascon J. Shaping of ZSM-5-Based Catalysts via Spray Drying: Effect on Methanol-to-Olefins Performance. ACS Appl Mater Interfaces. 2019;11:44133–43.

[130] Herbert R, Wang D, Schomäcker R, Schlögl R, Hess C. Stabilization of Mesoporous Silica SBA-15 by Surface Functionalization. ChemPhysChem. 2009;10:2230–33.

[131] Schüth F. Engineered porous catalytic materials. Annu Rev Mater Res. 2005;35:209–38.

[132] Baldovino-Medrano VG, Farin B, Gaigneaux EM. Establishing the Role of Graphite as a Shaping Agent of Vanadium-Aluminum Mixed (Hydr)oxides and Their Physicochemical Properties and Catalytic Functionalities. ACS Catal. 2012:322–36.

[133] Buurmans ILC, Weckhuysen BM. Heterogeneities of individual catalyst particles in space and time as monitored by spectroscopy. Nat Chem. 2012;4:873–86.

[134] Bluhm H, Hävecker M, Kleimenov E, Knop-Gericke A, Liskowski A, Schlögl R, Su DS. In Situ Surface Analysis in Selective Oxidation Catalysis: n-Butane Conversion Over VPP. Top Catal. 2003;23:99–107.

[135] Hävecker M, Mayer RW, Knop-Gericke A, Bluhm H, Kleimenov E, Liskowski A, Su D, Follath R, Requejo FG, Ogletree DF, Salmeron M, Lopez-Sanchez JA, Bartley JK, Hutchings GJ, Schlögl R.

In Situ Investigation of the Nature of the Active Surface of a Vanadyl Pyrophosphate Catalyst during n-Butane Oxidation to Maleic Anhydride. J Phys Chem B. 2003;107:4587–96.
[136] Hutchings GJ. Vanadium Phosphate: A New Look at the Active Components of Catalysts for the Oxidation of Butane to Maleic Anhydride. J Mater Chem. 2004;14:3385–95.
[137] Carreon MA, Guliants VV. Chapter 6 Selective Oxidation of n-Butane over Vanadium-Phosphorous Oxide. In: Nanostructured Catalysts: Selective Oxidations. The Royal Society of Chemistry; 2011, 141–68.
[138] Lintz H-G, Reitzmann A. Alternative Reaction Engineering Concepts in Partial Oxidations on Oxidic Catalysts. Catal Rev Sci Eng. 2007;49:1–32.
[139] Centi G. Vanadyl Pyrophosphate – A Critical Overview. Catal Today. 1993;16:5–26.
[140] Litster S, McLean G. PEM fuel cell electrodes. J Power Sources. 2004;130:61–76.
[141] Kleijn SEF, Lai SCS, Koper MTM, Unwin PR. Electrochemistry of Nanoparticles. Angew Chem Int Ed. 2014;53:3558–86.
[142] Goodenough JB, Kim Y. Challenges for Rechargeable Li Batteries. Chem. Mater. 2010;22: 587–603.
[143] Mehta V, Cooper JS. Review and analysis of PEM fuel cell design and manufacturing. J Power Sources. 2003;114:32–53.
[144] De Las Casas C, Li W. A review of application of carbon nanotubes for lithium ion battery anode material. J Power Sources. 2012;208:74–85.
[145] Winter M, Barnett B, Xu K. Before Li Ion Batteries. Chem Rev. 2018;118:11433–56.
[146] Sharma S, Pollet BG. Support materials for PEMFC and DMFC electrocatalysts – A review. J Power Sources. 2012;208:96–119.
[147] Sun C, Stimming U. Recent anode advances in solid oxide fuel cells. J Power Sources. 2007;171:247–60.
[148] Tucker MC. Progress in metal-supported solid oxide fuel cells: A review. J Power Sources. 2010;195:4570–82.
[149] Ralphs K, Hardacre C, James SL. Application of heterogeneous catalysts prepared by mechanochemical synthesis. Chem Soc Rev. 2013;42:7701–18.
[150] Soerijanto H, Rödel C, Wild U, Lerch M, Schomäcker R, Schlögl R, Ressler T. The impact of nitrogen mobility on the activity of zirconium oxynitride catalysts for ammonia decomposition. J Catal. 2007;250:19–24.
[151] Wang B. Recent development of non-platinum catalysts for oxygen reduction reaction. J Power Sources. 2005;152:1–15.
[152] Schlögl R. Fused Catalysts. In: Ertl G, Knözinger H, Weitkamp J, editors. Handbook of Heterogeneous Catalysis. 2008, 81–92.
[153] Oyama ST. Crystal structure and chemical reactivity of transition metal carbides and nitrides. J Solid State Chem. 1992;96:442–45.
[154] Cotter T, Frank B, Zhang W, Schlögl R, Trunschke A. The Impact of V Doping on the Carbothermal Synthesis of Mesoporous Mo Carbides. Chem. Mater. 2013;25:3124–36.
[155] Koch G, Hävecker M, Teschner D, Carey SJ, Wang Y, Kube P, Hetaba W, Lunkenbein T, Auffermann G, Timpe O, Rosowski F, Schlögl R, Trunschke A. Surface Conditions That Constrain Alkane Oxidation on Perovskites. ACS Catal. 2020;10:7007–20.
[156] Djenadic R, Winterer M. Chemical Vapor Synthesis of Nanocrystalline Oxides. In: Lorke A, Winterer M, Schmechel R, Schulz C, editors. Nanoparticles from the Gasphase: Formation, Structure, Properties. Heidelberg: Springer Berlin Heidelberg: Berlin, 2012, 49–76.
[157] Niedermaier M, Schwab T, Kube P, Zickler GA, Trunschke A, Diwald O. Catalytic activity, water formation, and sintering: Methane activation over Co- and Fe-doped MgO nanocrystals. J Chem Phys. 2020;152:074713.
[158] Oses C, Toher C, Curtarolo S. High-entropy ceramics. Nat Rev Mater. 2020;5:295–309.

[159] Strobel R, Baiker A, Pratsinis SE. Aerosol flame synthesis of catalysts. Adv Powder Technol. 2006;17:457–80.

[160] Pokhrel S, Mädler L. Flame-made Particles for Sensors, Catalysis, and Energy Storage Applications. Energy Fuels. 2020;34:13209–24.

[161] Djenadic R, Botros M, Benel C, Clemens O, Indris S, Choudhary A, Bergfeldt T, Hahn H. Nebulized spray pyrolysis of Al-doped Li7La3Zr2O12 solid electrolyte for battery applications. Solid State Ionics. 2014;263:49–56.

[162] Tessonnier J-P, Su DS. Recent Progress on the Growth Mechanism of Carbon Nanotubes: A Review. ChemSusChem. 2011;4:824–47.

[163] Tessonnier J-P, Rosenthal D, Hansen TW, Hess C, Schuster ME, Blume R, Girgsdies F, Pfänder N, Timpe O, Su DS, Schlögl R. Analysis of the structure and chemical properties of some commercial carbon nanostructures. Carbon. 2009;47:1779–98.

[164] Schüth F, Hesse M, Unger KK. Precipitation and Coprecipitation. In: Handbook of Heterogeneous Catalysis. Wiley-VCH Verlag GmbH & Co. KGaA; 2008.

[165] Behrens M, Schlögl R. How to Prepare a Good Cu/ZnO Catalyst or the Role of Solid State Chemistry for the Synthesis of Nanostructured Catalysts. Zeitschrift für anorganische und allgemeine Chemie. 2013;639:2683–95.

[166] Geus JW, Van Dillen AJ. Preparation of Supported Catalysts by Deposition–Precipitation. In: Handbook of Heterogeneous Catalysis. Wiley-VCH Verlag GmbH & Co. KGaA; 2008.

[167] Jolivet J-P. Metal Oxide Chemistry and Synthesis – From Solution to Solid State. John Wiley & Sons Ltd: Chichester; 2000.

[168] Cavani F, Trifiró F, Vaccari A. Hydrotalcite-type anionic clays: Preparation, properties and applications. Catal Today. 1991;11:173–301.

[169] Behrens M, Brennecke D, Girgsdies F, Kissner S, Trunschke A, Nasrudin N, Zakaria S, Idris NF, Hamid SBA, Kniep B, Fischer R, Busser W, Muhler M, Schlögl R. Understanding the complexity of a catalyst synthesis: Co-precipitation of mixed Cu,Zn,Al hydroxycarbonate precursors for Cu/ZnO/Al2O3 catalysts investigated by titration experiments. Appl Catal A Gen. 2011;392:93–102.

[170] Baltes C, Vukojevic S, Schüth F. Correlations between synthesis, precursor, and catalyst structure and activity of a large set of CuO/ZnO/Al2O3 catalysts for methanol synthesis. J Catal. 2008;258:334–44.

[171] Thanh NTK, Maclean N, Mahiddine S. Mechanisms of Nucleation and Growth of Nanoparticles in Solution. Chem Rev. 2014.

[172] Jung S Speciation of molybdenum- and vanadium-based polyoxometalate species in aqueous medium and gas-phase and its consequences for M1 structured MoV oxide synthesis. Doctoral Thesis, Technische Universität Berlin, Berlin, 2018.

[173] Potter ME, Light ME, Irving DJM, Oakley AE, Chapman S, Chater P, Cutts G, Watts A, Wharmby M, Vandegehuchte BD, Schreiber MW, Raja R. Exploring the origins of crystallisation kinetics in hierarchical materials using in situ X-ray diffraction and pair distribution function analysis. Phys Chem Chem Phys. 2020;22:18860–67.

[174] Norby P. In-situ XRD as a tool to understanding zeolite crystallization. Curr Opin Colloid Interface Sci. 2006;11:118–25.

[175] Michailovski A, Kiebach R, Bensch W, Grunwaldt J-D, Baiker A, Komarneni S, Patzke GR. Morphological and Kinetic Studies on Hexagonal Tungstates. Chem. Mater. 2006;19: 185–97.

[176] Beale AM, Sankar G. In Situ Study of the Formation of Crystalline Bismuth Molybdate Materials under Hydrothermal Conditions. Chem. Mater. 2002;15:146–53.

[177] Beale AM, Sankar G. Understanding the Crystallization of Nanosized Cobalt Aluminate Spinel from Ion-Exchanged Zeolites Using Combined in Situ QEXAFS/XRD. Chem. Mater. 2005;18:263–72.
[178] Kiebach R, Pienack N, Bensch W, Grunwaldt J-D, Michailovski A, Baiker A, Fox T, Zhou Y, Patzke GR. Hydrothermal Formation of W/Mo-Oxides: A Multidisciplinary Study of Growth and Shape. Chem. Mater. 2008;20:3022–33.
[179] O'Brien MG, Beale AM, Weckhuysen BM. The role of synchrotron radiation in examining the self-assembly of crystalline nanoporous framework materials: from zeolites and aluminophosphates to metal organic hybrids. Chem Soc Rev. 2010;39:4767–82.
[180] Aerts A, Kirschhock CEA, Martens JA. Methods for in situ spectroscopic probing of the synthesis of a zeolite. Chem Soc Rev. 2010;39:4626–42.
[181] Goesten M, Stavitski E, Pidko EA, Gücüyener C, Boshuizen B, Ehrlich SN, Hensen EJM, Kapteijn F, Gascon J. The Molecular Pathway to ZIF-7 Microrods Revealed by In Situ Time-Resolved Small- and Wide-Angle X-Ray Scattering, Quick-Scanning Extended X-Ray Absorption Spectroscopy, and DFT Calculations. Chem Eur J. 2013;19:7809–16.
[182] Aerts A, Haouas M, Caremans TP, Follens LRA, Van erp TS, Taulelle F, Vermant J, Martens JA, Kirschhock CEA. Investigation of the Mechanism of Colloidal Silicalite-1 Crystallization by Using DLS, SAXS, and 29Si NMR Spectroscopy. Chem Eur J. 2010;16:2764–74.
[183] Häne J, Brühwiler D, Ecker A, Hass R. Real-time inline monitoring of zeolite synthesis by Photon Density Wave spectroscopy. Microporous Mesoporous Mater. 2019;288:109580.
[184] Baumgartner J, Dey A, Bomans PHH, Le Coadou C, Fratzl P, Sommerdijk NAJM, Faivre D. Nucleation and growth of magnetite from solution. Nat Mater. 2013;12:310–14.
[185] Woehl TJ. Metal Nanocrystal Formation during Liquid Phase Transmission Electron Microscopy: Thermodynamics and Kinetics of Precursor Conversion, Nucleation, and Growth. Chem. Mater. 2020;32:7569–81.
[186] Perego C, Villa P. Catalyst preparation methods. Catal Today. 1997;34:281–305.
[187] Khalid S, Caliebe W, Siddons P, So I, Clay B, Lenhard T, Hanson J, Wang Q, Frenkel AI, Marinkovic N, Hould N, Ginder-Vogel M, Landrot GL, Sparks DL, Ganjoo A. Quick extended x-ray absorption fine structure instrument with millisecond time scale, optimized for in situ applications. Rev Sci Instrum. 2010;81:015105.
[188] Ivanova II, Kolyagin YG, Kasyanov IA, Yakimov AV, Bok TO, Zarubin DN. Time-Resolved In Situ MAS NMR Monitoring of the Nucleation and Growth of Zeolite BEA Catalysts under Hydrothermal Conditions. Angew Chem Int Ed. 2017;56:15344–47.
[189] Haouas M, Taulelle F, Martineau C. Recent advances in application of 27Al NMR spectroscopy to materials science. Prog Nucl Magn Reson Spectrosc. 2016;94-95:11–36.
[190] Maksimovskaya RI, Bondareva VM, Aleshina GI, Spectroscopic NMR. Studies of Interactions in Solution during the Synthesis of MoVTeNb Oxide Catalysts. Eur J Inorg Chem. 2008;2008:4906–14.
[191] Hu JZ, Hu MY, Zhao Z, Xu S, Vjunov A, Shi H, Camaioni DM, Peden CHF, Lercher JA. Sealed rotors for in situ high temperature high pressure MAS NMR. ChemComm. 2015;51:13458–61.
[192] Kongmark C, Martis V, Rubbens A, Pirovano C, Lofberg A, Sankar G, Bordes-Richard E, Vannier R-N, Van Beek W. Elucidating the genesis of Bi2MoO6 catalyst by combination of synchrotron radiation experiments and Raman scattering. ChemComm. 2009:4850–52.
[193] Celaya Sanfiz A, Hansen TW, Girgsdies F, Timpe O, Rödel E, Ressler T, Trunschke A, Schlögl R. Preparation of Phase-Pure M1 MoVTeNb Oxide Catalysts by Hydrothermal Synthesis-Influence of Reaction Parameters on Structure and Morphology. Top Catal. 2008;50:19–32.
[194] Fan F, Feng Z, Li G, Sun K, Ying P, Li C. In Situ UV Raman Spectroscopic Studies on the Synthesis Mechanism of Zeolite X. Chem Eur J. 2008;14:5125–29.

[195] Sanchez Sanchez M, Girgsdies F, Jastak M, Kube P, Schlögl R, Trunschke A. Aiding the Self-Assembly of Supramolecular Polyoxometalates under Hydrothermal Conditions To Give Precursors of Complex Functional Oxides. Angew Chem Int Ed. 2012;51:7194–97.
[196] Noack J, Rosowski F, Schlögl R, Trunschke A. Speciation of Molybdates under Hydrothermal Conditions. Zeitschrift für anorganische und allgemeine Chemie. 2014;640:2730–36.
[197] Zaera F. Probing Liquid/Solid Interfaces at the Molecular Level. Chem Rev. 2012;112: 2920–86.
[198] Falcone Jr JS, Bass JL, Krumrine PH, Brensinger K, Schenk ER. Characterizing the Infrared Bands of Aqueous Soluble Silicates. J Phys Chem A. 2010;114:2438–46.
[199] Patis A, Dracopoulos V, Nikolakis V. Investigation of Silicalite-1 Crystallization Using Attenuated Total Reflection/Fourier Transform Infrared Spectroscopy. J Phys Chem C. 2007;111:17478–84.
[200] Groen JC, Hamminga GM, Moulijn JA, Pérez-Ramírez J. In situ monitoring of desilication of MFI-type zeolites in alkaline medium. Phys Chem Chem Phys. 2007;9:4822–30.
[201] Pelster SA, Schrader W, Schuth F. Monitoring temporal evolution of silicate species during hydrolysis and condensation of silicates using mass spectrometry. J Am Chem Soc. 2006;128:4310–17.
[202] Wilson EF, Miras HN, Rosnes MH, Cronin L. Real-Time Observation of the Self-Assembly of Hybrid Polyoxometalates Using Mass Spectrometry. Angew Chem Int Ed. 2011;50:3720–24.
[203] Marianski M, Seo J, Mucha E, Thomas DA, Jung S, Schlögl R, Meijer G, Trunschke A, Von Helden G. Structural Characterization of Molybdenum Oxide Nanoclusters Using Ion Mobility Spectrometry–Mass Spectrometry and Infrared Action Spectroscopy. J Phys Chem C. 2019;123:7845–53.
[204] Kashchiev D, Van Rosmalen GM. Review: Nucleation in solutions revisited. Cryst Res Technol. 2003;38:555–74.
[205] Fitz E. Dynamische Lichtstreuung zur Partikelgrößenbestimmung von CuZnAl Hydroxykarbonaten. Berlin: Beuth Hochschule für Technik Berlin; 2008.
[206] Galarneau A, Nader M, Guenneau F, Di Renzo F, Gedeon A. Understanding the Stability in Water of Mesoporous SBA-15 and MCM-41. J Phys Chem C. 2007;111:8268–77.
[207] Carrier X, Marceau E, Lambert J-Fß, Che M. Transformations of gamma-alumina in aqueous suspensions: 1 Alumina chemical weathering studied as a function of pH. J Colloid Interface Sci. 2007;308:429–37.
[208] Tichit D, Layrac G, Gérardin C. Synthesis of layered double hydroxides through continuous flow processes: A review. Chem Eng J. 2019;369:302–32.
[209] Landau MV. Sol–Gel Process. In: Handbook of Heterogeneous Catalysis. Wiley-VCH Verlag GmbH & Co. KGaA; 2008.
[210] Hench LL, West JK. The sol-gel process. Chem Rev. 1990;90:33–72.
[211] Sui R, Charpentier P. Synthesis of Metal Oxide Nanostructures by Direct Sol-Gel Chemistry in Supercritical Fluids. Chem Rev. 2012.
[212] Fröschl T, Hörmann U, Kubiak P, Kučerová G, Pfanzelt M, Weiss CK, Behm RJ, Hüsing N, Kaiser U, Landfester K, Wohlfahrt-Mehrens M. High surface area crystalline titanium dioxide: potential and limits in electrochemical energy storage and catalysis. Chem Soc Rev. 2012;41:5313–60.
[213] Debecker DP, Hulea V, Mutin PH. Mesoporous mixed oxide catalysts via non-hydrolytic sol–gel: A review. Appl Catal A Gen. 2013;451:192–206.
[214] Ciriminna R, Fidalgo A, Pandarus V, Béland F, Ilharco LM, Pagliaro M. The Sol–Gel Route to Advanced Silica-Based Materials and Recent Applications. Chem Rev. 2013;113:6592–620.
[215] Stöber W, Fink A, Bohn E. Controlled growth of monodisperse silica spheres in the micron size range. J Colloid Interface Sci. 1968;26:62–69.

[216] Cushing BL, Kolesnichenko VL, O'Connor CJ. Recent advances in the liquid-phase syntheses of inorganic nanoparticles. Chem Rev. 2004;104:3893–946.
[217] Niederberger M. Nonaqueous Sol-Gel Routes to Metal Oxide Nanoparticles. Acc Chem Res. 2007;40:793–800.
[218] Kolen'ko YV, Amakawa K, d'Alnoncourt RN, Girgsdies F, Weinberg G, Schlögl R, Trunschke A. Unusual Phase Evolution in MoVTeNb Oxide Catalysts Prepared by a Novel Acrylamide-Gelation Route. ChemCatChem. 2012;4:495–503.
[219] Guo Y, Bae J, Fang Z, Li P, Zhao F, Yu G. Hydrogels and Hydrogel-Derived Materials for Energy and Water Sustainability. Chem Rev. 2020;120:7642–707.
[220] Wainwright SG, Parlett CMA, Blackley RA, Zhou W, Lee AF, Wilson K, Bruce DW. True liquid crystal templating of SBA-15 with reduced microporosity. Microporous Mesoporous Mater. 2013;172:112–17.
[221] Arnal PM, Comotti M, Schüth F. High-Temperature-Stable Catalysts by Hollow Sphere Encapsulation. Angew Chem Int Ed. 2006;45:8224–27.
[222] Aymonier C, Loppinet-Serani A, Reveron H, Garrabos Y, Cansell F. Review of supercritical fluids in inorganic materials science. J Supercrit Fluids. 2006;38:242–51.
[223] Cheetham AK, Rao CNR, Feller RK. Structural diversity and chemical trends in hybrid inorganic-organic framework materials. ChemComm. 2006:4780–95.
[224] Cundy CS, Cox PA. The Hydrothermal Synthesis of Zeolites: History and Development from the Earliest Days to the Present Time. Chem Rev. 2003;103:663–702.
[225] Cundy CS, Cox PA. The hydrothermal synthesis of zeolites: Precursors, intermediates and reaction mechanism. Microporous Mesoporous Mater. 2005;82:1–78.
[226] Demazeau G. Solvothermal reactions: an original route for the synthesis of novel materials. J Mater Sci. 2008;43:2104–14.
[227] Parnham ER, Morris RE. Ionothermal synthesis of zeolites, metal-organic frameworks, and inorganic-organic hybrids. Acc Chem Res. 2007;40:1005–13.
[228] Wakihara T, Okubo T. Hydrothermal synthesis and characterization of zeolites. Chem Lett. 2005;34:276–81.
[229] Kirschhock CEA, Feijen EJP, Jacobs PA, Martens JA. Hydrothermal Zeolite Synthesis. In: Handbook of Heterogeneous Catalysis. Wiley-VCH Verlag GmbH & Co. KGaA; 2008.
[230] Rabenau A. The Role of Hydrothermal Synthesis in Preparative Chemistry. Angew Chem Int Ed Engl. 1985;24:1026–40.
[231] Rajamathi M, Seshadri R. Oxide and chalcogenide nanoparticles from hydrothermal/solvothermal reactions. Curr Opin Solid State Mater Sci. 2002;6:337–45.
[232] Soler-Illia GJAA, Sanchez C, Lebeau B, Patarin J. Chemical Strategies To Design Textured Materials: from Microporous and Mesoporous Oxides to Nanonetworks and Hierarchical Structures. Chem Rev. 2002;102:4093–138.
[233] Schueth F. Non-siliceous Mesostructured and Mesoporous Materials. Chem. Mater. 2001;13:3184–95.
[234] D. A. A. Soler-illia GJ, Crepaldi EL, Grosso D, Sanchez C. Block copolymer-templated mesoporous oxides. Curr Opin Colloid Interface Sci. 2003;8:109–26.
[235] Taguchi A, Schü th F. Ordered mesoporous materials in catalysis. Microporous Mesoporous Mater. 2005;77:1–45.
[236] Corr SA, Shoemaker DP, Toberer ES, Seshadri R. Spontaneously formed porous and composite materials. J Mater Chem. 2010;20:1413–22.
[237] Shiju NR, Guliants VV. Microwave-assisted hydrothermal synthesis of monophasic Mo-V-Te-Nb-O mixed oxide catalyst for the selective ammoxidation of propane. ChemPhysChem. 2007;8:1615–17.

[238] Tompsett GA, Conner WC, Yngvesson KS. Microwave synthesis of nanoporous materials. Chemphyschem. 2006;7:296–319.
[239] Vanetsev AS, Tretyakov YD. Microwave synthesis of individual and multicomponent oxides. Usp Khim. 2007;76:435–53.
[240] Xie H, Shen D, Wang X, Shen G. Microwave hydrothermal synthesis and visible-light photocatalytic activity of [gamma]-Bi2MoO6 nanoplates. Mater Chem Phys. 2008;110:332–36.
[241] Patzke GR, Zhou Y, Kontic R, Conrad F. Oxide Nanomaterials: Synthetic Developments, Mechanistic Studies, and Technological Innovations. Angew Chem Int Ed. 2011;50:826–59.
[242] Baranchikov AY, Ivanov VK, Tretyakov YD. Sonochemical synthesis of inorganic materials. Usp Khim. 2007;76:147–68.
[243] Shchukin DG, Möhwald H. Sonochemical nanosynthesis at the engineered interface of a cavitation microbubble. Phys Chem Chem Phys. 2006;8:3496–506.
[244] Adschiri T. Supercritical hydrothermal synthesis of organic-inorganic hybrid nanoparticles. Chem Lett. 2007;36:1188–93.
[245] Aymonier C, Cansell F. Supercritical fluid techniques. Annales De Chimie-Science Des Materiaux. 2006;31:317–37.
[246] Galkin AA, Lunin VV. Water in sub- and supercritical states as a universal medium for chemical reactions. Usp Khim. 2005;74:24–40.
[247] Lam UT, Mammucari R, Suzuki K, Foster NR. Processing of iron oxide nanoparticles by supercritical fluids. Ind Eng Chem Res. 2008;47:599–614.
[248] Reverchon E, Adami R. Nanomaterials and supercritical fluids. J Supercrit Fluids. 2006;37:1–22.
[249] Darr JA, Zhang J, Makwana NM, Weng X. Continuous Hydrothermal Synthesis of Inorganic Nanoparticles: Applications and Future Directions. Chem Rev. 2017;117:11125–238.
[250] Galkin AA, Kostyuk BG, Lunin VV, Poliakoff M. Continuous Reactions in Supercritical Water: A New Route to La2CuO4 with a High Surface Area and Enhanced Oxygen Mobility. Angew Chem. 2000;112:2850–52.
[251] Hellstern HL, Becker J, Hald P, Bremholm M, Mamakhel A, Iversen BB. Development of a Dual-Stage Continuous Flow Reactor for Hydrothermal Synthesis of Hybrid Nanoparticles. Ind Eng Chem Res. 2015;54:8500–08.
[252] Munnik P, De Jongh PE, De Jong KP. Recent Developments in the Synthesis of Supported Catalysts. Chem Rev. 2015;115:6687–718.
[253] Stein A, Melde BJ, Schroden RC. Hybrid Inorganic–Organic Mesoporous Silicates – Nanoscopic Reactors Coming of Age. Adv Mater. 2000;12:1403–19.
[254] Carvalho WA, Wallau M, Schuchardt U. Iron and copper immobilised on mesoporous MCM-41 molecular sieves as catalysts for the oxidation of cyclohexane. J Mol Catal A Chem. 1999;144:91–99.
[255] Hess C, Wild U, Schlögl R. The mechanism for the controlled synthesis of highly dispersed vanadia supported on silica SBA-15. Microporous Mesoporous Mater. 2006;95:339–49.
[256] Amakawa K, Sun L, Guo C, Hävecker M, Kube P, Wachs IE, Lwin S, Frenkel AI, Patlolla A, Hermann K, Schlögl R, Trunschke A. How Strain Affects the Reactivity of Surface Metal Oxide Catalysts. Angew Chem Int Ed. 2013;52:13553–57.
[257] Amakawa K, Wang Y, Kröhnert J, Schlögl R, Trunschke A. Acid sites on silica-supported molybdenum oxides probed by ammonia adsorption: Experiment and theory. Mol Catal. 2019;478:110580.
[258] Amakawa K, Kröhnert J, Wrabetz S, Frank B, Hemmann F, Jäger C, Schlögl R, Trunschke A. Active Sites in Olefin Metathesis over Supported Molybdena Catalysts. ChemCatChem. 2015;7:4059–65.
[259] Parks GA, Bruyn PLD. The zero point of charge of oxides. J Phys Chem. 1962;66:967–73.

[260] Llorente I, Fajardo S, Bastidas JM. Applications of electrokinetic phenomena in materials science. J Solid State Electrochem. 2014;18:293–307.
[261] Parks GA. The Isoelectric Points of Solid Oxides, Solid Hydroxides, and Aqueous Hydroxo Complex Systems. Chem Rev. 1965;65:177–98.
[262] Sun K-Q, Marceau E, Che M. Evolution of nickel speciation during preparation of Ni–SiO2 catalysts: effect of the number of chelating ligands in [Ni(en)x(H2O)6−2x]2+ precursor complexes. Phys Chem Chem Phys. 2006;8:1731–38.
[263] Espinosa-Alonso L, Beale AM, Weckhuysen BM. Profiling Physicochemical Changes within Catalyst Bodies during Preparation: New Insights from Invasive and Noninvasive Microspectroscopic Studies. Acc Chem Res. 2010;43:1279–88.
[264] Eggenhuisen TM, Friedrich H, Nudelman F, Zečević J, Sommerdijk NAJM, De Jongh PE, De Jong KP. Controlling the Distribution of Supported Nanoparticles by Aqueous Synthesis. Chem. Mater. 2013;25:890–96.
[265] Budevski E, Staikov G, Lorenz WJ. Electrocrystallization: Nucleation and growth phenomena. Electrochim Acta. 2000;45:2559–74.
[266] Penner RM. Mesoscopic Metal Particles and Wires by Electrodeposition. J Phys Chem B. 2002;106:3339–53.
[267] Kang D, Kim TW, Kubota SR, Cardiel AC, Cha HG, Choi K-S. Electrochemical Synthesis of Photoelectrodes and Catalysts for Use in Solar Water Splitting. Chem Rev. 2015;115: 12839–87.
[268] Reichenberger S, Marzun G, Muhler M, Barcikowski S. Perspective of Surfactant-Free Colloidal Nanoparticles in Heterogeneous Catalysis. ChemCatChem. 2019;11:4489–518.
[269] Quinson J, Neumann S, Wannmacher T, Kacenauskaite L, Inaba M, Bucher J, Bizzotto F, Simonsen SB, Theil Kuhn L, Bujak D, Zana A, Arenz M, Kunz S. Colloids for Catalysts: A Concept for the Preparation of Superior Catalysts of Industrial Relevance. Angew Chem Int Ed. 2018;57:12338–41.
[270] Bond GC, Tahir SF. Vanadium oxide monolayer catalysts Preparation, characterization and catalytic activity. Appl Catal. 1991;71:1–31.
[271] Wegener SL, Marks TJ, Stair PC. Design Strategies for the Molecular Level Synthesis of Supported Catalysts. Acc Chem Res. 2011;45:206–14.
[272] Santacesaria E, Cozzolino M, Di Serio M, Venezia AM, Tesser R. Vanadium based catalysts prepared by grafting: preparation, properties and performances in the ODH of butane. Appl Catal A Gen. 2004;270:177–92.
[273] Love AM, Carrero CA, Chieregato A, Grant JT, Conrad S, Verel R, Hermans I. Elucidation of Anchoring and Restructuring Steps during Synthesis of Silica-Supported Vanadium Oxide Catalysts. Chem. Mater. 2016;28:5495–504.
[274] Fu T, Wang Y, Wernbacher A, Schlögl R, Trunschke A. Single-Site Vanadyl Species Isolated within Molybdenum Oxide Monolayers in Propane Oxidation. ACS Catal. 2019;9:4875–86.
[275] Copéret C. Fuels and energy carriers from single-site catalysts prepared via surface organometallic chemistry. Nat Energy. 2019;4:1018–24.
[276] O'Neill BJ, Jackson DHK, Lee J, Canlas C, Stair PC, Marshall CL, Elam JW, Kuech TF, Dumesic JA, Huber GW. Catalyst Design with Atomic Layer Deposition. ACS Catal. 2015;5:1804–25.
[277] Knemeyer K, Piernavieja Hermida M, Ingale P, Schmidt J, Kröhnert J, Naumann d'Alnoncourt R, Driess M, Rosowski F. Mechanistic studies of atomic layer deposition on oxidation catalysts – AlOx and POx deposition. Phys Chem Chem Phys. 2020;22:17999–8006.
[278] Strempel VE, Löffler D, Kröhnert J, Skorupska K, Johnson B, d'Alnoncourt RN, Driess M, Rosowski F. Enhancing of catalytic properties of vanadia via surface doping with phosphorus using atomic layer deposition. J Vac Sci Technol. 2016;34:01A135.

[279] Zander S, Kunkes EL, Schuster ME, Schumann J, Weinberg G, Teschner D, Jacobsen N, Schlögl R, Behrens M. The Role of the Oxide Component in the Development of Copper Composite Catalysts for Methanol Synthesis. Angew Chem Int Ed. 2013;52:6536–40.

[280] Zhao Q, Yan Z, Chen C, Chen J. Spinels: Controlled Preparation, Oxygen Reduction/Evolution Reaction Application, and Beyond. Chem Rev. 2017;117:10121–211.

[281] Wachs IE, Routray K. Catalysis Science of Bulk Mixed Oxides. ACS Catal. 2012;2:1235–46.

[282] Popova GY, Andrushkevich TV, Dovlitova LS, Aleshina GA, Chesalov YA, Ishenko AV, Ishenko EV, Plyasova LM, Malakhov VV, Khramov MI. The investigation of chemical and phase composition of solid precursor of MoVTeNb oxide catalyst and its transformation during the thermal treatment. Appl Catal A Gen. 2009;353:249–57.

[283] Girgsdies F, Schlögl R, Trunschke A. In-situ X-ray diffraction study of phase crystallization from an amorphous MoVTeNb oxide catalyst precursor. Catal Commun. 2012;18:60–62.

[284] Zenkovets GA, Kryukova GN, Gavrilov VY, Tsybulya SV, Anufrienko VA, Larina TA, Khabibulin DF, Lapina OB, Rödel E, Trunschke A, Ressler T, Schlögl R. The structural genesis of a complex (MoVW)5O14 oxide during thermal treatments and its redox behavior at elevated temperatures. Mater Chem Phys. 2007;103:295–304.

[285] Bron M, Teschner D, Knop-Gericke A, Steinhauer B, Scheybal A, Hävecker M, Wang D, Fodisch R, Honicke D, Wootsch A, Schlögl R, Claus P. Bridging the pressure and materials gap: in-depth characterisation and reaction studies of silver-catalysed acrolein hydrogenation. J Catal. 2005;234:37–47.

[286] Tessonnier JP, Ersen O, Weinberg G, Pham-Huu C, Su DS, Schlögl R. Selective Deposition of Metal Nanoparticles Inside or Outside Multiwalled Carbon Nanotubes. ACS Nano. 2009;3:2081–89.

[287] Behrens M, Furche A, Kasatkin I, Trunschke A, Busser W, Muhler M, Kniep B, Fischer R, Schlögl R. The Potential of Microstructural Optimization in Metal/Oxide Catalysts: Higher Intrinsic Activity of Copper by Partial Embedding of Copper Nanoparticles. ChemCatChem. 2010;2:816–18.

[288] Kasatkin I, Kurr P, Kniep B, Trunschke A, Schlögl R. Role of lattice strain and defects in copper particles on the activity of Cu/ZnO/Al2O3 catalysts for methanol synthesis. Angew Chem Int Ed. 2007;46:7324–27.

[289] Zhang J, Gao M-R, Luo J-L. In Situ Exsolved Metal Nanoparticles: A Smart Approach for Optimization of Catalysts. Chem. Mater. 2020;32:5424–41.

[290] Feng Z, Hong WT, Fong DD, Lee Y-L, Yacoby Y, Morgan D, Shao-Horn Y. Catalytic Activity and Stability of Oxides: The Role of Near-Surface Atomic Structures and Compositions. Acc Chem Res. 2016;49:966–73.

[291] Hansen TW, DeLaRiva AT, Challa SR, Datye AK. Sintering of Catalytic Nanoparticles: Particle Migration or Ostwald Ripening?. Acc Chem Res. 2013;46:1720–30.

[292] Götsch T, Schlicker L, Bekheet MF, Doran A, Grünbacher M, Praty C, Tada M, Matsui H, Ishiguro N, Gurlo A, Klötzer B, Penner S. tructural investigations of La0.6Sr0.4FeO3–δ under reducing conditions: kinetic and thermodynamic limitations for phase transformations and iron exsolution phenomena. RSC Adv. 2018;8:3120–31.

[293] Delmon B. Reactions During Catalyst Activation. In: Handbook of Heterogeneous Catalysis. Wiley-VCH Verlag GmbH & Co. KGaA; 2008.

[294] Kalz KF, Kraehnert R, Dvoyashkin M, Dittmeyer R, Gläser R, Krewer U, Reuter K, Grunwaldt J-D. Future Challenges in Heterogeneous Catalysis: Understanding Catalysts under Dynamic Reaction Conditions. ChemCatChem. 2017;9:17–29.

[295] Choi YW, Mistry H, Roldan Cuenya B. New insights into working nanostructured electrocatalysts through operando spectroscopy and microscopy. Curr Opin Electrochem. 2017;1:95–103.

[296] Amakawa K, Wrabetz S, Kröhnert J, Tzolova-Müller G, Schlögl R, Trunschke A. In Situ Generation of Active Sites in Olefin Metathesis. J Am Chem Soc. 2012;134:11462–73.

[297] Kube P, Frank B, Wrabetz S, Kröhnert J, Hävecker M, Velasco-Vélez J, Noack J, Schlögl R, Trunschke A. Functional Analysis of Catalysts for Lower Alkane Oxidation. ChemCatChem. 2017;9:573–85.

[298] Ishikawa S, Ueda W. Microporous crystalline Mo-V mixed oxides for selective oxidations. Catal Sci Technol. 2016;6:617–29.

[299] Bruix A, Margraf JT, Andersen M, Reuter K. First-principles-based multiscale modelling of heterogeneous catalysis. Nat Catal. 2019;2:659–70.

[300] Andersen M, Levchenko SV, Scheffler M, Reuter K. Beyond Scaling Relations for the Description of Catalytic Materials. ACS Catal. 2019;9:2752–59.

[301] Freeze JG, Kelly HR, Batista VS. Search for Catalysts by Inverse Design: Artificial Intelligence, Mountain Climbers, and Alchemists. Chem Rev. 2019;119:6595–612.

[302] Pacchioni G, Freund H-J. Controlling the charge state of supported nanoparticles in catalysis: lessons from model systems. Chem Soc Rev. 2018;47:8474–502.

[303] Kang J, Mahapatra M, Rui N, Orozco I, Shi R, Senanayake SD, Rodriguez JA. Growth and structural studies of In/Au(111) alloys and InOx/Au(111) inverse oxide/metal model catalysts. J Chem Phys. 2020;152:054702.

[304] Jung K, Corrigan N, Ciftci M, Xu J, Seo SE, Hawker CJ, Boyer C. Designing with Light: Advanced 2D, 3D, and 4D Materials. Adv Mater. 2020;32:1903850.

[305] Huang Z, Geyer N, Werner P, De Boor J, Gösele U. Metal-Assisted Chemical Etching of Silicon: A Review. Adv Mater. 2011;23:285–308.

[306] Wang Y, Mao J, Meng X, Yu L, Deng D, Bao X. Catalysis with Two-Dimensional Materials Confining Single Atoms: Concept, Design, and Applications. Chem Rev. 2019;119:1806–54.

[307] Wang W, Tuci G, Duong-Viet C, Liu Y, Rossin A, Luconi L, Nhut J-M, Nguyen-Dinh L, Pham-Huu C, Giambastiani G. Induction heating: an enabling technology for the heat management in catalytic processes. ACS Catal. 2019;9:7921–35.

[308] Roldan Cuenya B. Metal Nanoparticle Catalysts Beginning to Shape-up. Acc Chem Res. 2013;46:1682–91.

[309] Bañares MA. Operando methodology: combination of in situ spectroscopy and simultaneous activity measurements under catalytic reaction conditions. Catal Today. 2005;100:71–77.

Dmitry E. Doronkin, Jan-Dierk Grunwaldt

12 In Situ and Operando Analysis of Heterogeneous Catalysts in Chemical Energy Conversion

This chapter gives an overview of in situ catalyst characterization in general, with a focus on materials in chemical energy conversion. The introduction (Section 12.1) is followed by an overview on the different characterization techniques available and their application in energy-related catalysis (Section 12.2); the importance of in situ and operando studies (Section 12.3); factors to be considered during engineering of in situ cells and experiments (Section 12.4); case studies from gas phase, liquid phase, and high pressure including industrially relevant conditions (Section 12.5); and spatially resolved and spatiotemporal studies (Section 12.6). Finally, an outlook is given on the opportunities and limitations of in situ and operando characterization in chemical energy conversion (Section 12.7).

12.1 Setting the Scene for Catalyst Characterization in Energy-Related Catalysis and Energy Storage

Catalyst characterization requires nowadays an approach that spans all the length and complexity scales from atomic resolution to the reactor scale, as well as stationary and transient processes. Electronic and physical properties on an atomic scale

Acknowledgments: Authors would like to thank all coworkers at the Institute for Chemical Technology and Polymer Chemistry (ITCP) and Institute for Catalysis Research and Technology (IKFT) at Karlsruhe Institute of Technology (KIT) and within the collaboration research centre "TrackAct" (CRC1441, funded by DFG), former co-workers and present collaboration partners at DTU, A. Baiker (ETH Zurich), R. Frahm (University Wuppertal), C.G. Schroer (DESY) and their group members for stimulating discussions and support as well as HASYLAB at DESY (beamlines at DORIS and PETRA rings, Hamburg, Germany), the APS (ID-1, Chicago, USA), ESRF (ID26 and SNBL, Grenoble, France), KIT synchrotron (ANKA-XAS, CAT-ACT), DIAMOND (I18, Didcot, UK) and SLS (superXAS and microXAS, Villigen, Switzerland) for granting beamtime for the different projects shown in this chapter. Financial support by KIT, by BMBF (e.g., "Materials in Action" and "X-ray microscopy for operando investigations of chemical reactions"), the European Community – Research Infrastructure Action under the FP6: "Structuring the European Research Area" ("Integrating Activity on Synchrotron and Free Electron Laser Science" (IA-SFS) RII3-CT-2004-506008), by the DFG priority program SPP2080 and many more initiatives are gratefully acknowledged.

Dmitry E. Doronkin, Jan-Dierk Grunwaldt, Karlsruhe Institute of Technology, Karlsruhe, Germany

https://doi.org/10.1515/9783110608458-012

are at the origin of the catalytic properties of surfaces and are needed to predict and model the performance of catalysts. Equally important are the meso- and the macroscales, which include spillover effects or mass and heat transport in heterogeneous catalysis. They are crucial for the catalyst design (e.g., shaping) but also for modeling and designing the reactor as the final end of the complexity chain. Also, in reactors themselves, catalyst characterization is vital since concentration and temperature gradients or transients can result in different structures of the active site or catalyst particles at the different locations.

The catalyst structure is very dependent on the reaction conditions (see, e.g., [1–6] and references therein). For a long period of time it has been well known that CO-containing atmospheres lead to the reconstruction of Rh surfaces and reshaping of Rh particles [7]. Also, in energy-related catalysis and processes for chemical energy storage, materials undergo such structural changes while they are "working" [8]. This can be in the gas phase, such as during the activation of methane or syngas conversion (e.g., [9–13]), but it can also be in the liquid phase (e.g., the deoxygenation of biomass platform molecules). In addition, structural changes can occur on different timescales [8, 14, 15]: Aging is usually a process that occurs over several hours, days, or even years.

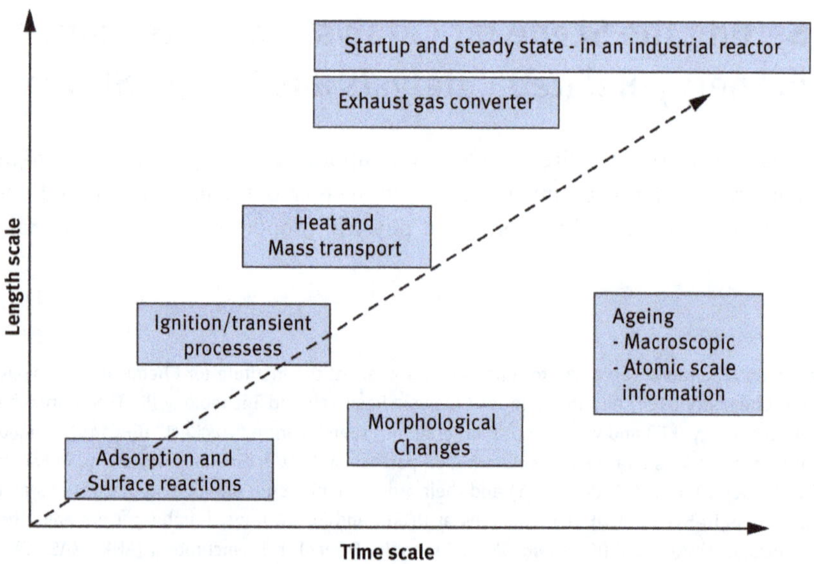

Figure 12.1: Catalyst characterization under reaction conditions spans all length scales from real reactors to the atomic structure with the catalytically active site as well as many orders in timescales considering long-term behavior in a real process and rapid changes of the structure and extremely fast surface processes.

On the other hand, morphological changes can occur on the minute scale [9], or transformations during activation of a catalyst (temperature-programmed reaction/reduction/sulfidation), ignition of a reaction, or oscillations can even occur on the subsecond timescale [12, 16, 17].

Hence, the task is not only to cover the different length scales from the atomic scale to the macroscale but also to study the structure at different timescales as highlighted by Figure 12.1. This will allow establishing a bridge between surface science and theoretical studies on the atomic scale and real-world catalysts in catalytic converters where reactor simulation and modeling on the macroscale, including description of transient processes, is required. This approach to monitoring the atomic scale, the mesoscale, and the macroscale is analogous to the approach in reactor modeling from first principles [18, 19] and required for a bottom-up design [20]. Atomic-scale information is required for modeling the surface reactions, the mesoscale for spillover and similar effects, and the macroscale to account for heat and mass transport.

12.2 The Bench of Complementary Characterization Methods

The scheme for catalyst characterization in general is given in Figure 12.2 (for a more detailed description of methods, cf. [21–23]). In order to properly describe the catalysts on an atomic scale, a mesoscale, and a macroscale, a number of different and complementary methods need to be applied. The overall bulk composition (elemental analysis, X-ray fluorescence [XRF]), the nature of crystalline phases by X-ray diffraction [XRD]), information on the total surface area (Brunauer, Emmett, Teller [BET] theory), and the specific surface area (e.g., chemisorption for supported metal particles) are required. Especially, the latter point is a very important dimension in catalysis, which can also be derived from other measurements giving information about the dispersion. Among them are the particle size from transmission electron microscopy (TEM, and related methods, like scanning transmission electron microscopy [STEM]), pair distribution function (PDF), small-angle X-ray scattering (SAXS), line broadening in XRD (Scherrer equation), the coordination number of the absorbing metal atom of interest extracted from extended X-ray absorption fine structure (EXAFS), and the distribution of the metal atoms by X-ray photoelectron spectroscopy (XPS). XPS together with ion-scattering spectroscopy (ISS), ion mass spectroscopies (e.g. secondary ion mass spectroscopy, SIMS) and scanning electron microscopy (SEM) can further give direct information on the surface composition. Not all catalysts in chemical energy conversion are metal-based catalysts; for example, Co–Mo–S catalysts are important in hydrodesulfurization and are potential candidates for hydrodeoxygenation of biomass, or titanium-, vanadium-, molybdenum-, or bismuth-based catalysts are well known for selective oxidation processes. To uncover the oxidation state and the coordination X-ray absorption near-edge structure (XANES), EXAFS, infrared (IR) spectroscopy, Raman spectroscopy, solid-state nuclear magnetic resonance (NMR), and electron paramagnetic resonance (EPR) are very powerful methods, and the reader is referred to other books for a detailed description of the mentioned methods [21–23].

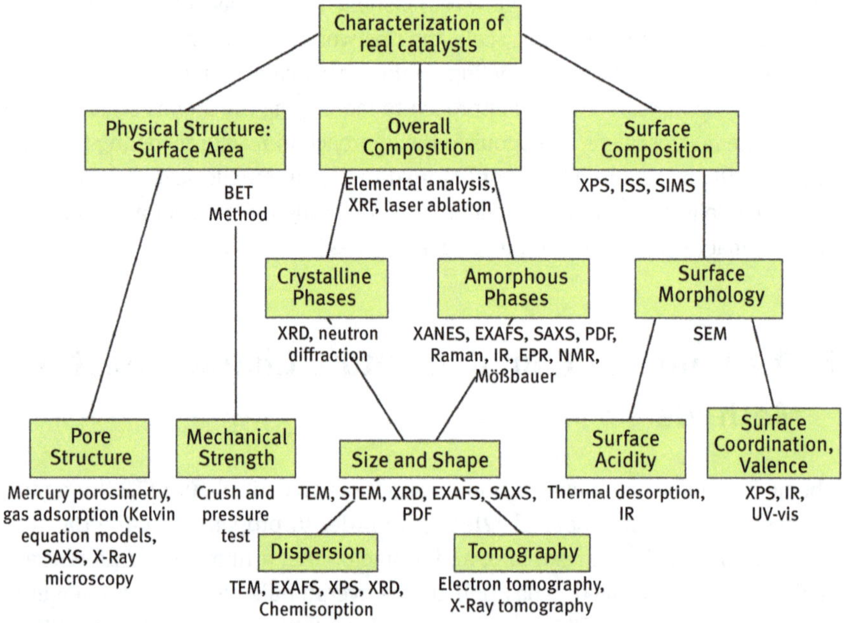

Figure 12.2: Schematic strategy for catalyst characterization; analysis by a number of different and complementary techniques is required. The full names of the respective characterization techniques are given in the text.

In order to gain insight into the meso- and macrostructure, similar methods can be applied. One of the most important aspects is the porosity that can either be measured by physisorption or by mercury porosimetry. In addition small angle X-ray scattering, pair distribution function, tomographic imaging are of great importance; X-ray and NMR tomography are among the few methods that do not require destruction of the catalyst bed and even allow imaging of reactors [6, 14, 24–30]. X-ray contrast exploiting diffraction, fluorescence, or absorption spectroscopy even allows distinguishing different species and phases [31]. Finally, the shaped catalysts can also be cut and analyzed by microspectroscopic methods like micro-IR, micro-Raman, and micro–ultraviolet–visible (UV–vis) spectroscopy [6, 28].

12.3 Importance of In Situ and Operando Studies

Until recently, the focus has been on catalyst characterization after preparation or after reaction. This is the first step toward an understanding of the structure and the functioning of the catalyst, but during recent years, it has been found that this is by far not sufficient and the catalyst structure may change under reaction conditions. Therefore "in situ," or sometimes also called "operando" studies have gained a great deal of attention [1–6]. The increasing significance is demonstrated by the strongly increasing number of such studies since the mid-1990s as shown in Figure 12.3.

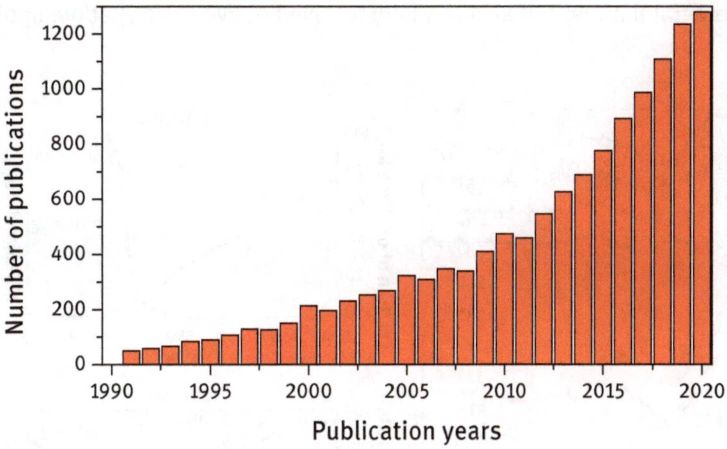

Figure 12.3: Number of papers published containing the words *in situ* or *operando* in connection with *catalysts* and *spectroscopy* (ISI Web of Knowledge).

Why is it important? An illustrative example is the catalytic partial oxidation of methane over Rh and Pt–Rh/Al_2O_3 catalysts that also will be looked at in later stages of this chapter. If a wet-chemically prepared Rh/Al_2O_3 catalyst is heated up to the reaction temperature, one observes a strong restructuring and the formation of $Rh^I(CO)_2$ species by IR spectroscopy [32] and a decrease of the Rh-coordination number in EXAFS [33] due to this formation of rhodium carbonyls. Even stronger are the structural changes if a catalyst is prepared by the so-called flame spray pyrolysis method. In this case, very small noble metal particles in the range of 2–4 nm can be supported on the alumina surface [15]. The noble metal particles are in an oxidized state after preparation and are even in an oxidized state after catalytic partial oxidation of methane at 500 °C [12]. This ex situ characterization may have led to the wrong conclusion that in this case oxidized Pt and Rh are responsible for the catalytic partial oxidation of methane, which is surprising since methane is better activated on metallic surfaces. In addition, in situ XANES (Figure 12.4) and EXAFS uncovered that at the ignition temperature of the catalytic partial oxidation of methane at approximately 330 °C the catalysts changed to a reduced state. However, when cooling down, the catalyst transformed back to oxidized rhodium and platinum at the moment when the catalytic partial oxidation of methane extinguished. A closer look at the ignition temperature uncovered that a recording time of 1 min was not sufficient to grab the reduction of either Pt or Rh, and furthermore, scanning along the catalytic microreactor with a 0.5 mm × 0.5 mm beam uncovered that the structure in the beginning of the catalytic reactor (oxidized) was different from the end of the catalyst bed (reduced state, cf. [12, 34]). This shows the requirement to (a) measure in certain cases in a time-resolved way and (b) watch the catalytic reactor not only in an integral manner but also spatially resolved or even in a spatiotemporal manner [35].

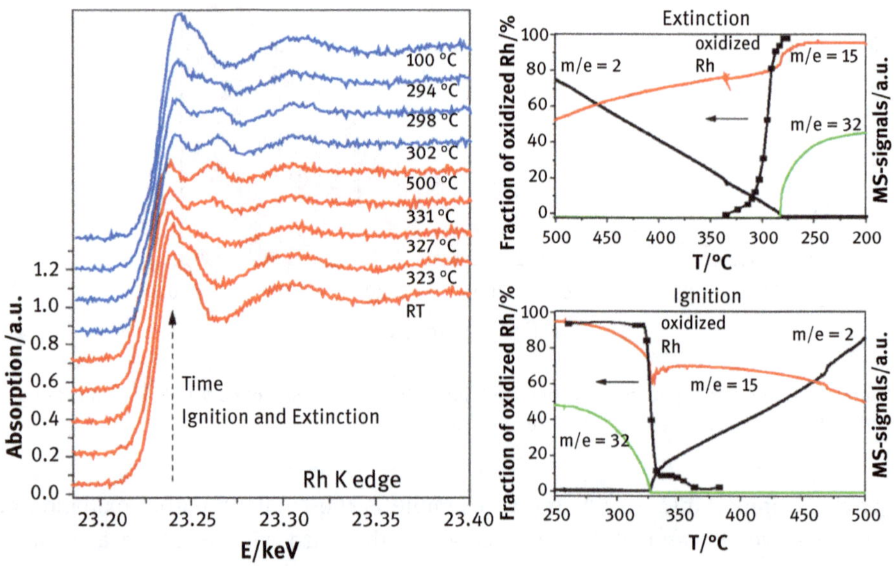

Figure 12.4: Example of the importance of in situ studies; the structural changes unraveled by selected XANES spectra at the Rh K edge during ignition and extinction of the catalytic partial oxidation of methane over 2.5% Pt–2.5% Rh/Al_2O_3 in 6% CH_4/3% O_2/He demonstrate that the structure changes during heating up and cooling down; relative concentrations for hydrogen (m/e = 2), methane (m/e = 15), and oxygen (m/e = 32) and the fraction of oxidized Rh (by linear combination analysis) are given on the right, cf. refs. [12, 34]. Reproduced from ref. [34]

12.4 In Situ Cell Design: A Challenge Between Engineering and Spectroscopy for Dynamic Experiments and Structure Performance Relationships

In situ or operando studies preferentially combine the identification of the structure and the catalytic performance. This is a demanding task and cannot be realized without compromises [36–40]. For X-ray-based techniques and UV–vis, Raman, or IR spectroscopy, the reactor needs windows where the light can shine through or a design so that NMR/EPR studies can become possible. The principle is given in Figure 12.5.

The catalyst must be as homogeneous as possible to get good spectroscopic data. On the other hand, basic engineering rules such as flow patterns through the reactor, heat- and mass-transport properties, dead volume, and catalytic measurements need to be fulfilled. Therefore, preferentially, a thin layer of a catalyst or a sieved catalyst fraction should be applied, especially if the reactions are rapid [37]. Moreover, such studies should be performed under realistic conditions (i.e., in gas phase, liquid phase [including catalyst preparation], or even at high pressure) and at realistic timescales in order to observe the relevant phenomena.

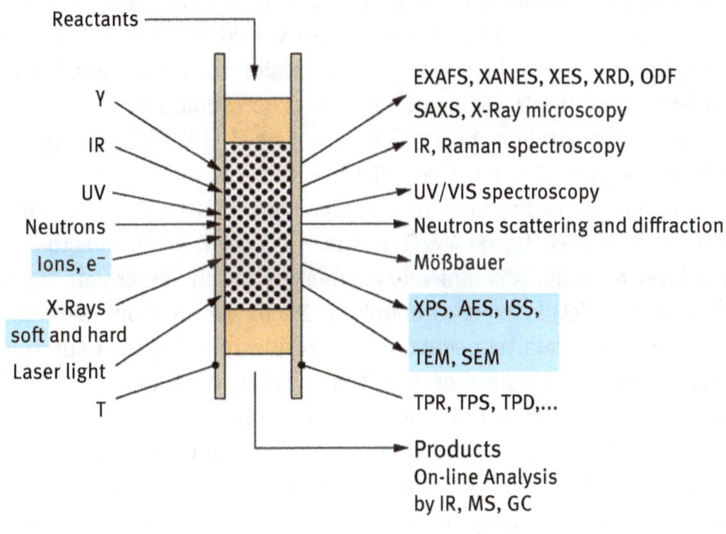

Figure 12.5: The task for in situ/operando studies; techniques that are blue shaded are difficult to realize under reaction conditions since they deal with ions or electrons; extremely thin window materials and differential pumping are presently designed for these techniques.

These challenges are underlined by two extreme cases in Figure 12.6. In the first part, a catalyst pellet is embedded in a catalytic reactor that allows optimal transmission of X-rays. However, the dead volume is large and dynamic studies are not so easy, whereas a fixed-bed reactor allows minimizing it. Further, using a sieved fraction allows establishing optimal conditions for reactions and dynamic changes of the reaction conditions. In fact, analyzing the reduction of Cu/ZnO diluted with boron nitride as pellet (2 mm thickness, left reactor) and as sieved fraction (80 µm particle diameter) uncovered the reduction at more than 40 °C lower temperature in the first case [37] and the presence of a much higher Cu(I) concentration than in the second case (ca. 15 min reduction time at a ramp rate of 3 °C/min). Calculations (details, c. f. [37]) on the basis of an effective diffusion coefficient of $D_e = 10^{-6}–10^{-8}$ $m^2 s^{-1}$ lead to the conclusion that internal mass transport is much more limited in the pellet (0.5–60 min) compared to the sieved fraction (0.02–2 s). In the case of gas-phase reactions, the external mass transport is usually negligible but may become very important in the case of liquid-phase reactions. Furthermore, the previous calculations demonstrate that under optimal conditions dynamic studies with subsecond time resolution are still reasonable; using powder catalysts, it is, however, not necessary to go below the subsecond timescale, except if special designs are used.

Not all catalytic reactions embrace rapid structural changes. Often one may only be interested in stationary conditions. Still, while measuring a catalytic reaction, one has to be very careful with the mass- and heat-transport effects. An illustrative example is the total oxidation of hydrocarbons over Pd particles [37, 41]. In catalytic converters, thin wash-coat layers are used to exploit the noble metal's catalytic activity efficiently, and the Thiele modulus is used for estimation of the effectiveness factor. What happens when the catalyst is not used efficiently enough is demonstrated in 1(c) of Figure 12.6. For the pellet-like catalyst, methane is rapidly consumed on the outer part since the catalytic reaction is much faster than the diffusion in the catalyst pellet. If 1% CH_4 and 4% O_2 in Helium is used, this means that in the top 50–100 µm layer methane is completely combusted and the inner part (more than 90% of the catalyst pellet) is exposed mainly to 2% oxygen without methane. Shining the X-rays through the catalyst pellet therefore results in reality in monitoring most of the catalyst in 2% oxygen instead of 1% CH_4 and 4% O_2 in Helium.

Despite the catalytic reactor in part 2 of Figure 12.6 being excellently suited for gas-phase reactions, it still has its limitations: quartz–glass wall thicknesses of and below 20 µm should be used if X-rays with photon energy below 8 keV are applied. Other designs may be required not only at lower energy but also if reactions are followed at significantly higher pressure, in liquid phase, or even when the preparation of catalysts is studied. X-rays are best suited for this purpose since they can penetrate a number of materials (Be, Kapton, graphite, aluminum foil, silicon nitride, quartz, etc.) rather well – but it remains a challenge to find the best compromise of the in situ cell for the specific scientific questions, and therefore new areas

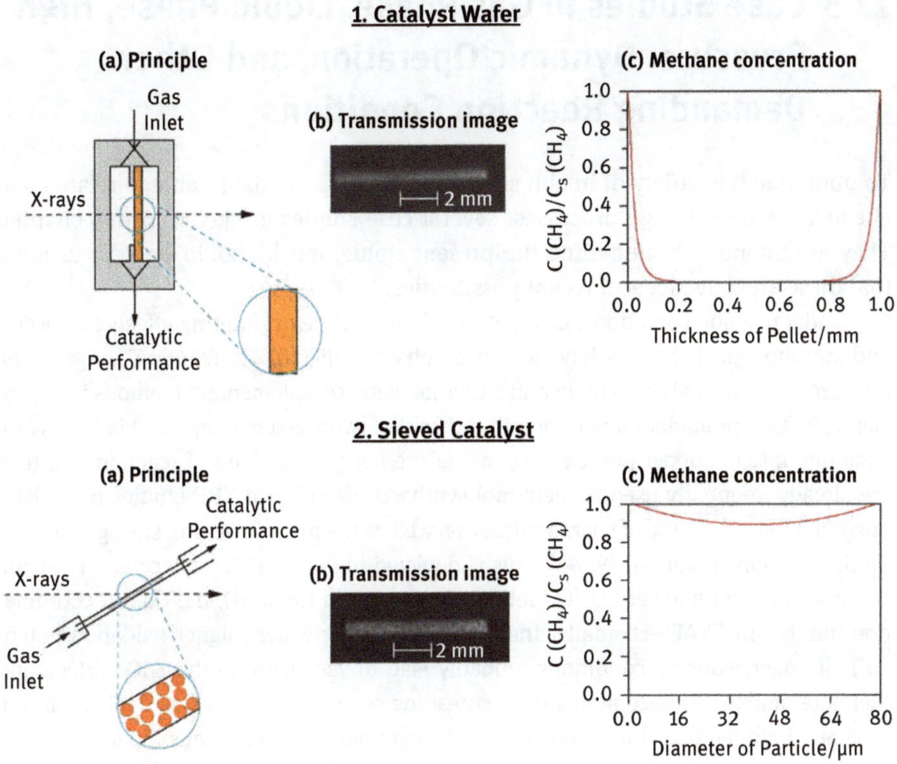

Figure 12.6: Principle of two in situ cells that are optimized for the spectroscopic studies and the catalytic experiments; the X-ray transmission image is recorded by an X-ray eye, and the calculated profile of the methane concentration is estimated for 1% CH_4/4% O_2/He on a pellet of 1 mm thickness and catalyst particles of 100 μm at 500 °C (adapted from ref. [37] with permission from the PCCP Owner Societies).

require the design of new in situ cells [36–38, 40]. The same is true for spectroscopic studies based on IR, UV light, and laser light for Raman spectroscopy [42]; the methods are outlined in Figure 12.5. The application of methods based on the use or detection of electrons or ions is difficult. This is especially true for TEM, as well as surface-sensitive techniques like XPS, ISS, SIMS, and so forth, which are preferentially performed in ultrahigh vacuum. Thanks to the development of thin window materials, based on silicon nitride membranes, as well as efficient differential pumping and high-intensity beams, nowadays samples can be exposed to atmospheres in the mbar and even bar regime [11, 43–45].

12.5 Case Studies in Gas Phase, Liquid Phase, High Pressure, Dynamic Operation, and Other Demanding Reaction Conditions

To underline the variety of in situ studies and their potential in energy-related and chemical energy storage processes, several case studies are given in this chapter. They are intended to underline the present status, but it should be kept in mind that these are only facets of recent possibilities.

Under reaction conditions, changes in oxidation state or sintering effects can occur, and morphological changes have also been observed [9, 10, 41, 42, 46, 47]. Again, the direct relation of catalytic activity and structure using complementary methods is beneficial [2, 9]. One prominent illustration in the field of hydrogen management is the hydrogenation of CO_2. Copper particles are one of the keys in this kind of reaction and they are already frequently used in methanol synthesis. EXAFS and XRD studies have demonstrated that the shape of such particles, as well as the microstructure, strongly depend on the reaction conditions [9, 46]. This is depicted in Figure 12.7. Under more reducing conditions (either a lower CO_2/CO ratio or less water in the feed), the Cu–Cu coordination number in EXAFS is smaller than when including water (higher oxidation potential). Stronger reducing conditions probably lead to vacancies on the ZnO surface and therefore more Cu-surface atoms (or a spreading out of the particles as depicted in the scheme). Both methods indicated at even stronger reducing conditions the formation of Cu–Zn alloys, also evidenced by IR spectroscopy [48]. The model derived from these studies (Figure 12.7b), was supported by in situ electron microscopy where specific reaction conditions could be mimicked in the mbar region [10]. Further in situ electron microscopy studies (e.g., [49]), however, showed that the structural changes are even more complex (Figure 12.7c). Despite the fact that real reaction conditions cannot be used (yet) in the electron microscope and particles can be investigated only very locally, the model could be further refined. The study also demonstrates the complementarity of these local imaging methods under idealized conditions and the integral spectroscopic and scattering methods under real reaction conditions. Note that further effects like strain in crystals may additionally play an important role [13] and there is presently a vivid discussion in literature on the role of ZnO.

When it comes to energy storage or biomass transformation reactions, elevated pressures also play an important role. This requires the design of appropriate in situ cells for flow-through reactors as well as batch reactors [50]. Designs that have adapted IR spectroscopy (e.g., using attenuated total reflection [ATR]) to monitor chemical intermediates and the surface as well as the catalyst structure by EXAFS, XRD, and small angle X-ray scatterings SAXS have been reported and successfully used. Obtaining information about catalyst deactivation when working under relevant conditions also requires long-term studies. To illustrate this, a commercial Co-Ni-Re/γ-Al_2O_3 Fischer–Tropsch synthesis (FTS) catalyst was studied by XAS for a period of >300 h on stream

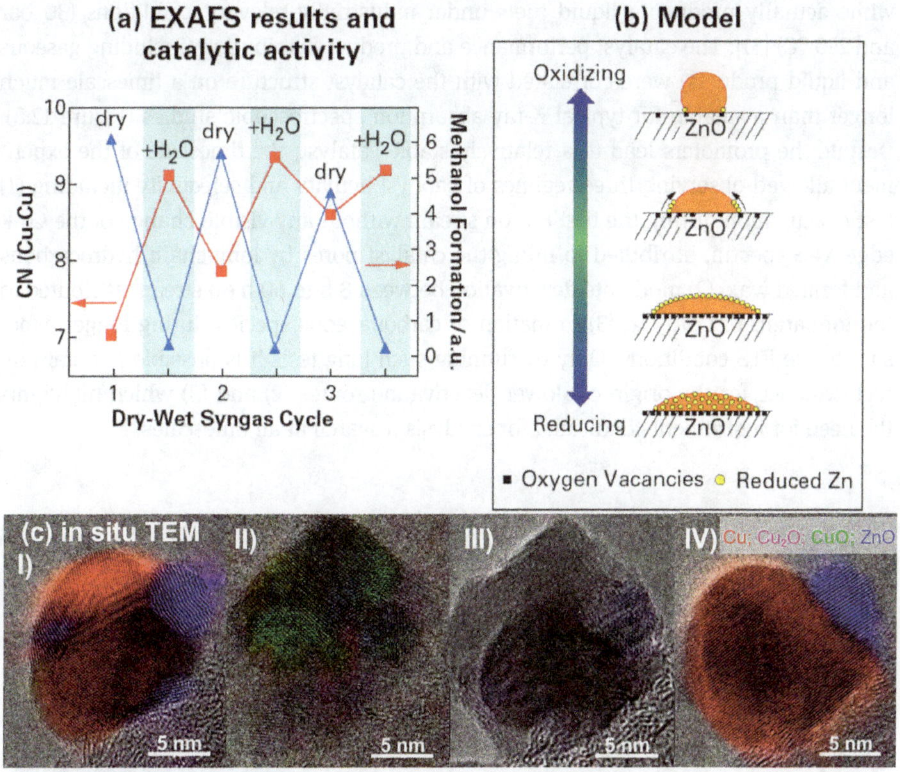

Figure 12.7: Dynamics of Cu/ZnO particles shown by in situ EXAFS (a) and the model derived from EXAFS; (b) synthesis gas with more CO_2 or H_2O in the feed is more oxidizing and leads to a higher Cu–Cu coordination number and more round-shaped particles (adapted from ref. [9], copyright 2000, with permission from Elsevier) supported by ETEM [10]; (c) in situ TEM images during an oxidation/reduction cycle probing the same Cu/ZnO nanoparticle: I) reduced after 5 h in 1 mbar H_2 at 300 °C; II) oxidized after 1 h in 1 mbar O_2 at 300 °C; III) partly reduced after 0.5 h in 1 mbar H_2 at 300 °C; IV) reduced after 1.5 h in 1 mbar H_2 at 300 °C. The degree of reduction/oxidation is based on the observation of lattice spacings from Cu metal or from any Cu oxide. Adapted from [49] with permission from ACS. Copyright ACS 2014.

while actually producing liquid fuels under industrially relevant conditions (30 bar and 250 °C) [51]. The catalyst performance and product distribution (including gaseous and liquid products) were correlated with the catalyst structure on a timescale much longer than available for typical X-ray absorption spectroscopic studies (Figure 12.8). Despite the promoters lead to a relatively stable catalyst, the timescale of the experiment allowed observing three regimes of catalyst activity and selectivity including (1) fast deactivation during the first 8 h on stream without any visible change of the Co K edge XAS spectra, attributed to filling the catalyst pores by long-chain hydrocarbons and formed wax, (2) moderate deactivation between 8 h to 60 h on stream attributed to the formation of $CoAl_2O_4$, (3) formation of carbonaceous species during longer exposure to the FTS conditions. Only by running such long tests it is possible to obtain direct evidence for the origin of slower deactivation routes (2) and (3) which highlights the need for instruments dedicated for catalysis research at all timescales.

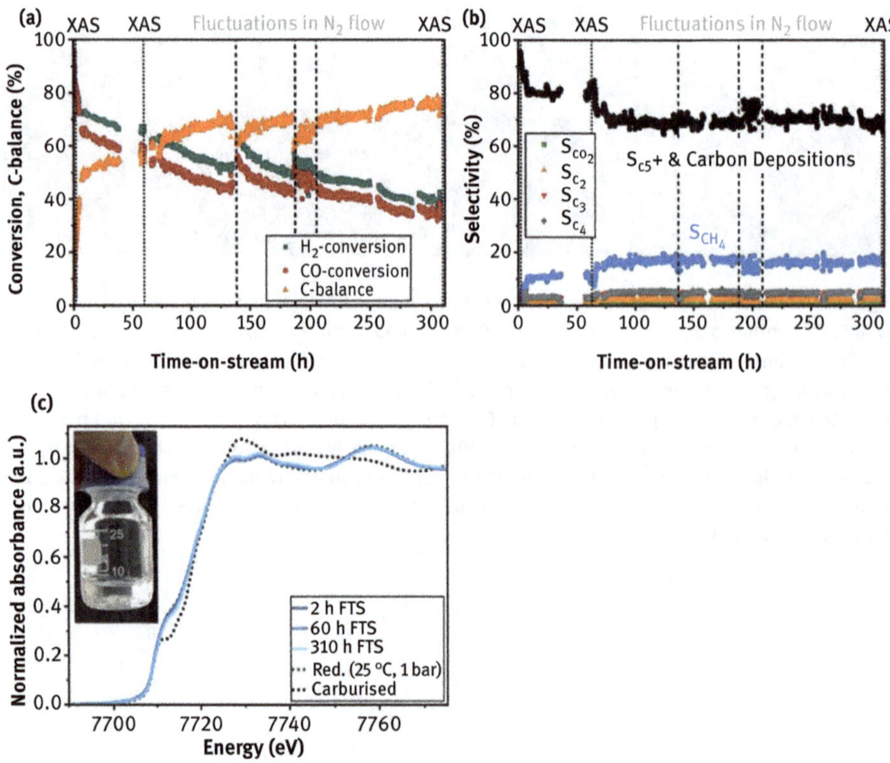

Figure 12.8: (a) Conversion of the reactants and carbon balance for the gaseous products, (b) selectivities, and (c) the corresponding Co K edge XANES spectra obtained during a long term XAS study of a commercial Co–Ni–Re/γ-Al_2O_3 catalyst during FTS (synthetic fuel produced during the spectroscopic study at the synchrotron radiation source in inset). Adapted from [51], used under CC BY 3.0 license, Copyright © 2020.

Another important question is to follow fast transient changes, and several methods are feasible for this area of interest. To name a few, IR spectroscopy, Raman spectroscopy, X-ray absorption spectroscopy (in terms of quick-EXAFS [QEXAFS] and dispersive EXAFS [DEXAFS]), XRD, and UV–vis spectroscopy are able to record structural changes or variations in the surface species in the subsecond regime. For example, structural changes during the ignition of the catalytic partial oxidation of methane to synthesis gas, which is given in Figure 12.4, lie in the subsecond regime and could be followed with subsecond time resolution [35].

A distinctive feature of renewable energy sources is their fluctuating availability, for example, of wind and solar energy [8] or changing biomass composition depending on the crops type, region of origin, weather, and so on. Catalysts designed for storing renewable energy (e.g., in Power-to-X concept) need to be stable also in operation regimes with transient energy or reactant supply. Knowledge-based development of such catalysts requires highly transient and highly sensitive operando characterization techniques. To illustrate this approach methanation of CO_2 on Ni-based catalysts was studied. Renewable hydrogen for this process originates from electrolysis of water. Temporary fluctuations in renewable energy may cause variations and even dropouts of hydrogen supply. To understand the influence of such short-term hydrogen dropouts on on the oxidation state and the atomic structure of Ni active sites in Ni/Al_2O_3 catalysts during CO_2 methanation operando QEXAFS studies were performed as depicted in Figure 12.9 [52]. To simulate intermittent hydrogen supply the feed was switched every 30 s between 5% CO_2/He and methanation conditions 5% CO_2, 20% H_2/He at 400 °C (temperature of highest activity and selectivity). Repeated short H_2 dropouts lead to loss of CH_4 production, which could not be fully recovered with hydrogen in the feed. After six consecutive H_2 dropouts QEXAFS revealed slight oxidation of a fraction of Ni sites manifested as increased white line (first peak above the absorption edge, Figure 12.9). Fast and significant oxidation of Ni during the following H_2 dropouts were observed with further H_2 drop outs. In this experiment, in principle, the most active Ni-sites were titrated away by small concentration of residual impurities of oxygen in the technical CO_2 used for methanation. Oxidized Ni sites could be only partially regenerated under methanation conditions with H_2 in the gas mixture. Also the CH_4 production rate could only partially be regained, confirming that the undercoordinated Ni sites, which are prone to oxidation, are most active for methane formation. Two strategies to overcome the observed irreversible deactivation were found: reactivation at 500 °C in hydrogen (undesired in practical application) and alloying Ni with Fe which was found to protect Ni sites from oxidation [53].

In order to isolate the small contribution of active sites in the total spectral response a so called modulation excitation spectroscopy (MES) approach was used [54, 55]. The approach can be universally combined with transient spectroscopic and scattering techniques (most explored with IR, Raman, XAS, and XRD) and allows isolating only spectral features which respond to periodic external perturbation of the system (e.g., gas concentration, temperature, and light). XAS coupled

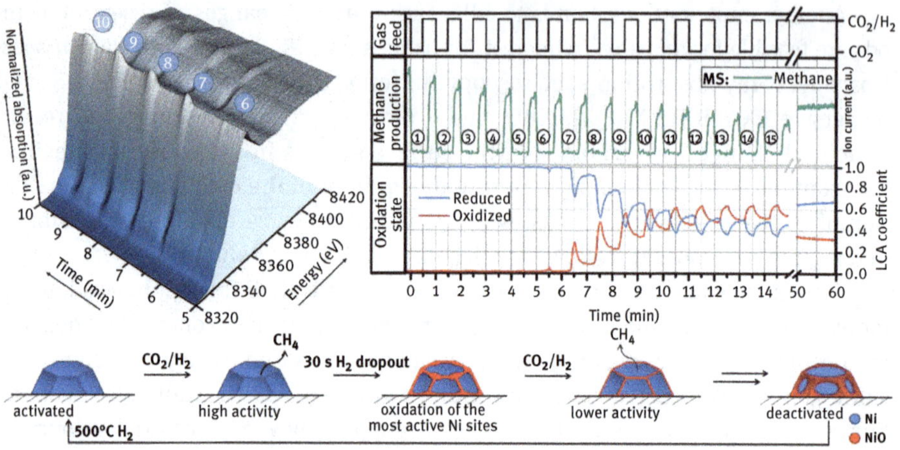

Figure 12.9: XANES spectra (left) and average Ni oxidation states alongside catalytic activity data (right) measured during CO_2 methanation with periodic 30 s H_2 dropouts. Conditions: $H_2/CO_2 = 4$; 75% He, 400 °C. Below is a schematic representation of the structural changes of supported Ni particles during methanation with continuous short H_2 dropouts. Adapted from [52], used under CC BY 4.0, Copyright © 2017.

with MES could isolate spectral features of redox-active Fe atoms on the interface between NiFe alloy and FeO_x clusters and relate those interface Fe sites with promoted CO_2 dissociation during methanation [55].

As outlined already in the examples in Figures 12.5, 12.7, and 12.8, the combination of complementary methods is also important under reaction conditions. This in situ characterization with several spectroscopic/imaging/scattering techniques has become a hot topic in recent years. In order to show the opportunities, Table 12.1 gives an overview of the combination of different spectroscopic methods while measuring the catalytic performance of the entitled heterogeneous catalyst system online (cf. [2, 56–58]). In the early stages, EXAFS and XRD were combined. Nowadays, also UV–vis, IR and diffuse reflectance infrared Fourier transform spectroscopy (DRIFTS), Raman spectroscopy, EPR, NMR, wide-angle X-ray scattering (WAXS), and other methods have been adapted to the same experiment to be sure that the conditions are really the same. In many cases, the structure is so dynamic that the results from the different techniques cannot just be connected via the online catalytic measurements. In addition, this allows verification of whether one of the methods has a destructive effect (e.g., due to the high photon flux).

Table 12.1: Examples of combining complementary spectroscopic methods while the catalytic performance was monitored online over a heterogeneous catalyst.

Methods	Reaction studied	Reference
EXAFS/XRD	Methanol synthesis, gas phase	[2, 9]
UV–vis/EPR EPR/UV–vis/Raman	Propane dehydrogenation	[59, 60]
UV–vis/Raman/XAFS	Propane dehydrogenation	[61]
XANES/XRD XANES/XRD/Raman	Total combustion of methane	[41, 42]
NMR/UV–vis	Conversion of methanol to cyclic compounds on weakly dealuminated zeolite H-ZSM-5	[62]
EXAFS/DRIFTS	CO and NO conversion over Rh catalysts	[63–65]
WAXS/XANES and UV–vis	Methanol to formaldehyde over iron-molybdate catalysts	[66]
XAFS/WAXS	Fischer–Tropsch synthesis over Fe-based catalysts	[67]
EPR/UV–vis	Selective catalytic reduction of NO_x over zeolites	[68]
Raman/IR/EPR	Oxidation of methanol over Mo/Al_2O_3	[69]
ATR-IR/UV–vis	Alcohol oxidation over Pd/Al_2O_3	[70]
EXAFS/DRIFTS	CO and NO conversion over Pd-based catalysts	[71]

12.6 Watching Ensembles and Reactors at Work: Spatially Resolved Studies

In the previous sections, the importance of identifying the structure at an atomic scale under operating conditions was discussed. As Figure 12.7 demonstrates, this can be achieved by microscopic methods but also spectroscopic or scattering methods. In fact, electron microscopy is the by far most used technique if local variations on an atomic scale are to be studied [10, 22, 28, 44, 72]. However, imaging is important not only on an atomic scale but also on the meso- and macroscales [6, 14, 28, 73]. Obviously, the catalysts are fine-tuned with respect to efficiency, physical strength, and minimization of pressure drops in continuous fixed-bed reactor systems. This requires atomic-scale information also at larger length scales and operating conditions as depicted in Figure 12.1. In fact, all the length scales up to several centimeters need to be covered if considering a real reactor with temperature and concentration gradients. For this purpose, microspectroscopic techniques have been widely applied (e.g., X-ray imaging (radiography), micro-Raman, IR, and UV–vis, as well as NMR tomography) [6, 28, 73]. Most demanding is the mesoscale under operating conditions. An illustrative recent example [74] is the study of a Fe-SiO_2–based Fischer–Tropsch catalyst using soft X-rays and an in situ cell based on 10 nm thin Si_3N_4 membranes (Figure 12.10). Probing with 15 nm resolution, 35 nm spatial variations in the iron valence and its metal/carbide/oxide nature were observed. On the basis of Fe $L_{2,3}$ XAS spectra, iron carbides were especially associated with those regions where α-Fe was formed after reduction apart from Fe(II) silicates and iron oxide [74]. Monitoring in the 10–100 nm regime appears important to grab the reactivity of different phases, spillover, aging, redispersion, and support effects. While X-rays with a focus of 20 nm and less have been available in the soft X-ray regime since the 1990s [75, 76], similar spot sizes have now also been achieved with hard X-rays [77–79]. Combination of sub-100 nm resolution and X-ray absorption spectroscopy now allows producing tomograms with hard X-rays on the

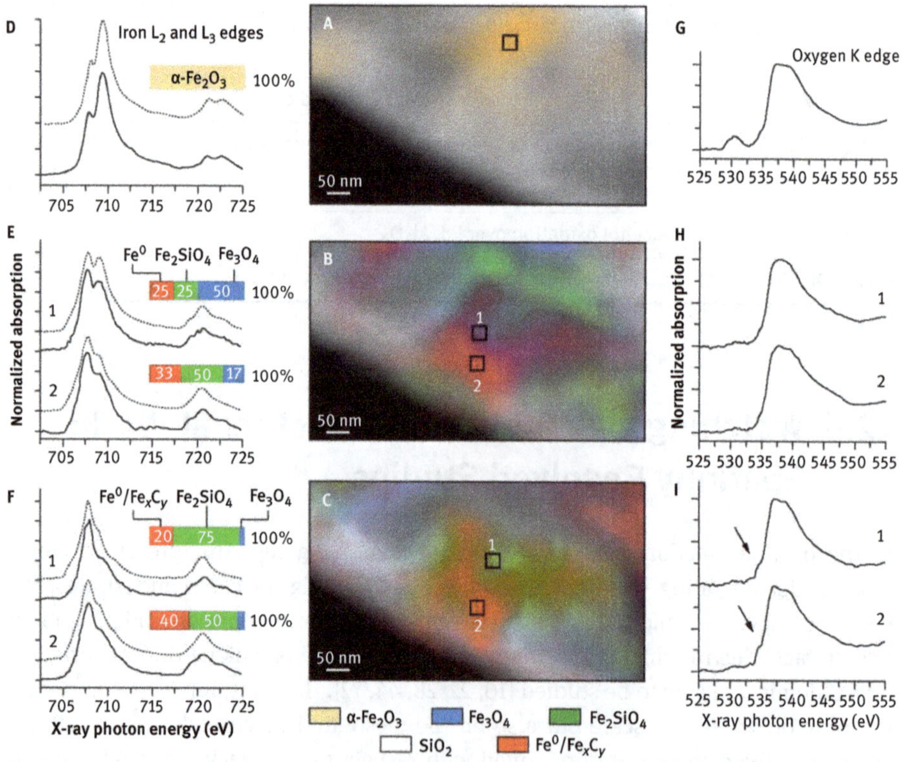

Figure 12.10: Chemical contour maps (A–C) and X-ray absorption spectra (D–I) of a Fischer–Tropsch catalyst during the following different stages of reaction: before treatment (top), after 2 h in H_2 at 350 °C (middle), after 4 h in synthesis gas at 250 °C (bottom). Chemical contour maps (A–C) of a 400 nm × 3,750 nm region and corresponding iron L_2 and L_3 edge (D–F) and oxygen K edge (G–I). Specific sampling regions and the corresponding X-ray absorption spectra are indicated in the figures (for details, cf. ref. [74], reproduced with permission, Copyright Nature Publishing Group, 2008).

nanoscale with chemical contrast as demonstrated by imaging oxygen diffusion pathways in ceria–zirconia-based catalysts [79].

Equally important are structural changes on the micrometer and the millimeter length scales, and eventually real reactors on a centimeter or even meter scale. In the first cases, full-field XAS is a well-matched method for in situ monitoring [6, 14]. An illustrative example is the catalytic partial oxidation (CPO) of methane to synthesis gas, a relevant reaction in the future's solid-oxide fuel cells. In Figure 12.4, it was demonstrated that the noble metal atoms are in a reduced state under the reaction conditions. However, full-field and scanning X-ray microscopy studies with XAS contrast on Pt-, Rh-, and Pd-based catalysts uncovered that the noble metals are in an oxidized state in the beginning of the catalyst bed where the total oxidation of methane occurs, and in the reduced state in the second part of the reactor where reforming of the remaining methane with the formed water and carbon dioxide occurs [80, 81]. Some of the results are shown in Figure 12.11. This is in line with spatially resolved product analysis together with temperature profiles [82, 83]. As in integral approaches, the aspect of dynamic variations is also important during spatially resolved studies. The ignition process of the catalytic partial oxidation of methane as depicted in Figure 12.4 has been recently monitored by full-field X-ray microscopy unraveling changes on the micrometer scale on the subsecond timescale [17, 84].

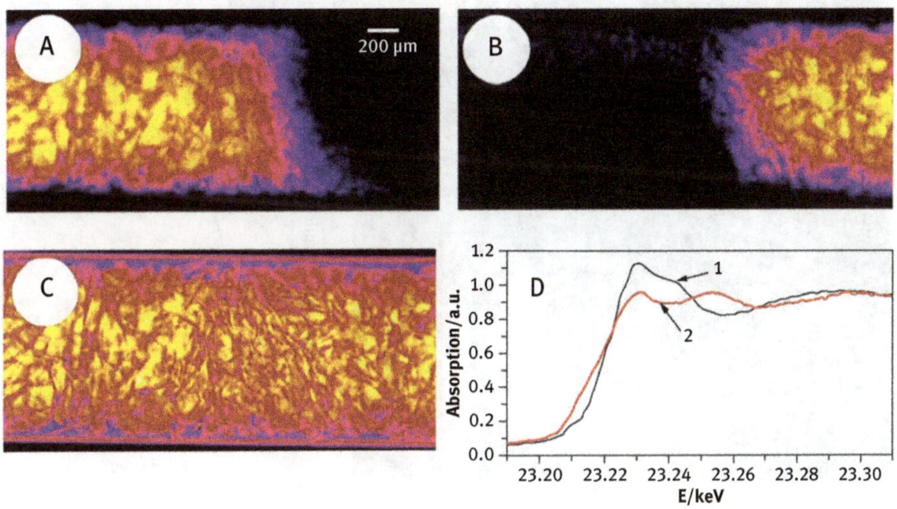

Figure 12.11: Full-field X-ray microscopy on a 5 wt% Rh/Al_2O_3 catalyst during catalytic partial oxidation of methane: (A) amount of oxidized Rh species (corresponds to XANES species 1 in [D]), (B) reduced Rh species (reduced species 2 in [D]), (C) the distribution of other elements that show a featureless absorption spectrum in the given energy range, and (D) spectra used for X-ray absorption contrast (original image taken by X-ray camera was 3.0 mm × 1.5 mm; the reaction gas mixture 6% CH_4/3% O_2/He enters from the left) (reproduced with permission from ref. [81], Copyright ACS, 2006).

Structured catalysts comprising several spatially separated phases such as e.g. bifunctional $Cu/ZnO/Al_2O_3$@ZSM-5 core@shell catalyst for dimethyl ether (DME) synthesis especially benefit from spatially resolved imaging techniques. Only by using simultaneous in situ scanning micro X-ray fluorescence (μ-XRF), X-ray diffraction (μ-XRD), and scanning transmission X-ray microscopy (STXM) computed tomography (CT) with micrometer spatial resolution it was possible to reveal the coexistence of Cu^0 in the active catalyst core together with partially oxidized metastable Cu^+ phase at the core–shell interface under model. DME synthesis conditions [31]. These studies demonstrate that in certain cases atomic-scale information on a macroscale can be obtained and is needed [28, 85]. As highlighted recently for the Cu–zeolite catalyst for the selective catalytic reduction NO_x with ammonia (NH_3-SCR) [30], the combination of XAS with full field or scanning X-ray imaging (spectrotomography) and a novel microreactor setup (Figure 12.12.) allows now to obtain operando spatially resolved information on the chemical state of catalytically active species in 3D space. Hence, one can observe the electronic state and local structure of any given microscopic volume

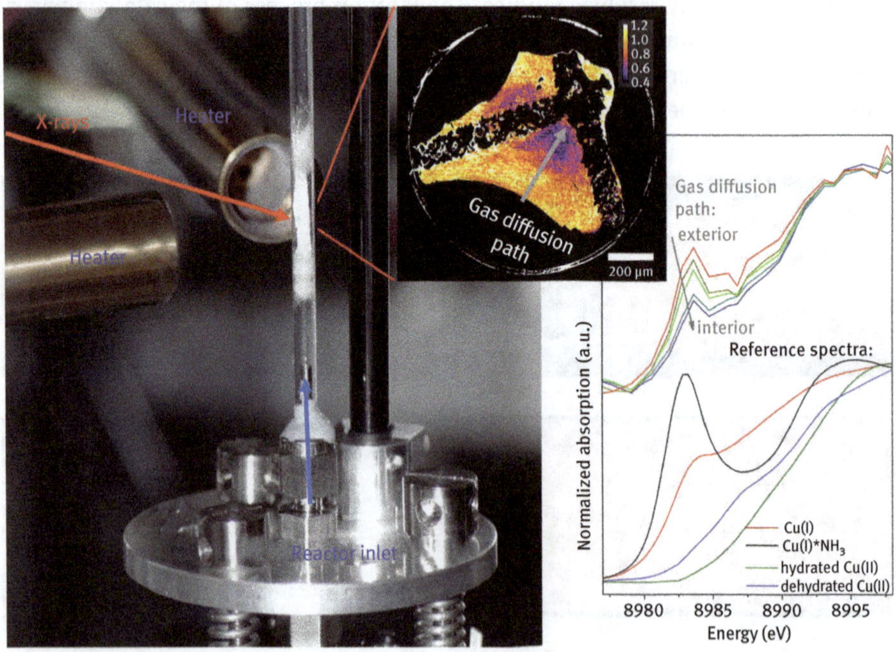

Figure 12.12: Operando X-ray spectrotomography on a structured industrial catalyst for automotive exhaust gas aftertreatment (Cu-SSZ-13 washcoated on a cordierite honeycomb monolith for NH_3-SCR). Left part shows the plug flow reactor cell based on the quartz capillary now applicable for operando X-ray tomographic studies. Right part demonstrates an ortogonal slice (color code stands for absorbance at 8983 eV) of a tomogram measured in $NH_3 + NO + O_2$/He gas feed at 300 °C and Cu K edge XANES spectra extracted from different depths of the washcoat layer together with reference spectra corresponding to different states of Cu active sites. For further details cf. [30].

of a structured catalyst in 3D in a reactor. This allows direct visualization of different oxidation states or phases including chemical gradients that may stem from occurring mass transfer limitations under relevant reaction conditions while monitoring and controlling integral catalyst activity as shown in Figure 12.12.

12.7 Conclusions and Outlook

This chapter has given an overview with selected examples and techniques on the potential and opportunities but also limitations for in situ and operando characterization of heterogeneous catalysts. A multitude of different characterization techniques are available nowadays, and more and more ways are being found to further adapt them to studying catalytic materials in their working state. This is important since each study with an in situ cell that tries to combine real-world catalysis with the most advanced characterization tools is a compromise. In certain cases, the reaction rates are high, and internal mass transport is important to consider during the spectroscopic study; in other cases, the reaction conditions with high temperature or pressure are demanding, the structure of the catalyst is very dynamic, or it is very difficult to grab the structure by only one characterization method. Hence, also with new materials and/or reactions in chemical energy conversion, new approaches have to be developed to come up with the best design for combining one or several spectroscopic techniques with the right timescale and length scale, as well as online measurement of the catalytic activity. Here, we focused on thermal catalysis, similar developments are found in electrocatalysis with sophisticated in situ cells [86–88]. The last section of this chapter has shown that spatially resolved and spatiotemporal studies pave the way to understand the dynamics on the atomic-scale structure in catalytic reactors (see overview and perspective in ref. [85]), which will have an important impact on the modeling of the catalytic reactions and reactors. Finally, with broader application of highly time and spatially resolved techniques the "big data" analysis approaches are becoming relevant for this area. High time and spatial resolution (e.g., 3D) lead to data in the TB range on a day basis and need not only clever data analysis but also data reduction codes calling for advanced spectral deconvolution approaches [89], to machine learning and for on-the-fly analysis of large datasets [90], but also reconstruction of missing information pieces [91].

References

[1] Thomas JM Design, synthesis and in situ characterization of new solid catalysts. Angew Chem Int Ed. 1999;38:3588.

[2] Grunwaldt J-D, Clausen BS Combining XRD and EXAFS with on-line catalytic studies for the in situ characterization of catalysts. Top Catal. 2002;18:37.
[3] Weckhuysen BM Determining the active site in a catalytic process: operando spectroscopy is more than a buzzword. Phys Chem Chem Phys. 2003;5:4351.
[4] Topsøe H Developments in operando studies and in situ characterization of heterogeneous catalysts. J Catal. 2003;216:155.
[5] Banares MA Operando methodology: combination of in situ spectroscopy and simultaneous activity measurements under catalytic reaction conditions. Catal Today. 2005;100:71.
[6] Grunwaldt J-D, Kimmerle B, Baiker A et al. Catalysts at work: from integral to spatially resolved X-ray absorption spectroscopy. Catal Today. 2009;145:267.
[7] Van't Blik HFJ, van Zon JBA, Huiziga T, Vis JC, Koningsberger DC, Prins R An extended x-ray absorption fine structure spectroscopy study of a highly dispersed rhodium/aluminum oxide catalyst: the influence of carbon monoxide chemisorption on the topology of rhodium. J Phys Chem. 1983;87:2264.
[8] Kalz KF, Kraehnert R, Dvoyashkin M, Dittmeyer R, Gläser R, Krewer U, Reuter K, Grunwaldt J-D Future challenges in heterogeneous catalysis: understanding catalysts under dynamic reaction conditions. Chem Cat Chem. 2017;9:17.
[9] Grunwaldt J-D, Molenbroek AM, Topsøe N-Y, Topsøe H, Clausen BS In situ investigations of structural changes in Cu/ZnO catalysts. J Catal. 2000;194:452.
[10] Hansen PL, Wagner JB, Helveg S, Rostrup-Nielsen JR, Clausen BS, Topsøe H Atom-resolved imaging of dynamic shape changes in supported copper nanocrystals. Science. 2002;295:2053.
[11] Helveg S, Lopez-Cartes C, Sehested J et al. Atomic-scale imaging of carbon nanofibre growth. Nature. 2004;427:426.
[12] Grunwaldt J-D, Baiker A Axial variation of the oxidation state of Pt-Rh/Al_2O_3 during partial methane oxidation in a fixed-bed reactor: an in situ X-ray absorption spectroscopy study. Catal Lett. 2005;99:5.
[13] Behrens M Meso- and nano-structuring of industrial Cu/ZnO/(Al_2O_3) catalysts. J Catal. 2009;267:24.
[14] Grunwaldt J-D, Schroer CG Hard and soft X-ray microscopy and tomography in catalysis: bridging the different time and length scales. Chem Social rev. 2010;39:4741.
[15] Schlögl R Heterogeneous Catalysis. Angew Chem Int Ed. 2015;54:3465.
[16] Ressler T, Hagelstein M, Hatje U, Metz W In situ X-ray absorption spectroscopy studies on chemical oscillations in the CO/O_2 system on supported Pd catalysts. J Phys Chem B. 1997;101:6680.
[17] Kimmerle B, Baiker A, Grunwaldt J-D Oscillatory behaviour of catalytic properties, structure and temperature during the catalytic partial oxidation of methane on Pd/Al_2O_3. Phys Chem Chem Phys. 2010;12:2288.
[18] Maestri M Atomic-scale understanding of complex chemical kinetics. In: Pignataro B editor, New Strategies in Chemical Synthesis and Catalysis, Wiley VCH, 2011.
[19] Deutschmann O editor. Modeling and Simulation of Heterogeneous catalytic Reactions: From the Molecular Process to the Technical System. Weinheim: Wiley-VCH; 2012.
[20] Sharapa DI, Doronkin DE, Studt F, Grunwaldt J-D, Behrens S Moving frontiers in transition metal catalysis: synthesis, characterization and modeling. Adv Mater. 2019;31:1807381.
[21] Baiker A, Kohler M Characterization of heterogeneous catalysts. In: Cheremisinoff PN editor, Handbook of Heat and Mass Transfer, Houston: Gulf Publishing, vol. 3, 1989:3.
[22] Niemantsverdriet JW Spectroscopy in Catalysis. Weinheim: Wiley-VCH;2007.
[23] Ertl G, Knözinger H, Schüth F, Weitkamp J editors. Handbook of Heterogeneous Catalysis. Weinheim: Wiley-VCH; 2008.

[24] Schroer CG, Kuhlmann M, Günzler TF et al. Mapping the chemical states of an element inside a sample using tomographic X-ray absorption spectroscopy. Appl Phys Lett. 2003;82:3360.

[25] Gladden LF Recent advances in MRI studies of chemical reactors: ultrafast imaging of multiphase flows. Top Catal. 2003;24:19.

[26] Koptyug IV, Lysova AA, Sagdeev RZ, Kirillov VA, Kulikov AV, Parmon VN In situ MRI of the structure and function of multiphase catalytic reactors. Catal Today. 2005;105:464.

[27] Beale AM, Jacques SDM, Bergwerff JA, Barnes P, Weckhuysen BM Tomographic energy dispersive diffraction imaging as a tool to profile in three dimensions the distribution and composition of metal oxide species in catalyst bodies. Angew Chem Int Ed. 2007;46:8832.

[28] Weckhuysen BM Chemical imaging of spatial heterogeneities in catalytic solids at different length and time scales. Angew Chem Int Ed. 2009;48:4910.

[29] Grunwaldt J-D, Wagner JB, Dunin-Borkowski RE Imaging catalysts at work: a hierarchical approach from the macro- to the meso- and nano-scale. Chem Cat Chem. 2013;5:62.

[30] Becher J, Ferreira Sanchez D, Doronkin DE, Zengel D, Motta Meira D, Pascarelli S, Grunwaldt J-D, Sheppard TL Chemical gradients in automotive Cu-SSZ-13 catalysts for NOx removal revealed by operando X-ray spectrotomography. Nat Catal. 2020;4:46.

[31] Sheppard TL, Price SWT, Benzi F, Baier S, Klumpp M, Dittmeyer R, Schwieger W, Grunwaldt J-D In situ multimodal 3D chemical imaging of a hierarchically-structured core@shell catalyst. J Am Chem Soc. 2017;139:7855.

[32] Basini L, Guarinoni A, Aragno A Molecular and temperature aspects in catalytic partial oxidation of methane. J Catal. 2000;190:284.

[33] Grunwaldt J-D, Basini L, Clausen BS In situ EXAFS study of Rh/Al_2O_3 catalysts for catalytic partial oxidation of methane. J Catal. 2001;200:321.

[34] Grunwaldt J-D Shining X-rays on catalysts at work. J Phys Conf Ser. 2009;190:012151.

[35] Grunwaldt J-D, Beier M, Kimmerle B et al. Structural changes of noble metal catalysts during ignition and extinction of the partial oxidation of methane studied by advanced QEXAFS techniques. Phys Chem Chem Phys. 2009;11:8779.

[36] Doronkin DE, Lichtenberg H, Grunwaldt J-D Cell designs for in situ and operando studies. In: Iwasawa Y, Asakura K, Tada M editors, XAFS Techniques for Catalysts, Nanomaterials, and Surfaces, Cham: Springer International Publishing Switzerland, 2017.

[37] Grunwaldt J-D, Caravati M, Hannemann S, Baiker A X-ray absorption spectroscopy under reaction conditions – opportunities and limitations of in situ monitoring and time-resolved studies of heterogeneous catalysts. Phys Chem Chem Phys. 2004;6:3037.

[38] Bare SR, Ressler T Characterization of catalysts in reactive atmospheres by X-ray absorption spectroscopy. Adv Catal. 2009;52:339.

[39] Grunwaldt J-D, Baiker A Time-resolved and operando XAS studies on heterogeneous catalysts – from the gas phase towards reactions in supercritical fluids. In: Hedman B, Oianetta P editors, AIP Conference Proceedings X-ray Absorption Fine Structure – XAFS13, vols CP 882, 2007:577–81.

[40] Meunier FC The design and testing of kinetically-appropriate operando spectroscopic cells for investigating heterogeneous catalytic reactions. Chem Social rev. 2010;39:4602.

[41] Grunwaldt J-D, Van Vegten N, Baiker A Insight into the structure of supported palladium catalysts during the total oxidation of methane. Chem Commun. 2007;4635.

[42] Grunwaldt J-D, van Vegten N, Baiker A, van Beek W Insight into the structure of Pd/ZrO_2 during the total oxidation of methane using combined in situ XRD, X-ray absorption and Raman spectroscopy. J Phys Conf Ser. 2009;190:012160.

[43] Creemer JF, Helveg S, Hoveling GH et al. Atomic-scale electron microscopy at ambient pressure. Ultramicroscopy. 2008;108:993.

[44] Plodinec M, Nerl HC, Farra R, Willinger MG, Stotz E, Schlögl R, Lunkenbein T Versatile homebuilt gas feed and analysis system for operando TEM of catalysts at work. Microsco Microanal. 2020;26:220.
[45] Bluhm H, Hävecker M, Knop-Gericke A, Kiskinova M, Schlögl R, Salmeron M In situ X-ray photoelectron spectroscopy studies of gas–solid interfaces at near-ambient conditions. MRS bull. 2007;32:1022.
[46] Clausen BS, Schiøtz J, Gråbek L et al. Wetting/non-wetting phenomena during catalysis: evidence from in situ on-line EXAFS studies of Cu-based catalysts. Top Catal. 1994;1:367.
[47] Simonsen SB, Dahl S, Johnson E, Helveg S Ceria-catalyzed soot oxidation studied by environmental transmission electron microscopy. J Catal. 2008;255:1.
[48] Fujitani T, Nakamura J The effect of ZnO in methanol synthesis catalysts on Cu dispersion and the specific activity. Catal Lett. 1998;56:119.
[49] Holse C, Elkjær CF, Nierhoff A, Sehested J, Chorkendorff I, Helveg S, Nielsen JH Dynamic Behavior of CuZn Nanoparticles under Oxidizing and Reducing Conditions. J Phys Chem C. 2015;119:2804.
[50] Grunwaldt J-D, Ramin M, Rohr M, Michailowski A, Patzke GR, Baiker A High pressure in situ x-ray absorption spectroscopy cell for studying simultaneously the liquid phase and the solid/liquid interface. Rev Sci Instrum. 2005;76:054104.
[51] Loewert M, Serrer M-A, Carambia T, Stehle M, Zimina A, Kalz KF, Lichtenberg H, Saraçi E, Pfeifer P, Grunwaldt J-D Bridging the gap between industry and iynchrotron: an operando study at 30 bar over 300 h during Fischer–Tropsch synthesis. React Chem Eng. 2020;5:1071.
[52] Mutz B, Gänzler AM, Nachtegaal M, Müller O, Frahm R, Kleist W, Grunwaldt J-D Surface oxidation of supported Ni particles and its impact on the catalytic performance during dynamically operated methanation of CO_2. Catalysts. 2017;7:279.
[53] Serrer M-A, Kalz KF, Saraçi E, Lichtenberg H, Grunwaldt J-D Role of iron on the structure and stability of $Ni_{3.2}Fe/Al_2O_3$ during dynamic CO_2 methanation for P2X applications. Chem Cat Chem. 2019;11:5018.
[54] Urakawa A, Bürgi T, Baiker A Sensitivity enhancement and dynamic behavior analysis by modulation excitation spectroscopy: principle and application in heterogeneous catalysis. Chem Eng Sci. 2008;63:4902.
[55] Serrer M-A, Gaur A, Jelic J, Weber S, Fritsch C, Clark AH, Saraçi E, Studt F, Grunwaldt J-D Structural dynamics in Ni–Fe catalysts during CO_2 methanation – role of iron oxide clusters. Catal Sci Technol. 2020;10:7542.
[56] Clausen BS, Topsøe H, Frahm R Application of combined X-ray diffraction and absorption techniques for in situ catalyst characterization. Adv Catal. 1998;42:315.
[57] Brückner A, Kondratenko E Simultaneous operando EPR/UV-vis/laser-Raman spectroscopy – a powerful tool for monitoring transition metal oxide catalysts during reaction. Catal Today. 2006;113:16.
[58] Bentrup U Combining in situ characterization methods in one set-up: looking with more eyes into the intricate chemistry of the synthesis and working of heterogeneous catalysts. Chem Social Rev. 2010;39:4718.
[59] Brückner A Simultaneous combination of in situ-EPR/UV-VIS/on line GC: a novel setup for investigating transition metal oxide catalysts under working conditions. Chem Commun. 2001;2122.
[60] Brückner A Killing three birds with one stone – simultaneous operando EPR/UV-vis/Raman spectroscopy for monitoring catalytic reactions. Chem Commun. 2005;1761.
[61] Beale AM, Van Der Eerden AMJ, Kervinen K, Newton MA, Weckhuysen BM Adding a third dimension to operando spectroscopy: a combined UV-Vis, Raman and XAFS setup to study heterogeneous catalysts under working conditions. Chem Commun. 2005;3015.

[62] Hunger M, Wang W Formation of cyclic compounds and carbenium ions by conversion of methanol on weakly dealuminated zeolite H-ZSM-5 investigated via a novel in situ CF MAS NMR/UV-Vis technique. Chem Commun. 2004;584.

[63] Newton MA, Jyoti B, Dent AJ, Fiddy SG, Evans J Synchronous, time resolved, diffuse reflectance FT-IR, energy dispersive EXAFS (EDE) and mass spectrometric investigation of the behaviour of Rh catalysts during NO reduction by CO. Chem Commun. 2004;2382.

[64] Newton MA, Dent AJ, Fiddy SG, Jyoti B, Evans J Combining diffuse reflectance infrared spectroscopy (DRIFTS), dispersive EXAFS, and mass spectrometry with high time resolution: potential, limitations, and application to the study of NO interaction with supported Rh catalysts. Catal Today. 2007;126:64.

[65] Ferri D, Kumar MS, Wirz R et al. First steps in combining modulation excitation spectroscopy with synchronous dispersive EXAFS/DRIFTS/mass spectrometry for in situ time resolved study of heterogeneous catalysts. Phys Chem Chem Phys. 2010;12:5634.

[66] O'Brien MG, Beale AM, Jacques SDM, Weckhuysen BM A combined multi-technique in situ approach used to probe the stability of iron molybdate catalysts during redox cycling. Top Catal. 2009;52:1400.

[67] De Smit E, Beale AM, Nikitenko S, Weckhuysen BM Local and long range order in promoted iron-based Fischer-Tropsch catalysts: a combined in situ X-ray absorption spectroscopy/wide angle X-ray scattering study. J Catal. 2009;262:244.

[68] Kumar MS, Schwidder M, Grünert W, Brückner A On the nature of different iron sites and their catalytic role in Fe-ZSM-5 DeNO(x) catalysts: new insights by a combined EPR and UV/VIS spectroscopic approach. J Catal. 2004;227:384.

[69] Brandhorst M, Cristol S, Capron M et al. Catalytic oxidation of methanol on Mo/Al_2O_3 catalyst: an EPR and Raman/infrared operando spectroscopies study. Catal Today. 2006;113:34.

[70] Bürgi T Combined in situ attenuated total reflection infrared and UV-vis spectroscopic study of alcohol oxidation over Pd/Al_2O_3. J Catal. 2005;229:55.

[71] Kubacka A, Martínez-Arias A, Fernández-García M, Newton MA Dynamic "operando" observation of 10 wt% Pd-based TWCs: simultaneous XAS/DRIFTS/mass spectrometry analysis of the effects of $Ce_{0.5}Zr_{0.5}O_2$ loading on structure, reactivity and performance. Catal Today. 2009;145::288.

[72] Thomas JM, Hernandez-Garrido JC Probing solid catalysts under operating conditions: electrons or X-rays?. Angew Chem Int Ed. 2009;48:3904.

[73] Urakawa A, Baiker A Space-resolved profiling relevant in heterogeneous catalysis. Top Catal. 2009;52:1312.

[74] De Smit E, Swart I, Creemer JF et al. Nanoscale chemical imaging of a working scanning transmission X-ray microscopy. Nature. 2008;456:222.

[75] Chao WL, Harteneck BD, Liddle JA, Anderson EH, Attwood DT Soft X-ray microscopy at a spatial resolution better than 15nm. Nature. 2005;435:1210.

[76] Vila-Comamala J, Jefimovs K, Raabe J et al. Advanced thin film technology for ultrahigh resolution X-ray microscopy. Ultramicroscopy. 2009;109:1360.

[77] Mimura H, Handa S, Kimura T et al. Breaking the 10 nm barrier in hard-X-ray focusing. Nat Phys. 2010;6:57.

[78] Schroer CG, Lengeler B Focusing hard x rays to nanometer dimensions by adiabatically focusing lenses. Phys Rev Lett. 2005;94:054802.

[79] Hirose M, Ishiguro N, Shimomura K, Nguyen D-N, Matsui H, Dam HC, Tada M, Takahashi Y Oxygen-diffusion-driven Oxidation Behavior and Tracking Areas Visualized by X-ray Spectro-ptychography with Unsupervised Learning. Commun Chem. 2019;2:50.

[80] Hannemann S, Grunwaldt JD, Kimmerle B, Baiker A, Boye P, Schroer C Axial changes of catalyst structure and temperature in a fixed-bed microreactor during noble metal catalysed partial oxidation of methane. Top Catal. 2009;52:1360.

[81] Grunwaldt J-D, Hannemann S, Schroer CG, Baiker A 2D-Mapping of the catalyst structure inside a catalytic microreactor at work: partial oxidation of methane over Rh/Al_2O_3. J Phys Chem B. 2006;110:8674.
[82] Horn R, Degenstein NJ, Williams KA, Schmidt LD Spatial and temporal profiles in millisecond partial oxidation processes. Catal Lett. 2006;110:169.
[83] Basile F, Fornasari G, Trifirò F, Vaccari A Effect of reaction parameters and catalyst composition on the thermal profile and heat distribution. Catal Today. 2001;64:21.
[84] Kimmerle B, Grunwaldt J-D, Baiker A et al. Visualizing the ignition of catalytic partial oxidation of methane. J Phys Chem C. 2009;113:3037.
[85] Schroer CG, Boye P, Feldkamp JM et al. Hard X-ray microscopy with elemental, chemical, and structural contrast. Acta Physica Polonica A. 2010;117:357.
[86] Binninger T, Fabbri E, Patru A et al. Electrochemical flow-cell setup for in situ X-ray investigations I. Cell for SAXS and XAS at synchrotron facilities. J Electrochem Soc. 2016;163: H906.
[87] Escudero-Escribano M, Pedersen AF, Paoli EA, Frydendal R, Friebel D, Malacrida P, Rossmeisl J, Stephens IEL, Chorkendorff I Importance of surface IrO_x in stabilizing RuO_2 for oxygen evolution. J Phys Chem B. 2018;122:947.
[88] Velasco-Velez -J-J, Mom RV, Sandoval-Diaz LE et al. Revealing the active phase of copper during the electroreduction of CO_2 in aqueous electrolyte by correlating *in situ* X-ray spectroscopy and *in situ* electron microscopy. ACS Energy Lett. 2020;5:2106.
[89] Voronov A, Urakawa A, Van Beek W, Tsakoumis NE, Emerich H, Ronning M Multivariate curve resolution applied to in situ X-ray absorption spectroscopy data: an efficient tool for data processing and analysis. Analytica chimica Acta. 2014;840:20.
[90] Timoshenko J, Frenkel A "Inverting" X-ray absorption spectra of catalysts by machine learning in search for activity descriptors. ACS Catal. 2019;9:10192.
[91] Yang X, Kahnt M, Brückner D, Schropp A, Fam Y, Becher J, Grunwaldt J-D, Sheppard TL, Schroer CG Tomographic reconstruction with a generative adversarial network. J Synchrotron Rad. 2020;27:486.

N. Nilius, M. Sterrer, S. Shaikhutdinov, D. Menzel, M.J. Prieto, Th. Schmidt, H.-J. Freund

13 Model Systems in Catalysis for Energy Economy

Abstract: The investigation of model systems for heterogeneous catalysts catching some of the complexity of real catalyst systems may add substantially to an attempt to rationally design catalysts and catalytic materials to effectively harvest and store energy. Five case studies discuss an approach to model disperse metal catalysts and its use in thermal and photo chemical processes.

13.1 Introduction

Harvesting, saving, and storing energy are key issues in today's general energy discussions [1]. Harvesting energy through photochemistry as well as storing energy in compounds using thermal energy-efficient processes are attractive solutions to some problems within the grand picture [2].

Heterogeneous thermal and photocatalysis are bound to play a key role in those solutions. However, while thermal heterogeneous catalysis is wide spread in chemical industries and photocatalysis not yet, there is still a strong demand in both areas for research in order to understand the process and its elementary steps as well as to rationally design the catalytic material.

Investigation of model catalysts can play a decisive role in a rational approach to understand heterogeneous catalysis and this is the topic of this chapter [3–6].

In the introduction, the model systems will be defined to familiarize the reader with the approach and to appreciate the connection to "real world" catalysis. Following the introduction we will demonstrate via five case studies various fundamental aspects in thermal and photocatalysis whereby studies on model systems might become important to unravel the foundations of reaction mechanisms.

The model systems have been created under the premises that it is possible to investigate them at the atomic scale using the toolbox that has been developed in surface science during the second half of the last century up to now [7].

The model systems are based on the idea to grow well-ordered oxide layers or thin films, representing the bulk material, and as such the catalyst support, on single crystal metal surfaces using the concepts of epitaxial growth[8]. This, in turn,

N. Nilius, M. Sterrer, S. Shaikhutdinov, D. Menzel, M.J. Prieto, Th. Schmidt
H.-J. Freund, Fritz-Haber-Institut der Max-Planck-Gesellschaft, Faradayweg 4-6, 14195 Berlin, Germany, email: freund@fhi-berlin.mpg.de

https://doi.org/10.1515/9783110608458-013

allows one to investigate bulk insulators without having to worry about charging when charged information carriers, such as electrons or ions, are used to investigate those systems. One thus avoids one of the key difficulties hampering the detailed study of real catalysts. Also the application of scanning probe techniques including electron tunneling is possible for thin film systems. The fact that the oxides are grown on a metal support also ensures the easy applicability of infrared reflection absorption spectroscopy at such systems. Moreover, controlling the growth parameters of the oxide film, also, allows one to vary the density of defects at the surface of the oxide films, a factor of importance when nanoparticles are grown on the oxide support.

The first case study presented in this chapter is related to a model system which is schematically represented in Figure 13.1a [9]. An oxide film representing the bulk material is grown on a metal surface. The growth of metal nanoparticles, and in particular their morphologies which may distinctively influence the catalytic performance are compared on pristine oxide films and films that have been doped with metal ions that may assume higher valences than the metal of the host material. Using the model system even the effect of doping as a function of the depth (Figure 13.1a) at which the dopant is deposited below the surface may be studied, and thus questions of the presence of dopants in materials used to prepare supports in real catalysis may be addressed. Since the effect of doping may be modeled to some extent by a metal substrate underneath a thin oxide film (Figure 13.1b), such that electrons may tunnel from the oxide/metal substrate interface to a metal nanoparticle, the chemistry triggered by electrons on the metal particle may be studied at higher resolution by an STM tip as the tunneling conditions are more favorable. As an example, adsorption of molecules and reactions are presented in the last section of the first case study.

Real dispersed metal catalysts are prepared by precipitation from solution. Investigating processes involving oxide surfaces in solution is a challenging topic. Figure 13.1c schematically shows the type of system investigated [10]. Particles are deposited from an aqueous solution of a metal chloride as a function of the pH of the solution. The pH strongly influences the outcome of the impregnation process. Understanding the details of such processes is of utmost importance for preparation of technical catalysts and first steps in this direction are being taken. This represents the second case study.

In a third case study the effect of a reducible support material, such as magnetite, on Pt nanoparticles anchored onto the magnetite surface after a heat treatment is investigated. In this system a typical strong metal support effect (SMSI), that is, the migration of a thin oxide film from the support onto the metal nanoparticles influencing catalytic performance, has been observed. An SMSI effect usually renders the system inactive in oxidation reactions such as CO oxidation. Here it is shown that depending on the chemical potential of oxygen in the gas phase the SMSI system schematically shown in Figure 13.1d may become more active than the

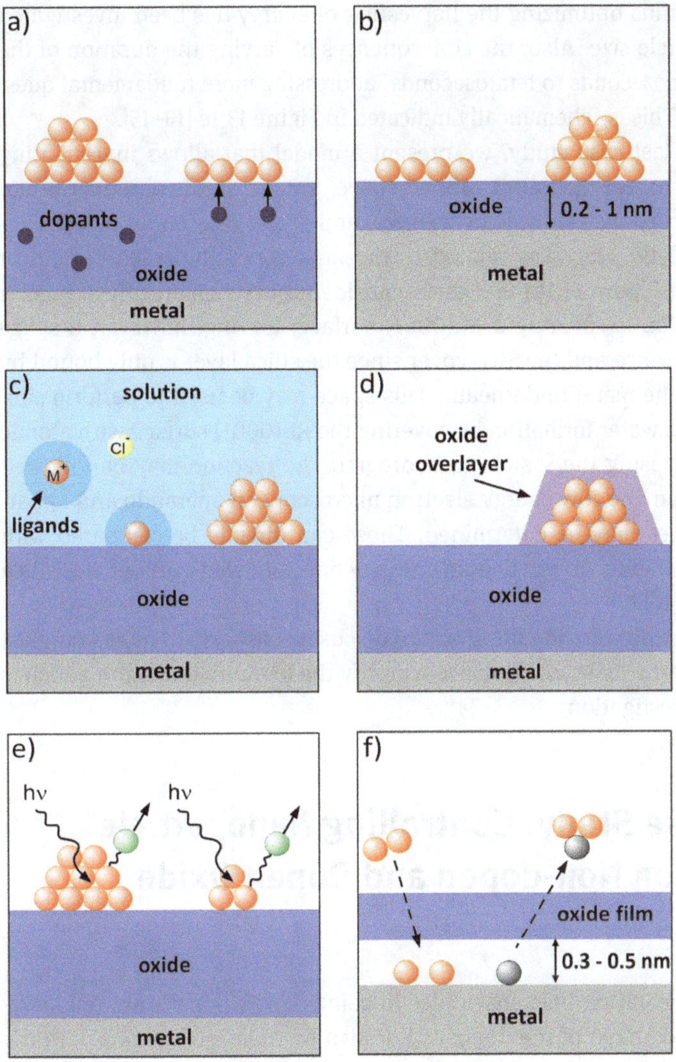

Figure 13.1: Schematic representation of the model systems discussed within the chapter. (a) Nanoparticle growth influence by dopants in the support, (b) nanoparticles on ultrathin oxide films (c) nanoparticle deposition from solution, (d) strong metal support interaction, (d) photochemistry at supported nanoparticles as a function of size, and (e) reactions in confined space underneath ultrathin oxide films.

clean metal at the given temperature. Some general design principles may be deduced from those studies [11, 12].

In the fourth case study the interaction of laser light with alumina supported Ag nanoparticles and the consequences for the desorption of molecules were studied. In particular, the question of tuning the light frequency into the plasmon frequency of the particles and thus optimizing the harvesting of energy has been investigated as a function of particle size. Also, the consequences of varying the duration of the laser pulses from nanoseconds to femtoseconds, addressing more fundamental questions, are discussed. This is schematically indicated in Figure 13.1e [13–15].

In the fifth and last case study, we present a model that allows investigating fundamental aspects of reactions in confined space, a topic that is of immense importance with respect to the catalysis by zeolites, and is also thought to play an important role in catalytic processes related to the area generally called "origin of life." As indicated in Figure 13.1f the system considered is schematically shown: a double layer silica film is covering a Ru(0001) surface, leaving, however, a space between the metal surface and the SiO_2 cover since the silica layer is only bound by dispersion forces to the metal underneath. This space may be used to perform simple reactions, such as water formation by covering the Ru(0001) surface with atomic oxygen and then exposing the system to hydrogen. The reaction may be followed using a photoemission and low energy electron microscope in operando and apparent activation energies may be determined. Those can directly be compared with the equivalent information deduced from the reaction followed without the silica cover being present [16].

This chapter does not provide the space to discuss all aspects of model catalyst studies in detail, but the five case studies exemplify the breadth and depth which is provided by such investigations.

13.2 First Case Study: Controlling Nanoparticle Shapes on Non-doped and Doped Oxide Supports

The performance of metal-particle ensembles in catalytic reactions is not only governed by the size and shape of the deposits but also by their interactions with the oxide support. The occurrence of certain active sites on a catalyst surface is often related to distinct particle geometries. For example, perimeter sites that enable a molecule to interact simultaneously with the metal and the oxide surface are most abundant on flat metal islands, while low-coordinated metal atoms are usually found on ultra-small aggregates [17–19]. Moreover, the geometry affects the electronic properties of metal particles, and many phenomena such as metal-to-insulator transitions are closely related to the atom arrangement within the aggregate [20]. Also in photo-catalysis, particle shapes are of utmost importance as they determine the energy position of plasmonic excitations, being the initial step in several photon-mediated processes [21]. The dependence of the chemical reactivity of a supported metal catalyst on size and morphology of the active metal species has been explored in detail for gold [22]. Using transmission electron microscopy, raft-shaped Au islands on a Fe_2O_3 support have been identified as the active species for the low-temperature CO oxidation [23, 24]. A similar conclusion was drawn from scanning tunnelling microscopy (STM) measurements on the Au/TiO_2 system, where again bilayer high deposits turned out to be the most active ones [25, 26]. Naturally, a close structure–reactivity relationship is not restricted to Au particles but can be found for other metal species in a multitude of chemical reactions [27].

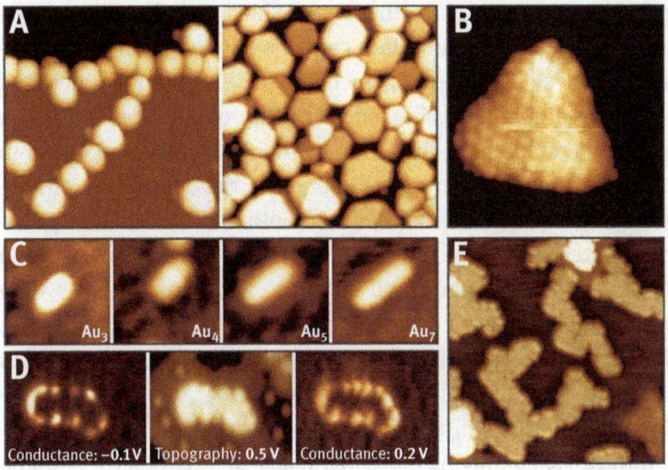

Figure 13.2: (A) STM images of Ag particles on pristine and sputtered Al_2O_3 films grown on NiAl(110) (1 V, 100 × 100 nm^2). Note the different particle density and shape on the defect-rich oxide surface. (B) Atomically resolved image of the top facet of a Pd particle on Al_2O_3/NiAl(110) (0.05 V, 4 × 4 nm^2). The facet is well ordered and corresponds to a Pd(111) plane. (C) One-dimensional Au clusters on ultra-thin alumina films that develop as a result of a charge transfer out of the support (−1.0 V, 5 × 5 nm^2). (D) Two-dimensional Au island on a 2 ML MgO/Ag(001) film. Whereas the central panel displays the island morphology (0.2 V, 25 × 25 nm^2), the left and right panels are conductance images and reflect the state density at the given bias voltage. (E) Large monolayer high Au sheets on MgO thin films with characteristic meandering shapes (0.2 V, 25 × 25 nm^2).

Given the importance of the particle geometry, a good fraction of catalysis research has been devoted to explore different means to optimize this parameter. In principle, particle morphologies are governed by the strength of the metal-support interaction with respect to the surface-free energies of the separated systems. For oxide supports, as commonly used in heterogeneous catalysis, this usually leads to the formation of 3D deposits, as both the free-energy of the oxide surface and the interface adhesion of the ad-metal are small (Figure 13.2a). Moreover, the metal particles tend to mimic the symmetry of the oxide lattice in order to establish interfacial bonds. This causes for instance the Ni lattice to switch from fcc in the bulk to hcp in small deposits grown on the MgO(001) surface [28]. To overcome these growth restrictions, several procedures have been developed that enable a certain control of particle equilibrium shapes on oxide supports. Most of them are based on a targeted modification of the substrate morphology, for example, by introducing defects [29] and suitable ad-species, e.g. hydroxyl groups [30]. Their function is to anchor the metal particles on the inert oxide surface, inhibiting sintering and ripening processes (Figure 13.2a). Alternatively, oxide surfaces with polar character might be used, because they feature high surface-free energies and therefore an enhanced adhesion of the ad-metal [31, 32]. This approach has been demonstrated for the polar ZnO(0001) surface, where Cu was found to grow in a layer-by-layer fashion [33]. Finally, electron transfer processes across the metal–oxide interface can be exploited to alter the equilibrium shape of the ad-particles [34]. The charge exchange between metal deposits and support is stimulated by different mechanisms. On thin oxide films grown on metal substrates, it often occurs spontaneously due to the different Fermi levels in support and ad-metal. The accumulation of excess charges in the deposit thereby activates new charge-mediated coupling schemes between both systems, such as polaronic and Coulomb type of interactions. The enhanced interface adhesion promotes a spreading of the metal on the oxide support and might, in an extreme case, switch the growth regime from 3D to 2D when going from neutral to charged ad-species [35, 36]. The charge transfer is however restricted to ultra-thin oxide films, as it requires electron tunnelling through the insulating layer (see below).

A charge-induced formation of low-dimensional metal aggregates has been identified on various oxide thin-films, employing mostly STM as imaging tool. On alumina films grown on NiAl(110), for example, the formation of 1D gold chains was observed (Figure 13.2c) [37]. The linear nature of the aggregates enables good spatial separation of the extra electrons inside the particle electronic system, which in turn lowers the internal Coulomb repulsion. With increasing atom number, the equilibrium shape changes to planar 2D islands, as observed for gold on magnesia thin films [35]. By analyzing the electronic structure of the 2D islands, their charge state could be quantified directly [38]. For a planar Au_{18} aggregate on 2 ML MgO/Ag(001) for example, four extra electrons were found to reside on the deposit, providing a rough idea on the efficiency of charge transfer processes across the metal–oxide interface. The relationship between the preferred growth geometry in the presence of excess charges and the need of the system to minimize the internal electron–electron repulsion has been visualized for even larger Au islands on the MgO support [39]. The transferred charges from the support localize exclusively at the perimeter of the metal island, where electron-rich edge states with Au 6s character become visible in tunnelling spectroscopy (Figure 13.2d). As the electron-storage capacity of those edge states is limited, the gold islands develop a peculiar meandering shape that keeps the ratio between inner and edge atoms constant (Figure 13.2e). The electron-rich nature of their perimeter renders charged monolayer Au islands particularly attractive for adsorbates. Not surprisingly, CO molecules dosed onto the Au/MgO thin-film system were found to bind exclusively to the low-coordinated edge atoms while no adsorption takes place in the island interior [40]. According to DFT calculations, the island perimeter is also prone to bind and dissociate O_2 molecules, being the initial step to form gold oxide at the island perimeter [41].

Although charge-mediated growth schemes are ideally suited to alter the equilibrium shape of metal particles on thin oxide films, the mechanism breaks down on thick, bulk-like oxide supports as used in real catalysts. The reason is the limited tunnelling length of electrons through an insulating oxide, being of the order of 1–2 nm. An obvious approach to overcome this restriction is to insert the charge donors directly into the oxide material, preferentially into a near-surface region. By this means, charge transfer into the ad-metal can be sustained even for arbitrarily thick oxide slabs, preserving the above-discussed possibility to tailor the particle shapes. There are two types of donor-like impurities in an oxide material. The first one comprises intrinsic lattice defects that are able to trap and release electrons [42]. Prominent examples are >single O vacancies that contain one or two excess electrons (F^o or F+ centers) and structural lattice defect with electron-trapping abilities. For example, grain boundaries that are present in most oxide films were found to trap large quantities of extra electrons (up to 5 per nm defect line) [43, 44]. The subsequent transfer of electrons out of these defects into suitable metal deposits has been demonstrated for the Au/MgO_{1-x} system, employing infrared absorption spectroscopy with CO as probe molecule [40, 45]. The main disadvantage of using intrinsic lattice defects as electron donors is the strong variation of their concentration with the oxide preparation conditions. Whereas the abundance of oxygen vacancies depends on the O_2 chemical potential during preparation or reaction conditions, the density of structural defects is given by the mechanical strain exerted on the oxide lattice during growth. Both influences cannot easily be controlled in real materials used for catalysis and energy storage, and are hence unsuitable to tailor the equilibrium geometry of ad-particles.

The second approach exploits new charge degrees of freedom that can be inserted into an oxide lattice by replacing some of the intrinsic ions with foreign species [46–51]. Whereas doping has proven to be extremely successful for tailoring the properties of semiconductors; a transfer of this technique to oxides is challenging due to the higher complexity of the latter. Two general approaches can be distinguished, which are overvalent doping, that is, replacing a lattice ion with a species of higher charge state, and undervalent doping where an electron-poor ion substitutes the original species. While the dopants have donor character in the first case, i.e. they release electrons to suitable adsorbates, they act as electron acceptors in the latter by creating hole states in the oxide that stimulate electron transfer out of the adsorbates. In both scenarios, the localization of excess charges in the ad-metal might lead to distinct changes of its geometric and electronic properties, in a similar fashion as described for thin oxide films before.

The impact of doping on the growth geometry of metal particles has been demonstrated for CaO films doped with Mo impurities in the 1–2% range [9]. After dosing gold onto the atomically flat CaO(001) surface, a perfectly 2D growth regime was

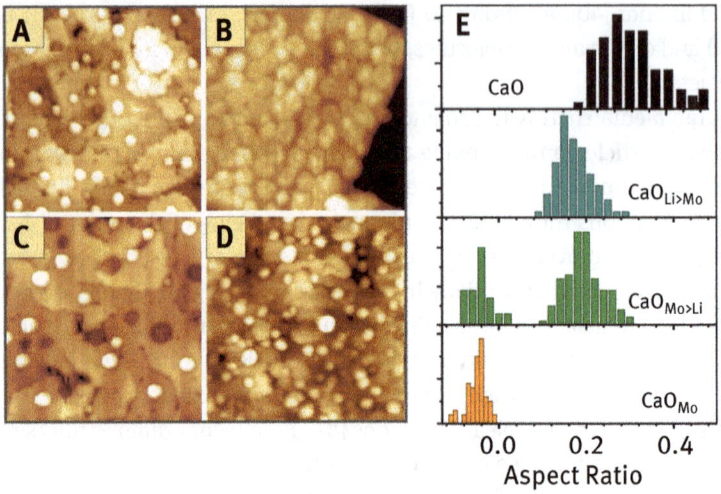

Figure 13.3: STM images of 0.5 ML Au deposited onto (A) pristine CaO, (B) CaO doped with 4% Mo; (C) CaO doped with 4% Mo plus 2% Li and (D) CaO doped with 4% Mo plus 8% Li (6.0 V, 50 × 50 nm^2). Whereas the presence of Mo donors changes the Au morphology to 2D, co-doping with Li reinstalls the original 3D growth mode. Note that monolayer Au islands in (C) appear as depressions, as their conductance is lower than the one of bare CaO(001). (E) Histogram of aspect ratios for Au particles grown on the differently doped CaO films. The low aspect-ratios observed for Mo-rich samples correspond to a wetting growth of gold, as induced by electron transfer from the donor-type impurities.

detected with the metal spreading into large monolayer high islands on the oxide support (Figure 13.3). This unusual growth behavior is not observed on non-doped films, where tall 3D Au deposits develop as expected from the weak interaction between the wide-gap insulator and the ad-metal. The observed cross-over in growth morphology is in full agreement with the donor character of the Mo impurities. Without gold, the Mo ions adopt a 2+ charge state that is typical for cations in the rocksalt lattice. This implies that the Mo *3d* electrons remain localized in the dopant, occupying t_{2g}-type defect states inside the CaO band gap [9]. However, some of the *d*-electrons are close to the onset of the CaO conduction band and hence in an energetically unfavourable situation, which renders a transfer into suitable acceptor states likely. Exactly those states are now provided by the Au adatoms, and indeed a spontaneous electron crossover takes place from a high-lying Mo *d*-state into the Au 6s affinity level. The anionic gold exhibits a reinforced bonding to the CaO surface, stimulating the 2D growth regime revealed in the experiment (Figure 13.3a, b).

The interrelation between electron transfer and realized Au particle shape can directly be proven by blocking the electron flow between the dopants and the ad-metal, for example by introducing a defined number of electron traps into the oxide lattice. As discussed before, hole-states with electron-trapping abilities are generated by undervalent dopants, for example, by monovalent Li^+ ions sitting in Ca^{2+} substitutional sites.

Indeed, a co-doping of CaO with both Mo and Li leads to an internal charge compensation of the two dopants, in which the excess electrons of the donors are trapped by the hole-states inserted by the acceptor species. As a consequence, no electron transfer into the Au ad-metal takes place anymore for the co-doped CaO films and the Au deposits adopt the original 3D equilibrium shape that is typical for pristine CaO (Figure 3c–e). Apparently, the desired doping effect gets cancelled out if dopants with opposite charge state coexist in the oxide host.

Interestingly, not every overvalent dopant is able to donate charges into an ad-metal. Cr ions embedded in an MgO matrix, for instance, are inappropriate electron donors and the Au growth morphology is hardly affected in this case [52]. The reason is that Cr in the MgO lattice cannot be stabilized in the 2+ charge state, because it prefers a high-spin configuration and consequently puts electrons into both the low-lying t_{2g} and the high-lying e_g state manifold. The latter already overlaps with the MgO conduction band, so that the electrons move away from the dopant and become trapped in other oxide defects, preferentially in Mg vacancies sites [53]. The Cr adopts a 3+ charge state right from the beginning and cannot donate further charges to surface species. An active dopant therefore needs to fulfil two requirements. On the one hand, it should occur in two stable charge configurations in the oxide host. On the other, the top-most occupied state needs to be high enough in energy to render charge transfer into an adsorbate affinity level energetically feasible. The same considerations also hold for acceptor-like dopants, only that the hole-state produced by doping localizes in an adjacent oxygen ion. It should be noted that holes in the oxygen *2p* states are intrinsically unstable against charge transfer from competing electron sources, such as electron-rich O defects (F^0 centers) or donor-type molecules that are always present on oxide surfaces (water and hydrogen) [54, 55]. Hole doping as a means to tailor the properties of metal ad-particles is hence more difficult to realize than electron doping with donor-type impurities.

Let us next consider chemical reactions as a function of the supported metal particle morphology, and specifically address the reactivity at the metal–support interface. In a series of studies we have addressed this area, realizing that the problem is difficult to tackle experimentally. So far, only evidence from model calculations have been presented indicating a particular reactivity of the metal/oxide interface at the rim of the particle, while all experimental evidence was indirect, that is, there have been no images of molecules in a reacting system at the rim of particles.

First attempts to image molecules at the rim were undertaken by Lin et al. [40] using CO as a probe molecule as alluded to above. While those studies suggest the localization of molecules at the rim of the particle they do not provide direct images of adsorbed molecules, which is the next challenge we will tackle. Perhaps a larger molecule, which is less mobile than CO, could be used to allow for imaging of individual molecules at the rim. Stiehler et al. [56] used isophorone in such an attempt. The molecule (structure shown in the inset) was identified via its vibrational fingerprint measured with IR spectroscopy. Figure 13.4 collects STM images of the same Au particle with (left) and without (right) molecules adsorbed at the rim. This was only possible, because isophorone is weakly adsorbed, and may be manipulated with the STM tip. By this means, images with adsorbed molecules could be taken directly after dosing, while those were removed with the tip afterwards to record images of the naked particle. This system provided another interesting option, as it allowed us to record the quantum well states of the nanoparticles with and without the adsorbed molecules. Quantum well states arise from the finite size of the object similar to the well-known particle in a box problem solved in undergraduate courses of physical chemistry. The quantum well states as measured through scanning tunneling spectroscopy are shown as spectra (blue and red) taken at two positions within the particle. In addition, the conductance maps of the system are shown, clearly indicating the nature of the states involved.

The states shift to lower energy as the molecules are adsorbed and also their mutual separation changes. In fact, the states are closer spaced for the system with adsorbed molecules as compared with the naked particle, indicating that the potential well get less steep and the gold wave functions are allowed to effectively spill out of the Au particle. The effect is similar to the situation encountered for a particle in a box with a wider potential well. Details have been analyzed in ref [57]. This knowledge may now be used to proceed to a situation where molecules at the rim are actually involved in a chemical reaction. Here we combine the above discussed ability of the Au particles to accumulate negative charge and consider a reaction of carbon dioxide [58, 59]. To transfer an electron to an isolated CO_2 molecule affords 0.6 eV and is thermodynamically unfavorable [15]. However, a CO_2-dimer has a negative electron-affinity (0.9 eV), as has been shown through electron attachment experiments in molecular beams and corroborated by calculations already decades ago [17, 18, 60]. Therefore, electron transfer to a dimer, which might be encountered at a surface at sufficiently high coverage, should be thermodynamically favorable.

On this basis, we propose the following scenario [21, 22]: The Au nanoparticles on a thin MgO(100) film are negatively charged and may be used as a source to trigger electron transfer to CO_2 dimers possibly inducing the formation of an oxalate species, this is nothing but a CO_2–dimer anion with a new carbon–carbon bond and an extra electron transferred, yielding $C_2O_4^{2-}$. Of course, since the Au particles, according to the paper by Pacchioni and Landman [36], only accumulate negative charge in the raft-like flat geometry, it is conceivable that upon heating the reaction is shut off, as with increasing atom mobility the 2D particles turn 3D and the electron transfer to the Au nanoparticle becomes unfavorable. This scenario is depicted schematically in Figure 13.5.

Here individual CO_2-anions are also shown, formed at defects of the MgO film. The latter is quantitatively and qualitatively consistent with earlier studies through EPR titration of oxygen anions and IR measurements on CO_2 on other oxide films [23, 24]. However, at those defects, only a single electron transfer may be achieved

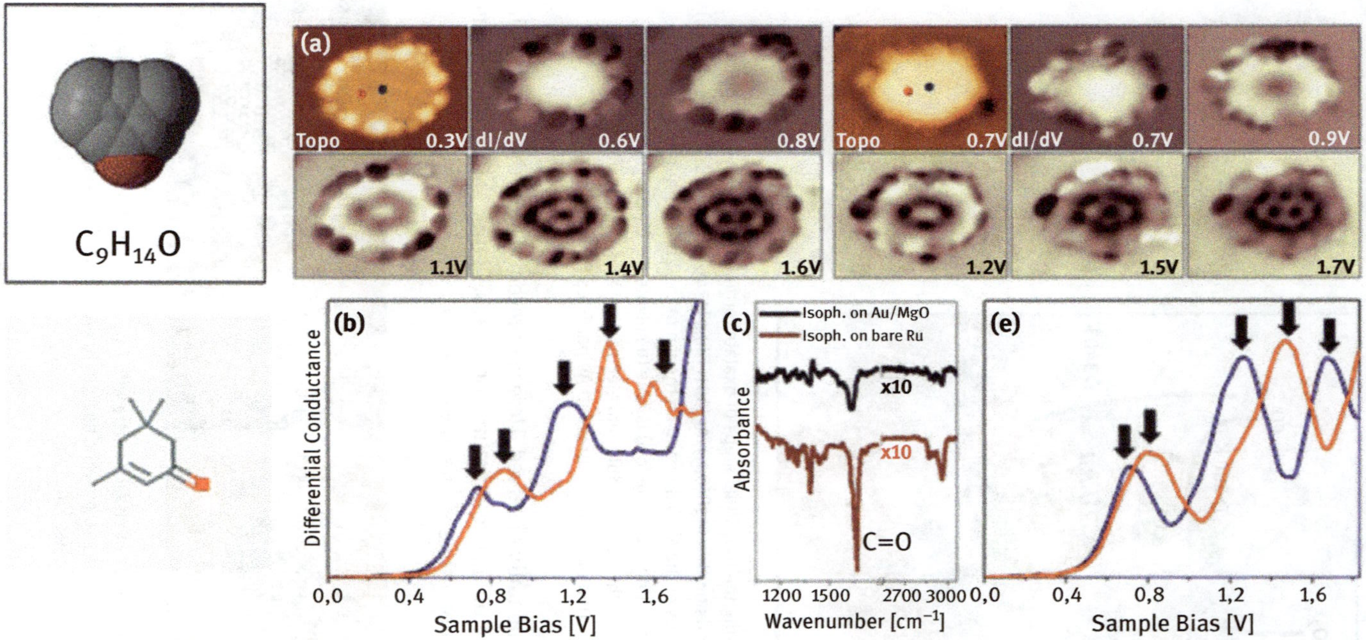

Figure 13.4: (color online). TopographicanddI=dVmaps of an Au island (a) with and (d) without isophoronemolecules bound to its perimeter (11×8nm2, 70 pA). (b) and (e) associated conductance spectra taken in the center (blue) and the left part of the island (red) as indicated in (a) and (d). Gaussian fitting revealed five QWSs that are marked by arrows.Their bias positions match the ones used for dI=dVimaging in (a) and (d). (c) Infrared absorption spectra of isophorone adsorbed on Au=MgO (black) and Ru(0001) (gray). Similar vibrational modes are resolved in both cases, indicating that the molecule adsorbs associatively on the two surfaces. [56]

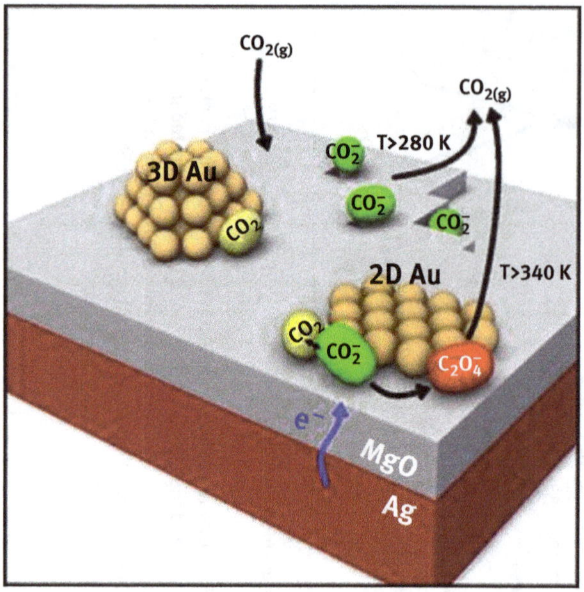

Figure 13.5: Schematic showing the individual steps of oxalate formation upon chemisorption of CO_2 on the rim of 2D Au islands on thin MgO(001)/Ag(001) films [134]. Carboxylate species (green) are formed both on defect sites of the MgO film and on the rim of the 2D Au islands by electron transfer. Only on the latter, additional CO_2 (yellow) is able to solvate the carboxylate species yielding a $(CO_2)^{2-}$dimer ion, which, after an additional electron transfer, results in adsorbed oxalate $C_2O_4{}^{2-}$ (red). This reaction does not occur on 3D Au particles [135, 136].

which is, in fact experimentally observed. Here we present part of the experimental evidence and refer to the published literature for details. Figure 13.6 shows STM images of a raft-like Au nanoparticle after exposure to CO_2.

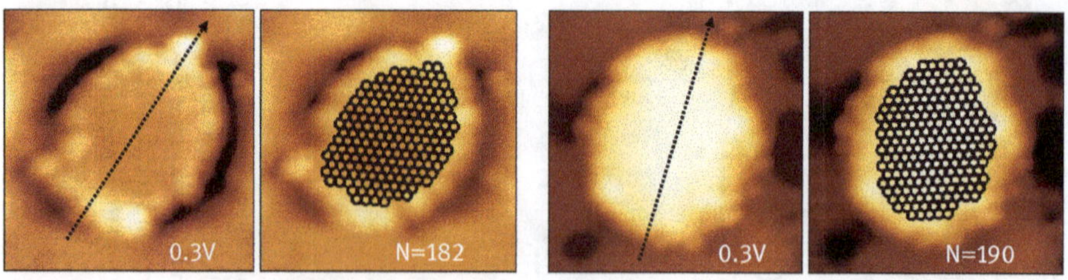

Figure 13.6: STM topographic images of an Au particle with CO_2 molecules at the rim (left panel), the Au island is schematically shown as a model. The right hand panels show a similarly shaped Au particle without adsorbed CO_2 molecules [56, 134].

Clearly, molecular protrusions are observed at the rim [21]. Infrared spectra (Figure 13.7a), taken as ensemble-averaging data, are also plotted as a function of temperature. The top trace shows the clean MgO film after exposure to CO_2. The band at 1295 cm^{-1} is due to carboxylate (CO_2^-) formed at defects. This was verified through isotopic labeling experiments [136]. The second trace has been recorded after Au deposition, the formation of the raft-like particles, and exposure to CO_2 at room temperature. The two extra bands are due to the formation of the oxalate. Again, detailed isotopic labeling experiments confirm the formation of a C–C bond in oxalate. Here we benefit from the possibility to not only label the CO_2 molecules but also the oxide film by growing it with either $^{18}O_2$ or $^{16}O_2$. This provides the unique way to differentiate the species formed from carbonate, which would need the involvement of oxygen from the substrate. For the subsequent spectra in Figure 13.7 the temperature of the system has been systematically increased. Clearly, upon heating above 400 K prior to exposure to CO_2, the oxalate bands attenuate [138], and at 600 K they have practically disappeared. If this information is combined with the STM data shown in Figure 13.7b, we may conclude that upon heating above 500 K the particles tend to assume a three-dimensional morphology, thus corroborating the idea, that those particles do not provide electrons to induce oxalate formation.

It should be noted that thermal desorption studies on the system [136, 138] revealed that the formed oxalate may desorb as CO_2 upon heating the system to a temperature larger than 340 K. This is interesting, since there is also the possibility to form CO and carbonate (CO_3^{2-}) upon heating, a typical observation made for oxalates in the presence of alkali co-adsorbates [137–139]. It seems that the low oxophilicity of Au prevents the formation of carbonate thus rendering the formation and decomposition of oxalate reversible in this system.

In summary, the investigation of metal nanoparticles on thin oxide film systems provides a conceptual basis to study the connections between charge transfer from the support, particle morphology and the concomitant chemistry driven by localization of charges as a function of morphology. Those concepts can be transferred to bulk systems by doping of oxide materials and opens promising new routes to modify the morphology and electronic properties of supported metal particles as used in heterogeneous catalysis. The identified mechanisms can be transferred to real catalysts, as the doping approach is not based on specific thin-film effects.

13.3 Second Case Study: Preparation of Oxide-Supported Palladium Model Catalysts by Pd Deposition from Solution

In typical surface science experiments as presented above, oxide-supported metal nanoparticles are deposited onto a clean oxide surface by physical vapor deposition. The precursors in this process are metal atoms in the gas phase, which impinge on the surface, diffuse until they eventually get trapped (either at surface

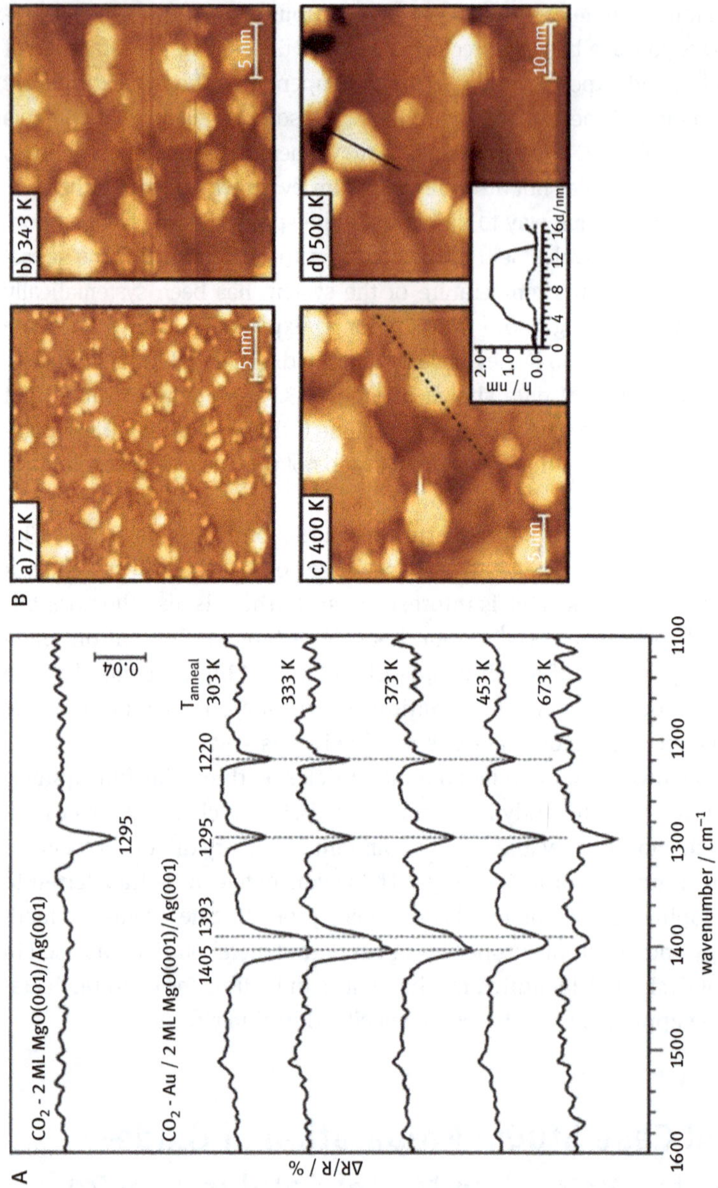

Figure 13.7: (a) IRA spectra of bare (top) and Au-loaded 2 ML MgO(001)/Ag(001) samples recorded after a saturation dose of CO_2 at 223 K. Au was deposited at 100 K and the samples subsequently annealed to the indicated temperature prior to CO_2 adsorption [138]. (b) STM images of (a) Au deposited on 2 ML MgO(001)/Ag(001) at 77 K, and after subsequent annealing of the system to 343 K (b), 400 K (c) and 500 K (d). All images were taken at 77 K. Scan area: (a)–(c) 25 nm × 25 nm; (d) 50 nm × 50 nm. $U_B = +(0.5–0.75)$ V. $I_T = 30$ pA. (e) Height profiles of representative islands/particles from the 400 K (dashed line) and 500 K (dotted line) annealed samples, highlighting the transition from 2D to 3D geometries [135, 136].

defects or by dimer formation) and then act as nuclei for the growth of larger particles. These processes are well understood for ideal model systems under UHV conditions [61, 62]. In contrast, most "realistic" supported metal catalyst preparation procedures, e.g. impregnation, make use of metal complexes as precursor, primarily in the form of salts dissolved in aqueous solution, which adsorb at the support-solution interface in the initial step [63, 64]. The pH of the solution is one of the important parameters, as it controls the interaction of precursor complexes with the support by determining the speciation of the solution complexes as well as the surface charge of the support. Based on the knowledge of the pH dependent precursor speciation in solution on one hand, and the properties of the support (point of zero charge, PZC, density of hydroxyl groups) on the other, this interaction can phenomenologically be described by complexation models [65]. The transformation of the adsorbed precursor complex into the catalytically active compound then usually involves a calcination step followed by reduction. The performance in a catalytic test reaction is the ultimate criterion for the usefulness of a certain preparation procedure. While pre, in situ, and post reaction characterization of the catalyst provides relevant information about the active sites in the catalytic reaction, initial nucleation and decomposition of the adsorbed precursor into metal nanoparticles are difficult to be tracked experimentally with powder catalysts and remain largely unknown aspects of catalyst preparation. As will be shown in the following, single-crystalline oxide thin films are, provided they are stable at typical impregnation conditions, ideally suited for studying these processes.

We have chosen to use Fe_3O_4(111) films grown on Pt(111) for studying the deposition of palladium from a $PdCl_2$ precursor. Fe_3O_4(111) films with a thickness of ~10 nm were prepared in a UHV chamber following a recipe described in the literature [66]. The surface of the films was subsequently covered by precursor solution, which was prepared from an acidic (5 M HCl) $PdCl_2$ stock solution by dilution to the desired concentration and pH adjustment with NaOH. Following washing with water and drying under a stream of He the sample was then subjected to thermal treatment in UHV to transform the adsorbed Pd precursor into nanoparticles.

Shown in Figure 13.8 are STM images (taken in air) of Fe_3O_4(111)-supported Pd particles formed by impregnation with 5 mM Pd precursor solutions of different pH ranging from 1.3 to 10 and after a final thermal treatment at 600 K in UHV [67]. In the acidic pH range (pH 1.3 to pH 4.7) we observe a substantial variation of Pd coverage, particle size, and particle dispersion. At the lowest pH only a few Pd particles with a diameter of 1 nm are present (note that after impregnation the surface of the Fe_3O_4(111) film is affected by heating leading to the observation of meander like features, most prominently seen for pH 1.3; on top of this surface the Pd particles appear as small spots). As the pH of the impregnation solution is changed to 1.6 and 2.5, higher Pd coverage is obtained with a slight increase of particle diameter, however, with still uniform size distribution. The behavior in this pH range reflects the results of typical uptake curves [68] for negatively charged precursor complexes (here $PdCl_4^{2-}$) on positively charged support surfaces (PZC(Fe_3O_4) ≈ 8). The retardation of Pd uptake in strongly acidic medium may be ascribed to (i) competitive adsorption of chloride [69] or (ii) a decrease of the adsorption equilibrium constant at high ionic strength [70]. As the pH is increased to 4.7, the very unfavorable situation of uncontrolled deposition of large and non-uniformly sized Pd particles occurs. Such deposition characteristic has been observed with various near neutral pH precursor solutions and can be explained by deposition of solution-precipitated Pd particles. At alkaline pH (pH 10, Figure 13.8) the Pd particle size distribution in the final catalyst is again uniform. However, compared to the acidic pH range, a significantly higher Pd loading is achieved.

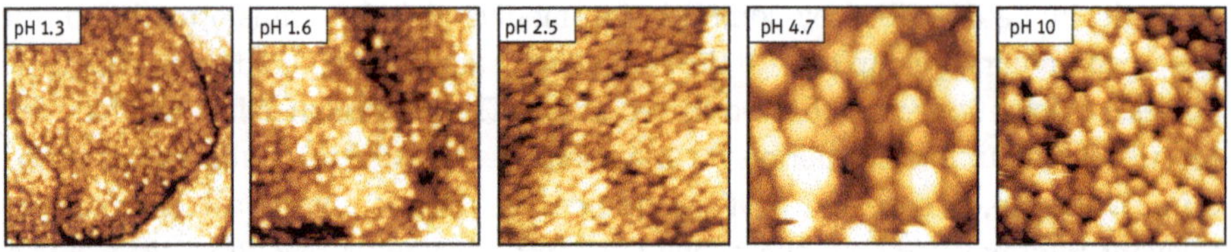

Figure 13.8: STM images (100 × 100 nm^2) of Pd particles formed on Fe_3O_4(111)/Pt(111) thin films by impregnation with Pd precursor solution of different pH prepared from $PdCl_2$,aq and subsequent annealing in UHV to 600 K.

A more detailed account on the nucleation process of Pd nanoparticles from adsorbed precursor complexes is given in Figure 13.9, which compares STM and XPS results obtained at different annealing temperatures after impregnation of Fe_3O_4 (111)/Pt(111) with precursor solutions of pH 1.3 (15 mM Pd^{2+}) and pH 10 (2 mM Pd^{2+}), respectively. The combination of STM and XPS allows the thermal evolution of morphological features with the chemical nature of the precursor species to be directly correlated. In the case of acidic precursor solution (Figure 13.9a) [10] the STM image obtained directly after removing the precursor solution without additional heating (RT) shows a low density of particles on a seemingly clean Fe_3O_4 substrate. As the temperature is increased to 390 K, the particle density is increased and at a final temperature of 600 K, particles uniformly cover the Fe_3O_4 surface (Figure 13.9a). Note that the particle size (2 nm) does not change from RT to 600 K showing that 600 K is well below the onset of Ostwald ripening under the present experimental conditions. The corresponding Pd 3d XPS spectra (Figure 13.9b) reveal the presence of two Pd species with binding energies (BE) of 337.8 and 336.2 eV (Pd $3d_{5/2}$) on the Fe_3O_4 sample directly after removing the pH 1 solution. With increasing temperature, the Pd component at higher BE is gradually transformed into the lower BE component. This behavior is perfectly in line with the STM observations and allows the high BE component to be assigned to adsorbed Pd precursor complexes, which cannot be resolved with STM. The low BE component, on the other hand, correlates with the emergence of nucleated particles at increasing annealing temperature. The chemical identity of the surface species may be inferred from knowledge of the solution speciation of Pd. It is safe to assume that in pH 1.3 (HCl) solution Pd is present as $PdCl_4^{2-}$ complex. The high Pd $3d_{5/2}$ BE obtained for the adsorbed species at RT (337.8 eV, compared to 335.2 eV for bulk Pd) is in line with previous observations [71, 72] and can, therefore, be attributed to adsorbed $PdCl_4^{2-}$ or aquochloro complexes of the form $Pd(H_2O)_xCl_y^{n-}$. According to STM results shown in Figure 13.9a, these adsorbed precursor complexes are thermally decomposed into Pd particles. At a final annealing temperature of 600 K the BE of the particles is 335.7 eV, which is significantly higher than expected for metallic Pd particles. XPS analysis suggests that this BE shift results from remaining chlorine adsorbed on the Pd particles rather than from a particle size effect [10].

A slightly different behavior is observed on the Fe_3O_4 sample contacted with pH 10 precursor solution (Figure 9c, d) [73]. In the initial state, no particles are observed suggesting that the surface is uniformly covered by adsorbed precursor. Only moderate drying at 390 K leads to the formation of small particles in the size range 1–2 nm covering the entire surface. The corresponding XPS spectrum shows the dominant abundance of a single Pd species with a Pd $3d_{5/2}$ BE of 335.9 eV. Significant particle sintering (average diameter 4–5 nm) occurs upon further annealing with a concomitant shift of the Pd $3d_{5/2}$ component to 335.4 eV. The different nucleation behavior observed on the pH 10 sample as compared to the pH 1.3 case is a result of the different speciation of Pd complexes. At pH 10, the adsorbed Pd species are hydroxo complexes. Their thermal decomposition into Pd nanoparticles proceeds via the formation of PdO particles as an intermediate step (observed at 390 K).

Based on the results presented in Figures 13.8 and 13.9 the deposition of Pd on single crystalline $Fe_3O_4(111)$ thin films from aqueous precursor solutions can be divided into three different regimes: electrostatic adsorption of $PdCl_4^{2-}$, precipitation, and adsorption of Pd-hydroxide. In the acidic pH range (<pH 3) the interaction is controlled by electrostatic adsorption of the negatively charged precursor complex ($PdCl_4^{2-}$) and the positively charged surface. Applied to the present case, this usually as strong electrostatic adsorption termed mechanism predicts a minimum of uptake at the PZC of the oxide, a maximum below the PZC, and again little uptake at strongly acidic pH [65]. The latter case, which is usually ascribed to the effect of high ionic strength, has been observed here. In this regime, small and uniformly sized Pd particles are formed by thermal decomposition of the adsorbed precursor. The regime of maximum adsorption due to strong electrostatic interaction could not be reached because uncontrolled deposition of precipitated Pd (most probably $PdCl_2$) species sets in at near-neutral pH. Shifting the solution pH into the basic range (pH 10) leads again to the formation of Pd particles with uniform size distribution after decomposition of the precursor. Strong electrostatic adsorption is not possible in this regime because both the solution complexes ($Pd(OH)_4^{2-}$) and the oxide surface are negatively charged. A chemical interaction between precursor and support, for example, hydrolytic

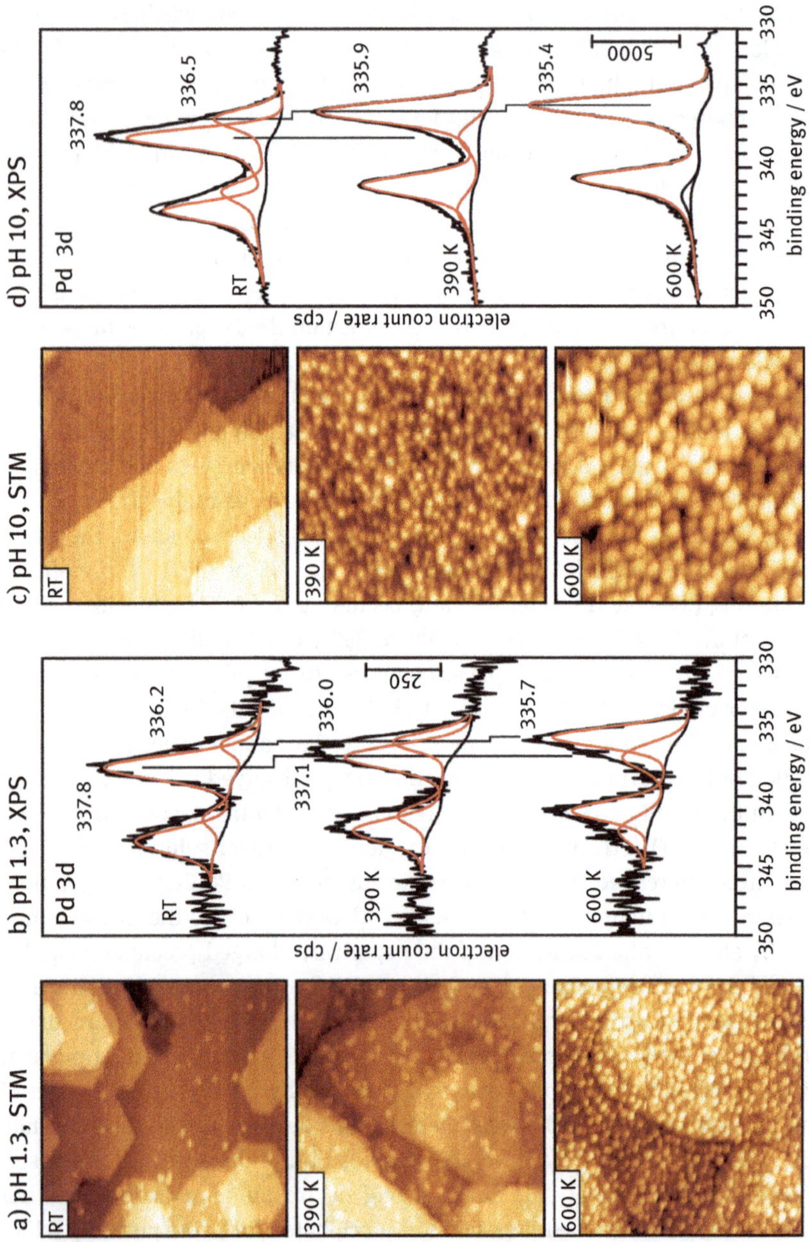

Figure 13.9: STM and XPS results for Pd deposited on $Fe_3O_4(111)/Pt(111)$ surfaces from (a and b) pH 1 precursor solution and (c and d) pH 10 precursor solution at different stages of pretreatment: directly after removing the precursor solution at room temperature (top), after drying in UHV at 390 K (middle) and 600 K (bottom).

adsorption of Pd-hydroxide, is more likely the dominant interaction at the support-solution interface at basic pH.

Summarizing, the way is paved to look at oxide supported metal nanoparticles, prepared in solution and to understand the formation of metal nanoparticles through calcination and reduction. However, there is still a way to go to identify the elementary steps in the interaction of the species from solution at the solid–liquid interface. Of course, this is what we really want.

13.4 Third Case Study: Strong Metal/Support Interaction Effects

SMSI effects have been known for many years to occur when a metal catalyst supported on a reducible oxide is subjected to high-temperature treatments in a reducing atmosphere (typically, hydrogen) [74–78]. In principle, the SMSI state depends on the metal, the oxide and the reaction, and may include both structural and electronic effects. The most prominent example is Pt supported on TiO_2 which undergoes SMSI via the formation of a thin titania film overgrowing the Pt particles, albeit of a very complex structure [79–81]. Certainly, the oxide overlayer suppresses adsorption of many molecules otherwise readily adsorbing on a metal surface and, as a result, renders metal catalysts inactive. However, when the encapsulated oxide film is very thin ("ultra-thin"), the situation is not straightforward since structural and chemical properties of an ultra-thin film are often considerably different from those of a bulk counterpart [82]. In some cases, the encapsulation may lead to the promotional effects. Here, we illustrate this phenomenon for Pt particles supported on iron oxide, $Fe_3O_4(111)$.

The Pt nanoparticles were formed by physical vapour deposition of Pt onto ~10 nm-thick $Fe_3O_4(111)$ films grown on Pt(111). The particles become more uniform in size and well-faceted upon heating to elevated temperatures. Figure 13.10 shows the typical morphology of Pt nanoparticles vacuum-annealed at temperatures above 800 K. The particles, about 1 nm in height and 5 nm in width, expose atomically flat top facets, which are Pt(111) in nature, owing to the epitaxial relationships between $Fe_3O_4(111)$ and Pt(111) [83]. It has turned out, however, that high-temperature annealing leads to a dramatic attenuation of the CO uptake compared to the samples annealed at 600 K, which cannot be assigned solely to Pt sintering [83, 84]. This behavior is, in fact, a "classical" manifestation of the SMSI [74].

High resolution STM study showed that the top facets exhibit the hexagonal lattice of protrusions with a ~3 Å periodicity, in turn forming a superstructure with a ~25 Å periodicity (see inset in Figure 13.10). This structure is well-documented in the literature for thin iron oxide films grown on Pt(111) [85] and can unambiguously be assigned to a FeO(111) film which consists of close-packed layers of iron and

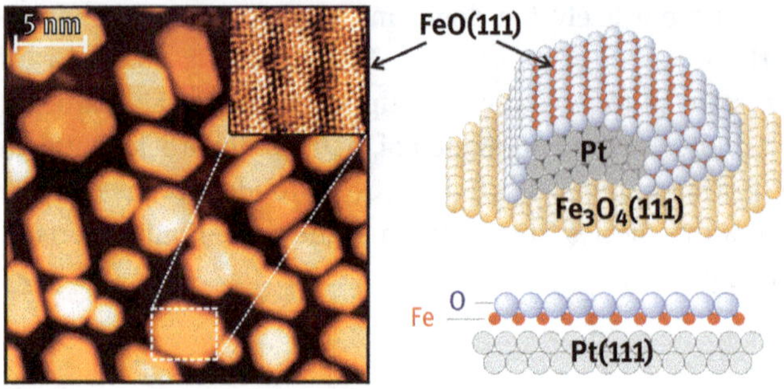

Figure 13.10: (a) STM image of Pt particles deposited onto an $Fe_3O_4(111)$ film at 300 K and annealed in UHV at 850 K. Atomically resolved STM image of the top facet is shown in the inset. (b) The cross view of the FeO(111)/Pt(111) interface and schematic presentation of the Pt particles encapsulated by the FeO(111) overlayer.

oxygen stacked as O–Fe–Pt(111). The Moiré superstructure originates from a mismatch between the Pt(111) and FeO(111) lattices. Since FeO(111) films can be grown on the Pt(100) surface as well [86], it seems plausible that the encapsulation by the FeO(111) layer also occurs on the side facets, which, according to the particles' habitus, expose (111) and (100) surfaces.

The mechanism of the encapsulation is poorly understood. The encapsulation implies high adhesion energy between Pt and iron oxide, which could, in principle, be derived from the structural information, obtained by STM on the particle size and shape, using the modified Wulff construction [87]. The analysis yielded an energy in the range of 3.8–4.2 J m^{-2}, which is, indeed, considerably larger than those obtained for Pd particles on $Fe_3O_4(111)$ and alumina films (i.e., 3.1–3.3 J m^{-2}), for which the encapsulation has not been observed [83]. Note also that CO adsorption experiments indicated Fe–Pt surface intermixing with the onset at ca. 600 K [88], probably as the first step in the encapsulation. However, high-resolution TEM studies performed on the model systems discussed here by Willinger et al. [89] indicate convincingly that the oxide encapsulates the particle like a carpet resorting to material located close to the nanoparticle. This situation is illustrated in Figure 13.11, where we show on the left side a cross-sectional view of the system, i.e. the Pt(111) support, the grown magnetite film and the Pt nanoparticle on top. The EELS (electron energy loss spectroscopy) traces next to the cross sectional view indicate that the particle is encapsulated by an iron oxide film as proven by monitoring the Fe L-edge. The three images on the right side show in one case a high-resolution image of two particles on top of the magnetite film. The arrows indicate the region with the loss of material possibly used to encapsulate the particle. The two images below in false color indicate that the interface underneath the encapsulated particles is still intact.

The well-defined $Pt/Fe_3O_4(111)$ systems were examined in the CO oxidation reaction at near-atmospheric pressures and relatively low temperatures (~450 K) [90]. Figure 13.12

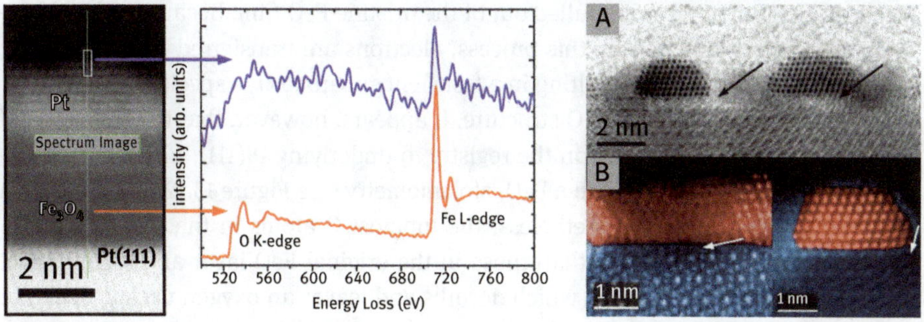

Figure 13.11: Left: TEM image and EEL spectra taken in the ranges of the OK-edge and the Fe L-edge in the oxide film (red) and at the region of the Pt–particle surface (blue). Right: (a) STEM-ABF image and (b) false color HAADF images. Arrows indicate a reduced contrast and a reduced structural order in the $Fe_3O_4(111)$ film near the Pt particle.

shows kinetic curves of CO_2 production under O-rich conditions (e.g., 10 mbar CO + 50 mbar O_2, He balance to 1 bar) over two samples possessing the same amount of Pt, but annealed either at 600 K (i.e., exposing clean Pt surface) or 850 K (i.e., encapsulated by FeO) prior to the reaction. The results for pristine $Fe_3O_4(111)$ films and clean Pt(111) under the same conditions are also shown for comparison.

It is clear that the encapsulated Pt particles exhibit higher reactivity than the clean Pt particles. The difference must be even higher if one normalizes the reactivity to the particles surface area, which obviously decreases at 850 K due to particle sintering, albeit not measured in those experiments. The same effect was observed also at the stoichiometric ratio (40 mbar CO + 20 mbar O_2) [90]. Such promotional effect of encapsulation seems counterintuitive since the FeO film covering Pt particles and exposing a close-packed O layer must be essentially inert. In order to rationalize these findings on highly dispersed systems we have to address the structure–reactivity relationships observed for extended, well-ordered FeO(111) films on Pt(111).

The FeO(111) film is, indeed, extremely stable and chemically inert under conditions typically used in UHV-based experiments. However, the situation changes dramatically in the mbar range of pressures. At low temperatures studied here (400–450 K), Pt(111) is inactive in CO oxidation due to the well-known blocking effect of CO on O_2 dissociation. The nm-thick $Fe_3O_4(111)$ film shows some activity, but negligible as compared to ultrathin FeO(111) films which showed an order of magnitude higher reaction rate under the same conditions (Figure 13.12a). Therefore, it is the thin FeO overlayer on Pt that is responsible for the enhanced reactivity of encapsulated Pt particles in CO oxidation.

The experimental results in combination with DFT calculations provided compelling evidence that at elevated pressures the FeO(111) film transforms into a different structure containing an Fe layer sandwiched between two oxygen layers like O–Fe–O film (Figure 13.12b) [11, 12]. The mechanism for this transformation starts by

O_2 adsorption on an Fe atom pulled out of the pristine FeO film. Because of local lowering of the work function by this process, electrons are transferred from the oxide/metal interface to oxygen, resulting in a transient superoxo O_2^- species, which dissociates, thus forming the O–Fe–O structure. It appears, however, that the formation of the trilayer structure depends on the registry to underlying Pt(111), ultimately resulting in close-packed islands with a FeO_2 stoichiometry (see Figure 13.12b) rather than a continuous O–Fe–O film. Nonetheless, the topmost O atoms in the resulted FeO_{2-x} films are more weakly bound than those in the original FeO layer and readily react with incoming CO to form CO_2, which desorbs and leaves an oxygen vacancy behind. The overall activation barrier for CO_2 formation on the ideal O–Fe–O overlayer, as determined by DFT (~0.3 eV), is considerably lower than the computed barrier (~1 eV) for the CO oxidation reaction on Pt(111), and as such may explain higher reactivity of FeO(111)/Pt(111) than pure Pt(111) [12]. Certainly, to end the catalytic cycle, the oxygen vacancies must be replenished via the reaction with O_2 from the gas phase. STM studies provided strong evidence for this mechanism of Mars-van Krevelen type [91]. Interestingly, NO transforms the FeO film into the trilayer film in the same way as O_2.

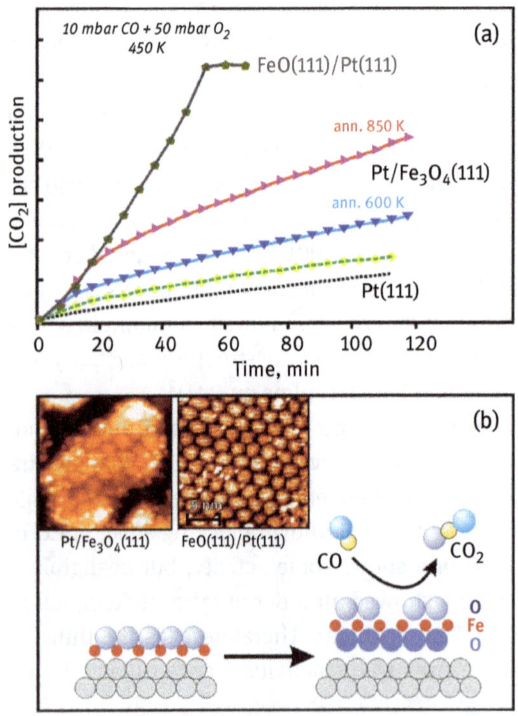

Figure 13.12: (a) CO_2 production over Pt(111), iron oxide films on Pt(111), and Pt/Fe_3O_4(111) annealed to 600 and 850 K, as indicated. (b) STM images of the encapsulated Pt particle on Fe_3O_4(111) and, for comparison, of the FeO(111)/Pt(111) film, both exposed to 20 mbar O_2 at 450 K. The scheme illustrates the originally proposed reaction mechanism (see the text).

Comparison of the $CO + O_2$ and $CO + NO$ reactions over the FeO(111)/Pt(111) surface showed that the replenishment of oxygen vacancies is the rate-limiting step that proceeds much faster with O_2 than NO [92].

It is important to note that both the transformation of the FeO into FeO_2-like film and the oxygen vacancy replenishment under the reaction conditions, involve the charge transfer accompanied by a lattice distortion. Both factors favor the reaction on ultra-thin films.

Apparently, the same scenario holds true for the encapsulated Pt particles. Figure 13.12b shows STM images of the encapsulated Pt particle on Fe_3O_4(111) and, for comparison, of the FeO(111)/Pt(111) film, both exposed to 20 mbar O_2 at 450 K. The close similarities between these two systems with respect to the surface morphology and reactivity indicate the absence of the "material gap", suggesting that the results and conclusions drawn for extended surfaces can be transferred to the supported nanoparticles.

In the experiments, presented in Figure 13.12a, the initial reaction rate over the encapsulated particles is almost identical to that of measured on the FeO(111) film, because of the particle surface area (including both top and side facets) at the high Pt loadings studied here is close to the surface area of the FeO film. Whereas the rate over the FeO film is almost constant until all CO in the ambient is consumed and the reaction stops, the reaction slows down over the encapsulated particles indicating catalyst deactivation. The latter may include carbon deposition, but this issue is beyond the scope of the present paper.

The situation becomes more complex at sub-monolayer (sub-ML) film coverages as the surface additionally exposes oxide/metal interfacial cites which are commonly considered to be active. Bao and coworkers [93] addressed reactivity of FeO(111) islands on Pt(111) by scanning tunneling microscopy (STM) and ultraviolet photoelectron spectroscopy (UPS) under ultrahigh vacuum (UHV) conditions and concluded that the reaction takes place at the FeO(111)/Pt(111) boundary [93; 95]. Although this mechanism may be operative under UHV-based conditions [94], it hardly holds true at the realistic pressures, which force the formation of the FeO_{2-x} phase [96]. Our follow-up study [97] provided strong evidence that it is the FeO_{2-x}/Pt interface that catalyzes the reaction under realistic pressure conditions.

Even though certain progress has recently been reached towards understanding the reactivity of the FeO(111)/Pt(111) systems (see also recent studies [98, 99], the details of the reaction mechanism remains poorly understood. Recently, we provided experimental evidence that the CO oxidation reaction takes place not at oxide/metal interface, but at sites provided by the boundary between reduced and oxidized FeO_x phases in the film, whose formation is promoted by the metal support at the initial stages of the reaction [100]. We investigated FeO_{2-x} islands on Pt(111) after exposure to 10^{-6} mbar of CO at 400 K and determined the area representing the tri-layers as a function of exposure time. The STM images in Figure 13.13A show that the protrusions (each representing the area with a FeO_2 structure within the Moire pattern of the initial FeO film) disappear starting from the side that is closer to the island edge. The interior region remains intact, indicating that no reaction occurs on the oxidized FeO_{2-x} surface. The progressive reduction (see the curve in Figure 13.13B) takes place at the interface between the compact FeO_{2-x} domain and the reduced surface formed by the reaction with CO, and is typical for etching the islands from the side following first order kinetics. A model based on DFT calculation [101] for the FeO/FeO_{2-x} interface is shown in Figure 13.13C. Extraction of oxygen at this interface only costs 1.31 eV [97], thus rendering the reaction with CO feasible. Other characteristic oxygen removal energies range between 1.53 eV at the FeO_{2-x}/Pt interface and 2.15 eV at the FeO/Pt interface [97].

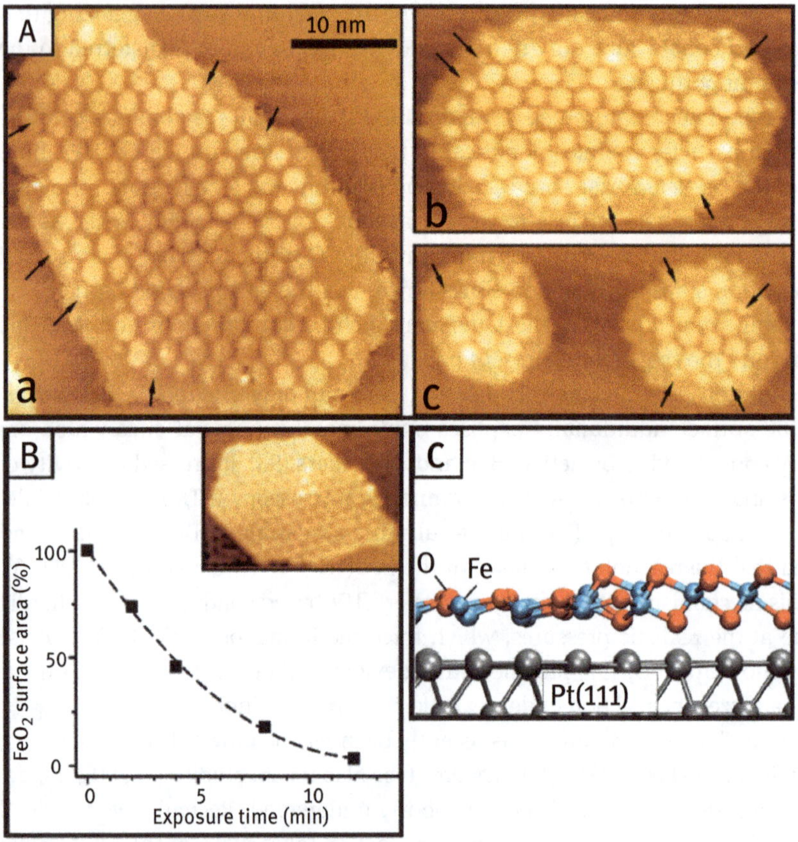

Figure 13.13: (A) STM images of the FeO_{2-x} islands after exposure to 10^{-6} mbar CO at 400 K for 4 min. The arrows highlight non-uniform disappearance of the Moiré spots close to the island edge. (B) The FeO_{2-x} surface area normalized to the area in the “as-prepared” sample as a function of the accumulative exposure time. (C) Model of the FeO/FeO_{2-x} interface based on DFT calculations [97].

We also studied morphological changes induced by mixtures of CO and O_2 to link the observed structure and reactivity. We carried out reaction tests in another UHV chamber using a quadrupole mass spectrometer (QMS). Figure 13.14 (red curve) depicts the CO_2 signal recorded on the 0.5 ML FeO_{2-x}/Pt(111) sample in an O-rich mixture consisting of 1×10^{-6} mbar of CO and 5×10^{-6} mbar of O_2. After the crystal was heated up to the reaction temperature (450 K), sustained CO_2 production is observed, suggesting that the prepared surface is catalytically active. The reaction rate stays fairly constant, at least within the first 10 min. In the next experiment, we stopped the oxygen flow after 2 min of the reaction at 450 K (Figure 13.14b, black curve). As expected, the CO_2 production drops to zero and recovers after re-introducing O_2. However, the CO_2 signal does not reach the original level. This is more clearly seen by switching oxygen off and on. The steady-state activity almost linearly decreases with increasing the number of oxygen "pulses". Bearing in mind that FeO_{2-x} islands in pure CO ambient (i.e., when O_2 is off) exhibit progressive reduction propagating from the edge (see Figure 13.13C), which cannot be recovered even in oxygen-rich ambient (not shown here), the observed rate attenuation can readily be explained by decreasing the total perimeter length of the boundary between the oxidized, FeO_{2-x} and reduced, FeO-like phases. Therefore, the results provide compelling evidence that this interface provides the most active sites.

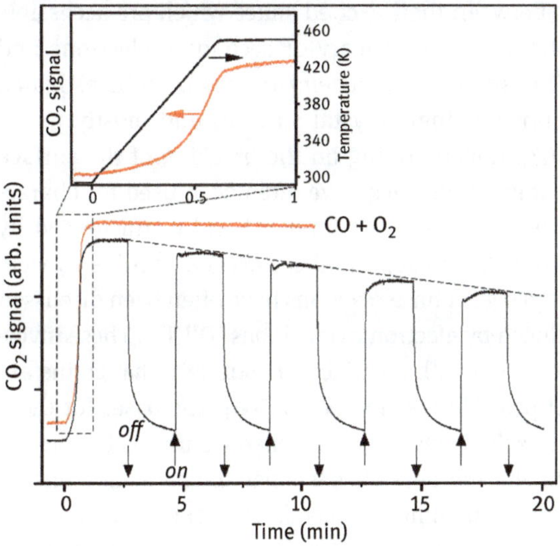

Figure 13.14: CO_2 (44 amu) signal measured by QMS in front of the submonolayer FeO_{2-x} film. The reaction ($CO/O_2 = 1/5$) mixture consisting of 10^{-6} mbar of CO and 5×10^{-6} mbar of O_2. At time zero, the sample was heated up to the reaction temperature 450 K with a rate 2 K s^{-1}. The red curve shows CO_2 production under steady state conditions. Black curve shows CO_2 kinetics upon switching O_2 flow in the mixture off and on, as indicated by the arrows, while keeping the CO pressure constant.

Interestingly, a similar picture has previously been observed by STM on continuous FeO_{2-x} films on Pt(111) treated in the mbar pressure range, where compact domains of FeO_{2-x} were surrounded by areas formed upon reduction with CO [91]. This finding suggests the similar reaction mechanism for the case of a dense FeO film.

This case study demonstrates that ultra-thin oxide films may enhance reactivity of metal catalysts, particularly in oxidation reactions in the low temperature regime, where pure metal catalysts may suffer from site blocking effects and strong chemisorption of reactants. Approach presented here allows us not only to identify the involved structures forming under reaction conditions but also to pinpoint the active site of the reaction for CO oxidation under oxygen rich conditions. A continuously growing body of studies on reactivity of ultrathin oxide films makes us to believe in a rational design of monolayer oxidation catalysts by combining different ultrathin films oxides with different metals, thereby controlling the charge transfer.

13.5 Fourth Case Study: Photochemistry at Nanoparticles

The coupling of molecules to surfaces not only strongly affects their thermal reaction paths, but also the reactions between their excited states which are accessible selectively by irradiation with photons (also primary or secondary electrons and other excitations can be used). This surface photochemistry has been investigated in detail for model systems comprising single crystal surfaces and mostly rather simple adsorbate systems [102, 103]. Concentrating on the breaking of the surface bond or/and internal bonds, desorbing molecules have often been used for characterization. Also the changing coverages have been followed and characterized by some spectroscopy, which makes the observation of other effects accessible as well. The relevant differences to comparable gas phase reactions have often been discussed under the heading "desorption induced by electronic transitions (DIET)". These investigations have shown that the interaction with a surface can strongly change the absolute and relative probabilities of possible reaction paths [104]. The bases for these effects are the changes induced in excited state potential energy surfaces by interaction with the surface, and in particular the possible exchange of charge and energy with the substrate during the excitation lifetime. On metals and semiconductors this leads often to strong, selective quenching of the excited adsorbate states which, because of its sensitive dependence on several adsorbate properties, can select certain pathways and suppress others. But also new channels can open up; in the range of low excitation energies (visible to near ultraviolet light) such channels often consist of transient negative ion (TNI) states [103].

It is a very interesting question which additional modifications are introduced if one uses nanoparticles instead of surfaces of bulk materials. As our earlier review

[13] has summarized, for metallic nanoparticles (MNPs) new electronic excitations – notably the Mie or particle plasmon [21] can be expected to feed photochemical reactions on the metallic nanoparticles. But also the lifetimes and thus the reactivity of excited states may be changed by going to nanoparticles; the main expected effect here is the confinement of excitations in the small particles which would quickly disperse in the volume of bulk materials [13].

In the past 7 years we have examined this range of questions for an apparently simple (though in fact quite complex) system. Since one of the focal points was the influence of Mie plasmons the chosen material was silver. The deposition of silver nanoparticles (Ag NPs) on ultrathin alumina films (on AlNi(110) alloy surfaces) has been studied and their properties have been characterized in detail [105]. The preparation of narrowly defined particle sizes in the range 2–10 nm is possible. These particles possess a strong plasmonic mode at ~3.5 eV, polarized perpendicular to the surface, as probed with photon STM [105] (see Figure 13.15a) and two-photon photoemission (2PPE) [106]. The lower energy mode parallel to the surface is screened by the metal substrate below. The thin alumina film decouples the Ag NPs quite efficiently electronically from the metal substrate without leading to charge-up, so that confinement effects might be expected. This is in contrast to the situation of Ag NPs on strongly reduced TiO_2 surfaces, where plasmon excitation leads to electron-hole pairs in the TiO_2 which decay radiatively [107]. NO was chosen as adsorbate because of the ease of its state-selective detection. Its adsorption on silver surfaces is, however, more complex than usually observed on (transition) metal surfaces. Because of its weak interaction with the noble metal silver the adsorption has to be done below 80 K. In this range, NO dimers are formed on the surface. On Ag(111), the resulting adsorption layers have been characterized in great detail [108] including their photochemistry [109, 110], so that we were able to start on a good basis. The formation of dimers makes the photochemical reaction channels more complex: besides breaking of the ON–NO bond and the bond to the surface simultaneously which leads to NO desorption into the gas phase (and some NO left on the surface), the dimers can also react to $N_2O + O$ which stay on the surface (N_2O can be detected by subsequent thermal desorption; adsorbed O leads to a stronger bond of the NO monomer which desorbs thermally at much higher temperature and has a much smaller photoreaction cross section than the dimer). Even the desorption of (extremely fast) N_2 molecules has been observed under certain conditions [111]. However, NO desorption is the strongest channel and it has been investigated in great detail. Besides measuring yields and desorption cross sections, the energy distributions over the translational [14] and internal modes (rotation, vibration) [112] of the desorbing NO have been measured. The influences of photon energy and polarization, particle size, and laser pulse duration have been investigated as well. In all cases direct comparison to Ag(111) in the same experimental system and with the same methods has been made.

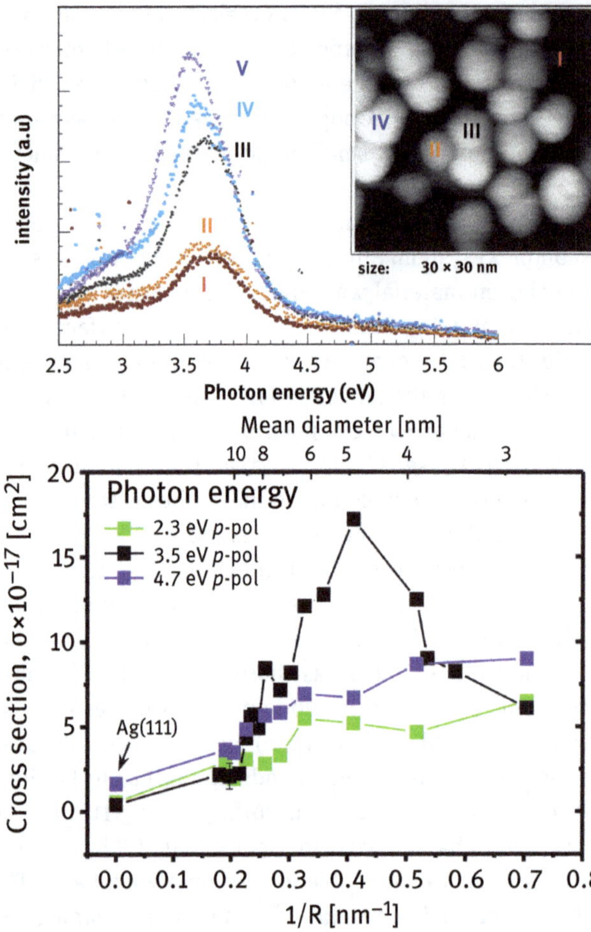

Figure 13.15: (a) Radiation emitted from individual Ag NPs of various sizes, observed with the Photon STM. Insert: STM micrograph of the NP ensemble investigated.[105]. (b) Variation of desorption cross sections, σ, of NO from $(NO)_2$ monolayers on Ag NPs as a function of mean particle size (top abscissa), for the three photon energies given. The plasmon is excited at 3.5 eV. The bottom abscissa gives the inverse radius and emphasizes the approximate scaling off the plasmon resonance [14].

In the following some of the main results are listed:

(1) We found that the cross section, σ, for NO desorption from $(NO)_2$ monolayers on Ag NPs is indeed strongly enhanced by excitation of the plasmon which is known to lie around 3.5 eV (with a weak dependence on NP size [105, 106]). Compared to Ag(111) an enhancement of σ by up to a factor 40 (depending on NP size; see point (3) below) has been found [14].

(2) There is also an enhancement of σ off the plasmon resonance, albeit weaker, which we interpret as due to confinement of excitations, here of hot electrons

in the Ag NP. The photochemical mechanism of all the processes seen is believed to involve TNI states [103, 110] which is consistent with detailed characterizations of the desorbing NO (see (4) below [112]). This means that a hot electron of sufficient energy which is excited in an NP by light absoption tunnels from the NP into an adsorbed $(NO)_2$. In the formed TNI the N-N and N-O bonds are weakened, and the bond to the surface is strengthened by the charge–image charge interaction, so that the corresponding wave packet is put into areas with potential gradients in all dimensions. When the electron jumps back into the NP after a very short residence time a considerable fraction of the molecules has evolved sufficiently along the gradients to lead to breaking of the N–N and the NO–Ag bonds and to translational, rotational, and vibrational excitation of the NO that appears in the gas phase.

(3) This cross section enhancement is also size-dependent (Figure 13.15b [14]). It generally increases with decreasing NP size (diameter d) which we ascribe to confinement of the primary hot electrons in the NP ($\sigma \sim 1/d$, going with the surface/volume ratio S/V; see (2)). For excitation in the plasmon resonance a strong maximum of σ at a certain NP size (here at $d \sim 5$ nm) is found which is explained by two counteracting influences. On the one hand the conversion of plasmons into hot electrons increases with decreasing NP size against the competing radiative decay; on the other hand the total number of photons collected decays with the number of atoms in the particle [14, 112].

(4) On the other hand, the dynamics of the bond dissociations, that is, the motion of the representative wave packet on the potential energy surfaces of ground and excited states for the various molecular entities, turns out to be always the same. This is concluded from the fact that the final state of the desorbing NO stays constant: all energy distributions, translational, rotational, and vibrational, as well as their correlations are identical despite the strong variations of cross sections [14, 112]. These characteristics are well compatible with the proposed TNI mechanism (see (2)). The only exception to this general mechanism is found for excitation with the highest photon energy used (4.7 eV) and the smallest NPs ($d < 4$ nm). For these conditions the mean translational energy of NO was considerably enhanced – in fact a new much faster component was found in addition to the normal one [14, 112]. We have explained this new path by excitation of a transient positive ion of the NO dimer which becomes accessible at an excitation energy sufficient to produce holes in the Ag d-band or direct excitations from a filled adsorbate state to empty states of the NP. That this path becomes observable only for very small NPs is due to the fact that its contribution is proportional to the surface area while the TNI contribution goes down with the volume [14, 112].

(5) All these experiments have been carried out with nanosecond lasers under fluences which lead to linear behavior of the photochemical yield with fluence, that is, the determined cross sections are independent of fluence. This means that the

excitations do not interact, even when they are confined in the NP. Very different behavior is found with femtosecond (~100 fs) lasers of even smaller fluences, for which the photons of a pulse arrive in a much shorter time so that the spatio-temporal photon density is increased by at least a factor 10^4 [15]. Here the cross section is considerably enhanced and increases with fluence, that is, the yield has nonlinear dependence on the fluence. With two-pulse measurements, in which one pulse is split into two and the two are delayed with respect to each other, it is found that the memory of the system is confined to the time of overlapping pulses. So there is a very short-lived confinement of interacting hot electrons. We have explained this by a re-pumping of hot electrons within the same pulse [15]. However, despite this dramatic change of reactivity it is found again that the mechanism of desorption, as indicated by NO final state energies, is unchanged.

To summarize, we find that nanoparticles induce strong increases of photochemical cross sections if they possess strong new excitations, in particular of plasmon type. Also confinement of excitations leads to (more gradual) enhancement of cross sections. If this confinement is also temporally constrained additional, nonlinear effects can be caused. The time scales of the electronic processes (including excitation and decay of plasmons) are very short, probably in the range below 10 fs (our results can only impose an upper limit of 100 fs). The dynamics of evolving molecular states, however, and thus the mechanism, appear essentially unchanged. This makes sense since these processes occur on much longer time scales. It cannot be excluded that there are systems in which the time scales can mix and e.g. the plasmon excitation has a direct photochemical influence. However, we have rarely observed examples for this [113]. We thus consider that these main conclusions will be fairly general and do not depend on any particular system properties.

In order to make a bridge to practical photochemical processes, the main emphasis of the use of NPs should lie on the improved light harvesting made possible by plasmon excitations. The fact that plasmon resonances are fairly narrow, even if ensembles with broad particle size distribution are employed, limits the usable part that can be caught of a broadband source like sun light. If one aims at photochemical reactions which require a certain energy it will be important to use NPs with plasmons in the required energy range. The additional effects offered by confinement will depend more strongly on the particular system since it is strongly influenced by the photochemical mechanism which will not always be describable by the TNI process. Generally, we do not expect that particular new mechanisms open with using NPs. The subject of photochemistry at nanoparticles is still in its infancy and deserves to be investigated more fully on a broader front, both for model systems and practical setups.

13.6 Fifth Case Study: Water Formation in Confined Space

Carrying out chemical reactions by performing them in confined spaces has fascinated chemists for a long time [114–118]. Confinement occurs, for example, inside the voids of zeolite materials that are well-known to be active for some chemical reactions due to their acidic reactive sites [119–121]. Another important field, where reactions in confined space may have played a decisive role is the area of origin of life. In this context, it is important to consider the mechanism by which confinement can influence a reaction in order to reach a rational understanding. In general, the physical confinement of reactants and products can affect the reaction kinetics by changing the transition state or the reaction path, particularly of the rate-determining step.

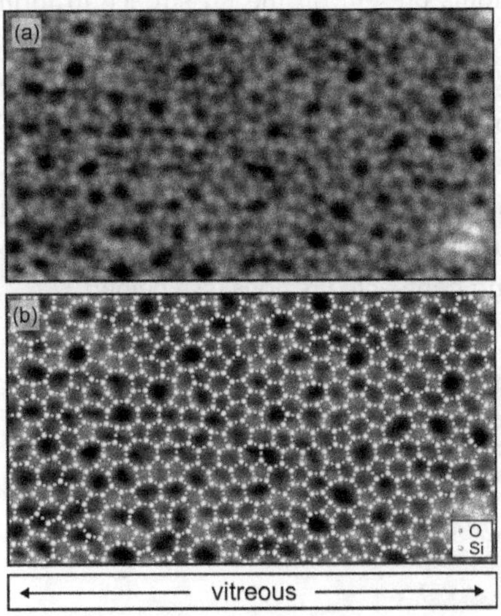

Figure 13.16: STM image of a silica film grown on Ru(0001) (a) image as recorded 12.3 × 7 nm^2, $V_s = 2$ V, $I_T = 100$ pA (b) atomic model indicating ring size in color.

This model system is based upon a thin silica film only attached via dispersive forces to a Ru(0001) single crystal surface. This setup leaves a confined space of nanometer scale between the film and the metal surface as indicated in Figure 13.1f. Figure 13.16 provides an STM image of the film as viewed from the top indicating that the film comes in a crystalline form, where SiO_4 tetrahedra are edge connected to form two hexagonal ring structures, which are directly connected via Si–O–Si bonds, shared by a tetrahedral edge in the top and the bottom layer. The film also comes in a vitreous form, in which the ring sizes vary between five- and nine-membered rings. We are currently looking into the transformation between the crystalline and the vitreous morphologies. This, however, is not the topic of the case study considered here. The scenario for the confined reaction is schematically represented in Figure 13.17. The figure schematically shows at the top the Ru(0001) surface decorated with atomic oxygen up to a well-defined coverage, depending on the preparation conditions. Molecular oxygen intercalates, and spontaneously dissociates to form a well-ordered oxygen adsorbate state on Ru, achieving different final coverages depending on the exposure time. On top of such an oxygen-covered Ru(0001) surface, the silica layer as described above is located as indicated schematically. Here the crystalline structure has been chosen for illustration and one recognizes the above-mentioned Si–O–Si bonds between the two hexagonal edge-shared SiO_4 layers. The stoichiometry is perfect SiO_2. We note that the layer is extremely stable against water and hydroxylation can only be achieved by electron bombardment in the presence of an ice film, and even in this case, the surface concentration of –OH groups has proven to be quite low [122, 123]. In the next step of the reaction, H_2 is dosed, intercalates into the confined space and, after dissociation on Ru(0001)-free sites, reacts with oxygen atoms present on the Ru surface to form water, which desorbs from Ru under the experimental conditions used. In the experiments we follow the exposure to molecular hydrogen at 10^{-6} mbar at different temperatures, using a low energy electron microscope (LEEM), which is part of the SMART setup operating at the UE49-PGM beam line at the synchrotron light source BESSY II. This setup allows us not only to follow with LEEM in real-time chemical processes on surfaces but also to record local x-ray photoelectron spectra on different areas of inhomogeneous samples with coexisting reacted and nonreacted regions (see below).

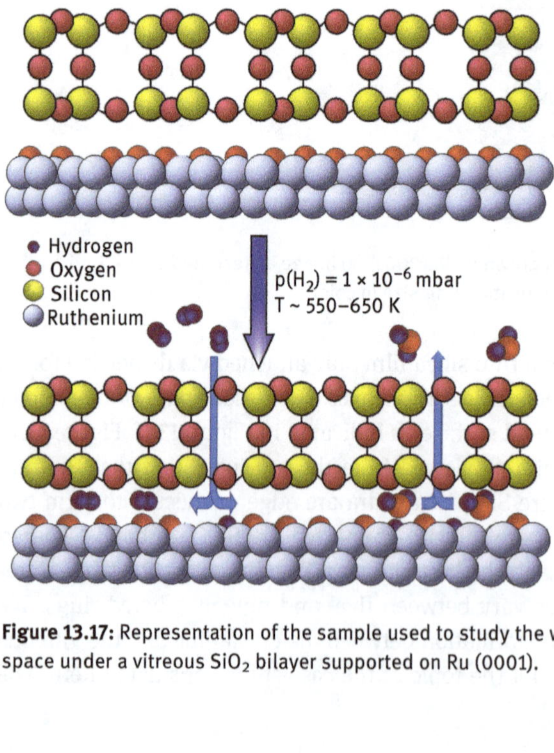

Figure 13.17: Representation of the sample used to study the water formation reaction in confined space under a vitreous SiO_2 bilayer supported on Ru (0001).

During the annealing in hydrogen atmosphere, a reaction front characterized by an abrupt change in intensity was seen moving across the sample surface. Figure 13.18 displays the time evolution of the front propagation. Real time LEEM measurements performed on different areas of the sample (not shown) suggest that the front propagation does not follow a preferential direction on the surface, though the presence of step bunches on the Ru(0001) substrate can affect its movement (see arrows in Figure 13.18).

By stopping the progression of the reaction front it was possible to characterize the nature of the intensity change by means of local LEEM-IV and x-ray photoemission electron spectroscopy (XPS) measurements. Results are presented in Figure 13.18 (for details, see Prieto et al.) [16]. A detailed description of the working principle of LEEM-IV is outside the scope of this publication, so the readers are referred to specific literature on this topic [124–126].

Local photoemission spectra of Si 2p and O1s lines, collected on both sides of the reaction front revealed the changes produced by the treatment. For instance, analysis of the O1s line reveals (Figure 13.18) that the component at lower binding energies, ascribed to interfacial oxygen adsorbed on the Ru substrate (O_{Ru}) under the silica film, follows the change in the work function observed in LEEM-IV measurements, both indicating the removal of O_{Ru} during the annealing treatment. In this context, the work function change due to the removal of O_{Ru} is the main of contrast variation across the front. The results in the confined space situation may be directly compared with the situation with the silica film removed, where the formation of the a similar reaction front is observed, thus allowing us to exclude any contribution in contrast formation due to changes in the SiO_2–Ru(0001) distance. Koch et al. [127] concluded from kinetic analysis based on integral work function measurements that, in the water formation reaction, the first $O_{ads} + H_{ads}$ combination step is rate limiting, with very low H_2 and H_2O transient coverages on bare Ru under reaction conditions (H_2 co-adsorbed with O desorbs completely at 400 K [128] and H_2O at 220 K [129]). Moreover, the authors were able to determine that the reaction propagates by the emptying O sites, which are subsequently required for further H_2 dissociative adsorption. This finding, derived from an unexpected dependence on starting O_{Ru} coverage and non-equilibrium effects, gives a reasonable explanation for the formation of reaction fronts on both silica-covered and -uncovered Ru(0001).

The evaluation of the temperature dependence of the front velocity (v_{front}) reveals the apparent activation energy (E_a^{app}) of the water formation reaction. The values are presented in the Arrhenius plot in Figure 13.19. The Arrhenius analysis yields an E_a^{app} for the confined reaction under SiO_2 of 0.27 ± 0.02 eV. On the other hand, the corresponding value for the non-confined water formation on a bare Ru (0001) surface yields a value of 0.59 ± 0.04 eV for the E_a^{app}, in excellent agreement with the value reported by Koch et al. [127] for the reaction performed on basal and stepped Ru surfaces, although with a lower starting O_{Ru} coverage.

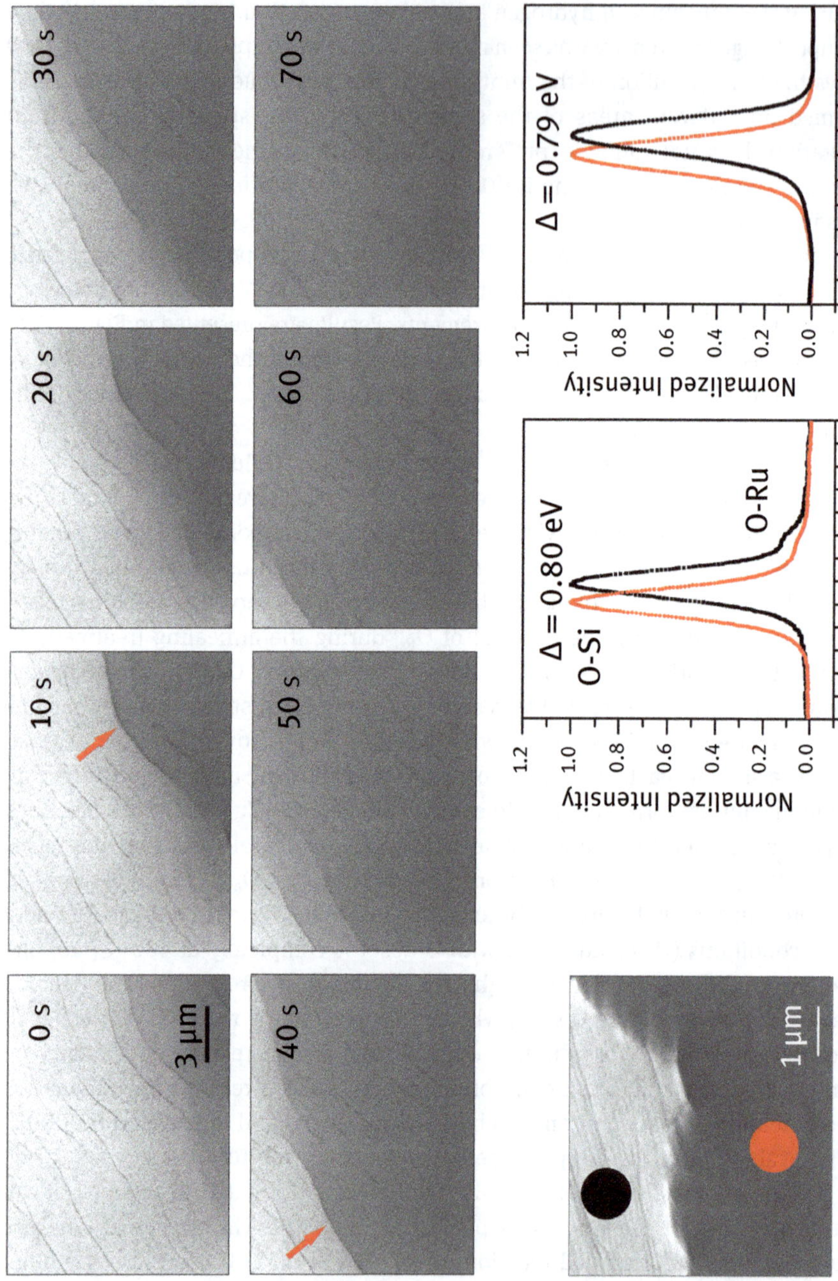

Figure 13.18: Snapshots of the reaction front observed during annealing in 1×10^{-6} mbar H_2 at 550 K. $E_{kin} = 14$ eV.

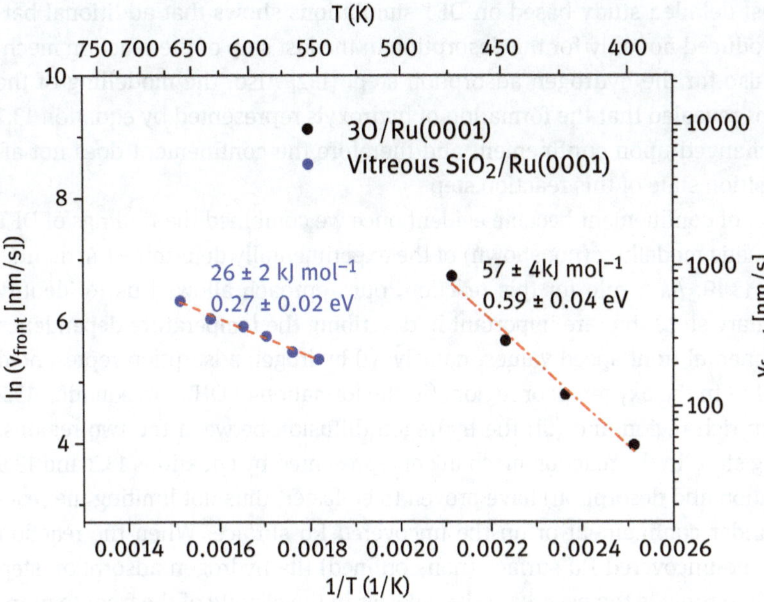

Figure 13.19: Arrhenius plots of the temperature-dependent velocities of the H_2 oxidation front on SiO_2/Ru(0001) (blue dots) and 3O/Ru(0001) (black dots).

In general, the following elementary reaction steps can lead to water formation on Ru(0001), with both reaction direction being rather fast under our experimental conditions.

$$H_{2(g)} \rightleftharpoons 2\,H_{(ads)} \tag{13.1}$$

$$H_{(ads)} + O_{(ads)} \rightleftharpoons OH_{(ads)} \tag{13.2}$$

$$OH_{(ads)} + H_{(ads)} \rightleftharpoons H_2O_{(ads)} \tag{13.3}$$

$$2\,OH_{(ads)} \rightleftharpoons H_2O_{(ads)} + O_{(ads)} \tag{13.4}$$

$$H_2O_{(ads)} \rightleftharpoons H_2O_{(g)} \tag{13.5}$$

We emphasize that, under our experimental conditions, both directions of each step are fast.

On bare Ru(0001), reactions (13.1)–(13.3) and 13.5 were identified as part of the mechanism, with the first H + O combination being the rate determining step [127, 130]. Although a SiO2 BL can be penetrated both by molecular hydrogen (see [131]) and water molecules, permeation is a thermally activated process. Therefore, it might be reasonable to assume that their permeation to or away from the confined space can be strongly affected by the presence of the silica cover.

Our latest detailed study based on DFT simulations shows that additional barriers are introduced not only for the desorption in the last step of the reaction mechanism, but also for the hydrogen adsorption step. [132] Also, the modelling of the reaction steps revealed that the formation of hydroxyls represented by equation 13.2 remains unchanged upon confinement and therefore the confinement does not affect the transition state of this reaction step.

The effect of confinement became evident once we combined the findings of DFT with microkinetic modelling (not shown) of the experimentally determined Arrhenius plots in Fig. 13.19. As a rule for this reaction, our approach allowed us to identify three elementary steps that are important in describing the temperature dependence of the experimental front speed values, namely: (i) hydrogen adsorption represented by equation 13.1 in the oxygen poor region, (ii) the formation of OH_{ads} in equation 13.2 in the oxygen rich region and (iii) the hydrogen diffusion between the two regions. The following steps in the reaction mechanism represented by equations 13.3 and 13.5 (water formation and desorption) have proven to be faster, thus not limiting the reaction speed under confinement or on the uncovered Ru surface. When the reaction proceeds on the uncovered Ru surface (non-confined) the hydrogen adsorption step is fast enough to provide the necessary H_{ads} species in the vicinity of the front to propagate it. Under these circumstances, the reaction rate is determined solely by the formation of OH on bare ruthenium, in agreement with previous findings. On the other hand, under equivalent experimental conditions, when the silica lid is added to the system its presence does not affect the equilibrium concentration of adsorbed hydrogen but affects how fast this coverage is attained at a given temperature and hydrogen pressure. Therefore, if this process happens to be slower than the formation of OH_{ads}, the hydrogen density (coverage) must be the limiting factor for the propagation of the reaction front because then H needs to be transported from a larger distance towards the O-front. As for water accumulation in the confined space, our calculations show that even though water permeation through the silica film is indeed a thermally activated process, its coverage on the Ru surface remains significantly too low to contribute through the comproportionation path (back reaction in 13.4) to the reaction kinetics. A detailed description of the phenomenon is not presented here but readers are referred to our complete study of the important variables [132].

In conclusion, we have shown that confinement can influence the apparent activation energy of a given chemical reaction by either modifying existing elementary steps of the reaction or by adding new ones due to the confinement of the reaction product. Our simulation studies indicate that confinement under a silica bilayer can act in two of the three ways described by Csicsery [133], either by preventing that reactant can diffuse and access the active site or by preventing products from leaving the reaction site. The balance between these two effects will ultimately determine the kinetics of the water formation reaction under confinement. In this sense, the scarcity of adsorbed hydrogen at low H2 pressures (10^{-6} mbar range) becomes relevant for the spatiotemporal propagation of the reaction cascade, whereas at high H2 pressures (near ambient pressure – NAP – conditions) the entrapment of water molecules H_2O blocks the active sites necessary for H adsorption and diffusion.

13.7 Synopsis

Via five case studies this chapter has demonstrated how model studies on complex materials related to heterogeneous catalysis may help to unravel an atomistic view at processes at solid–vacuum, solid–gas, and solid–liquid interfaces. Going beyond metal single crystal surfaces towards modelling some of the real complexity of catalytic materials is an important step forward to help designing energy-efficient catalysts by providing information on principles more than on specific systems. We are approaching a situation where interplay between work on real catalytic materials and work on specifically designed model catalysts may lead to a detailed understanding of the relation between structure-morphology and catalytic activity and selectivity.

References

[1] Renn J, Schlögl R, Zenner H-P. Herausforderung Energie. Max Planck Research Library for the History and Development of Knowledge, Proceedings 1. Berlin, 2011.

[2] Behrens M, Schlögl R. Energie ist Chemie – Katalyse als Schlüsseltechnik. Energie von Morgen: Eine Momentaufnahme. In: Herausforderung Energie Renn J, Schlögl R, Zenner H-P, Eds. Max Planck Research Library for the History and Development of Knowledge, Proceedings 1, Berlin: 2011.

[3] Freund HJ. Model Studies in Heterogeneous Catalysis. Chem Eur J. 2010;16(31):9384–97.

[4] Shaikhutdinov S, Freund H-J. Ultrathin Oxide Films on Metal Supports: Structure-Reactivity Relations. Annu Rev Phys Chem. 2012;63(1):619–33.

[5] Freund H-J, Pacchioni G. Oxide Ultra-Thin Films on Metals: New Materials for the Design of Supported Metal Catalysts. Chem Soc Rev. 2008;37(10):2224–42.

[6] Freund HJ. Clusters and Islands on Oxides: From Catalysis via Electronics and Magnetism to Optics. Surf Sci. 2002;500(1-3):271–99.

[7] Duke CB. Surface science: the first thirty years. Surf Sci. 1994;299/300:vii–viii.

[8] Freund H-J, Goodman DW. Ultrathin Oxide Films. In: Handbook of Heterogeneous Catalysis Ertl G, Knözinger H, Schüth F, Weitkamp J, Eds. 2. Weinheim: Wiley-VCH Verlagsgesellschaft mbH; 2008. Vol. 3, 1309–38.

[9] Shao X, Prada S, Giordano L, Pacchioni G, Nilius N, Freund H-J. Tailoring the shape of Metal Ad-Particles by doping the oxide support. Angew Chem Int Ed. 2011;50(48):11525–27.

[10] Wang H-F, Ariga H, Dowler R, Sterrer M, Freund H-J. Surface Science Approach to Catalyst Preparation – Pd Deposition onto thin Fe3O4(111) Films from PdCl2 Precursor. J Catal. 2012;286:1–5.

[11] Sun YN, Qin ZH, Lewandowski M, Carrasco E, Sterrer M, Shaikhutdinov S, Freund HJ. Monolayer iron oxide film on platinum promotes low temperature CO oxidation. J Catal. 2009;266(2):359–68.

[12] Sun Y-N, Giordano L, Goniakowski J, Lewandowski M, Qin Z-H, Noguera C, Shaikhutdinov S, Pacchioni G, Freund H-J. The Interplay between Structure and CO Oxidation Catalysis on Metal-Supported Ultrathin Oxide Films. Angew Chem Int Ed. 2010;49(26):4418–21.
[13] Watanabe K, Menzel D, Nilius N, Freund H-J. Photochemistry on metal nanoparticles. Chem Rev. 2006;106(10):4301–20.
[14] Mulugeta D, Kim KH, Watanabe K, Menzel D, Freund HJ. Size effects in thermal and photochemistry of (NO)(2) on Ag nanoparticles. Phys Rev Lett. 2008;101(14):146103.
[15] Kim KH, Watanabe K, Mulugeta D, Freund H-J, Menzel D. Enhanced photoinduced desorption from metal nanoparticles by photoexcitation of confined hot electrons using femtosecond laser pulses. Phys Rev Lett. 2011;107(4):047401.
[16] Prieto MJ, Klemm HW, Xiong F, Gottlob DM, Menzel D, Schmidt T, Freund HJ. Water formation under silica thin films: real-time observation of a chemical reaction in a physically confined space. Angew Chem Int Ed Engl. 2018;57(28):8749–53.
[17] Molina LM, Hammer B. Some recent theoretical advances in the understanding of the catalytic activity of Au. Appl Catal A-Gen. 2005;291(1-2):21–31.
[18] Green IX, Tang WJ, Neurock M, Yates JT. Spectroscopic observation of dual catalytic sites during oxidation of CO on a Au/TiO2 catalyst. Science. 2011;333(6043):736–39.
[19] Abbet S, Heiz U, Hakkinen H, Landman U. CO oxidation on a single Pd atom supported on magnesia. Phys Rev Lett. 2001;86(26):5950–53.
[20] Ekardt W. Metal Clusters. John Wiley; Chichester: 1999.
[21] Kreibig U, Vollmer W. Optical Properties of Metal Clusters. Berlin-New York: Springer-Verlag; 1995, Vol. 25.
[22] Risse T, Shaikhutdinov S, Nilius N, Sterrer M, Freund HJ. Gold supported on thin oxide films: from single atoms to nanoparticles. Acc Chem Res. 2008;41(8):949–56.
[23] Hashmi AS, Hutchings GJ. Gold catalysis. Angew Chem Int Ed Engl. 2006;45(47):7896–936.
[24] Herzing AA, Kiely CJ, Carley AF, Landon P, Hutchings GJ. Identification of active gold nanoclusters on iron oxide supports for CO oxidation. Science. 2008;321(5894):1331–35.
[25] Valden M, Lai X, Goodman DW. Onset of catalytic activity of gold clusters on titania with the appearance of nonmetallic properties. Science. 1998;281(5383):1647–50.
[26] Chen MS, Goodman DW. The structure of catalytically active Au on titania. Science. 2004;306:252–55.
[27] Mostafa S, Behafarid F, Croy JR, Ono LK, Li L, Yang JC, Frenkel AI, Cuenya BR. Shape-dependent catalytic properties of Pt nanoparticles. J Am Chem Soc. 2010;132(44):15714–19.
[28] Benedetti S, Myrach P, Di Bona A, Valeri S, Nilius N, Freund H-J. Growth and morphology of metal particles on MgO/Mo(001): A comparative STM and diffraction study. Phys Rev B. 2011;83(12):125423-1-10.
[29] Benia HM, Nilius N, Freund HJ. Effect of electromagnetic interactions on plasmon excitations in silver particle ensembles. Surf Sci. 2006;600(10):128–33.
[30] Brown MA, Fujimori Y, Ringleb F, Shao X, Stavale F, Nilius N, Sterrer M, Freund H-J. Oxidation of Au by Surface OH: Nucleation and Electronic Structure of Gold on Hydrohylated MgO(001). J Amer Chem Soc. 2011;133(27):10668–76.
[31] Goniakowski J, Noguera C. Electronic states and schottky barrier height at Metal/MgO(100) Interfaces. Interface Sci. 2004;12(1):93–103.
[32] Goniakowski J, Finocchi C, Noguera C. Polarity of oxide surfaces and nanostructures. Rep Progr Phys. 2008;71:016501-1-55.
[33] Koplitz LV, Dulub O, Diebold U, Xe STM. Study of Copper Growth on ZnO(0001)–Zn and ZnO (0001¯)–O Surfaces. J Phys Chem B. 2003;107(38):10583–90.
[34] Pacchioni G, Giordano L, Baistrocchi M. Charging of Metal Atoms on Ultrathin MgO/Mo(100) Films. Phys Rev Lett. 2005;94(22):226104.

[35] Sterrer M, Risse T, Heyde M, Rust H-P, Freund H-J. Crossover from three-dimensional to two-dimensional geometries of au nanostructures on thin MgO(001) Films: A confirmation of theoretical predictions. Phys Rev Lett. 2007;98(20):206103.

[36] Ricci D, Bongiorno A, Pacchioni G, Landman U. Bonding trends and dimensionality crossover of gold nanoclusters on Metal-supported MgO thin films. Phys Rev Lett. 2006;97(3):036106.

[37] Nilius N, Ganduglia-Pirovano MV, Brázdová V, Kulawik M, Sauer J, Freund HJ. Counting electrons transferred through a thin alumina film into Au chains. Phys Rev Lett. 2008;100 (9):096802.

[38] Lin X, Nilius N, Freund HJ, Walter M, Frondelius P, Honkala K, Häkkinen H. Quantum well states in two-dimensional gold clusters on MgO thin films. Phys Rev Lett. 2009;102 (20):206801.

[39] Lin X, Nilius N, Sterrer M, Koskinen P, Häkkinen H, Freund H-J. Characterizing low-coordinated atoms at the periphery of MgO-supported Au islands using scanning tunneling microscopy and electronic structure calculations. Phys Rev B. 2010;81(15):153406.

[40] Lin X, Yang B, Benia HM, Myrach P, Yulikov M, Aumer A, Brown M, Sterrer M, Bondarchuk O, Kieseritzky E, Rocker J, Risse T, Gao H, Nilius N, Freund HJ. Charge-mediated adsorption behavior of CO on MgO-Supported Au clusters. J Am Chem Soc. 2010;132(22):7745–49.

[41] Frondelius P, Häkkinen H, Honkala K. Formation of Gold(I) edge oxide at flat gold nanoclusters on an ultrathin MgO film under ambient conditions. Angew Chem Int Ed. 2010;49(43):7913–16.

[42] Pacchioni G. Ab initio theory of point defects in oxide materials: structure, properties, chemical reactivity. Solid State Sci. 2000;2(2):161–79.

[43] McKenna KP, Shluger A. Electron-trapping polycrystalline materials with negative electron affinity. Nat Mater. 2008;7:859–62.

[44] Benia HM, Myrach P, Gonchar A, Risse T, Nilius M, Freund HJ. Electron trapping in misfit dislocations of MgO thin films. Phys Rev B. 2010;81:241415.

[45] Sterrer M, Yulikov M, Fischbach E, Heyde M, Rust H-P, Pacchioni G, Risse T, Freund H-J. Interaction of Gold Clusters with Color Centers on MgO(001) Films. Angew Chem Int Ed. 2006;45(16):2630–32.

[46] Wang JX, Lunsford JH. Characterization of [Li+O] centers in lithium-doped magnesium oxide catalysts. J Phys Chem. 1986;90(22):5883–87.

[47] Ito T, Wang J, Lin CH, Lunsford JH. Oxidative dimerization of methane over a lithium-promoted magnesium oxide catalyst. J Am Chem Soc. 1985;107(18):5062–68.

[48] Rodriguez JA, Hanson JC, Kim J-Y, Liu G, Iglesias-Juez A, Fernández-García M. Properties of CeO_2 and $Ce_{1-x}ZrxO_2$ Nanoparticles: X-ray Absorption Near-Edge Spectroscopy, Density Functional, and Time-Resolved X-ray Diffraction Studies. J Phys Chem B. 2003;107(15): 3535–43.

[49] Nambu A, Graciani J, Rodriguez JA, Wu Q, Fujita E, Sanz JF. N doping of TiO_2(110): Photoemission and density-functional studies. J Chem Phys. 2006;125(9):094706.

[50] Ganesh R, Pala S, Metiu H. The Structure and Energy of Oxygen Vacancy Formation in Clean and Doped, Very Thin Films of ZnO. J Phys Chem C. 2007;111(34):12715–22.

[51] Nolan M, Verdugo VS, Metiu H. Vacancy formation and CO adsorption on gold-doped ceria surfaces. Surf Sci. 2008;602(16):2734–42.

[52] Stavale F, Nilius N, Freund HJ. Cathodoluminescence of near-surface centres in Cr-doped MgO (001) thin films probed by scanning tunnelling microscopy. New J Phys. 2012;14.

[53] Henderson B, Imbusch GF. Optical Spectroscopy of Inorganic Solids. Oxford: Oxford University Press: 1989.

[54] Hu Z, Li B, Sun X, Metiu H. Chemistry of Doped Oxides: The Activation of Surface Oxygen and the Chemical Compensation Effect. J Phys Chem C. 2011;115(7):3065–74.

[55] Myrach P, Nilius N, Levchenko SV, Gonchar A, Risse T, Dinse K-P, Boatner LA, Frandsen W, Horn R, Freund H-J, Schlögl R, Scheffler M. Temperature-Dependent Morphology, Magnetic and Optical Properties of Li-Doped MgO. Chemcatchem. 2010;2(7):854–62.
[56] Stiehler C, Calaza F, Schneider W-D, Nilius N, Freund H-J. Molecular Adsorption Changes the Quantum Structure of Oxide-Supported Gold Nanoparticles: Chemisorption versus Physisorption. Phys Rev Lett. 2015;115(3):036804.
[57] Stiehler C, Calaza F, Sterrer M, Schneider W-D, Nilius N, Freund H-J, to be published.
[58] Freund H-J, Roberts MW. Surface Chemistry of Carbon Dioxide. Surf Sci Rep. 1996;25(8): 225–73.
[59] Solymosi F. The bonding, structure and reactions of CO_2 adsorbed on clean and promoted metal surfaces. J Mol Catal. 1991;65(3):337–58.
[60] Abbet S, Riedo E, Brune H, Heiz U, Ferrari AM, Giordano L, Pacchioni G. Identification of defect sites on MgO(100) thin films by decoration with Pd atoms and studying CO adsorption properties. J Am Chem Soc. 2001;123(25):6172–78.
[61] Campbell CT. Ultrathin metal films and particles on oxide surfaces: structural, electronic and chemisorption properties. Surf Sci Rep. 1997;27:1–111.
[62] Fu Q, Wagner T. Interaction of nanostructured metal overlayers with oxide surfaces. Surf Sci Rep. 2007;62:431–98.
[63] Ertl G, Knözinger H, Schueth F, Weitkamp J. Handbook of Heterogeneous Catalysis, 2., compl. rev. and enlarged ed. VCH: Weinheim, 2008.
[64] De Jong KP. Synthesis of Solid Catalysts. Wiley-VCH; Weinheim: 2009.
[65] Regalbuto JR, Navada A, Shadid S, Bricker ML, Chen Q. An experimental Verification of the physical nature of Pt adsorption onto alumina. J Catal. 1999;184:335–48.
[66] Weiss W, Ritter M. Metal oxide heteroepitaxy: Stranski-Krastanov growth of iron oxides on Pt (111). Phys Rev B. 1999;59:5201–13.
[67] Sterrer M, Freund HJ. Surface Science Approach to Catalyst Preparation Using Thin Oxide Films as Substrates. In: Encyclopedia of Interfacial Chemistry Wandelt K, Ed. Elsevier: Oxford; 2018. 632–42.
[68] Hao X, Spieker WA, Regalbuto JR. A further simplification of the revised physical adsorption (RPA) model. J Coll Interface Sci. 2003;267:259–64.
[69] Olsbye U, Wendelbo R, Akporiaye D. Stucy of Pt/alumina catalyst preparation. Appl Catal A. 1997;152:127–41.
[70] Spieker WA, Regalbuto JR. A fundamental model of platinum impregnation onto alumina. Chem Eng Sci. 2001;56:3491–504.
[71] Bozon-Verduraz F, Omar A, Escard J, Pontvianne B. J Catal. 1978;53:126.
[72] Fleisch TH, Hicks RF, Bell AT. J Catal. 1984;87:398.
[73] Wang H-F, Kaden WE, Dowler R, Sterrer M, Freund H-J. Model oxide-supported metal catalysts – comparison of ultrahigh vacuum and solution based preparation of Pd nanoparticles on a single-crystalline oxide substrate. Phys Chem Chem Phys. 2012;14(32): 11525–33.
[74] Tauster SJ. Strong metal-support interactions. Acc Chem Res. 1987;20(11):389–94.
[75] Tauster SJ, Fung SC, Garten RL. Strong metal-support interactions. Group 8 noble metals supported on titanium dioxide. J Am Chem Soc. 1978;100(1):170–75.
[76] Ko EI, Garten RL. Ethane hydrogenolysis studies of TiO2-supported group VIII metal catalysts. J Catal. 1981;68(1):233–36.
[77] Haller GL, Resasco DE. Metal-support interactions between Group VIII metals and reducible oxides. Adv Catal. 1989;36:173.

[78] Bernal S, Calvino JJ, Cauqui MA, Gatica JM, Larese C, Pérez Omil JA, Pintado JM. Some recent results on metal/support interaction effects in NM/CeO2 (NM: noble metal) catalysts. Catal Today. 1999;50(2):175–206.

[79] Dulub O, Hebenstreit W, Diebold U. Imaging Cluster Surfaces with Atomic Resolution: The Strong Metal-Support Interaction State of Pt Supported on TiO_2(110). Phys Rev Lett. 2000;84 (16):3646–49.

[80] Bowker M, Stone P, Morrall P, Smith R, Bennett R, Perkins N, Kvon R, Pang C, Fourre E, Hall M. Model catalyst studies of the strong metal–support interaction: Surface structure identified by STM on Pd nanoparticles on TiO2(110). J Catal. 2005;234(1):172–81.

[81] Barcaro G, Agnoli S, Sedona F, Rizzi GA, Fortunelli A, Granozzi G. Structure of Reduced Ultrathin TiOx Polar Films on Pt(111). J Phys Chem C. 2009;113(14):5721–29.

[82] Netzer FP, Allegretti F, Surnev S. Low-dimensional oxide nanostructures on metals: Hybrid systems with novel properties. J Vac Sci Technol B. 2010;28(1):1–16.

[83] Qin ZH, Lewandowski M, Sun YN, Shaikhutdinov S, Freund HJ. Encapsulation of Pt Nanoparticles as a Result of Strong Metal-Support Interaction with Fe_3O_4(111). J Phys Chem C. 2008;112(27):10209–13.

[84] Qin ZH, Lewandowski M, Sun YN, Shaikhutdinov S, Freund HJ. Morphology and CO adsorption on platinum supported on thin Fe_3O_4(111) films. J Phys Condens Matter. 2009;21(13):134019.

[85] Weiss W, Ranke W. Surface chemistry and catalysis on well-defined epitaxial iron-oxide layers. Prog Surf Sci. 2002;70(1–3):1–151.

[86] Shaikhutdinov S, Ritter M, Weiss W. Hexagonal heterolayers on a square lattice: Acombined STM and LEED study of FeO(111) on Pt(100). Phys Rev B. 2000;62:7535–41.

[87] Hansen KH, Worren T, Stempel S, Lægsgaard E, Bäumer M, Freund HJ, Besenbacher F, Stensgaard I. Palladium Nanocrystals on Al_2O_3: Structure and Adhesion Energy. Phys Rev Lett. 1999;83(20):4120–23.

[88] Sun YN, Qin ZH, Lewandowski M, Shaikhutdinov S, Freund HJ. CO adsorption and dissociation on iron oxide supported Pt particles. Surf Sci. 2009;603(20):3099–103.

[89] Willinger MG, Zhang W, Bondarchuk O, Shaikhutdinov S, Freund H-J, Schlögl R. A Case of Strong Metal–Support Interactions: Combining Advanced Microscopy and Model Systems to Elucidate the Atomic Structure of Interfaces. Angew Chem Int Ed. 2014;53(23):5998–6001.

[90] Lewandowski M, Sun YN, Qin ZH, Shaikhutdinov S, Freund HJ. Promotional effect of metal encapsulation on reactivity of iron oxide supported Pt catalysts. Appl Catal, A. 2011;391(1): 407–10.

[91] Lewandowski M, Groot IMN, Shaikhutdinov S, Freund H-J. Scanning tunneling microscopy evidence for the Mars-van Krevelen type mechanism of low temperature CO oxidation on an FeO(111) film on Pt(111). Catal Today. 2012;181(1):52–55.

[92] Lei Y, Lewandowski M, Sun Y-N, Fujimori Y, Martynova Y, Groot IMN, Meyer RJ, Giordano L, Pacchioni G, Goniakowski J, Noguera C, Shaikhutdinov S, Freund H-J. CO+NO versus CO+O_2 Reaction on Monolayer FeO(111) Films on Pt(111). Chemcatchem. 2011;3(4):671–74.

[93] Fu Q, Li W-X, Yao Y, Hongyang L, Fu Q, Su H-Y, Ma D, Gu X-K, Chen L, Wang Z, Zhang H, Wang B, Bao X. Interface-Confined Ferrous Centers for Catalytic Oxidation. Science. 2010;328 (5982):1141–44.

[94] Kudernatsch W, Peng G, Zeuthen H, Bai Y, Merte LR, Lammich L, Besenbacher F, Mavrikakis M, Wendt S. Direct Visualization of Catalytically Active Sites at the FeO–Pt(111) Interface. ACS Nano. 2015;9(8):7804–14.

[95] Sun D, Gu X-K, Ouyang R, Su H-Y, Fu Q, Bao X, Li W-X. Theoretical study of the role of a metal-cation ensemble at the oxide-metal boundary on CO oxidation. J Phys Chem C. 2012;116: 7491–98.

[96] Fu Q, Yao Y, Guo X, Wei M, Ning Y, Liu H, Yang F, Liu Z, Bao X. Reversible structural transformation of FeOx nanostructures on Pt under cycling redox conditions and its effect on oxidation catalysis. Phys Chem Chem Phys. 2013;15(35):14708–14.
[97] Pan Q, Weng X, Chen M, Giordano L, Pacchioni G, Noguera C, Goniakowski J, Shaikhutdinov S, Freund H-J. Enhanced CO Oxidation on the Oxide/Metal Interface: From Ultra-High Vacuum to Near-Atmospheric Pressures. Chemcatchem. 2015;7(17):2620–27.
[98] Merte LM, Heard CJ, Zhang F, Choi J, Gustafson J, Weaver JF, Grönbeck H, Lundgren E. Angew Chem Int Ed. 2016;55:9267–71.
[99] Johansson N, Merte LR, Grånäs E, Wendt S, Andersen JN, Schnadt J, Knudsen J. Top Catal. 2016;59:506–15.
[100] Zhang K, Li L, Shaikhutdinov S, Freund H-J. Carbon Monoxide Oxidation on Metal-Supported Monolayer Oxide Films: Establishing Which Interface is Active. Angew Chem Int Ed. 2018;57: 1261–65.
[101] Giordano L, Lewandowski M, Groot IMN, Sun YN, Goniakowski J, Noguera C, Shaikhutdinov S, Pacchioni G, Freund HJ. Oxygen-Induced Transformations of an FeO(111) Film on Pt(111): A Combined DFT and STM. J Phys Chem C. 2010;114(49):21504–09.
[102] Dai H-L, Ho W. Laser Spectroscopy and Photochemistry on Metal Surfaces. Singapore: World Scientific: 1995.
[103] Zimmermann FM, Ho W. State resolved studies of photochemical dynamics at surfaces. Surf Sci Rep. 1995;22(4-6):127–247.
[104] Menzel D. Electronically induced surface chemistry: localised bond breaking versus delocalisation. Surf Interface Anal. 2006;38(12-13):1702–11.
[105] Nilius N, Ernst N, Freund HJ. Photon emission spectroscopy of individual oxide-supported silver clusters in a scanning tunneling microscope. Phys Rev Lett. 2000;84(17):3994–97.
[106] Evers F, Rakete C, Watanabe K, Menzel D, Freund H-J. Two-photon photoemission from silver nanoparticles on thin alumina films: Role of plasmon excitation. Surf Sci. 2005;593(1-3): 43–48.
[107] Nilius N, Ernst N, Freund H-J. On energy transfer processes at cluster-oxide interfaces: silver on titania. Chem Phys Lett. 2001;349:351–57.
[108] Carlisle CI, King DA. Direct Molecular Imaging of NO Monomers and Dimers and a Surface Reaction on Ag(111). J Phys Chem B. 2001;105(18):3886–93.
[109] Vondrak T, Burke DJ, Meech SR. The dynamics and origin of NO photodesorbed from NO/Ag (111). Chem Phys Lett. 2000;327(3-4):137–42.
[110] So SK, Franchy R, Ho W. Photodesorption of NO from Ag(111) and Cu(111). J Chem Phys. 1991;95(2):1385–99.
[111] Kim KH, Watanabe K, Menzel D, Freund H-J. Photoinduced Abstraction Reactions within NO Dimers on Ag(111). J Amer Chem Soc. 2009;131(5):1660–61.
[112] Mulugeta D, Watanabe K, Menzel D, Freund H-J. State-resolved investigation of the photodesorption dynamics of NO from $(NO)_2$ on Ag nanoparticles of various sizes in comparison with Ag(111). J Chem Phys. 2011;134(16):164702 ((1-11)).
[113] Watanabe K, Kim KH, Menzel D, Freund H-J. Hyperthermal chaotic photodesorption of xenon from alumina-supported silver nanoparticles: Plasmonic coupling and plasmon-induced desorption. Phys Rev Lett. 2007;99(22):225501.
[114] Li H, Xiao J, Fu Q, Bao X. Confined catalysis under two-dimensional materials. PNAS. 2017;114 (23):5930–34.
[115] Emmez E, Anibal Boscoboinik J, Tenney S, Sutter P, Shaikhutdinov S, Freund H-J. Oxidation of the Ru(0001) surface covered by weakly bound, ultrathin silicate films. Surf Sci. 2016;646: 19–25.

[116] Emmez E, Yang B, Shaikhutdinov S, Freund H-J. Permeation of a Single-Layer SiO_2 Membrane and Chemistry in Confined Space. J Phys Chem C. 2014;118(50):29034–42.
[117] Küchler A, Yoshimoto M, Luginbühl S, Mavelli F, Walde P. Enzymatic reactions in confined environments. Nat Nanotechnol. 2016;11:409.
[118] Hansma HG. The Power of Crowding for the Origins of Life. Orig Life Evol Biosph. 2014;44:307.
[119] Sauer J. Bronsted activity of two-dimensional zeolites compared to bulk materials. Faraday Discuss. 2016;188:227–34.
[120] Boronat M, Concepción P, Corma A, Navarro MT, Renz M, Valencia S. Reactivity in the confined spaces of zeolites: The interplay between spectroscopy and theory to develop structure–activity relationships for catalysis. Phys Chem Chem Phys. 2009;11(16):2876–84.
[121] Jenny J, Krishnakanth C, Aditya G, Jeremiah JG. The Chemistry of Confined Spaces. Curr Org Chem. 2014;18(15):2002–09.
[122] Kaden WE, Pomp S, Sterrer M, Freund H-J. Insights into silica bilayer hydroxylation and dissolution. Top Catal. 2016. DOI: 10.1007/s11244-016-0715-7.
[123] Yu X, Emmez E, Pan Q, Yang B, Pomp S, Kaden WE, Sterrer M, Shaikhutdinov S, Freund H-J, Goikoetxea I, Wlodarczyk R, Sauer J. Electron stimulated hydroxylation of a metal supported silicate film. Phys Chem Chem Phys. 2016;18(5):3755–64.
[124] Bauer E, Surface Microscopy with Low Energy Electrons. Springer Science+Business Media New York 2014 2014, Print ISBN 978-1-4939-0934-6, Online ISBN 978-1-4939-0935-3.
[125] Bauer E. Low energy electron microscopy. Rep Progr Phys. 1994;57(9):895–938.
[126] Flege JI, Krasovskii EE. Intensity–voltage low-energy electron microscopy for functional materials characterization. Physica Status Solidi (RRL). 2014;8(6):463–77.
[127] Koch MH, Jakob P, Menzel D. The influence of steps on the water-formation reaction on Ru (001). Surf Sci. 1996;367(3):293–306.
[128] Schiffer A, Jakob P, Menzel D. Structure and lateral interactions in binary and ternary coadsorbate layers of O, H and CO on Ru(001). Surf Sci. 2000;465(3):198–210.
[129] Faradzhev NS, Kostov KL, Feulner P, Madey TE, Menzel D. Stability of water monolayers on Ru (0001): Thermal and electronically induced dissociation. Chem Phys Lett. 2005;415(1):165–71.
[130] Schick M, Xie J, Mitchell WJ, Weinberg WH. Interaction of gas-phase atomic deuterium with the Ru(001)–p(1×2)–O surface: Kinetics of hydroxyl and water formation. J Chem Phys. 1996;104(19):7713–18.
[131] Hacarlioglu P, Lee D, Gibbs GV, Oyama ST. Activation energies for permeation of He and H2 through silica membranes: An ab initio calculation study. J Memb Sci. 2008;313(1):277–83.
[132] Prieto MJ, Mullan Th, Schlutow M, Gottlob DM, Tănase LC, Menzel D, Sauer J, Usvyat D, Schmidt Th, Freund HJ. Insights into Reaction Kinetics in Confined Space: Real Time Observation of Water Formation under a Silica Cover. J. Am. Chem. Soc. 2021; 43(23): 8780–8790.
[133] Csicsery S M. Shape-selective catalysis in zeolites. Zeolites. 1984; 4 (3):202–213.
[134] Calaza F, Stiehler C, Fujimori Y, Sterrer M, Beeg S, Ruiz-Oses M, Nilius N, Heyde M, Parviainen T, Honkala K, Häkkinen H, Freund H-J. Carbon dioxide activation and reaction induced by electron transfer at an oxide–metal interface. Angew Chem Int Ed. 2015;54(42):12484–87.
[135] Freund H-J. Models for heterogeneous catalysts: studies at the atomic level. Rend Fis Acc Lincei. 2016:1–14.
[136] O'Brien CP, Dostert KH, Hollerer M, Stiehler C, Calaza F, Schauermann S, Shaikhutdinov S, Sterrer M, Freund HJ. Supports and modified nanoparticles for designing model catalysts. Faraday Discuss. 2016;188:309–21.
[137] J. Paul, F. M. Hoffmann, L. L. Robbins, J. Phys. Chem. 1988, 92, 6967.
[138] F. M. Hoffmann, M. D. Weisel, J. Paul, Surf. Sci. 1994, 316, 277.
[139] R. L. Toomes, D. A. King, Surf. Sci. 1996, 349, 65.

Cornelius M. Berger, Martin Muhler, Jan Van De Loosdrecht, (Hans) J.W. Niemantsverdriet, Edward Kunkes, Klaus Friedel Ortega, Malte Behrens

14 CO_2 Utilization: Methane, Methanol, and Synthetic Fuels

The switch to an efficient large-scale chemical industry with a neutral CO_2 footprint is a major scientific and engineering challenge, mainly because the current industrial processes have not been developed and optimized under the boundary conditions of conversion of anthropogenic CO_2 or coupling to the de-centralized generation of renewable electricity. In fact, most chemical production plants today are not sustainable, because they are part of a petrochemical route and rely strongly on fossil feedstocks.

To improve this situation, non-fossil feedstocks like biomass and carbon-rich sources – for example, waste gas streams from fossil-fueled power plants or cement factories – should be used as well. However, to fulfill the requirements of a sustainable system, the needed hydrogen should also originate from non-fossil sources. Therefore, the use of regenerative power sources is a very important pillar in the future chemical industry.

The electricity-assisted generation of hydrogen for the subsequent production of synthetic natural gas, base chemicals like methanol, or synthetic fuels from carbon-rich feeds (power-to-X) is quite promising, because it can fulfill more purposes than the mere provision of valuable chemicals. It is also a way of storing energy, given that electrical excess energy is converted into bond energies in chemical products. Depending on the product, it can be stored and re-electrified (e.g., methane), used as a base material for further synthesis (methanol) or used as a fuel (hydrocarbons). Even if the conversion back to electricity does not take place, the initial conversion of electricity into chemical energy carriers can be considered as energy storage and helps enhancing robustness and flexibility of the utility network by coupling the electricity sector with the gas grid and transportation.

Cornelius M. Berger, Martin Muhler, Laboratory of Industrial Chemistry, Ruhr-Universität Bochum, Universitätsstr. 150, 44801 Bochum, Germany, e-mail: muhler@techem.rub.de
Jan Van De Loosdrecht, Council for Scientific and Industrial, Mering Naude Road, 0001 Pretoria, South Africa
(Hans) J.W. Niemantsverdriet, Syngaschem BV, SynCat@DIFFER, P.O. Box 6336, 5600 HH Eindhoven, The Netherlands
Edward Kunkes, BASF Corp., Iselin, NJ 08830, USA
Klaus Friedel Ortega, Malte Behrens, Kiel University, Institute of Inorganic Chemistry, Max-Eyth-Strasse 2, 24118 Kiel, e-mail: mbehrens@ac.uni-kiel.de

https://doi.org/10.1515/9783110608458-014

As all these chemical syntheses predominantly rely on the conversion of syngas, a mixture of hydrogen, carbon monoxide, and carbon dioxide, they could be part of a circular carbon economy or at least a means to reduce net carbon emissions [1]. Clearly, the source of the syngas plays a very important role and decides if a scenario is truly sustainable or just "light green."

Hence, in this chapter we describe the current state of syngas chemistry and the complex interplay of the different available options for syngas generation (Section 14.1). The synthesis of chemicals like methane, methanol, or hydrocarbons follow well-established large-scale industrial processes (Sections 14.2, 14.3, and 14.4). Yet, in a future decentralized scenario of power-to-X these processes must be capable of handling inherent fluctuations in the electricity grid related to the intermittent nature of renewable energy [2]. This is currently not given by conventional large-scale plants [3]. Also, the feeds may shift from purified mixtures of CO and H_2 to streams with a higher content of CO_2 and possibly impurities. As a result, new catalysts, novel reactor concepts, and innovative operation strategies are needed as well as understanding of the effects of a dynamic operation on a catalytic system at various micro- and macroscopic scales.

14.1 Syngas and Hydrogen Generation

Sustainable production routes for chemicals like methanol or hydrocarbons are particularly interesting, because in principle these substances can be produced by hydrogenation of the greenhouse gas CO_2. Thus, anthropogenic CO_2 could be used as feed. This is especially feasible in combination with high-emission industries like fossil-fueled power plants, cement factories, or blast furnaces.

Obviously, the other reactant for CO_2 hydrogenation is H_2. Hence, all considerations on the use of chemicals as sustainable energy carriers are only useful, if H_2 is available from regenerative sources.

Starting from the conventional fossil feedstock-based production routes for hydrogen and syngas we highlight some improvements to the current situation and some possible alternatives. Yet, the main focus lies on the conversion of syngas into value-added chemicals and fuels.

14.1.1 Conventional Pathways

Industrially, syngas is mainly produced by steam reforming of natural gas and to a minor extent by partial oxidation of coal and gasification of biomass or other hydrocarbon feedstocks. Syngas is a versatile intermediate for the production of bulk chemicals such as hydrogen, methanol, aldehydes and alcohols, and liquid hydrocarbon fuels. As such, syngas is a key component in present and future sustainable energy systems. Figure 14.1 summarizes the conventional routes of syngas production and some of the applications. All these technologies are well established and performed economically at large scales.

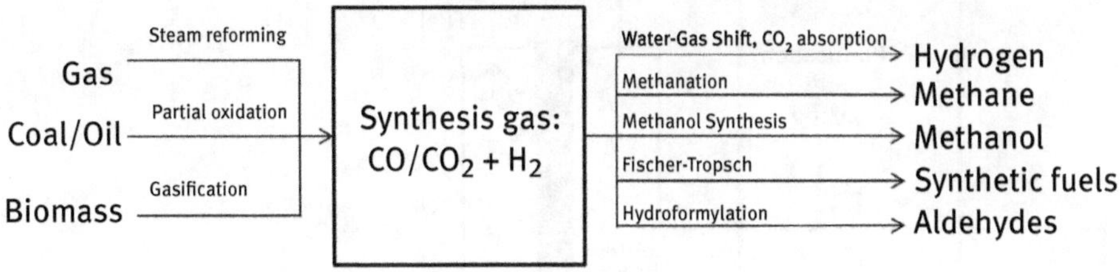

Figure 14.1: Conventional production routes of synthesis gas and a selection of applications.

Table 14.1 lists the important reactions in the reaction network of syngas production [4]. The reforming of methane or other hydrocarbons to yield syngas is endothermic and requires high temperatures. Indeed, only at temperatures above 973 K the steam reforming becomes thermodynamically favored over the competing water–gas shift (WGS) and carbon formation reactions.

Table 14.1: Reactions in syngas production [4]*.

Reaction	Equation	Enthalpy	No.
Steam reforming of methane (SRM)	$CH_4 + H_2O \rightleftharpoons CO + 3H_2$	+206 kJ mol^{-1}	(14.1)
	$C_nH_m + n\ H_2O \rightleftharpoons nCO + \left(n + \frac{1}{2}m\right) H_2$	endothermic	(14.2)
Dry reforming of methane (DRM)	$CH_4 + CO_2 \rightleftharpoons 2CO + 2H_2$	+247 kJ mol^{-1}	(14.3)
Water–gas shift (WGS)	$CO + H_2O \rightleftharpoons CO_2 + H_2$	−41 kJ mol^{-1}	(14.4)
Partial oxidation	$CH_4 + \frac{1}{2}O_2 \rightleftharpoons CO + 2H_2$	−36 kJ mol^{-1}	(14.5)
	$CH_4 + O_2 \rightleftharpoons CO + H_2O + H_2$	−278 kJ mol^{-1}	(14.6)
	$CH_4 + 1\frac{1}{2}\ O_2 \rightleftharpoons CO + 2H_2O$	−519 kJ mol^{-1}	(14.7)
Total oxidation	$CH_4 + 2O_2 \rightleftharpoons CO_2 + 2H_2O$	−802 kJ mol^{-1}	(14.8)
Methane pyrolysis	$CH_4 \rightleftharpoons C + 2H_2$	+ 75 kJ mol^{-1}	(14.9)
Boudouard reaction	$2CO \rightleftharpoons C + CO_2$	−172 kJ mol^{-1}	(14.10)
Reverse coal gasification	$CO + H_2 \rightleftharpoons C + H_2O$	−131 kJ mol^{-1}	(14.11)
	$C_nH_m \rightleftharpoons nC + \frac{1}{2}m\ H_2$	endothermic	(14.12)

* The enthalpies are valid for standard conditions (298.15 K, 1 bar).

Steam reforming of methane (SRM) is predominantly carried out in tubular reactors that are externally heated by burners to achieve the necessary high temperatures (Figure 14.2). Inside, supported nickel catalysts are used which are designed to cope with the harsh conditions at high temperatures and high-pressure steam. The waste heat is recovered to preheat the gas feed and to generate steam.

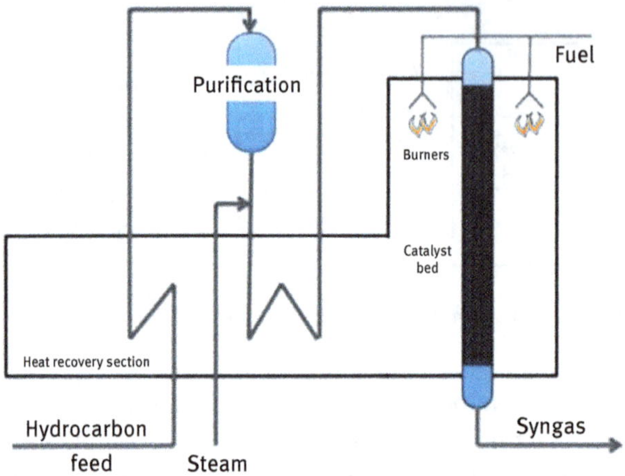

Figure 14.2: Flow scheme of an SRM plant. Simple SRM is often used for operation at moderate scale, such as relatively small methanol synthesis plants or H_2 production. It produces syngas with H_2/CO ratios in the range of 3–4 [4].

Recently, a new development has been proposed which aims at suppression of CO_2 emissions [5]. This is accomplished by replacing the large reactor-heater section by a much smaller design in which the catalyst/reactor assembly is heated directly through electric current. This interesting example of process intensification offers attractive opportunities to integrate the process with electricity from renewable sources and to utilize the methane feedstock exclusively for the steam reforming reaction itself, without the need to burn a considerable fraction for heat generation.

To optimize energy efficiency and suppress unwanted side reactions, a number of alternative processes to produce syngas are available. These processes differ in cost, tolerance for contaminants, and syngas composition. Figure 14.3 summarizes the syngas production processes from natural gas. One option is the partial oxidation of methane (or hydrocarbons). This process can be operated with or without a catalyst. It needs high temperatures for high CH_4 conversion and to suppress soot formation. The process yields syngas with a H_2/CO ratio of about 2. Moreover, it can handle other feedstocks such as coal, heavy oil fractions, or biomass. Therefore, the process is eminently suitable for large-scale production of syngas, for example, for gas-to-liquid plants.

The catalytic version of partial oxidation is based on short-contact time conversion of methane, hydrocarbons, or biomass over rhodium and other catalysts [6, 7]. This process is suitable for small-scale applications.

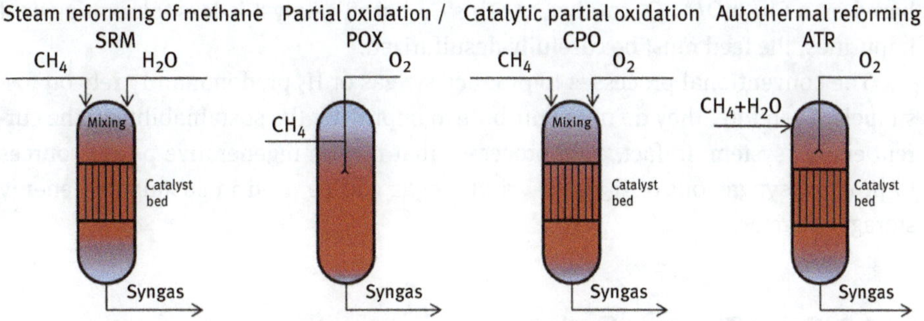

Figure 14.3: Schematic production processes for synthesis gas out of natural gas. Left: Steam reforming of methane, middle: Partial oxidation with and without catalyst for large-scale syngas production, right: autothermal reforming.

The combination of endothermic steam reforming and exothermic partial oxidation of methane results in the autothermal reforming process [4, 8]. The necessary heat for the endothermic reforming reaction is provided inside the reactor by partial oxidation. The H_2/CO ratio can be varied between 2 and 3.5. This process is also well suited for large-scale production of syngas for gas-to-liquids or larger-scale methanol synthesis processes (see Section 14.3.1.1). CO_2-containing syngas is usually classified by a modulus value M (14.13). Values around 2 satisfy the requirements for most syngas reactions:

$$M = \frac{[H_2] - [CO_2]}{[CO] + [CO_2]} \tag{14.13}$$

In all processes, avoiding carbon deposition on the catalyst is a major challenge [9, 10]. Carbon can be present as graphite-like coke and in the form of whiskers, or carbon nanofibers. The latter lead to detachment of the nickel crystallites from the support and the breakup of catalyst pellets. This may cause blockage of the reformer reactor tubes and the formation of hotspots. Higher hydrocarbons exhibit a larger tendency to form carbon than methane. Thus, reforming of naphtha imposes more complex requirements on the catalysts and the process design.

The WGS reaction (eq. (14.4)) is not only implicitly present in the production of syngas, but can also be applied separately to adjust the syngas modulus value or to even produce high-purity hydrogen [11, 12]. CO acts as reducing agent converting water into H_2. Being mildly exothermic, the reaction runs at much lower temperatures than the reforming and oxidation reactions described above.

Usually, the WGS reaction is performed at two different temperature levels. The high-temperature WGS runs at 573–773 K and uses an iron oxide-based catalyst stabilized by alumina. This catalyst is tolerant against impurities such as sulfur and resistant against sintering. For higher conversion and purity of H_2 the high-temperature stage is followed by a more delicate, low-temperature process around 473 K based on a $Cu/ZnO/Al_2O_3$ catalyst. As this Cu-based catalyst is less tolerant against impurities, the feed must be carefully desulfurized.

The conventional processes to produce syngas or H_2 predominantly rely on fossil fuels. Therefore, they do not contribute to improving the sustainability of the current energy system. In fact, only processes that rely on regenerative power sources to produce syngas out of H_2O, CO_2, and biogas can be used in sustainable energy storage schemes.

14.1.2 Solar-Thermal Cycles

Concentrated solar power (CSP) can provide extremely high temperatures. Point-focusing CSP consists of a multitude of sun-tracking mirrors (heliostats), which reflect the incident sunlight onto a small area where the receiver is placed. Conventionally, in the receiver a heat transfer fluid such as air is heated to up to 1273 K and then used to generate steam for a steam turbine. Alternatively, the high temperatures can directly be used to drive endothermic chemical reactions which require high temperatures.

14.1.2.1 Dry Reforming of Methane

Dry reforming of methane (DRM) by CO_2 is one possibility to utilize excess power because of the conversion of two potent greenhouse gases into high-quality syngas [13]. This process may be suitable for the conversion of biogas, which already consists mainly of CH_4 and CO_2. DRM yields a syngas composition with a H_2/CO ratio close to unity as opposed to the H_2/CO ratio in SRM of around 3. This makes it suitable for synthesis reactions like hydroformylation that require this ratio. Also, the combination of DRM and SRM could lead to syngas compositions with a H_2/CO ratio of 2 required by Fischer–Tropsch synthesis (FTS).

Due to the strong endothermicity of the reaction (eq. (14.3)), high temperatures above 973 K are required [13]. Full conversion is obtained at temperatures around 1,173 K. Under these conditions unwanted side reactions like the exothermic Boudouard reaction (eq. (14.10)) are effectively suppressed. Yet, these conditions impose strong requirements on the nickel catalyst, which needs to be protected against sintering by ceramic supports like Al_2O_3 or $MgAl_2O_4$. Furthermore, carbon deposits on the catalyst are likely to occur and thus increase the difficulty of a large-scale commercialization.

14.1.2.2 Thermal Splitting of H_2O/CO_2

An elegant way to provide syngas would be the direct thermal CO_2 splitting and water splitting:

$$CO_2 \rightleftharpoons CO + \tfrac{1}{2}\,O_2 \tag{14.14}$$

$$H_2O \rightleftharpoons H_2 + \tfrac{1}{2}\,O_2 \tag{14.15}$$

However, this thermolysis requires extremely high temperatures of above 2500 K for even minimal conversion. Also O_2 would need to be removed in order not to recombine at lower temperatures in the downstream section. Thus, direct thermolysis is unfeasible [14]. Still, in a two-step process involving redox-active metal oxides as oxygen carrier, the relevant temperatures can be significantly lower and oxygen would be bound to the metal oxide [15].

Therefore, the unwanted back reaction and the challenging separation of the reaction products are circumvented. The proposed process for the splitting of H_2O (14.18) follows the reactions (14.16) and (14.17). The splitting of CO_2 is analogous but proceeds at different temperatures and potentially at different rates. For CO_2 splitting, the Boudouard (14.10) reaction has to be considered and for the combined process the WGS equilibrium (14.4) also plays a role.

$$MO_x \rightarrow MO_{x-y} + \tfrac{y}{2}\,O_2 \tag{14.16}$$

$$MO_{x-y} + yH_2O \rightarrow MO_x + yH_2 \tag{14.17}$$

$$H_2O \rightarrow H_2 + \tfrac{1}{2}\,O_2 \tag{14.18}$$

Instead of thermally reducing steam to H_2, in this approach the metal oxide is thermally reduced (14.16) in an inert atmosphere with low oxygen partial pressure. In the second step, the reduced species is cooled down and reoxidized by H_2O/CO_2 resulting in the formation of H_2/CO and the initial metal oxide (eq. (14.17)). The reduced species can either be a metal, a stoichiometric or a nonstoichiometric oxide. Depending on the oxygen carrier, realistic reduction temperatures can be as low as 1,773–1,973 K [16].

These temperature levels can be achieved in receivers of point-focusing concentrated solar power plants. While the general concept seems to be promising due to its simplicity, the practical realization of the cyclic process in the receiver suffers from problems caused by the severe reaction conditions.

Generally, the reduction of any metal oxide according to eq. (14.16) is endothermic and has a positive reaction entropy. As a consequence, high temperatures are needed for the release of O_2 yielding the reduced state. The reaction of the reduced species with H_2O/CO_2 is slightly exothermic (14.17). Obviously, the individual reactions require quite different temperatures levels and atmospheres. Hence, the process must be performed dynamically with respect to both reaction temperature and atmosphere. The reaction rate r of an oxide reacting with a surrounding gas phase to form another oxide/metal depends on many parameters:

$$r = f\left(T, A, p_{O_2}, D, \dot{V}, X, \epsilon\right) \tag{14.19}$$

Fast kinetics demands temperatures as high as possible and a high surface area A. Naturally, the oxygen partial pressure p_{O_2} of the surrounding atmosphere also influences the reaction rate. The reduction of the solid entails the formation of vacancies at the surface. The diffusion of the vacancies into the particles and the outward diffusion of oxygen anions to the surface can be rate-limiting. Mass transfer limitations can lead to high local O_2 concentrations and are partially resolved by increasing the flow rate $\dot{V}$ of the inert carrier gas and the porosity ϵ of the solids. High flow rates also result in a thinner stagnant boundary layer, in which transport only occurs via molecular diffusion.

It may be regarded as advantageous to perform short cycles limiting the reduction to surface layers, which are more easily reduced, thus avoiding bulk diffusion limitations. Yet, high conversion X is possible only with longer cycle duration with eventually slow rates. While the former strategy operates at high reaction rates, it only uses a small fraction of the active metal oxide and leads to high thermal stress in the material due to fast heating and cooling. The latter uses most of the active mass and therefore has a higher yield per cycle, but the mean reaction rates and productivity may be lower.

Apart from process design aspects and challenges of heat transfer in the receiver, the actual oxygen carrier needs to fulfill a set of requirements [16]. First, the material should provide the possibility of thermal reduction within a reasonable temperature range. Also, it needs to be active for reoxidation by H_2O/CO_2. The performance in these reactions and the storage capacity for oxygen anions determines the overall productivity.

For efficient solar-thermal cycles the oxygen carrier must effectively absorb solar radiation and convert it into heat without substantial reradiation. Therefore, bright materials are unfavorable, because they always cause reradiation.

Ideally, the cycle possesses unlimited repeatability without loss of performance and capacity [17]. Hence, a crucial point is the stability of the material. As stated above, the reaction rate for the reduction of the oxygen carrier increases with high surface area and temperature. Additionally, the reaction rates of the thermal reduction and the reoxidation by H_2O/CO_2 benefit from high oxygen anion diffusivity. Oxygen anion conductivity depends on crystal structure and usually increases with temperature.

Yet, high temperatures may lead to sintering or even volatilization of the solid. Clearly, it is of limited use to employ a high surface area oxide with excellent reducibility in terms of oxygen release capacity and reaction rate, if it undergoes extreme sintering, phase segregation, and layer formation within the first cycles.

As a result, research on solar splitting of CO_2 and H_2O focuses on the development of oxides with high stability and high oxygen mobility resulting in improved reducibility and high oxygen storage/release capacities. Typical examples for suitable materials are the oxide/metal system ZnO/Zn or multivalent metal oxide systems like Fe_3O_4/FeO or CeO_2/Ce_2O_3 [15].

14.1.2.2.1 The ZnO/Zn System

The system ZnO/Zn has attracted considerable attention for solar-thermal splitting of CO_2 or H_2O:

$$ZnO_{(s)} \rightarrow Zn_{(g)} + 1/2\ O_{2,g} \tag{14.20}$$

$$Zn_{(s)} + CO_{2,g} \rightarrow ZnO_{(s)} + C\,O_{(g)} \tag{14.21}$$

The reduction is experimentally performed at approximately 1,873 K in inert atmosphere [18]. The melting and boiling points of elemental zinc are at 693 and 1,180 K at 1 bar, respectively. Accordingly, the reaction products of ZnO dissociation are gaseous Zn and O_2, which have to be quenched at high temperatures to prevent recombination.

Obviously, the ZnO/Zn system does not technically meet the requirements for material stability of the working oxide and hence extended processing to handle the multiple material streams is necessary. For example, a moving-front reactor for solar-thermal ZnO dissociation has been proposed [18]. This concept enables the recovery of approximately 30% of the volatilized product as a powder consisting of nanoparticles containing up to 50% of reduced Zn species.

For rapid reaction rates and complete conversion of Zn during the splitting reaction temperatures below 973 K are mandatory, which are still far above the melting point and lead to considerable vaporization of Zn [19–22]. Due to the rather low temperatures and the exothermic reaction this step does not necessarily involve the need for solar heating. Thus, the solar reactor for oxide dissociation can be operated continuously without temperature cycling and reproducibly yields Zn metal with high surface area and favorable properties for subsequent CO_2 splitting, avoiding the degradation of the working oxide.

Thermodynamic considerations predict the formation of some carbon for temperatures up to 973 K according to the Boudouard reaction (14.10). However, this was not observed experimentally. Although the ZnO/Zn thermochemical cycle does not meet the initially formulated criteria with respect to material stability, its properties render it attractive and promising.

The volatilization of Zn (14.20) imposes difficulties in terms of product separation and aerosol processing and is accompanied with an energy penalty for quenching leading to low thermal efficiencies. Reducing zinc oxide with methane using CSP at temperatures of above 1273 K directly yields syngas and elemental zinc for separate use or application in the CDS/WS step [23]:

$$ZnO_{(s)} + CH_4 \rightarrow Zn_{(g)} + CO + 2H_2 \tag{14.22}$$

14.1.2.2.2 The Fe_3O_4/FeO System

The basic ferrite cycle operates within the iron oxides magnetite (Fe_3O_4) and wustite (FeO:

$$Fe_3O_{4(s)} \rightarrow 3FeO + 1/2\ O_{2,g} \tag{14.23}$$

$$3FeO + CO_{2,g} \rightarrow Fe_3O_4 + CO_{(g)} \tag{14.24}$$

In a batch system the reduction of magnetite becomes thermodynamically favorable at temperatures far above 2,200 K. Reduction was demonstrated in an inert atmosphere at ca. 300 K above the melting point of 1,811 K, resulting in a low surface area of the reduced product which makes mechanical activation necessary to achieve high reaction rates in the subsequent step. This is as a major obstacle for successful implementation [24]. Nevertheless, several ferrite systems $M^{II}Fe_2O_4$ (M: Mn, Ni, Co, Zn) have been proposed as alternatives exhibiting lower reduction temperatures while still offering favorable thermodynamics for H_2O and CO_2 splitting [24, 25].

In the temperature range from 973 to 1,273 K the initial reaction rates of FeO oxidation using H_2O/CO_2 increase with high H_2O concentration and with temperature [20, 21]. A FeO conversion of up to 80% is obtained in the diffusion-controlled reaction with full conversion appearing to be within reach. As sintering takes place at these temperatures, a transition from a fast interface-controlled to a slow diffusion-controlled regime was observed. It was found that utilization and reduction of Fe oxide was especially effective, when it was dispersed on yttria-stabilized zirconia (YSZ) to mitigate sintering and to enhance oxygen transport [24].

Multiple cycles of thermal reduction and CO_2 splitting were demonstrated without a significant loss of activity using a cobalt ferrite $Co_{0.67}Fe_{2.33}O_4$ supported on YSZ [17]. The results indicate that a favorable temperature range for performing the reoxidation is from 1,373 to 1,473 K, because slow kinetics limit below 1,373 K and unfavorable thermodynamics limit above 1,473 K.

A solar-to-chemical system efficiency of 29% to 36% represents the theoretical limit for this process without the recuperation of heat [17, 19]. To increase this efficiency a reactor/receiver design consisting of counterrotating discs enabling efficient heat transfer in the temperature cycling of the working oxide was described yielding an increase of efficiency to up to 76% [17].

14.1.2.2.3 The CeO_2/Ce_2O_3 System

The redox properties of ceria and ceria-based materials have been studied in great detail because of their oxygen storage and release capacities [26, 27]. Some case studies of ceria-based systems in solar-thermal cycles are published in the literature [17, 28–30]:

$$2CeO_2 \rightarrow Ce_2O_3 + \frac{1}{2}\,O_{2,g} \tag{14.25}$$

$$Ce_2O_3 + CO_2 \rightarrow 2CeO_2 + CO \tag{14.26}$$

Thermodynamic calculations show that the reduction of CeO_2 to Ce_2O_3 (14.25) takes place spontaneously at temperatures above 2,623 K, which is in between the melting points of the reactant (2,873 K) and the product (2,503 K). Nevertheless, partial reduction of ceria to suboxides can be performed at lower temperatures [17]. Typical drawbacks of ceria for the solar-thermal cycling are its tendency to sinter and its surface-limited oxygen storage/release capacity. Modification with dopants such as Zr helps to overcome these drawbacks [17]. Solar-thermal cycles with Zr-doped ceria $(CeO_2)_{0.25}(ZrO_2)_{0.75}$ were successfully performed and the rate of CO_2 splitting was limited by surface processes. Higher ceria contents lead to increased sintering of the materials and therefore to a loss in activity during the reduction step. For pure ceria, thermodynamic analysis results in solar-to-fuel energy conversion efficiencies of maximum 16% to 19% without heat recovery [28]. However, in the same study efficiencies of only 0.8% were achieved.

14.1.3 Electrolysis and Co-electrolysis

Instead of thermally reducing mixtures of steam and CO_2 at very high temperatures, the same reactions (14.14) and (14.15) can be performed using electric currents at lower temperatures. For water electrolysis several technologies are available which differ in the applied temperature range, the used catalyst, and the cell design [31]. Resulting H_2 can be mixed with purified CO_2 from industry flue gases to obtain syngas.

At low temperatures, CO is a potent reversible poison for catalysts like platinum blocking adsorption sites strongly. Therefore, CO_2 electrolysis and co-electrolysis (simultaneous electrolysis of CO_2 and H_2O) is preferably performed in solid oxide electrolyzers at temperatures around 973–1,273 K [32]. These electrolyzers mostly consist of an oxygen-anion conducting membrane such as yttria-stabilized zirconia (YSZ) and two electrodes. At the fuel electrode (e.g., Ni + YSZ) the mixture of steam and CO_2 is electrocatalytically reduced into syngas. For co-electrolysis under these conditions reverse WGS plays an important role and is in fact the major route to CO generation [33]. The separated oxygen anions diffuse through the membrane and are oxidized at the oxygen electrode (e.g., perovskite) recombining to gaseous O_2, which itself is a valuable chemical. Thus, integration of electrolysis units at sites where O_2 is needed may yield substantial advantages.

Several reactor designs have been published, among them planar and tubular system architectures. Whereas planar design allows for higher current densities due to thinner electrolytes, the tubular systems are claimed to be more robust. For both configurations successful direct co-electrolysis has been reported [34–37]. Major research activities focus on improving degradation stability and reducing temperatures by developing new electrode and electrolyte materials [33]. Lowering temperatures is especially important because the produced syngas can directly be converted into methane (Section 14.2), methanol (Section 14.3), or synthetic fuels (Section 14.4).

14.2 Methanation

In a general energy storage concept, methanation of CO and CO_2 is a very promising option, because it provides the opportunity for seasonal storage [38–40]. Moreover, the generated synthetic natural gas (SNG) can be easily stored in the existing gas grid without modifications of the infrastructure. Most current methanation schemes rely on the use of electrolytic H_2 and captured CO_2 from industry off-gases or from biomass gasification. The main reactions during methanation (14.27) and (14.28) are quite exothermic:

$$\text{CO methanation} \quad CO + 3H_2 \rightleftharpoons CH_4 + H_2O, \quad \Delta_R H = -206\,\text{kJ mol}^{-1} \tag{14.27}$$

$$CO_2\,\text{methanation} \quad CO_2 + 4H_2 \rightleftharpoons CH_4 + 2H_2O, \quad \Delta_R H = -165\,\text{kJ mol}^{-1} \tag{14.28}$$

$$\text{Reverse water-gas shift} \quad CO_2 + H_2 \rightleftharpoons CO + H_2O, \quad \Delta_R H = +41\,\text{kJ mol}^{-1} \tag{14.29}$$

$$\text{Reverse DRM} \quad 2CO + 2H_2 \rightleftharpoons CH_4 + CO_2, \quad \Delta_R H = -246\,\text{kJ mol}^{-1} \tag{14.30}$$

Therefore, methanation plants need powerful heat removal at temperature levels between 498 and 673 K, often achieved by a combination of active cooling between separate reactors, multiple feed inlets, and product recycling [41]. Figure 14.4 depicts a simplified scheme for the methanation of CO_2.

14.2.1 Catalysts for Methanation

Several catalysts have been investigated for methanation of CO and CO_2 [42–44]. Among the known active metals for methanation (Ni, Fe, Co, Ru, Pd, Mo, W), mainly nickel has been used commercially because of the high methane selectivity and relatively low cost. For low temperatures ruthenium is more active but considerably more expensive than nickel. In general, the catalyst has to be active, selective, and resistant against degradation. Deactivation may be due to sintering at elevated temperatures in hotspots or due to loss of active sites by means of catalyst poisoning (e.g., H_2S) or catalyst fouling by solid deposits like coke or coke precursors.

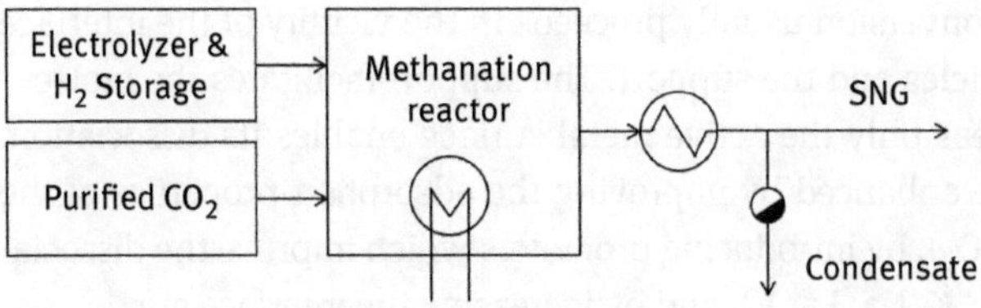

Figure 14.4: Simplified scheme for CO_2 methanation. Methanation may take place in a sequence of fixed-bed reactors with interstage cooling or using tube bundles filled with catalyst and immersed in a cooling medium (e.g., salt melt) [41].

These processes demand a careful purification step of the involved gases and a process design that enables the removal of potential carbon deposits, for example, by dynamically changing the gas compositions, in particular the H_2O concentration [40]. To provide a stable and active catalyst, the active metal nanoparticles are supported on metal oxides such as Al_2O_3, SiO_2, TiO_2, ZrO_2, or CeO_2. In turn, these supports may be promoted by dopants such as MgO, La_2O_3, or V_2O_3. The promoters can increase thermal stability and coking resistance or even increase adsorption properties and thus the activity [40, 42, 43].

14.2.2 Mechanism

The mechanisms of CO and CO_2 conversion to methane are quite similar. In fact, one possible route of CO_2 hydrogenation advances via the CO_2 adsorption step followed by dissociation into adsorbed CO and adsorbed O [45]. Similarly, CO adsorbs either molecularly or dissociatively depending on the exact temperature and catalyst surface properties. In either case the methanation reaction proceeds via the sequential hydrogenation of adsorbed C_{ads} species. A simplified Langmuir–Hinshelwood reaction mechanism is described in the following.

Step 1: $H_2 + 2^* \rightleftharpoons 2H - ^*$ (14.31)
Step 2a: $CO_2 + 2^* \rightleftharpoons CO - ^* + O - ^*$ (14.32)
Step 2b: $CO - ^* + ^* \rightleftharpoons C - ^* + O - ^*$ (14.33)
Step 3a: $CO + 2^* \rightleftharpoons CO - ^{**}$ (14.34)
Step 3b: $CO + 2^* \rightleftharpoons C - ^* + O - ^*$ (14.35)
Step 3c: $CO - ^{**} \rightleftharpoons C - ^* + O - ^*$ (14.36)
Step 4a: $O - ^* + H - ^* \rightleftharpoons OH - ^* + ^*$ (14.37)
Step 4b: $OH - ^* + H - ^* \rightarrow H_2O + 2^*$ (14.38)
Step 5a: $C - ^* + H - ^* \rightleftharpoons CH - ^* + ^*$ (14.39)
Step 5b: $CH - ^* + H - ^* \rightarrow CH_2 - ^* + ^*$ (14.40)
Step 5c: $CH_2 - ^* + H - ^* \rightarrow CH_3 - ^* + ^*$ (14.41)
Step 5d: $CH_3 - ^* + H - ^* \rightarrow CH_4 + 2^*$ (14.42)

In supported catalysts CO_2 conversion usually proceeds in the vicinity of the interface between the metal nanoparticles and the support. The support facilitates the molecular adsorption of CO_2, whereas only the active metal surface enables its dissociation. Thus, CO_2 conversion can be enhanced by improving the adsorption properties of the supporting material (e.g., Al_2O_3), by introducing promoters which improve the dissociation of adsorbed CO/CO_2 (e.g., K, Na, La, V), and by increasing the interface area.

CO methanation proceeds via adsorption of CO at the active metal surface. After dissociation the reaction is analogous to CO_2 methanation. The catalyst composition and morphology as well as the process conditions (pressure, temperature, composition of atmosphere) determine the steady-state concentrations of the adsorbed species. If, for example, the feed contains CO, CO_2 conversion is hindered because of preferred adsorption of CO on the metal surface.

The rate-determining steps in CO and CO_2 methanation are still unclear. While hydrogenation reactions of adsorbed carbon species are generally more facile, the dissociation of adsorbed CO or the removal of O-atoms may be limiting [40, 46–49]. This also depends on the exact reaction conditions as the dissociation step may be hydrogen-assisted.

14.3 Methanol Synthesis and Steam Reforming

Methanol is the smallest alcohol and liquid at ambient temperature. Today, methanol is an important platform molecule for many syntheses in chemical industry [50–52]. It can be upgraded into other chemical intermediates like formaldehyde or acetic acid. Zeolite-catalyzed methanol-to-olefins (MTO) and methanol to gasoline (MTG) processes open a pathway for direct conversion of methanol into fuels and chemicals currently derived from mineral oils.

However, these are only expected to become economically viable in the case of high oil prices. Also, more research and intensified catalyst and process optimization is urgently needed to realize the idea of a sustainable "methanol economy" [50, 51]. Methanol can be used directly or as fuel additive in combustion engines. Thus, it is also regarded as a medium for energy storage. Currently existing gasoline distribution and storage infrastructure like pipelines, road tankers, and filling stations require little modification to operate with methanol [50].

Direct utilization of methanol in combustion engines is attractive due to its good combustion properties. Methanol has a high octane number (RON = 133), comparable to currently available gasoline blends. The emissions of NO_x and hydrocarbons (not of CO and formaldehyde) are lower compared with gasoline, while thermal efficiency and motor power are higher by ca. 15% and 10%, respectively [53]. Yet, the energy density of methanol is only ca. 50% of that of gasoline, leading to an increased volume-based consumption.

Currently, up to 3% methanol is used as a blend in gasoline (M3). Pure methanol (M100) or mixtures of methanol and gasoline (e.g., M85, 85% methanol, 15% gasoline) were shown to be applicable in large-scale studies with several hundred cars in the USA and Germany [54–56]. Among the direct products of methanol conversion, methyl-*tert*-butyl-ether (MTBE), dimethylether (DME), and oxymethylene ethers (OMEs) are of relevance for fuel applications. MTBE is used as an octane-booster to improve the anti-knocking properties of gasoline, but recently has been banned by some states due to environmental concerns. DME and OMEs have been proposed as potentials substitute for diesel fuel (Section 14.5.3).

Instead of combustion, methanol can be applied in fuel cells, either in direct methanol fuel cells (DMFC), or as onboard hydrogen storage for proton exchange membrane fuel cells (PEMFC) circumventing the efficiency limitations of the Carnot cycle. The use of methanol for hydrogen storage is convenient because of the simple storage as a liquid at ambient conditions, the high H:C ratio of 4:1, and the relatively facile methanol steam reforming for recovering hydrogen (MSR, see Section 14.3.3).

14.3.1 Methanol Synthesis

The current primary feed stock for industrial methanol synthesis is syngas [57]. The interconversion of carbon oxides and methanol, central to methanol synthesis and steam reforming, is defined by three equilibrium equations:

$$CO_2 + 3H_2 \rightleftharpoons CH_3OH + H_2O, \qquad \Delta_R H = -50\,\mathrm{kJ\,mol^{-1}} \tag{14.43}$$

$$CO + 2H_2 \rightleftharpoons CH_3OH, \qquad \Delta_R H = -91\,\mathrm{kJ\,mol^{-1}} \tag{14.44}$$

$$CO_2 + H_2 \rightleftharpoons CO + H_2O, \qquad \Delta_R H = 41\,\mathrm{kJ\,mol^{-1}} \tag{14.45}$$

Methanol synthesis from CO_2 (14.43) and CO (14.44) is mildly exothermic under standard conditions and results in volumetric contraction. The slightly endothermic reverse WGS (rWGS) reaction (14.45) occurs as a side reaction to methanol synthesis from CO_2. According to Le Chatelier's principle, high pressures and low temperatures favor methanol synthesis.

Thus, the methanol yield displays a positive dependence on pressure and an inverse dependence on temperature, whereas the rWGS yield shows weak pressure dependence and increases with temperature. The addition of CO to the feed mixture has a positive effect on the equilibrium yield of methanol, as is discussed in detail in Section 14.3.1.4.

The current low-pressure synthesis process (see Section 14.3.1.1) yields methanol with greater than 99% selectivity on a CO-free basis. The equilibrium constants (per CO_2 molecule reacted) for the formation of common by-products at 523 K are shown in Figure 14.5.

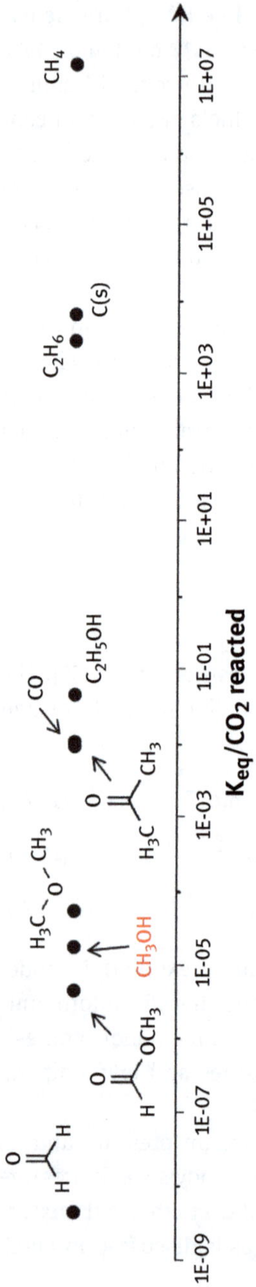

Figure 14.5: The equilibrium constants of the formation of methanol synthesis by-products from a 3:1 H_2:CO_2 mixtures at 523 K.

The high methanol selectivity is remarkable considering that the formation of ethers, ketones, and alkane impurities found in industrial methanol is thermodynamically more favored than the formation of methanol. Similarly, coke, another thermodynamically favored product and common cause of catalyst deactivation, is not observed under methanol synthesis conditions [58].

Thus, the aforementioned products are kinetically inaccessible using the Cu/ZnO-based catalyst described in more detail in Section 14.3.1.2. From these thermodynamic considerations the need for more active CO_2 hydrogenation catalysts becomes obvious. Even at temperatures associated with the low pressure process, K_2 lies between 10^{-5} and 10^{-6} allowing for a single pass methanol yield of only 15–25% necessitating the implementation of costly recycling loops.

14.3.1.1 Industrial Methanol Synthesis

Currently, the worldwide production capacity of methanol amounts to approximately 110 million metric tons. Although this amount represents only 0.01% of the worldwide gasoline production, it is nearly equivalent to the total biodiesel and bioethanol production [59]. From this number it is clear that a large-scale replacement of gasoline by methanol as fuel requires an enormous increase of the world-wide methanol synthesis capacities. Today, chemical intermediates dominate methanol consumption. Formaldehyde – a platform molecule for the synthesis of polymer resins – is responsible for nearly half of the total demand. Acetic acid, MTBE, and methyl methacrylate (MMA) constitute another 25% [60, 61]. Direct fuel and additive usage accounts for 15% of demand, but is expected to rise.

Until the commercialization of the first heterogeneous catalytic process for methanol synthesis by BASF in the 1920s, methanol was produced exclusively from the dry distillation of wood. The BASF process utilized sulfur-containing coal- or coke-derived synthesis gas and a ZnO/Cr_2O_3 catalyst operating at 573–723 K [62]. High pressures (100–300 bar) were required to counteract these thermodynamically unfavorable temperatures. Although the superior activity of Cu-based methanol synthesis catalysts was reported shortly afterward [63], only the advent of natural gas–derived sulfur-free syngas allowed for industrial application.

The commercialization of more active $Cu/ZnO/Al_2O_3$-based catalysts (see Section 14.3.1.2) by ICI in the 1960s led to milder reaction conditions of 513–533 K and 50–100 bar in the low-pressure process [64]. Since its inception, this process has been optimized to yield methanol with a selectivity above 99% and a 75% energy efficiency and has thus become the exclusive means of methanol production [57]. Catalysts based on supported palladium and other noble metals have also been shown to yield methanol at current process conditions, but due to their price and insufficient methanol selectivity, these materials are so far only of academic interest. Recent catalyst developments beyond copper include reports on Ni-Ga intermetallics [65], mixed oxides like MnO_x/Co_3O_4 [66], and In_2O_3/ZrO_2 [67].

Methanol synthesis plants utilizing the low pressure process currently operate at capacities of 2×10^5 to 2×10^6 metric tons per year [64]. Such installations consist of a synthesis gas production unit, the actual methanol synthesis reactor, and a separation and purification section. The production and purification of synthesis gas accounts for 50–80% of the total cost of methanol production, with the remaining cost associated with the actual synthesis and purification of methanol [57, 60]. Although a variety of carbonaceous feedstocks can be transformed into synthesis gas, the steam reforming of natural gas and the partial oxidation of hydrocarbons/coal are by far the most common option, especially for large plants [57, 64, 68]. Countries with large domestic coal reserves, such as China and South Africa rely primarily on coal gasification to produce syngas. This syngas is hydrogen-deficient and undergoes a WGS step to yield mixtures of more suitable H_2 and CO_2 content [57]. Methanol synthesis from CO_2 and CO_2-rich mixtures requires special catalysts and poses additional reactor design challenges, which will be further discussed in more detail.

Industrial methanol synthesis is carried out in fixed-bed reactors, which are designed to achieve effective removal or dissipation of the heat generated during this exothermic reaction. A simplified schematic of a methanol synthesis plant is shown in Figure 14.6.

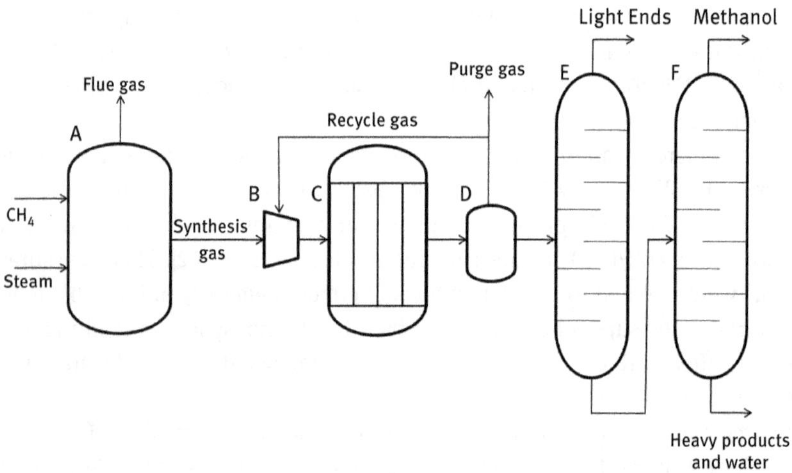

Figure 14.6: A simplified scheme of a methanol synthesis plant. (A) Gasifier, (B) compressor, (C) methanol reactor, (D) flash drum, (E) light-ends column, and (F) methanol column.

The original adiabatic quench reactor developed by ICI consists of a single catalyst bed with cold syngas being injected at several points along the axial direction of the bed. The Kellogg and Haldor–Topsøe reactor designs consist of a series of catalyst beds with inter-stage cooling of the products.

A quasi-isothermal reactor developed by LURGI consists of several tubular catalyst beds surrounded by an outer shell of boiling water [57]. The pressure in the shell is used to control the reaction temperature. Another rather novel reactor type is the radial-flow reactor which has benefits in terms of pressure drop and manufacturing costs [69]. Still, the ICI and LURGI configurations account for approximately 60% and 30% of global methanol production, respectively. Crude methanol leaving the reactor contains volatile impurities such as dissolved gases, light hydrocarbons, esters, and ketones that are removed in an initial stripping step. Less volatile impurities such as water and heavier alkanes (C_{8+}) are removed as bottoms in subsequent distillation steps [57, 60, 64].

14.3.1.2 Cu/ZnO-Based Methanol Synthesis Catalysts

Low-pressure methanol synthesis relies almost exclusively on catalysts based on copper, zinc oxide and alumina. The catalysts are produced by Johnson Matthey (formerly ICI), Clariant (formerly Süd-Chemie), Haldor–Topsøe, BASF, and other chemical enterprises and contain 50–70% CuO, 20–50% ZnO and 5–20% Al_2O_3. Instead of alumina, also chromia, zirconia, and rare earth oxides have been used. The mixed oxide catalysts are usually prepared as 4–6 mm cylindrical pellets with specific surface areas between 60 and 100 m^2/g. The catalysts are activated in situ with dilute H_2, often derived from off-gases from syngas production. The activation procedure takes place at 463–503 K, completely reducing CuO to metallic Cu^0 crystallites interspersed by a ZnO-Al_2O_3 matrix.

Commercial Cu/ZnO/Al_2O_3 methanol synthesis catalysts are characterized by a molar Cu:Zn ratio close to 70:30, while the amount of Al_2O_3 is significantly lower than that of ZnO. This Cu-rich composition manifests itself in a peculiar microstructure of the industrial Cu/ZnO/Al_2O_3 catalyst (Figure 14.7).

The catalyst is composed of spherical Cu nanoparticles of ca. 10 nm in size (Figure 14.7B,C) and often even smaller ZnO nanoparticles arranged in an alternating fashion. Thus, porous aggregates are formed (Figure 14.7A) in which the oxide particles act as spacers between Cu particles (Figure 14.7B[71]). The presence of inter-particle pores as seen in the HRTEM image allows some access to the inner surface of larger Cu/ZnO aggregates.

This unique microstructure can be described as an intermediate stage between a supported catalyst and a bulk metallic sponge or skeletal Raney-type catalyst. It enables a reasonably high dispersion of Cu and exposure of many Cu-ZnO interface sites at a high total Cu content. The specific Cu surface area (SA_{Cu}) of methanol catalysts can be determined by H_2 adsorption [73, 74] or reactive N_2O titration [75, 76], which causes surface oxidation of the Cu particles and allows calculation of SA_{Cu} from the amount of evolved N_2. The SA_{Cu} of state-of-the-art methanol synthesis catalysts measured by this method amounts to 25–35 $m^2\,g^{-1}$.

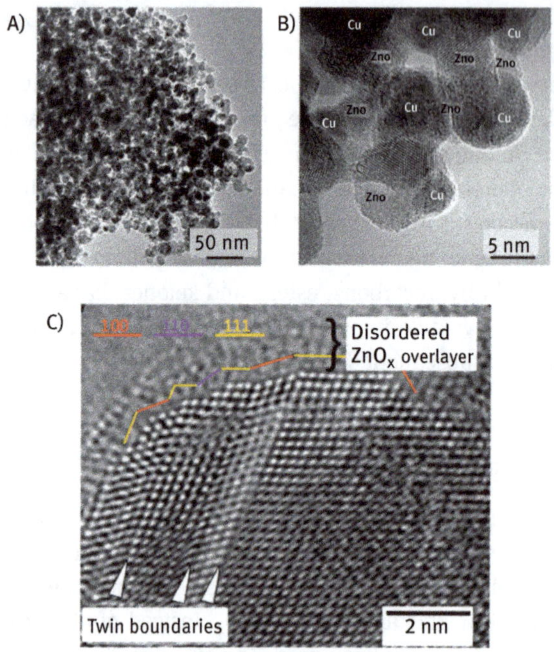

Figure 14.7: (High-resolution) TEM images of a $Cu/ZnO/Al_2O_3$ methanol synthesis catalyst consisting of porous aggregates (A) of metallic Cu and ZnO nanoparticles (B [71]) showing details of the surface faceting, decoration and defect structure (C [72]), which is discussed in detail in the text.

If reliable data of the average Cu particle size are available, for example, by TEM observations [71], the degree of oxide coverage of the Cu particles, that is, the average ratio of interface area to surface area, can be calculated [77]. For industrial $Cu/ZnO/Al_2O_3$ catalysts, this value is around 35%. The favorable microstructure and the proper balance of Cu dispersion and loading in this type of $Cu/ZnO/Al_2O_3$ catalysts lead to a large SA_{Cu}, which is probably the most important property of a methanol synthesis catalyst.

The SA_{Cu} has been observed to scale linearly with the activity for sample families with a similar preparation history [78]. However, between these families considerably different intrinsic activities, that is, activities normalized by SA_{Cu}, can be found [77]. Thus, in agreement with the structure sensitivity of methanol synthesis over Cu [79], different Cu surfaces can be prepared, which vary in the activity of their exposed sites and/or in the concentration of these sites. A thorough study of the particle size effect has shown that the TOF of methanol synthesis decreases if the Cu particles do not exceed a minimum size of approximately 10 nm [80].

Differences in intrinsic activity of the exposed SA_{Cu} can be related to defects and disorder in the Cu nanoparticles and to the role of ZnO. One role of ZnO is to act as spacer and stabilizer avoiding direct contact of the Cu nanoparticles thus preventing them from sintering (Figure 14.7B) [81].

Yet, it is generally agreed that the role of ZnO in Cu-based methanol synthesis catalysts exceeds the function of a mere physical stabilizer. In addition to this geometrical function, a so-called Cu-ZnO synergy is described in literature for methanol synthesis [60, 82, 83]. The nature of this synergy and the contribution of ZnO to the active site of methanol synthesis are strongly debated and several models have been proposed, for example, Cu^+ ions in the ZnO matrix [84], ZnO segregated on Cu^+ [85], electron-rich Cu at the Cu-ZnO heterojunction [86], CuZn surface alloy formation [87, 88], Cu metal supported on ZnO [89], or perimeter sites between the two components [90]. Strong metal-oxide interactions (SMSI) between Cu and ZnO were observed under highly reducing conditions [91, 92] and it was suggested that partially reduced ZnO_x migrates onto the surface of the Cu particles under methanol synthesis conditions [93]. On a supported Cu/ZnO model catalyst, reversible wetting/de-wetting was observed as the reduction potential of the gas phase was varied [94], an observation not made on Cu/SiO_2.

Another contribution to variations of intrinsic activity is the different amount of defects and disorder in the metallic Cu phase. This disorder can manifest itself in form of lattice strain detectable, for example, by line profile analysis of XRD reflections [63, 95], Cu-NMR lines [96], or as an increased disorder parameter (Debye–Waller factor) derived from extended X-ray absorption fine structure (EXAFS) spectroscopy [97]. Strained Cu^0 has been shown theoretically [98] and experimentally [99] to have different adsorptive properties compared with unstrained surfaces. Strain, that is, local variation in the lattice parameter, is known to shift the center of the d-band and alter the interactions of metal surface and absorbate [100].

The origin of strain and defects in Cu/ZnO is probably related to the crystallization of kinetically trapped non-ideal Cu^0 in close interfacial contact to the oxide during catalyst activation under mild conditions. A correlation of the concentration of planar defects in the Cu^0 particles with the catalytic activity in methanol synthesis was observed in a series of industrial $Cu/ZnO/Al_2O_3$ catalysts [71]. Planar defects like stacking faults and twin boundaries can also be observed by HRTEM and are marked with arrows in Figure 14.7C [72]. They lead to changes in surface faceting of the Cu^0 nanoparticles (Figure 14.7C) associated with the formation of steps and kinks that were assumed to represent high-energy surface sites of special catalytic activity [72].

For a series of Cu/ZnO-based catalysts a linear correlation of the defect concentration with the intrinsic activity of the exposed Cu surface was observed. In addition, (partial) surface decoration of Cu with ZnO_x by SMSI has been confirmed by HRTEM (Figure 14.7C) and in situ XPS. The high catalytic activity of surface steps containing Zn was supported by DFT calculations. Thus, the active site of industrial methanol synthesis was identified as a complex surface ensemble requiring a high-energy site due to defects in Cu and the presence of Zn in the close vicinity due to in situ adjustment of Zn surface decoration.

In the technical catalyst, the three requirements large SA_{Cu}, defective Cu nanoparticles and many reactive interfaces to ZnO are efficiently established by the nanoparticulate and porous Cu/ZnO arrangement shown in Figure 14.7. Preparation of this microstructure requires a homogeneous and maximized intermixing of the Cu and Zn species in order to stabilize the alternating arrangement of small Cu and ZnO nanoparticles. Different methods of Cu/ZnO catalyst preparation can be found in literature [101], but by far the most important route is co-precipitation.

The main goal of catalyst synthesis is to preserve and maintain the perfectly homogeneous cation distribution in the starting solution to a maximum extent in the final catalyst [102]. Figure 14.8 gives a schematic overview of the multistep synthesis route of Cu/ZnO catalysts introduced by ICI in the 1960s [81, 103].

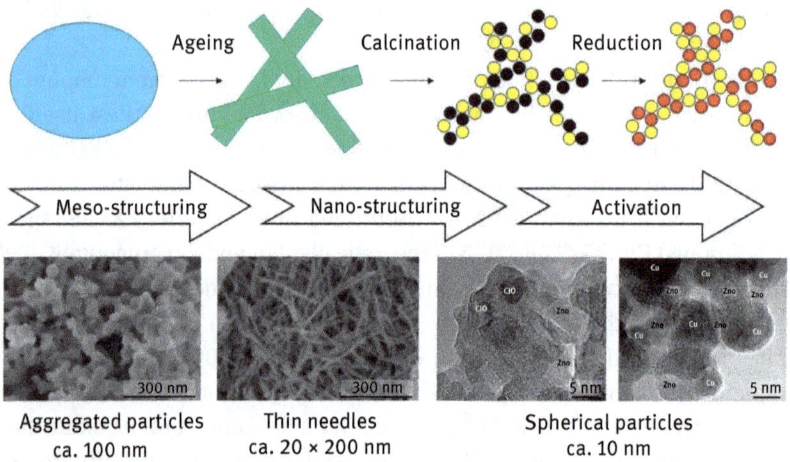

Figure 14.8: Simplified geometrical model [104, 105] for the preparation of industrial Cu/ZnO catalysts comprising subsequent meso- and nano-structuring of the material from [70]. In the first microstructure-directing step (meso-structuring), the Cu,Zn co-precipitate crystallizes in the form of thin needles of the zincian malachite precursor $(Cu,Zn)_2(OH)CO_3$. In the second step, the individual needles are decomposed and de-mix into CuO and ZnO. The effectiveness of this nanostructuring step depends critically on a high Zn content in the precursor, which in zincian malachite is limited to Cu:Zn ca. 70:30 due to solid-state chemical constraints [97]. Finally, inter-dispersed CuO/ZnO is reduced to yield active Cu/ZnO.

The synthesis route comprises co-precipitation and aging [96, 106] of a mixed Cu, Zn,(Al) hydroxycarbonate precursor material, thermal decomposition yielding an intimate mixture of the oxides and finally activation of the catalyst by reduction of the Cu component [107]. The synthesis parameters of this route have been studied in many academic and industrial groups and a high degree of optimization has been achieved over the last decades by mostly empirical fine-tuning of the conditions [102, 108–110], The delicate nanoparticulate and porous microstructure of the industrial methanol synthesis catalyst (see above) can be obtained only if the optimized parameters are strictly obeyed during synthesis.

Especially the synthesis conditions during the early co-precipitation and aging steps are crucial for the catalytic properties of the resulting methanol synthesis catalyst. This phenomenon, sometimes termed “chemical memory” of the Cu/ZnO system, indicates the critical role and sensitivity of the parameters pH and temperature for the final product [95, 109, 111].

Typical $Cu/ZnO/Al_2O_3$ catalyst lifetimes are about 2 years, with one third of the total activity loss occurring during the first 1000 hours of operation. The loss of activity with time is compensated by increasing the reaction temperature. Deactivation occurs through loss of copper dispersion during particle growth or poisoning by impurities. Recent studies of deactivation kinetics [112] and structural consequences [113] of aging showed that ZnO and $ZnAl_2O_4$ crystallize and grow during the first 60 days on stream. Also, high partial pressures of water associated with CO_2 rich gas mixtures lead to an increased particle growth. However, the complete removal of CO_2 leads to an even faster deactivation by dispersion loss [114].

Sulfur is a potent poison for Cu catalysts but sulfur poisoning is seldom a problem as syngas feeds are desulfurized to less than 0.5 ppm. The ZnO component provides some protection against sulfur poisoning by scavenging sulfur compounds irreversibly as ZnS, thereby preserving a large fraction of catalyst activity [60, 114].

14.3.1.3 The Reaction Mechanism of Methanol Synthesis

Gas feeds for industrial methanol synthesis usually contain both CO and CO_2. Which carbon oxide serves as the primary source for methanol formation has been arguably the most important question pertaining to the reaction mechanism. Early work by Klier and co-workers assumed that CO was the primary source, and that the active site consisted of Cu^+ species dissolved in ZnO [84]. However, this model predicted a zero rate of methanol production in the absence of CO_2, and it was assumed that CO_2 and water prevented over-reduction of Cu and thus helped maintaining a population of active Cu^+ species [115, 116].

In the 1980s, experiments conducted by Rozovskii and later by Chinchen et al. involving the use of ^{14}CO or ^{14}CO tracers in methanol synthesis from $CO_2/CO/H_2$ mixtures over commercial catalysts proved conclusively that CO_2 was the primary methanol source [117, 118]. Chinchen et al. measured the radioactivity of reaction products at the outlet of a reactor operating with a $^{14}CO/^{12}CO$ feed. As shown in Figure 14.9, at lower space velocities, scrambling of carbon isotopes between CO and CO_2 through the WGS reaction resulted in the incorporation of both isotopes into methanol.

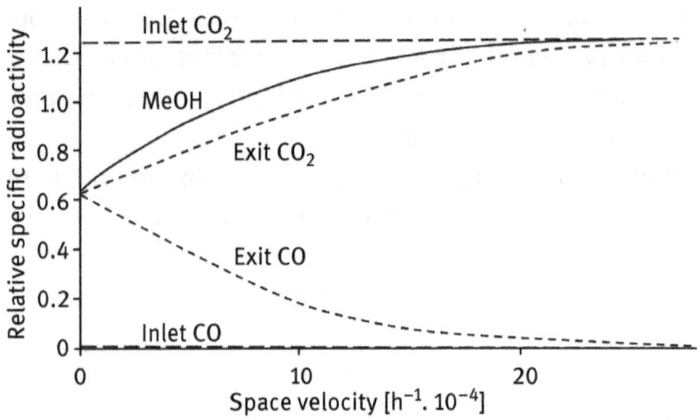

Figure 14.9: Effect of space velocity on ^{14}C distribution in methanol synthesis products (reproduced from ref. [117]).

At high space velocities, where the rate of scrambling is negligible and conversion is low, methanol retained the specific radioactivity of the $^{14}CO_2$, indicating that only carbon from CO_2 was incorporated. Chinchen et al. showed that CO_2 was the primary carbon source for methanol even when present at very low concentrations (100 ppm).

In situ spectroscopic studies identified a variety of species such as formate, carbonate, and methoxide to co-exist under methanol synthesis conditions on Cu/ZnO-based catalysts [89, 119]. FTIR studies of Cu/ZnO based catalysts under H_2/CO_2 identified the presence of formate bound to both Cu and ZnO, whereas methoxy was found on ZnO only. Carbonates form via CO_2 adsorption on ZnO [120] and partially oxidized Cu [89] and quickly convert into formate via Cu-activated hydrogen. Upon exposure to CO mixtures, only zinc-bound formate is observed [119].

The hydrogenation of these formates to methoxy is thought to be rate-determining in methanol synthesis. However, this hydrogenation may not be direct, as – even in the presence of H_2 – the rate of methanol synthesis on Cu/SiO_2 from formate was negligible compared with the rates of formate decomposition into CO_2 and H_2 [121]. The presence of water and/or OH groups is critical for methanol formation [122] and the absence of direct coverage by reducible oxygen species (and therefore of Cu^+) was confirmed by CO pulse experiments on catalysts in the working state [123].

Surface science studies on Cu single crystals also identified formate as an abundant surface species, but reported a wide range of activities for methanol synthesis. The intrinsic reaction rates reported for the hydrogenation of CO/CO_2 mixtures on Cu(100) were four orders of magnitude below those reported for high surface area Cu/ZnO catalysts [124, 125].

In comparison, using CO_2/H_2 mixtures the rates for Cu(100) increased up to 2–3 orders of magnitude. Similar rates to those on high surface area catalysts were achieved on polycrystalline copper [126], and even three times higher rates on the more open Cu(110) surface [79]. As there is no oxygen on the Cu surface, metallic Cu is assumed to be the active phase [79, 127]. Furthermore, ZnO is considered to maintain metallic Cu in a form that more closely resembles an open surface, thereby acting as a promoter [79]. Direct deposition of up to 0.19 ML of Zn on a polycrystalline Cu surface has been shown to increase CO_2 hydrogenation activity by a factor of 6, thus suggesting a more direct role for ZnO [128].

The structure sensitivity of the rWGS reaction accompanying methanol synthesis is even more pronounced, with Cu(110) being an order of magnitude more active than polycrystalline Cu. Furthermore, kinetic studies of CO_2 hydrogenation on Cu/ZnO catalysts yielded a temperature-dependent non-zero CO selectivity at the limit of zero CO_2 conversion [79, 119, 129]. These findings suggest that both reactions do not share a common intermediate. Whereas methanol formation is associated with formate hydrogenation, the rate of the rWGS reaction is controlled by direct dissociation of CO_2, a direct redox process that is more favorable on open surfaces [79].

Theoretical studies confirm the importance of formate as a surface intermediate and its sequential hydrogenation as being rate-determining. A highly simplified version of the mechanism proposed by Askgaard et al. for methanol synthesis from CO_2 on Cu(111) is given below [130]:

$$\text{Step 1:}\quad H_2 + 2^* \rightleftharpoons 2H-^* \tag{14.46}$$

$$\text{Step 2:}\quad H_2O + 2^* \rightleftharpoons H-^* + OH-^* \tag{14.47}$$

$$\text{Step 3:}\quad CO_2 + 2^* \rightleftharpoons CO-^* + O-^* \tag{14.48}$$

$$\text{Step 4:}\quad CO_2 + H^* \rightleftharpoons HCOO-^* \tag{14.49}$$

$$\text{Step 5:}\quad HCOO-^* + 3H-^* \rightarrow CH_3O-^* + OH-^* + 2^* \tag{14.50}$$

$$\text{Step 6:}\quad H_3CO-^* + H-^* \rightleftharpoons CH_3OH + 2^* \tag{14.51}$$

Therein, steps 3–6 may each represent a series of reaction events. The formation of formate from CO_2 with adsorbed atomic hydrogen is a fast process (step 4), compared with forming carbonate by binding to surface oxygen species [131, 132]. Askgaard et al. propose that the hydrogenation of HCOO* is rate-limiting (step 5), whereas Yang et al., Grabow, Hu et al., and Mavrikakis identify the hydrogenation of HCOO* and CH_3O^*, respectively, as rate-determining [131–133].

The presence of adsorbed Zn may promote methanol synthesis on Cu(111) by stabilizing formate and associated transition states [134]. It was shown that methanol cannot form directly through formate hydrogenation on Cu(111), but instead forms by a hydrogen transfer reaction requiring the presence of water [135]. According to Grabow and Mavrikakis, the main role of CO in promoting the hydrogenation of CO_2 is the removal of site-blocking OH species, although the direct hydrogenation of CO is said to account for 1/3 of the methanol produced under industrial conditions [132]. Studt et al. discuss the role of Zn at the surface for the promotion of the hydrogenation of the key intermediate formate [136].

14.3.1.4 Methanol Synthesis from CO_2 – Thermodynamic and Kinetic Considerations

Methanol synthesis from waste CO_2 streams has potential to contribute to the limitation of worldwide CO_2 emissions and to serve as an alternative carbon source to fossil fuels, provided that a renewable source of H_2 is available (Section 14.3) [137]. The main obstacle to methanol synthesis from CO_2-rich streams is thermodynamics. Figure 14.10 depicts the equilibrium yield of methanol from mixtures with 75% H_2 and 25% CO/CO_2 with varying CO/CO_2 ratio.

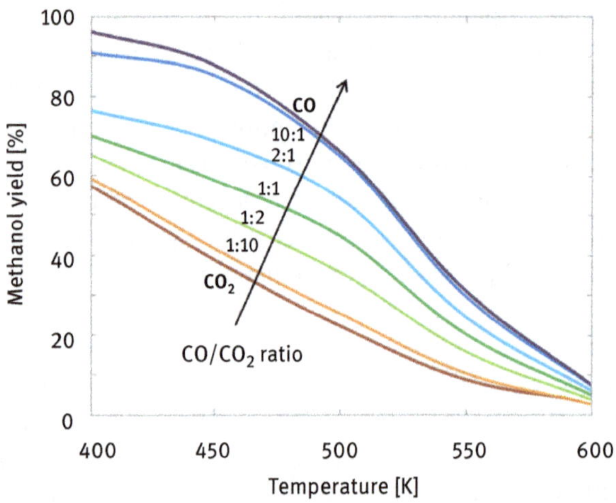

Figure 14.10: Equilibrium yields of methanol as a function of the temperature and the $CO:CO_2$ ratio in 3:1 $(CO:CO_2)/H_2$ mixtures at 50 bar.

For pure CO_2, a one-pass methanol yield of 18% is realized at 525 K. For pure CO nearly 55% can be obtained. Thus, the equilibrium limitation can be overcome by the addition of CO to CO_2. Operating at lower temperatures would also be an option, which, however, calls for more active catalysts. Furthermore, higher recycle ratios or product extraction would improve the yield but result in higher capital investment [60]. The thermodynamic limitations of methanol synthesis from CO_2 can be circumvented by innovative reactor design, which involves efficient product removal without excessive recycling of feed gases.

One strategy involves a liquid-phase reactor using a high-boiling hydrocarbon oil solvent [138]. Methanol and water can be separated from the solvent upon cooling and the solvent is recycled back into the reactor. Liquid-phase methanol synthesis has the advantage of higher capacities and better temperature control. However, diffusion limitations may be encountered with solvent present in the pores of the catalyst. Furthermore, catalysts that are stable in the solvent environment must be developed. Other strategies involve the use of alkane and alcohol solvents under supercritical conditions. A series of gas-phase reactors with inter-stage absorption of methanol and water by a polar liquid solvent or an in situ desorption with a solid such as alumina have also been proposed [60, 139].

Although CO_2 has been shown to undergo hydrogenation faster than CO, kinetic limitations would arise in any industrial process operating at high conversion. Several investigators have shown that a maximum in methanol production with varying CO_2/CO ratio occurs at CO_2 concentrations of 2–5 mol% of total carbon [116, 129, 140]. Further investigations showed that such a maximum in the synthesis rate is only exhibited at high reactant conversion, thus indicating that product inhibition plays an important role [129, 140–142]. Sahibzada et al. studied the methanol synthesis rate as a function of the CO_2/CO ratio in an integral internal recycle reactor compared with a differentially operated down flow micro reactor [129]. Their results (Figure 14.11) show that a maximum exists at 2% CO_2 in the integral reactor, while in the differential reactor operated at <0.3% methanol yield an almost linear relation between methanol synthesis rate and CO_2 concentration was observed.

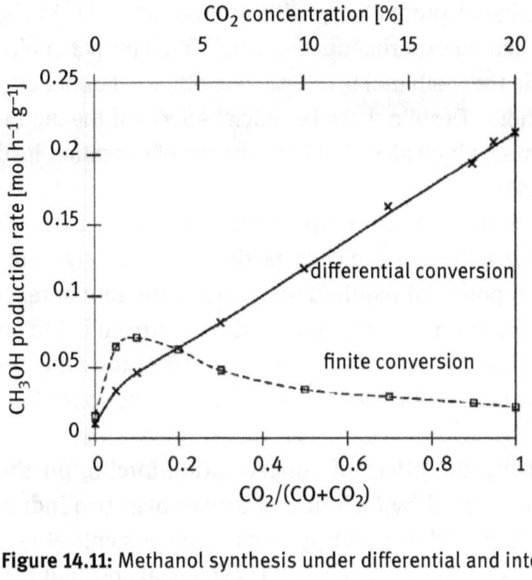

Figure 14.11: Methanol synthesis under differential and integral conditions at 523°K and 50 bar from ref. [129].

These investigators also correlated the methanol synthesis rate under integral conditions with the concentration of water in the product stream during CO_2 hydrogenation. Furthermore, they showed that co-feeding of water to a differential reactor in concentrations similar to those formed during integral operation resulting in a 10-fold decrease of the methanol synthesis rate [129]. These results show conclusively that water inhibits methanol synthesis from CO_2 and that the promotional effects of CO at integral conversion arise from its ability to convert excess water via the WGS reaction [89, 143].

14.3.2 Decentralized Methanol Synthesis Under Dynamic Reaction Conditions

First experimental studies dealing with a forced variation of reactant concentrations in methanol synthesis were carried out already during the 1980s [144–147] to identify optimal process parameters. Nappi et al. [146] observed a beneficial response for periodically switching between feeds with and without 2% CO_2 at constant H_2/CO ratio. Methanol production increased by 27% compared with steady state. This maximum in terms of methanol production was obtained for a cycle split of 0.01, which represents the fraction of the period during which the catalyst was exposed to the CO_2-containing pulse.

In view of the impact that CO_2 had on the methanol synthesis rate during the periodic modulation of reactant concentrations, Chanchlani et al. investigated the effect of cycling parameters such as cycle period, the amplitude of the composition change, and the cycle split on the catalytic activity in greater detail [144, 145]. Steady-state measurements performed with different H_2:CO_2 mixtures showed that Cu/ZnO and Cu/ZnO/Al_2O_3 reach maximum methanol productivity with around 20% CO_2 in the feed [144]. Periodic composition switching experiments indicated that a proper choice of the cycling parameters can double the methanol synthesis rate obtained under otherwise identical steady-state conditions. Despite these beneficial effects of the modulation approach, stability measurements indicated that both modes of operation lead to catalyst deactivation over time [145].

Evaluation of the data suggests that forced composition modulation between CO_2 and H_2 mixtures modified the structure or the composition of the catalyst surface so that it changed the chemical potential required for an optimum rate of reaction in comparison to steady state. Furthermore, the modulation approach led to an increased methanol yield, which might be related to an adjustment of the ratio between the reactants at the catalyst surface that meets the stoichiometric conditions necessary to maximize the rate.

Similar investigations addressing the effect of concentration-forcing on the rate of methanol synthesis were performed by McNeil and Rinker over two industrial catalysts (BASF S 3-85 and ICI 51-2) [147]. For both systems, improvements of 15% (ICI 51-2) and 25% (BASF S 3-85) were achieved during H_2/CO cycling at constant CO_2 concentrations of around 2–3%. Some discrepancies are found compared with the results reported by Chanchlani et al. [144], which may be related to differences in the feed compositions, the nature of the employed catalysts, and probably the time scale of the cycle periods. However, these studies show the potential of improving the methanol synthesis rate if proper values for the cycling parameters are chosen. A theoretical investigation that addresses the impact of a square-wave concentration cycling of reactants on the performance of a dynamically operated isothermal catalytic reactor reveals that the degree of improvement strongly depends on whether sorption or reaction nonlinearities dominate during the modulation process [148].

Another type of forced unsteady-state operation that has been applied to methanol synthesis is the periodic flow reversal approach, which consists of cycling the direction of the gas flow that passes through the catalyst bed [149–151]. The idea behind the dual concept is that the catalyst acts both as an accelerator of the chemical reaction and as a heat regenerator [152]. At the beginning of the process, the gas fed into the reactor exhibits a lower temperature compared with the reaction zone, resulting in a negligible rate. A gradual increase in conversion is accompanied by heat release due to the exothermic character of the process. Thus, the temperature and concentration profiles that are formed move parallel to one another in flow direction. After flow reversal, the accumulated heat is transferred to the colder gas front thereby lowering the temperature inside of the reactor, giving rise to favorable conditions for an equilibrium-limited reaction like the synthesis of methanol [149, 152].

An alternative to overcome the sharp decrease of product concentrations upon switching while maintaining the same strategy is to perform the reverse flow using three reactors arranged in a so-called STAR configuration [151]. Each bed is operated in a cyclic mode starting with an *inlet*, followed by a *blow out* and ending with an *exit* phase. Although the "dip" phenomenon was still present, the amplitude was reduced by more than 80% using the STAR configuration. As a consequence, the time average conversion is increased to an extent that it exceeds the values of the steady-state process by 10% and that of the conventional reverse flow process by around 5%.

In addition to the use of CO_2 for direct hydrogenation to methanol, the thermochemical cleavage of CO_2 and H_2O on redox-active metal oxides using sunlight has proven to be a highly advantageous approach for the sustainable generation of synthesis gas [153] (Section 14.1.2). Using a solar-powered reactor, about 700 standard liters of syngas (H_2:CO:CO_2 = 33.7%: 19.2%: 30.5%) were generated after 290 redox cycles on a CeO_2 catalyst. The Pérez-Ramírez group used a syngas mixture of similar composition to highlight the effect of the "daily startup-shutdown" (DSS) routine on methanol synthesis on a commercial Cu/ZnO/Al_2O_3 catalyst (see Figure 14.12) [154].

Contrarily to the forced variation of reactant concentrations and the reversed flow operation discussed above, the focus of this investigation was to study the robustness of a sunlight-driven process against periodic variations arising from the natural fluctuations of daylight. Pure CO/H_2 or CO_2/H_2 mixtures and stationary conditions were used as reference states. As expected, pure CO hydrogenation leads to a rapid decrease of the initial catalytic activity, whereas the deactivating effect is barely noticeable in the presence of pure CO_2. According to DFT calculations, active sites for CO hydrogenation are blocked due to the low interaction of carbon atoms with Zn [136]. In contrast, the adsorption of oxygen species on the surface in the presence of Zn atoms is improved to accelerate CO_2 hydrogenation.

In view of the inherent fluctuations discussed for renewable energy sources, the synthesis of methanol was carried out under non-stationary conditions. The effects of the dynamic operation on the catalyst were systematically tested in "wet DSS"

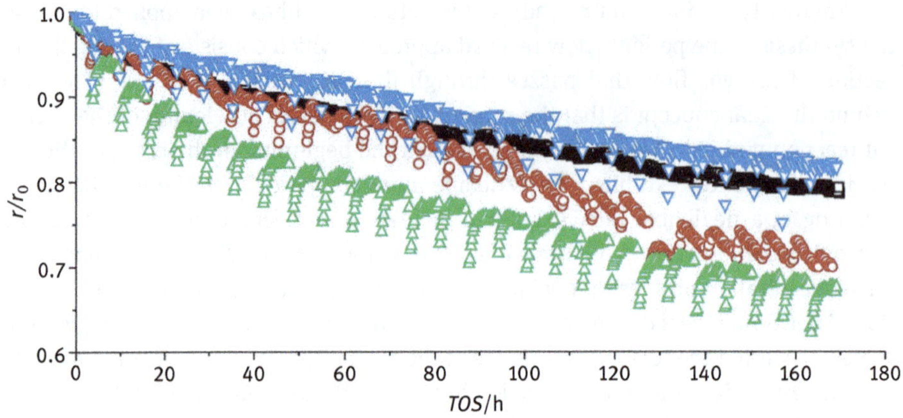

Figure 14.12: Evolution of the relative methanol production rate during steady-state (□), wet DSS (○), dry DSS(△), and CO_x DSS (▽) measurements over Cu–ZnO–Al_2O_3 [154]..

(feed off, reactor outlet closed, entire system cooled down), "dry DSS" (initial pressure drop from 50 to 2 bar, reaction chamber purged with nitrogen at 230 °C for 1 h, system cooled under static inert conditions), and "CO_x DSS" (dry DSS protocol, but switching off and on the reactor in mixture of CO/CO_2 and in the absence of H_2). Monitoring more than 25 cycles of the different DSS strategies, no significant difference in deactivation between "CO_x DSS" and steady-state conditions was observed. After approximately 70 h of operation, catalyst deactivation becomes more pronounced for the "wet DSS" operation mode compared to steady state and the rate of "wet DSS" converges toward that of dry DSS, which does not seem suitable to retain catalyst performance.

14.3.3 Methanol Steam Reforming (MSR)

Methanol steam reforming (MSR) refers to the reverse methanol formation reaction out of CO_2-rich syngas (eq. (14.43)). MSR is the preferred way to liberate H_2 from methanol as it generates a high H_2 concentration in the product stream, runs at relatively low temperatures of 473–573 K, and unlike methanol decomposition (reverse of eq. (14.44)) does not directly produce CO, which would act as a poison for the downstream PEM-FC anode catalyst. MSR was first described in 1921 by Christiansen [155] and research on its application for H_2 production has a long history [156]. The recently renewed interest was triggered by the development of fuel cell technology requiring clean and preferably renewable hydrogen. A number of overview articles and reviews are available addressing the role of MSR in this context [61, 156–161]. In areas where steam reforming of natural gas is not an option, MSR is also applied in the methanol-to-hydrogen (MTH) process to produce H_2 in relatively small-sized units.

MSR is an endothermic reaction and requires external heating. It is sometimes used in combination with exothermic partial oxidation of methanol (autothermal reforming or oxidative steam reforming) [162, 163] or combustion of methanol [159] in order to generate the necessary heat. The endothermicity of MSR is much weaker compared with steam reforming of other hydrocarbons or higher alcohols [159]. Therefore, reformer units can be relatively small enabling the onboard combination with PEM-FCs. A comparison between methanol and other molecules as reactants for onboard H_2 production can be found in the comprehensive review by Palo et al. [61].

In technical MSR applications, a gas purification step has to be introduced between reformer and fuel cell because of the rWGS equilibrium yielding CO. CO chemisorbs irreversibly on the Pt-based fuel cell catalysts and causes irreversible site blocking. Its concentration has to be below ca. 20 ppm in order to prevent poisoning, which is usually not achieved in the reformer outlet gas.

The CO concentration in the gas stream can lowered by means of the WGS reaction, preferential oxidation (PROX) or using palladium membranes, which in all cases complicates the setup and generates costs [164]. Generally, in addition to high activity and stability, a low selectivity to CO is a major and particular requirement for a successful MSR catalyst to be used for onboard H_2 production. Lower temperature operation would thermodynamically hinder CO formation via methanol decomposition and rWGS. Also, the methanol reforming catalysts should be particularly stable toward abrupt changes of the conditions of reforming, that is, work reliably in transient situations like on-off operations as well as at steady state to produce sufficient amounts of H_2 on demand.

14.3.3.1 Catalysts for Methanol Steam Reforming

Commercial industrial Cu/ZnO-based catalyst for methanol synthesis (see Section 14.3.1) or WGS are also active in MSR. $Cu/ZnO/Al_2O_3$ catalysts or the unpromoted binary Cu/ZnO model system were thus employed in many studies of MSR [70]. Modifications of the Cu/ZnO/X system, in particular, choosing another second oxide phase X like rare earths [165] or ZrO_2 [166–170], employing new catalyst precursors like layered double hydroxides [171–174], or even changing to ZnO-free samples and using ZrO_2 and/or CeO_2 [175–179] were reported to lead to interesting MSR performance.

There are many similarities between methanol synthesis and MSR [180], but due to different working conditions, the application of the concept of microscopic reversibility is limited as has been pointed out by Spencer for WGS and rWGS [181] and is valid accordingly also for methanol synthesis and MSR [58]. However, the optimized methanol synthesis catalyst will also be active in MSR due to its generally large SA_{Cu} and represents a powerful reference system, but it does not necessarily represent the optimal catalyst for this reaction [58]. Furthermore, once activated it is highly pyrophoric. Finding Cu/ZnO/X systems with a composition and microstructure optimized for the MSR reaction is thus the major current challenge in development of a MSR catalyst for energy applications. In the case of MSR the Cu-ZnO synergy (see Section 14.3.1.2) does not seem to be as critical as compared to methanol synthesis [182] or this synergistic effect is not as strictly limited to ZnO. Highly active Cu-based MSR catalysts can also be prepared without ZnO, for example, as Cu/ZrO_2.

Also commercial WGS catalysts have been tested in the MSR reaction and $Cu/ZnO/Al_2O_3$ formulations were found to be most active but also very prone to deactivation by sintering [183] and decrease in intrinsic activity [184]. In addition to pyrophoricity and instability, the other major problem of Cu-based MSR catalysts is the formation of CO during MSR, typically in the low percent range. However, this could be attenuated by increasing the steam-to-methanol ratio or by the addition of O_2 or air (oxidative MSR) [185].

Pt- and Pd-based intermetallic catalysts have been suggested as an alternative for Cu [70, 186]. In particular, the ordered intermetallic compound PdZn shows interesting MSR activity and CO_2 selectivity, which are higher compared with non-alloyed pure Pd which is a methanol decomposition catalyst [187–190]. With regard to application, these and other new catalysts have to prove their competitiveness in comparison to the well-established Cu/ZnO system.

14.3.3.2 Mechanism and Active Sites of Methanol Steam Reforming

Several studies are available addressing the mechanism and kinetics of the MSR reaction over Cu-based catalysts [180, 191–196]. There is agreement that CO_2 is a direct product of the MSR reaction and not of a sequence of methanol decomposition and WGS reactions. The main source of CO is the rWGS reaction taking place as a secondary reaction after MSR. Frank et al. [196] presented a comprehensive microkinetic analysis of the MSR reaction based on the work of Peppley et al. [193]. They investigated several Cu-based catalysts with various oxide components showing considerably different activities. Similar activation energies support the idea that the surface chemistry is independent of the oxide material with the exception of $Cu/Cr_2O_3/Fe_2O_3$, which behaved differently. Dehydrogenation of methoxy groups is the rate-limiting step and by means of DRIFTS experiments methoxy and formate species were found as the dominating species at the surface. Two reaction pathways of the methoxy intermediate are discussed via dioxomethylene/formate, intermediates of the reverse methanol synthesis reaction, or methyl formate. Theoretical studies favor the former pathway [197, 198].

The active site of Cu-based catalysts in MSR is still unclear and debated. Two distinct kinds of active sites were considered, one for the adsorption and desorption of oxygenates and one for hydrogen. Hydrogen is dissociatively adsorbed at the metallic Cu surface. Similar to methanol synthesis, metallic Cu^0 and Cu^+ sites dispersed on the oxide component or at the Cu-oxide interface are discussed to contribute to the active ensembles at the Cu surface. In addition to the available surface area SA_{Cu}, the variable oxidation state of Cu, in particular the Cu^0/Cu^+ ratio at the catalyst surface has a great impact on MSR activity [167, 172, 199–205], Furthermore, the defect structure and varying amount of disorder in metallic Cu depending on the microstructure and preparation history of the catalyst seem to be important [96, 97, 206, 207]. Finally, the oxidic surface of the refractory component may take part in the catalytic reaction and provide adsorption sites [208].

14.4 Fischer–Tropsch Synthesis

Fischer-Tropsch synthesis (FTS) converts syngas (CO + x H_2, where x can range between 1 and 3) into mostly liquid hydrocarbons [47, 209–211], Depending on the origin of the syngas, the overall process from carbon feedstock to liquid product is nowadays termed "gas to liquids" (GTL), "coal-to-liquids" (CTL), "biomass-to-liquids" (BTL) or "anything-to-liquids" (XTL). Also the recently adopted term "power-to-liquids" often refers to the FTS, although other liquids such as methanol and DME also come under this denominator. The product spectrum of FTS, however, is broader than liquid hydrocarbons alone, and includes gaseous products like C_1–C_4 hydrocarbons, liquid alkanes, or alkenes, and – to a lesser extent – oxygenated products such as alcohols. Hence, the FTS offers the opportunity to convert gas, coal, or biomass-derived syngas into transportation fuels, such as gasoline, jet fuel and diesel oil, and chemicals such as olefins, naphtha, and waxes.

While FTS technology relies in principle on CO_2-free syngas, the applicability of CO_2-rich mixtures, obviously of interest in the context of the present book, is presently actively being researched. We address this in Section 14.4.5.

Catalysts for FTS are based on iron and cobalt. Ruthenium also shows good performance, but is considered too expensive, while nickel produces mainly methane and can be used for methanation, for example, in biomass or coal to substitute natural gas processes. At present, FTS is applied commercially in South Africa (CTL, GTL), Malaysia (GTL), Qatar (GTL), Nigeria (GTL), and China (CTL), while another GTL plant in Uzbekistan as well as CTL plants in China are under construction. Recent reviews of FTS, emphasizing industrial application, are available [212, 213].

The formal overall reaction equations are straightforward, but hide a tremendous amount of mechanistic detail as we discuss later on in this chapter:

$$n\,CO + (2n+1)\,H_2 \rightarrow C_nH_{2n+2} + n\,H_2O, \quad \Delta H_R = -50\,kJ\,mol^{-1} \tag{14.52}$$

$$n\,CO + 2n\,H_2 \rightarrow C_nH_{2n} + n\,H_2O, \quad \Delta H_R = -91\,kJ\,mol^{-1} \tag{14.53}$$

$$n\,CO + 2n\,H_2 \rightarrow C_nH_{2n+1}OH + (n-1)\,H_2O, \quad \Delta H_R = 41\,kJ\,mol^{-1} \tag{14.54}$$

14.4.1 Elementary Reaction Steps

The mechanism of FTS is still a subject of scientific debate. In the simplest mechanism [214], CO adsorbs and dissociates, carbon atoms are hydrogenated to CH_x species, which then couple to form longer hydrocarbons. We illustrate this with the most straightforward – but not necessarily correct! – sequence, in which * denotes an active site at the catalyst surface:

Adsorption and dissociation:

$$CO + * \rightleftharpoons CO - * \tag{14.55}$$

$$CO - * + * \rightleftharpoons C - * + O - * \tag{14.56}$$

$$H_2 + 2^* \rightleftharpoons 2H - * \tag{14.57}$$

O-rejection by water formation

$$O - * + H - * \rightleftharpoons OH - * + * \tag{14.58}$$

$$OH - * + H - * \rightleftharpoons H_2O + 2^* \tag{14.59}$$

C-hydrogenation:

$$C - * + H - * \rightleftharpoons CH - * + * \tag{14.60}$$

$$CH - * + H - * \rightleftharpoons CH_2 - * + * \tag{14.61}$$

$$CH_2 - * + H - * \rightleftharpoons CH_3 - * + * \tag{14.62}$$

Chain growth (for example):

$$CH_3 - * + CH_2 - * \rightleftharpoons C_2H_5 - * + * \tag{14.63}$$

$$C_nH_{2n+1} - * + CH_2 - * \rightleftharpoons C_{n+1}H_{2(n+1)+1} - * + * \tag{14.64}$$

Termination:

$$CH_3 - * + H - * \rightleftharpoons CH_4 + 2^* \tag{14.65}$$

$$C_2H_5 - * + H - * \rightleftharpoons C_2H_6 + 2^* \tag{14.66}$$

$$C_2H_5 - * \rightleftharpoons C_2H_4 + H - * \tag{14.67}$$

Other schemes exist as well, for example, where the initial CO dissociation step is replaced by a hydrogen-assisted step (Figure 14.13) [46, 48, 49, 214], or where chain growth is proposed to occur by insertion of CO, while the C–O bond breaks after incorporation into the growing chain. Also, for the chain growth several alternatives have been proposed [212].

Computational modeling can investigate the likelihood of the reaction steps by predicting the energies of the adsorbed intermediates and the barriers for the transition states. In some cases, elementary steps have been studied by surface science, notably by in situ high-resolution XPS at a synchrotron, in combination with vibrational spectroscopy and temperature-programmed reaction spectroscopy [215].

On cobalt metal, for example, CO dissociates directly at step sites [216], and removal of oxygen atoms by hydrogen has the highest barrier [217]. The monomer species is the methylidyne (CH), and chain growth is believed to occur via the coupling reaction between methylidyne and alkylidyne, for example, CH + CCH_3 yields adsorbed propyne, HCC–CH_3, which subsequently reacts with a hydrogen atom to propylidyne, CCH_2CH_3. Due to the coupling step between adjacent species, the growing chain is not fixed to a single site, but switches to the site of the monomer and is mobile on the surface [218].

14.4.2 Hydrocarbon Product Distribution

Irrespective of what the exact elementary steps at the molecular level may look like, FTS is a polymerization process. And even without knowing the exact nature of the monomer and the initiating species, we can derive a general expression for the distribution of chain lengths in the products [210]. We will assume that the monomer is a species containing one carbon atom, and the chain initiator is a one-carbon species as well. They may be the same, but do not have to be.

For convenience we follow the scheme of reactions (14.55)–(14.67) above, and consider the initiating species as an adsorbed methyl, CH_3, and the monomer as adsorbed CH_2. Chain growth would then correspond to the insertion of the methylene species into the metal–alkyl bond, as expressed in the reaction steps (14.63) and (14.64). The derivation, however, does not need that we specify the mechanism in detail.

Any alkyl species on the surface can either grow or can terminate by hydrogenation to an alkane or by dehydrogenation to an alkene. The chain-growth probability α determines the average chain length of the products. Accordingly, the probability that a hydrocarbon terminates as a molecule with one carbon atom equals $C_1=1-\alpha$, for two carbon atoms it amounts to $C_2=(1-\alpha)\alpha$, for three carbon atoms $C_3=(1-\alpha)\alpha^2$ and so on. Obviously, a high value of α results in a higher share of long hydrocarbon products.

The expected molar fraction of a hydrocarbon with n carbon atoms can thus be determined only using the chain growth probability:

$$C_n = (1-\alpha)\cdot\alpha^{n-1} \tag{14.68}$$

The total amount of carbon in the product spectrum can be expressed as the convergent sum

$$\sum_1^\infty n\cdot C_n = \sum_1^\infty n\cdot(1-\alpha)\cdot\alpha^{n-1} = \frac{1}{1-\alpha} \tag{14.69}$$

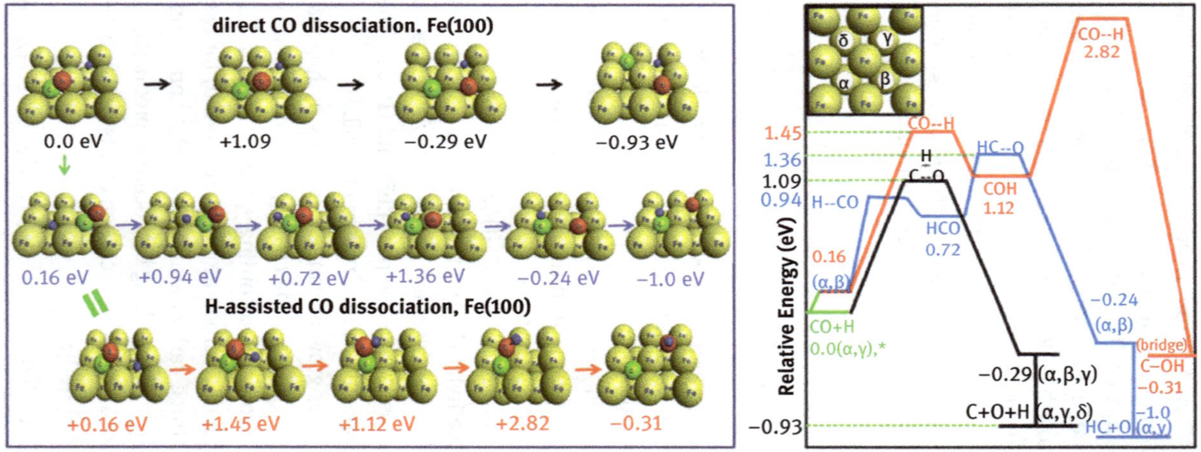

Figure 14.13: Molecular modeling simulation of different CO dissociation pathways on an iron (100) surface, along with relative energies in electron volt (1 eV≈ 96 kJ mol^{-1}). The direct dissociation of CO yields the lowest activation barrier, but the two-step mechanism, in which CO first reacts with H to an HCO intermediate, yields an overall barrier which is not much higher. Hence under conditions where CO is coadsorbed with hydrogen atoms, this route might compete with the direct dissociation (adapted from Elahifard et al. [48]). However, at surfaces containing steps, the direct route is usually energetically the most favorable.

Therefore, the selectivity toward a product with a chain length of n equals

$$S_n = \frac{nC_n}{\sum_1^\infty n \cdot C_n} = n \cdot (1-\alpha)^2 \alpha^{n-1} \tag{14.70}$$

or in logarithmic form:

$$\ln\left(\frac{S_n}{n}\right) = n \cdot \ln \alpha + \ln\left(\frac{(1-\alpha)^2}{\alpha}\right) \tag{14.71}$$

Figure 14.14 depicts the dependence of the Fischer–Tropsch product distribution on the chain-growth probability. The product distribution follows the well-known Anderson–Schulz–Flory (ASF) distribution.

Although the ASF distribution is usually a quite good description of the chain-length distribution, deviations occur as well. Often the methane yield is higher and the C_2 yield lower than predicted. Also, the chain-growth probability is not constant and may be higher for longer hydrocarbons than for shorter ones [219].

14.4.3 Fischer–Tropsch Reactors

Fischer–Tropsch reactions are strongly exothermic and dealing with the released heat is an important engineering consideration when designing reactors for the process. FTS can be carried out in several different reactor types, for example, fixed bed, fluidized bed, or even slurry phase reactors (Figure 14.15).

The slurry reactor, where the catalyst is suspended in the liquid product wax, offers advantages in terms of efficient heat removal and temperature homogeneity. Yet, an efficient solid–liquid separation step is required to recover the product. Also, the catalyst needs to be highly mechanically stable to offer sufficient resistance against attrition.

Typical reaction conditions in non-slurry reactors include temperatures between 473 and 623 K and pressures between 20 and 50 bar. Depending on the temperature the FT process is divided into low-, medium- and high-temperature modes. The high-temperature Fischer–Tropsch (HTFT) synthesis runs at 593–623 K, where typically all products are in the gas phase [220].

HTFT synthesis is operated in fluidized bed reactors with iron catalysts. The product distribution is well suited for gasoline production (C_4–C_{12}) and also contains some olefins and oxygenates. Low-temperature Fischer–Tropsch (LTFT) synthesis runs at temperatures between 473 and 523 K [221–223]. Under these conditions the product distribution extends well into the liquid waxes. LTFT reactors are thus three-phase systems: solid catalysts, gaseous reactants, and gaseous and liquid products. Both cobalt and iron catalysts are applied, although cobalt catalysts are generally used in all modern GTL plants. The LTFT synthesis is ideally suited for the production of high-

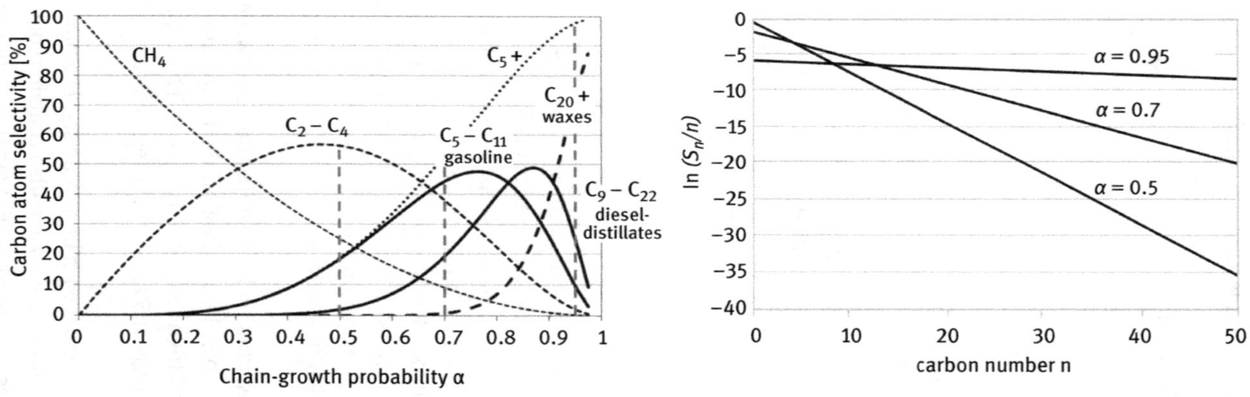

Figure 14.14: Left: Fischer–Tropsch product distributions in terms of hydrocarbon fractions as a function of the chain-growth probability α (14.70). Right: Anderson–Schulz–Flory plots according to the logarithmic formula (14.71).

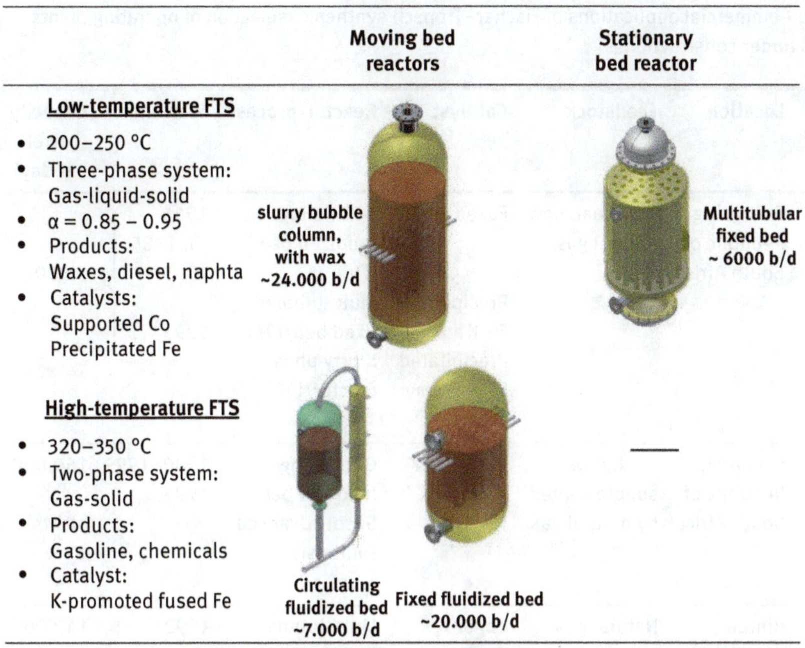

Figure 14.15: Overview of Fischer–Tropsch technologies and common reactor types (figure adapted from [212]). There is also a variation known as medium temperature FTS, which runs typically at 275 °C, and is based on iron catalysts applied in a slurry bubble column [213].

quality middle distillates (diesel and jet fuel, C_{12}–C_{20}) after hydrocracking of the long chain waxes. The waxes can be used as specialty waxes. Finally, the product also contains naphtha, which is a feedstock for further chemical synthesis.

Medium-temperature Fischer–Tropsch (MTFT) synthesis is performed in slurry phase at temperatures around 550–565 K and yields products in the liquids and waxes regime. It relies exclusively on iron catalysts, as cobalt catalysts produce too many short hydrocarbons and methane under these conditions. MTFT is typically used in CTL plants in China, where the world's largest single CTL plant, Shenhua Ningmei, produces about 100,000 barrels of synthetic fuel per day in an MTFT slurry phase process developed by Synfuels China Technology [213].

Nowadays, FTS is regarded as proven technology and an important factor in the monetization of natural gas. Table 14.2 shows an overview of the current and planned commercial Fischer–Tropsch plants anno 2019. Note that capacity of the large FT plants totals to only some 650,000 barrels per day in 2018, compared to a total crude oil production of about 85 million barrels per day.

Table 14.2: Commercial applications of Fischer–Tropsch synthesis: selection of operating plants and plants under construction.

Company	Location	Feedstock	Catalyst	Reactor/process	Start-up	Capacity (barrel per day)
Sasol	Sasolburg, Republic of South Africa	First coal, now natural gas	Fused Fe/K Precipitated Fe/K Precipitated Fe/K (spray dried)	Circulating fluidized bed/ HTFT Multi-tubular fixed bed/LTFT Slurry phase reactor/LTFT	1955 – ca. 1985 1955 1993	5,000
Sasol	Secunda, Republic of South Africa	Coal, now supplemented by natural gas	Fused Fe/K	Circulating fluidized bed Sasol advanced synthesis reactor/HTFT	1980–1999 1995	160,000 (two plants)
Shell	Bintulu, Malaysia	Natural gas	Co/SiO_2 Co/TiO_2	Multi-tubular fixed bed/LTFT	1992	14,500
Petro SA	Mosselbay, Republic of South Africa	Natural gas	Fused Fe/K	Circulating fluidized bed (Sasol technology)/ HTFT	1993	22,000
Sasol-QP (Oryx)	Ras Laffan, Qatar	Natural gas	Co/Al_2O_3	Slurry phase/ LTFT	2007	34,000
Inner Mongolia Yitai	Erdos, Inner Mongolia, China	Coal	Fe catalyst	Slurry bed (Synfuels China technology)/ MTFT	2009	32,000 (260,000 by 2020)
Shell (Pearl)	Ras Laffan, Qatar	Natural gas	Co/TiO_2	Multi-tubular fixed bed/LTFT	2011	140,000
Chevron-Sasol	Escravos, Nigeria	Natural gas	Co/Al_2O_3	Slurry phase/ LTFT	2013	34,000
Shenhua	Yinchuan, Ningxia, China	Coal	Fe catalyst	Slurry bed (Synfuels China technology)/ MTFT	2016	100,000

Table 14.2 (continued)

Company	Location	Feedstock	Catalyst	Reactor/process	Start-up	Capacity (barrel per day)
Shanxi Lu'an	Xiangyuan, Shanxi, China	Coal	Fe catalyst	Slurry bed (Synfuels China technology)/ MTFT	2017	36,000
Oltin Yo'L GTL	Karshi, Uzbekistan	Natural gas	Co/Al_2O_3	Slurry phase (Sasol technology)/ LTFT	2020	37,000
Yankuang-Air Products	Yulin, Shaanxi, China	Coal	Fe catalyst	Slurry bed (Yankuang technology)/ MTFT	2021	80,000

Note: Capacity given in barrels (159 L) per day.

14.4.4 Fischer–Tropsch Catalysts

Metals that are capable of dissociating CO are iron, cobalt, ruthenium, and nickel [224]. Ruthenium is scarce and expensive, whereas nickel converts syngas mostly into methane (see Section 14.2). As a result, only iron and cobalt are used in commercial applications of FTS.

For the HTFT process iron-based catalysts are used, which are promoted with alkali metals to improve activity and selectivity [211, 220]. They are normally prepared by fusion at temperatures in excess of 1273 K, during which the mill scale (iron oxide) and the promoters are fused together, followed by casting into ingots and cooling, milling (to obtain the correct particle sizes) and reduction/activation in H_2.

Iron catalysts can also be used for the LTFT and MTFT processes [44, 211, 212, 221, 225]. Iron-based LTFT catalysts contain promoters like potassium and sometimes copper, as well as a structural modifier such as silica. These catalysts are normally prepared by co-precipitation of iron and copper, followed by impregnation with potassium and the structural promoters. Thereafter, the catalyst has to be shaped for use in either a fixed bed (e.g., extrudates) or a slurry-phase FT reactor (e.g., spray-dried spheres), followed by calcination. Prior to use in the FT process, the oxidized iron catalysts need to be activated or conditioned in CO, H_2 or synthesis gas. During the FT process the iron metal is converted to iron carbide, and, depending on the conditions, oxidized again to an iron oxide [212, 226, 227].

While iron catalysts change under Fischer–Tropsch conditions to a complex mixture of iron carbides and oxides [211], cobalt is active mainly in the metallic state.

Although it is significantly more expensive than iron, cobalt is predominantly used in modern LTFT processes [47, 212, 222, 228–231]. Cobalt is normally supported on silica, alumina, titania, or zinc oxide. Supported cobalt catalysts often contain structural (e.g., Zr, Si) and reduction promoters (e.g., Re, Pt, Ru). A silica-supported cobalt catalyst containing cobalt crystallites of about 10 nm is shown in Figure 14.16.

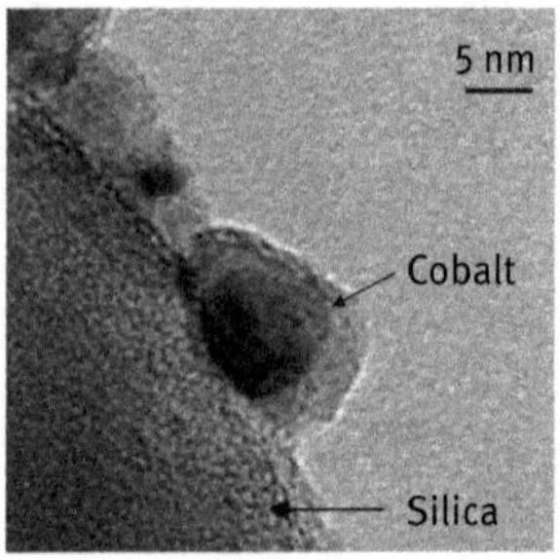

Figure 14.16: Silica-supported cobalt catalyst with a 10 nm cobalt crystallite (adapted from ref. [232]).

Cobalt catalysts for slurry-phase FTS are normally prepared on pre-shaped supports by means of impregnation techniques, followed by calcination and reduction. For fixed-bed applications the cobalt and support can be mixed together (by, for example, co-precipitation or simply preparing a slurry of solids) followed by shaping (e.g., extrusion), calcination, and reduction.

The cobalt crystallite size is of crucial importance for the performance of the catalyst. There seems to be an optimum cobalt crystallite size of about 8–10 nm, which combines the highest activity with the highest selectivity toward longer hydrocarbons [233]. Long-term catalyst stability is also very important, and the following deactivation mechanisms are at play: poisoning by sulfur- and nitrogen-containing compounds, sintering, carbon deposition, and surface reconstruction [234].

Water is an important coupled product of FTS. With cobalt catalysts, essentially all oxygen from CO dissociation (typically around 99%) is discarded as water (14.72). In contrast, iron catalysts possess activity for the WGS reaction, and hence particularly at higher temperatures part of the oxygen evolves as CO_2 (14.73):

$$CO + 2H_2 \rightleftharpoons -CH_2- + H_2O \tag{14.72}$$

$$2CO + H_2 \rightleftharpoons -CH_2- + CO_2 \tag{14.73}$$

The important implication is that processes based on iron can handle syngas with low H_2/CO ratio as is obtained from partial oxidation of coal. In fact, the H_2:CO ratio may be as low as 0.5. Iron generally produces more olefins and oxygenates than cobalt. The low WGS activity of cobalt makes it the preferred catalyst for gas-to-liquid technology, since the H_2/CO ratio of syngas from steam reforming of natural gas is close to the desired composition for hydrocarbon formation (Section 14.1). In this case, FTS needs a H_2:CO ratio of at least 2 (eqs. (14.52)–(14.54)).

14.5 Future Perspectives

The synthesis of methanol or hydrocarbons out of hydrogen and CO/CO_2 (syngas) can be one way to reduce the need for fossil raw materials such as crude oil, coal, or natural gas provided that the syngas originates from non-fossil sources. As discussed, syngas generation could be realized in solar-thermal cycles (Section 14.1.2), by electrolysis of water (Section 14.1.3) combined with capture of anthropogenic CO_2 from industrial processes, or by gasification of biomass. This concept is especially feasible in a power system with a very high share of volatile power sources requiring large seasonal storage capacities.

Whereas the conversion of renewable syngas into methane is already in the pilot stage, almost all methanol and FTS plants operate with syngas derived from coal or natural gas. This is mainly due to cost and capacity reasons. Nevertheless, on smaller scale (200–5,000 barrels per day) FT plants are in operation that use biomass, municipal waste, landfill- or flare gases as a source for syngas in addition to natural gas [235, 236].

For all schemes the future wide-spread use of anthropogenic CO_2 in combination with renewable H_2 and other non-conventional sources of synthesis gas will lead to sub-optimal syngas compositions. Efficient incorporation of these syngas mixtures into the current synthesis infrastructure will necessitate the re-development of catalysts to perform stably under high concentrations of CO_2, water and impurities.

14.5.1 Methanol

As one among other energy storage strategies, methanol has great potential as a sustainable synthetic fuel and seems particularly promising for the transportation sector. The industrial methanol synthesis process with Cu/ZnO-based catalysts represents a good starting point for the implementation of methanol in a future energy scenario. Although mature technologies and long-lasting experience already exist, the thermodynamic constraints of CO_2 hydrogenation require further research. The key challenge remains the development of catalysts with better activity at low temperatures. Despite the long-lasting experience with the industrial processes, methanol chemistry is scientifically not yet mature. Elaboration of a firm scientific basis for effective catalyst design by resolving the open questions concerning the mechanism and the nature of the active sites of the relevant reactions is still a major challenge.

Although most researchers agree that copper-based catalysts will remain the industrial standard, numerous possibilities exist for the improvement of the activity, selectivity, and stability of these materials through the incorporation of promoters, more efficient exploitation of the Cu-ZnO synergy and implementation of new synthesis strategies. Therefore, a deeper understanding of the active sites for methanol synthesis and rWGS on Cu-based catalysts is desirable and essential to further knowledge-based catalyst development.

Here, combined efforts from the fields of theory, material science, model catalysis, and reaction kinetics are needed. Advanced characterization methods must be implemented to discriminate between the individual degradation effects, surface area loss by sintering, loss of active sites by defect annealing and poisoning by impurities. Current high-throughput testing methods for methanol synthesis catalysts need to be modified to take product inhibition from water into account. With that in mind, complementary in situ structural und surface studies on working industrial catalysts would be beneficial to validate the conclusions drawn from model studies and promise further progress in the understanding and optimization of methanol catalysts.

In that respect, properties such as particle size, interaction with oxide matrix, and identity of exposed Cu crystal planes should be investigated in more detail with respect to CO selectivity. Furthermore, targeted synthesis strategies can be developed to take these factors into account and in the end allow for tailoring the CO_2 and CO reaction channels on Cu surfaces.

14.5.2 Fischer–Tropsch Synthesis

The Fischer–Tropsch process dates back to the early 1920s when Franz Fischer and Hans Tropsch demonstrated the conversion of syngas into a mixture of higher hydrocarbons, over cobalt and iron as catalysts [237, 238]. Some 20 years earlier Sabatier had already discovered the reaction from syngas to methane catalyzed by nickel [239]. A hundred years later FTS still relies on cobalt and iron as catalysts and for the most part syngas from fossil sources. Nevertheless, the plant design and the capacities have improved tremendously and the understanding of the process has evolved considerably (see the Section 14.4).

FTS is now commercially applied on large scale by Shell and Sasol in the shape of improved versions of gas-to-liquids processes, which are mainly based on cobalt catalysts and production of long hydrocarbon chains, which can, for example, be hydrocracked to diesel fuel [47, 211, 223, 240]. Synfuels China medium temperature MTFT process results in a similar product slate but starts from coal and is based on iron catalysts [213].

The challenge for any Fischer–Tropsch process is to minimize CO_2 emissions and cost. Replacing the considerable electric power supply from fossil sources by renewable electricity and integrating water electrolysis to H_2 and O_2 in the process are the most obvious measures that would diminish CO_2 emissions from these plants. Applications of Fischer–Tropsch technology in the utilization of biomass are being explored as well [241–243]. Here, the challenge is to cope with higher impurity levels of biomass-derived syngas. Nevertheless, demonstration plants have been built and it is expected that applications will grow in the future.

In situations where both H_2 from renewable sources (e.g., electrolysis) and large amounts of CO_2 are available (e.g., at power stations, steel or cement plant), the direct conversion of CO_2 into hydrocarbons might be interesting. Hence it is relevant to know whether the FTS is applicable using syngas consisting of CO_2 and H_2. Table 14.3 presents a comparison of several catalysts applied in CO_2 hydrogenation under identical conditions taken from the work of Song and coworkers [244].

Table 14.3: Test results of several catalysts during CO_2 hydrogenation at 573 K, 1.1 MPa, 3600 mL g^{-1} h^{-1} (data from [244]).

Catalyst	X_{CO2} [%]	Space–time–yield [µmol · g^{-1} · s^{-1}]			C-atom selectivity [%]		O/P in C_2–C_4
		CH_4	C2–C7	CO	CH_4	C2–C7	
Cu	22.6	–	–	2.33	–	–	–
Pd	12.5	0.22	0.0	1.18	100	0	–
Ni	70.1	6.57	0.0	0.01	100	0	–
Co	48.8	3.61	0.01	0.07	99	1	0.0
Fe	12.1	0.54	0.13	0.65	62	38	0.0
Fe/K	27.0	0.46	0.34	0.63	29	71	1.1
FeCo	25.2	1.19	0.36	0.39	55	45	0.0
FeCo/K	33.7	0.94	0.49	0.46	36	64	0.7
FeCu	27.8	0.78	0.37	0.66	41	59	0.0
FeCu/K	31.5	0.55	0.42	0.52	29	71	0.4

Catalysts contain 15 wt% metal on Al_2O_3 (Pd: 5 wt%); K/Fe= 0.3 atom basis.

The data illustrates the general characteristics of the metals in this reaction. CO_2 hydrogenation on copper yields CO, owing to the rWGS at a conversion which is close to equilibrium under these conditions. On palladium the CO selectivity is high, too. As expected, nickel produces exclusively methane. According to the authors, 70% CO_2 conversion under these conditions is close to equilibrium, confirming that Ni/Al_2O_3 is a good catalyst for synthetic natural gas production. Iron catalysts are less active but do produce higher hydrocarbons, which, however, are paraffinic in nature. Addition of potassium is effective in promoting hydrocarbons with a significant contribution of olefins. Combining iron with a second metal is also effective in increasing chain growth and adding potassium promotes the activity further and shifts the selectivity of the hydrocarbons toward olefins [244].

The data of Table 14.3 can be interpreted on the basis of a two-step mechanism, in which first CO_2 undergoes reduction to CO in the rWGS reaction, followed by CO hydrogenation as in the Fischer–Tropsch process [245].

Visconti et al. [246] reported encouraging results based on a K-promoted, unsupported iron catalyst, consisting mainly of Fe_3O_4 (for rWGS) and iron carbides (for FTS), which produced a product distribution with about 65% of the carbon ending up in C_2+ hydrocarbons, with 55% being olefins, and 10% paraffins. Methane make remains relatively high at 20% C-selectivity while the rest is mostly unconverted CO (12%).

The chain growth parameter α reached a maximum of about 0.65, while LTFT and MTFT processes operate on α values in excess of 0.9. According to Song and coworkers [247], the underlying reason for the lower chain growth probability and saturated nature of the hydrocarbons is the relatively high H-to-C coverage ratio on the surface due to the intrinsically weak interaction of CO_2 with the surface. Hence, they recommend strategies to increase the adsorption strength of CO_2 or weaken the interaction of the surface with hydrogen atoms to arrive at more favorable surface concentrations for chain growth and olefin production.

Another factor that is likely to be important is that the gas mixture contains both water and CO_2, which are both known to negatively affect the FTS reaction above certain percentages. Hence, FTS starting from CO_2 and H_2 as the syngas appears feasible, but is a compromise with a product selectivity that is less attractive and more difficult to tune than FTS based on CO and H_2.

14.5.3 Next-Generation Synthetic Fuels

Diesel- and gasoline-fueled transportation has been criticized recently due to high emissions of NO_x and particulate matter. Thus, new fuels with improved combustion behavior and ideally facile synthesis are sought after. A promising surrogate or blend for diesel fuels may be (poly)oxymethylenethers (OMEs) [248]. This class of chemical does not contain any C–C bonds, which leads to a very clean combustion behavior with almost no emission of particulates. This enables an effective exhaust gas recirculation, which reduces NO_x emissions without increasing emissions of particulate matter as it would be the case using conventional diesel or FT-diesel fuels [249].

The synthesis of OMEs involves two platform chemicals: methanol and formaldehyde (Figure 14.17). Formaldehyde, paraformaldehyde, or trioxane provide oxymethylene groups; and methanol/DME acts as a capping agent by providing methoxy groups.

Successful routes are frequently carried out under non-aqueous conditions converting, for example, trioxane and methylal (OME_1). The attractiveness of this route strongly depends on the cost of the raw materials representing up to 62% of the total production cost [250]. In view of the feedstock cost, a more economic approach

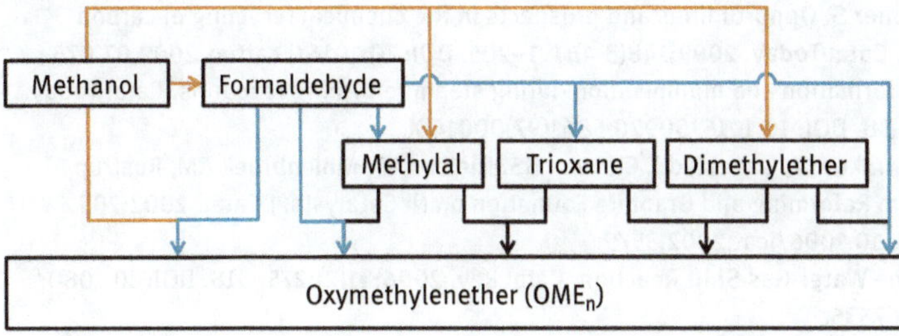

Figure 14.17: Selection of possible routes for the synthesis of OMEs.

is based on the conversion of aqueous solutions of formaldehyde or paraformaldehyde with either methanol or OME_1 [249–251].

Whereas OMEs are considered to be an alternative fuel for diesel engines, conventional gasoline can be blended with methanol, ethanol, higher alcohols, or ethers like MTBE. Recently, the addition of dimethyl carbonate and methylformate has attracted attention because of the high potential to reduce particulate matter emissions from compression-ignited (diesel) and spark-ignited (gasoline) combustion engines [252]. Both compounds can be synthesized using methanol and CO_2 or CO [253, 254].

For a widespread sustainable market introduction, large-scale H_2 production from renewable sources has to be established and the corresponding production capacities have to be expanded. Finally, it remains a societal and political decision to introduce these high-price fuels to the existing system.

References

[1] Koytsoumpa EI, Bergins C, Kakaras E. The CO_2 economy: Review of CO_2 capture and reuse technologies, J Supercrit Fluids, 2018, 132, 3–16, DOI: 10.1016/j.supflu.2017.07.029.

[2] Kalz KF, Kraehnert R, Dvoyashkin M, Dittmeyer R, Gläser R, Krewer U, Reuter K, Grunwaldt J-D. Future challenges in heterogeneous catalysis: understanding catalysts under dynamic reaction conditions. ChemCatChem. 2017;9(1):17–29. DOI: 10.1002/cctc.201600996.

[3] Stadler I. Energiespeicher: Bedarf, Technologien, Integration. Berlin: Springer Vieweg; 2014.

[4] Rostrup-Nielsen JR, Christiansen LJ. Concepts in Syngas Manufacture: Catalytic Science Series Vol. 10, 10th ed., London: Imperial College Press; 2011.

[5] Wismann ST, Engbæk JS, Vendelbo SB, Bendixen FB, Eriksen WL, Aasberg-Petersen K, Frandsen C, Chorkendorff I, Mortensen PM. Electrified methane reforming: A compact approach to greener industrial hydrogen production. Science. 2019;364(6442):756–59. DOI: 10.1126/science.aaw8775.

[6] Bharadwaj SS, Schmidt LD. Catalytic partial oxidation of natural gas to syngas. Fuel Process Technol. 1995;42(2-3):109–27. DOI: 10.1016/0378-3820(94)00098-E.

[7] Hu YH, Ruckenstein E. Catalytic Conversion of Methane to Synthesis Gas by Partial Oxidation and CO_2 Reforming', in 'Advances in Catalysis. (eds. Gates BC, Knüpffer H), 1st ed. Elsevier textbooks; 2004. 297–345.

[8] Centi G, Perathoner S. Opportunities and prospects in the chemical recycling of carbon dioxide to fuels. Catal Today. 2009;148(3-4):191–205. DOI: 10.1016/j.cattod.2009.07.075.

[9] Trimm DL. Coke formation and minimisation during steam reforming reactions. Catal Today. 1997;37(3):233–38. DOI: 10.1016/S0920-5861(97)00014-X.

[10] Bengaard HS, Nørskov JK, Sehested J, Clausen BS, Nielsen LP, Molenbroek AM, Rostrup-Nielsen JR. Steam Reforming and Graphite Formation on Ni Catalysts. J Catal. 2002;209 (2):365–84. DOI: 10.1006/jcat.2002.3579.

[11] Newsome DS. The Water-Gas Shift Reaction. Catal Rev. 2006;21(2):275–318. DOI: 10.1080/03602458008067535.

[12] Armor JN. The multiple roles for catalysis in the production of H_2. Appl Catal A: General. 1999;176(2):159–76. DOI: 10.1016/S0926-860X(98)00244-0.

[13] Agrafiotis C, Von Storch H, Roeb M, Sattler C. Solar thermal reforming of methane feedstocks for hydrogen and syngas production – A review. Renewable Sustainable Energy Rev. 2014;29:656–82. DOI: 10.1016/j.rser.2013.08.050.

[14] Perkins C. Likely near-term solar-thermal water splitting technologies. Int J Hydrogen Energy. 2004;29(15):1587–99. DOI: 10.1016/j.ijhydene.2004.02.019.

[15] Agrafiotis C, Roeb M, Sattler C. A review on solar thermal syngas production via redox pair-based water/carbon dioxide splitting thermochemical cycles. Renewable Sustainable Energy Rev. 2015;42:254–85. DOI: 10.1016/j.rser.2014.09.039.

[16] Carrillo RJ, Scheffe JR. Advances and trends in redox materials for solar thermochemical fuel production. Solar Energy. 2017;156:3–20. DOI: 10.1016/j.solener.2017.05.032.

[17] Miller JE, Allendorf MD, Diver RB, Evans LR, Siegel NP, Stuecker JN. Metal oxide composites and structures for ultra-high temperature solar thermochemical cycles. J Mater Sci. 2008;43 (14):4714–28. DOI: 10.1007/s10853-007-2354-7.

[18] Chambon M, Abanades S, Flamant G. Thermal dissociation of compressed ZnO and SnO_2 powders in a moving-front solar thermochemical reactor. AIChE J. 2011;57(8):2264–73. DOI: 10.1002/aic.12432.

[19] Gálvez ME, Loutzenhiser PG, Hischier I, Steinfeld A. CO_2 Splitting via Two-Step Solar Thermochemical Cycles with Zn/ZnO and FeO/Fe_3O_4 Redox Reactions: Thermodynamic Analysis. Energy Fuels. 2008;22(5):3544–50. DOI: 10.1021/ef800230b.

[20] Stamatiou A, Loutzenhiser PG, Steinfeld A. Solar Syngas Production from H_2O and CO_2 via Two-Step Thermochemical Cycles Based on Zn/ZnO and FeO/Fe_3O_4 Redox Reactions: Kinetic Analysis. Energy Fuels. 2010;24(4):2716–22. DOI: 10.1021/ef901544v.

[21] Stamatiou A, Loutzenhiser PG, Steinfeld A. Solar Syngas Production via H_2O/CO_2-Splitting Thermochemical Cycles with Zn/ZnO and FeO/Fe_3O_4 Redox Reactions. Chem Mater. 2010;22 (3):851–59. DOI: 10.1021/cm9016529.

[22] Abanades S, Chambon M. CO_2 Dissociation and Upgrading from Two-Step Solar Thermochemical Processes Based on ZnO/Zn and SnO_2/SnO Redox Pairs. Energy Fuels. 2010;24(12):6667–74. DOI: 10.1021/ef101092u.

[23] Steinfeld A, Brack M, Meier A, Weidenkaff A, Wuillemin D. A solar chemical reactor for co-production of zinc and synthesis gas. Energy. 1998;23(10):803–14. DOI: 10.1016/S0360-5442 (98)00026-7.

[24] Coker EN, Ambrosini A, Rodriguez MA, Miller JE. Ferrite-YSZ composites for solar thermochemical production of synthetic fuels: Operando characterization of CO_2 reduction. J Mater Chem. 2011;21(29):10767. DOI: 10.1039/c1jm11053e.

[25] Kodama T, Gokon N. Thermochemical cycles for high-temperature solar hydrogen production. Chem Rev. 2007;107(10):4048–77. DOI: 10.1021/cr050188a.

[26] Reddy BM, Bharali P, Saikia P, Khan A, Loridant S, Muhler M, Grünert W. Hafnium Doped Ceria Nanocomposite Oxide as a Novel Redox Additive for Three-Way Catalysts. J Phys Chem C. 2007;111(5):1878–81. DOI: 10.1021/jp068531i.

[27] Reddy BM, Katta L, Thrimurthulu G. Novel Nanocrystalline $Ce_{1-x}La_xO_{2-\delta}$ (x = 0.2) Solid Solutions: Structural Characteristics and Catalytic Performance. Chem Mater. 2010;22 (2):467–75. DOI: 10.1021/cm903282w.

[28] Chueh WC, Falter C, Abbott M, Scipio D, Furler P, Haile SM, Steinfeld A. High-flux solar-driven thermochemical dissociation of CO_2 and H_2O using nonstoichiometric ceria. Science. 2010;330(6012):1797–801. DOI: 10.1126/science.1197834.

[29] Le Gal A, Abanades S, Flamant G. CO_2 and H_2O Splitting for Thermochemical Production of Solar Fuels Using Nonstoichiometric Ceria and Ceria/Zirconia Solid Solutions. Energy Fuels. 2011;25(10):4836–45. DOI: 10.1021/ef200972r.

[30] Furler P, Scheffe JR, Steinfeld A. Syngas production by simultaneous splitting of H_2O and CO_2 via ceria redox reactions in a high-temperature solar reactor. Energy Environ Sci. 2012;5 (3):6098–103. DOI: 10.1039/c1ee02620h.

[31] Sapountzi FM, Gracia JM, Weststrate CJ, Fredriksson HO, Niemantsverdriet JW. Electrocatalysts for the generation of hydrogen, oxygen and synthesis gas. Prog Energy Combust Sci. 2017;58:1–35. DOI: 10.1016/j.pecs.2016.09.001.

[32] Zheng Y, Wang J, Yu B, Zhang W, Chen J, Qiao J, Zhang J. A review of high temperature co-electrolysis of H_2O and CO_2 to produce sustainable fuels using solid oxide electrolysis cells (SOECs): advanced materials and technology. Chem Soc Rev. 2017;46(5):1427–63. DOI: 10.1039/C6CS00403B.

[33] Zhang X, Song Y, Wang G, Bao X. Co-electrolysis of CO_2 and H_2O in high-temperature solid oxide electrolysis cells: Recent advance in cathodes. J Energy Chem. 2017;26(5):839–53. DOI: 10.1016/j.jechem.2017.07.003.

[34] Chen L, Chen F, Xia C. Direct synthesis of methane from $CO_2 - H_2O$ co-electrolysis in tubular solid oxide electrolysis cells. Energy Environ Sci. 2014;7(12):4018–22. DOI: 10.1039/C4EE02786H.

[35] Chen M, Høgh JVT, Nielsen JU, Bentzen JJ, Ebbesen SD, Hendriksen PV. High Temperature Co-Electrolysis of Steam and CO_2 in an SOC Stack: Performance and Durability. Fuel Cells. 2013;13(4):638–45. DOI: 10.1002/fuce.201200169.

[36] Graves C, Ebbesen SD, Mogensen M. Co-electrolysis of CO_2 and H_2O in solid oxide cells: Performance and durability. Solid State Ionics. 2011;192(1):398–403. DOI: 10.1016/j.ssi.2010.06.014.

[37] Jensen SH, Sun X, Ebbesen SD, Knibbe R, Mogensen M. Hydrogen and synthetic fuel production using pressurized solid oxide electrolysis cells. Int J Hydrogen Energy. 2010;35 (18):9544–49. DOI: 10.1016/j.ijhydene.2010.06.065.

[38] Götz M, Lefebvre J, Mörs F, McDaniel Koch A, Graf F, Bajohr S, Reimert R, Kolb T. Renewable Power-to-Gas: A technological and economic review. Renewable Energy. 2016;85:1371–90. DOI: 10.1016/j.renene.2015.07.066.

[39] Blanco H, Faaij A. A review at the role of storage in energy systems with a focus on Power to Gas and long-term storage. Renewable Sustainable Energy Rev. 2018;81:1049–86. DOI: 10.1016/j.rser.2017.07.062.

[40] Rönsch S, Schneider J, Matthischke S, Schlüter M, Götz M, Lefebvre J, Prabhakaran P, Bajohr S. Review on methanation – From fundamentals to current projects. Fuel. 2016;166:276–96. DOI: 10.1016/j.fuel.2015.10.111.

[41] Ghaib K, Nitz K, Ben-Fares F-Z. Chemical Methanation of CO_2: A Review. ChemBioEng Rev. 2016;3(6):266–75. DOI: 10.1002/cben.201600022.

[42] Gao J, Liu Q, Gu F, Liu B, Zhong Z, Su F. Recent advances in methanation catalysts for the production of synthetic natural gas. RSC Adv. 2015;5(29):22759–76. DOI: 10.1039/c4ra16114a.
[43] Frontera P, Macario A, Ferraro M, Antonucci P. Supported Catalysts for CO_2 Methanation: A Review. Catalysts. 2017;7(12):59. DOI: 10.3390/catal7020059.
[44] Aziz MAA, Jalil AA, Triwahyono S, Ahmad A. CO_2 methanation over heterogeneous catalysts: Recent progress and future prospects. Green Chem. 2015;17(5):2647–63. DOI: 10.1039/c5gc00119f.
[45] Fujita S, Terunuma H, Nakamura M, Takezawa N. Mechanisms of methanation of carbon monoxide and carbon dioxide over nickel. Ind Eng Chem Res. 1991;30(6):1146–51. DOI: 10.1021/ie00054a012.
[46] Ojeda M, Nabar R, Nilekar AU, Ishikawa A, Mavrikakis M, Iglesia E. CO activation pathways and the mechanism of Fischer–Tropsch synthesis. J Catal. 2010;272(2):287–97. DOI: 10.1016/j.jcat.2010.04.012.
[47] Iglesia E, Reyes SC, Madon RJ, Soled SL. Selectivity Control and Catalyst Design in the Fischer-Tropsch Synthesis: Sites, Pellets, and Reactors. Elsevier; 1993. 221–302.
[48] Elahifard MR, Jigato MP, Niemantsverdriet JWH. Direct versus hydrogen-assisted CO dissociation on the Fe (100) surface: A DFT study. ChemPhysChem. 2012;13(1):89–91. DOI: 10.1002/cphc.201100759.
[49] Inderwildi OR, Jenkins SJ, King DA. Fischer–Tropsch Mechanism Revisited: Alternative Pathways for the Production of Higher Hydrocarbons from Synthesis Gas. J Phys Chem C. 2008;112(5):1305–07. DOI: 10.1021/jp710674q.
[50] Olah GA. Beyond oil and gas: the methanol economy. Angewandte Chemie (International ed in English). 2005;44(18):2636–39. DOI: 10.1002/anie.200462121.
[51] Olah GA. Towards oil independence through renewable methanol chemistry. Angewandte Chemie (International ed in English). 2013;52(1):104–07. DOI: 10.1002/anie.201204995.
[52] Rihko-Struckmann LK, Peschel A, Hanke-Rauschenbach R, Sundmacher K. Assessment of Methanol Synthesis Utilizing Exhaust CO_2 for Chemical Storage of Electrical Energy. Ind Eng Chem Res. 2010;49(21):11073–78. DOI: 10.1021/ie100508w.
[53] Aldrich B 'ABC's of AFV's: a guide to alternative fuel vehicles'; 1995.
[54] Moffat AS. Methanol-powered cars get ready to hit the road. Science. 1991;251(4993):514–16.
[55] 'Methanol: the clear alternative for transportation', 2011; available at http://www.methanol.org.
[56] Reed TB, Lerner RM. Methanol: A Versatile Fuel for Immediate Use: Methanol can be made from gas, coal, or wood. It is stored and used in existing equipment. Science. 1973;182(4119):1299–304. DOI: 10.1126/science.182.4119.1299.
[57] Olah GA, Goeppert A, Prakash GKS. Beyond oil and gas:: The methanol economy. Weinheim an der Bergstrasse, Germany: Wiley-VCH; 2006
[58] Twigg MV, Spencer MS. Deactivation of Copper Metal Catalysts for Methanol Decomposition, Methanol Steam Reforming and Methanol Synthesis. Top Catal. 2003;22(3):191–203. DOI: 10.1023/A: 1023567718303.
[59] Global B Statistical Review of World Energy, BP Global, 2011.
[60] Hansen JB, Nielsen PEH. Methanol synthesis. In: Ertl, et al. editors. Handbook of heterogeneous catalysis. Weinheim, Germany: Wiley-VCH: 2008.
[61] Palo DR, Dagle RA, Holladay JD. Methanol steam reforming for hydrogen production. Chem Rev. 2007;107(10):3992–4021. DOI: 10.1021/cr050198b.
[62] Stiles AB. Methanol, past, present, and speculation on the future. AIChE J. 1977;23(3):362–75. DOI: 10.1002/aic.690230321.

[63] Lewis WK, Frolich PK. Synthesis of Methanol from Carbon Monoxide and Hydrogen. Ind Eng Chem. 1928;20(3):285–90.

[64] Weissermel K, Arpe: H-J. Industrial organic chemistry. 4th ed, Weinheim, Germany: Wiley-VCH; 2003.

[65] Studt F, Sharafutdinov I, Abild-Pedersen F, Elkjær CF, Hummelshøj JS, Dahl S, Chorkendorff I, Nørskov JK. Discovery of a Ni-Ga catalyst for carbon dioxide reduction to methanol. Nat Chem. 2014;6(4):320–24. DOI: 10.1038/nchem.1873.

[66] Li C-S, Melaet G, Ralston WT, An K, Brooks C, Ye Y, Liu Y-S, Zhu J, Guo J, Alayoglu S, Somorjai GA. High-performance hybrid oxide catalyst of manganese and cobalt for low-pressure methanol synthesis. Nat Commun. 2015;6:6538. DOI: 10.1038/ncomms7538.

[67] Martin O, Martín AJ, Mondelli C, Mitchell S, Segawa TF, Hauert R, Drouilly C, Curulla-Ferré D, Pérez-Ramírez J. Indium Oxide as a Superior Catalyst for Methanol Synthesis by CO_2 Hydrogenation. Angewandte Chemie (International ed in English). 2016;55(21):6261–65. DOI: 10.1002/anie.201600943.

[68] Fiedler E, et al. Methanol in Ullmann's Encyclopedia of Industrial Chemistry. Weinheim, Germany: Wiley-VCH; 2011.

[69] Rahimpour MR, Abbasloo A, Sayyad Amin J. A Novel Radial-Flow, Spherical-Bed Reactor Concept for Methanol Synthesis in the Presence of Catalyst Deactivation. Chem Eng Technol. 2008;31(11):1615–29. DOI: 10.1002/ceat.200800284.

[70] Behrens M. Methanol Steam Reforming. In: Guczi L, Erdohelyi A, editors. Catalysis for alternative energy generation. New York: Springer; 2012.

[71] Kasatkin I, Kurr P, Kniep B, Trunschke A, Schlögl R. Role of lattice strain and defects in copper particles on the activity of $Cu/ZnO/Al_2O_3$ catalysts for methanol synthesis. Angew Chem Int Ed. 2007;46(38):7324–27. DOI: 10.1002/anie.200702600.

[72] Behrens M, Studt F, Kasatkin I, Kühl S, Hävecker M, Abild-Pedersen F, Zander S, Girgsdies F, Kurr P, Kniep B-L, Tovar M, Fischer RW, Nørskov JK, Schlögl R. The active site of methanol synthesis over $Cu/ZnO/Al_2O_3$ industrial catalysts. Science. 2012;336(6083):893–97. DOI: 10.1126/science.1219831.

[73] Kuld S, Conradsen C, Moses PG, Chorkendorff I, Sehested J. Quantification of zinc atoms in a surface alloy on copper in an industrial-type methanol synthesis catalyst. Angewandte Chemie (International ed in English). 2014;53(23):5941–45. DOI: 10.1002/anie.201311073.

[74] Fichtl MB, Schumann J, Kasatkin I, Jacobsen N, Behrens M, Schlögl R, Muhler M, Hinrichsen O. Counting of oxygen defects versus metal surface sites in methanol synthesis catalysts by different probe molecules. Angewandte Chemie (International ed in English). 2014;53 (27):7043–47. DOI: 10.1002/anie.201400575.

[75] Chinchen G. The measurement of copper surface areas by reactive frontal chromatography 1. J Catal. 1987;103(1):79–86. DOI: 10.1016/0021-9517(87)90094-7.

[76] Hinrichsen O, Genger T, Muhler M. Chemisorption of N_2O and H_2 for the Surface Determination of Copper Catalysts. Chem Eng Technol. 2000;23(11):956–59. DOI: 10.1002/1521-4125(200011)23:11<956::AID-CEAT956>3.0.CO;2-L.

[77] Behrens M, Furche A, Kasatkin I, Trunschke A, Busser W, Muhler M, Kniep B, Fischer R, Schlögl R. The Potential of Microstructural Optimization in Metal/Oxide Catalysts: Higher Intrinsic Activity of Copper by Partial Embedding of Copper Nanoparticles. ChemCatChem. 2010;2(7):816–18. DOI: 10.1002/cctc.201000017.

[78] Kurtz M, Bauer N, Büscher C, Wilmer H, Hinrichsen O, Becker R, Rabe S, Merz K, Driess M, Fischer RA, Muhler M. New Synthetic Routes to More Active Cu/ZnO Catalysts Used for Methanol Synthesis. Catal Lett. 2004;92(1):49–52. DOI: 10.1023/B:CATL.0000011085.88267.a6.

[79] Yoshihara J, Campbell CT. Methanol Synthesis and Reverse Water–Gas Shift Kinetics over Cu (110) Model Catalysts: Structural Sensitivity. J Catal. 1996;161(2):776–82. DOI: 10.1006/jcat.1996.0240.
[80] Van Den Berg R, Prieto G, Korpershoek G, Van Der Wal LI, Van Bunningen AJ, Lægsgaard-Jørgensen S, De Jongh PE, De Jong KP. Structure sensitivity of Cu and CuZn catalysts relevant to industrial methanol synthesis. Nat Commun. 2016;7:13057. DOI: 10.1038/ncomms13057.
[81] Spencer MS. The role of zinc oxide in Cu/ZnO catalysts for methanol synthesis and the water–gas shift reaction. Top Catal. 1999;8(3-4):259–66. DOI: 10.1023/A: 1019181715731.
[82] Burch R, Chappell RJ, Golunski SE. Synergy between copper and zinc oxide during methanol synthesis. Transfer of activating species. J Chem Soc, Faraday Trans 1. 1989;85(10):3569–78. DOI: 10.1039/f19898503569.
[83] Kanai Y, Watanabe T, Fujitani T, Uchijima T, Nakamura J. The synergy between Cu and ZnO in methanol synthesis catalysts. Catal Lett. 1996;38(3):157–63. DOI: 10.1007/BF00806562.
[84] Klier K, Methanol Synthesis. In: Advances in catalysis and related subjects. New York: Academic Press; 1948, 243–313.
[85] Jansen W. Dynamic Behavior of the Surface Structure of Cu/ZnO/SiO_2 Catalysts. J Catal. 2002;210(1):229–36. DOI: 10.1006/jcat.2002.3679.
[86] Frost JC. Junction effect interactions in methanol synthesis catalysts. Nature. 1988;334 (6183):577–80. DOI: 10.1038/334577a0.
[87] Nakamura J, Choi Y, Fujitani T. On the Issue of the Active Site and the Role of ZnO in Cu/ZnO Methanol Synthesis Catalysts. Top Catal. 2003;22(3):277–85. DOI: 10.1023/A: 1023588322846.
[88] Kuld S, Thorhauge M, Falsig H, Elkjær CF, Helveg S, Chorkendorff I, Sehested J. Quantifying the promotion of Cu catalysts by ZnO for methanol synthesis. Science. 2016;352(6288):969–74. DOI: 10.1126/science.aaf0718.
[89] Waugh KC. Methanol Synthesis. Catal Today. 1992;15(1):51–75. DOI: 10.1016/0920-5861(92) 80122-4.
[90] Kattel S, Ramírez PJ, Chen JG, Rodriguez JA, Liu P. Active sites for CO_2 hydrogenation to methanol on Cu/ZnO catalysts. Science (New York, N Y). 2017;355(6331):1296–99. DOI: 10.1126/science.aal3573.
[91] Topsøe N-Y, Topsøe H. On the nature of surface structural changes in Cu/ZnO methanol synthesis catalysts. Top Catal. 1999;8(3):267–70. DOI: 10.1023/A: 1019133832569.
[92] Naumann d'Alnoncourt R, Xia X, Strunk J, Löffler E, Hinrichsen O, Muhler M. The influence of strongly reducing conditions on strong metal-support interactions in Cu/ZnO catalysts used for methanol synthesis. Phys Chem Chem Phys. 2006;8(13):1525–38. DOI: 10.1039/b515487a.
[93] Grunwaldt J-D, Molenbroek A, Topsøe N-Y, Topsøe H, Clausen BS. In Situ Investigations of Structural Changes in Cu/ZnO Catalysts. J Catal. 2000;194(2):452–60. DOI: 10.1006/jcat.2000.2930.
[94] Hansen PL, Wagner JB, Helveg S, Rostrup-Nielsen JR, Clausen BS, Topsøe H. Atom-resolved imaging of dynamic shape changes in supported copper nanocrystals. Science. 2002;295 (5562):2053–55. DOI: 10.1126/science.1069325.
[95] Kurr P, Kasatkin I, Girgsdies F, Trunschke A, Schlögl R, Ressler T. Microstructural characterization of Cu/ZnO/Al_2O_3 catalysts for methanol steam reforming – A comparative study. Appl Catal A: General. 2008;348(2):153–64. DOI: 10.1016/j.apcata.2008.06.020.
[96] Kniep BL, Ressler T, Rabis A, Girgsdies F, Baenitz M, Steglich F, Schlögl R. Rational design of nanostructured copper-zinc oxide catalysts for the steam reforming of methanol. Angew Chem Int Ed. 2004;43(1):112–15. DOI: 10.1002/anie.200352148.

[97] Günter MM, Ressler T, Jentoft RE, Bems B. Redox Behavior of Copper Oxide/Zinc Oxide Catalysts in the Steam Reforming of Methanol Studied by in Situ X-Ray Diffraction and Absorption Spectroscopy. J Catal. 2001;203(1):133–49. DOI: 10.1006/jcat.2001.3322.

[98] Sakong S, Groß A. Dissociative adsorption of hydrogen on strained Cu surfaces. Surf Sci. 2003;525(1-3):107–18. DOI: 10.1016/S0039-6028(02)02550-5.

[99] Girgsdies F, Ressler T, Wild U, Wübben T, Balk TJ, Dehm G, Zhou L, Günther S, Arzt E, Imbihl R, Schlögl R. Strained thin copper films as model catalysts in the materials gap. Catal Lett. 2005;102(1):91–97. DOI: 10.1007/s10562-005-5208-4.

[100] Hammer B, Nørskov JK. Electronic factors determining the reactivity of metal surfaces. Surf Sci. 1995;343(3):211–20. DOI: 10.1016/0039-6028(96)80007-0.

[101] Schimpf S, Muhler M, De Jong KP, editor. Synthesis of Solid Catalysts: Methanol catalysts. Weinheim, Germany: Wiley-VCH; 2009.

[102] Behrens M, Brennecke D, Girgsdies F, Kißner S, Trunschke A, Nasrudin N, Zakaria S, Idris NF, Hamid SBA, Kniep B, Fischer R, Busser W, Muhler M, Schlögl R. Understanding the complexity of a catalyst synthesis: Co-precipitation of mixed Cu,Zn,Al hydroxycarbonate precursors for $Cu/ZnO/Al_2O_3$ catalysts investigated by titration experiments. Appl Catal A: General. 2011;392(1-2):93–102. DOI: 10.1016/j.apcata.2010.10.031.

[103] Waller D, Stirling D, Stone FS, Spencer MS. Copper–zinc oxide catalysts. Activity in relation to precursor structure and morphology. Faraday Discuss Chem Soc. 1989;87(0):107–20. DOI: 10.1039/DC9898700107.

[104] Behrens M. Meso- and nano-structuring of industrial $Cu/ZnO/(Al_2O_3)$ catalysts. J Catal. 2009;267(1):24–29. DOI: 10.1016/j.jcat.2009.07.009.

[105] Behrens M, Kissner S, Girgsdies F, Kasatkin I, Hermerschmidt F, Mette K, Ruland H, Muhler M, Schlögl R. Knowledge-based development of a nitrate-free synthesis route for Cu/ZnO methanol synthesis catalysts via formate precursors. Chem Commun. 2011;47(6):1701–03. DOI: 10.1039/c0cc04933f.

[106] Whittle DM, Mirzaei AA, Hargreaves JSJ, Joyner RW, Kiely CJ, Taylor SH, Hutchings GJ. Co-precipitated copper zinc oxide catalysts for ambient temperature carbon monoxide oxidation: Effect of precipitate ageing on catalyst activity. Phys Chem Chem Phys. 2002;4(23):5915–20. DOI: 10.1039/b207691h.

[107] Fierro G, Lo Jacono M, Inversi M, Porta P, Cioci F, Lavecchia R. Study of the reducibility of copper in CuO ZnO catalysts by temperature-programmed reduction. Appl Catal A: General. 1996;137(2):327–48. DOI: 10.1016/0926-860X(95)00311-8.

[108] Günter MM, Ressler T, Bems B, Büscher C, Genger T, Hinrichsen O, Muhler M, Schlögl R. Implication of the microstructure of binary Cu/ZnO catalysts for their catalytic activity in methanol synthesis. Catal Lett. 2001;71(1):37–44. DOI: 10.1023/A: 1016696022840.

[109] Baltes C, Vukojevic S, Schüth F. Correlations between synthesis, precursor, and catalyst structure and activity of a large set of $CuO/ZnO/Al_2O_3$ catalysts for methanol synthesis. J Catal. 2008;258(2):334–44. DOI: 10.1016/j.jcat.2008.07.004.

[110] Li J-L, Inui T. Characterization of precursors of methanol synthesis catalysts, copper/zinc/aluminum oxides, precipitated at different pHs and temperatures. Appl Catal A: General. 1996;137(1):105–17. DOI: 10.1016/0926-860X(95)00284-7.

[111] Shen G. Preparation of precursors for the Cu/ZnO methanol synthesis catalysts by coprecipitation methods: Effects of the preparation conditions upon the structures of the precursors. J Catal. 1992;138(2):754–58. DOI: 10.1016/0021-9517(92)90323-A.

[112] Fichtl MB, Schlereth D, Jacobsen N, Kasatkin I, Schumann J, Behrens M, Schlögl R, Hinrichsen O. Kinetics of deactivation on $Cu/ZnO/Al_2O_3$ methanol synthesis catalysts. Appl Catal A Gen. 2015;502:262–70. DOI: 10.1016/j.apcata.2015.06.014.

[113] Lunkenbein T, Girgsdies F, Kandemir T, Thomas N, Behrens M, Schlögl R, Frei E. Bridging the Time Gap: A Copper/Zinc Oxide/Aluminum Oxide Catalyst for Methanol Synthesis Studied under Industrially Relevant Conditions and Time Scales. Angewandte Chemie (International ed in English). 2016;55(41):12708–12. DOI: 10.1002/anie.201603368.
[114] Kung HH. Deactivation of methanol synthesis catalysts – a review. Catal Today. 1992;11 (4):443–53. DOI: 10.1016/0920-5861(92)80037-N.
[115] Kung HH. Methanol Synthesis. Catal Rev -Sci Eng. 1980;22(2):235–59. DOI: 10.1080/03602458008066535.
[116] Klier K. Catalytic synthesis of methanol from CO/H_2:: IV. The effects of carbon dioxide. J Catal. 1982;74(2):343–60. DOI: 10.1016/0021-9517(82)90040-9.
[117] Chinchen GC, Denny PJ, Parker DG, Spencer MS, Whan DA. Mechanism of methanol synthesis from $CO_2/CO/H_2$ mixtures over copper/zinc oxide/alumina catalysts: Use of ^{14}C-labelled reactants. Appl Catal. 1987;30(2):333–38. DOI: 10.1016/S0166-9834(00)84123-8.
[118] Rozovskii AY. Kinetika i Kataliz. 1980, 21, 87.
[119] Fujita S-I, Satriyo AM, Shen GC, Takezawa N. Mechanism of the formation of precursors for the Cu/ZnO methanol synthesis catalysts by a coprecipitation method. Catal Lett. 1995;34 (1):85–92. DOI: 10.1007/BF00808325.
[120] Saussey J, Lavalley J-C, Bovet C. Infrared study of CO_2 adsorption on ZnO. Adsorption sites. J Chem Soc, Faraday Trans 1. 1982;78(5):1457. DOI: 10.1039/f19827801457.
[121] Yang Y, Mims CA, Disselkamp RS, Mei D, Kwak J-H, Szanyi J, Peden CH, Campbell CT. Isotope Effects in Methanol Synthesis and the Reactivity of Copper Formates on a Cu/SiO_2 Catalyst. Catal Lett. 2008;125(3):201–08. DOI: 10.1007/s10562-008-9592-4.
[122] Yang Y, Mims CA, Disselkamp RS, Kwak J-H, Peden CH, Campbell CT. (Non)formation of Methanol by Direct Hydrogenation of Formate on Copper Catalysts. J Phys Chem C. 2010;114 (40):17205–11. DOI: 10.1021/jp104068k.
[123] Muhler M, Törnqvist E, Nielsen LP, Clausen BS, Topsøe H. On the role of adsorbed atomic oxygen and CO_2 in copper based methanol synthesis catalysts. Catal Lett. 1994;25(1):1–10. DOI: 10.1007/BF00815409.
[124] Nakamura J, Campbell JM, Campbell CT. Kinetics and mechanism of the water-gas shift reaction catalysed by the clean and Cs-promoted Cu(110) surface: A comparison with Cu(111). J Chem Soc Faraday Trans. 1990;86(15):2725–34. DOI: 10.1039/ft9908602725.
[125] Szanyi J, Goodman DW. Methanol synthesis on a Cu(100) catalyst. Catal Lett. 1991;10(5):383–90. DOI: 10.1007/BF00769173.
[126] Yoshihara J, Parker SC, Schafer A, Campbell CT. Methanol synthesis and reverse water-gas shift kinetics over clean polycrystalline copper. Catal Lett. 1995;31(4):313–24. DOI: 10.1007/BF00808595.
[127] Rasmussen PB, Holmblad PM, Askgaard T, Ovesen CV, Stoltze P, Nørskov JK, Chorkendorff I. Methanol synthesis on Cu(100) from a binary gas mixture of CO_2 and H_2. Catal Lett. 1994;26 (3):373–81. DOI: 10.1007/BF00810611.
[128] Nakamura J, Nakamura I, Uchijima T, Kanai Y, Watanabe T, Saito M, Fujitani T. Methanol synthesis over a Zn-deposited copper model catalyst. Catal Lett. 1995;31(4):325–31. DOI: 10.1007/BF00808596.
[129] Sahibzada M, Metcalfe IS, Chadwick D. Methanol Synthesis from $CO/CO_2/H_2$ over Cu/ZnO/Al_2O_3 at Differential and Finite Conversions. J Catal. 1998;174(2):111–18. DOI: 10.1006/jcat.1998.1964.
[130] Askgaard TS, Norskov JK, Ovesen CV, Stoltze P. A Kinetic Model of Methanol Synthesis. J Catal. 1995;156(2):229–42. DOI: 10.1006/jcat.1995.1250.

[131] Yang Y, Evans J, Rodriguez JA, White MG, Liu P. Fundamental studies of methanol synthesis from CO_2 hydrogenation on Cu(111), Cu clusters, and Cu/ZnO(0001). Phys Chem Chem Phys. 2010;12(33):9909–17. DOI: 10.1039/c001484b.

[132] Grabow LC, Mavrikakis M. Mechanism of Methanol Synthesis on Cu through CO_2 and CO Hydrogenation. ACS Catal. 2011;1(4):365–84. DOI: 10.1021/cs200055d.

[133] Hu Z-M, Takahashi K, Nakatsuji H. Mechanism of the hydrogenation of CO_2 to methanol on a Cu(100) surface: Dipped adcluster model study. Surf Sci. 1999;442(1):90–106. DOI: 10.1016/S0039-6028(99)00900-0.

[134] Morikawa Y, Iwata K, Terakura K. Theoretical study of hydrogenation process of formate on clean and Zn deposited Cu(111) surfaces. Appl Surf Sci. 2001;169-170:11–15. DOI: 10.1016/S0169-4332(00)00631-0.

[135] Zhao Y-F, Yang Y, Mims C, Peden CH, Li J, Mei D. Insight into methanol synthesis from CO_2 hydrogenation on Cu(111): Complex reaction network and the effects of H_2O. J Catal. 2011;281(2):199–211. DOI: 10.1016/j.jcat.2011.04.012.

[136] Studt F, Behrens M, Kunkes EL, Thomas N, Zander S, Tarasov A, Schumann J, Frei E, Varley JB, Abild-Pedersen F, Nørskov JK, Schlögl R. The Mechanism of CO and CO_2 Hydrogenation to Methanol over Cu-Based Catalysts. ChemCatChem. 2015;7(7):1105–11. DOI: 10.1002/cctc.201500123.

[137] Meunier N, Chauvy R, Mouhoubi S, Thomas D, De Weireld G. Alternative production of methanol from industrial CO_2'. Renewable Energy. 2020;146:1192–203. DOI: 10.1016/j.renene.2019.07.010.

[138] Lee S, Sardesai A. Liquid phase methanol and dimethyl ether synthesis from syngas. Top Catal. 2005;32(3-4):197–207. DOI: 10.1007/s11244-005-2891-8.

[139] Saito M. R&D activities in Japan on methanol synthesis from CO_2 and H_2'. Catalysis Surveys from Asia. 1998;2(2):175–84. DOI: 10.1023/A: 1019082525994.

[140] Lee JS, Lee KH, Lee SY, Kim YG. A Comparative Study of Methanol Synthesis from CO_2/H_2 and CO/H_2 over a $Cu/ZnO/Al_2O_3$ Catalyst. J Catal. 1993;144(2):414–24. DOI: 10.1006/jcat.1993.1342.

[141] Chanchlani K. Methanol synthesis from H_2, CO, and CO_2 over Cu/ZnO catalysts. J Catal. 1992;136(1):59–75. DOI: 10.1016/0021-9517(92)90106-R.

[142] Liu G. The rate of methanol production on a copper-zinc oxide catalyst: The dependence on the feed composition. J Catal. 1984;90(1):139–46. DOI: 10.1016/0021-9517(84)90094-0.

[143] Bart JCJ, Sneeden RPA. Copper-zinc oxide-alumina methanol catalysts revisited. Catal Today. 1987;2(1):1–124. DOI: 10.1016/0920-5861(87)80001-9.

[144] Chanchlani KG, Hudgins RR, Silveston PL 'Activation of Cu/ZnO methanol catalysts by periodic composition switching, in Progress in Catalysis, Proceedings of the 12th Canadian Symposium on Catalysis, 331–33; 1992, Elsevier.

[145] Chanchlani KG, Hudgins RR, Silveston PL. Methanol synthesis under periodic operation: An experimental investigation. Can J Chem Eng. 1994;72(4):657–71. DOI: 10.1002/cjce.5450720416.

[146] Nappi A, Fabbricino L, Hudgins RR, Silveston PL. Influence of forced feed composition cycling on catalytic methanol synthesis. Can J Chem Eng. 1985;63(6):963–70. DOI: 10.1002/cjce.5450630613.

[147] McNeil MA, Rinker RG. An Experimental Study of Concentration Forcing Applied to the Methanol Synthesis Reaction. Chem Eng Commun. 2007;127(1):137–49. DOI: 10.1080/00986449408936229.

[148] Gutsche R, Lange R, Witt W. The effect of process nonlinearities on the performance of a periodically operated isothermal catalytic reactor. Chem Eng Sci. 2003;58(22):5055–68. DOI: 10.1016/j.ces.2003.08.011.

[149] Neophytides SG, Froment GF. A bench scale study of reversed flow methanol synthesis. Ind Eng Chem Res. 1992;31(7):1583–89. DOI: 10.1021/ie00007a001.

[150] Vanden Bussche KM, Neophytides SN, Zolotarskii IA, Froment GF. Modelling and simulation of the reversed flow operation of a fixed-bed reactor for methanol synthesis. Chem Eng Sci. 1993;48(19):3335–45. DOI: 10.1016/0009-2509(93)80150-O.

[151] Vanden Bussche KM, Froment GF. The STAR configuration for methanol synthesis in reversed flow reactors. Can J Chem Eng. 1996;74(5):729–34. DOI: 10.1002/cjce.5450740524.

[152] Matros YS. Forced unsteady-state processes in heterogeneous catalytic reactors. Can J Chem Eng. 1996;74(5):566–79. DOI: 10.1002/cjce.5450740504.

[153] Marxer D, Furler P, Scheffe J, Geerlings H, Falter C, Batteiger V, Sizmann A, Steinfeld A. Demonstration of the Entire Production Chain to Renewable Kerosene via Solar Thermochemical Splitting of H_2O and CO_2'. Energy Fuels. 2015;29(5):3241–50. DOI: 10.1021/acs.energyfuels.5b00351.

[154] Ash-Kurlander U, Martin O, Fontana LD, Patil VR, Bernegger M, Mondelli C, Pérez-Ramírez J, Steinfeld A. Impact of Daily Startup-Shutdown Conditions on the Production of Solar Methanol over a Commercial Cu-ZnO-Al_2O_3 Catalyst. Energy Technol. 2016;4(5):565–72. DOI: 10.1002/ente.201600022.

[155] Christiansen JA A reaction between methyl alcohol and water and some related reactions., J Am Chem Soc, 1921, (43), 1670–1672.

[156] Prigent: M. On Board Hydrogen Generation for Fuel Cell Powered Electric Cars. a Review of Various Available Techniques. Rev Inst Fr Pét. 1997;52(3):349–60. DOI: 10.2516/ogst:1997045.

[157] Navarro RM, Peña MA, Fierro JLG. Hydrogen production reactions from carbon feedstocks: Fossil fuels and biomass. Chem Rev. 2007;107(10):3952–91. DOI: 10.1021/cr0501994.

[158] Sá S, Silva H, Brandão L, Sousa JM, Mendes A. Catalysts for methanol steam reforming – A review. Appl Catal B: Env. 2010;99(1-2):43–57. DOI: 10.1016/j.apcatb.2010.06.015.

[159] De Wild P, Verhaak MJFM. Catalytic production of hydrogen from methanol. Catal Today. 2000;60(1-2):3–10. DOI: 10.1016/S0920-5861(00)00311-4.

[160] Joensen F, Rostrup-Nielsen JR. Conversion of hydrocarbons and alcohols for fuel cells. J Power Sour. 2002;105(2):195–201. DOI: 10.1016/S0378-7753(01)00939-9.

[161] Cheekatamarla PK, Finnerty CM. Reforming catalysts for hydrogen generation in fuel cell applications. J Power Sour. 2006;160(1):490–99. DOI: 10.1016/j.jpowsour.2006.04.078.

[162] Velu S, Suzuki K, Osaki T Oxidative steam reforming of methanol over CuZnAl(Zr)-oxide catalysts; a new and efficient method for the production of CO-free hydrogen for fuel cells, Chem Commun, 1999, (23), 2341–42, DOI: 10.1039/a907047h.

[163] Lattner JR, Harold MP. Autothermal reforming of methanol: Experiments and modeling. Catal Today. 2007;120(1):78–89. DOI: 10.1016/j.cattod.2006.07.005.

[164] Park ED, Lee D, Lee HC. Recent progress in selective CO removal in a H_2-rich stream. Catal Today. 2009;139(4):280–90. DOI: 10.1016/j.cattod.2008.06.027.

[165] Tsai M-C, Wang J-H, Shen -C-C, Yeh C-T. Promotion of a copper–zinc catalyst with rare earth for the steam reforming of methanol at low temperatures. J Catal. 2011;279(2):241–45. DOI: 10.1016/j.jcat.2010.12.018.

[166] Breen JP, Ross JRH. Methanol reforming for fuel-cell applications: Development of zirconia-containing Cu–Zn–Al catalysts. Catal Today. 1999;51(3-4):521–33. DOI: 10.1016/S0920-5861(99)00038-3.

[167] Agrell J, Birgersson H, Boutnonnet M, Melián-Cabrera I, Navarro RM, Fierro JLG. Production of hydrogen from methanol over Cu/ZnO catalysts promoted by ZrO_2 and Al_2O_3. J Catal. 2003;219(2):389–403. DOI: 10.1016/S0021-9517(03)00221-5.

[168] Matsumura Y, Ishibe H. High temperature steam reforming of methanol over $Cu/ZnO/ZrO_2$ catalysts. Appl Catal B: Env. 2009;91(1-2):524–32. DOI: 10.1016/j.apcatb.2009.06.023.
[169] Jones SD, Hagelin-Weaver HE. Steam reforming of methanol over CeO_2- and ZrO_2-promoted Cu-ZnO catalysts supported on nanoparticle Al_2O_3. Appl Catal B: Env. 2009;90(1-2):195–204. DOI: 10.1016/j.apcatb.2009.03.013.
[170] Velu S, Suzuki K. Selective Production of Hydrogen for Fuel Cells Via Oxidative Steam Reforming of Methanol Over CuZnAl Oxide Catalysts: Effect of Substitution of Zirconium and Cerium on the Catalytic Performance. Top Catal. 2003;22(3-4):235–44. DOI: 10.1023/A: 1023576020120.
[171] Velu S, Suzuki K, Okazaki M, Kapoor MP, Osaki T, Ohashi F. Oxidative Steam Reforming of Methanol over CuZnAl(Zr)-Oxide Catalysts for the Selective Production of Hydrogen for Fuel Cells: Catalyst Characterization and Performance Evaluation. J Catal. 2000;194(2):373–84. DOI: 10.1006/jcat.2000.2940.
[172] Constantino U, Marmottini F, Sisani M, Montanari T, Ramis G, Busca G, Turco M, Bagnasco G. Cu–Zn–Al hydrotalcites as precursors of catalysts for the production of hydrogen from methanol. Solid State Ionics. 2005;176(39-40):2917–22. DOI: 10.1016/j.ssi.2005.09.051.
[173] Turco M, Bagnasco G, Constantino U, Marmottini F, Montanari T, Ramis G, Busca G. Production of hydrogen from oxidative steam reforming of methanol: I. Preparation and characterization of $Cu/ZnO/Al_2O_3$ catalysts from a hydrotalcite-like LDH precuros. J Catal. 2004;228(1):43–55. DOI: 10.1016/j.jcat.2004.08.026.
[174] Turco M, Bagnasco G, Constantino U, Marmottini F, Montanari T, Ramis G, Busca G. Production of hydrogen from oxidative steam reforming of methanol: II. Catalytic activity and reaction mechanism on $Cu/ZnO/Al_2O_3$ hydrotalcite-derived catalysts. J Catal. 2004;228 (1):56–65. DOI: 10.1016/j.jcat.2004.08.029.
[175] Ritzkopf I, Vukojević S, Weidenthaler C, Grunwaldt J-D, Schüth F. Decreased CO production in methanol steam reforming over Cu/ZrO_2 catalysts prepared by the microemulsion technique. Appl Catal A: General. 2006;302(2):215–23. DOI: 10.1016/j.apcata.2006.01.014.
[176] Purnama H, Girgsdies F, Ressler T, Schattka JH, Caruso RA, Schomäcker R, Schlögl R. Activity and Selectivity of a Nanostructured CuO/ZrO_2 Catalyst in the Steam Reforming of Methanol. Catal Lett. 2004;94(1-2):61–68. DOI: 10.1023/B:CATL.0000019332.80287.6b.
[177] Liu Y, Hayakawa T, Suzuki K, Hamakawa S, Tsunoda T, Ishii T, Kumagai M. Highly active copper/ceria catalysts for steam reforming of methanol. Appl Catal A: General. 2002;223(1-2):137–45. DOI: 10.1016/S0926-860X(01)00733-5.
[178] Szizybalski A, Girgsdies F, Rabis A, Wang Y, Niederberger M, Ressler T. In situ investigations of structure–activity relationships of a Cu/ZrO_2 catalyst for the steam reforming of methanol. J Catal. 2005;233(2):297–307. DOI: 10.1016/j.jcat.2005.04.024.
[179] Mastalir A, Frank B, Szizybalski A, Soerijanto H, Deshpande A, Niederberger M, Schomäcker R, Schlögl R, Ressler T. Steam reforming of methanol over $Cu/ZrO_2/CeO_2$ catalysts: A kinetic study. J Catal. 2005;230(2):464–75. DOI: 10.1016/j.jcat.2004.12.020.
[180] Rozovskii AY, Lin GI. Fundamentals of Methanol Synthesis and Decomposition. Top Catal. 2003;22(3-4):137–50. DOI: 10.1023/A: 1023555415577.
[181] Spencer MS. On the activation energies of the forward and reverse water-gas shift reaction. Catal Lett. 1995;32(1-2):9–13. DOI: 10.1007/BF00806097.
[182] Saito M, Wu J, Tomoda K, Takahara I, Murata K. Effects of ZnO Contained in Supported Cu-Based Catalysts on Their Activities for Several Reactions. Catal Lett. 2002;83(1):1–4. DOI: 10.1023/A: 1020693226903.
[183] Löffler DG, McDermott SD, Renn: CN. Activity and durability of water-gas shift catalysts used for the steam reforming of methanol. J Power Sour. 2003;114(1):15–20. DOI: 10.1016/S0378-7753(02)00589-X.

[184] Thurgood CP, Amphlett JC, Mann RF, Peppley BA. Deactivation of $Cu/ZnO/Al_2O_3$ Catalyst: Evolution of Site Concentrations with Time. Top Catal. 2003;22(3-4):253–59. DOI: 10.1023/A:1023680121028.

[185] Agrell J, Birgersson H, Boutonnet M. Steam reforming of methanol over a $Cu/ZnO/Al_2O_3$ catalyst: A kinetic analysis and strategies for suppression of CO formation. J Power Sour. 2002;106(1-2):249–57. DOI: 10.1016/S0378-7753(01)01027-8.

[186] Iwasa N, Masuda S, Takezawa N. Steam reforming of methanol over Ni, Co, Pd and Pt supported on ZnO. React Kinet Catal Lett. 1995;55(2):349–53. DOI: 10.1007/BF02073070.

[187] Penner S, Jenewein B, Gabasch H, Klotzer B, Wang D, Knopgericke A, Schlögl R, Hayek K. Growth and structural stability of well-ordered PdZn alloy nanoparticles. J Catal. 2006;241(1):14–19. DOI: 10.1016/j.jcat.2006.04.005.

[188] Karim A, Conant T, Datye A. The role of PdZn alloy formation and particle size on the selectivity for steam reforming of methanol. J Catal. 2006;243(2):420–27. DOI: 10.1016/j.jcat.2006.07.024.

[189] Liu S, Takahashi K, Eguchi H, Uematsu K. Hydrogen production by oxidative methanol reforming on Pd/ZnO: Catalyst preparation and supporting materials. Catal Today. 2007;129(3-4):287–92. DOI: 10.1016/j.cattod.2006.08.069.

[190] Föttinger K, Van Bokhoven JA, Nachtegaal M, Rupprechter G. Dynamic Structure of a Working Methanol Steam Reforming Catalyst: In Situ Quick-EXAFS on Pd/ZnO Nanoparticles. J Phys Chem Lett. 2011;2(5):428–33. DOI: 10.1021/jz101751s.

[191] Jiang CJ, Trimm DL, Wainwright MS, Cant NW. Kinetic mechanism for the reaction between methanol and water over a $Cu-ZnO-Al_2O_3$ catalyst. Appl Catal A: General. 1993;97(2):145–58. DOI: 10.1016/0926-860X(93)80081-Z.

[192] Takezawa N, Iwasa N. Steam reforming and dehydrogenation of methanol: Difference in the catalytic functions of copper and group VIII metals. Catal Today. 1997;36(1):45–56. DOI: 10.1016/S0920-5861(96)00195-2.

[193] Peppley BA, Amphlett JC, Kearns LM, Mann RF. Methanol–steam reforming on $Cu/ZnO/Al_2O_3$ catalysts.: Part 2. A comprehensive kinetic model. Appl Catal A: General. 1999;179(1-2):31–49. DOI: 10.1016/S0926-860X(98)00299-3.

[194] Lee JK, Ko JB, Kim DH. Methanol steam reforming over $Cu/ZnO/Al_2O_3$ catalyst: Kinetics and effectiveness factor. Appl Catal A: General. 2004;278(1):25–35. DOI: 10.1016/j.apcata.2004.09.022.

[195] Peppley BA, Amphlett JC, Kearns LM, Mann RF. Methanol–steam reforming on $Cu/ZnO/Al_2O_3$ catalysts.: Part 1. The reaction network. Appl Catal A: General. 1999;179(1-2):21–29. DOI: 10.1016/S0926-860X(98)00298-1.

[196] Frank B, Jentoft F, Soerijanto H, Krohnert J, Schlögl R, Schomäcker R. Steam reforming of methanol over copper-containing catalysts: Influence of support material on microkinetics. J Catal. 2007;246(1):177–92. DOI: 10.1016/j.jcat.2006.11.031.

[197] Lin S, Johnson RS, Smith GK, Xie D, Guo H. Pathways for methanol steam reforming involving adsorbed formaldehyde and hydroxyl intermediates on Cu(111): Density functional theory studies. Phys Chem Chem Phys. 2011;13(20):9622–31. DOI: 10.1039/c1cp20067d.

[198] Lin S, Xie D, Guo H. Methyl Formate Pathway in Methanol Steam Reforming on Copper: Density Functional Calculations. ACS Catal. 2011;1(10):1263–71. DOI: 10.1021/cs200311t.

[199] Goodby BE, Pemberton JE. XPS Characterization of a Commercial $Cu/ZnO/Al_2O_3$ Catalyst: Effects of Oxidation, Reduction, and the Steam Reformation of Methanol. Appl Spectrosc. 1988;42(5):754–60. DOI: 10.1366/0003702884429148.

[200] Raimondi F, Schnyder B, Kötz R, Schelldorfer R, Jung T, Wambach J, Wokaun A. Structural changes of model Cu/ZnO catalysts during exposure to methanol reforming conditions. Surf Sci. 2003;532-535:383–89. DOI: 10.1016/S0039-6028(03)00236-X.

[201] Raimondi F, Wambach J, Wokaun A. Structural properties of Cu/ZnO/Si methanol reforming catalysts: Influence of the composition of the reactant mixture and of the Cu island size. Phys Chem Chem Phys. 2003;5(18):4015–24. DOI: 10.1039/b305714c.
[202] Reitz TL, Lee P, Czaplewski KF, Lang J, Popp KE, Kung HH. Time-Resolved XANES Investigation of CuO/ZnO in the Oxidative Methanol Reforming Reaction. J Catal. 2001;199(2):193–201. DOI: 10.1006/jcat.2000.3141.
[203] Busca G, Constantino U, Marmottini F, Montanari T, Patrono P, Pinzari F, Ramis G. Methanol steam reforming over ex-hydrotalcite Cu–Zn–Al catalysts. Appl Catal A: General. 2006;310:70–78. DOI: 10.1016/j.apcata.2006.05.028.
[204] Busca G, Montanari T, Resini C, Ramis G, Costantino U. Hydrogen from alcohols: IR and flow reactor studies. Catal Today. 2009;143(1-2):2–8. DOI: 10.1016/j.cattod.2008.09.010.
[205] Larrubia Vargas MA, Busca G, Constantino U, Marmottini F, Montanari T, Patrono P, Pinzari F, Ramis G. An IR study of methanol steam reforming over ex-hydrotalcite Cu–Zn–Al catalysts. J Mol Catal A Chem. 2007;266(1-2):188–97. DOI: 10.1016/j.molcata.2006.08.085.
[206] Zhang X-R, Wang L-C, Cao Y, Dai W-L, He H-Y, Fan K-N A unique microwave effect on the microstructural modification of Cu/ZnO/Al_2O_3 catalysts for steam reforming of methanol, Chem Commun, 2005, (32), 4104–06, DOI: 10.1039/b502997j.
[207] Wang L, Liu Y, Chen M, Cao Y, He H, Wu G, Dai W, Fan K. Production of hydrogen by steam reforming of methanol over Cu/ZnO catalysts prepared via a practical soft reactive grinding route based on dry oxalate-precursor synthesis. J Catal. 2007;246(1):193–204. DOI: 10.1016/j.jcat.2006.12.006.
[208] Noei H, Qiu H, Wang Y, Löffler E, Wöll C, Muhler M. The identification of hydroxyl groups on ZnO nanoparticles by infrared spectroscopy. Phys Chem Chem Phys. 2008;10(47):7092–97. DOI: 10.1039/b811029h.
[209] Storch H, Golumbic N, Anderson RB. The Fischer-Tropsch and Related Syntheses. New York: Wiley; 1951.
[210] Van Der Laan GP, Beenackers AACM. Kinetics and Selectivity of the Fischer–Tropsch Synthesis: A Literature Review. Catal Rev. 1999;41(3-4):255–318. DOI: 10.1081/CR-100101170.
[211] Steynberg A, Dry M (eds.). Fischer-Tropsch Technology. Amsterdam: Elsevier; 2004.
[212] Van De Loosdrecht J, Botes FG, Ciobica IM, Ferreira A, Gibson P, Moodley DJ, Saib AM, Visagie JL, Weststrate CJ, Niemantsverdriet JW. Fischer–Tropsch Synthesis: Catalysts and Chemistry, in Comprehensive Inorganic Chemistry II, (ed. Reedijk KPJ). Elsevier; 2013, 525–57.
[213] Xu J, Yang Y, Li Y-W. Recent development in converting coal to clean fuels in China. Fuel. 2015;152:122–30. DOI: 10.1016/j.fuel.2014.11.059.
[214] Biloen P, Sachtler WMH. Mechanism of Hydrocarbon Synthesis over Fischer-Tropsch Catalysts. In: Eley DD, Pines H, Weisz PB, editor. Advances in Catalysis. vol. 30, New York: Academic Press; 1981.
[215] Weststrate CJ, Mahmoodinia M, Farstad MH, Svenum I-H, Strømsheim MD, Niemantsverdriet JW, Venvik HJ Interaction of hydrogen with flat (0001) and corrugated (11–20) and (10–12) cobalt surfaces: Insights from experiment and theory, Catal Today, 2019, DOI: 10.1016/j.cattod.2019.04.002.
[216] Van Santen RA, Markvoort AJ, Filot IAW, Ghouri MM, Hensen EJM. Mechanism and microkinetics of the Fischer-Tropsch reaction. Phys Chem Chem Phys. 2013;15(40):17038–63. DOI: 10.1039/C3CP52506F.
[217] Kizilkaya AC, Niemantsverdriet JW, Weststrate CJ. Oxygen Adsorption and Water Formation on Co(0001). J Phys Chem C. 2016;120(9):4833–42. DOI: 10.1021/acs.jpcc.5b08959.
[218] Weststrate CJ, Van Helden P, Niemantsverdriet JW. Reflections on the Fischer-Tropsch synthesis: Mechanistic issues from a surface science perspective. Catal Today. 2016;275:100–10. DOI: 10.1016/j.cattod.2016.04.004.

[219] Donnelly TJ, Yates IC, Satterfiled CN. Analysis and Prediction of Product Distributions of the Fischer-Tropsch Synthesis. Energy Fuels. 1988;2(6):734–39.
[220] Steynberg AP, Espinoza RL, Jager B, Vosloo AC. High temperature Fischer–Tropsch synthesis in commercial practice. Appl Catal A: General. 1999;186(1-2):41–54. DOI: 10.1016/S0926-860X(99)00163-5.
[221] Espinoza RL, Steynberg AP, Jager B, Vosloo AC. Low temperature Fischer–Tropsch synthesis from a Sasol perspective. Appl Catal A: General. 1999;186(1-2):13–26. DOI: 10.1016/S0926-860X(99)00161-1.
[222] Geerlings JJC, Wilson JH, Kramer GJ, Kuipers HPCE, Hoek A, Huisman HM. Fischer–Tropsch technology – from active site to commercial process. Appl Catal A: General. 1999;186(1-2):27–40. DOI: 10.1016/S0926-860X(99)00162-3.
[223] Sie ST. Process Development and Scales Up: IV. Case History of the Development of a Fischer-Tropsch Synthesis Process. Rev Chem Eng. 1998;14(2):109–57. DOI: 10.1515/REVCE.1998.14.2.109.
[224] Dry ME, Hoogendoorn JC. Technology of the Fischer-Tropsch Process. Catal Rev -Sci Eng. 1981;23(1-2):265–78. DOI: 10.1080/03602458108068078.
[225] Davis BH. Fischer-Tropsch synthesis: Relationship between iron catalyst composition and process variables. Catal Today. 2003;84(1-2):83–98. DOI: 10.1016/S0920-5861(03)00304-3.
[226] Niemantsverdriet JW, Van Der Kraan AM, Van Dijk WL, Van Der Baan HS. Behavior of metallic iron catalysts during Fischer-Tropsch synthesis studied with Moessbauer spectroscopy, x-ray diffraction, carbon content determination, and reaction kinetic measurements. J Phys Chem. 1980;84(25):3363–70.
[227] De Smit E, Weckhuysen: BM. The renaissance of iron-based Fischer-Tropsch synthesis: On the multifaceted catalyst deactivation behaviour. Chem Soc Rev. 2008;37(12):2758–81. DOI: 10.1039/b805427d.
[228] Khodakov AY, Chu W, Fongarland P. Advances in the development of novel cobalt Fischer-Tropsch catalysts for synthesis of long-chain hydrocarbons and clean fuels. Chem Rev. 2007;107(5):1692–744. DOI: 10.1021/cr050972v.
[229] Tsakoumis NE, Rønning M, Borg Ø, Rytter E, Holmen A. Deactivation of cobalt based Fischer–Tropsch catalysts: A review. Catal Today. 2010;154(3-4):162–82. DOI: 10.1016/j.cattod.2010.02.077.
[230] Van De Loosdrecht J, Balzhinimaev B, Dalmon J-A, Niemantsverdriet JWH, Tsybulya SV, Saib AM, Van Berge PJ, Visagie JL. Cobalt Fischer-Tropsch synthesis: Deactivation by oxidation?. Catal Today. 2007;123(1-4):293–302. DOI: 10.1016/j.cattod.2007.02.032.
[231] Reedijk KPJ (ed.). Comprehensive Inorganic Chemistry II. Elsevier; 2013.
[232] Saib A, Borgna A, Vandeloosdrecht J, Vanberge P, Geus J, Niemantsverdriet JWH. Preparation and characterisation of spherical Co/SiO_2 model catalysts with well-defined nano-sized cobalt crystallites and a comparison of their stability against oxidation with water. J Catal. 2006;239(2):326–39. DOI: 10.1016/j.jcat.2006.02.004.
[233] Bezemer GL, Bitter JH, Kuipers HPCE, Oosterbeek H, Holewijn JE, Xu X, Kapteijn F, Van Dillen AJ, De Jong KP. Cobalt particle size effects in the Fischer-Tropsch reaction studied with carbon nanofiber supported catalysts. J Am Chem Soc. 2006;128(12):3956–64. DOI: 10.1021/ja058282w.
[234] Saib AM, Moodley DJ, Ciobîcă IM, Hauman MM, Sigwebela BH, Weststrate CJ, Niemantsverdriet JWH, Van De Loosdrecht J. Fundamental understanding of deactivation and regeneration of cobalt Fischer–Tropsch synthesis catalysts. Catal Today. 2010;154(3-4):271–82. DOI: 10.1016/j.cattod.2010.02.008.

[235] LeViness S, Deshmukh SR, Richard LA, Robota HJ. Velocys Fischer–Tropsch Synthesis Technology – New Advances on State-of-the-Art. Top Catal. 2014;57(6-9):518–25. DOI: 10.1007/s11244-013-0208-x.

[236] Menon M. Comparative Study of Gas to Liquids (GTL) Technology for Monetising Marginal Reserves and Wasted Flared Gas, Kuala Lumpur, Malaysia, 2016- 03-22.

[237] Fischer F, Tropsch H. The Preparation of synthetic oil mixtures (synthol) from carbon monoxide and hydrogen. Brenst Chem. 1923;4:276–85.

[238] Fischer F, Tropsch H. The synthesis of petroleum at atmospheric pressures from gasification products of coal. Brenst Chem. 1926;7:97–104.

[239] Sabatier P, Senderens JB. Hydrogenation of CO over Nickel to Produce Methane. Hebd Seances Acad Sci. 1902;134:514.

[240] Oukaci R, Singleton AH, Goodwin JG. Comparison of patented Co F–T catalysts using fixed-bed and slurry bubble column reactors. Appl Catal A: General. 1999;186(1-2):129–44. DOI: 10.1016/S0926-860X(99)00169-6.

[241] Van Steen E, Claeys M. Fischer-Tropsch Catalysts for the Biomass-to-Liquid (BTL)-Process. Chem Eng Technol. 2008;31(5):655–66. DOI: 10.1002/ceat.200800067.

[242] Tristantini D, Lögdberg S, Gevert B, Borg Ø, Holmen A. The effect of synthesis gas composition on the Fischer–Tropsch synthesis over Co/γ-Al_2O_3 and Co–Re/γ-Al_2O_3 catalysts. Fuel Process Technol. 2007;88(7):643–49. DOI: 10.1016/j.fuproc.2007.01.012.

[243] Tijmensen M. Exploration of the possibilities for production of Fischer Tropsch liquids and power via biomass gasification. Biomass Bioenergy. 2002;23(2):129–52. DOI: 10.1016/S0961-9534(02)00037-5.

[244] Satthawong R, Koizumi N, Song C, Prasassarakich P. Comparative Study on CO_2 Hydrogenation to Higher Hydrocarbons over Fe-Based Bimetallic Catalysts. Top Catal. 2014;57(6-9):588–94. DOI: 10.1007/s11244-013-0215-y.

[245] Landau MV, Meiri N, Utsis N, Vidruk Nehemya R, Herskowitz M. Conversion of CO_2, CO, and H_2 in CO_2 Hydrogenation to Fungible Liquid Fuels on Fe-Based Catalysts. Ind Eng Chem Res. 2017;56(45):13334–55. DOI: 10.1021/acs.iecr.7b01817.

[246] Visconti CG, Martinelli M, Falbo L, Infantes-Molina A, Lietti L, Forzatti P, Iaquaniello G, Palo E, Picutti B, Brignoli F. CO_2 hydrogenation to lower olefins on a high surface area K-promoted bulk Fe-catalyst. Appl Catal B. 2017;200:530–42. DOI: 10.1016/j.apcatb.2016.07.047.

[247] Li W, Wang H, Jiang X, Zhu J, Liu Z, Guo X, Song C. A short review of recent advances in CO_2 hydrogenation to hydrocarbons over heterogeneous catalysts. RSC Adv. 2018;8(14):7651–69. DOI: 10.1039/C7RA13546G.

[248] Deutz S, Bongartz D, Heuser B, Kätelhön A, Schulze Langenhorst L, Omari A, Walters M, Klankermayer J, Leitner W, Mitsos A, Pischinger S, Bardow A. Cleaner production of cleaner fuels: Wind-to-wheel – environmental assessment of CO_2 -based oxymethylene ether as a drop-in fuel. Energy Environ Sci. 2018;11(2):331–43. DOI: 10.1039/C7EE01657C.

[249] Baranowski CJ, Bahmanpour AM, Kröcher O. Catalytic synthesis of polyoxymethylene dimethyl ethers (OME): A review. Appl Catal B. 2017;217:407–20. DOI: 10.1016/j.apcatb.2017.06.007.

[250] Schmitz N, Burger J, Ströfer E, Hasse H. From methanol to the oxygenated diesel fuel poly (oxymethylene) dimethyl ether: An assessment of the production costs. Fuel. 2016;185:67–72. DOI: 10.1016/j.fuel.2016.07.085.

[251] Lautenschütz L, Oestreich D, Haltenort P, Arnold U, Dinjus E, Sauer J. Efficient synthesis of oxymethylene dimethyl ethers (OME) from dimethoxymethane and trioxane over zeolites. Fuel Process Technol. 2017;165:27–33. DOI: 10.1016/j.fuproc.2017.05.005.

[252] Härtl M, Pélerin D, Dworschak P, Maier T, Stadler A, Blochum S, Gaukel K, Jacob E, Wachtmeister G. Potential of the sustainable C1 fuels OME, DMC, and MeFo for particle-free

combustion in SI and CI engines. In: J. Liebl et al. editor. Internationaler Motorenkongress 2018. Wiesbaden: Springer Fachmedien Wiesbaden; 2018, 459–78.

[253] Liu Z, Zhang R, Wang S, Li N, Sima R, Liu G, Wu P, Zeng G, Li S, Sun Y. Highly Efficient and Stable Vanadia–Titania–Sulfate Catalysts for Methanol Oxidation to Methyl Formate: Synthesis and Mechanistic Study. J Phys Chem C. 2016;120(12):6591–600. DOI: 10.1021/acs.jpcc.5b12621.

[254] Kongpanna P, Pavarajarn V, Gani R, Assabumrungrat S. Techno-economic evaluation of different CO_2-based processes for dimethyl carbonate production. Chem Eng Res Des. 2015;93:496–510. DOI: 10.1016/j.cherd.2014.07.013.

Gabriele Centi, Siglinda Perathoner

15 Photoelectrocatalytic CO_2 Activation Toward Artificial Leaves: Outlooks and Needs of a System Approach

15.1 Introduction

The concept of developing artificial leaf (AL) devices which can directly realize CO_2 conversion to valuable fuels/chemicals using sunlight and water [1] is one of the more challenging and fascinating to realize in full the ongoing energy transition. They are a missing element to transform energy (and chemistry) infrastructure from centralized to delocalized and a key technology for implementing a fully renewable energy system and overcome the use of fossil fuels (FFs). They are a crucial factor to implement chemical energy storage (CES) vectors as the backbone for energy, equivalent to what play today by FFs, overcoming the limits and constraints deriving from the grid dependence of an energy system based on electrical energy. They are the next step for a green H_2 world to overcome hydrogen limits in storage and transport. Thus, AL devices are a necessary technology for a future sustainable society.

The concept of AL is applied to different cases (production of H_2, NH_3, or products of CO_2 reduction), the first case (hydrogen) being the most studied. However, in terms of mimicking natural leaves and technological perspectives, the CO_2 case is the most appropriate. Chemicals and fuels produced from CO_2 by ALs can be readily integrated in the actual infrastructure, minimizing the cost of transformation, and thus greatly accelerating the conversion. By integrating functionalities which allow to directly use low-concentration CO_2 emissions, if not directly CO_2 from the air, it is possible to close the carbon cycle disbalance present, without the massive investments otherwise necessary for using other solutions, including H_2.

Acknowledgments: This work has been realized in the frame of the PRIN2017 projects 2017WR2LRS "CO_2 as only source of carbons for monomers and polymers: a step forwards circular economy" and 20179337R7 "Multielectron transfer for the conversion of small molecules: an enabling technology for the chemical use of renewable energy" which are gratefully acknowledged. The FET ProActive A-LEAF "An Artificial Leaf: a photoelectrocatalytic cell from earth-abundant materials for sustainable solar production of CO_2-based chemicals and fuels" project (grant no. 732840) and the ERC Synergy Project SCOPE "Surface-COnfined fast-modulated Plasma for process and Energy intensification in small molecules conversion" (grant no. 810182) are also gratefully acknowledged.

Gabriele Centi, Siglinda Perathoner, Dept. ChiBioFarAm – Industrial Chemistry, ERIC aisbl and INSTM/CASPE, University of Messina, V.le F. Stagno D'Alcontres 31 – 98166, Messina, Italy

https://doi.org/10.1515/9783110608458-015

There are significant differences to account with respect to natural leaves:

i) The function of CO_2 capture directly from the air (or diluted CO_2 streams), a key element in natural leaves, is not present, neither actually investigated, in AL devices, although a key challenge as commented above; it can be realized, for example, by incorporating functional membranes in the electrode design.
ii) The target products of CO_2 reduction in AL differ from those present in natural leaves, and still too limited knowledge exists on how to control the selectivity and especially realize C2 + (products with two or more carbon atoms) products which involve a multielectron/proton reduction.
iii) Natural leaves contain crucial functionalities for their life times, such as self-reparation and self-limitation in light absorption, which cannot be present in ALs, because they are also factors limiting productivity and solar-to-fuel (STF) efficiency. To make ALs economic, STF should be over one order of magnitude higher than in natural leaf, but this means to use a radical different, and more robust, design with respect to a natural leaf.

Thus, going from natural to ALs requires the use of a radical different system design taking inspiration from nature, but developing conceptually new and robust devices that overcome the limits of natural leaves. Often this concept is not well present in literature studies and approaches. In addition, as outlined above, some key aspects for AL are not investigated. AL are complex devices, still largely underperforming, and it is thus reasonable that simplifications in the study are necessary, as Einstein stated "made as simple as possible, but no simpler."

A relevant example is given by the materials used to realize AL. To mimic the complex machinery present in natural leaves, complex supramolecular systems are often constructed to realize artificial photosynthesis devices, another way to indicate AL. These supramolecular systems are based on organic complexes, molecular or biomaterials [2–10] which have limited thermal stability. In addition, they typically need to operate in contact with a liquid electrolyte. They cannot operate above 100 °C.

From an operational perspective, the efficiency of the systems could be low-cost increased by adding Fresnel lenses acting as a sun concentrator. Depending on the design, sun concentration factors up to three orders of magnitude can be reached [11], but even a simple (not-costly) design can allow to increase light intensity by a factor 10–20 at a minimal additional cost. However, without using a cooling system, these concentration factors can easily increase the temperature of the cell even above 100 °C. Most of the current cell designs, and related materials, cannot run in a temperature range of around 100–200 °C. This is a current challenge, but it indicates that these aspects, together with others, have to be accounted for already at the stage of development of the materials. "Turning perspective in photoelectrocatalytic cells for solar fuels" is the title of a viewpoint contribution [12] where this and other relevant aspect for design, such as product recovery, are analyzed to remark the need to focus research on AL on aspects relevant for their future practical implementation. Other aspects relevant for the design of AL devices for CO_2 conversion are discussed also in refs [13, 14].

An abundant literature on artificial photosynthesis or leafs is available, with selected recent reviews being reported in refs. [15–30]. The terms photosynthesis and leaf are used as synonymous (the first is the most used), although correctly they describe different aspects. The first refers more properly to the material, while the second to the device. The cited reviews and related works discuss important aspects of the light harvesting mechanism, of the charge separation and transfer and of the reaction mechanism and the nature of the active elements, but often the studies on artificial photosynthesis are limited to H_2O conversion to H_2 and O_2 rather than to consider the more challenging conversion of CO_2. These studies advance the knowledge in the field, but many of the proposed devices cannot operate in CO_2 conversion rather than in water splitting [1, 12–14]. However, there is an increasing effort on devices for CO_2 conversion recently.

An AL device for CO_2 conversion (CO_2RR) should be composed by elements which adsorb light, generate electrons and holes, the latter which oxidize water to generate O_2 and protons, and elements able to adsorb and convert CO_2 with H^+/e^- to generate the target products. It should be thus present a photoactive element (for light absorption generation of photocurrent and water oxidation) and an electrocatalytic element for CO_2 reduction. These two typologies of processes should operate in physically separate zones, although with closure of the electro-ionic circuit (realized by using a membrane). The term photoelectrocatalytic (PEC) indicates this device where the photoactivated processes occur in one section while the electrocatalytic processes in a different section operating (typically) in dark (except for tandem cells, not discussed here for conciseness; although potentially interesting, these tandem cells have severe limitations from a practical perspective, being necessary to illuminate both cell sites, or develop photoanodes and membranes transparent to the light).

Often the term "photoelectrochemical" is used instead of "photoelectrocatalytic" (although the term PEC indicates both), but the second is preferable, because a process limiting the overall performances is the catalytic redox reaction occurring both at the photoanode section (water oxidation) and at the cathode section (CO_2 reduction). The electrodes should be thus modified typically by adding proper catalysts to accelerate these reactions. They are typically indicated as co-catalysts, although the term is not fully correct, because it is not present an effective primary catalyst.

This contribution is a largely revised and updated version of the original version present in the first edition of the book *Chemical Energy Storage* (De Gruyter, 2013).

15.2 Working Principles of PEC Devices, Questions, and Strategies in Their Design

A PEC solar cell device (Figure 15.1) comprises three key elements:

1. A *photoanode*, composed of a *semiconductor* (photovoltaic device, PV), which generates a photocurrent by absorbing sunlight (promoting electrons from the valence to the conduction bands, preferably using the entire solar light spectrum) and a *catalyst* able to use the photogenerated holes to oxidize water to O_2 and protons (in the most common typology of n-type semiconductors).
2. A *cell system* (membrane, electrodes, electronic connections, etc.) that transports the protons and electrons produced in the anodic compartment to the cathodic compartment.
3. A *cathodic electrocatalyst*, which uses the protons and electrons produced in the anodic section for electrocatalytic reduction of CO_2 (CO_2RR), which typically occurs with the side reaction of H_2 evolution (HER).

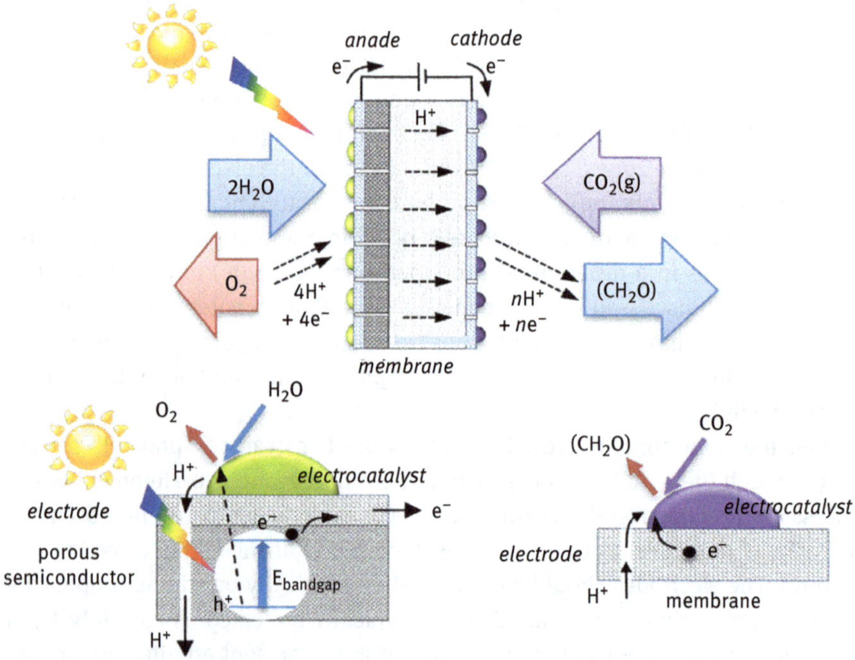

Figure 15.1: PEC solar cell device with a scheme of the photoelectrolytic process at the anodic side and the electrocatalytic process of CO_2 reduction at the cathodic side. (CH_2O) indicates in general the products of CO_2 hydrogenation. Reproduced from the contribution present in the first edition of the book *Chemical Energy Storage*.

The first two elements could either be integrated in a single photocatalytic element able to absorb the light and oxidize water or instead be physically separated (i.e., using an external PV cell, which provides the current to an electrolytic cell). In the latter configuration, that we can indicate as PV/EC where PV indicates the photovoltaic element and EC the electrochemical cell, the solar PV cell is connected in series to an electrolyzer (and a counter electrode): the redox reactions are carried out by the electrolyzer (and the counter electrode), using the charge carriers photogenerated by the PV component. With this architecture, the solar PV cell could be integrated within the EC cell, or externally, for example, not in direct contact with the electrolyte. This solution prevents corrosion effects which limits up to now the development of these cells from an application perspective, but introduces limitations in the performances, due to not efficient coupling of the two components. Since both photovoltaic and electrolyzer technologies have reached high efficiencies, the PV/EC approach is the most promising solution for production of H_2 and O_2 by water splitting [21], with the current record of solar-to-hydrogen efficiency of about 30% [31], even if a fast deactivation is present. While the PV/EC approach appears as preferable to the PEC approach for water splitting based on current data [21], the situation is different for CO_2 reduction. The preference between PV/EC and PEC architectures is an open question and there are no specific studies which can clearly indicate the preferable approach. It is often assumed that the same typology of devices used for water splitting could be used for CO_2 reduction, but this is not true. There are large differences between the two cases indicating that distinct devices should be used in the two cases. Many solutions used to improve performances for water splitting are not effective for CO_2 reduction, for various reasons, among which selectivity and the consequent impossibility to operate at the optimal current density.

In H_2 evolution, recombination of H^+ and e^- occurs, while in CO_2 reduction their recombination should be avoided and instead favored their reaction with CO_2. There are problems of CO_2 solubility in the electrolyte (except for electrolyte-less operations, see later), diffusion of CO_2 through the double-layer to access to electrode surface, surface coverage of the electrolyte by CO_2, needs to avoid back-diffusion of CO_2 after the first electron transfer (the $^{\bullet}CO_2^-$ anion radical which form move away from the electrode negatively charged). H_2, forming from H^+/e^- recombination, generates gas bubbles remaining sticked to the electrode, while the products of CO_2 reduction remain in the electrolyte (when present), except for the case of CO, and may be further transformed. In CO_2RR the optimal current density is determined from the need to limit the side reaction of hydrogen evolution and thus solar-to-fuel (η_{STF}) efficiencies are lower. In addition, the catalytic process undergoes reaction intermediates where the light absorption process may be coupled to the catalytic mechanism, precluding a PV/EC strategy. In CO_2RR the further complexity derives from the many products of CO_2 reduction forming at similar potentials (see later).

There is an additional relevant aspect to consider. The hydrogen evolution is a two-electron reaction which could also occurs in sequential steps, while CO_2RR involves a higher number of electrons, except for CO or HCOOH production (two-electron reduction). For all other cases, it requires a multielectron/proton simultaneous transfer, rather than a sequential transfer [32, 33]. For two-electron reduction of CO_2, high Faradaic efficiencies could be obtained, above 90%, with high current densities (but only using an external applied potential, that is, in electrochemical reactions, not PV/EC or PEC approaches). For example, a Faradaic efficiency of ~96% and a current density of ~41 $mA\cdot cm^{-2}$ have been reported for the electroreduction of CO_2 to HCOOH in an ionic liquid electrolyte on a PbO_2 electrode [34]. However, selectivity is largely lower when products of CO_2 reduction involving >2 electrons are targeted, such as methanol (a 6e– reduction) or products also involving C–C bond formation (indicated as C2+ products).

The aspects remarked above are some of the issue indicating that the cell design should be different, not only the electrode/electrocatalyst, differently from what often supposed. While PV/EC is likely preferable over other PEC strategies for water splitting, the situation is much less defined, and still open to discussion for CO_2RR. In these cases, an integrated PEC device, instead of the sequential PV/EC approach, may be likely preferable, even if reliable strategies to obtain high performances for CO_2RR in PEC devices have not been proposed yet. This intrinsic difficulty has greatly limited the results in literature regarding the optimal design for the artificial-leaf type device. Despite the many reviews present on the topic (a selection was cited before), the focus of discussion has been mainly centered to aspects dealing on materials separately considered from the cell design without considering the implications of cell design on material (electrocatalyst, electrodes, membranes, etc.) requirements.

Materials and cell design are instead strictly correlated in artificial photosynthesis/leaf devices, but this is a concept scarcely recognized in literature. Most of the studies are interesting from the scientific perspective but have limited impact in terms of effective practical implementation of these devices for CO_2 reduction in relevant industrial emissions. This is not only related to aspects such as presence of impurities and scale-up, as typically supposed, but to the lack of a strategic vision which clearly identify the needs from an application perspective and the S&T (science and technology) inputs to address the requirements. The problem of thermal stability, product recovery, current density, target products in CO_2RR, optimal cell design for industrial operations is among the open questions which have received a scarce attention in the abundant literature on the topic.

In terms of industrial exploitation, a compact cell design as illustrated in Figure 15.1 is preferable for a practical development and to minimize resistances and efficiency losses, while maintaining an easier and more robust approach regarding the development of a single catalytic element able to be effective in the three steps: light absorption, charge separation, and the multielectron CO_2 reduction. However, most of the studies used a PEC configuration based on two electrodes immersed in a large volume electrolyte, sometimes even without physical separation of the anodic and cathodic zones obtained by putting a membrane between the two compartments. Often this is considered a more convenient solution for laboratory studies, for which the development to cells of more practical interest is a consecutive stage. This is not a correct approach, because the cell configuration largely determines the interfaces present and the slow steps in the process. It is a chain-type process where each step influences the others and understanding properly this aspect is of fundamental importance. Optimal electrodes observed for the conventional PEC cell differ from those for the compact design in Figure 15.1. To accelerate the process of application of PEC devices for CO_2 reduction it is necessary to minimize the time of discovery of novel solutions, which can be achieved only with a system approach integrating materials development with cell design and engineering, driven from the clear identification of the needs from an application perspective. This system approach is largely lacking.

Given the already high η_{STF} performances in water splitting in PV/EC devices (apart the questions of costs and stability, which can be substantially improved by adequate research and development effort), a strategy is to produce H_2 by the PV/EC approach, and then use H_2 in catalytic (thermal) processes of CO_2 conversion, also well established. This was made, in a slightly different version, in large demo plants. Renewable electrical energy is produced by wind, PV, or other methods, then H_2 is produced in electrolyzers (which already have been scaled-up to large dimensions), H_2 is then compressed and fed with CO_2 to be converted to methane or methanol, the power-to-gas or power-to-liquid (P2G or P2L, respectively) technologies, originally developed to seasonal/yearly storage of excess renewable energy produced [35–39]. This solution, however, does not well adapt to development of AL (distributed) devices. In addition, direct CO_2RR in PEC (or analogous) devices remain a preferable strategy for several reasons: (i) avoiding the overpotentials necessary in H_2 formation and its use (need to activate and compress H_2), due to the instead direct use of protons/electrons for CO_2 reduction, (ii) operating in mild conditions, and (iii) using an integrated device rather than two units (electrolyzer, catalytic unit). A potential reduction of both fixed and operative costs can be thus achieved, if improved electrocatalysts are developed.

15.3 Target Products in CO_2 Conversion in PEC Devices

The products obtained in CO_2 reduction in PEC devices depend not only from the electrocatalyst (as typically supposed), but also on the choice of the operative conditions (electrolyte and its presence or not, for example, electrolyte-less conditions, see later) which in electrochemical cells can be imposed externally, but in the PEC device are determined both from the photoanode characteristics and the chemical potential determined from the eventual differences in the composition/pH of the electrolytes in the anodic and cathodic cell compartments. Note that operating for example at strong basic pH (12–14) at the anode and strong acid (1–2) pH at the cathode the generated potential would be nearly half of the potential of 1.23 V necessary for water splitting (V_0 = –1.23 V at 25 °C for pH 0). By adding the overpotential necessary for the redox reactions of water oxidation and H^+/e^- recombination, typically 0.4–0.6 V, the minimum energy gap required to split water is 1.6–1.8 V, which reduces to around 1.0 when the chemical potential is present, nearly doubling the theoretical maximum current density (and thus productivity) which can be obtained without the chemical potential. However, besides the problem of stability in the materials, this approach is not well suited for CO_2 reduction, rather than H_2 evolution, for several reasons: (i) dependence of the nature of CO_2 species and CO_2 solubility from the pH, and surface coverage on the electrocatalyst, which strongly determine the Faradaic selectivity, (ii) the crossover through the membrane which depends strongly on the pH difference between the compartments. This is another element showing why concepts for water electrolysis cannot be translated to the CO_2 case, explaining the large differences in efficiency between water splitting and CO_2RR.

The generation of hydrogen (HER) always competes with CO_2RR, because the protons/ electrons in the cathodic side should reduce CO_2 instead of generating H_2. The potential to apply varies depending on the reaction, going from 1.281 V when CO_2 is reduced to CO to higher values for other products of CO_2 reduction that involve multielectron reductions. Due to the overpotential, the potential to apply could be higher. However, the CO_2 chemisorption at the electrocatalyst surface, which depends on many parameters, including the electrolyte nature and pressure, affects largely this overpotential and the difference in overpotential between HER and CO_2RR, which is perhaps the most relevant aspect. A systematic analysis of these issues, however, is lacking.

The main issue is to realize a selective conversion of CO_2, because its potential of reduction to different type of products is similar, as shown in Figure 15.2 [40]. The reduction of CO_2 to CO or to $HCOO^-$ is a 2-electron reduction, while it is a 6-electron reduction to methanol and an 8-electron reduction to methane. The reduction to C2+ products is an even more difficult reaction (isopropanol formation, for example, is an 18 e^- reduction). While difficulty in achieving a multielectron transfer increases exponentially with the number of electrons involved, the energy density of the products increases, and they become more valuable as fuels/chemicals and avoid costly further processing. For example, CO should be further converted (in the presence of H_2) to methanol, for example, following the well-established catalytic routes of syngas (CO/H_2) conversion. However, the direct electrocatalytic reduction of CO_2 to methanol is more interesting because it avoids the consecutive steps with related cost and energy efficiency loss. Formic acid production with respect to methanol from CO_2 requires less electrons and avoids the formation of H_2O as by-product, but the energy density of formic acid is lower than that of methanol. Isopropanol formation from CO_2 is a difficult reaction, even if demonstrated to be feasible [41], but in terms of practical use, it is a preferable solution over methanol synthesis.

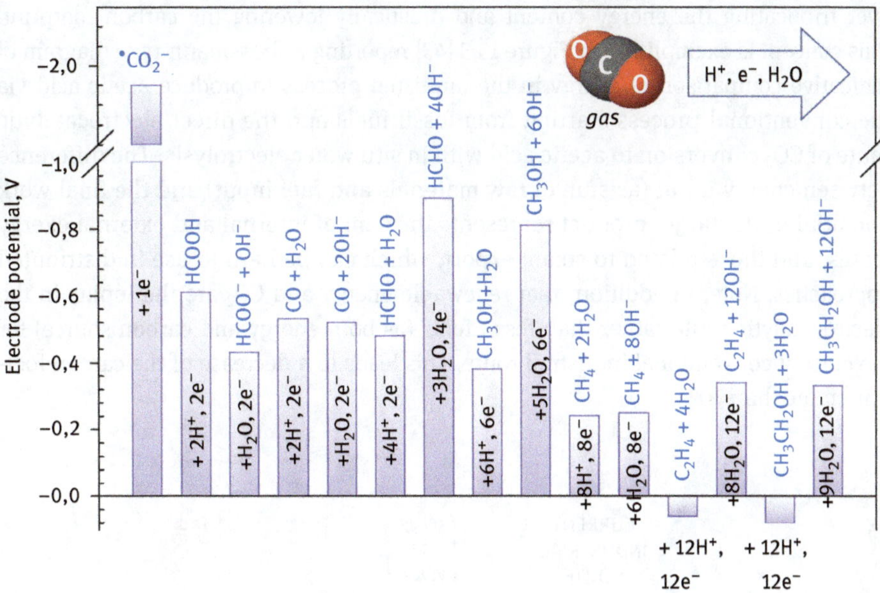

Figure 15.2: Electrode potential at pH 7 (versus standard hydrogen electrode) for half-reactions of electrochemical CO_2 reduction.

There are advantages in terms of both process intensification and cost reduction due to the decreased number of process steps. A good example is the production of acetic acid or acetate from CO_2 reduction, an eight-electron conversion. The industrial synthesis of acetic acid is a multistep process, first the production of syngas from methane, then the catalytic conversion of syngas to methanol, and finally the carbonylation of the latter. The maximum overall energy efficiency, that is, the minimum energy required for the process (based on chemical reactions, standard conditions and 100% yield) divided by the total process energy input, is about 27% for acetic acid [33]. The practical overall energy efficiency is lower, but this value of maximum energy efficiency remarks that most of the energy content in fossil fuel raw materials is lost in production of chemicals using the current processing technologies. By a direct electrocatalytic route, it is possible to synthetize acetic acid from CO_2 and H_2O with higher energy efficiency and avoiding the need of multisteps. To raise the performances to a possible exploitation, it is necessary to understand better the reaction mechanism and particularly the crucial step of C–C bond formation [42].

In terms of energy content of the final molecule, by considering the full value chain, only less than about 25% of the overall fossil fuel used (in the production steps) is found in the final product with the conventional production of acetic acid taken as an example. Direct synthesis of acetic acid from CO_2, even considering a Faradaic selectivity of 60%, could reach an energy content of 50% or more, thus over triplicating the energy content and drastically lowering the carbon footprint. This concept is exemplified in Figure 15.3 [43] reporting a Grassmann-type diagram of indicative comparison of exergy in the multistep process to produce acetic acid via the conventional process starting from fossil fuels and the direct electrocatalytic route of CO_2 conversion to acetic acid with in situ water electrolysis. The difference between energy input (as sum of raw materials and fuel input) and the final work potential of the target product represents the sum of internal and external exergy losses, and those related to steam export, which can be hard to use in distributed approaches. Note, in addition, that renewable energy and CO_2 are the inputs in the electrocatalytic route rather than fossil fuels (as both energy and carbon source) for the current conventional industrial route. This leads to a decrease of the carbon footprint more than 70%,

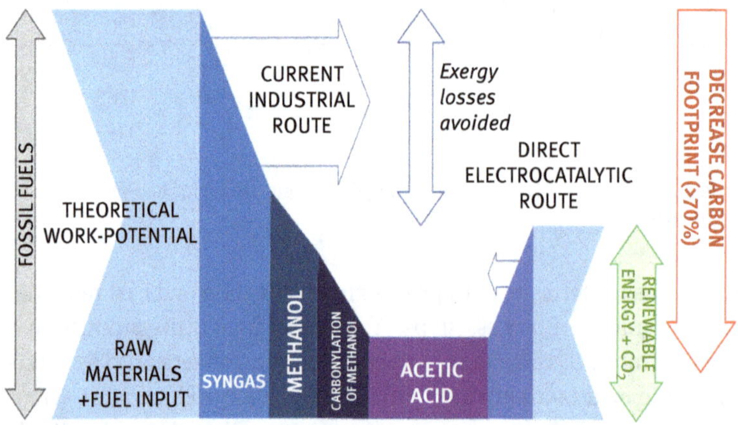

Figure 15.3: Grassmann-type diagram of indicative comparison of exergy in the multistep process to produce acetic acid via the conventional process starting from fossil fuels and the direct electrocatalytic route of CO_2 conversion to acetic acid with in situ water electrolysis. Reproduced with permission from ref. [43]. Copyright The Catalyst Group Inc.@2020.

The further step is to integrate all these knowledges to address the distributed production of chemicals/fuels directly using solar energy, water and small molecules as CO_2. Realize an artificial leaf type device able to use sunlight to convert CO_2 and H_2O (eventually captured directly from the air, as commented before) and produce chemicals (as acetic acid) opens new possibilities for delocalized production.

A better understanding would lead the way for improvement of other relevant reactions, such as the reductive coupling of CO_2 to oxalic acid [33], which selective reduction could also be used to synthetize ethylene glycol or the other C2 chemicals.

There are thus pros and cons for the different products obtained from CO_2 electrocatalytic reduction, and the preferable route has not been established, which will greatly depend on the identification of efficient electrocatalysts for the reaction. But the above discussion remarks that, notwithstanding the higher difficulties, PEC devices should make more complex molecules than those which can be obtained from a two-electron reduction. The future is in this direction, but requires a breakthrough change in the way how these reactions are approached, considering the development of electrocatalysts able to make simultaneously multielectron/proton reductions, rather than sequential ones. This is a true biomimicking approach, rather than other just based on the replica of metal complexes as those present in enzymes [32], which thus suffer of the many limitations present there.

In water, the reduction of protons to H_2 (a 2-electron reduction) always competes with the reduction of CO_2. Increasing the potential difference against the reference electrode (e.g., the standard hydrogen electrode, SHE), the multielectron oxidation is favored and thus increases the selectivity in the reduction of CO_2 with respect to the formation of H_2, and the selectivity toward the formation of products such as CH_4 or C2+. This is a general trend, because the selectivity, Faradaic efficiency (the energy efficiency for which a species is electrolyzed at a given charge), and type of species formed in the reduction of CO_2 greatly depend on several parameters, including the electrode (composition, morphology, etc.) and the operative conditions (electrolyte, pressure, and temperature, etc.). Reviews discussing these aspects were published by Lvov et al. [44], Gattrell et al. [45], and Whipple and Kenis [46].

High overpotentials are needed to convert CO_2 because the first step is the formation of a $^{\bullet}CO_2^-$ (radical anion type) intermediate [47] with negative equilibrium potential of formation in water and in most common solvents. It is necessary to charge the cathode with a very negative potential (i.e., use a high overpotential) to make possible that the reaction occurs. However, by using suitable solvents or catalytic pathways, it is possible to lower the potential for formation of the $^{\bullet}CO_2^-$-type intermediate, which then reacts with H^+ on the cathode to produce CO.

It is possible to electrocatalytically reduce CO_2 to carbon monoxide (CO) at overpotentials below 0.2 V using an ionic liquid electrolyte to lower the energy of the $^{\bullet}CO_2^-$-type intermediate, most likely by complexation, and lower the initial reduction barrier [48]. There are two main approaches, depending on whether the conversion of CO_2 is studied in aqueous or nonaqueous solutions. Formic acid is the main reaction product in electrolysis of aqueous solutions of CO_2. A problem in the utilization of CO_2 in aqueous solution derives from its low solubility in water at standard temperature and pressure. Therefore, at the electrode surface, only very small amounts of CO_2 to be reduced are available. For aqueous solutions, to speed up the reaction rate, the pressure must be increased.

Numerous studies have been made on the electrochemical reduction of CO_2 under high pressure on various electrodes in an aqueous electrolyte. Productivity increases substantially at 30 bar of CO_2 with respect that at one bar of CO_2. However, the total cathodic current barely increases with increasing CO_2 pressure. In terms of products, CO, H_2, and formic acid are mainly observed. A fast deactivation is also typically present.

Solvents with high solubility for CO_2 are used in the nonaqueous electrochemical reduction of CO_2. Carbon-dioxide concentration in dimethyl-formamide is about 20 times higher than in aqueous solutions, in propylene carbonate and methanol, the CO_2 solubility is about 8 and 5 times higher, respectively. However, a high CO_2 solubility requires a larger current density, but low electrolytic conductivity leads to high ohmic losses. Methanol is often used to balance these two aspects. Cu-based foils give the best results as electrodes. Another problem is that high current densities at the copper cathode are necessary to maximize the formation of hydrocarbons, leading to a fast deactivation in these conditions [49].

The direct electrochemical reduction to hydrocarbons (methane and ethylene) is interesting but suffers from a low energy efficiency due to high overpotentials, even if progresses have been made recently. The electrochemical reduction to CO has high energy efficiencies, and the main side product, hydrogen, is included in the syngas product. However, because hydrogen evolution has a higher (less negative) theoretical reduction potential than carbon monoxide, it is more efficient to optimize the electrolysis cell for CO production and supply the hydrogen from a different renewable source.

The reaction rate, as measured by the current density, is also an important parameter as it determines the reactor size and thus capital cost of the process. In the electrocatalytic reduction of CO_2, that is, when an external potential and current is applied to the cell, moderate to high current densities (200–600 $mA{\cdot}cm^{-2}$) have been reported using gas diffusion electrodes (GDEs). Instead in PEC devices, where the potential and current is determined from the photoactive element incorporated in the cell (typically a photoanode), the current densities are limited. Current values are below 10 $mA{\cdot}cm^{-2}$ (see later) and thus about two order of magnitude lower. It is thus evident that this is the main technological limitation for PEC devices for converting CO_2.

15.4 Optimizing the Complex Machinery in PEC Devices

The last observations of the previous section remark the need to rethink from a more systemic approach the design of PEC devices for CO_2 conversion, to understand what the critical elements are to optimize the complex machinery present, composed of many sequential elements and interfaces limiting the overall performances, but also influencing each other. The overall optimization, including selection of electrocatalysts, is not the result of the independent optimization of each specific element, and that does not account for the constraints imposed from the cell architecture. While many advances have been made over the last years on single specific aspects, what is missing is the understanding of the whole process and the interdependence of each specific element. Without this identification, the valid scientific advances made in the last years remain largely an academic work which has a minor impact on developing these cells. Accelerating the development of PEC devices for CO_2 conversion means to clearly identify the bottlenecks and factors limiting the overall behavior to focus research. This does not mean to have an applied (only) perspective, because fundamental research is needed, but to understand that a complex and highly interconnected systems is not the sum of the single components, and the optimization of the whole system is not coincident with the optimization of the single components.

It is also necessary to identify global behavior indicators which can be easy to use and provide values which can allow to compare different approaches and typologies of cell design, as the comparison between PV/EC and PEC approaches discussed before. System efficiency is often used to assess the overall performances. It is defined as the ratio of the total output power (in electricity and/or energy stored in the products, often called also solar fuels – SFs) to the total input power (i.e., electricity and/or solar energy) [50]. For water splitting, the solar-to-hydrogen conversion efficiency (η_{STH}) is often used to quantify the behavior, but there is no equivalent, at least widely used, for CO_2 conversion. Given the variety of products which can be obtained, differently from the hydrogen case, this efficiency cannot be used. Also, the system efficiency shows limitations because solar light is not a reactant as others. It is formally for free; the cost is associated not to its inefficient use (natural leaves have efficiencies lower than 1%). Increasing the efficiency of its use implies a higher productivity of the system, but if this is realized by devices in which the increase of the performances is much less pronounced than the increase of the costs (which depend on the material used, the cell manufacture and operations, the lifetime), the improvement is not worthy, if not from only a scientific perspective. In addition, the energy stored in the products does not account for the quality and value of what is formed, the costs (also energetic) of their recovery and separation, and so on. System efficiencies, such as η_{STH}, are values that may be suited for hydrogen, but not for CO_2, where the complexity of the products that are formed remarks the need of a different approach.

In addition, often the parameters proposed to indicate in a single number the overall behavior (other proposals include an ideal regenerative cell efficiency, the ratiometric power-saved and the applied bias photon-to-current component component [50]) require elaborating the experimental data, while direct experimental data, which can be also of relevance to understand industrial exploitability, would be preferable. Current density (J) meets these requirements [51]. Among the advantages: (i) J can extract much information about the physicochemical processes during the device operation; (ii) it is readily obtained from simple electrical or electrochemical characterizations (thus, it is easily accessible); and (iii) it represents a useful parameter to compare with existing technologies (e.g., commercial electrolyzers work with J ~0.5–2 A cm^{-2}) [52, 53].

Several steps are involved in the complex machinery to transform simple molecules (CO_2, H_2O) to valuable fuels/chemicals using sunlight. They can be schematically divided in these series of processes, which in PEC devices occur in different cell zones:

a) At the photoanode
   a. Photon absorption and generation of free charge carriers
   b. Separation and transport of charge carriers
   c. Redox (catalytic) conversion (water oxidation)
   d. Evolution of O_2 (in the form of gaseous bubbles, avoiding that they remain sticked on the semiconductor)
   e. Collection of electrons, transport at the cathode through an external wire
   f. Diffusion of protons to the membrane
b) At the membrane
   a. Diffusion of protons through the membrane (this when n-type semiconductors are used as photoanode, the most common case)
   b. Inhibition of cross-over from the cathodic zone
c) At the cathode
   a. Mass transfer/diffusion of protons from the membrane to the electrocatalyst (here, significant differences exist if the electrocatalyst operates in the presence or not of a bulk liquid electrolyte, see later)
   b. Transport of electrons to the electrocatalyst sites
   c. Diffusion of CO_2 through the electrolyte and double layer (if a liquid electrolyte is present), adsorption on the electrocatalyst surface
   d. Electrocatalytic reduction of CO_2 by electron/protons
   e. Inhibition of the side reaction of proton/hydrogen recombination to form H_2
   f. Diffusion of the products of CO_2 conversion out of the electrocatalyst

The semiconductor absorbs photons whose energy equals or exceeds that of its band gap (E_g). Upon absorption of sunlight that satisfy this condition, electron–hole pairs are generated which (depending on the properties of the semiconductor) can form a stable bound state (exciton) or dissociate into free charge carriers. The photogenerated carriers are responsible to trigger the reactions for CO_2 reduction. The reactions are endergonic (i.e., they require energy to take place). The Gibbs free energy ($\Delta G°$) for CO_2 reduction into CH_3OH or CH_4, for example, require 689 and 800 kJ·mol^{-1} respectively (at 25 °C and 1 bar conditions). The $\Delta G°$ values can be readily converted into potential difference, which are 1.19 and 1.03 eV, respectively, that correspond to the minimum photon energy needed to trigger these solar-to-fuel reactions, respectively, in the absence of a bias, external or internal (due to chemical potential between cathodic and anodic zones), respectively. These are the thermodynamic limits, but an overpotential is necessary due to ohmic and charge transport losses, kinetic overpotentials and reactor characteristics [54]. Increasing the potential regarding RHE (reversible hydrogen electrode), increases the rate of formation and thus the electrode productivity.

However, in CO_2RR, higher potentials also favor the side reaction of H^+/e^- recombination to form H_2. Thus, to maintain high the selectivity to the target product, the overall overpotential is kept low (from 0.1–0.2 to 0.4–0.5 V) which implies low reaction rates and electrode productivities. Thus, semiconductors with such band gaps must efficiently produce enough photovoltage to conduct the desired reactions. The band edge absorption lies at wavelengths higher than 600 nm, which reduces drastically the utilization of sunlight.

Once electrons and holes are generated in semiconductors, they must be kept separated with the holes driven toward the interfaces formed with the electrolyte, while the electrons collected and transported to the cathode. For these processes to take place favorably, charge carriers inside the semiconductor must lie into proper energy levels, that is, the conduction band minimum (CBM, for electrons) and valence band maximum (VBM, for holes) [55]. When both oxidation and reduction reactions occur on the same semiconductor particle, electrons can reduce a species if the CBM lies above the reduction potential while holes can drive oxidation of a reactant if the VBM lies below the oxidation potential. The quasi-Fermi levels should be the energies at which carriers lie (the position of these level changes upon light absorption) but in practice these are approximated by the CBM and VBM [55]. In the photoanode, however, the electrons should be transported through an external wire to the cathodic zone, where these react. Thus, the electrical losses related to this transport, and related interfaces present, should be also accounted. Finally, the circuit is electrically closed by the motion of the ions within the electrolyte. This step is crucial to obtain high reaction rates because it is responsible to bring "new" active species from the electrolyte to the particle surface where they can be used for the redox processes.

In PEC devices, differently from semiconductor particles (on which most of theoretical studies focus), the closure of the circuit by ion transport is critical. In conventional PEC reactors, in the presence of large volumes of electrolyte, the distance between the electrodes could determine the performance, becoming the limiting factor. Often this limitation does not emerge, because there is a buffering effect given by the large volume of the electrolyte that with the low reaction rates masks the effective presence of limitations by mass diffusion. However, in practical devices, with large electrodes, that should operate for long-term, this issue becomes crucial. In an electrolyte-less configuration (see later), the transport occurs through the membrane on which two sides the photoanode and the cathode are located. The interface between them, but also the characteristics of both the membrane and the electrode (including the substrate elements to collect/distribute electrons) becomes crucial. Note that protons on the cathodic side, in the electrolyte-less configuration, should also diffuse on the surface of the electron-transport element (usually carbon paper or cloth treated with Nafion, for the gas-diffusion electrodes used in this PEC cell configuration) to arrive to the electrocatalytic element. The mechanism of transport largely determines the availability of protons at the electrocatalytic surface, the reaction rate, and the Faradaic selectivity. Note that the situation differs from that present when the electrocatalyst is in contact with the bulk liquid electrolyte. Although there are still limited studies to understand this aspect, the evidence is that not only the reaction rate, but also the Faradaic selectivity change, with also different types of products observed when the same electrocatalysts are used in conventional situation (in the presence of a liquid bulk electrolyte) or in the electrolyte-less conditions [56, 57], that is, when the membrane itself acts as the electrolyte to close the circuit, as shown in Figure 15.1. A fast mass transfer is fundamental to enhance the kinetics of the reactions and keep high J values.

High J implies fast charge carrier transport during each step of device operation. However, there is a significant difference in the timescales (τ) involved during light harvesting, charge separation and transport, redox reactions, and electrolyte transport. Charge generation and separation are fast processes (τ ~ fs) while charge transport, redox reactions and mass transfer take place at slower timescales (τ ~ µs, µs/ms and ms/s, respectively) as reported by Domen and collaborators [58] and summarized in Figure 15.4. There is thus a large (various order of magnitude) differences in the time scale of the process, with the redox (catalytic) processes being largely the slower process. This situation is even worse in heterojunction systems, where each interface contributes with additional resistances. Using a "cocatalysts" (see the previous comment on this term) usually improves the performances, but typically the mechanism is not properly of catalyzing these reactions, but to avoid that the charges recombine. They act thus mainly as hetero- and phase-junctions, rather than as a truly catalyst, and the effect observed is typically limited. Literature studies focused especially on this and analogous aspects, such as Schottky junction, interface engineering, facets and defects, and surface functionalization [59–61], rather than investigating from a fundamental viewpoint how to increase the redox (charge transfer) reaction rates. This is another of the aspects on which literature studies miss to address the focal elements to improve the performances.

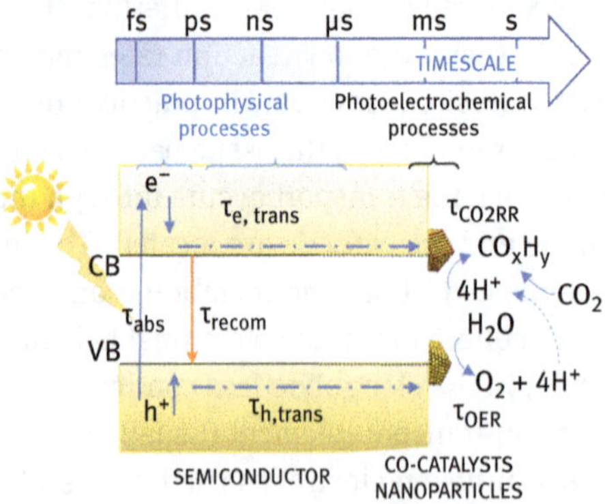

Figure 15.4: Indicative timescale (τ) of the photophysical and (photo)electrochemical processes in water splitting. CB: conduction band; VB: valence band; trans: transfer. Based on indications in ref. [58].

Motivations of the “slow” reactivity (and turnover) in redox (charge transfer) reactions, and in the identification of the key elements to promote them, is far to be understood, given the potentially high reactivity of electrons/holes. Localization of these charges to make possible the electron/energy transfer, difficulties by incoming reacting molecules to reach a charged surface, need of reducing rotational modes of incoming molecule to favor the transfer are among the possible reasons, which are far to be understood. What is evident, however, is that only by proper combination of catalysis to concepts of electrochemistry (in photocatalysis, besides the photophysical processes, the aspects of reactivity are like those present in electrocatalysis) can be likely make the step forward in controlling these aspects and accelerate the performances in redox (charge transfer) reactions. However, still limited advances are present in this direction.

15.5 Designing the Electrocatalytic Cathode for CO_2 Reduction

Several advances have been made recently in the understanding of the catalysts able to selectively activate CO_2 [62–65]. Another key issue is the need to optimize the multielectron transfer necessary for the reduction of CO_2, as commented in Section 15.3. Most electrochemical and photoelectrochemical systems for the reduction of CO_2 produce only the 2 e^- reduction products of CO and formate, while few reported the formation of methane and methanol and even less authors reported C2+ products, as briefly discussed before. The main limitation in achieving such products formation has been attributed to the inability of most catalysts to affect multielectron transfers along with the required multiproton transfers.

One key point regarding this aspect is whether a sequential mechanism of electron and proton transfer should occur (as most of the proposed mechanisms) or whether a simultaneous electron/proton transfer should occur, as happens in enzymes [32, 33]. It should be a key focal point of discussion because an intrinsic difference like the active sites and mechanism is evident. However, a discussion on these aspects is still limited in literature.

An example of the sequential mechanism is that proposed by pyridinium-modified electrodes [66]. Although they received significant attention when these results were proposed, no effective application turns out after a decade, pointing out the limit in the approach, even if interesting from a fundamental perspective. Pyridinium and its substituted derivatives are effective and stable homogeneous electrocatalysts for the aqueous multiple-electron, multiple-proton reduction of carbon dioxide to products such as formic acid, formaldehyde, and methanol. High Faradaic yields for methanol have been reported in both electrochemical and photoelectrochemical systems at low reaction overpotentials. At hydrogenated Pd electrodes, Faradaic efficiencies for the reduction of CO_2 to methanol of ~30% were observed at overpotentials of only ~200 mV [67]. Instead a p-GaP photoelectrochemical system yielded nearly 100% in faradaic efficiency for methanol at underpotentials as high as 300 mV (below the thermodynamic potential of −0.52 V vs. saturated calomel electrode [SCE] at pH of 5.2) [68].

The pyridinium cation is a simple $1e^-$ electrocatalyst and thus it is intrinsically a sequential electron transfer catalyst [66–68]. The mechanism of the reduction proceeds through various coordinative interactions between the pyridinium radical and carbon dioxide, formaldehyde, and related species. This suggests an inner-sphere-type electron transfer from the pyridinium radical to the substrate for various mechanistic steps where the pyridinium radical covalently binds to intermediates and radical species.

Transition metal-based electrocatalysts containing multiple metal centers may instead give a multielectron transfers indicated as necessary to produce highly reduced species [69]. However, while multielectron charge-transfer catalysts have been demonstrated to affect the $2e^-$ reduction of CO_2 to CO and formate, more highly reduced products are only sporadically observed. This is apparently in contrast with the results of pyridinium ions. Likely they thus act as an *electron shuttle*, and this property probably explains high Faradaic yields. However, they are unstable and not well suited for robust electrocatalysts.

In terms of energy barrier of the intermediates, proton-assisted electron transfer is typically a lower energy path regarding sequential electron transfer followed by protonation. However, as the case of sequential versus multielectron mechanism, also for proton-assisted electron-transfer mechanisms in CO_2 reduction the studies are limited and not specific. Frustrated Lewis Pairs accelerate CO_2 reduction, which well follows the proton-assisted electron transfer [70]. HCO_3^- ions in the electrolyte were indicated to induce a proton-coupled electron transfer in CO_2 reduction on a $Cu_2O@SnO_x$ hybrid electrocatalyst [71].

Polyoxometalates were indicated as efficient electron and proton reservoir to assist electrochemical CO_2 reduction [72]. These, among others, are indications of the relevance of a proton-assisted electron transfer, but a design of how to realize efficient electrocatalysts for this scope is missing. Note, in addition, that enzymatic mechanism triggered a simultaneous multielectron and multiproton transfer [32, 33].

A further issue in CO_2 electrocatalytic reduction is how to favor the formation of C2+ versus C1 products, with thus the key step of formation of C–C bonds. Discordant indications on the mechanism exist in literature, and thus on the design of electrocatalysts to favor a selective path to C2+ products.

While past studies evidenced the possibility to form only ethylene (in low amounts) during the electrocatalytic reduction at high voltages of CO_2 and only using Cu as the active metal [73], heavier alcohols and hydrocarbons could be obtained on iron-oxide supported on carbon nanotubes (Fe_2O_3/CNT), particularly isopropanol [74]. Still many studies, particularly by theoretical methods, indicate that the formation of C–C bonds during the electrocatalytic reduction of CO_2 is associated to only specific sites on copper. For example that C2 products occurs through the dimerization of two chemisorbed CO* molecules on specific sites of the Cu(100) plane [75–77] or that the C–C bond formation involves the reaction between adsorbed CHO and CO species over a Cu(100) surface [78]. However, the selective formation of isopropanol on Fe_2O_3/CNT remarks that also different mechanisms of formation may exist. The possibility to shift the path from C1 to C2+ products formation by forcing a higher surface concentration of CO_2-derived adspecies was also shown [57, 79]. This introduces the more general question of whether these theoretical approaches can demonstrate a true mechanism in electrocatalysis or rather a possible mechanism, which not necessarily is that operative during electrocatalysis. This question was analyzed for the mechanism of electrocatalytic synthesis of ammonia from N_2, concluding that current methods cannot demonstrate what the true working mechanism is [80].

Many factors determine the formation of C–C bonds during CO_2 electrocatalytic conversion: (i) aspects related to the surface structure (at the nano- and meso-scale), (ii) reactions conditions (pH, buffer strength, ion effects), and (iii) mass transport aspects [81]. Disordered copper structures were indicated as responsible for C–C bond formation from CO_2 by Li et al. [82], while Chan et al. [83] by a theoretical approach indicate surface strains in the Cu (100) crystalline plane the centers responsible for the C2 path. Copper is postulated by both groups as the unique active metal. Lum et al. [84], on the contrary, indicated the surface coverage by CO_2 as the crucial element, thus only indirectly related to a specific metal. The coverage by adspecies depends on realizing a current density sufficiently high enough to achieve a moderately high pH near the catalyst surface but not so high as to significantly reduce the local concentration of CO_2. Independently, Marepally et al. [79] demonstrated that even electrocatalysts which are indicated by theoretical studies as unable to form C2+ products from CO_2 electrocatalytic conversion, can instead give up to about 60% selectivity to C2+ products, where the conditions to have a high surface coverage by CO_2 are realized, not by increasing operating pressure, but working with gas-diffusion electrodes – GDE (thus without the presence of a bulk liquid electrolyte) and adding in the GDE a component (metal-organic framework – MOF) favoring capture of CO_2. As earlier commented, thus experiments evidence a different picture from what emerges from theoretical indications.

Genovese et al. [85] proposed a mechanism of C–C bond formation, based mainly on reactivity tests of various intermediates and co-feeding experiments, in relation to the synthesis of acetic acid by electrocatalytic reduction of CO_2. The experimental results indicate that the key step is the reaction of $^{\bullet}CO_2^-$ anion radicals with surface adsorbed $-CH_3$ like species. A key aspect to favor this path over the C1 path is to keep the $^{\bullet}CO_2^-$ species (formed by one-electron transport) near to the metal nanoparticles surface, while the tendency of this negatively charged species is to move away from the negatively charged metal nanoparticles. Chang et al. [86] suggested that multicarbon (C3+) products derive from the coupling mechanism between adsorbed CO and acetaldehyde. Zhuang et al. [87] evidence that instead of catalyst structural and electronic aspects, morphological one are determining, by introducing the concept that the nanoconfinement in the electrocatalyst would promote the C2:C1 coupling to form C3 alcohols in the electrocatalytic reduction of CO_2. Zhang et al. [88] also reported that Au nanoparticles in Cu_2O result very active for CO_2 reduction to ethanol due to the confinement of the intermediates. Nanofoams, by an analogous mechanism, can induce an enhanced formation of C2+ products in CO_2 electroreduction with respect to the analogous, flat-type metal [89, 90]. Also, in the CO_2 photoreduction the C–C coupling occurs when a high photon energy and flux determines a simultaneous harvesting of more than one electron from the photocharged Au nanoparticle catalyst [91]. Thus, also in this case the C–C bond will depend on the concentration of adspecies rather than on the presence of an unique type of active sites.

Munir et al. [92] studying CO_2 electrocatalytic reduction on Cu-Zn composite electrodes, proposed a mechanistic pathway helping to form 2-propanol based on a first step of coupling between two surface HCO adspecies to form an adsorbed ethanol-like species which further react with another HCO adspecies. Chan et al. [83] used Cu(100) with surface strain as model to determine by a theoretical approach the conditions to determine C–C bond formation by CO_2 electrochemical reduction. A surface strain with one compressed axis and one elongated axis is geometrically beneficial for C2 product formation. The surface strain stabilizes the CO binding on the bridge sites ($^*CO_{bridge}$) and the C2 intermediates – *OCCOH and *OCCO and maintains a low activation energy of CO–CO coupling. High compressing strains would favor the further coupling of CO with C2 intermediates to form C3 products. The atomic arrangement with an adjacent rectangle and parallelogram is found to play an important role in producing C3 products. Thus, a specific strain arrangement in Cu(100) is responsible of the formation of C3 products as a consecutive step of C2 intermediate formation. Chemisorbed CO is a key element in this mechanism based on theoretical simulation, but for example Genovese et al. [85] experimentally found that by adding small amounts of CO during the electrocatalytic reduction of CO_2 on copper nanoparticles supported on CNTs, the C2 path is inhibited rather than promoted.

Han et al. [93] indicate that isopropanol form in the CO_2 electrocatalytic reduction on Cu nanoparticles supported on N-doped porous carbons by reaction of CO (formed at pyridinic sites related to N-doped carbon) with C2 species formed on Cu nanoparticles. Zhuang et al. [87] related instead the formation of C3 regarding C2 or C1 products in CO_2 electrocatalytic reduction on Cu to the geometric characteristics of Cu nanocavities, and thus to the nanoconfinement of C2 intermediates. Chang et al. [86] indicate instead that propanol forms by a coupling mechanism between CO and acetaldehyde on copper electrodes. A short-living adsorbed methylcarbonyl is a key intermediate. Its hydrogenation leads to C2 products (acetaldehyde and ethanol), while its reaction with CO helps to form *n*-propanol.

There is thus a wide range of concepts and mechanisms proposed to explain the formation of C2+ products in the electrocatalytic reduction of CO_2. Each proposed mechanism can explain the specific results, but it is missing an overall view of the problem and capability to derive general indications which can be the basis for a design of tailored electrocatalysts. A large discrepancy between theoretical and experimental results is evident. The proposed mechanistic models cannot explain relevant experimental observations, among others, that different C2+ products form in CO_2 reduction up to relatively high Faradaic selectivities. For example, acetic and oxalic acids, isopropanol and methyl formate are obtained on Cu_2O/TiO_2 [94], and that other metals, such as Fe, are also equally active to Cu in forming C2+ products, when the proper reaction conditions are investigated [57]. Thus, a system approach addressing all the results rather than specific ones would be necessary.

Another key limiting factor for the conversion of CO_2 is the low mass-transfer rates of CO_2 at the electrode/catalyst surface, which is exacerbated by the low solubility of CO_2 in many electrolytes. Ionic liquids [95] may help to solve this problem, as Zhao et al. [96] originally proposed.

In electrodes, these mass-transfer limitations could be overcome using gas-diffusion electrodes (GDEs), which create a three-phase interface between the gaseous reactants, the solid catalyst, and the electrolyte. Thus, optimization of the electrodes will be the key to improve current densities. The extensive work on GDE optimization for fuel cells over the past decades will greatly accelerate progress. As seen in fuel-cell works, temperature is the key to enhance the performance of fuel cells, indicating that optimizing the reaction temperature will yield significant improvements for CO_2 reduction. This is another further possibility to increase productivity, which is impossible in the photochemical approach.

In addition, to find a suitable mechanism for an energy-efficient electron transfer that avoids overpotentials, the technical feasibility of CO_2 conversion to fuels from the exploitation perspective depends on the possibility of forming under solventless conditions liquid fuels such as long-chain hydrocarbons and/or alcohols, which can be easily collected without the need to distillate from liquid solutions (a energy-intensive process). The key point is thus to pass from a conventional electrochemical conversion of CO_2 in aqueous or nonaqueous solutions to the electrocatalytic conversion of CO_2 in gaseous phase over solid electrodes. There are two possibilities, related to operate at high or low temperatures. High-temperature CO_2 conversions are typically used in the solid oxide fuel cell (SOFC), whereas low-temperature systems largely utilize a variation of polymer electrolyte membrane (PEM) fuel cells. Spinner et al. [97] has discussed the use of SOFC-type cells for the electrochemical CO_2 conversion in comparison with low-temperature systems based on transition-metal electrodes in aqueous or nonaqueous electrolytes. However, the low-temperature approach avoiding the use of a liquid electrolyte was not discussed, even if it was shown that new breakthrough possibilities in terms of type of products (long-carbon-chain hydrocarbons and alcohols) exist deriving from this approach [74, 98].

There are very few studies based on this approach, but it was demonstrated that using nanostructured carbon-based electrodes, it is possible to reduce electrocatalytically CO_2 in the gas phase using the protons flowing through a membrane [74, 99]. Long-chain hydrocarbons and alcohols up to C9–C10 are formed, with preferential formation of isopropanol using carbon-nanotube-based electrodes [41, 74]. Productivities are still limited, but these results demonstrate the concept of a new approach to recycle CO_2 back to fuels.

The features of the electrode used in this gas-phase electrocatalytic reduction of CO_2 are close to those used in PEM fuel cells [98, 100, 101] (e.g., a carbon cloth/Pt or Fe on carbon black/Nafion assembled electrodes, GDE). The electrocatalysts are Pt or Fe nanoparticles supported on nanocarbons (doped carbon nanotubes), which are then deposited on a conductive carbon cloth to allow the electrical contact and the diffusion of gas phase CO_2 to the electrocatalyst. The metal nanoparticles are at the contact of Nafion, through which protons diffuse. On the metal nanoparticles, the gas-phase CO_2 reacts with the electrons and protons to be reduced to longer-chain hydrocarbons and alcohols, the relative distributions of which depend on the reaction temperature and type of metal nanoparticles. Isopropanol forms selectively from the electrocatalytic reduction of CO_2 using a gas diffusion electrode based on an Fe/N carbon nanotube (Fe/N-CNT) [41, 101, 102]. Not only the nature of carbon is relevant, but also the presence of nanocavities, which could favor the consecutive conversion of intermediates with formation of C–C bonds.

The SOFC cell approach is instead used for methane wet reforming with CO_2 to produce syngas [103]. These high-temperature cells typically utilize a yttria-stabilized zirconia (YSZ) electrolyte or tube with various metal and/or mixed-metal catalysts at the anode and cathode. When CO_2 is the fuel without CH_4, CO alone can be synthesized at the cathode along with oxygen at the anode [104]. Current densities of over 1 $A \cdot cm^{-2}$ could be obtained by this high-temperature CO_2 electrolysis approach, but the high temperature necessary (over 900 °C to have enough ionic conductivity in the YSZ membrane) and the possibility of obtaining only CO or syngas, which should be thus further converted, greatly limit the applicability of this approach.

15.6 Designing the Photoanode

There are two main possibilities for designing the photoanode: (1) integrate a "conventional" PV cell with a catalyst for water oxidation (also called oxygen evolution catalyst, OEC) and (2) develop a specifically designed photoanode.

The first approach corresponds to the well-advertised (as the first "practical" artificial leaf) device proposed by Nocera and coworkers [105]. The cells consist of a triple junction, amorphous silicon photovoltaic cell interfaced to HEC and OEC components. The OEC is based on a cobalt catalyst that self-assembles upon oxidation of Co^{2+} in phosphate- or borate-buffered solutions at neutral pH [106]. It is the structural relative of Mn_3CaO_4–Mn cubane (the oxygen evolution center of PSII), with Co replacing Mn and the cubane extending in a corner-sharing head-to-tail dimer. The H_2-evolving catalyst is a ternary alloy of NiMoZn. In the wireless device, the NiMoZn alloy was deposited directly on the opposing stainless-steel surface of the 3jn-a-Si solar cell. The device is a photoelectrochemical cell, but without the membrane and without the physical separation of the evolving H_2 and O_2. From the application perspective, this is a limit, in terms of avoiding recombination of H_2 and O_2 (intrinsic productivity), the cost of downstream separation of O_2 from H_2, and safety. The application with the PEC reduction of CO_2 is even more difficult, due to the competition of dissolved O_2 in the reduction of CO_2, besides the problems of solubility of CO_2. This cell design is thus not suitable for a practical development of artificial leaves able to convert CO_2 to fuels using sunlight.

The device in the wired configuration is essentially a variation of the conventional photoelectrochemical cells. Lewerenz et al. [107] reviewed the main aspects of photoelectrocatalysis for light-induced fuel generation, such as water photoelectrolysis or carbon-dioxide reduction. The photoelectrochemical approach involves the separation of the oxidation and reduction processes into two half-cell reactions [108]. Three approaches are possible: (1) the use of photovoltaic (PV) cells, (2) semiconductor-liquid junctions (SCLJ), or (3) a combination of the two (PV/SCLJ). The approach based on solid-state photovoltaics is to couple a PV system and an electrolyzer into a single system. Semiconductor layers are connected in series, one behind the other, in a single monolithic device capable of generating the potential needed to split water. These so-called tandem cells or multijunction cells are modified with, or connected to, H_2- and O_2-producing electrodes, like Pt and RuO_2-modified Pt acting as cathode and anode, respectively. Several aspects of the operating principles of the photoelectrochemical devices, the materials requirements, main bottlenecks, and the device concepts (in relation to H_2 production), were discussed in a chapter of a recent book on sustainable energy technologies [109]. However, this approach may be adapted with difficulty to the reduction of CO_2, and up to now, the attempts in this direction were limited.

Tandem or multijunction cells are interesting since the absorption efficiency can be enhanced by complementary spectrum absorption, while a corrosion-resistant material can be the top junction (e.g., WO_3 on $GaInP_2$). For the development of efficient tandem cell structures that are based on non-oxidic semiconductors, stabilization strategies are the key challenge because semiconductors are not thermodynamically stable at the reactive electrolyte interface. Physical surface passivation could be realized using the nano-emitter concept [107], first developed for n-Si for operation in the photovoltaic mode of an electrochemical solar cell [110] yielding efficiencies higher than 10%.

When in contact with redox electrolytes, the Fermi level of the metallic nano-emitters adjusts to that of the solution. If the metal nano-islands (size less than approximately 50 nm) form rectifying junctions without Fermi-level pinning, the system is characterized by a rectifying junction between the electrolyte and the semiconductor absorber. The decoupling of the metal/semiconductor junction, which is often impeded by metal-induced gap states, can be achieved by spatial separation introducing, for example, an ultrathin interfacial film that inhibits interpenetration of evanescent metal states into the semiconductor energy gap. Upon illumination, the transparent passive layer transmits the sunlight, which is absorbed in the underlying semiconductor. Using p-Si and p-InP thin films, solar-to-hydrogen efficiencies reach 12.1% for homoepitaxial InP thin films covered with Rh nano-islands [107]. There are other strategies for solving stability of the photoanode in PEC cells, such as (a) chemical passivation of the surface, as, for instance [107], by H-termination or methylization, and (b) stabilization by energy band structure properties. The latter has been found in transition-metal chalcogenide materials where the upper valence band and the bottom of the conduction band are composed predominantly of metallic d-states.

The hydrogen production efficiency in a photoelectrochemical cell was already demonstrated a decade ago to be over 12% by Khaselev and Turner [111] (i.e., two to three times higher than that reported by the recent cited approach by Nocera and coworkers [106]). However, performances of the Turner cell fast degrade with time. However, the approach described will be a good way to have more stable performances.

These strategies are interesting for the production of H_2 and O_2 from water by a photoelectrochemical approach, but with CO_2, it is necessary to (1) avoid the use of a liquid electrolyte (to eliminate problems of CO_2 solubility, diffusion limitation due to double layer formation, solvent competition, and to simplify cell sealing and facilitate product recovery eliminating the solvent), (2) have the anodic and cathodic reactions in separated compartments (reduce separation costs and eliminate safety issues), and (3) have a cell design suitable for ease of scaling-up and application. Most of the artificial-leaf-type solar cells proposed in literature show similar issues [1].

A new PEC design was originally proposed by the authors of this chapter [74, 100, 102] as a further development of the original concept proposed by "Hitachi Green Centre" researchers [53]. The design fits the mentioned requirements and can be used for the reduction of carbon dioxide using the protons/electrons generated at a photoanode. The advantage is that this solar-fuel-cell has a configuration close to commercial PEM fuel cells, which are already optimized in terms of engineering and mass/charge transfer. A significant difference in this new design with respect to the PEC cell proposed "Hitachi Green Centre" [112] is present: the CO_2 electrocatalyst operates in gas phase, instead of liquid phase. The difference in cell operations leads not only to an easier product recovery, but also to a different type of electrode. It is a flat-type copper electrode in the PEC cell proposed by "Hitachi Green Centre" [112], while metallic nanoparticles supported on nanocarbon materials in the new PEC design proposed by the authors of this chapter [98, 100, 113].

The scheme of this solar fuel cell is discussed in the following section, but similarly to some other recent advanced solar leaf cell designs [1], it is necessary to have a porous photoanode that combines high effectiveness in sunlight harvesting with fast collection of the electrons produced and an effective transport of the protons generated to the proton-diffusion membrane on which the photoanode is integrated. This is a key property determining the cell performances and that is not present in conventional PEC schemes, including variations of the dye-sensitized solar cells originally proposed by Grätzel [114] and expected for combination with hematite (α-Fe_2O_3) photoelectrode materials for producing H_2/O_2 by water photoelectrolysis [115].

A porous photoanode for PEC solar cells must fit several requirements:

- It must be cost-effective and easily scaled up to larger sizes.
- It must be robust for stable operations.
- It must have an optimal nanostructure allowing (1) an enhanced light harvesting (possible over the entire sunlight spectrum, with effective use of the radiation for creating a photocurrent); (2) low rate of charge recombination and reduced interfaces/grain boundaries, which favors the charge recombination; (3) negligible defects and centers that favor thermal or radiative pathways (which reduce quantum efficiency); and (4) fast transport of the electrons to a conductive substrate.
- It must have a porous nanostructure that allows a fast transport of protons (generated from water oxidation) to the underlying proton-conductive membrane, avoid surface recombination between protons and electrons (which should have different paths of transport), and presenting an optimal interface with the membrane.

It is thus necessary to use knowledge in the synthesis and mastering of semiconductor nano-objects to realize these demanding characteristics in the photoanode [116]. It is necessary to realize a semiconductor film characterized by an array of vertically aligned nanotubes that allow good light harvesting, limit charge recombination at the grain boundaries regarding an assembly of nanoparticles while maintaining a high geometrical surface area necessary to improve the photoresponse, have in principle separate paths for electron and proton movement, and have a good porosity and interface with the membrane to allow a good transport of the protons [123, 124]. This concept is schematically presented in Figure 15.5.

A good method to achieve these characteristics and suitable also for scale-up is the anodic oxidation of thin titanium foils to form ordered arrays of vertically aligned titania nanotubes [117–122].

The process of anodic anodization used to prepare the nanostructured titania thin films may be summarized as follows. When the Ti foil, after the preliminary cleaning treatment, is immersed in the electrolyte for the anodization process, a fast surface oxidation occurs with the formation of a thin TiO_2 layer. This process may be monitored by a fast decrease of the current, the TiO_2 layer being not conductive. Due to the presence of F^- anions in the electrolyte, the solubilization of Ti^{4+} ions and/or of small TiO_2 particles starts simultaneously with the formation of the oxide layer. These processes help to form holes, which locally modify the electrical field and induce from one side the acceleration of the process of dissolution, due to field-enhanced effects, and from the other side the oriented growing of the one-dimensional structure (nanotubes, nanorods, etc.). The electrolyte and conditions of anodization have a relevant influence not only on the film thickness but also on the specific nanostructure obtained [117–121], which strongly influence both the stationary and transient photoresponse. The thicker walls prepared in the organic electrolyte allow a fast electron transport, but it is also necessary to avoid defects in the titania, which trap electrons and reduce the photocurrent.

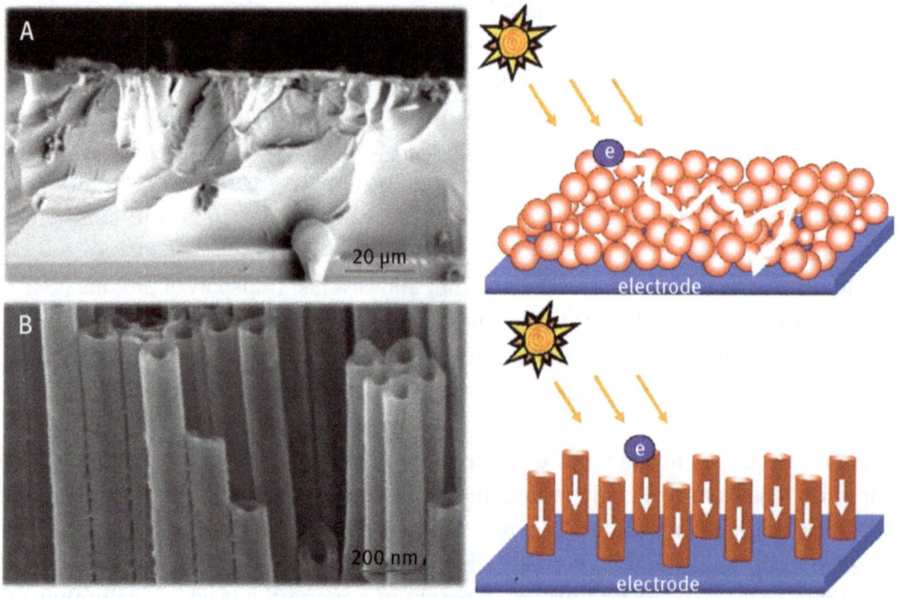

Figure 15.5: Scanning electron microscopy (SEM) images (cross section) of a TiO_2 film deposited over a conductive support by the sol–gel method (A) or produced by anodic oxidation of a Ti foil in ethylene glycol containing 0.3 wt% NH_4F and 2 vol% H_2O applying a 50 V potential for 6 h. The thickness of the titania nanostructured film is approximately 14 µm; tube internal diameter is approximately 40 nm, while tube external diameter is approximately 100 nm. (The latter SEM image is adapted from [41].) On the right side of the figure, a schematic model shows the differences between TiO_2 thin films and the aligned nanotube array of TiO_2. Adapted from [117]. Reproduced from the contribution present in the first edition of the book *Chemical Energy Storage*.

15.7 PEC Cells for CO_2 Conversion

Several proposals have been made recently for solar cells, although often at a conceptual level without proving the feasibility of realization [1]. Most are designed only for producing H_2 from water, but as discussed in the previous sections, developing AL

PEC solar cells for converting CO_2 using sunlight, requires a different design. Galan-Mascaros [125] recently reviewed the PEC systems for CO_2RR commenting the low number of examples of unbiased solar CO_2 reduction. Some PV/EC sequential approaches reported η_{STF} up to 13% [125]. but only toward CO [126] with one exception (based on a Si PV cell powering an electrolyzer equipped with a selective Bi anode) where formic acid production with 95% Faradaic efficiency (η_{STF} =8.5%) was reported [127].

Lower η_{STF} values have been reported in producing chemicals/fuels involving a higher number of electrons for the reduction than the $2e^-$ for CO and HCOOH. For example, ethylene was obtained at 2.9% STF efficiency combining Si solar panels with an electrolyzer [128]. Membrane-less electrolyzers powered by Si PV cells reached a >3% η_{STF} to CO [129] using earth-abundant catalysts.

The integrated PEC approach for CO_2RR results in lower efficiencies, with η_{STF} typically below 1% [125]. Zhou et al. [130] reported η_{STF}>10, but for a PEC device incorporating a photoanode made of a TiO_2/InGaP/GaAs heterostructure decorated with a Ni OER catalyst which drive a dark Pd/C/Ti cathode producing formic acid at >94% Faradaic efficiency.

Few results have been reported for integrated PEC devices in producing C2+ products [131]. Combined with an external PV cell, an electrolyzer based on a Cu/Ag/TiO_2 photocathode results in 60% Faradaic efficiency to form a mixture of C2/C3 products. A tandem device was developed by coupling a Si photocathode.

By using a compact PEC cell design, on earth-abundant materials for both photoanode (CuO electrodeposited on an array of ordered TiO_2 nanotubes) and cathode (Cu_2O deposited on GDE by electrodeposition), see Figure 15.6, without an external bias and use of sacrificial agents, the total carbon Faradaic selectivity was about 90%, with about 75% Faradaic selectivity to acetate, the other product being formate [132]. A stable current density of 0.2 mA was observed for 24 h operation.

There is an in situ transformation of the Cu_2O/GDE which helps to form a hybrid Cu_2O-Cu/GDE system. The same electrode gives, in electrocatalytic tests of CO_2 reduction at −1.5 V, high total Faradaic selectivity (>95%), forming selectively formate (about 80% selectivity), rather than acetate. The difference between the electrocatalytic tests and those in the PEC cell derives from the different potential applied to the cathode, being −1.5 V in the electrocatalytic tests and lower, determined from the photoanode, for the PEC tests. This result also remarks how the performances of the same electrodes may change when tested separately or integrated within a full PEC device.

The compact design of the PEC cell in Figure 15.6, and that shown in Figure 15.1, is close to that of commercial PEM fuel cells, to take advantage of the large knowledge portfolio on their engineering and mass/charge-transfer optimization. Scale-up is thus easier. GDE also allows operations in the absence of a bulk liquid electrolyte (indicated as "gas phase" or electrolyte-less operations [12, 56]). This concept is presented

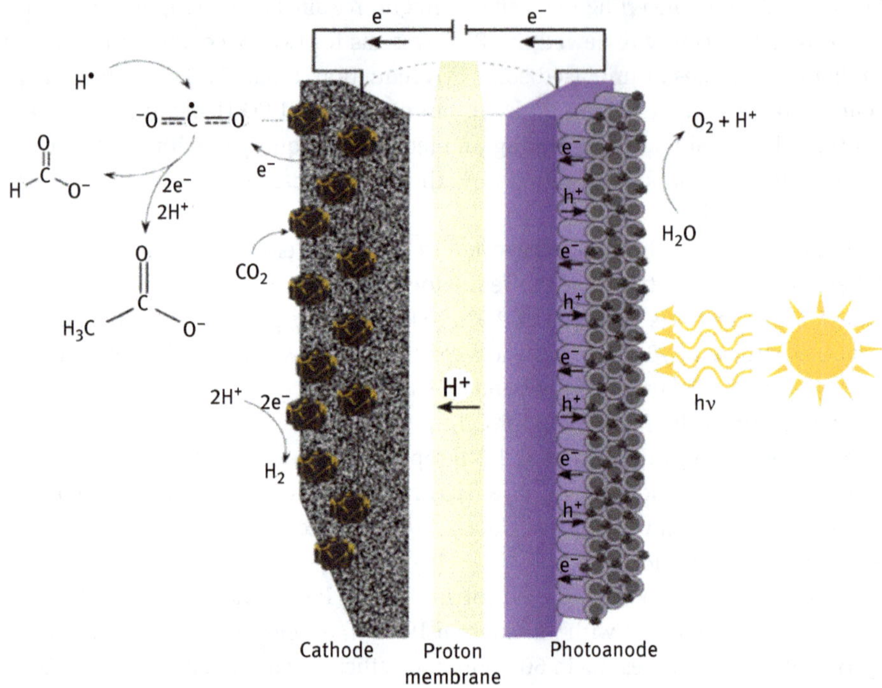

Figure 15.6: Scheme of a PEC full cell reactor with compact design used for water oxidation on the photoanode side and CO_2 reduction on the cathode side without external bias and use of sacrificial agents. The electrodes are based on earth-abundant materials. Graphical abstract in ref. [132]. Copyright: Wiley VCH, 2019.

schematically in Figure 15.7, where the architectures with (a) and without (b) a bulk liquid electrolyte is presented [12].

In the PEC cell configuration as in Figure 15.7b, the cathodic part operates in the gas phase because this (1) simplifies the recovery of the reaction products (they can be collected by cooling the gas outlet from the flowcell), (2) allows continuous operations using large concentrations of CO_2, (3) greatly reduces mass-transfer limitations, and especially (4) changes the nature of the products formed. Gas-phase operations enhance the formation of C2+ products.

This cell design could be transferred to an application stage as artificial leaf, but productivity of both electrodes must be improved in terms of (1) the response to visible light of the photoanode, (2) reducing surface self-quenching during reaction, (3) the presence of several interfaces that limit the mass/charge-transfer and cell efficiency, and (4) the rate of CO_2 reduction.

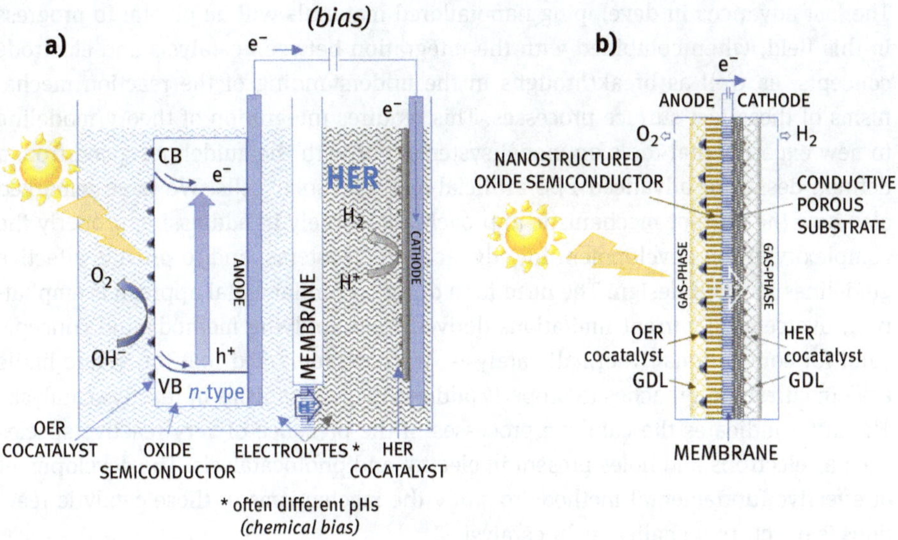

Figure 15.7: (a) Scheme of a conventional PEC cell operating with the two electrodes immersed in a liquid electrolyte. (b) Scheme of a compact (electrolyte-less) design for PEC operations.

15.8 Conclusions

Photoelectrochemical activation of CO_2 in artificial-leaf-type PEC cells is still a long-term goal for the practical, widespread use of the great potential of sunlight and to collect energy naturally, but we have shown here how this represents potentially the solution to make sustainable energy an approach compatible with the large investments made for the energy chain and infrastructure. To avoid intermittency of solar energy, it is necessary to design systems that directly capture CO_2 and convert it into liquid solar fuels, which can be easily stored. Chemical energy storage is the key for sustainable use of solar energy.

Even if there is great interest in research on these topics, we recommend the need for a radically different approach, particularly in terms of system design. This is because often current studies, although scientifically valuable, are not focused on the key issues, particularly regarding the transfer of the results to a practical use. We have remarked in this contribution several points that need to focus better on critical issues for the development, going beyond the mainstream. Only by addressing from a different perspective, and by using a system approach, it is possible to accelerate the progress.

Due to the complexity of the problems, a fundamental understanding is the key for the advancement considering the system engineering and integration and the following three keywords to designing artificial leaves: smart, cheap, and robust.

The fast advances in developing nanotailored materials will be pivotal to progress in this field, when combined with the integration between catalysis and electrode concepts, as well as breakthroughs in the understanding of the reaction mechanisms of these fast surface processes. This requires integration of theory/modeling to new experimental tools on model systems, but with the guidelines given from a defined design of advanced PEC artificial-leaf-type solar cells. We have remarked also how the current mechanistic approach fails largely in addressing properly the complexity of the development for advanced PEC systems, and to provide effective guidelines for their design. The need for a different fundamental approch is emphatized, overcoming current limitations derived from applying methods and concepts valid for conventional (thermal) catalysis. We have remarked that too severe limits exist in current approaches to properly address the mechanisms of reactive catalysis. The latter indicates the catalytic processes in the presence of very reactive species such as electrons and holes present in electro- and photocatalysis. The development of effective fundamental methods to study the mechanisms in these catalytic reactions is the current challenge for catalysis.

References

[1] Bensaid S, Centi G, Garrone E, Perathoner S, Saracco G. Artificial leaves for solar fuels from CO_2. ChemSusChem. 2012;5:500–21.

[2] Chen L, Chen G, Leung C-F, Cometto C, Robert M, Lau T-C. Molecular quaterpyridine-based metal complexes for small molecule activation: Water splitting and CO_2 reduction. Chem Soc Rev. 2020;49:7271–83.

[3] Shukla J, Singh VP, Mukhopadhyay P. Molecular and supramolecular multiredox systems. Chem Open. 2020;9:304–24.

[4] Smith PT, Nichols EM, Cao Z, Chang CJ. Hybrid catalysts for artificial photosynthesis: Merging approaches from molecular, materials, and biological catalysis. Acc Chem Res. 2020;53: 575–87.

[5] Zahran ZN, Tsubonouchi Y, Mohamed EA, Yagi M. Recent advances in the development of molecular catalyst-based anodes for water oxidation toward artificial photosynthesis. ChemSusChem. 2019;12:1775–93.

[6] Bae S, Jang JE, Lee H-W, Ryu J. Tailored assembly of molecular water oxidation catalysts on photoelectrodes for artificial photosynthesis. Eur J Inorg Chem. 2019;15:2040–57.

[7] Whang DR, Apaydin DH. Artificial photosynthesis: Learning from nature. Chem Photo Chem. 2018;2:148–60.

[8] Garrido-Barros P, Gimbert-Surinach C, Matheu R, Sala X, Llobet A. How to make an efficient and robust molecular catalyst for water oxidation. Chem Soc Rev. 2017;46:6088–98.

[9] Rudolf M, Kirner SV, Guldi DM. A multicomponent molecular approach to artificial photosynthesis – the role of fullerenes and endohedral metallofullerenes. Chem Soc Rev. 2016;45:612–30.

[10] Zhang B, Sun L. Artificial photosynthesis: Opportunities and challenges of molecular catalysts. Chem Soc Rev. 2019;48:2216–64.

[11] Shanks K, Senthilarasu S, Mallick TK. Optics for concentrating photovoltaics: Trends, limits and opportunities for materials and design. Renew Sustain Energy Rev. 2016;60:394–407.

[12] Perathoner S, Centi G, Su D. Turning perspective in photoelectrocatalytic cells for solar fuels. ChemSusChem. 2016;9:345–57.

[13] Perathoner S, Centi G. Artificial leaves using sunlight to produce fuels. Stud Surf Sci Catal. 2019;179:415–30.

[14] Centi G, Perathoner S. Artificial Leaves, in Kirk-Othmer Encyclopedia. Hoboken, NJ: Wiley & Son, 2013.

[15] Wang J-W, Zhong DC, Lu T-B. Artificial photosynthesis: Catalytic water oxidation and CO_2 reduction by dinuclear non-noble-metal molecular catalysts. Coord Chem Rev. 2018;377: 225–36.

[16] Djokic M, Soo HS. Artificial photosynthesis by light absorption, charge separation, and multielectron catalysis. Chem Comm. 2018;54:6554–72.

[17] Wang Y, Suzuki H, Xie J, Tomita O, Martin DJ, Higashi M, Kong D, Abe R, Tang J. Mimicking natural photosynthesis: Solar to renewable H_2 fuel synthesis by Z-scheme water splitting systems. Chem Rev. 2018;118:5201–41.

[18] Xu Y, Li A, Yao T, Ma C, Zhang X, Shah. JH, Han H. Strategies for efficient charge separation and transfer in artificial photosynthesis of solar fuels. ChemSusChem. 2017;10:4277–305.

[19] Jiang C, Moniz SJA, Wang A, Zhang T, Tang J. Photoelectrochemical devices for solar water splitting – materials and challenges. Chem Soc Rev. 2017;46:4645–60.

[20] Zhou H, Xiao C, Yang Z, Du Y. Three-dimensional structured materials and devices for artificial photosynthesis. Nanotechnology. 2020;31:282001.

[21] Kim JH, Hansora D, Sharma P, Jang J-W, Lee JS. Toward practical solar hydrogen production – an artificial photosynthetic leaf-to-farm challenge. Chem Soc Rev. 2019;48:1908–71.

[22] Dogutan DK, Nocera DG. Artificial photosynthesis at efficiencies greatly exceeding that of natural photosynthesis. Acc Chem Res. 2019;52:3143–48.

[23] Li X-B, Xin Z-K, Xia S-G, Gao X-Y, Tung C-H, Wu L-Z. Semiconductor nanocrystals for small molecule activation via artificial photosynthesis. Chem Soc Rev. 2020;49:9028–56.

[24] Wang H, Liu W, Jin S, Zhang X, Xie Y. Low-dimensional semiconductors in artificial photosynthesis: An outlook for the interactions between particles/quasiparticles. ACS Cent Sci. 2020;6:1058–69.

[25] Butburee T, Chakthranont P, Phawa C, Faungnawakij K. Beyond artificial photosynthesis: Prospects on photobiorefinery. ChemCatChem. 2020;12:1873–90.

[26] Ng AYR, Boruah B, Chin KF, Modak JM, Soo HS. Photoelectrochemical cells for artificial photosynthesis: Alternatives to water oxidation. Chem Nano Mat. 2020;6:185–203.

[27] Feng J, Huang H, Yan S, Luo W, Yu T, Li Z, Zou Z. Non-oxide semiconductors for artificial photosynthesis: Progress on photoelectrochemical water splitting and carbon dioxide reduction. Nano Today. 2020;30:100830.

[28] Zhang JZ, Reisner E. Advancing photosystem II photoelectrochemistry for semi-artificial photosynthesis. Nature Rev Chem. 2020;4:6–21.

[29] Abas N, Kalair E, Kalair A, Hasan Q, Khan N. Nature inspired artificial photosynthesis technologies for hydrogen production: Barriers and challenges. Int J Hydrogen Energy. 2020;45:20787–99.

[30] Roy N, Suzuki N, Terashima C, Fujishima A. Recent improvements in the production of solar fuels: From CO_2 reduction to water splitting and artificial photosynthesis. Bull Chem Soc Jpn. 2019;92:178–92.

[31] Jia J, Seitz LC, Benck JD, Huo Y, Chen Y, Ng JWD, Bilir T, Harris JS, Jaramillo TF. Solar water splitting by photovoltaic-electrolysis with a solar-to-hydrogen efficiency over 30. Nat Commun. 2016;7:13237.

[32] Mallamace D, Papanikolaou G, Perathoner S, Centi G, Lanzafame P. Comparing molecular mechanisms in solar NH_3 production and relations with CO_2 reduction. Int J Mol Sci. 2021;22:139.

[33] Perathoner S, Centi G. Catalysis for solar-driven chemistry: The role of electrocatalysis. Catal Today. 2019;330:157–70.

[34] Wu H, Song J, Xie C, Hu Y, Han B. Highly efficient electrochemical reduction of CO_2 into formic acid over lead dioxide in an ionic liquid–catholyte mixture. Green Chem. 2018;20:1765–69.

[35] Centi G, Perathoner S. Chemistry and energy beyond fossil fuels. A perspective view on the role of syngas from waste sources. Catal Today. 2020;342:4–12.

[36] Centi G, Perathoner S. Towards solar fuels from water and CO_2. ChemSusChem. 2010;3:195–208.

[37] Buttler A, Spliethoff H. Current status of water electrolysis for energy storage, grid balancing and sector coupling via power-to-gas and power-to-liquids: A review. Renewable Sustain Energy Rev. 2018;82:2440–54.

[38] Dieterich V, Buttler A, Hanel A, Spliethoff H, Fendt S. Power-to-liquid via synthesis of methanol, DME or Fischer-Tropsch-fuels: A review. Energy Env Sci. 2020;13:3207–52.

[39] Thema M, Bauer F, Sterner M. Power-to-Gas: Electrolysis and methanation status review. Renewable Sustain Energy Rev. 2019;112:775–87.

[40] Zhang W, Hu Y, Ma L, Zhu G, Wang Y, Xue X, Chen R, Yang S, Jin Z. Progress and perspective of electrocatalytic CO_2 reduction for renewable carbonaceous fuels and chemicals. Adv Sci. 2017;5:1700275.

[41] Ampelli C, Centi G, Passalacqua R, Perathoner S. Synthesis of solar fuels by a novel photoelectrocatalytic approach. Energy Environ Sci. 2010;3:292–301.

[42] Genovese C, Ampelli C, Perathoner S, Centi G. Mechanism of C–C bond formation in the electrocatalytic reduction of CO_2 to acetic acid. A challenging reaction to use renewable energy with chemistry. Green Chem. 2017;19:2406–15.

[43] Centi G. Catalysis to go beyond the use of fossil fuels: A dream or an effective possibility?. Cata Rev. 2020;10:6–14.

[44] Lvov SN, Beck JR, LaBarbera MS. Electrochemical reduction of CO_2 to fuels. In: Muradov NZ, Veziroglu TN, editors. Carbon-Neutral Fuels and Energy Carriers, Boca Raton, FL: CRC Press (Taylor & Francis Group); 2012. 363–400.

[45] Gattrell M, Gupta N, Co A. A review of the aqueous electrochemical reduction of CO_2 to hydrocarbons at copper. J Electroanal Chem. 2006;594:1–19.

[46] Whipple DT, Kenis PJA. Prospects of CO_2 utilization via direct heterogeneous electrochemical reduction. J Phys Chem Lett. 2010;1:3451–58.

[47] Bockris JOM, Wass JC. The photoelectrocatalytic reduction of carbon dioxide. J Electrochem Soc. 1989;136:2521–28.

[48] Rosen BA, Salehi-Khojin A, Thorson MR et al.. Ionic liquid–mediated selective conversion of CO_2 to CO at low overpotentials. Science. 2011;334:643–44.

[49] Hori Y, Konishi H, Futamura T et al.. Deactivation of copper electrode" in electrochemical reduction of CO_2. Electrochim Acta. 2005;50:5354–69.

[50] Coridan RH, Nielander AC, Francis SA, McDowell MT, Dix V, Chatman SM, Lewis NS. Methods for comparing the performance of energy-conversion systems for use in solar fuels and solar electricity generation. Energy Environ Sci. 2015;8:2886–901.

[51] Romano V, D'Angelo G, Perathoner S, Centi G. Current density in solar fuel technologies. Env Sci Techn. 2021 submitted.

[52] Roger I, Shipman MA, Symes MD. Earth-abundant catalysts for electrochemical and photoelectrochemical water splitting. Nat Rev Chem. 2017;1:0003.

[53] Carmo M, Fritz DL, Mergel J, Stolten D. A comprehensive review on PEM water electrolysis. Int J Hydrogen Energy. 2013;38:4901–34.

[54] Smith WA. Photoelectrochemical Solar Fuel Production: From Basic Principles to Advanced Devices. Giménez S, Bisquert J. Eds.. 2016.

[55] Stolarczyk JK, Bhattacharyya S, Polavarapu L, Feldmann J. Challenges and prospects in solar water splitting and CO_2 reduction with inorganic and hybrid nanostructures. ACS Catal. 2018;8:3602–35.

[56] Ampelli C, Centi G, Passalacqua R, Perathoner S. Electrolyte-less design of PEC cells for solar fuels: Prospects and open issues in the development of cells and related catalytic electrodes. Catal Today. 2016;259:246–58.

[57] Ampelli C, Genovese C, Marepally BC, Papanikolaou G, Perathoner S, Centi G. Electrocatalytic conversion of CO_2 to produce solar fuels in electrolyte or electrolyte-less configurations of PEC cells. Faraday Disvuss. 2015;183:125–45.

[58] Hisatomi T, Takanabe K, Domen K. Photocatalytic water-splitting reaction from catalytic and kinetic perspectives. Catal Lett. 2015;145:95–108.

[59] Chen S, Qi Y, Li C, Domen K, Zhang F. Surface strategies for particulate photocatalysts toward artificial photosynthesis. Joule. 2018;2:2260–88.

[60] Kment S, Riboni F, Pausova S, Wang L, Wang L, Han H, Hubicka Z, Krysa J, Schmuki P, Zboril R. Photoanodes based on TiO_2 and α-Fe_2O_3 for solar water splitting – superior role of 1D nanoarchitectures and of combined heterostructures. Chem Soc Rev. 2017;46:3716–69.

[61] Ma Y, Wang X, Jia Y, Chen X, Han H, Li C. Titanium dioxide-based nanomaterials for photocatalytic fuel generations. Chem Rev. 2014;114:9987–10043.

[62] Cokoja M, Bruckmeier C, Rieger B, Herrmann WA, Kühn FE. Transformation of carbon dioxide with homogeneous transition-metal catalysts: A molecular solution to a global challenge?. Angew Chem Int Ed. 2011;50:8510–37.

[63] Quadrelli EA, Centi G, Duplan JL, Perathoner S. Carbon dioxide recycling: Emerging large-scale technologies with industrial potential. ChemSusChem. 2011;4:1194–215.

[64] Centi G, Perathoner S. Opportunities and prospects in the chemical recycling of carbon dioxide to fuels. Catal Today. 2009;148:191–205.

[65] Sakakura T, Choi JC, Yasuda H. Transformation of carbon dioxide. Chemical Rev. 2007;107: 2365–87.

[66] Morris AJ, McGibbon RT, Bocarsly AB. Electrocatalytic carbon dioxide activation: The rate-determining step of pyridinium-catalyzed CO_2 reduction. ChemSusChem. 2011;4:191–96.

[67] Seshadri G, Chao L, Bocarsly AB. A new homogeneous electrocatalyst for the reduction of carbon dioxide to methanol at low overpotential. J Electroanal Chem. 1994;372:145–50.

[68] Barton EE, Rampulla DM, Bocarsly AB. Selective solar-driven reduction of CO_2 to methanol using a catalyzed p-GaP based photoelectrochemical cell. J Am Chem Soc. 2008;130:6342–44.

[69] Tanaka K, Ooyama D. Multi-electron reduction of CO_2 via Ru-CO_2, -C(O)OH, -CO, -CHO, and -CH_2OH species. Coordination Chem Rev. 2002;226:211–18.

[70] Wang X, Lu L, Wang B, Xu Z, Xin Z, Yan S, Geng Z, Zou Z. Frustrated lewis pairs accelerating CO_2 reduction on oxyhydroxide photocatalysts with surface lattice hydroxyls as a solid-state proton donor. Adv Funct Mater. 2018;28:1804191.

[71] Li T, Yang C, Luo J-L, Zheng G. Electrolyte driven highly selective CO_2 electroreduction at low overpotentials. ACS Catal. 2019;9:10440–47.

[72] Lang Z, Miao J, Lan Y, Cheng J, Xu X, Cheng C. Polyoxometalates as electron and proton reservoir assist electrochemical CO_2 reduction. APL Mater. 2020;8:120702.

[73] Gattrell M, Gupta N, Co A. A review of the aqueous electrochemical reduction of CO_2 to hydrocarbons at copper. J Electroanal Chem. 2006;594:1–19.

[74] Centi G, Perathoner S, Winè G, Gangeri M. Electrocatalytic conversion of CO_2 to long carbon-chain hydrocarbon. Green Chem. 2007;9:671–78.

[75] Huang Y Y, Handoko AD, Hirunsit P, Yeo BS. Electrochemical reduction of CO_2 using copper single-crystal surfaces: Effects of CO* coverage on the selective formation of ethylene. ACS Catal. 2017;7:1749–56.
[76] Schouten KJP, Qin Z, Gallent EP, Koper MTM. Two pathways for the formation of ethylene in CO reduction on single-crystal copper electrodes. J Am Chem Soc. 2012;134:9864–67.
[77] Montoya JH, Shi C, Chan K, Nørskov JK. Theoretical insights into a CO dimerization mechanism in CO_2 electroreduction. J Phys Chem Lett. 2015;6:2032–37.
[78] Goodpaster JD, Bell AT, Head-Gordon M. Identification of possible pathways for C–C bond formation during electrochemical reduction of CO_2: New theoretical insights from an improved electrochemical model. J Phys Chem Lett. 2016;7:1471–77.
[79] Marepally BC, Ampelli C, Genovese C, Saboo T, Perathoner S, Wisser FM, Veyre L, Canivet J, Quadrelli EA, Centi G. Enhanced formation of >C1 products in electroreduction of CO_2 by adding a CO_2 adsorption component to a gas-diffusion layer-type catalytic electrode. ChemSusChem. 2017;10:4442–46.
[80] Centi G, Perathoner S. Nanocarbon for energy material applications: N_2 reduction reaction. Smalll. 2021;in press. DOI: 10.1002/smll.202007055.
[81] Birdja YY, Pérez-Gallent E, Figueiredo MC, Göttle AJ, Calle-Vallejo F, Koper MTM. Advances and challenges in understanding the electrocatalytic conversion of carbon dioxide to fuels. Nat Energy. 2019;4:732–45.
[82] Li Y, Kim D, Louisia S, Xie C, Kong Q, Yu S, Lin T, Aloni S, Fakra SC, Yang P. Electrochemically scrambled nanocrystals are catalytically active for CO_2-to-multicarbons. PNAS. 2020;117: 9194–201.
[83] Chan Y-T, Huang I-S, Tsai M-K. Phys Chem Chem Phys. 2019;21:22704–10.
[84] Lum Y, Yue B, Lobaccaro P, Bell AT, Ager JW. J Phys Chem C. 2017;121:14191–203.
[85] Genovese C, Ampelli C, Perathoner S, Centi G. Green Chem. 2017;19:2406–15.
[86] Chang X, Malkani A, Yang X, Xu BJ. Am Chem Soc. 2020;142:2975–83.
[87] Zhuang -T-T, Pang Y, Liang Z-Q, Wang Z, Li Y, Tan C-S, Li J, Dinh CT, De Luna P, Hsieh P-L et al.. Nature Catal. 2018;1:946–51.
[88] Zhang -B-B, Wang Y-H, Xu S-M, Chen K, Yang Y-G, Kong Q-H. RSC Adv. 2020;10:19192–98.
[89] Sen S, Liu D, Palmore GTR. ACS Catal. 2014;4:3091–95.
[90] Marepally BC, Ampelli C, Genovese C, Tavella F, Quadrelli EA, Perathoner S, Centi GJ. CO_2 Utiliz. 2020;35:195–204.
[91] Yu S, Wilson AJ, Heo J, Jain PK. Nano Lett. 2018;18:2189–94.
[92] Munir S, Varzeghani AR, Kaya S. Sustain Energy Fuels. 2018;2:2532–41.
[93] Han H, Noh Y, Kim Y, Park S, Yoon W, Jang D, Choi SM, Kim WB. Green Chem. 2020;22:71.
[94] Giusi D, Ampelli C, Genovese C, Perathoner S, Centi G. A novel gas flow-through photocatalytic reactor based on copper-functionalized nanomembranes for the photoreduction of CO_2 to C1-C2 carboxylic acids and C1-C3 alcohols. Chem Eng J. 2021;408:127250.
[95] Brennecke JF, Gurkan BE. Ionic liquids for CO_2 capture and emission reduction. J Phys Chem Lett. 2010;1:3459–64.
[96] Zhao GY, Jiang T, Han BX et al.. Electrochemical reduction of supercritical carbon dioxide in ionic liquid 1-n-butyl-3-methylimidazolium hexafluorophosphate. J Supercrit Fluids. 2004;32: 287–91.
[97] Spinner NS, Vega JA, Mustain WE. Recent progress in the electrochemical conversion and utilization of CO_2. Catal Sci Technol. 2012;2:19–28.
[98] Centi G, Perathoner S. Carbon nanotubes for sustainable energy applications. ChemSusChem. 2011;4:913–25.
[99] Genovese C, Ampelli C, Perathoner S, Centi G. Electrocatalytic conversion of CO_2 on carbon nanotube-based electrodes for producing solar fuels. J Catal. 2013;308:237–49.

[100] Centi G, Perathoner S. Nanostructured electrodes and devices for converting carbon dioxide back to fuels: Advances and perspectives. In: Zang L, editor. Energy efficiency and Renewable Energy through Nanotechnology, London, UK: Springer-Verlag; 2011. 561–84.
[101] Centi G, Perathoner S. The role of nanostructure in improving the performance of electrodes for energy storage and conversion. Eur J Inorg Chem. 2009;26:3851–78.
[102] Gangeri M, Perathoner S, Caudo S et al.. Fe and Pt carbon nanotubes for the electrocatalytic conversion of carbon dioxide to oxygenates. Catal Today. 2009;143:57–63.
[103] Kim T, Moon S, Hong S-I. Internal carbon dioxide reforming by methane over Ni-YSZ-CeO_2 catalyst electrode in electrochemical cell. Appl Catal A Gen. 2002;224:111–20.
[104] Bidrawn F, Kim G, Corre G, Irvine J, Vohs J, Gorte R. Efficient reduction of CO_2 in a solid oxide electrolyzer. Electrochem Solid-State Lett. 2008;11:B167–70.
[105] Reece SY, Hamel JA, Sung K et al.. Wireless solar water splitting using silicon-based semiconductors and earth-abundant catalysts. Science. 2011;334:645–48.
[106] Pijpers JJH, Winkler MT, Surendranath Y, Buonassisi T, Nocera DG. Light-induced water oxidation at silicon electrodes functionalized with a cobalt oxygen-evolving catalyst. PNAS. 2011;108:10056–61.
[107] Lewerenz HJ, Heine C, Skorupska K et al.. Photoelectrocatalysis: Principles, nanoemitter applications and routes to bio-inspired systems. Energy Environ Sci. 2010;3:748–60.
[108] Currao A. Photoelectrochemical water splitting. Chimia. 2007;61:815–19.
[109] Van De Krol R, Schoonman J. Photo-electrochemical production of hydrogen. In: Hanjalić K, Van De Krol R, Lekić A, editors. Sustainable Energy Technologies. Dordrecht, Netherlands: Springer; 2008. 121–42.
[110] Stempel T, Aggour M, Skorupska K, Munoz A, Lewerenz H-J. Efficient photoelectrochemical nanoemitter solar cell. Electro Chem Commun. 2008;10:1184–86.
[111] Khaselev O, Turner JA. A monolithic photovoltaic-photoelectrochemical device for hydrogen production via water splitting. Science. 1998;280:425–27.
[112] Ichikawa S, Doi R. Hydrogen production from water and conversion of carbon dioxide to useful chemicals by room temperature photoelectrocatalysis. Catal Today. 1996;27:271–77.
[113] Su D, Centi G. Carbon nanotubes for energy applications. In: Rios G, Centi G, Kanellopoulos N, editors. Nanoporous Materials for Energy and the Environment, Singapore: Pan Stanford Pub; 2012. 173–202.
[114] Grätzel M. Dye-sensitized solar cells. J Photochem Photobiol C Photochem Rev. 2003;4: 145–53.
[115] Sivula K, Le Formal F, Grätzel M. Solar water splitting: Progress using hematite (α-Fe_2O_3) photoelectrodes. ChemSusChem. 2011;4:432–49.
[116] Centi G, Perathoner S. Creating and mastering nano-objects to design advanced catalytic materials. Coordination Chem Rev. 2011;255:1480–98.
[117] Centi G, Perathoner S. Nano-architecture and reactivity of titania catalytic materials. Part 2. Bidimensional nanostructured films. Catalysis. 2009;21:82–130.
[118] Ampelli C, Passalacqua R, Perathoner S, Centi G, Su DS, Weinberg G. Synthesis of TiO_2 thin films: Relationship between preparation conditions and nanostructure. Top Catal. 2008;50:133–44.
[119] Perathoner S, Passalacqua R, Centi G, Su DS, Weinberg G. Photoactive titania nanostructured thin films. Synthesis and characteristics of ordered helical nanocoil array. Catal Today. 2007;122:3–13.
[120] Grimes CA, Mor GK. TiO_2 Nanotube Arrays: Synthesis, Properties, and Applications. Heidelberg: Springer, 2009.
[121] Schmuki P. Self-organized oxide nanotube layers on titanium and other transition metals. In: Schmuki P, Virtanen S, editors. Electrochemistry at the Nanoscale. New York: Springer Science; 2009. 435–66.

[122] Centi G, Perathoner S. Nanostructured titania thin films for solar use in energy applications. In: Rios G, Centi G, Kanellopoulos N, editors. Nanoporous Materials for Energy and the Environment. Singapore: Pan Stanford Pub; 2012. 257–82.
[123] Centi G, Passalacqua R, Perathoner S, Su DS, Weinberg G, Schlögl R. Oxide thin films based on ordered arrays of one-dimensional nanostructure. A possible approach toward bridging material gap in catalysis. Phys Chem Chem Phys. 2007;9:4930–38.
[124] Centi G, Perathoner S. Nano-architecture and reactivity of titania catalytic materials. Part 2. Bidimensional nanostructured films. Catalysis. 2009;21:82–130.
[125] Galan-Mascaros JR. Photoelectrochemical solar fuels from carbon dioxide, water and sunlight. Catal Sci Technol. 2020;10:1967–74.
[126] Schreier M, Héroguel F, Steier L, Ahmad S, Luterbacher JS, Mayer MT, Luo J, Grätzel M. Solar conversion of CO_2 to CO using Earth-abundant electrocatalysts prepared by atomic layer modification of CuO. Nat Energy. 2017;2:17087.
[127] Piao I, Yoon SH, Han DS, Park H. On-enhanced conversion of CO_2 into formate on porous dendritic bismuth electrodes with high efficiency and durability. ChemSusChem. 2019;13: 698–706.
[128] Ren D, Loo NWX, Gong L, Yeo BS. Continuous production of ethylene from carbon dioxide and water using intermittent sunlight. ACS Sustain Chem Eng. 2017;5:9191–99.
[129] Arai T, Sato S, Sekizawa K, Suzuki TM, Morikawa T. Solar-driven CO_2 to CO reduction utilizing H_2O as an electron donor by earth-abundant Mn–bipyridine complex and Ni-modified Fe-oxyhydroxide catalysts activated in a single-compartment reactor. Chem Commun. 2019;55: 237–40.
[130] Zhou Y, Shin D, Ngaboyamahina E, Han Q, Parker CB, Mitzi DB, Glass JT. Efficient and Stable $Pt/TiO_2/CdS/Cu_2BaSn(S,Se)_4$ Photocathode for Water Electrolysis Applications. ACS Energy Lett. 2018;3:177–83.
[131] Gurudayal, Beeman JW, Bullock J, Wang H, Eichhorn J, Towle C, Javey A, Toma FM, Mathews N, Ager JW. Si photocathode with Ag-supported dendritic Cu catalyst for CO_2 reduction. Energy Environ Sci. 2019;12:1068–77.
[132] Ferreira De Brito J, Genovese C, Tavella F, Ampelli C, Boldrin Zanoni MV, Centi G, Perathoner S. CO_2 reduction of hybrid Cu_2O–Cu/Gas diffusion layer electrodes and their integration in a Cu-based photoelectrocatalytic cell. ChemSusChem. 2019;12:4274–84.

ACRONYMS

AL	Artificial Leaf
C2+	Products with two or more carbon atoms
CBM	Conduction band Minimum
CES	Chemical Energy Storage
CNT	Carbon Nanotubes
CO_2RR	CO2 Reductive Reaction
FF	Fossil Fuel
GDE	Gas Diffusion Electrode
HER	Hydrogen Evolution Reaction
J	Current Density
MOF	Metal Organic Framework
OEC	Oxygen Evolution Reaction
P2G	Power to Gas

P2L	Power to Liquid
PEC	Photoelectrocatalytic
PEM	Polymer Electrolyte Membrane
PV	Photovoltaic
PV/EC	Photovoltaic cell/Electrolyzer
SHE	Standard Hydrogen Electrode
SOFC	Solid Oxide Fuel Cell
SCLJ	Semiconductor-Liquid Junctions
SEM	Scanning Electron Microscopy
STF	Solar to Fuel
VBM	Valence Band Maximum
YSZ	Yttria-Stabilized Zirconia

Serena DeBeer, Maurice van Gastel, Eckhard Bill, Shengfa Ye, Taras Petrenko, Dimitrios A. Pantazis, Frank Neese

16 Challenges in Molecular Energy Research

16.1 Introduction

The leading theme of this volume is to document the state of energy research in chemistry. It is evident that the challenges associated with renewable, sustainable, and scalable energy reactions touch all fields of chemistry, ranging from the most fundamental aspects of mechanistic catalysis to intricate questions of large-scale chemical engineering and process design. The efficient and reversible storage and release of energy in chemical bonds is indeed an essential aspect, and a great challenge, in energy research. Ideally, such energy-conserving chemistry follows the capture of photon energy provided by sunlight. In this respect, the most impressive chemistry of renewable energy occurs in green plants, algae, and photosynthetic bacteria, all of which convert CO_2 and H_2O into O_2 and energy-rich sugar molecules under the action of sunlight [1].

The relevant processes all involve the activation of small, largely inert molecules. The most important elementary processes are summarized by the following six reactions:

$$2\,H^{+} + 2e^{-} \rightarrow H_2 \tag{16.1}$$

$$2\,H_2O \rightarrow O_2 + 4H^{+} + 4e^{-} \tag{16.2}$$

$$O_2 + 4H^{+} + 4e^{-} \rightarrow 2H_2O \tag{16.3}$$

Serena DeBeer, Max Planck Institute for Chemical Energy Conversion, Department of Inorganic Spectroscopy, Stiftstraße 34-36, 45470 Mülheim an der Ruhr, Germany, e-mail: serena.debeer@cec.mpg.de
Maurice van Gastel, Max-Planck-Institut für Kohlenforschung, Department for Molecular and Theoretical Spectroscopy, Kaiser-Wilhelm-Platz 1, 45470 Mülheim an der Ruhr, Germany, e-mail: maurice.van-gastel@kofo.mpg.de
Eckhard Bill, Max Planck Institute for Chemical Energy Conversion, Department of Inorganic Spectroscopy, Stiftstraße 34-36, 45470 Mülheim an der Ruhr, Germany, e-mail: eckhard.bill@cec.mpg.de
Shengfa Ye, Max-Planck-Institut für Kohlenforschung, Department for Molecular and Theoretical Spectroscopy, Kaiser-Wilhelm-Platz 1, 45470 Mülheim an der Ruhr, Germany, e-mail: shengfa.ye@kofo.mpg.de
Taras Petrenko, High Performance Computing Center Stuttgart (HLRS), Nobelstr. 19, 70569 Stuttgart, Germany, e-mail: taras.petrenko@hlrs.de
Dimitrios A. Pantazis, Max-Planck-Institut für Kohlenforschung, Department for Molecular and Theoretical Spectroscopy, Kaiser-Wilhelm-Platz 1, 45470 Mülheim an der Ruhr, Germany, e-mail: dimitrios.pantazis@kofo.mpg.de
Frank Neese, Max-Planck-Institut für Kohlenforschung, Department for Molecular and Theoretical Spectroscopy, Kaiser-Wilhelm-Platz 1, 45470 Mülheim an der Ruhr, Germany, e-mail: frank.neese@kofo.mpg.de

https://doi.org/10.1515/9783110608458-016

$$CO_2 + 2|H^+ + 2e^- \rightarrow HCOOH \tag{16.4}$$

$$CH_4 + \frac{1}{2}O_2 \rightarrow H_3COH \tag{16.5}$$

$$N_2 + 6\,H^+ + 6e^- \rightarrow 2\,NH_3 \tag{16.6}$$

Reaction (16.1) is one of the most elementary chemical reactions and is of central importance for energy research as it provides H_2, which either serves as a fuel by itself or acts as a precursor for further reduction reactions (e.g., in the activation of O_2, CO_2, and N_2 reactions [3, 4, 6]). Ideally, the protons and electrons required for reaction (16.1) are provided by reaction (16.2), the oxidation of water. The latter process has been found to be very difficult to achieve in chemical catalysis, and its technological realization by homogenous, heterogeneous, or electrochemical means is of fundamental importance for energy research. The oxidation chemistry of methane (or higher reduced hydrocarbons, reaction (16.5)) is a paradigm for the conversion of biomass to fuel. It involves the activation of nonactivated C–H bonds in a controlled manner, again a process that is exceedingly difficult to achieve in catalysis. Finally, the activation of the triple bond of dinitrogen (reaction (16.6)) is a pivotal step of fundamental importance for the large-scale production of fertilizer and consequently feedstock, and thus it is also an integral part of energy research.

The chemistry underlying reactions (16.1)–(16.6) has, of course, been intensely studied in various fields of chemistry. Most of these reactions can be achieved with varying degrees of efficiency in biological, homogenous, or heterogeneous catalysis. Of these fields, nature is unsurpassed in terms of catalyzing all of these reactions with stunning efficiency and selectivity in enzymatic catalysis. In addition, in all cases, nature utilizes highly abundant transition metals in the enzyme active site in order to carry out these processes. Any chemistry that aims at scalability must also be based on abundant elements, which basically restricts the relevant transition metals to the first transition row and molybdenum (which is also sufficiently abundant in nature that it could be used for large-scale chemical processes). Many heterogeneous or homogenous processes are based on noble metals from the second or third row. While many impressive chemical conversions have been achieved on the basis of these metals, they will not be covered in this chapter, as these processes will eventually fall short of the requirement of scalability.

It seems clear that progress toward the efficient, selective, scalable, sustainable, and cost-effective catalysis of reactions (16.1)–(16.6) will require a concerted effort by the chemical community, as well as strong synergy with neighboring disciplines. The fundamental basis for rational process design is a thorough understanding of the chemical principles that involve the key reactions of energy research.

The aim of this book is to summarize the state of the art of the efforts within individual chemical subdisciplines toward achieving the common goals associated with energy research. By bringing together the various communities, this will hopefully develop into an internal joint research effort that benefits from cross-fertilization. In this chapter, we will focus on the biological and molecular aspects of energy research.

Obtaining a basic understanding of the chemistry of natural systems has certainly been a beneficial outcome of research in the field of biological catalysis. However, to date, it has not been possible to design biomimetic catalysts that resemble the natural active sites and have similar activity. In fact, enzyme active sites have evolved to a very high degree of sophistication in which the placement of individual hydrogen bonds or active-site water molecules is absolutely essential for catalytic activity and selectivity. If one aims at the construction of low-molecular-weight catalysts that reproduce *all* of the geometric features of enzyme active sites, the result would very likely be an extremely complex synthesis that would produce molecules that are as fragile, or even more fragile, than the enzymes themselves. The design of chemically highly complex ligands counters the requirement of scalability, as eventually these catalysts must be produced in kilogram quantities. Thus, it appears to be important to take mechanistic inspiration from studying enzyme active sites, while judiciously searching for reactions that utilize readily available base metals within simple ligand frameworks.

Understanding reactions (16.1)–(16.6), catalyzed by first-row transition metals, is a complex undertaking. The chemistry of the first transition row is particularly difficult as most of the ions involved can exist in multiple redox and spin states. Hence, in understanding these reactions, input from theoretical chemistry is essential, and there are ample opportunities for the interaction between theory and experiment. In the next section, we describe a research strategy that emphasizes the interaction between high-level spectroscopy and quantum chemistry. This approach has been found to be particularly fruitful in bioinorganic chemistry and catalysis, and we expect that it will prove equally useful in facing the many challenges associated with energy research.

16.2 Modern Spectroscopy and Quantum Chemistry as a Means to Decipher Reaction Mechanisms

Unraveling the mechanisms of chemical reactions, as well as the structures of various thermodynamically unstable transient species that are involved in molecular transformations, constitutes a significant challenge for both experiment and theory. Due to the "fleeting existence" of transitions states, which correspond to saddle points on the potential energy surface of molecules, and the typically short-lived nature of the transient products, their chemical structures cannot be determined using conventional X-ray, neutron, or electron-diffraction techniques. Nevertheless, specifically designed time-resolved spectroscopic methods can probe molecular structure extremely close to the transition point. The kinetics of a reaction can be studied by various experimental methods using stopped-flow instrumentation. Quench-flow techniques enable the capture of reaction intermediates, which thus can be characterized by spectroscopic methods [2–8]. In particular, the local geometric structure of the intermediates can be directly assessed by extended X-ray absorption fine structure (EXAFS) spectroscopy [9]. In general, however, the elucidation of the chemical information on the basis of spectroscopic data cannot be accomplished without recourse to electronic structure calculations.

One of the main advantages of the quantum chemical calculations is that they can, in principle, be used on their own to study chemical reactions and, in particular, to explore reactive species, which may be difficult to study experimentally. Because an understanding of the reaction mechanisms cannot be deduced from geometric parameters alone, an important aspect of applying first-principles methods for studying chemical reactions is that they provide an insight into the electronic structure of reacting species. In this case, it is important to elucidate the details of the electron (and spin) distribution in molecules and to explain mechanisms of bond formation and cleavage in reactions. In particular, understanding a catalytic reaction at the atomic level requires a detailed analysis of the electronic structure of the entire system, involving the substrate, product, and catalyst. Studies of photochemical reactions represent another problem that requires a detailed understanding of the electronically excited states in molecules.

In recent decades, the development of quantum chemical methods, together with advances in computer hardware and the development of efficient computer codes, has enabled quantum chemistry to make significant contributions to the study of complex chemical systems. The best approximations within the ab initio wave-function-based methods provide energies within <1 kcal mol^{-1} of their exact energies, which is sufficient accuracy for predictive studies of chemical reactions. However, because of the steep scaling of computational effort with system size, the associated methods can only be applied in a rigorous way to molecules with not more than roughly 10 atoms and in approximated form to systems of up to about 100 atoms. In contrast, density functional theory (DFT) methods feature significantly better scaling and can readily be applied to systems with up to 200–1,000 atoms, which makes DFT the method of choice for most large-scale chemical applications. However, the intrinsic deficiencies of DFT, related to the approximate nature of present-day density functionals, result in a noticeable loss of accuracy as compared to ab initio methods. Although the metric parameters associated with weaker metal-ligand and hydrogen bonds are typically poorly predicted by DFT, the gradient-corrected density functionals tend to yield reliable geometric structures [10–13]. It is known that DFT methods do not yet provide uniform accuracy over broad ranges of compounds. When applied to the extended G2 set of molecules, which mainly consists of main group elements and involves closed-shell molecules, the energetic predictions with the hybrid B3LYP functional are accurate to within 3 kcal mol^{-1} [14]. However, even this level of accuracy is scarcely achieved in transition-metal chemistry, in particular for open-shell species [15]. Thus, when only relative energies along the reaction coordinate are taken into account, and no reference is made to experimental data, the interpretation of reaction mechanisms can suffer significantly from the deficiencies of approximate theoretical methods. Another significant problem is that theoretical modeling of the reaction kinetics by means of quantum chemical methods requires testing of many possible reaction pathways to determine the one with the lowest transition-state energy. These procedures are quite cumbersome and, by and large, are based on guessing an approximate transition-state structure guided by chemical intuition. The studies of multistep chemical reactions with unknown intermediates, as in catalytic cycles, are immensely complex [16, 17].

In this respect, the combination of quantum chemistry and spectroscopic methods has proved to be an indispensable tool for the elucidation of reaction mechanisms. Different captured chemical intermediates can be characterized by a variety of spectroscopic methods. The principles of each spectroscopic technique are based on certain physical phenomena, which involve the interactions of atoms and molecules with external electromagnetic fields. These interactions can induce certain transitions in the energy spectrum of the system if the applied photon energy matches the corresponding energy differences. The parameters characterizing transition energies and transition rates, which are deduced from spectroscopic measurements, very often provide a unique fingerprint of the compound and corresponding electronic structure. The spectral intensity pattern as well as the behavior of certain characteristic spectral lines can be very sensitive to the structural parameters, bonding features, chemical environment, spin, and oxidation states. Together with spectroscopic techniques commonly used in chemical analysis, such as nuclear magnetic resonance, infrared, and Raman, a number of methods have proved to be particularly valuable due to their special selectivity. The element-specific techniques, such as Mössbauer spectroscopy, nuclear resonance vibrational spectroscopy, x-ray absorption spectroscopy, and x-ray emission spectroscopy, allow one to identify and characterize species within complex chemical environments. Electron paramagnetic resonance (EPR), electron nuclear double resonance (ENDOR), ultraviolet-visible spectroscopy absorption in combination with resonance Raman spectroscopy, and magnetic circular dichroism also possess enhanced selectivity [18, 19]. Thus, if the spectroscopic data available for the reaction intermediates are included as constraints in the theoretical modeling, a number of proposed structures and mechanisms may be excluded from consideration.

Considerable progress has been made in recent years in the development of quantum chemical methods for predicting molecular properties, thus allowing for the efficient and accurate calculations of various spectroscopic parameters [20, 21]. Inasmuch as any approximate theoretical method needs to be validated, the comparison of calculated and experimental spectroscopic parameters for model systems is very useful for calibrating the calculations. By correlating properly calibrated calculations to the results of multiple spectroscopic methods, one is able to have a greater degree of confidence in identifying a given chemical species. In many experiments, even a qualitative assignment and interpretation of spectroscopic data presents a significant challenge (and in some cases is even impossible) without an initial guess provided by first-principles modeling [22–24]. Thus, for many spectroscopic studies, quantum chemical procedures provide essential input into experimental data analysis. Moreover, in certain cases, theoretical calculations have helped in the identification of artifacts in spectroscopic measurements. Hence, there is a clear synergistic relationship between theory and experiment, which allows for a more detailed understanding of fundamental chemical processes and, in our view, will form an essential component of understanding small-molecule activation in the broader context of energy research. In the sections that follow, an overview of the six fundamental reactions in energy research and the current understanding of these processes from the perspective of both biological and small-molecule catalysis is provided. Throughout this chapter, the contribution that both experiment and theory have had in current understanding is highlighted, as well as continued opportunities for future work in energy-related research.

16.3 Fundamental Chemistry of Energy Conversion

16.3.1 Hydrogen Production

The hydrogen economy represents a means for fulfilling energy demands by using molecular hydrogen as an energy carrier and using reactions in which polluting products like greenhouse gases are avoided. Establishing such an economy requires the development of clean and efficient ways of producing and storing molecular hydrogen.

Presently, the most common method for the production of molecular hydrogen on an industrial basis is steam reforming, a process in which steam is allowed to react with fossil fuels at high temperatures, according to the following reaction:

$$CH_4 + H_2O \rightarrow CO + 3H_2$$

The energy required (i.e., the enthalpy change ΔH) for methane steam reforming, for example, is +49 kcal mol^{-1} [25]. A second well-established and widely applied method for hydrogen production, introduced during the early days of electrochemistry in the 1800s and that has recently become commercially available, concerns electrolysis of water,

$$2\ H_2O \rightarrow 2H_2 + O_2$$

which requires +116 kcal mol^{-1}. For comparison, the bond dissociation energy of H_2 is +104 kcal mol^{-1} [25]. These numbers indicate that the production of molecular hydrogen, the key ingredient of the hydrogen economy, is by no means a trivial task, and that the presently used industrial processes to produce molecular hydrogen will in the long term likely not be sustainable.

16.3.1.1 Hydrogen Production in Nature

Fortunately, nature has found a way to produce molecular hydrogen very efficiently, by using metalloenzymes as catalysts for this reaction. The proteins that evolve hydrogen are called hydrogenases, and they catalyze both the formation and the decomposition of molecular hydrogen:

$$H_2 \leftrightharpoons 2H^+ + 2e^-$$

The family of hydrogenases is divided into three classes, depending on the metal content of the active site [26]. The classes comprise [NiFe], [FeFe], and [Fe] hydrogenases. The turnover numbers for hydrogen production of the [FeFe] hydrogenases amount to 9,000 molecules per second [27]. A disadvantage of all the enzyme systems is their oxygen sensitivity; this is particularly problematic for the [FeFe] hydrogenases. The active sites of the [NiFe] and [FeFe] hydrogenases display an unusual arrangement, which includes inorganic CO and CN ligands (Figure 16.1). In the case of the [FeFe] hydrogenase, the enzyme contains a [4Fe4S] cluster coupled to a [2Fe2S] cluster, both of which are linked to the protein by the cysteine thiolate ligands.

In general, the active sites of all hydrogenases have one open coordination position, which is most likely where substrate interacts [26]. The [NiFe] hydrogenases display a rich redox structure in which the nickel atom cycles between the 3+ and 2+ redox state and the iron is 2+, low spin. Using Hyperfine Sublevel Correlation (HYSCORE) spectroscopy, a hydride ligand has been detected [26] providing direct evidence that coordination of the substrate occurs at the nickel. Currently, there is much interest in improving the issue of oxygen sensitivity. In this respect, the hydrogenases from extremophile organisms are promising candidates since these enzymes are much more robust, oxygen insensitive, and even function at elevated temperatures.

In a broader sense, in addition to the H-H bond, nature often stores energy in the chemical bonds of reduced molecules. Well-known examples are nicotinamide adenine dinucleotide (NADH) or nicotinamide adenine dinucleotide phosphate. As with molecular hydrogen, the energy stored in the respective C-H bonds may be released in an oxidizing environment. Moreover, by storing energy in chemical bonds as opposed to utilizing charge separation, nature has greatly simplified the issue of energy storage. Besides NADH, another common molecule is adenosine triphosphate (ATP), in which energy (approx. -7 kcal mol^{-1}) is stored in the P-O bond. These molecules are common "fuels" employed by nature to store energy. Additionally, in plant photosynthesis, light energy is converted and stored as sugars (see Section 16.3.2).

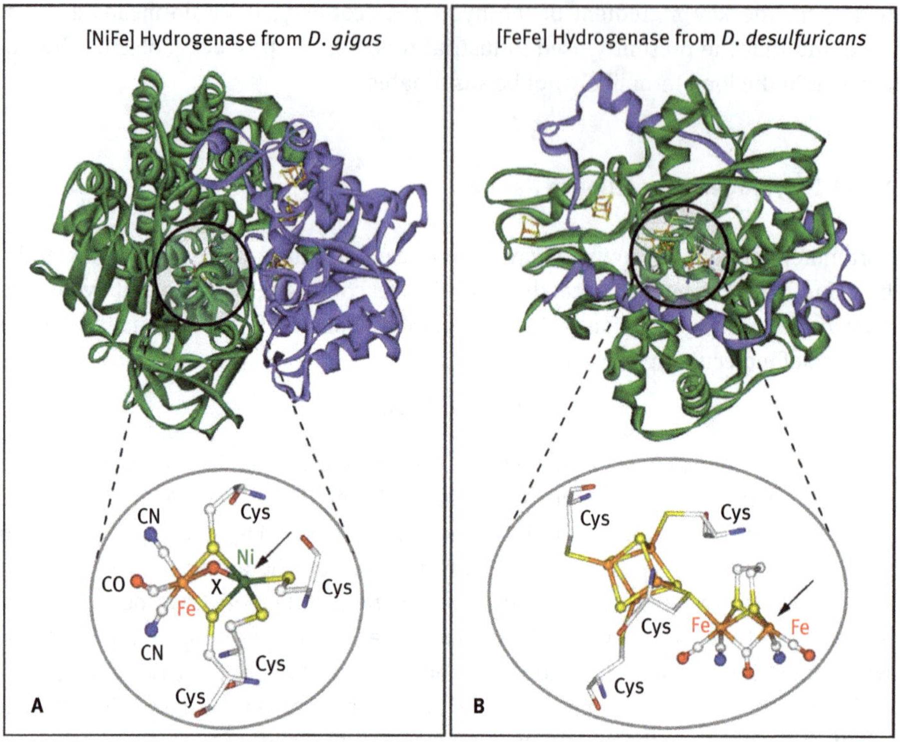

Figure 16.1: (A) Structure of [NiFe] hydrogenase from *D. gigas* (PDB 1FRV). The two subunits are indicated in blue (small subunit) and green (large subunit). The structure of the active site is shown enlarged at the bottom (see text). The arrow indicates the sixth coordination site at Ni, which is found to be unoccupied. (B) Structure of [FeFe] hydrogenase from *Desulfovibrio desulfuricans* (PDB 1HFE). The H cluster (hydrogen-activating cluster) and the two additional [4Fe-4S] clusters are all located in the large subunit. The molecular structure of the H cluster with the cubane $[4Fe\text{-}4S]_H$ and the dinuclear $[2Fe]_H$ subclusters is shown enlarged at the bottom. The arrow indicates the free coordination site at the distal iron Fe_d. The figure is reproduced from Lubitz W, Reijerse EJ, van Gastel M. Chem Rev. 2007;107:4331–65.

16.3.1.2 Molecular Models for Hydrogen Production

The active sites of hydrogenases have been a focal point for inorganic chemists with the aim of preparing biomimetic inorganic molecules that possess catalytic activity. One of the first molecules, which showed catalytic activity for hydrogen splitting in aqueous solution, was reported by Ogo et al. [28]. This molecule features a nickel-ruthenium dinuclear metal center, which is bridged by a hydride (as demonstrated by neutron-diffraction studies). The hydride was additionally shown to be the product of the heterolytic cleavage of molecular hydrogen. Other systems that show catalytic activity in organic solvents involve homonuclear metal centers featuring ruthenium or iridium. Yet, despite intense efforts, it has proved to be very difficult to synthesize catalytically active complexes featuring nickel and iron, or only iron.

The challenges for the future are thus significant in number. The oxygen sensitivity of the enzyme systems remains an issue of crucial importance. While the synthesis of inorganic complexes has recently resulted in catalytic activity toward hydrogen splitting, no model systems that catalyze hydrogen production currently exist. This feat so far remains limited to hydrogenases. Research efforts are also being directed toward storing energy in the form of other reduced molecules (i.e., sugars and biomass) in a broader sense. It is clear that in the general area of hydrogen splitting and production significant research opportunities and challenges exist as we move toward an understanding and broader utilization of this fundamental chemical reaction.

16.3.2 Water Oxidation

Among the possible sources of hydrogen production discussed in the preceding section, water is one of the most desirable potential sources for sustainable large-scale production of hydrogen in the future. The abundance of water virtually guarantees that a hydrogen-based economy can have a practically endless supply of clean basic material for fuel production. The central chemical process for extracting hydrogen from water is the oxidation of water. The products of complete oxidation of two water molecules are molecular oxygen, four electrons, and four protons. It is from the coupling of these water-derived protons and electrons that molecular hydrogen can be obtained, in a process that in total is endergonic by 4.92 eV (113 kcal mol^{-1}). The elementary electrochemical reactions involved in water oxidation [29] are one-, two-, or four-electron processes:

$$E^0/V(\text{vs.Standard hydrogen electrode})$$

$$OH + e^- \rightarrow OH^- 2.0$$

$$H_2O_2 + 2\ H^+ + 2e^- \rightarrow 2\ H_2O\ 1.77(pH\,0)$$

$$H_2O_2 + 2e^- \rightarrow 2\,OH^-\,0.88$$

$$O_2 + 4\ H^+ + 4e^- \rightarrow 2\ H_2O\ 1.23(pH\,0)$$

$$O_2 + 2\ H^+ + 4e^- \rightarrow 2\,OH^-\ 0.41(pH\,14)$$

16.3.2.1 Water Oxidation in Nature

In Nature, the deceptively simple reaction $2\ H_2O \rightarrow O_2 + 4\ H^+ + 4e^-$ is carried out by long-evolved biochemical systems in photosynthetic organisms. Due to the photochemical nature of the reaction, the four oxidation equivalents need to be accumulated one at a time, in contrast to electrochemical processes. At large-scale quantities, the reaction can at present be achieved electrolytically only at substantial cost due to the excess energy required to overcome the "over-potential" arising from various activation barriers encountered in the electrolytic process. A cheap and efficient synthetic catalytic system for water oxidation must become available if hydrogen is to be produced sustainably from water with low energy input and in the vast quantities required for a fuel in the role of a major energy vector. Since biology offers the only current example of efficient water oxidation, it forms the starting point for most considerations regarding the design of artificial systems.

The oxidation of water is the first step in the chain of photosynthetic process that results in the biosynthesis of carbohydrates. The reaction takes place in the multiunit enzyme photosystem II (PSII), which is embedded in the membranes of the thylakoids in the leaves of plants and in algae and cyanobacteria [1]. For photosynthetic organisms, sunlight is the only source of energy – or "oxidative power" – to achieve the oxidation of water, making this process unique on the planet. However, sunlight is of course not nearly enough to induce the splitting of water by itself. Elaborately organized excitation energy and electron transfer systems, as well as an inorganic clusters unique in biology, are integral parts for the catalytic activity of PSII. The locus of water oxidation is the inorganic Mn_4O_5Ca cluster known as the oxygen-evolving complex (OEC) or water-oxidizing complex (WOC). Its composition has been known for some time, but its structure has only recently been resolved at a sufficiently high resolution of 1.9 Å [30]. The optimized geometry of the cluster for the S_2 state of the Kok cycle is shown in Figure 16.2 [31].

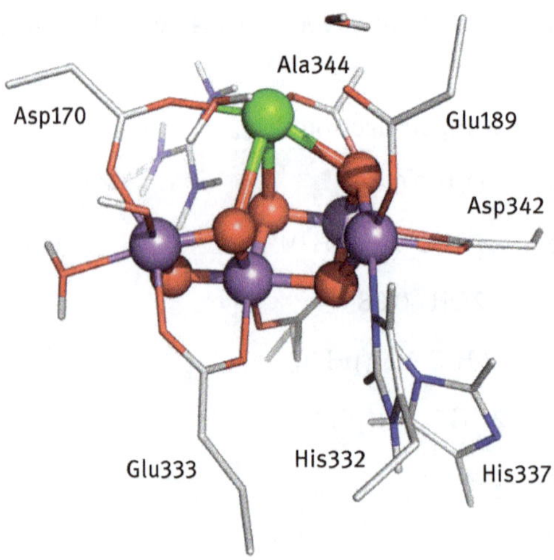

Figure 16.2: Structure of the Mn_4O_5Ca core and the immediate protein surrounding the WOC of PSII in its S_2 state, as obtained from DFT geometry optimization of the 1.9 Å crystal structure (Mn: purple; Ca: green; C: gray; N: blue; O: red).

The WOC is oxidized stepwise by a nearby tyrosine residue (Tyr_Z), which is itself oxidized by the chlorophyll cation radical $P680^{+\cdot}$ (formed by light-induced charge separation). The electrons are eventually used by PSII for the reduction of plastoquinone. After the WOC has lost four electrons, the accumulated oxidizing power drives the formation of molecular oxygen from two substrate water molecules, and the catalytic system is reset. The sequence of the four electron-transfer steps is summarized in the Kok cycle [32] of Figure 16.3, where the most probable spectroscopically derived oxidation states of the Mn ions [33] are shown for each of the five redox state intermediates S_n ($n = 0–4$).

A prominent feature of the natural system is the separation of components and their finely controlled spatial organization. This distribution is coupled to efficient electron- and proton-transfer pathways, which are defined, for example, by extended hydrogen-bond networks or accurately spaced donors and acceptors. The whole of the protein is organized in such a way that the direction of several events becomes practically irreversible; for example, there is a very low probability of charge recombination after the initial charge separation step at P680. The separation of primary (P680) and secondary (Tyr_Z) oxidants and the WOC cluster are critical for regulating the efficiency of each catalytic step. Of particular importance is the functional role of Tyr_Z as a proton-coupled redox active link. It has been shown that slight variations in proton-transfer distance and hydrogen-bond dynamics can change the rate of proton-coupled electron transfer by orders of magnitude [34]. Clearly, if this design is to be carried over to synthetic systems, it is necessary to have simultaneously efficient dedicated components and delicate control over their spatial distribution and interconnection even at the molecular scale. It remains to be seen whether this can be achieved through a supramolecular approach or through the combination of different phases.

In terms of the specific components, advancement to each step of the cycle is possible because the absorption of light and the transfer of excitation energy to P680 leads to formation of an extremely strong oxidant. The preparation of an analogous durable synthetic system that can be selectively switched into a strong oxidizing form

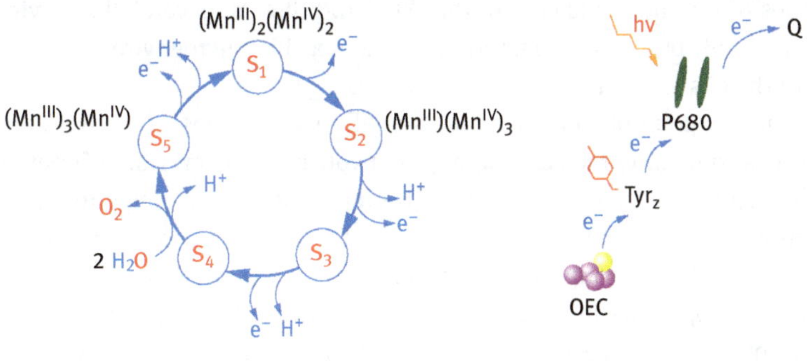

Figure 16.3: The Kok cycle showing the sequence of redox and proton- and electron-transfer steps that lead to the oxidation of water in the WOC (left), and the electron-transfer steps in PSII (right).

by visible-light activation would be an interesting development toward artificial articulated water-splitting systems. However, the catalyst can also be driven electrochemically to reduce the complexity at the molecular level.

An important observation regarding the order of events in the catalytic cycle is that the loss of electrons and protons occurs in a strictly alternating way [35, 36] that leads to a "redox-leveling" effect [37]. This enables the accumulation of four oxidizing equivalents within a narrow range of potential. The reduction potentials for each S-state transition, $\Delta G^0(S_{i+1}/S_i)$ have been approximately determined as 0.85, 1.10, 1.15, and 1.0 eV for $i = 0$, 1, 2, and 3, respectively, while the barriers for the oxidation of the OEC by Tyr_Z at each S-state starting from S_0 are ca. 1, 3, 9, and 7 kcal mol^{-1}, respectively [38]. Given the relatively large number of electrons that is necessary for the oxidation of water, controlling the redox potential range should be central in constructing artificial systems to ensure low barriers at each step.

16.3.2.2 Molecular Models for Water Oxidation

The inorganic cluster itself satisfies a number of requirements necessary for its role: it can accommodate the four necessary oxidation equivalents, it remains intact in all states of the Kok cycle, and it avoids at all steps the Mn(II) oxidation state that is labile. Unsurprisingly, there have been many attempts to produce structural mimics of the WOC (i.e., complexes containing manganese ions and possibly calcium) [39]. One of the most recent ones is a Mn_3Ca cubane model [40], which reproduces the bonding of part of the WOC quite successfully. Until now, however, the success of such systems in terms of catalytic efficiency, stability, and turnover has not been impressive. This is rather unsurprising since it is not only the metal stoichiometry and connectivity of the WOC core that determine its catalytic properties, but also a functionally important protein matrix that stabilizes the various states of the core and offers specific transfer pathways. Even the presence of extrinsic regulatory proteins affects the stability of the Mn_4O_5Ca cluster [38]. The ways that the surrounding protein modifies the intrinsic properties of the inorganic core of the WOC and directs its catalytic activity are not known in detail, but an understanding of the specific microenvironment effects is crucial for the design of artificial water-oxidation systems.

From a mechanistic point of view, the details of all reaction steps in the WOC are not certain, but most probably, radical coupling is involved at the critical O-O bond-formation step [41]. Alternative mechanisms have been proposed, including direct coupling of two hydroxo groups and nucleophilic attack of a free or loosely bound water molecule onto a terminal oxo group [38]. Even though only a radical mechanism appears at this point to be operative in the WOC, nucleophilic attack has been shown to be the dominant, if not exclusive, mechanism in some synthetic catalysts, notably many ruthenium-based ones [42, 43]. This suggests that no pathway can be excluded a priori: different mechanisms are operative in different systems, depending on the nature of the metal and other aspects of the electronic structure or the setup of components. Therefore, there is more than one way to achieve water oxidation, and it is crucial to recognize which one should be optimized for any given catalytic system.

In conclusion, the challenge in terms of fundamental chemistry is broadly understood; there are multiple potential ways to address it, but the best way is by no means obvious. It might be naïve to expect that a synthetic catalyst will mimic the WOC both structurally and functionally. It is quite probable that the WOC will offer insight into the fundamental chemical and physical aspects of water-oxidation catalysis but will not be a synthetic blueprint. For instance, the requirement for a four-electron oxidation-state span and the avoidance of labile oxidation states can be satisfied in multiple ways. Thus, the goal can be achieved with metals other than manganese and with different chemistry, as has been shown, for example, by ruthenium, platinum, and iridium complexes [42–45]. It should be borne in mind, though, that it is uncertain whether systems based on rare or precious metals can be developed into practical platforms for applications. Cobalt-based systems are an attractive alternative in this respect [46]. Besides, an appealing approach for large-scale applications is the development of catalytic surfaces or nanoclusters, relying on the carefully controlled composition and the properties of the surface rather than on the homogeneous chemistry of an expensive molecular catalyst. Recent demonstrations of activity by, among others, cobalt polyoxometalate [47] and cobalt-phosphate catalytic systems [48, 49] have shown that this might be a promising way forward in terms of ease of development and implementation.

16.3.3 Oxygen Activation

The reverse reaction of water oxidation is the reductive activation of dioxygen by four electrons and four protons to produce two molecules of water. This reaction is of fundamental importance for energy research as it is the chemistry underlying fuel cells. There, the reductive activation of dioxygen occurs on electrode surfaces, often involving platinum or other noble metals. Despite the fact that the reaction is highly exothermic, a significant overpotential is necessary to achieve oxygen activation, which consequently leads to energy loss. Hence, there is significant interest in understanding the mechanism of oxygen activation on the electrode surface in atomic detail in the hope that the emerging insight could be used to design more efficient fuel cells. However, despite its importance and the rapidly growing interest, there is relatively little mechanistic understanding of how dioxygen activation is actually happening on the electrode surface. Part of the problem in investigating this chemistry experimentally is the limited number of spectroscopic methods that can be used to probe elementary reaction steps under in situ conditions. Given that only a fraction of the surface is catalytically active, it is difficult to obtain spectroscopic signals that can be correlated to the active surface species with a high degree of confidence.

By contrast, there is a wealth of mechanistic information available in biochemistry and homogenous catalysis concerning oxygen activation mechanisms on first-row transition metals. It is therefore appropriate to provide a brief overview of this field here. However, this is a very large and active research field; hence, it is not possible to do full justice to this subject here. For more extensive overviews, see references [50–60] in Section 16.6.

16.3.3.1 Oxygen Activation in Nature

Nature is extremely efficient in activating dioxygen without releasing biologically harmful intermediates such as the hydroxyl radical. In fact, one can think of a biological cell as an elaborate fuel cell in which the energy released by oxygen reduction is stored in a proton gradient across the mitochondrial membrane that is in turn used to drive the synthesis of energy-rich ATP molecules – a paradigm for storing energy in chemical bonds. The terminal oxidase in this cell respiration process is the enzyme cytochrome *c* oxidase (COX), which is a large enzyme that spans the mitochondrial membrane. Using electrons provided by metabolic processes that oxidatively break down sugars or fatty acids, COX reduces dioxygen to water while pumping protons across the membrane. As found invariably throughout biology, oxygen activation is achieved by an active site containing first-row transition metals. In the case of COX, the active site consists of a mixed-metal heme-iron/copper center.

The reaction of COX starts with the fully reduced active site containing an Fe (II)/Cu(I) binuclear center. This unit serves as a two-electron donor. Hence, binding of O_2 leads to a simultaneous two-electron transfer creating an Fe(III)-(O_2^{2-})-Cu(I) intermediate (Figure 16.4). In this way, the sluggish first reduction step from dioxygen to superoxide is avoided. The strategy to use a simultaneous two-electron transfer to create a bridging peroxide species is widespread in biochemistry in enzymes with binuclear metal centers. The O-O bond is known to cleave after injection of the third electron. This is sensible from an electronic structure point of view because this electron has to enter a strongly antibonding sigma* orbital with respect to the O-O bond. Hence, stabilizing this orbital is a prerequisite for O-O bond cleavage. Obviously, coordination of peroxide to transition-metal cations is a good way to achieve this objective. A second strategy is protonation of the peroxide to yield a hydroperoxide, which is known to have a much deeper-lying sigma* orbital. In the case of COX, it is not known whether protonation of the peroxide occurs prior to or after O-O bond cleavage. However, it is agreed upon that the third electron is provided by the heme center, thus creating a high-valent heme-Fe(IV) = O intermediate together with a Cu(II)-OH configuration in the active site. The heme-Fe(IV) = O species is a powerful oxidant itself. Hence, the last step of the reaction consists of the oxidation of a nearby tyrosine residue in a proton-coupled electron transfer, which creates an Fe(III)-OH species. The final stages of the reaction involve the recharging of the active site with electrons concomitant with protonation of the hydroxide ligands and product release.

The COX mechanism serves as a paradigm for oxygen activation at binuclear active sites. A key feature is that the active site is constructed such that a peroxide molecule (O-O distance ~ 1.4 Å) fits comfortably between the two metal centers.

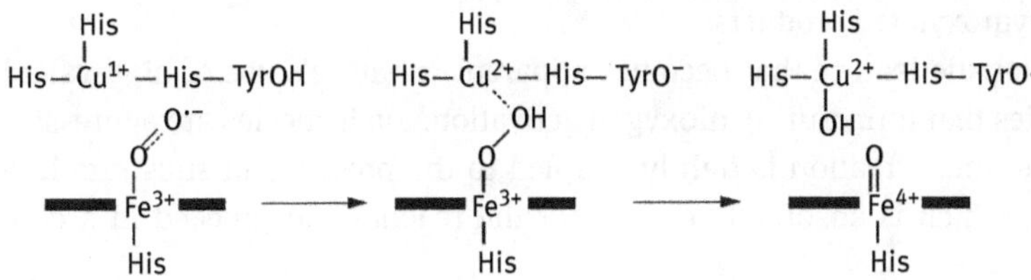

Figure 16.4: Dioxygen bond activation by cytochrome *c* oxidase (adapted from [60]).

One particular feature of this mechanism is that the two metals in the active site can provide three electrons, which means that a fourth electron has to come from a different source, in this case a redox active tyrosine residue. A different solution to the same problem has been found in the enzyme methane monooxygenase, which involves a binuclear iron center and will be discussed subsequently.

In summary, the metal centers serve the dual roles of providing electrons and activating the O-O bond for cleavage through lowering of the energy of the critical sigma* orbital. The second key feature of enzymatic O_2 activation concerns the role of protons. Protons represent essentially "naked" positive charges and hence are very aggressive chemical species that greatly influence the electronic structure of reactants. Thus, finely controlled acid/base chemistry is a key feature of enzyme reactions. Recalling that a change of the pK_a by one unit only involves an energy change of about 1.2 kcal mol^{-1}, it becomes evident how finely tuned nature orchestrates the flow of charge in enzyme active sites and how difficult it will be to try to achieve comparable subtlety in the design of low-molecular-weight catalysts.

Obviously, oxygen activation is more difficult to achieve at mononuclear transition-metal centers. Assuming that a given metal center cannot provide more than two electrons in the biologically accessible potential window of at most 1.27 V [61], it becomes clear that oxygen activation either has to stop at the peroxide level or has to rely on other electron sources to reach completion. The paradigm for such a reaction is provided by the family of cytochrome P450 (CYP) enzymes, for which the reaction mechanism has been investigated in great detail (Figure 16.5) [58, 60, 62–64]. The active site of CYP consists of a heme-iron center with a thiolate axial ligand. Binding of O_2 to the ferrous enzyme leads to a very characteristic intermediate in the chemistry of O_2 with first-row transition metals – namely, an end-on Fe(III)$O_2^{\bullet-}$ unit in an antiferromagnetic alignment. This intermediate cannot be further processed before another external electron and a proton enter the reaction, thus creating an Fe(III)-OOH intermediate. This intermediate spontaneously decays under heterolytic cleavage of the O-O bond to yield a very characteristic intermediate that is called “compound I” and consists of a high-valent Fe(IV) = O center coupled to a porphyrin radical. Thus, overall the metal and porphyrin together have contributed three electrons toward O-O cleavage, while a fourth electron must be provided externally. Compound I is known to be a “magical” oxidant and reacts with unactivated C-H bonds to yield, following a rebound step, hydroxylated products.

The characteristic feature that becomes apparent is nature’s use of high-valent metal-oxo species that form during dioxygen activation. Such species are aggressive oxidants. Hence, their creation is tightly coupled to the presence of substrate in a confined space, which is absolutely critical for the reactions to proceed in a controlled manner.

We should finally note the importance of the overall spin state in reactions involving dioxygen. The fact that both the open-shell transition metal and the dioxygen unit can exist as open-shell fragments leads to a significant variety of possible spin couplings. For example, a quintet high-spin Fe(II) center coupled to a triplet dioxygen fragment gives rise to triplet, quintet, or septet final states. Each spin state is characterized by a different electronic structure and hence different reactivity. In fact, the important concept of *two-state reactivity* states that a chemical reaction may proceed on several of those surfaces as the barriers for the elementary reaction steps may differ widely for each spin state. It is not possible to go into further detail here.

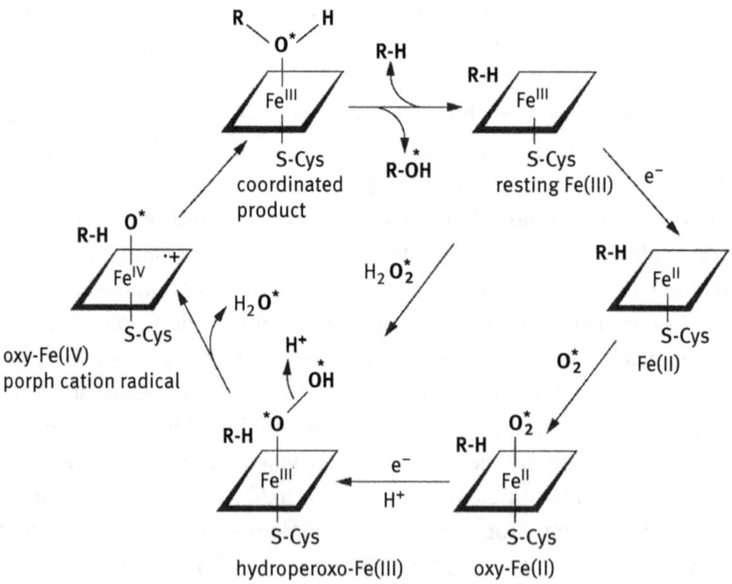

Figure 16.5: Reaction cycle of cytochrome P450 (adapted from [58]).

16.3.3.2 Molecular Models for Oxygen Activation

The design of oxygen-activating catalysts on the basis of first-row transition metals has been greatly inspired by the insights obtained in biology. In fact, there has been significant progress in the design of catalysts on the basis of relatively simple amine- and pyridine-based ligands (such as tripyridine-amine, triazacyclononane). Also, reaction mechanisms and reaction intermediates have been carefully investigated, leading to a variety of transition-metalcomplexes involving Fe(II) and Fe(III) peroxides and hydroperoxides as well as high-valent Fe(IV) = O and Fe(V) = O oxo units that could even be stabilized to the point where crystal structures could be obtained. In addition to iron, a widely studied motif is provided by dicopper complexes that bind O_2 in the Cu (II) state to produce either $Cu(II)_2(O_2^{2-})$ or $Cu(III)_2$(bis-μ-oxo) species, which can be thought of as a fascinating case of valence isomerism. Either oxidation state features distinct reactivity. The relative stability of the two states can be controlled via the supporting ligand framework.

Unfortunately, it is not possible here to go into further detail about the rich chemistry involving first-row transition metals. It, however, hopefully became apparent that there are a number of subtle electronic and geometric structure design criteria that must be met for efficiently catalyzing dioxygen activation. Many of those factors are fairly well understood. In the context of energy research, it remains to be seen how these insights could be applied to the fields of heterogeneous catalysis and electrochemistry. An attractive route could consist of anchoring oxygen-activating catalysts on electrode surfaces.

16.3.4 Methane Oxidation

Methane is the simplest alkane and the principal component (more than 80%) of natural gas. The high abundance of methane renders it a promising energy carrier. In fact, by mass, methane combustion releases the largest amount of energy due to its high gravimetric hydrogen content (20%):

$$CH_4(g) + 2\ O_2(g) \rightarrow CO_2(g) + 2\ H_2O(1), \quad \Delta H = -213\,\text{kcal/mol}$$

However, at normal temperature and pressure, methane is a gas; therefore, it is difficult to transport.

At present, the Fischer-Tropsch process is employed in industry to convert standard natural gas into synthetic gasoline, diesel, or jet fuel [65]. It involves a series of chemical reactions that lead to a variety of hydrocarbons:

$$H_2O + CH_4 \rightarrow CO + 3\ H_2$$

$$(2n+1)\ H_2 + n\ CO \rightarrow C_nH_{2n+2} + n\ H_2O$$

The reactions are catalyzed by transition metals (cobalt, iron, and ruthenium) on high-surface-area silica, alumina, or zeolite supports. However, the exact chemical identity of the catalysts is unknown, and their characterization presents challenges as these transformations are carried out under very harsh reaction conditions. Typically, the Fischer-Tropsch process is operated in the temperature range of 150–300 ° C and in the pressure range of one to several tens of atmospheres [66]. Thus, the entire process is costly and inefficient and even produces waste [67]. Hence, development of more economical and sustainable strategies for the gas-to-liquid conversion of methane is highly desirable.

16.3.4.1 Methane Oxidation in Nature

In nature, transformation of methane to methanol can be very efficiently carried out at ambient conditions. Methanotrophic bacteria perform a methane hydroxylation reaction using metalloenzymes called methane monooxygenases (MMOs) through dioxygen-activation mechanisms. There are two types of MMOs that have been discovered so far. Soluble MMO (sMMO) [56, 68] oxidizes methane with a well-characterized catalytic di-iron center (Figure 16.6A). The intermediate called Q that is kinetically competent for methane hydroxylation has been experimentally characterized using a variety of spectroscopic methods. The extended X-ray absorption spectroscopy suggests that intermediate Q likely contains an $[Fe^{IV}{}_2(\mu\text{-}O)_2]^{4+}$ diamond core [69]. The second type of MMO is particulate MMO (pMMO), which is an integral membrane metalloprotein that requires copper for catalytic activity. Despite the long-term intensive research work on pMMO [70, 71], only recently was the active site for methane oxidation shown to be a di-copper center [72]. In addition to di-iron enzymes, a range of mononuclear iron enzymes are capable of effectively modifying unactivated C-H bonds with Bond Dissociation Energies (BDEs) comparable to methane, such as taurine/α-ketoglutarate dioxygenase (TauD) (shown in Figure 16.6B). Experiments have revealed that the C-H bond cleaving intermediates intheir catalytic circles all involve a high-valent metal-oxo unit [73]. These metalloenzymes thus provide the optimal model for an efficient, environmentally friendly catalyst.

16.3.4.2 Molecular Models for Methane Oxidation

Recently, the synthetic precedent for the $[Fe^{IV}{}_2(\mu\text{-}O)_2]^{4+}$ diamond core proposed for intermediate Q has been reported [74, 75]. Surprisingly, this complex is a sluggish oxidant and is only able to oxidize weak C-H bonds (BDE <80 kcal mol^{-1}) [74]. However, its one-electron reduced species possessing an open $[HO\text{-}Fe^{III}\text{-}O\text{-}Fe^{IV}=O]^{2+}$ core structure displays much higher reactivity toward C-H bond activation, and the reaction rate is at least 105-fold faster than that determined for the diamond complex with the higher

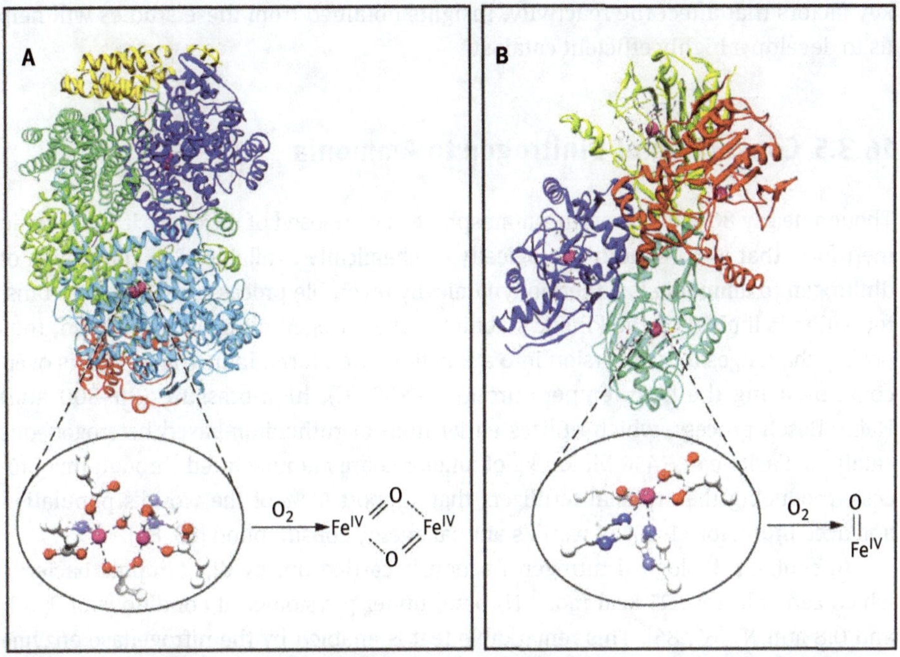

Figure 16.6: (A) Structure of sMMO (PDB code 1MTY) and (B) structure of TauD (PDB code 1OS7). The active sites and the key intermediates responsible for C-H bond oxidation are shown at the bottom.

oxidation states of the metal centers [76]. Interestingly, the oxidation state does not appear to dictate their relative reactivities; hence, there must be more critical factors, such as spin states of ferryl centers (see subsequent discussion), that need to be taken into account in order to fully understand high-valent di-iron chemistry.

In parallel, a series of mononuclear Fe^{IV}-oxo complexes have been prepared [59, 77]. Most of them show significant oxidizing power, which is comparable to that observed for the open-core di-iron complex. And the mechanism of alkane hydroxylation by ferryl intermediates has spurred considerable interest in theoretical studies as well [78–81]. Recently, the changes in the electronic structure along the reaction coordinate of alkane hydroxylation by Fe^{IV}-oxo intermediates with different spin ground states ($S = 1, 2$) have been carefully analyzed and revealed that the quintet ferryl species is intrinsically more reactive than its triplet congener [82]. This result may shed light on how to rationalize the different reactivity between the diamond-core model compound and intermediate Q. In comparison with intermediate Q, which consists of the two high-spin ferryl centers ($S = 2$), two intermediate-spin Fe^{IV} ($S = 1$) ions were identified in the model complex. More research work will be needed to clarify the reaction mechanism of both MMOs and the model complexes, and to pinpoint key factors that affect the reactivity. Insights obtained from these studies will help us to develop a highly efficient catalyst.

16.3.5 Conversion of Dinitrogen to Ammonia

Though nearly 80% of the earth's atmosphere is composed of nitrogen, it exists in an inert form that is not readily biologically or chemically available. The conversion of dinitrogen to ammonia is a thermodynamically favorable process; however, its transformation is limited by the kinetic stability of the nitrogen-nitrogen triple bond, thus posing challenges for conversion into a functionalized form. Industrially, this is overcome by using the high-temperature (300–400 °C), high-pressure (90–300 atm) Haber-Bosch process, which utilizes either iron- or ruthenium-based heterogeneous catalysts. Each year, ~ 450 billion kg of ammonia are manufactured through this process, producing the artificial fertilizers that support 40% of the world's population and accounting for ~1–2% of world's annual energy consumption [83, 84].

In contrast, biological nitrogen fixation is carried out by diazotrophic bacteria, which can split the 225 kcal mol^{-1} N_2 bond under physiological conditions of 290 K and 0.8 atm N_2 [85, 86]. This remarkable feat is enabled by the nitrogenase enzyme family utilizing ATP as an energy source. The lower reaction temperature for the enzyme system, as compared to the industrial process, implies a more efficient activation of dinitrogen and provides a thermodynamic advantage for ammonia synthesis. This has provided great impetus for better understanding the mechanism of biological nitrogen fixation and the fundamental insights it can provide into catalytic mechanisms. In this chapter, biological nitrogen fixation and the progress that has been made in modeling this process on the molecular level are briefly reviewed.

16.3.5.1 Nitrogen Fixation in Nature

The nitrogenase enzyme is a two-component protein that consists of an electron-transfer Fe protein and a catalytic protein [85]. Three different nitrogenase enzymes are known, which differ primarily in the nature of the putative active site within the catalytic protein. The most common form is the MoFe protein, in which the active site for nitrogen reduction, the so-called FeMo cofactor (FeMoco), is composed of seven irons, one molybdenum, and nine sulfides (Figure 16.7). A vanadium nitrogenase is also known in which the molybdenum is replaced by vanadium. In addition, iron-only forms of the enzyme have been identified. Though all three forms are capable of reducing dinitrogen to ammonia, the vanadium and iron-only forms are less active and also less well understood. However, recent studies of vanadium nitrogenase have shown that it is three orders of magnitude more efficient at catalyzing C–C bond coupling than its Mo counterpart, which raises interesting questions about the role of the heterometal in catalysis [87, 88].

The FeMoco factor of nitrogenase, as reported in a 2002 crystal structure by Einsle et al., is shown in Figure 16.7 [89]. The seven iron and nine sulfides form a prismane-like structure, with the capping molybdenum ligated to a homocitrate. At the center of the cluster is a single light atom, previously identified as single C^{4-}, N^{3-}, or O^{2-}. Detailed ENDOR, electron spin echo envelope modulation, nuclear resonance vibrational spectroscopy, EXAFS, and DFT studies were carried out in an effort to elucidate the structure, however, without a clear consensus as to the identity [90–94]. Recently, based on X-ray emission spectroscopy and correlated computational studies, the central atom has been assigned as C^{4-} [95]. In addition, very recent high-resolution crystallography and ^{13}C ENDOR have provided independent support for this assignment [96], thus identifying a unique example of an organometallic biological cluster. The identification of the central atom raises many new questions about the role of the carbide, the biosynthesis of the cluster, and the contributions a central carbon has to the electronic structure and reactivity.

With regard to catalysis, however, perhaps the more interesting questions focus on reactivity and the interaction of the cofactor with substrate. Electron paramagnetic resonance, ENDOR, and HYSCORE studies by Hoffman and coworkers suggest that an iron ion or multiple iron ions (within a single FeS face) are the site of substrate binding and reduction [97]. Such a mechanism is appealing, given that all known forms of nitrogenase contain iron. However, currently no synthetic iron models that are capable of catalytic dinitrogen reduction are known. The only homogeneous catalyst known to reduce dinitrogen is a molybdenum complex, again lending intrigue to the role of the heterometal.

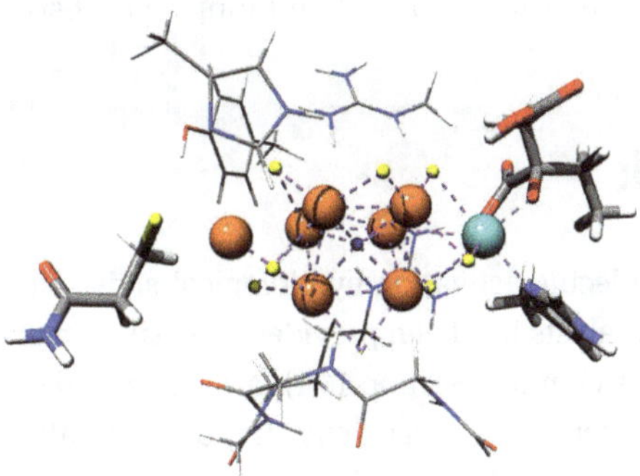

Figure 16.7: Structure of the FeMoco in the MoFe protein (PDB: 1M1N). Orange spheres correspond to iron atoms; cyan, to molybdenum; and yellow, to sulfur.

16.3.5.2 Molecular Model for Nitrogen Fixation

Since the first report of an N_2-bound transition-metal complex in 1965, chemists have been hopeful that molecular catalysts could be made to mimic the chemistry of nitrogenase. However, it was not until 2003 that Schrock and coworkers reported the first discrete transition-metal catalyst that reduces dinitrogen to ammonia in the presence of electrons and protons [98]. The complex is a monomeric Mo species that utilizes a triamidoamine ligand framework and cycles between oxidation states of Mo(III) to Mo(VI) during the course of catalysis. Several intermediates in the reaction cycle have been isolated and characterized, and the suggested mechanism parallels the original proposals of Chatt for low-valent Mo and W species.

Notable progress in ammonia synthesis has also been made by Chirik and coworkers, who have shown that ammonia can be evolved from a zirconium metallocene complex. Using a tetramethylated-bis-Cp-dichloride complex, they were able to first add N_2 to form a dimeric Zr complex and subsequently add excess H_2, with heating, to evolve ammonia [99]. Their work, though not catalytic, also provides basic mechanistic insights into transition-metal-mediated N-H bond catalysis.

In contrast, a molecular iron catalyst for conversion of dinitrogen into ammonia has remained elusive, despite its relevance to both the heterogeneous process and biological catalysis. A number of iron-dinitrogen complexes have been isolated and characterized with varying degrees N-N bond activation [100]. Both iron (0) and iron(I) complexes have been shown to mediate conversion of N_2 to ammonia, however, in low yields and not catalytically. Recent studies by Holland and coworkers demonstrate that a trimeric iron complex can evolve ammonia stoichiometrically, though not catalytically [101]. With much continued work in this area, there is promise that a molecular iron catalyst is on the horizon and that there is still much to learn at the interface of chemical and biological ammonia synthesis.

16.4 Summary and Outlook

In this chapter, a brief overview of small-molecule activation by biological active sites and related small-molecule homogeneous catalysts has been provided. The six primary reactions, outlined in the introduction of this chapter (Section 16.1), are all carried out by enzyme active sites with remarkable efficiency and specificity. Importantly, all of the enzymes discussed in the preceding sections utilize readily available base metals in order to activate small molecules. This is a key requirement as we look to the needs and challenges of our energy future, which will require not only efficient catalysts, but also scalable and sustainable catalysts. Here, it is clear that there is a gap between what biology can enable and what has been achieved with small-molecule catalysts. While remarkable transformations have been reported for small-molecule catalysts that use noble metals, the documented examples of highly active base-metal catalysts in small-molecule complexes is far more limited. As summarized in the preceding sections, there have been considerable synthetic efforts directed toward utilizing base-metal catalysts to carry out the same chemical transformations as enabled by enzyme active sites. In general, however, these have met with limited success, often resulting in low activity and limited (or no) catalytic activity. We emphasize, however, that the goal should not be to mimic these enzyme active sites, but rather to obtain fundamental chemical insights into the mechanisms that can then be more broadly applied. We feel it is at this interface between biological and chemical catalysis that there remains a wealth of fundamental chemical insights, which have yet to be fully understood. In our opinion, a research approach that integrates theory and experiment is the best way to approach this challenge and to ultimately obtain insights into fundamental chemical transformations.

References

[1] Blankenship RE. Molecular mechanisms of photosynthesis. Oxford: Blackwell; 2001.

[2] Ballou DP, Palmer GA. Practical rapid quenching instrument for study of reaction-mechanisms by electron-paramagnetic resonance spectroscopy. Anal Chem. 1974;46: 1248–53.

[3] Bollinger JM, Krebs C. Stalking intermediates in oxygen activation by iron enzymes: Motivation and method. J Inorg Biochem. 2006;100:586–605.

[4] Bollinger JM, Tong WH, Ravi N, Huynh BH, Edmondson DE, Stubbe J. Redox-active amino acids in biology. 1995, 258, 278.

[5] Cherepanov AV, De Vries S. Microsecond freeze-hyperquenching: development of a new ultrafast micro-mixing and sampling technology and application to enzyme catalysis. Biochim Biophys Acta Bioenerg. 2004;1656:1–31.

[6] Krebs C, Bollinger JM Jr. Freeze-quench $^{(57)}$Fe-Mössbauer spectroscopy: trapping reactive intermediates. Photosynth Res. 2009;102:295–304.

[7] Krebs C, Price JC, Baldwin J, Saleh L, Green MT, Bollinger JM. Rapid freeze-quench ^{57}Fe Mössbauer spectroscopy: monitoring changes of an iron-containing active site during a biochemical reaction. Inorg Chem. 2005;44:742–57.

[8] Mitic N, Saleh L, Schenk G, Bollinger JM, Solomon EI. Rapid-freeze-quench magnetic circular dichroism of intermediate X in ribonucleotide reductase: new structural insight. J Am Chem Soc. 2003;125:11200–01.

[9] Koeningsberger DC, Prins R. X-ray absorption: principles, applications and techniques of EXAFS, SEXAFS and XANES. New York: John Wiley & Sons; 1988.

[10] Buhl M, Reimann C, Pantazis DA, Bredow T, Neese F. Geometries of third-row transition metal complexes from density-functional theory. J Chem Theory Comput. 2008;4: 1449–59.

[11] Ray K, Begum A, Weyhermüller T, et al. The electronic structure of the isoelectronic, square-planar complexes $[FeII(L)_2]^{2-}$ and $[CoIII(L\ Bu)_2]^{-}$ (L^{2-} and $(L\ Bu)^{2-}$=benzene-1, 2-dithiolates): an experimental and density functional theoretical study. J Am Chem Soc. 2005;127:4403–15.

[12] Schoneboom JC, Neese F, Thiel W. Toward identification of the compound I reactive intermediate in cytochrome P450 chemistry: a QM/MM study of its EPR and Mössbauer parameters. J Am Chem Soc. 2005;127:5840–43.
[13] Berry JF, Bill E, Bothe E, et al. An Octahedral Coordination Complex of Iron(VI). Science. 2006;312:1937–41.
[14] Curtiss LA, Raghavachari K, Redfern PC, Pople JA. Assessment of Gaussian-2 and density functional theories for the computation of enthalpies of formation. J Chem Phys. 1997;106: 1063–79.
[15] Neese F. A critical evaluation of DFT, including time-dependent DFT, applied to bioinorganic chemistry. J Biol Inorg Chem. 2006;11:702–11.
[16] Shaik S, Cohen S, Wang Y, Chen H, Kumar D, Thiel W. P 450 enzymes: their structure, reactivity, and selectivity-modeled by QM/MM calculations. Chem Rev. 2010;110:949–1017.
[17] Siegbahn PEM, Eriksson L, Himo F, Pavlov M. Hydrogen Atom Transfer in Ribonucleotide Reductase (RNR). J Phys Chem B. 1998;102:10622–29.
[18] Holm RH, Kennepohl P, Solomon EI. Structural and Functional Aspects of Metal Sites in Biology. Chem Rev. 1996;96:2239–314.
[19] Solomon EI, Lever ABP, editors. Inorganic electronic structure and spectroscopy. New York: John Wiley & Sons; 1999. Volumes 1–2.
[20] Jensen F. Introduction to computational chemistry. New York: Wiley; 2009.
[21] Neese F. Prediction of molecular properties and molecular spectroscopy with density functional theory: From fundamental theory to exchange-coupling. Coord Chem Rev. 2009; 253:526–63.
[22] Petrenko T, DeBeer George S, Aliaga-Alcalde N, et al. Characterization of a genuine iron (V)-nitrido species by nuclear resonant vibrational spectroscopy coupled to density functional calculations. J Am Chem Soc. 2007;129:11053–60.
[23] Petrenko T, Krylova O, Neese F, Sokolowski M. Optical Absorption and Emission Properties of Rubrene: Insight by a Combined Experimental and Theoretical Study. New J Phys. 2009;11: 11053–60.
[24] Bratus VY, Petrenko TT, Okulov SM, Petrenko TL. Positively charged carbon, vacancy in three inequivalent lattice sites of 6H-SiC: Combined EPR and density functional theory study. Phys Rev B. 2005;71:1–23.
[25] Handbook of chemistry and physics. 81st ed. New York: Taylor and Francis Group; 2000.
[26] Michl, J. Chem Rev Hydrogen Issue. 2007;107:3899–4435.
[27] Cammack R. Hydrogenase sophistication. Nature. 1999;397:214–15.
[28] Ogo S, Kabe R, Uehara K, et al. A dinuclear Ni(mu-H)Ru complex derived from H_2. Science. 2007;316, 585–87.
[29] Inoue H, Shimada T, Kou Y, et al. The water oxidation bottleneck in artificial photosynthesis: how can we get through it? An alternative route involving a two-electron process. ChemSusChem 2011;4:173–79.
[30] Umena Y, Kawakami K, Shen J.-R, Kamiya N. Crystal structure of oxygen-evolving photosystem II at a resolution of 1.9 Å. Nature. 2011;473:55–60.
[31] Ames W, Pantazis DA, Krewald V, et al. Theoretical evaluation of structural models of the S2 state in the oxygen evolving complex of Photosystem II: protonation states and magnetic interactions. J Am Chem Soc. 2011;133:19743–57.
[32] Kok B, Forbush B, McGloin M. Cooperation of charges in photosynthetic O_2 evolution-I. A linear four step mechanism. Photochem Photobiol. 1970;11:457–75.
[33] Kulik LV, Epel B, Lubitz W, Messinger J. ^{55}Mn pulse ENDOR at 34 GHz of the S0 and S2 states of the oxygen-evolving complex in photosystem II. J Am Chem Soc. 2005; 127:2392–93.

[34] Hammarström L, Styring S. Proton-coupled electron transfer of tyrosines in Photosystem II and model systems for artificial photosynthesis: the role of a redox-active link between catalyst and photosensitizer. Energy Environ Sci. 2011;4:2379–88.

[35] Dau H, Haumann M. Eight steps preceding O-O bond formation in oxygenic photosynthesis–A basic reaction cycle of the Photosystem II manganese complex. Biochim Biophys Acta Bioenerg. 2007;1767:472–83.

[36] Dau H, Haumann M. The manganese complex of photosystem II in its reaction cycle–Basic framework and possible realization at the atomic level. Coord Chem Rev. 2008; 252:273–95.

[37] McEvoy JP, Brudvig GW. Water-splitting chemistry of photosystem II. Chem Rev. 2006; 106: 4455–83.

[38] Messinger J, Renger G. In: Renger G, editor. Primary processes of photosynthesis, part 2: Principles and apparatus. Cambridge: The Royal Society of Chemistry; 2008. Volume 9. p. 291.

[39] Mukhopadhyay S, Mandal SK, Bhaduri S, Armstrong WH. Manganese clusters with relevance to photosystem II. Chem Rev. 2004;104:3981–4026.

[40] Kanady JS, Tsui EY, Day MW, Agapie T. A synthetic model of the Mn_3Ca subsite of the oxygen-evolving complex in photosystem II. Science. 2011;333:733–36.

[41] Siegbahn PEM. Structures and energetics for O2 formation in photosystem II. Acc Chem Res. 2009;42:1871–80.

[42] Romain S, Vigara L, Llobet A. Oxygen-oxygen bond formation pathways promoted by ruthenium complexes. Acc Chem Res. 2009;42:1944–53.

[43] Concepcion JJ, Jurss JW, Brennaman MK, et al. Making oxygen with ruthenium complexes. Acc Chem Res. 2009;42:1954–65.

[44] Sala X, Romero I, Rodríguez M, Escriche L, Llobet A. Molecular catalysts that oxidize water to dioxygen. Angew Chem Int Ed. 2009;48:2842–52.

[45] McDaniel ND, Coughlin FJ, Tinker LL, Bernhard S. Cyclometalated iridium(III) Aquo complexes: efficient and tunable catalysts for the homogeneous oxidation of water. J Am Chem Soc. 2008;130:210–17.

[46] Artero V, Chavarot-Kerlidou M, Fontecave M. Splitting Water with Cobalt. Angew Chem Int Ed. 2011;50:7238–66.

[47] Yin Q, Tan JM, Besson C, et al. A fast soluble carbon-free molecular water oxidation catalyst based on abundant metals. Science. 2010;328:342–45.

[48] Kanan MW, Nocera DG. In situ formation of an oxygen-evolving catalyst in neutral water containing phosphate and Co^{2+}. Science. 2008;321:1072–75.

[49] Lutterman DA, Surendranath Y, Nocera DG. A self-healing oxygen-evolving catalyst. J Am Chem Soc 2009;131:3838–39.

[50] Costas M, Mehn MP, Jensen MP, Que L. Dioxygen activation at mononuclear nonheme iron active sites: enzymes, models, and intermediates. Chem Rev. 2004;104:939–86.

[51] Fox BG, Lyle KS, Rogge CE. Reactions of the diiron enzyme stearoyl-acyl carrier protein desaturase. Acc Chem Res. 2004;37:421–29.

[52] Friedle S, Reisner E, Lippard SJ. Current challenges of modeling diiron enzyme active sites for dioxygen activation by biomimetic synthetic complexes. Chem Soc Rev. 2010; 39:2768–79.

[53] Kovaleva EG, Lipscomb JD. Versatility of biological non-heme Fe(II) centers in oxygen activation reactions. Nat Chem Biol. 2008;4:186–93.

[54] Krebs C, Galonica Fujimori D, Walsh CT, Bollinger JM. Non-heme Fe(IV)-oxo intermediates. Acc Chem Res. 2007;40:484–92.

[55] Siegbahn PEM, Borowski T. Modeling enzymatic reactions involving transition metals. Acc Chem Res. 2006;39:729–38.

[56] Tinberg CE, Lippard SJ. Dioxygen activation in soluble methane monooxygenase. Acc Chem Res. 2011;44:280–88.

[57] Stone KL, Borovik AS. Lessons from nature: unraveling biological CH bond activation. Curr Opin Chem Biol. 2009;13:114–18.

[58] Groves JT. High-valent iron in chemical and biological oxidations. J Inorg Biochem. 2006;100: 434–47.

[59] Nam W. High-valent iron(IV)-oxo complexes of heme and non-heme ligands in oxygenation reactions. Acc Chem Res. 2007;40:522–31.

[60] Bassan A, Blomberg MRA, Borowski T, Siegbahn PEM. Theoretical studies of enzyme mechanisms involving high-valent iron intermediates. J Inorg Biochem. 2006;100:727–43.

[61] Bakac A. Oxygen activation with transition-metal complexes in aqueous solution. Inorg Chem. 2010;49:3584–93.

[62] Jung C. The mystery of cytochrome P450 Compound I: a mini-review dedicated to Klaus Ruckpaul. Biochim Biophys Acta Proteins Proteomics. 2011;1814:46–57.

[63] Ortiz De Montellano PR. Hydrocarbon hydroxylation by cytochrome P450 enzymes. Chem Rev. 2009;110:932–48.

[64] Shaik S, Cohen S, Wang Y, Chen H, Kumar D, Thiel W. P 450 enzymes: their structure, reactivity, and selectivity-modeled by QM/MM calculations. Chem Rev. 2009;110:949–1017.

[65] Hamelinck CN, Faaij APC, Den Uil H, Boerrigter H. Production of FT transportation fuels from biomass; technical options, process analysis and optimisation, and development potential. Energy. 2004;29:1743–71.

[66] Hermans I, Spier ES, Neuenschwander U, Turra N, Baiker A. Selective Oxidation Catalysis: Opportunities and Challenges. Top Catal. 2009;52:1162–74.

[67] Arakawa H, Aresta M, Armor JN, et al. Catalysis research of relevance to carbon management: progress, challenges, and opportunities. Chem Rev. 2001;101:953–96.

[68] Que L Jr, Dong YH. Modeling the oxygen activation chemistry of methane monooxygenase and ribonucleotide reductase. Acc Chem Res. 1996;29:190–96.

[69] Shu LJ, Nesheim JC, Kauffmann K, Münck E, Lipscomb JD, Que L. An Fe2IVO2 diamond core structure for the key intermediate Q of methane monooxygenase. Science. 1997;275:515–18.

[70] Lieberman RL, Rosenzweig AC. Crystal structure of a membrane-bound metalloenzyme that catalyses the biological oxidation of methane. Nature. 2005;434:177–82.

[71] Chan SI, Yu SSF. Controlled oxidation of hydrocarbons by the membrane-bound methane monooxygenase: the case for a tricopper cluster. Acc Chem Res. 2008;41:969–79.

[72] Balasubramanian R, Smith SM, Rawat S, Yatsunyk LA, Stemmler TL, Rosenzweig AC. Oxidation of methane by a biological dicopper centre. Nature. 2010;465:115–19.

[73] Krebs C, Fujimori DG, Walsh CT, Bollinger JM Jr. Non-heme Fe(IV)-oxo intermediates. Acc Chem Res. 2007;40:484–92.

[74] Xue G, Wang D, De Hont R, et al. A synthetic precedent for the [FeIV$_2$(mu-O)$_2$] diamond core proposed for methane monooxygenase intermediate Q. Proc Natl Acad Sci U S A. 2007;104: 20713–18.

[75] Wang D, Farquhar ER, Stubna A, Münck E, Que L Jr. A diiron(IV) complex that cleaves strong C-H and O-H bonds. Nat Chem. 2009;1:145–50.

[76] Xue G, De Hont R, Münck E, Que L Jr. Million-fold activation of the [Fe(2)(micro-O)(2)] diamond core for C-H bond cleavage. Nat Chem. 2010;2:400–05.

[77] Que L Jr. The road to non-heme oxoferryls and beyond. Acc Chem Res. 2007;40:493–500.

[78] Louwerse MJ, Baerends EJ. Oxidative properties of FeO^{2+}: electronic structure and solvation effects. Phys Chem Chem Phys. 2007;9:156–66.

[79] Shaik S, Hirao H, Kumar D. Reactivity of high-valent iron-oxo species in enzymes and synthetic reagents: a tale of many states. Acc Chem Res. 2007;40:532–42.

[80] Ye S, Neese F. Quantum chemical studies of C-H activation reactions by high-valent nonheme iron centers. Curr Opin Chem Biol. 2009;13:89–98.

[81] Solomon EI, Wong SD, Liu LV, Decker A, Chow MS. Peroxo and oxo intermediates in mononuclear nonheme iron enzymes and related active sites. Curr Opin Chem Biol. 2009;13: 99–113.

[82] Ye S, Neese F. Nonheme oxo-iron(IV) intermediates form an oxyl radical upon approaching the C-H bond activation transition state. Proc Natl Acad Sci U S A. 2011; 108:1228–33.

[83] Smith BE. Nitrogenase reveals its inner secrets. Science. 2002;297:1654–55.

[84] Fryzuk MD. Inorganic chemistry: ammonia transformed. Nature. 2004;427:498–99.

[85] Burgess BK, Lowe DL. Mechanism of Molybdenum Nitrogenase. Chem Rev. 1996;96: 2983–3012.

[86] Howard JB, Rees DC. Structural Basis of Biological Nitrogen Fixation. Chem Rev. 1996;96: 2965–82.

[87] Hu YL, Lee CC, Ribbe MW. Extending the carbon chain: hydrocarbon formation catalyzed by vanadium/molybdenum nitrogenases. Science. 2011;333:753–55.

[88] Lee CC, Hu YL, Ribbe MW. Vanadium nitrogenase reduces CO. Science. 2010;329:642.

[89] Einsle O, Tezcan FA, Andrade SLA, et al. Nitrogenase MoFe-protein at 1.16 A resolution: a central ligand in the FeMo-cofactor. Science. 2002;297:1696–700.

[90] George SJ, Igarashi RY, Xiao Y, et al. Extended X-ray absorption fine structure and nuclear resonance vibrational spectroscopy reveal that NifB-co, a FeMo-co precursor, comprises a 6Fe core with an interstitial light atom. J Am Chem Soc. 2008;130:5673–80.

[91] Xiao YM, Fisher K, Smith MC, et al. How nitrogenase shakes–initial information about P-cluster and FeMo-cofactor normal modes from nuclear resonance vibrational spectroscopy (NRVS). J Am Chem Soc. 2006;128:7608–12.

[92] Hoffman BM, Dean DR, Seefeldt LC. Climbing nitrogenase: toward a mechanism of enzymatic nitrogen fixation. Acc Chem Res. 2009;42:609–19.

[93] Lukoyanov D, Pelmenschikov V, Maeser N, et al. Testing if the interstitial atom, X, of the nitrogenase molybdenum-iron cofactor is N or C: ENDOR, ESEEM, and DFT studies of the S = 3/2 resting state in multiple environments. Inorg Chem. 2007;46:11437–49.

[94] Harris TV, Szilagyi RK. Comparative Assessment of the Composition and Charge State of Nitrogenase FeMo-Cofactor. Inorg Chem. 2011;50:4811–24.

[95] Lancaster KM, Roemelt M, Ettenhuber P, et al. X-ray emission spectroscopy evidences a central carbon in the nitrogenase iron-molybdenum cofactor. Science. 2011;334:974–77.

[96] Spatzal T, Aksoyoglu M, Zhang LM, et al. Evidence for interstitial carbon in nitrogenase FeMo cofactor. Science. 2011;334:940.

[97] Lukoyanov D, Dikanov SA, Yang ZY, et al. ENDOR/HYSCORE studies of the common intermediate trapped during nitrogenase reduction of N_2H_2, CH_3N_2H, and N_2H_4 support an alternating reaction pathway for N2 reduction. J Am Chem Soc. 2011:133: 11655–64.

[98] Yandulov DV, Schrock RR. Catalytic reduction of dinitrogen to ammonia at a single molybdenum center. Science. 2003;301:76–78.

[99] Pool JA, Lobkovsky E, Chirik PJ. Hydrogenation and cleavage of dinitrogen to ammonia with a zirconium complex. Nature. 2004;427:527–30.

[100] Hazari N. Homogeneous iron complexes for the conversion of dinitrogen into ammonia and hydrazine. Chem Soc Rev. 2010;39:4044–56.

[101] Rodriguez MM, Bill E, Brennessel WW, Holland PL. N(2) reduction and hydrogenation to ammonia by a molecular iron-potassium complex. Science. 2011;334:780–83.

4 Biobased

Regina Palkovits, William R.H. Wright

17 Biofuels Derived from Renewable Feedstocks

17.1 Introduction

Today's economy is dominated by fossil fuels including oil, natural gas, and coal as feedstocks for production of heat, energy, and chemicals. Various studies discussing the potential availability of remaining fossil resources present rather different results. Nevertheless, current studies estimate natural gas and coal can be utilized for another 100 years or even longer, while depletion of crude oil is expected to occur within 40 years [1]. Consequently, solutions for the sustainable production of energy in the future need to be established, and promising technologies that include water and wind power, solar thermal energy, and photovoltaic systems already exist. However, alternative carbon sources for the future production of chemicals and liquid fuels are indispensable. Although electromobility via high-performance batteries or fuel-cell systems is suitable for application in short-distance individual transportation, future air traffic and heavy duty vehicles will inevitably rely on liquid fuels, which possess the high energy density required to guarantee sufficient cruising range and performance.

Herein, an account of recent developments in the field of biofuel production is given, and technological challenges with regard to conventional biofuels such as biodiesel and bioethanol are discussed. Advances in the fields of syngas and Fischer-Tropsch technology are only illustrated briefly. Instead, major emphasis is given to novel approaches aimed at the controlled (chemo-)catalytic transformation of lignocellulose into potential alternative biofuel compounds as energy storage media for mobile applications.

Acknowledgment: We acknowledge financial support of the Robert Bosch Foundation in the frame of the Robert Bosch Junior Professorship for sustainable utilization of natural renewable resources.

Regina Palkovits, RWTH Aachen University, Institut für Technische und, Makromolekulare Chemie (ITMC), Worringerweg 1, 52074 Aachen, Germany, e-mail: palkovits@itmc.rwth-aachen.de
William R. H. Wright, RWTH Aachen University, Institut für Technische und, Makromolekulare Chemie (ITMC), Worringerweg 1, 52074 Aachen, Germany, e-mail: right@itmc.rwth-aachen.de

https://doi.org/10.1515/9783110608458-017

17.2 Sources of Biomass

Plant material presents a potential alternative carbon source for the production of liquid fuels and as feedstock for the whole chemical industry. Transformed into hydrocarbons, biofuels can achieve high energy storage densities. In terms of environmental impacts, biomass could allow a closed carbon cycle, thereby avoiding further anthropogenic increase of the atmospheric CO_2 concentration. Nevertheless, careful assessment of the available resources and their efficient utilization is highly essential for the facilitation of sustainable production and reduced CO_2 emissions. Comprehensive balances need to account for all inputs concerning cultivation, harvesting, transportation, and processing of plant material as well as release of greenhouse gases such as nitrous oxides due to application of fertilizers [2–5]. Additionally, prior changes in land use (e.g., forest clearance or monocropping with the related consequences for the respective ecosystem) have to be considered. Furthermore, current analysis even tries to account for socio-economic and ethical impacts [6, 7].

Initially, potential sources of biomass have to be evaluated, focusing on the availability of renewable feedstocks. However, the identification of suitable biomass sources represents a considerable challenge. Today's production of biofuels and biomass-derived chemicals predominantly relies on vegetable oils, animal fats, terpenes, and sugar or starch crops. Consequently, competition between food and fuel concerning available cropping areas may arise. This difficulty could be worsened by an increasing usage of such renewable feedstocks in energy and fuel production.

The most prominent examples of biomass utilization are arguably biodiesel and bioethanol. Biodiesel is produced via transesterification of oils and fats with methanol delivering fatty acid methyl esters together with crude glycerol as the by-product. In recent years, environmental considerations and political targets have caused a tremendous increase in biodiesel production accompanied by an inevitable overproduction of crude glycerol. This has caused a significant decrease in the value of this feedstock. With the aim of utilizing surplice glycerol, various recent investigations examine novel transformations of crude glycerol to value-added products [8, 9]. Additionally, alternative biodiesel processes have been presented that aim to produce glycerol in higher purity. New approaches also seek to treat glycerol with tert-butanol as the resulting etherification reactions generate additional blending fuels. Together, such strategies could avoid glycerol overproduction and enhance the overall fuel yields associated with biodiesel production [10, 11]. Novel strategies toward transesterification with bioethanol have been established, which aim at the production of fully biomass-derived biodiesel. Nevertheless, several technological challenges remain as well as the choice of appropriate sources of oils and fats, which will become increasingly essential as biodiesel demand further increases. The sources of feedstocks used for biodiesel production must receive special consideration due to the rather low fuel yield per hectare cultivation area compared to other biofuel candidates (see Table 17.1).

Table 17.1: Cruising range with biofuels derived from 1 ha cropping area. (Table adapted from Fachagentur Nachwachsende Rohstoffe e.V.).

Biofuel	Natural resource	Cruising range/km ha^{-1}
Biogas	Maize silage, etc.	67,600
BTL (biomass to liquid)	Straw	64,000
Rapeseed oil	Rapeseed	23,300 (17,600*)
Biodiesel	Rapeseed	23,300 (17,600*)
Bioethanol	Corn	22,400 (14,400*)

Car mileage: Otto, 7.4 l/100 km; diesel, 6.1 l/100 km.
*Biogas production from crop remaining.

With regard to bioethanol, conventional processes are based on enzymatic hydrolysis of starch, which yields glucose that can be converted to ethanol via fermentation. Significantly, ethanol is an inhibitor for the producing microorganisms, allowing only limited product concentrations. This results in diluted aqueous ethanol solutions. The low product concentration, together with azeotrope formation, means that product separation requires a significant energy input. With the aim of optimizing bioethanol production, numerous scientific investigations have been conducted focusing on alternative separation technologies, bioengineering to accelerate product formation, and increasing the accessible ethanol concentrations to reduce the overall energy demand. In terms of fuel properties, even dry ethanol has only roughly two-thirds of the energy content per unit volume found for gasoline. Nevertheless, oxygenates such as ethanol increase knock resistance and reduce exhaust emissions, making them suitable for integration into today's gasoline technology. Ethanol can be blended into gasoline in amounts up to 15% without the need for any engine modification. Alternatively, 85% ethanol blends can be used in modern flex cars that are now commercially available.

Starting from the 1980s, Brazil has pursued a national strategy for ethanolic fuels called the "Proàlcool program" to abate the country's strong dependence on fossil-oil imports. However, in the 1990s, increasing world market prices for sugar nearly disrupted Brazil's ethanol production, although in recent years a significant upturn in bioethanol production has again occurred. Until 2005, Brazil was the biggest producer and consumer of bioethanol derived from sugarcane. Combustion of residual plant material for energy and heat production results in a net positive energy and CO_2 balance for such bioethanol. Interestingly, in recent years, the United States has taken over as the largest bioethanol producer reaching production capacities around 26 billion liters (2009), mainly derived from corn and crops. However, this increase in production has been accompanied by governmental funding and has furthered competition between food and fuel production. This precipitated the

"Tortilla Crisis" in Mexico during 2007, when prices for corn-based ground staple food increased by up to 400%. Consequently, a careful choice of suitable feedstocks is essential with regard to biofuel yields, and competition between food and fuel production must be avoided.

The potential yields per hectare cropping area for different biomass-derived biofuels are illustrated in Table 17.1. Biogas, which is not further discussed in this overview, is produced via anaerobic digestion or fermentation of energy crops such as maize, silage, or biodegradable wastes, including sewage sludge, and food wastes. The composition of biogas may vary depending on feedstock and microorganism; nevertheless, it is mainly composed of methane (50–75%), followed by carbon dioxide (25–50%) and traces of other compounds such as N_2, O_2, H_2S, H_2, or NH_3.

The data presented in Table 17.1 clearly illustrate the advantage of transformation technologies utilizing the whole plant material and not only the oil or sugar and starch fraction.

An alternative to the utilization of food crops in biofuel production is lignocellulose, which can for instance be utilized in biomass-to-liquid (BTL) processes. That way, a direct competition between food and fuel production can be avoided. This feedstock is much more abundant than vegetable oils or sugar and starch crops. Moreover, lignocellulose is not edible and could theoretically be utilized without any impact on food production. The cellulose and hemicellulose fraction of lignocellulose may serve for the production of cellulosic ethanol, which could be produced via acid or enzymatic catalyzed hydrolysis of cellulose, followed by further fermentation to yield ethanol. Alternatively, the whole plant can be gasified to yield syngas, followed by methanol or dimethyl ether synthesis or Fischer-Tropsch technology that produces hydrocarbon fuels. Furthermore, controlled (bio-)chemical transformations to novel fuel compounds based on cellulose, hemicellulose, or lignin are possible, and numerous recent publications emphasize intense research in this direction.

17.3 Lignocellulose as Feedstock

Lignocellulose, which comprises the main construction material of plant biomass, accounts for up to 90% of all biomass and is formed in amounts of approximately 1.5 trillion tons per year [12]. Consequently, lignocellulose is much more abundant than available amounts of vegetable oils, starch, and sugar crops. In addition to the high abundance of lignocellulose, it is inedible, and its utilization as feedstock for production of biofuels and chemicals could drastically reduce challenges of food versus fuel production.

Focusing on the status of Germany with regard to renewable feedstocks, recent estimates discuss a possible additional annual production of dry matter of approximately 39.8–68.5 million tons, corresponding to an energy content of approximately

712–1,219 PJ year^{-1}. This material is mainly composed of lignocellulose. Consequently, efficient technologies for the conversion of lignocellulose will gain increasing importance and will be utilized in parallel with biodiesel and bioethanol. For example, based on BTL, which will be discussed in the following section, the available dry matter could yield ca. 15 million tons per year of fuel (42% efficiency factor), which would meet approximately 22% of the current fuel demand of Germany. Thus, biofuel production could contribute to a significant fraction of future mobility energy demands. Nevertheless, a 100% substitute of today's fossil fuel-based production is certainly not possible. Therefore, future mobility will require mixed solutions (e.g., combining electromobility for short distance traveling and conventional combustion technology for air traffic and long-range vehicles).

Lignocellulose itself is composed of cellulose (50–70%), a polymer of glucose units connected via β-1,4-glucosidic linkages; hemicellulose (10–40%), which is a polymer of various sugars mainly comprising xylose as well as arabinose; and lignin (10–30%), a polymer consisting of aromatic subunits (Figure 17.1).

Cellulose is the major component of lignocellulose and is the most abundant natural biopolymer. Thus, cellulose could serve as a future source of glucose, which could be fermented to bioethanol or converted to other potential platform chemicals. However, the hydrolysis of cellulose to release glucose is hindered by the high degree of crystallinity caused by hydrogen bonding between cellulose chains. Within wood cellulose, hemicellulose and lignin are closely packed and protected against chemical breakdown. This further complicates the controlled fractionation and depolymerization

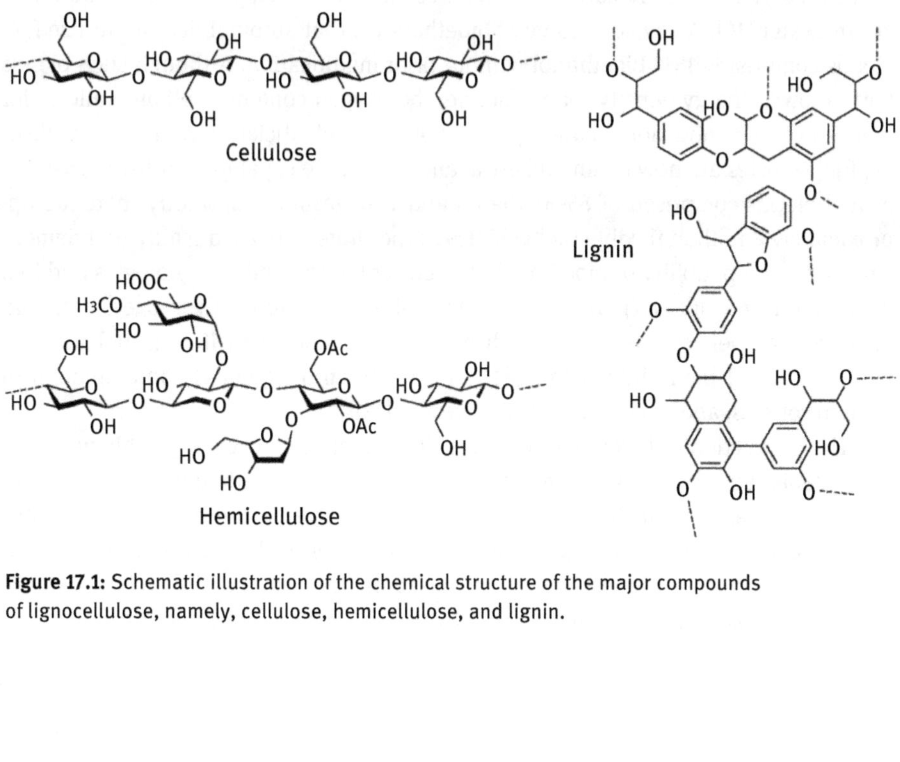

Figure 17.1: Schematic illustration of the chemical structure of the major compounds of lignocellulose, namely, cellulose, hemicellulose, and lignin.

of these biopolymers. Additionally, neither cellulose nor lignin is soluble in any conventional solvent system. Thus, enzymatic processes still suffer from rather low space-time-yields. There are also disadvantages associated with chemical hydrolysis, which like traditional wood saccharification requires harsh conditions utilizing concentrated mineral acids that have high corrosion potentials and the capacity to form undesirable salts.

Consequently, the controlled (bio-)chemical breakdown of lignocellulose, and particularly cellulose, requires the development of tailor-made catalysts. Furthermore, new solvent systems will be required to facilitate future refinery technologies based on lignocellulosic feedstocks [13].

17.4 Bioethanol as Sustainable Biofuel

Due to the inevitable depletion of petroleum supplies, governments around the world are seeking to develop new transportation fuels based on sustainable and renewable biomass. One fuel additive that is currently obtained from biomass is bioethanol, which is produced on an industrial scale in both Brazil and the United States from sugarcane and maize, respectively [14]. As a fuel additive, bioethanol has a number of advantages, such as a high octane number and a capacity to burn cleanly, with lower carbon monoxide emissions [15]. Furthermore, bioethanol is nontoxic and can be used instead of other oxygenated fuel additives such as methyl tertiary butyl ether, a hazardous substance that has been found to contaminate groundwater [16]. A key advantage of bioethanol is that automobiles can utilize fuel blends comprising 15% bioethanol without prior modification. Although such blends have a lower energy density than gasoline, the oxygen content of ethanol allows for more efficient combustion, enabling similar overall fuel efficiency. Additionally, flexible fuel vehicles are now commercially available and are capable of utilizing gasoline or fuel blends comprised of 85% ethanol and 15% gasoline. Currently there are approximately 8 million flexible fuel vehicles in the United States, a reality that demonstrates the compatibility of bioethanol with current automobile technologies and fuel distribution networks [17]. While bioethanol offers numerous advantages, it should be noted that there are also potential disadvantages associated with this fuel additive, with bioethanol being linked to a decrease in urban air quality, contamination of water resources, and a reduction in biodiversity [18].

The production of bioethanol requires the sequential processing of biomass into fermentable reducing sugars, fermentation of these sugars to ethanol, and then a separation of ethanol from the fermentation mixtures. Overall the net energy balance (i.e., the energy required during processing vs. the energy obtained from bioethanol) has been found to be highest in systems based on sugarcane and starch feedstocks [19]. This fact is arguably demonstrated by the success of bioethanol production in

Brazil from sugarcane. The capacity of bioethanol to partially replace gasoline was recognized by the Brazilian government in the 1970s, and a political decision was made to stimulate bioethanol production [20]. Although bioethanol production was initially subsidized, during the 1990s these subsidies were removed; in 2004, Brazilian bioethanol became commercially competitive with gasoline on international markets. This achievement can be attributed to both the rising cost of crude oil and efficiencies made in bioethanol production, which in 2007 reached 16 billion liters, requiring 5.6 million hectares of land. Overall, this only required 1% of the land available for agriculture in Brazil, meaning there is significant scope for expansion. This could prove to be desirable as analysis indicates Brazilian bioethanol has an energy yield ratio of 7.9 (bioenergy yield to fossil energy input), and each hectare of land farmed annually could mitigate CO_2 production by approximately 25,000 kg per annum [21]. In contrast to bioethanol obtained from sugarcane, bioethanol produced from corn in the United States offers lower overall net benefits [22]. This is problematic as in 2006, U.S. production of bioethanol reached 1.2 billion liters, with virtually all transportation ethanol being obtained from corn. With an associated energy yield ratio of just 1.3, it is clear that the energy obtained from corn-derived bioethanol is minimal. Furthermore, some high-profile analysis suggests that bioethanol production in the United States is leading to a substantial net increase in greenhouse gas emissions [23]. Alternative analysis indicates that the greenhouse gas emissions associated with maize-derived bioethanol can vary significantly based on plant design [24]. For example, processing based on using coal as a heat source results in a 3% net increase in greenhouse gas emissions. In contrast, utilization of woodchips as a fuel in processing results in a 52% decrease in net greenhouse gas emissions. Evidently, more research is needed to reconsider the different methodologies used to produce ethanol from corn, with the aim of developing more sustainable processes.

While production of bioethanol from tropical sugar-rich crops has been demonstrated to be efficient, the expansion of such industries could eventually impact the food supply. With this in mind, a substantial research effort has been focused on obtaining "second-generation" bioethanol from lignocellulosic materials [25–27]. This is particularly attractive as lignocellulosic materials are cheap, highly abundant, and infinitely renewable and can be exploited without compromising food production. Additionally, analysis of life-cycle greenhouse gas emissions indicates that bioethanol obtained from lignocellulosic materials could enable a 50–80% reduction in the emissions associated with gasoline. These figures compare favorably with bioethanol obtained from corn, which can currently only facilitate savings of 25–40% [28]. One disadvantage associated with deriving bioethanol from lignocellulosic materials is the need to utilize additional processing in order to break down plant fibers into fermentable reducing sugars. A number of different methodologies have been applied to the problem, including milling, pyrolysis in the presence of acids, steam explosion, ammonia fiber explosion, ozonolysis, acid/base hydrolysis, and oxidative delignification. Alternatively, organisms such as fungi can be used to convert waste materials into reducible sugars, and a wide variety of enzymes can be used to convert cellulosic materials into glucose. Although a wide range of techniques has been applied to the breaking down of lignocellulose, the financial and energy costs associated with this step are currently inhibiting the commercial production of ethanol from such materials. A second problem is the difficulties associated with fermenting the mixtures of pentose and hexose sugars obtained from crude lignocellulosic materials. In order for the overall system to have high efficiency, both types of sugars must be converted to bioethanol. However, this is problematic as pentose-utilizing microorganisms ferment pentose and hexose at a slower rate than microorganisms that only assimilate hexose. Furthermore, the microorganisms that ferment pentose tend to be more sensitive to ethanol, the result being that efficient pentose fermentation cannot be easily achieved in a single vessel alongside hexose fermentation. This difficulty can be overcome by using separate fermentation steps, although this is costly in energy and processing time. Overall, further research is needed to develop combinations of microorganisms that are capable of cofermenting both pentose and hexose sugars.

While a number of problems need to be overcome before second-generation bioethanol can be efficiently produced from lignocellulose materials, it should be noted that historically biobutanol has been produced from such routes on an industrial scale [29]. It is well established that butanol can be utilized as a solvent; however, this alcohol may also find application as a fuel. Compared to ethanol, butanol has a number of key advantages [30]. It can be mixed in any proportion with gasoline, in contrast to ethanol, which can only be blended up to 85%. Unlike ethanol, it can be used as a sole fuel, without modification of an internal combustion engine. It has a lower vapor pressure and is thus easier to handle, and it is immiscible with water. This means that less care is required to minimize water contamination during production and handling. Butanol is also less corrosive than ethanol and has an energy density similar to that of gasoline (ethanol, 21.2 MJ L^{-1} vs. butanol, 29.2 MJ L^{-1} vs. gasoline, 32.5 MJ L^{-1}). Although butanol is currently produced via petrochemicals in the oxo process, at the start of the twentieth century, butanol and acetone were obtained from the fermentation of starch [31, 32]. However, during the 1960s, such approaches became unfavorable due to the development of more competitive processes based on cheap petrochemicals. However, with the rising cost of petrochemicals, such technologies are being reconsidered. For instance, China has already restarted butanol production based on fermentation processes [33]. Evidently, the production of butanol from biomass could help meet the worldwide demand for sustainable fuels, a possibility that requires further consideration.

17.5 Biodiesel as Potential Biofuel

The utilization of biodiesel is not a new concept, as almost a century ago Rudolph Diesel invented engines that ran on peanut oil. However, in the second half of the twentieth century, humanity began the large-scale exploitation of reserves, enabling the production of cheap petroleum transport fuels that were more economic than their biodiesel counterparts. Recently, the rising cost of diminishing petroleum resources is beginning to reverse this situation, and biodiesel is again being utilized as transportation fuel. This is particularly convenient as biodiesel can be used without major modification of current diesel engines or delivery networks. Moreover, biodiesel has numerous advantages compared to petroleum-derived diesel. For instance, the combustion of biodiesel in prototype vehicles results in 94% less carcinogenic particle matter than that resulting from conventional diesel. Furthermore, biodiesel has substantially lower sulfur content than petroleum fuels, which means that utilization of biodiesel would render desulfurization strategies of fossil fuels redundant. Biodiesel also has relatively high lubricity compared to conventional fuels, which can extend engine lifetimes and enhance fuel economy. Additionally, biodiesel has a higher flash point (150 °C) than conventional fuels and is thus more easy to handle and less volatile. Finally, biodiesel degrades more rapidly than diesel fuel, meaning that any adverse spills will have lower environmental impact. Due to these advantages, biodiesel is presently the most widely utilized biofuel in Europe, and the consumption of biodiesel in the United States exceeded 300 million gallons in 2008.

First-generation biodiesel is currently produced via the modification of vegetable oils, which cannot themselves be utilized as transportation fuels due to their high viscosity. Instead, such oils are modified via treatment with methanol or ethanol, inducing transesterification, a chemical transformation that yields the fatty acid alkyl esters that make up biodiesel and the by-product glycerol. The transesterification reaction requires an acid or base catalyst, and the extent of transesterification can be influenced by variation of the catalyst type, reaction temperature, and the molar ratio of alcohol to vegetable oil, as well as the concentrations of water and free fatty acids (FFAs). When alkaline conditions are employed, care must be taken to ensure that FFAs do not react with the catalyst to form soaps, which can produce emulsions causing separation problems during processing. Although alkaline catalysts are more sensitive to the problem of saponification, homogeneous alkaline catalysts are routinely utilized instead of acidic catalysts, as basic conditions facilitate significantly faster transesterification rates. Current industrial processes rely on the utilization of catalysts such as sodium hydroxide or potassium hydroxide as these salts are cheap and readily available. Significantly, the type of vegetable oil employed as a feedstock can have a considerable impact on the physical properties of the resulting fuel. This is a consequence of different triglycerides and fatty acids with different chain lengths being present in the vegetable oils of alternative plants. For example, biodiesel fuels obtained from palm oil and soybeans have very different viscosities, cloud points, and combustion properties, making these fuels suitable for different applications. Furthermore, the vegetable oil feedstock has a significant impact on the economics of biodiesel production; as for first-generation biodiesel, the price of the feedstock is estimated as contributing up to 70–90% of the total operating costs. In 2007, it was reported that biodiesel could be produced from European oilseed rape or Malaysian palm oil at a cost of US$0.80 per liter gasoline equivalent, an amount that compares favorably with the US$0.80 per gallon production costs associated with petroleum diesel in 2008. One factor that reduces the economic attractiveness of biodiesel production is the low value of the by-product glycerol, which is typically obtained in low purities. Despite these difficulties, the sale of glycerol has been shown to result in a marginal decrease in biofuel production costs. Alternatively, glycerol could form the basis of a range of value-added chemicals including fuel additives. The development of such products could further enhance the economic viability of biofuel production, by adding value to an otherwise low-value commodity by-product and increasing the atom efficiency of the overall process.

Although biodiesel production is currently being achieved by the utilization of homogeneous base and acid catalysts, such approaches have a number of key disadvantages. For instance, systems based on homogeneous catalysts often require an excess of alcohol and are more suited to less economic batch processing and additional separation steps. Secondly, solubilized homogeneous catalysts often migrate into the glycerol by-product, which makes catalyst recovery and reuse problematic. Solubilized homogeneous base catalysts are also particularly susceptible to saponification. In contrast, solid catalysts are more easily separated from reaction mixtures and thus offer higher reusability and suitability for continuous processing. Solid catalysts also tend to have higher selectivity for biodiesel production and are more tolerant to the water and FFAs that are invariably present in the biodiesel streams. Lastly, heterogeneous catalysts facilitate the production of glycerol in higher purities than homogeneous systems (99% vs. 85%), enabling the production of a more utilizable by-product [34]. Although these advantages make heterogeneous catalysis highly attractive, it should be noted that in general solid catalysts tend to be less active, with systems based on heterogeneous catalysis often requiring the use of higher temperatures and pressures to achieve high conversions. One factor that can inhibit heterogeneous catalysis is the low extent of catalyst surface area available for the initiation of transesterification. For an effective solid catalyst to proceed, the catalyst surface cannot become saturated with polar molecules such as water or glycerol. With this in mind, the most effective catalysts tend to have hydrophobic surfaces that preferentially absorb triglyceride substrates. Overall, the solid catalyst must consist of a robust material that can resist decomposition. Examples of materials currently being considered for biodiesel production include basic zeolites, hydrotalcites, metal oxides, insoluble basic salts, immobilized organic bases, supported basic oxides, and alkali earth oxides.

One approach that could enable the production of cheaper biodiesel is to utilize low-value vegetable oils that contain a high content of FFA. This is particularly attractive as such materials can be obtained from either waste or the cultivation of alternative crops, something that would increase the biodiversity from which biofuels are obtained. Significantly, vegetable oils that contain mixtures of FFAs and triglycerides are incompatible with the base catalysts required for triglyceride transesterification, due to the propensity of FFAs to undergo saponification. Instead, such oils would first require treatment with alcohols in the presence of an acidic catalyst, with the aim of converting the FFAs to the corresponding esters. Once the FFA content of the oil has been removed, the triglyceride portion can then undergo base-catalyzed transesterification without the undesirable formation of soaps. Such a two-step process requires the combination of acid and base catalysts, which is most easily achieved when the initial esterification step is catalyzed by a solid acid. Furthermore, utilization of solid catalysts for both steps has been proposed to facilitate more efficient continuous processes that do not suffer from catalyst neutralization or the production of significant wastewater streams. Indeed, it has been suggested that such a continuous process could enable a 40–50% reduction in capital costs and a 30–60% reduction in energy compared to batch processes. With this in mind, the identification of suitable esterification solid acid catalysts is increasingly being viewed as an integral part of biodiesel development.

Esterification and transesterification are by no means the only methodologies by which fuels can be obtained from biologically derived oils. An alternative route by which the glycerol fragment of triglycerides can be removed is by thermocatalytic cracking, a process that can render diesel-like oils. This has the advantage of producing more stable nonester fuels that have higher energy contents, and desirable combustion properties, as well as physical properties that make them compatible for use as transportation fuels. Examples of alternative fuels that can be produced by this approach include nonesterified renewable diesel and hydrotreated renewable jet. The possibility of producing jet fuels from biomass is of interest to the U.S. military, which has explored cracking oils derived from biomass in the presence of hydrogen. Such processing enables deoxygenation and decarboxylation of triglyceride, producing less polar compounds more suited for use as jet fuels. In 2009, the U.S. military announced that this technology would be utilized in the EniEconfining™ process, which will supply 2.23 million liters of jet fuel to the U.S. military. Such processes cannot be viewed as being fully sustainable as they currently are reliant on hydrogen that is obtained from nonrenewable resources. However, the EniEconfining™ process clearly demonstrates that a range of methodologies can be employed to convert oils derived from biomass into utilizable fuels.

17.6 Production of Biofuel via Chemical Transformations of Lignocellulose

Recent attempts aim at the controlled transformation of cellulose, hemicellulose, and lignin to platform molecules for a potential future biorefinery scenario. In this regard, the U.S. Department of Energy has published studies on potential future platform molecules that could be derived from renewable resources [35, 36]. Tailored transformation of biomass to these platform chemicals could serve as a starting point for biofuel production. This would allow the development of comprehensive biorefinery approaches that incorporate both the production of biofuels and chemicals. The generation of such value-added commodities could optimize the overall process economics. Figure 17.2 illustrates promising platform chemicals derived from cellulose and hemicellulose that have been transformed into potential fuel compounds.

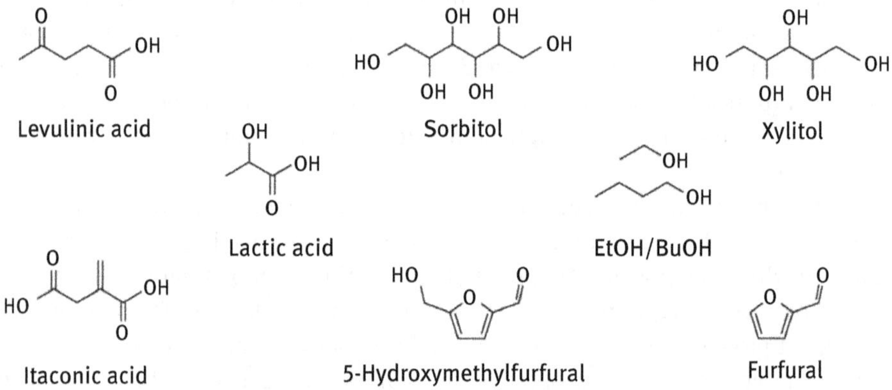

Figure 17.2: Selected biomass-derived platform chemicals that have been demonstrated to be feedstocks for the production of potential biofuels (EtOH, ethanol; BuOH, butanol).

A detailed account of reports concerning the synthesis of such platform chemicals can be found in these reviews [37–39]. Significantly, boundary conditions relating to the potential production of platform chemicals should be set, and the following rhetorical question should always be considered: What would future scenarios for the production of these molecules entail? With regard to sustainable chemical processes, Anastas et al. have published a clear definition of the requirements for sustainable chemical processes in the frame of green chemistry [40, 41]. Furthermore, aspects of health and safety, environmental protection, and economics have been considered and summarized by Poliakoff et al. (see Figure 17.3):

Green Chemistry is the design of chemical products and processes that reduce or eliminate the use and generation of hazardous substances [42].

Sustainable development can be broadly defined as those that "meet present needs without compromising the ability of future generations to meet their needs" [43]. Alongside the principles of green chemistry and sustainable processing, this definition provides a clear guideline for future technology developments. Many of the conditions associated with green chemistry and sustainable development can be met if catalytic processes and extensive process integration are utilized. Consequently, the development of suitable catalysts and efficient reaction systems will be indispensable in facilitating future sustainable transformation technologies. Thus, novel developments in the field of biomass processing clearly have to be benchmarked against these requirements. When designing new processes, emphasis must be placed on optimizing reaction temperature, solvent systems, hydrogen requirements, and carbon efficiencies. In general, optimized conversions and selectivity are desirable. However, the overall efficiency of a chemical transformation can be heavily influenced by product separation and workup, and so these processes must always be considered. These points should already be taken into account during basic research to facilitate implementation of efficient technical processes. Interestingly, recent studies emphasize that chemical disintegration of lignocellulose into the individual biopolymers, and the subsequent depolymerization into sugars and single aromatic motifs, will be of great importance to the overall process economics.

Today's refineries are based on utilization of crude oil. Consequently, technologies and catalyst systems for the treatment of oxygen-deficient molecules in

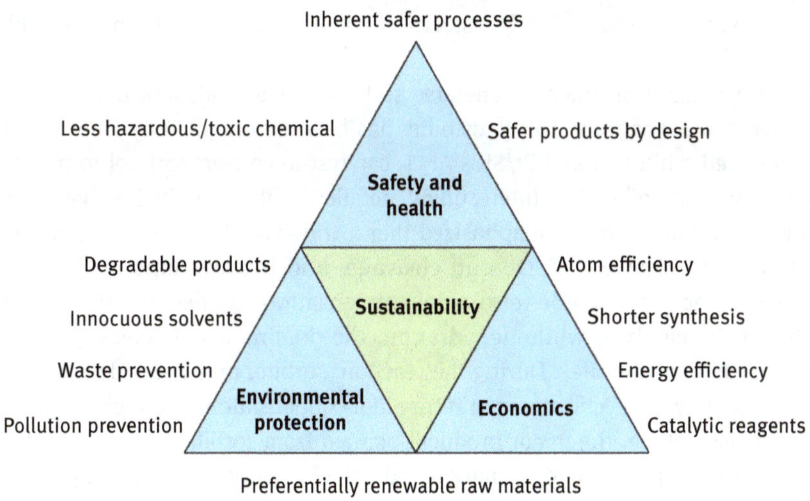

Figure 17.3: The essential aspects of sustainability in chemical processing as summarized by Poliakoff et al. [42].

high-temperature gas-phase processes have been established. In contrast, the main constituents of lignocellulose are rich in oxygen and require chemical transformations in liquid-phase processes at rather low temperature. Additionally, neither cellulose nor lignin is soluble in conventional solvents, which further complicates their controlled chemical breakdown [44].

These new technologies mostly serve to increase the carbon chain length and decrease the oxygen content of typical lignocellulose-based platform chemicals to deliver molecules of high energy content and suitable evaporation and combustion properties. Overall, three main approaches to the controlled low-temperature transformation of lignocellulose into biofuels can be distinguished (Figure 17.5). Besides the transformation of lignocellulose into bioalcohols, which has already been discussed, several studies focus on the production of simple hydrocarbons to feed conventional petrochemical supply chains and today's automotive combustion systems [45]. However, only a few approaches address the production of novel, potentially oxygen-containing biofuel candidates [46, 47].

17.7 Controlled Transformations of Carbohydrates into Hydrocarbon Fuels

The synthesis of hydrocarbons from carbohydrates requires the removal of oxygen, which is usually achieved by employing transformations that are demanding in hydrogen, such as hydrogenation and hydrodeoxygenation. To enable sustainable processes, hydrogen consumption must be minimized, and any hydrogen utilized should ideally be derived from renewable feedstocks or water splitting based on renewable energy sources [48].

The hydrogenation of glucose generates sorbitol, a sugar alcohol that can undergo further transformations. This possibility has been investigated by Huber et al. who have utilized a bifunctional Pt/SiO_2-Al_2O_3 catalyst to convert sorbitol to hexane via hydrodeoxygenation [49, 50]. Interestingly, detailed analyses of the involved reaction pathways and mechanisms emphasized that a wide variety of reactions occurs, such as C–C bond cleavage, C–O bond cleavage, and hydrogenations. The C–C bond cleavage reactions include retro-aldol condensation and decarbonylation on metal sites of the catalyst, while dehydration, the dominant C–O cleavage reaction, takes place on acidic sites. During the reaction, sorbitol undergoes dehydration and ring closure to yield cyclic ethers or retro-aldol condensations producing primarily C_3 polyols. Isosorbide, the major product obtained from sorbitol dehydration further undergoes ring opening hydrogenation and dehydration/hydrogenation yielding 1,2,6-hexanetriol, which is subsequently converted into hexanol and hexane by dehydration/hydrogenation. In addition to C_6 hydrocarbons, various smaller oxygenates

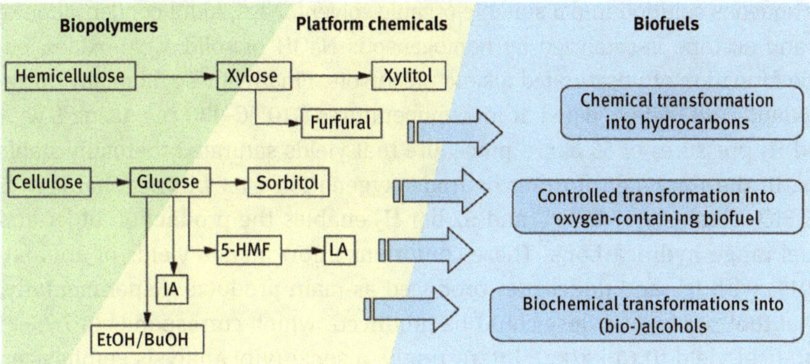

Figure 17.4: Potential platform chemicals considered for the controlled chemical transformation into fuel compounds and possible approaches to obtain novel biofuel motifs (5-HMF, 5-hydroxy methyl furfural; LA, levulinic acid; IA, itaconic acid; EtOH, ethanol).

and alkanes are formed. Overall, the yields of alkanes and hexane reached in this system were 56.5% and 30%, respectively.

However, six equivalents of hydrogen are necessary to convert one mole of glucose into hexane. Thus, such an approach requires efficient hydrogen management (e.g., via aqueous-phase reforming of sorbitol to yield the desired hydrogen) [51–53]. Combining both reactions, hydrodeoxygenation of sorbitol with its aqueous-phase reforming, a net exothermic process results requiring 1.5 mol of sorbitol to produce 1 mol of hexane:

$$C_6O_6H_{14} + 6\,H_2 \longrightarrow C_6H_{14} + 6\,H_2O$$

$$C_6O_6H_{14} + 6\,H_2O \longrightarrow 6CO_2 + 13\,H_2$$

$$\Sigma\, 19/13\, C_6O_6H_{14} \longrightarrow C_6H_{14} + 36/13\, CO_2 + 42/13\, H_2O$$

Based on these results and the disadvantageously high volatility of hexane, Huber et al. decided to develop another approach utilizing 5-hydroxymethylfurfural (5-HMF) or furfural, compounds that can be obtained from the dehydration of glucose or xylose, respectively [54–57]. The production of C_8-C_{15} alkenes from these furans is achievable by sequential aldol condensation with acetone followed by hydrogenation and hydrodeoxygenation. These products are in the desired range for diesel and jet fuel applications. Maximum yields were accomplished by applying different catalysts and reaction conditions in each reaction step. An overall process analysis starting from aqueous solutions of waste hemicellulose indicate that the utilization of a four-step process is optimal. Such a process would comprise (1) combined acid hydrolysis and xylose dehydration, (2) aldol condensation, (3) low-temperature hydrogenation, and (4) high-temperature hydrodeoxygenation. In the first step, xylose oligomers are hydrolyzed and further dehydrated to furfural, which is continuously extracted from the aqueous solution into a suitable organic solvent. Next, aldol condensation of furfural and acetone is catalyzed by homogeneous NaOH or solid MgAl-oxides, enabling the formation of unsaturated alkane precursors. Finally, these thermally unstable compounds are hydrogenated at low temperatures (110 °C–130 °C), using 5 wt% Ru/C and H_2 pressures of 55 bar, a procedure that yields saturated thermally stable products. In the final step, further hydrodeoxygenation over bifunctional 4 wt% Pt/Al_2O_3-SiO_2 catalysts at 260 °C and 62 bar H_2 enables the production of jet and diesel fuel range hydrocarbons. These conditions allow alkane yields of approximately 91% with tri- and dodecanes produced as main products. Experimentally, 0.46 kg jet fuel per kg of xylose could be produced, which corresponds to 76% of the theoretical yield (0.61 kg/kg). Interestingly, a sensitivity analysis emphasized that the price of the raw materials, the organic-to-aqueous mass ratio in the biphasic dehydration step, and the xylose concentration in the feed solution significantly affects the product cost, while investment and operation costs have less impact.

Recently, Corma et al. followed yet another approach aimed at producing branched C_{15}-alkanes [58]. In this investigation, 2-methylfuran served as the substrate undergoing solvent-free hydroalkylation and hydrodeoxygenation. 2-Methylfurfural can be derived from furfural [59–61], and can also be produced in the industrial production of furfuryl alcohol [62], with 93% selectivity simply by raising the reaction temperature from 135 °C to 250 °C [63]. Hydroalkylation of 2-methylfurfural with butanol can be catalyzed by soluble para-toluolsulfonic acid or solid Amberlyst-15. The hydrodeoxygenation of the obtained 1,1-bisylvylalkanes (1,1-bisylvylbutane) was performed without any additional solvents, using a continuous fixed-bed reactor alongside the catalysts Pt/C or Pt/Al_2O_3. Overall, 90% of all carbon ends up in the organic phase with 95% selectivity to alkanes. Furthermore, other approaches allow for the production of alkanes from 2-methylfuran, without the need for any additional reagents. In this system, 4-oxopentanal is generated in situ from 2-methylfuran via acid-initiated ring opening. This intermediate then reacts with two other molecules of 2-methylfuran to produce a trimer. Finally, hydrodeoxygenation of the intermediate yields pentadecane isomers as the main products, and overall fuel yields of up to 87% are generated. The obtained organic fraction exhibits an excellent pour point (90 °C) and cetane number (70.9), making this mixture suitable as a fuel. Alternatively, the product mixture could serve as a valuable blending agent, the utilization of which could improve both the cetane number and low-temperature flow properties of other fuels. Notably, all of these methods are demanding in hydrogen, a factor that must be considered before their widespread utilization (Figure 17.6).

The utilization of levulinic acid (LA) as a platform chemical could enable the implementation of routes to biofuels that demand less hydrogen. Furthermore, the formic acid released during LA formation may serve as a hydrogen source. This could be achieved by separate dissociation of formic acid to CO_2 and hydrogen or by employing formic acid directly in transfer hydrogenations [64–66]. Starting from γVl (γ-valerolactone), Dumesic and coworkers reported an integrated approach for the synthesis of C_{8+} alkenes without the need for external hydrogen. In this system, γVl is converted into pentenoic acid, which undergoes decarboxylation to butene and is subsequently oligomerized to yield C_{8+} alkenes [67]. When an aqueous solution of γVl is used as a feedstock, an isomeric mixture of pentenoic acids is formed, which is then decarboxylated to n-butenes and carbon dioxide. Both transformations are catalyzed by solid acids such as SiO_2/Al_2O_3 and can be carried out in a single fixed-bed reactor. While elevated pressures hinder the decarboxylation and shift the selectivity toward the formation of pentenoic acid, improved selectivity for butenes results from higher temperatures. However, this is accompanied by coke formation and associated catalyst deactivation. Additionally, the inhibitory effect of water necessitates its removal before the oligomerization reactor in which the butene/CO_2 mixture is oligomerized over solid acids such as Amberlyst-70. The reactions may be integrated into a single process, which runs at 36 bar pressure. In the first reactor unit, a temperature of 375 °C is maintained, while in the second unit a lower temperature of 170 °C is employed. This approach avoids the need for additional compression or thermal energy. Overall, the system gives a 77% yield for C_{8+}alkenes (based on the initial amount of γVl used), of which C_8-C_{16} alkenes represent the major fraction [68, 69]. The main advantages of this approach are (1) no requirement for external hydrogen or other substrates, (2) potential integration into a single continuous process, (3) moderate reaction conditions, and (4) the utilization of simple solid acid catalysts that do not contain any precious metals. Notably, this system releases CO_2 in pressures of up to 36 bar. This by-product could be used for sequestration or chemical synthesis without any need for additional compression energy.

Currently, the efficient dehydration of carbohydrates into LA limits overall process efficiencies. However, several investigations have identified solid acid catalysts that offer promising selectivities [70–72]. Despite these advances, the use of sulfuric acid catalysts for the hydrolysis and dehydration of cellulose remains superior with regard to space-time yields, despite the associated problem of salt formation [73]. Consequently, potential product streams resulting from cellulose dehydration would comprise sulfuric acid containing an aqueous solution of LA. Therefore, LA would have to be separated from the aqueous sulfuric acid solution before further transformation. With this in mind, Gürbüz et al. developed the extraction of LA via etherification with butane, which allows the development of an integrated process concept [74]. Alternatively, Braden et al. developed a sulfuric acid–resistant catalyst for the transformation of LA into γVl. Conventional catalysts such as Ru/C can facilitate the transfer hydrogenation of LA into γVl, utilizing formic acid as the hydrogen source. However, Ru/C suffers from fast deactivation in the presence of sulfuric acid, prompting the development of a bimetallic RuRe/C catalyst (4:3 molar ration of Re to Ru). This bimetallic catalyst offers superior activity for the simultaneous hydrogenation of LA and decomposition of formic acid, as well as displaying good stability in the presence of sulfuric acid [75]. Interestingly, a techno-economic analysis of a complete process utilizing lignocellulosic feedstocks to

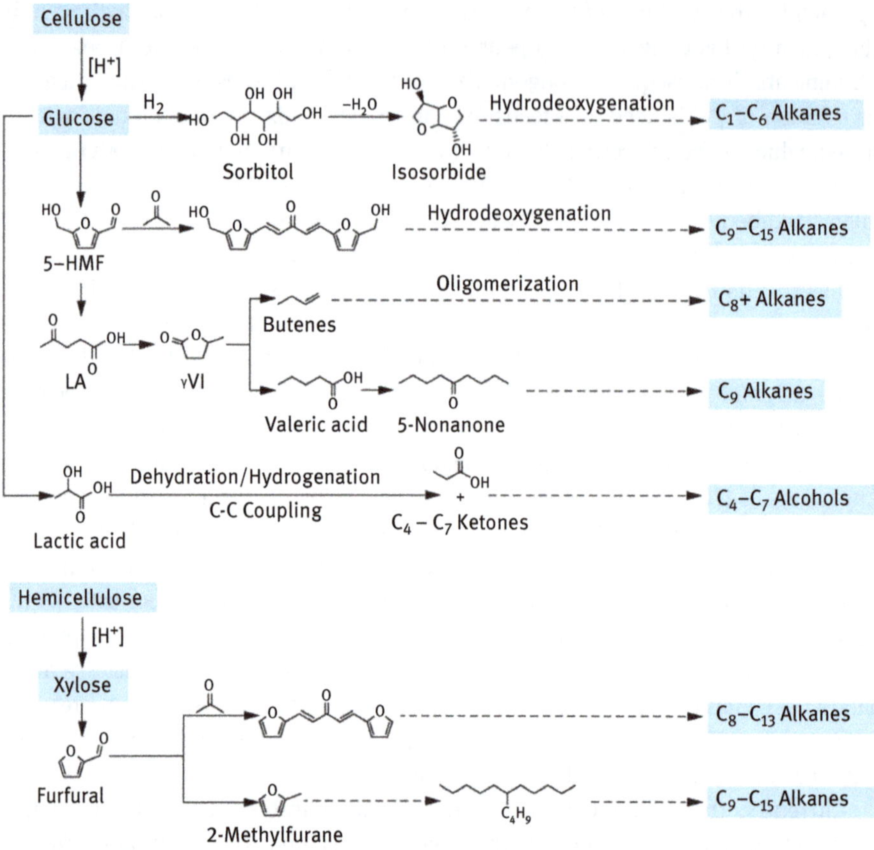

Figure 17.5: Graphic summary of alternative approaches to alkanes starting from cellulose and hemicellulose, respectively.

produce liquid hydrocarbons, in comparison to a state-of-the-art lignocellulosic ethanol production strategy, revealed that feedstocks and total installation costs are the most sensitive process parameters. Although the proposed process appears to be economically viable, further investigations are required to improve both the production of LA and its transformation into γVl.

Another means by which the oxygen content of LA can be reduced is conversion to valeric acid. This is achieved by transformation of LA to γVl followed by ring opening and hydrogenation over a water-stable bifunctional Pd/Nb_2O_5 catalyst at moderate temperatures and pressures. This transformation could prove to be advantageous, as it results in a hydrophobic and less reactive intermediate that is suitable for further processing [76]. One means by which valeric acid can be further modified is ketonization, which generates 5-nonane and stoichiometric amounts of CO_2 and water. The reaction conditions can even be adapted to higher temperatures and lower weight hour space velocity (WHSV), which enables both valeric acid production and ketonization to be initiated by the same Pd/Nb_2O_5 catalyst in a single fixed-bed reactor (350 °C, 35 bar, 0.1 h^{-1}). An organic layer that spontaneously separates from the aqueous 50 wt% γVl feed is produced, accounting for 81% of the reacted carbon with 60% carbon yield of C_9 ketones together with unreacted valeric acid, C_{4-5} ketones, and lower ketones. Although this strategy reduces the number of reaction steps and could reduce capital and operating costs in a potential technical application, higher yields are reached by separating both steps and individually optimizing catalyst and reaction conditions. In such systems, valeric acid is formed over Pd/Nb_2O_5 at 325 °C, 35 bar, and 1.2 h^{-1} with a 95% carbon yield in the organic phase and 92% selectivity to valeric acid. Next, valeric acid can be converted to 5 -nonane via decarboxylation, which is catalyzed by a ceria-zirconia catalyst at 425 °C, 20 bar, and 1.1 h^{-1} [77–81]. In this cascade approach, a 5-nonane yield of approximately 84% is obtained. The product, 5-nonanone, can be further converted via hydrogenation/dehydration over Pt/ Nb_2O_5 to linear nonane, which possesses a good cetane number and lubricity, making it suitable for use as a diesel fuel blender. Additionally, 5-nonanone can be hydrogenated over zeolites such as USY, yielding 5-nonanol. This alcohol can then undergo dehydration/isomerization to give hydrocarbons that are suitable for use as gas components [82, 83].

In addition to synthesis strategies based on sorbitol, 5-HMF, and LA, investigations have been made examining the conversion of lactic acid into hydrocarbons. Lactic acid can be catalytically upgraded via combined dehydration/hydrogenation and C–C coupling reactions [84, 85]. In such systems, an aqueous solution of lactic acid is processed over Pt supported on Nb_2O_5 causing the formation of an organic layer composed of propanoic acid and C_4-C_7 ketones, which spontaneously separates from the aqueous feed solution. Through subsequent hydrogenations, C_4-C_7 alcohols are generated, which are suitable for use as liquid fuels and have high energy densities. It has been demonstrated that the niobia support catalyzes both the dehydration and CC coupling reactions, while platinum provides catalytically active sites suitable for hydrogenation. Adjustment of the reaction conditions even allows further oxygen removal, again delivering hydrocarbon compounds as potential biomass-derived fuel compounds. Significantly, this example presents a transition between attempts purely targeting the production of hydrocarbons in controlled transformations and the utilization of similar approaches for the production of biomass-derived oxygenates suitable for use in novel biofuel compounds.

Certainly, various alternative approaches are possible to deliver liquid hydrocarbons based on cellulosic feedstocks. However, various factors including the overall yield, energy requirements, ease of product separation, and not least the demand for additional hydrogen will determine the economic and ecological competitiveness of these approaches. Nevertheless, as discussed previously, oxygenates exhibit several advantages as fuels (e.g., knock resistance and reduced exhaust emissions). These advantages have furthered the integration of ethanol and lately ethyl tert-butyl ether into today's gasoline chains.

17.8 Controlled Transformations of Carbohydrates into Novel Biofuels

As mentioned previously, only a few approaches address the potential of novel oxygen-containing biofuel candidates. Nevertheless, such oxygenates may exhibit several advantages compared to simple biomass-derived hydrocarbons. For instance, their production demands considerably less hydrogen, and many oxygenates exhibit superior performance in combustion reactions. Furthermore, oxygenates are known to increase knock resistance and reduce exhaust emissions, especially with regard to soot and NO_x formation.

17.8.1 Transformations Based on LA

Interestingly, research into the synthesis of novel oxygenates based on renewable feedstocks was only started in recent years. Currently, research has mainly been based on the aforementioned platform chemicals. Levulinic acid esters have already been identified as potential fuel candidates. These are not only compatible with conventional diesel fuels but also result in significantly reduced soot formation [86]. Additionally, ethyl levulinate has been utilized as a bio-based diluent for biodiesel and improves cold flow properties [87]. Interestingly, alkyl levulinates can be derived via dehydration of fructose or glucose in the presence of alcohols [88]. Moreover, even a direct transformation of cellulose into alkyl levulinates has been demonstrated. The sequential hydrolysis and dehydration of cellulose, in the presence of alcohols, reduces the formation of by-products such as humines. Hu et al. have demonstrated that the synthesis of alkyl levulinates starting from dehydration of sugars proceeds via etherification of the employed carbohydrate, transformation into the corresponding ethers of 5-HMF and subsequent rehydration to yield alkyl levulinate [88]. In this system, polymerization reactions of 5-HMF and sugars are suppressed, inhibiting the main route of humine formation. Thus, higher yields and selectivities for alkyl levulinates can be achieved than when the target product is LA. For example, a methyl levulinate yield of up to 90% can be obtained in a methanol-rich medium [88].

Apart from dehydration of glucose or fructose in the presence of alcohols and acid catalysts, synthesis strategies via 5-(chloromethyl)furfural and subsequent heating in the desired alcohol or utilizing furfuryl alcohol as feedstock have been demonstrated [89–98]. While 5-(chloromethyl)furfural may be derived in one step based on cellulose, a synthesis procedure starting from furfuryl alcohol utilizes hemicellulose as feedstock, bridging the C_5 and C_6 carbohydrate value chain.

Notably, ethyl or butyl levulinates could be entirely derived from cellulosic feed-stocks, without the need for additional hydrogen. However, the chemical properties of these levulinates, including their capacity to act as solvents, would require modification of the materials used in today's combustion systems. Furthermore, long-term road trials employing alkyl levulinate as blends or even as a pure biofuel would be necessary to assess the resulting overall engine efficiency and exhaust emissions.

Another biofuel candidate that can be derived from LA is 2-methyl tetrahydrofuran (2-MTHF), which is obtained via the intermediate γVl in a reaction pathway that includes both dehydration and hydrogenation steps [99, 100]. In a comparable reaction, itaconic acid, a platform chemical produced industrially by fermentation of carbohydrates such as glucose, can be converted into 3-methyl tetrahydrofuran [100]. The hydrogen required to produce γVl may be obtained from formic acid, via transfer hydrogenation, thus only one equivalent of additional hydrogen is required in the overall transformation [101]. Assessment of the combustion properties of a 70% 2-MTHF and 30% di-butylether blend indicated that the mixture gave almost soot- and NO_x-free combustion, without a significant reduction in engine efficiency. This clearly demonstrates the advantages associated with biomass-derived oxygenates [102].

Lange et al. presented pentenoic acid esters as potential biofuels for application in gasoline or diesel. They evaluated the process chain starting from cellulose and identified the transformation of cellulose into LA as well as the overall low space-time yields as major challenges [103, 104]. They revisited the complete value chain starting from acid-catalyzed hydrolysis of lignocellulose to LA, hydrogenation to γVl, and further hydrogenation to valeric acid, a new transformation that has not been reported yet. Finally, esterification generated alkyl valerate esters suitable as potential fuel compounds. With regard to the hydrogenation of LA to γVl, catalyst screening and stability tests show that Pt supported on TiO_2 or ZrO_2 is the most suitable. These catalysts result in greater than 95% selectivity to γVl at a differential productivity of 10 h^{-1} with negligible deactivation over 100 h at 473 K and 40 bar H_2pressure. The one-step transformation of γVl to valeric acid is catalyzed by bifunctional catalysts with acid and hydrogenation sites, a typical example being Pt-loaded SiO_2-bound H-ZSM-5. The reaction is not structure selective, but balancing the acid and hydrogenation activity of the catalyst is important in excluding the formation of pentenoic acid, incomplete hydrogenation, or further hydrogenation. Additional hydrogenation steps are most likely to occur for catalysts with high surface metal concentrations and can result in the formation of 2-MTHF, alcohols, and alkanes. The production of valeric acid appears to proceed by acid-catalyzed ring opening of γVl to pentenoic acid followed by hydrogenation. When a Pt/ZSM-5 catalyst is employed, a selectivity of 90% is obtained using a WHSV of 2 h^{-1}, 523 K, and 10 bar H_2. Furthermore, the catalyst can operate for more than 1,500 h when regenerated with H_2 at 673 K and 10 bar.

Ethyl esters are promising fuels for gasoline, and pentyl, ethylene, and propylene glycol esters could be used in diesel applications. These esters meet many of the requirements needed for fuel applications, including suitable energy density and polarity. Furthermore, these materials can be obtained in yields in excess of 95% by esterification of valeric acid with the appropriate alcohol over acidic ion-exchange resins. With regard to process integration, a one-step conversion of γVl into pentyl valerate can be achieved using Pt or Pd/TiO_2 catalysts, which give selectivities of between 20% and 50% at 275–300 °C. The metal content of the catalyst has to be enhanced to increase the formation of products obtained via more extensive hydrogenation, such as 2-MTHF or pentanol, which are reacted in situ with valeric acid to yield pentyl valerate. Assessment of fuel properties confirms the superior properties of alkyl valerates for gasoline and diesel applications. A 250,000 km road trial examining blends consisting of 15% vol. ethyl valerate in gasoline further substantiates the suitability of alkyl valerates for fuel applications. Remarkably, pentenoic acid esters also proved to have promising properties; hence, the hydrogenation to valeric acid could be dispensable (see Figure 17.7).

17.8.2 Biofuel Compounds Based on 5-HMF

In addition to LA, the alternative platform chemical HMF can also form the basis of synthesis aimed at producing compounds suited for use as biofuels. Casanova et al. have developed the etherification of 5-HMF and its oxidative esterification, which yield ethers of 5-HMF [105, 106]. However, the potential of such ethers as biofuel candidates has not yet been investigated. Lanyafame et al. have studied the etherification of 5-HMF with ethanol, a reaction that generates potential biodiesel components. The capacity of a range of solid catalysts to initiate this transformation was explored, with the activities of mesoporous silicas (Al-MCM-41) and supported zirconia compared with H_2SO_4 and Amberlyst-15 [107]. Reaction products included 5-(ethoxymethyl) furan-2-carbaldehyde (EMF), 1,1-dietoxy ethane, and ethyl 4-oxopentanoate (EOP), which may be obtained selectively by tuning the acidity of the catalyst. Notably, EMF is a valuable biodiesel component, while EOP could open interesting perspectives to produce valeric biofuels by subsequent selective hydrogenation.

In another approach, glucose is dehydrated to 5-HMF, which undergoes further hydrodeoxygenation or hydrodeoxygenation and hydrogenation to 2,5-dimethylfuran (2,5-DMF) or 2,5-dimethyltetrahydrofuran (2,5-DMTHF). The synthesis of DMF has been investigated by Yang and coworkers, who developed a multifunctional catalyst system for the production of DMF from biomass-derived carbohydrates, such as fructose, glucose, and cellulose. Furthermore, DMF was produced from raw lignocellulose in the form of corn stover. Notably, DMF exhibits an energy density of 31.8 MJ L^{-1} and a boiling point of 90–92 °C and is immiscible in water, properties that make DMF suitable for use in fuel applications. It has been demonstrated that 2,5-DMTHF can also be derived from carbohydrates obtained from biomass, including cellulose. Combinations of rhodium chloride, HCl, NaI, and benzene

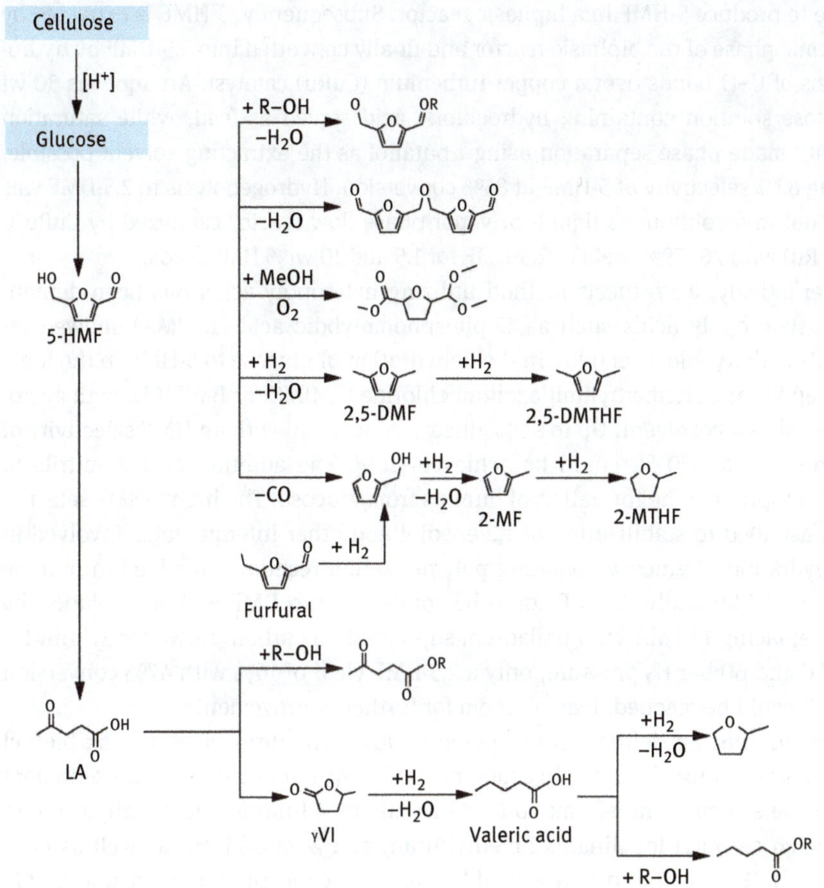

Figure 17.6: Schematic illustration of controlled chemical transformations of 5-HMF and LA into potential novel biofuels.

facilitated the conversion of cellulose into DMTHF, which was obtained in yields as high as 76% at 160 °C, 16 h, and 20 bar H_2. When this system is applied to untreated corn stover, DMTHF is again obtained, albeit in a lower yield of 41%.

Due to their good storage and transport stability, the utilization of saturated compounds as biofuels is receiving the most attention, while potential applications of furanes are being relatively neglected. Nevertheless, 2,5-DMF presents a promising fuel compound that exhibits 40% higher energy density compared to ethanol, a suitable boiling point of 93 °C, and immiscibility with water, which facilitates product recovery based on aqueous synthesis systems and subsequent fuel storage.

Roman-Leshkov et al. developed a concept for continuous production of 2,5-DMF based on fructose [108]. The first step involves the acid-catalyzed dehydration of fructose to produce 5-HMF in a biphasic reactor. Subsequently, 5-HMF is extracted by the organic phase of the biphasic reactor and finally converted into 2,5-DMF by hydrogenolysis of C–O bonds over a copper-ruthenium (CuRu) catalyst. An aqueous 30 wt % fructose solution containing hydrochloric acid served as feed, while saturation with NaCl made phase separation using 1-butanol as the extracting solvent possible, reaching 82% selectivity of 5-HMF at 85% conversion. Hydrogenolysis to 2,5-DMF was carried out in a continuous liquid- or vapor-phase flow reactor catalyzed by CuRu/C (3:2 Cu:Ru) with 76–79% yield of 2,5-DMF for 1.5 and 10 wt% HMF feeds.

Alternatively, a synthesis method utilizing heteropoly acids has been demonstrated. Heteropoly acids such as 12-phosphomolybdic acid (12-PMA) showed remarkable activity and selectivity in the dehydration of glucose to 5-HMF in the ionic liquids ethyl- or butylmethylimidazolium chloride (EMIMCl or BMIMCl), with acetonitrile used as a cosolvent. Up to 98% glucose conversion with an HMF selectivity of 99% after 3 h at 120 °C could be achieved [109]. The addition of acetonitrile to EMIMCl suppressed the formation of humins from glucose. The high 5-HMF selectivity was ascribed to stabilization of 1,2-enediol and other intermediates involved in the dehydration of glucose, avoiding polymerization reactions involved in humine formation. Additionally, 5-HMF could be converted into DMF without isolation by simply replacing 12-PMA with palladium supported on carbon. However, within 1 h at 100 °C and 60 bar H_2 pressure, only a 2,5-DMF yield of 16% with 47% conversion of 5-HMF could be reached, leaving room for further improvements.

Overall, one can define a set of potential lead structures of promising biofuel candidates currently discussed in academia. Thereby, tailored catalyst development allowed the selective transformation into levulinates, furane, and tetrahydrofuran motifs such as butyl levulinate, 2-methylfuran, and 2- or 3-MTHF as well as derivates of 5-HMF. Essential progress could be reached concerning flexible and highly selective catalysts for efficient transformation of levulinic and itaconic acid into these targets [100]. Novel catalysts for efficient C–O cleavage allowed synthesis of 2- and 2,5-furane, respectively. Nevertheless, most synthesis strategies still suffer from low space-time yields and low product concentrations. Additionally, low hydrogen demand and efficient product recovery will be essential to make the technical implementation of such technologies feasible. Consequently, future biorefinery concepts will strongly rely on the development of suitable catalyst systems and efficient reaction systems.

17.9 Controlled Transformations of Lignin into Potential Fuel Compounds

Compared to carbohydrate feedstocks, fewer studies have addressed the controlled transformation of lignin into potential fuel compounds. Zakzeski et al. published a comprehensive review summarizing the literature outlining the utilization of catalysts in the production of value-added products from lignin [110].

The directed degradation of lignin allows access to aromatic compounds, which are important structural motifs as fuel additives and complement the product range available from the carbohydrate fraction. Consequently, methods for the selective cleavage of the linkages in the highly branched lignin structure are a prerequisite to make these innovative fuel components accessible. With respect to biocatalytic transformations, many attempts for directed lignin conversion focus currently on oxidative processes. Various oxidoreductases (e.g., laccases and lignin peroxidases) have been used in enzymatic reactions [111, 112]. In nonoxidative bio-catalytic processes, mostly esterases were identified as effective ligninolytic enzymes [113, 114].

For chemical transformations of lignin, robust heterogeneous catalysts are used in hydrogenolysis reactions typically at elevated temperature [113]. These procedures usually result in a broad range of smaller fragments, although a complete chemical breakdown of the lignin structure is possible, rendering hydrocarbon molecules. Lercher et al. have demonstrated a new route of aqueous-phase depolymerization of lignin combining base hydrolysis and protecting agents to suppress repolymerization, which usually limits the efficient lignin depolymerization [114]. The resulting phenolic oil is further converted via hydrodeoxygenation utilizing bifunctional catalysts (Pd/C with H_3PO_4 or Raney Ni with Nafion/SiO_2), leading to the quantitative yield of hydrocarbons at 200–250 °C and 50 bar hydrogen pressure [116, 117]. In such systems, a suitable balance between metal and acid-catalyzed reactions is essential [118, 119].

More selective transformations can be expected from methods that allow milder conditions or the application of homogeneous catalytic systems with molecularly defined active sites. Initial examples with model compounds corroborated the expectations, and particularly vanadium- and ruthenium-catalyzed nonoxidative bond cleavage shows promising potential [120, 121]. Nevertheless, further investigations are necessary to develop suitable bio- or chemocatalysts that allow an efficient and selective hydrolysis of lignin toward defined chemical building blocks. With regard to applications for fuel production, hydrocarbons are certainly compatible with today's engines, but also aromatics and cyclic alcohols are effective as fuel additives.

17.10 Summary

Clearly, in the final analysis, market demand and market price will decide whether high-temperature processes aiming for syngas or bio-oil production, selective enzymatic transformation yielding methane or bioethanol, or tailored chemical conversion technologies via catalytic processes can deliver the required products. Based on today's diversity in the chemical industry, various technology platforms may be utilized to cover the required product range. Nevertheless, efficient utilization of available carbon sources both with respect to carbon and energy content and resource management will be decisive in developing a sustainable future scenario for carbon-based industry.

References

[1] BP Statistical Review of World Energy, June 2012. Available from: http://www.bp.com/statisticalreview.
[2] Hill J, Nelson E, Tilman D, Polasky S, Tiffany D. Proc Natl AcadSci U S A. 2006;103(30):11206–10.
[3] Demirbas A. Prog Energy Combustion Sci. 2007;33(1):1–18.
[4] Adler PR, Del Grosso SJ, Parton WJ. Ecol Appl. 2007;17(3):675–91.
[5] Crutzen PJ, Mosier AR, Smith KA, Winiwarter W. Atmos Chem Phys. 2008;8(2):389–95.
[6] Tilman D, Hill J, Lehman C. Science. 2006;314(5805):1598–600.
[7] Demirbas A. Energy Convers Manag. 2008;49(8):2106–16.
[8] Behr A, Eilting J, Irawadi K, Leschinski J, Lindner F. Green Chem. 2008;10(1):13–30.
[9] Zhou C-HC, Beltramini JN, Fan Y-X, Lu GQM Chem Soc Rev. 2008;37(3):527–49.
[10] Ma FR, Hanna MA Bioresour Technol. 1999;70(1):1–15.
[11] Meher C, Sagar DV, Naik SN Renewable Sustainable Energy Rev. 2006;10(3):248–68.
[12] Kamm B. *Biorefinery*.
[13] Palkovits R. Chem Ing Tech. 2011;83(4):1–10.
[14] Von Blottuitz H, Curran MA J Cleaner Prod. 2007;15:607–19.
[15] Available from: http://www.afdc.energy.gov/afdc/ethanol/e85.html.
[16] Browner CM EPA Administrator, Press Conference, March 20, 2000. Available from: http://www.epa.gov/otaq/consumer/fuels/mtbe/press34b.pdf.
[17] Available from: http://www.afdc.energy.gov/afdc/vehicles/flexible_fuel.html.
[18] Niven RK. Renewable Sustainable Energy Rev. 2005;9(6):535–55.
[19] Hayes DJ. Catal Today. 2008June http://www.afdc.energy.gov/afdc/vehicles/flexible_fuel.html.
[20] Goldemberg J. Science. 2007;315:808.
[21] De Carvalhomacedo I. Biomass Bioenergy. 1998;14:77–81.
[22] Shapouri H, Duffield J, Wang M The Energy Balance of Corn Ethanol: An Update, in Agricultural Economic Report No. 813. United States Department of Agriculture; 2002.
[23] Searchinger T, Heimlich R, Houghton RA et al. Science. 2008;319:1238–40.
[24] Wang M, Wu M, Huo H. Environ Res Lett. 2007;2:1–13.
[25] Sun Y, Cheng J. Bioresour Technol. 2002;83:1–11.
[26] Kumar P, Barrett DM, Delwiche MJ, Stroeve P. Ind Eng Chem Res. 2009;48:3713–29.
[27] Galbe M, Zacchi G. Appl Microbiol Biotechnol. 2002;59:618–28.
[28] Hayes DJ. Catal Today. 2009;145(1–2):138–51.
[29] García V, Päkkilä J, Ojamo H, Muurinen, Keiski RL. Renewable Sustainable Energy Rev. 2011;15:964–80.
[30] Dürre P. Biotechnol J. 2007;2:1525–34.
[31] Gabriel CL. Ind Eng Chem. 1928;20:1063–67.
[32] Jones DT, Woods DR. Microbiol Rev. 1986;50:484–524.
[33] Ni Y, Sun Z. Appl Microbiol Biotechnol. 2009;83:415–23.

[34] Centi G, Trifiró F, Perathoner S, Cavani F. Sustainable Industrial Chemistry. Weinheim: Wiley-VCH; 2009.
[35] Werpy T, Petersen G. Top Value Added Chemicals From Biomass. Vol. 1, Oak Ridge, TN: U.S. Department of Energy; 2004.
[36] Bozell JJ, Petersen GR. Green Chem. 2010;2(12):539–54.
[37] Huber GW, Iborra S, Corma A. Chem Rev. 2006;106(9):4044–98.
[38] Corma A, Iborra S, Velty A. Chem Rev. 2007;107:2411–502.
[39] Climent MJ, Corma A, Iborra S. Green Chem. 2011;13:520–40.
[40] Anastas PT, Warner JC. Green chemistry: theory and practice. New York: Oxford University Press USA; 2000.
[41] Anastas P, Eghbali N. Chem Soc Rev. 2010;39:301–12.
[42] Tang SLY, Smith RL, Poliakoff M. Green Chem. 2005;7:761.
[43] World Commission on Environment and Development. Our Common Future, Chapter 2: Towards Sustainable Development. Available from: www.un-documents.net/ocf-02.htm/.
[44] Chheda JN, Huber GW, Dumesic JA. Angew Chem Int Ed. 2007;46:7164–83.
[45] Serrano-Ruiz JC, Dumesic JA. Energy Environ Sci. 2011;4(1):83–99.
[46] Huber GW, Iborra S, Corma A. Chem Rev. 2006;106(9):4044–98.
[47] Alonso DM, Bond JQ, Dumesic JA. Green Chem. 2010;12:1493–513.
[48] Davda RR, Shabaker JW, Huber GW, Dumesic JA. Appl Catal B Environ (Special Issue on H2 Prod). 2005;56:171–86.
[49] Huber GW, Cortright RD, Dumesic JA. Angew Chem Int Ed. 2004;43:1549–51.
[50] Lin N, Huber GW. J Catal. 2010;270(1):48–59.
[51] Huber GW, Shabaker JW, Dumesic JA. Science. 2003;300(5628):2075–77.
[52] Wen G, Xu Y, Xu Z, Tian Z. Catal Commun. 2010;11(6):522–26.
[53] Shabakera JW, Hubera GW, Dumesic JA. J Catal. 2004;222(1):180–91.
[54] Huber GW, Chheda J, Barrett CB, Dumesic JA. Science. 2005;308:1446–2079.
[55] Xing R, Subrahmanyam AV, Olcay H et al. Green Chem. 2010;12:1933–46.
[56] Subrahmanyam AV, Thayumanavan S, Huber GW. Chem Sus Chem. 2010;3:1158–61.
[57] West RM, Liu ZY, Peter M, Gärtner CA, Dumesic JA. J Mol Catal A Chem. 2008;296(1–2):18–27.
[58] Corma A, De La Torre O, Renz M, Villandier N. Angew Chem Int Ed. 2011;50(10):2375–78.
[59] Burnett LW, Johns IB, Holdren RF, Hixon RM. Ind Eng Chem. 1948;40(3):502–05.
[60] Zhu Y-L, Xiang H-W, Li Y-W et al. New J Chem. 2003;27:208–21.
[61] Zeitsch KJ. The chemistry and technology of furfural and its many by-products. Dordrecht: Elsevier Science; 2000:36–74(Sugar series; vol. 13).
[62] Ibid., p. 229–30.
[63] Zheng H-Y, Zhu Y-L, Teng B-T et al. J Mol Catal. 2006;246:18–23.
[64] Mehdi H, Fábos V, Tuba R, Bodor A, Mika LT, Horváth IT. Top Catal. 2008;48(1–4):49–54.
[65] Heeres H, Handana R, Chunai D, Rasrendra CB, Girisuta B, Heeres HJ. Green Chem. 2009;11: 1247–55.
[66] Deng L, Zhao Y, Li J, Fu Y, Liao B, Guo Q-X. Chem Sus Chem. 2010;3(10):1172–75.
[67] Bond JQ, Alonso DM, Wang D, West RM, Dumesic JA. Science. 2010;327:1110–14.
[68] Palkovits R. Angew Chem. 2010;122(26):4434–36.
[69] Angew Chem Int Ed. 2010;49(26):4336–38.
[70] Rackemann DW, Doherty WOS. Biofuels Bioprod Biorefining. 2011;5(2):198–214.
[71] Van De Vyver S, Thomas J, Geboers J et al. Energy Environ Sci. 2011;4:3601–10.
[72] Schraufnagel RA, Rase HF. Ind Eng Chem Prod Res Dev. 1975;14(1):40–44.
[73] Rinaldi R, Schüth F. Energy Environ Sci. 2009;2:610–26.
[74] Gürbüz EI, Alonso DM, Bond JQ, Dumesic JA. Chem Sus Chem. 2011;4(3):357–61.
[75] Braden DJ, Henao CA, Heltzel J, Maravelias CC, Dumesic JA. Green Chem. 2011;13:1755–65.

[76] Serrano-Ruiz JC, Wang D, Dumesic JA. Green Chem. 2010;12:574–77.
[77] Serrano-Ruiz JC, Wang D, Dumesic JA. Green Chem. 2010;12:574–77.
[78] Renz M. Eur J Org Chem. 2005;6:979–88.
[79] Gärtner CA, Serrano-Ruiz JC, Braden DJ, Dumesic JA. Ind Eng Chem Res. 2010;49:6027–33.
[80] Gärtner A, Serrano-Ruiz JC, Braden DJ, Dumesic JA. J Catal. 2009;266(1):71–78.
[81] Gärtner CA, Serrano-Ruiz JC, Braden DJ, Dumesic JA. Chem Sus Chem. 2009;2(12):1121–24.
[82] Kunkes EL, Simonetti DA, West RM, Serrano-Ruiz JC, Gärtner CA, Dumesic JA. Science. 2008;322(5900):417–23.
[83] West RM, Liu ZY, Peter M, Dumesic JA. Chem Sus Chem. 2008;1:417.
[84] Serrano-Ruiz JC, Dumesic JA. Green Chem. 2009;11:1101.
[85] Serrano-Ruiz JC, Dumesic JA. Chem Sus Chem. 2009;2:581.
[86] Janssen A, Pischinger S, Muether M. SAE Int J Fuels Lubricants. 2010;3(1):70–84.
[87] Joshi H, Moser BR, Toler J, Smith WF, Walker T. Biomass Bioenergy. 2011;35:3262–66.
[88] Hu X, Li C-Z. Green Chem. 2011;13:1676–79.
[89] Mascal M, Nikitin EB. R Soc Chem Green Chem. 2010;12:370–73.
[90] Mascal M, Nikitin EB. Angew Chem Int Ed. 2008;47:7924.
[91] Mascal M, Nikitin EB. Chem Sus Chem. 2009;2:423.
[92] Mascal M, Nikitin EB. Chem Sus Chem. 2009;2:859.
[93] Mascal M, Nikitin EB. Green Chem. 2010;12:30.
[94] Mascal M, Nikitin EB. Energy Fuels. 2010;24:2170.
[95] Mascal M, Nikitin EB. Chem Sus Chem. 2010;3:1349.
[96] Mascal M, Dutta S. Green Chem. 2011;13:40.
[97] Mascal M, Nikitin EB. Green Chem. 2010;12:370–73.
[98] Lange J-P, Van De Graaf WD, Haan RJ. Chem Sus Chem. 2009;2(5):437–41.
[99] Mehdi H, Fábos V, Tuba R, Bodor A, Mika LT, Horváth IT. Top Catal. 2008;48(1–4):49–54.
[100] Geilen FMA, Engendahl B, Harwardt A, Marquardt W, Klankermayer J, Leitner W. Angew Chem Int Ed. 2010;122(32):5642–46.
[101] Deng L, Zhao Y, Li J, Fu Y, Liao B, Guo Q-X. Chem Sus Chem. 2010;3:1172–75.
[102] Janssen AJ, Kremer FW, Baron JH, Muether M, Pischinger S, Klankermayer J. Energy Fuels. 2011;25(10):4734–44.
[103] Lange J-P, Price R, Ayoub PM et al. Angew Chem Int Ed. 2010;122(26):4581–85.
[104] Palkovits R. Angew Chem Int Ed. 2010;49:4336.
[105] Casanova O, Iborra S, Corma A. J Catal. 2010;275:236–42.
[106] Casanova O, Iborra S, Corma A. J Catal. 2009;265:109–16.
[107] Lanzafame P, Temi DM, Perathoner S et al. Catal Today. 2011;175(1):435–41.
[108] Roman-Leshkov Y, Barrett CJ, Liu ZY, Dumesic JA. Nature. 2007;447:982–86.
[109] Chidambaram M, Bell AT. Green Chem. 2010;12:1253–62.
[110] Zakzeski J, Bruijnincx PCA, Jongerius AL, Weckhuysen BM. Chem Rev. 2010;110(6):3552–99.
[111] Singh Arora D, Chander M, Gill PK. Int Biodeter Biodegr. 2002;50:115–20.
[112] Dashtban M, Schraft H, Syed TA, Qin W. Int J Biochem Mol Biol. 2010;1:36–50.
[113] Ruiz-Dueñas FJ, Martínez AT. Microb Biotechnol. 2009;2:164–77.
[114] Gallezot P. Catal Today. 2007;121:76–91.
[115] Roberts VM, Stein V, Reiner T et al. Chem Eur J. 2011;17(21):5939–48.
[116] Zhao C, Kou Y, Lemonidou AA, Li X, Lercher JA. Angew Chem Int Ed. 2009;48:3987.
[117] Zhao C, Kou Y, Lemonidou AA, Li X, Lercher JA. Chem Commun. 2010;46:412.
[118] Zhao C, He J, Lemonidou AA, Li X, Lercher JA. J Catal. 2011;280(1):8–16.
[119] Wawrzetz A, Peng B, Hrabar A, Jentys A, Lemonidou AA, Lercher JA. J Catal. 2010;269:411.
[120] Son S, Toste FD. Angew Chem Int Ed. 2010;49:3791.
[121] Nichols JM, Bishop LM, Bergman RG, Ellman JA. J Am Chem Soc. 2010;132:12554.

Armin Brandner, Peter Claus, Marcus Rose

18 Biomass Conversion to Chemicals

18.1 Introduction

In the earliest times of humankind, those auxiliaries today known as chemicals could only be obtained from nature. This may have included the use of native coal, gas, or oil, but heavy exploitation may not have occurred. Since industrialization, the use of coal (since the nineteenth century), acetylene (1930–1950), and finally oil and gas (since the 1950s) served as a base for fuels and chemicals as well.

In Europe, approximately 69 million tons of oil was used as the raw material for the chemical industry in 2008 [1]. The total oil demand in Europe was 703 million tons in that year [2]. In contrast, only approximately 5% of all industry feedstock is of renewable origin [3]. Most of this reflects direct use of natural products like cotton for textiles, wood pulp for papermaking, or different oils for special applications and for oleochemistry in general (detergents, lubricants, etc.) [3]. The demand for bio-based chemicals and materials in the EU in 2018 was about 5.5 million tons. Low and high growth scenarios predict a growth to at least 8 and up to 16 million tonnes per year in 2030 [64].

In Germany, approximately 4.1% of all fossil raw materials and 14% of the crude oil demand were used in 2008 as feedstock for material used in the chemical industry, together 17 million tons [4]. In contrast, the total energy demand in Germany was 224 million tons [5]. According to a recent study the German chemical industry used overall 19.6 million tonnes fossil and biogenic resources with the latter amounting up to 2.6 million tonnes (13 %) in 2018 [65].

Therefore, the ecological and environmental effects of switching the base of chemicals to renewable sources will be diminished as long as fuels are produced from crude oil. Conversely, the relatively low amount of crude oil needed for the chemical industry will be accessible for a rather long period of time at moderate expenditures and exploitation cost.

As the production of fuels from biomass is described in Chapter 5, the authors want to mention that chemicals may also be classified as energy storage materials as all energy input during production is present in chemical bonds. As the formation of chemical bonds, especially the C–C bond and C–O bond synthesized by nature, is tedious, the structure given by nature should wherever possible be retained

Armin Brandner, Green Chemicals & Fuels Product Line, Air Liquide, Olof-Palme-Str. 35, 60439 Frankfurt am Main, Germany, e-mail: armin.brandner@airliquide.com
Peter Claus / Marcus Rose, Technische Universität Darmstadt, Technische Chemie II, Ernst-Berl-Institut für Technische und, Makromolekulare Chemie, Petersenstraße 20, 64287 Darmstadt, Germany, e-mail: claus@ct.chemie.tu-darmstadt.de / rose@tc2.tu-darmstadt.de

https://doi.org/10.1515/9783110608458-018

(Figure 18.1). This holds especially true for chemicals, in which it is not the chemical energy that is needed but the property of a special molecule.

Another point in terms of CO_2 reduction is the fact that binding carbon from biomass into chemicals is a more sustainable approach than converting CO_2 to fuel – which is burnt afterward – or capturing CO_2 into subterranean caverns.

The use of biomass as feedstock for fuels means a big challenge in that biomass in general has a significantly higher oxygen content than crude oil and common fuels (diesel, gasoline). They comprise more or less completely unfunctionalized hydrocarbons of variable chain length. In current oil refineries, crude oil is therefore mainly fractionated directly or only slightly modified.

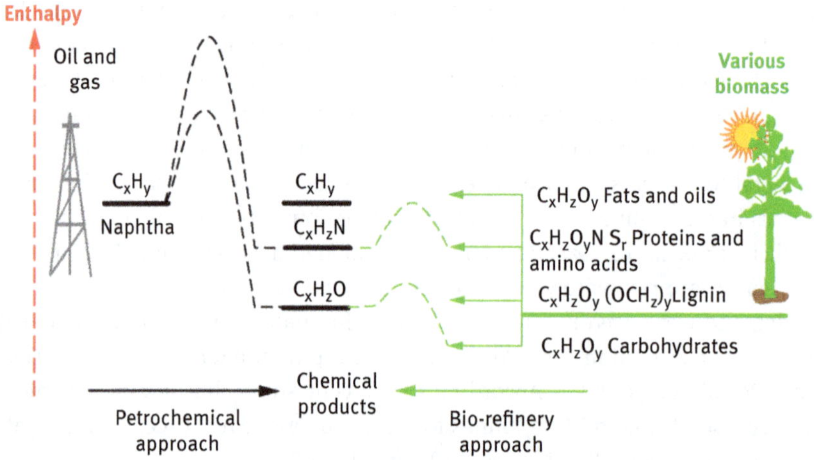

Figure 18.1: The production of functionalized chemicals from naphtha or from biomass [6].

With biomass as feedstock, completely new strategies are needed to reduce the oxygen content and the degree of functionalization in biomass to reach the levels of common fuels. Alternatively, the fuel basis can be switched to fuels more similar to the oxygen-rich biomass (e.g., bioethanol).

These points hold only partially for the production of chemicals from biomass. The smallest part of bulk chemicals consists of unfunctionalized hydrocarbons. Nevertheless, the oxygen content of biomass has to be reduced to the level needed for typical bulk chemicals. Therefore, strategies are strongly needed [7].

Politics has recognized the need for massive research. With the Biomass R&D Act of 2000, the U.S. Department of Energy has founded many studies regarding the use of biomass for energy and chemical use. Similar programs have also been conducted in Europe and other countries.

Funding the research on that topic has led to several major studies examining the potential of biomass as feedstock for chemical production. This includes the identification of possible intermediate chemicals, state-of-the-art processes for their production, and their current economic chances. Examples of these studies are the "Top Value-Added Chemicals from Biomass" [8] that have been revisited and updated later on [66] and, in Europe, the BREW project [9].

It is not the intention of this chapter to give an extensive review of the subject headlined but instead to discuss fundamentals, ways to select key chemicals, and, finally, examples for the current status of industrial chemical production from biomass.

18.2 Classification of Biomass

Biomass can be divided mainly into three compound classes – namely, lignocellulose, lipids, and proteins. These may come from different feedstocks and may differ in composition but should need only different steps in preprocessing to obtain a homogenous and constant feed (Figure 18.2). Not mentioned here are further carbohydrates such as free sugars and starch, because these are mainly used directly for human nutrition.

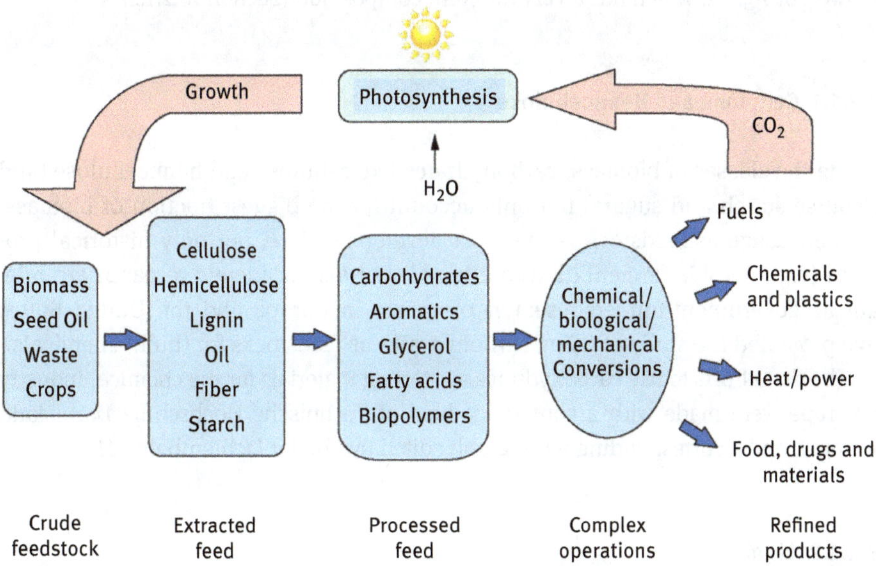

Figure 18.2: Overview of the processing of crude feedstocks to refined products in a sustainable biorefinery [10].

The direct use of biomass components may be subject to specialty and fine chemicals or pharmaceuticals. Processes are well established and mainly comprise extraction and simple transformation of biomass components. Examples are monoterpenes or fatty alcohols. They can be summarized as products with a very similar molecular structure compared to the deployed biomass.

When the structure of biomass components has to be quite drastically rearranged compared to the building blocks needed, synthesis gas or different sugars can serve as platform chemicals. The latter can be converted, for example, to sugar-derived building blocks (Figure 18.3) – that is, to glycerol, sorbitol, levulinic acid, and furfural.

Besides chemocatalysis, biotechnological processes are very important for converting biomass into valuable chemical compounds, which are summarized in Figure 18.4.

18.2.1 Lignocellulose

Lignocellulose denotes the mixture of the carbohydrate biopolymers cellulose and hemicellulose with the aromatic polymer lignin that is found in plants. Wooden raw materials consist mainly of cellulose (30–50 wt%), hemicellulose (10–40 wt%), and lignin (15–30 wt%). As the structure of cellulose (C6 carbohydrates) and hemicellulose (C5 carbohydrates) is quite similar, they will be discussed together in Section 18.2.1.1, followed by lignin, which has a very different composition (Section 18.2.1.2).

18.2.1.1 Cellulose and Hemicellulose

Among the classes of biomass, carbohydrates like cellulose and hemicellulose (and of course starch and sugars) not only account for the biggest fraction of biomass, but their usage as feedstock is the most developed. Here, not only historical processes (e.g., aerobic fermentation to ethanol) but also academic research are relevant as government-funded research programs in Europe and the United States have promoted the study of using carbohydrates as feedstocks for (bulk) chemicals.

The first efforts to use carbohydrates as starting materials for the chemical industry in Europe were made with a conference held at Technische Hochschule Darmstadt, Germany, and a corresponding monograph edited by Frieder Lichtenthaler [11].

18.2.1.2 Lignin

Lignin consists of randomly connected methoxypropylphenol units. It is obtained with 50 million tons/year [12] in paper pulp production as water soluble lignosulfonates from the sulfite pulping process or as Kraft lignin from the Kraft or sulfate process, where the Kraft process accounts for 95% of all produced lignin [13]. Further processes are more modern, such as organosolv processes (organic solvent) or so-called explosion processes with water steam or ammonia fiber expansion (AFEX).

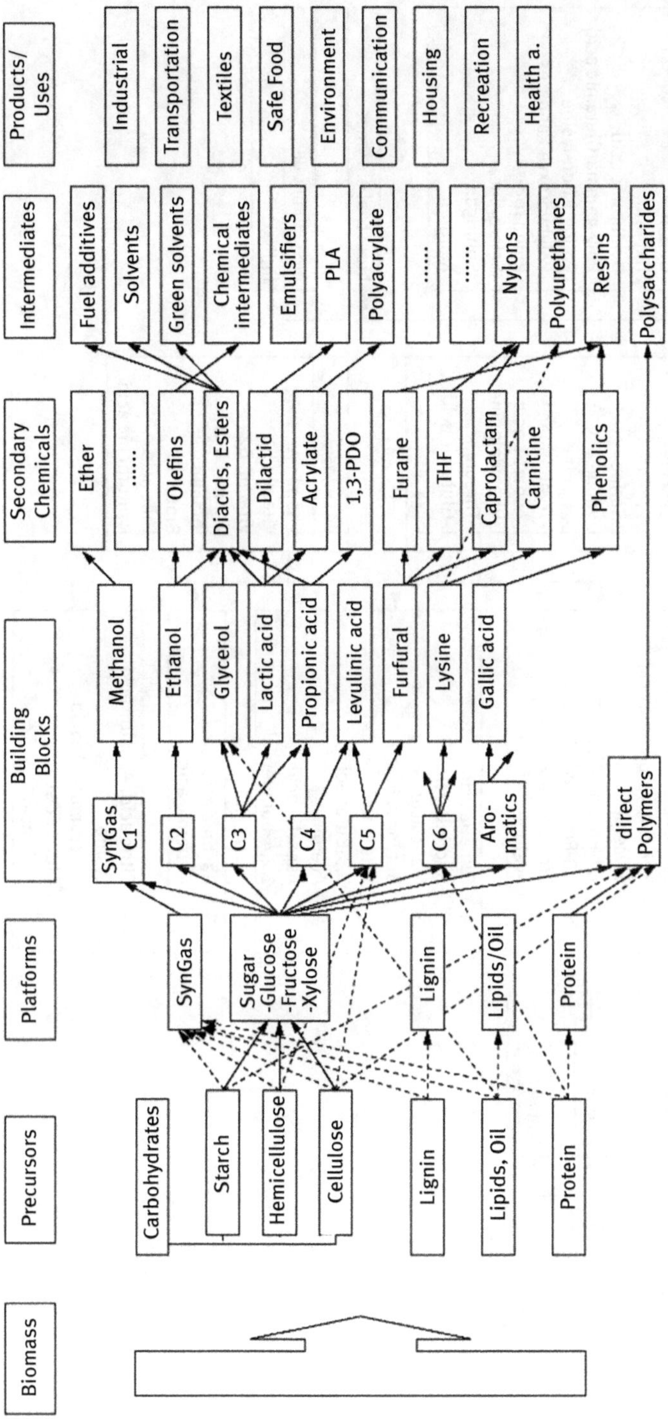

Figure 18.3: Bio-based product flowchart [8].

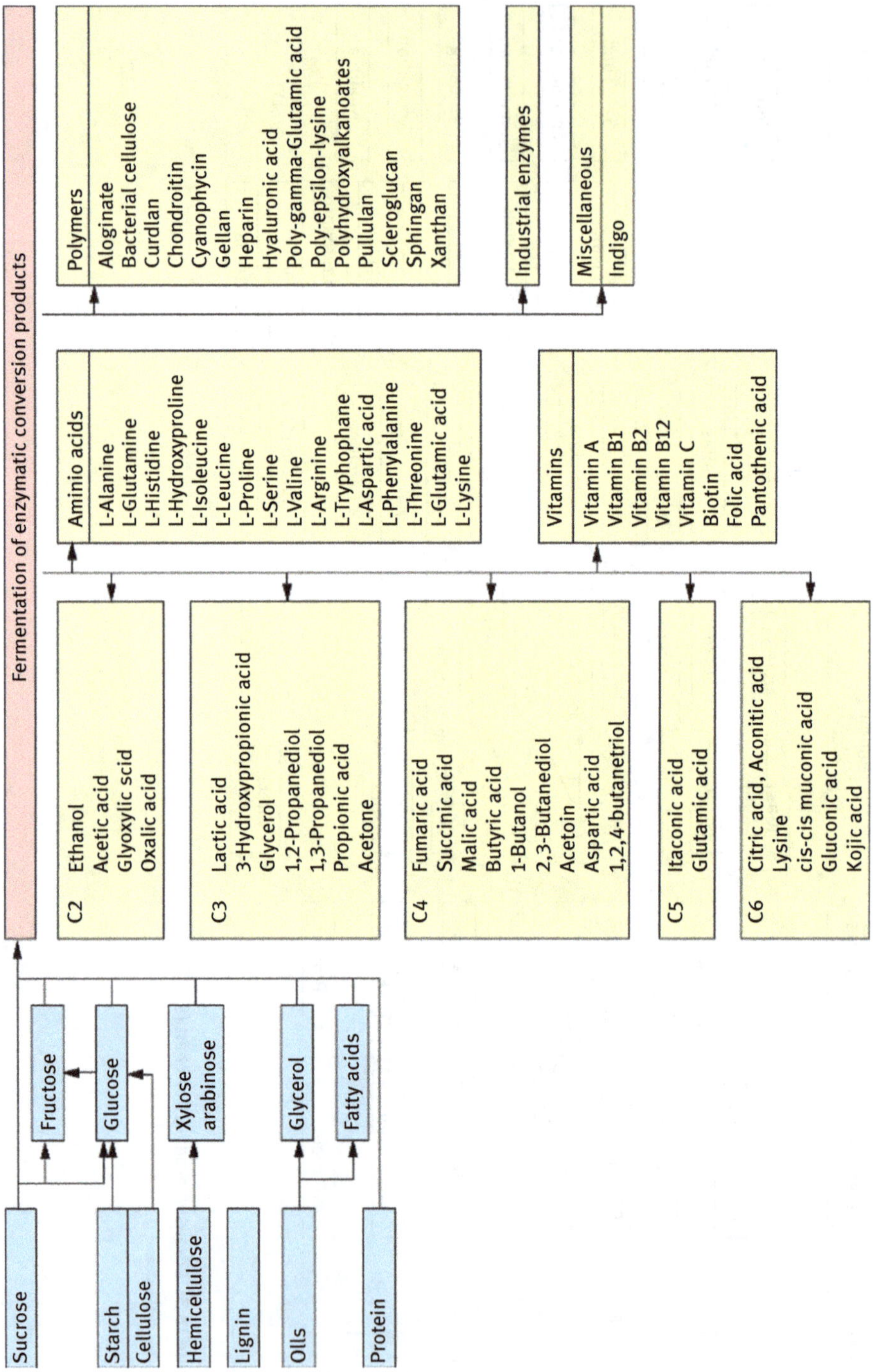

Figure 18.4: Overview of chemicals obtained from biomass by biotechnological processes [9].

Lignosulfonates (approximately 1 million tons/year) can be used directly from the so-called sulfite liquors (around 50% lignosulfonates) or further isolated. They exhibit many uses (e.g., as concrete additives and dispersion agents) [12].

Kraft lignin is soluble in alkaline media or some organic solvents but is not as easily degraded as cellulose. Ionic liquids have also been reviewed for their dissolution capability, but no processes are commercialized yet. This is mainly due to the high prices of ionic liquids compared to regular solvents and the challenging separation and recycling of ionic liquids [7].

Due to its high stability and poor degradability, Kraft lignin is mainly burnt to provide the energy for paper mills [12]. Without chemical modification, it can be used as a component in construction materials [14].

Another method is the derivatization to lignosulfonates with the aforementioned uses. Since the 1980s, bacteria and fungi have been under investigation to depolymerize lignin to valuable aromatic chemicals, but no commercialization has taken place yet [15].

In the past, processes for the production of vanillin from lignosulfonates and dimethyl sulfide and dimethyl sulfate were known, but these have been abandoned due to lower production costs with petrochemical raw materials [12].

A very comprehensive review regarding the use of lignin for chemicals has recently been published by Weckhuysen et al. [7]. They carved out three basic strategies to use Kraft lignins for chemicals (Figure 18.5).

One way is the complete or partial destruction of natural biomass by pyrolysis or gasification to yield so-called bio-oil or syngas, respectively. While the downstream processes for syngas (methanol, Fischer-Tropsch) are state of the art, smaller molecules generated by pyrolysis need a further pretreatment to be developed before they can be used in existing chemical process streams [16]. This is obviously a rather uneconomic way of cutting biomass down to C1 before rebuilding but is available on a short scale of time.

The second route involves the retrieval of current platform molecules (benzene, toluene, xylenes (BTX), phenols, etc.) by unselective catalytic processes, followed by familiar processes. Here, mainly hydrogenation takes place to remove excess oxygen, yielding aromatic hydrocarbons (BTX, phenols). This will need some time for commercialization as no results with high yields for BTX from genuine lignin are known. Instead, actual research is mainly performed on model components like p-coumaryl, coniferyl, and sinapyl alcohol [7].

A calculation for the U.S. market showed that the whole demand of aromatics (BTX; 20 million tons/year) could be optimally produced by an input of approximately 42 million tons/year. Due to the currently low effectiveness of only 10%, the needed amount would be 420 million tons/year of lignin [17].

More selective catalytic processes leading to desired chemicals directly have to be developed for a midterm period of time. This means mainly oxidation processes in which products such as aromatic aldehydes and acids can be obtained. Examples of the conversion of genuine lignins to syringaldehyde, vanillin, or corresponding

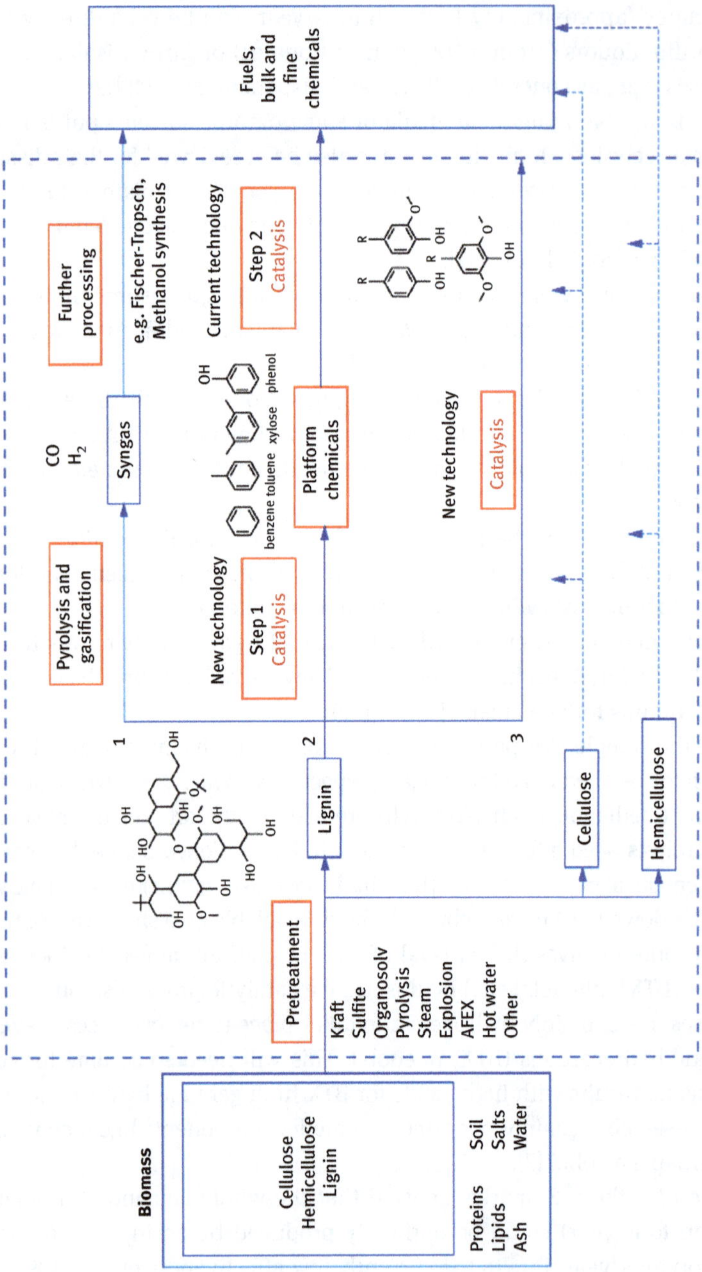

Figure 18.5: Lignocellulosic biorefinery scheme with particular emphasis on the lignin stream [7].

acids give yields around 10% [18]. For these reactions, mainly polyoxymetalate catalysts or simple metal salts were used [7]. The advances in the past decade cover all kinds of chemocatalytic conversion routes and have been significantly advanced towards the application of "real" lignin as obtained from pulp production and even the raw material lignocellulose. The comprehensive developments were recently summarized accordingly [67].

18.2.2 Lipids

Lipids include a variety of quite different substance classes, with a wider range than carbohydrates or proteins. The International Union of Pure and Applied Chemistry (IUPAC) definition is quite loose and says "substances of biological origin that are soluble in nonpolar solvents" [19]. Numerous further classification systems can be applied, but a recent initiative suggested eight different groups [20]. Industrially, the most important lipids are glycerides, fatty acids, and waxes. Further classes consist of functional plant and animal lipids such as glycerophospholipids, sphingolipids, prenol lipids, saccharolipids, and polyketides.

18.2.2.1 Glycerides

Glycerides or O-acylglycerols are esters of glycerol and fatty acids and are colloquially named oils and fats. In nature, monoglycerides, diglycerides, and triglycerides may occur, but 97% of all oils and fats are triesters of glycerol with different fatty acids. They are distinguished by their physical state at room temperature; fats are solid, whereas oils are liquid [21].

They comprise a major product in agriculture with worldwide production amounts of 173 million tons in 2009/2010 [22]. During the last decade, the global production rate increased by roughly 45%, which can be attributed to an increasing demand in developing countries and especially the increased biodiesel production. The shares in demand are 80% for human nutrition, 6% for animal nutrition, and approximately 14% for chemical industry [23]. Most of the raw materials come from vegetable oils (85%); only 15% are of animal origin [23].

The part of chemistry dealing with oils and fats is commonly known as oleochemistry. Figure 18.6 gives an overview of typical processes and products.

18.2.2.2 Fatty Acids and Fatty Alcohols

Fatty acids are traditionally meant as aliphatic unbranched monocarboxylic acids, either saturated or unsaturated, but with a chain length of 4–28 carbon atoms.

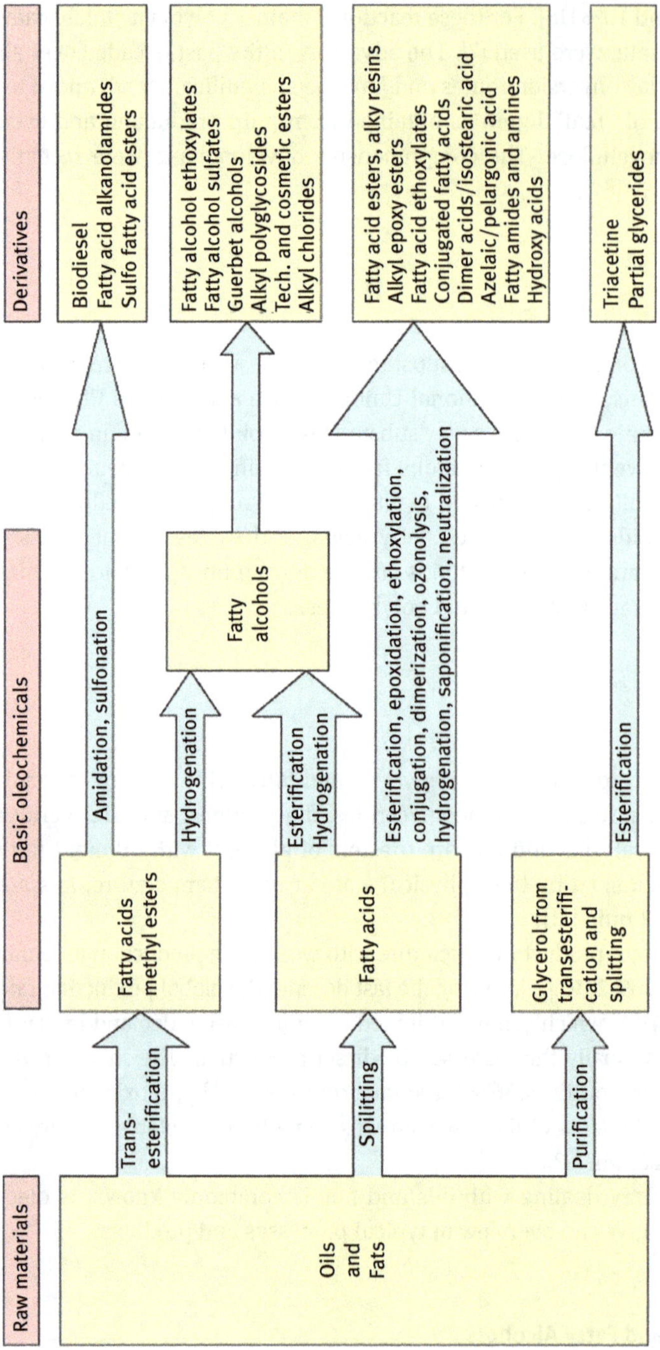

Figure 18.6: Oleochemical processes and products [23].

Sometimes even shorter acyclic aliphatic carboxylic acids like acetic acid are named fatty acids, although they are not found in oils and fats [19].

The most abundant fatty acids in vegetable oils and fats are palmitic acid (hexadecanoic acid or 16:0), oleic acid ([9Z]-octadec-9-enoic acid or 18:1 *cis*-9), and linoleic acid (*cis*, *cis*-9,12-octadecadienoic acid or 18:2 *cis*-9 *cis*-12) [21]. Other fatty acids are found in special oils (e.g., 80–87% ricinoleic acid in castor oil) [23], but these oils are quite rare. Castor oil, for example, has a production rate of 610,000 tons/year compared to the top four: palm oil (46 million tons/year), soya oil (40 million tons/year), rapeseed oil (24 million tons/year), and sunflower oil (12 million tons/year) [24]. Further sources of fatty acids are tall oils (2 million tons/year) [25] and to a lesser degree synthetic fatty acids derived by mainly hydroformylation and hydrocarboxylation of olefins [23]. The summed fatty acid production is estimated to be 8 million tons/year (2006) [23].

Fatty acids are obtained by fat splitting using water (hydrolysis), methanol (methanolysis), and base (saponification) of amines (aminolysis). Splitting with water or methanol can be considered transesterification because glycerol is liberated. The methanolysis is the reaction taking place in biodiesel production as the resulting product is called fatty acid methyl ester.

The hydrolysis is industrially conducted by acid or base catalysis (temperatures 210–260 °C) or by enzymatic hydrolysis in the case of sensitive fatty acids. The resulting crude product mixture is separated and purified mainly by means of distillation or crystallization, and more rarely by adsorption or extraction.

Pure fatty acids can be further treated (e.g., by hydrogenation, isomerization, and dimerization). The hydrogenation of fats and fatty acids leads to partially or fully unsaturated molecules under retention of the carboxylic group. This is known as hardening because the melting point is increased. Hardened products exhibit a higher stability to air and thermal treatment. Hydrogenation is mostly catalyzed by nickel catalysts and carried out under increased temperature and pressure [23].

Fatty alcohols are obtained by direct hydrogenation of fatty acids or by hydrogenation of fatty acid esters. Typically, this is performed over copper catalysts at elevated temperature (170–270 °C) and pressure (40–300 bar hydrogen) [26]. By this route, completely saturated fatty alcohols are produced. In the past, unsaturated fatty alcohols were produced via hydrolysis of whale oil (a natural wax occurring in whale blubber) or by reduction of waxes with sodium (Bouveault-Blanc reduction). Today, they can be obtained by selective hydrogenation at even higher temperatures (250–280 °C), but lower pressure up to 25 bar over metal oxides (zinc oxide, chromium oxide, iron oxide, or cadmium oxide) or partially deactivated copper chromite catalysts [26].

Approximately 75% of the fatty alcohols produced (around 3 million tons/year in 2008) are used in surfactants; further uses are cosmetics, lubricants, polymer ingredients, and flavors [26]. The same uses hold for fatty acids, either in pure form or as derivatives [23].

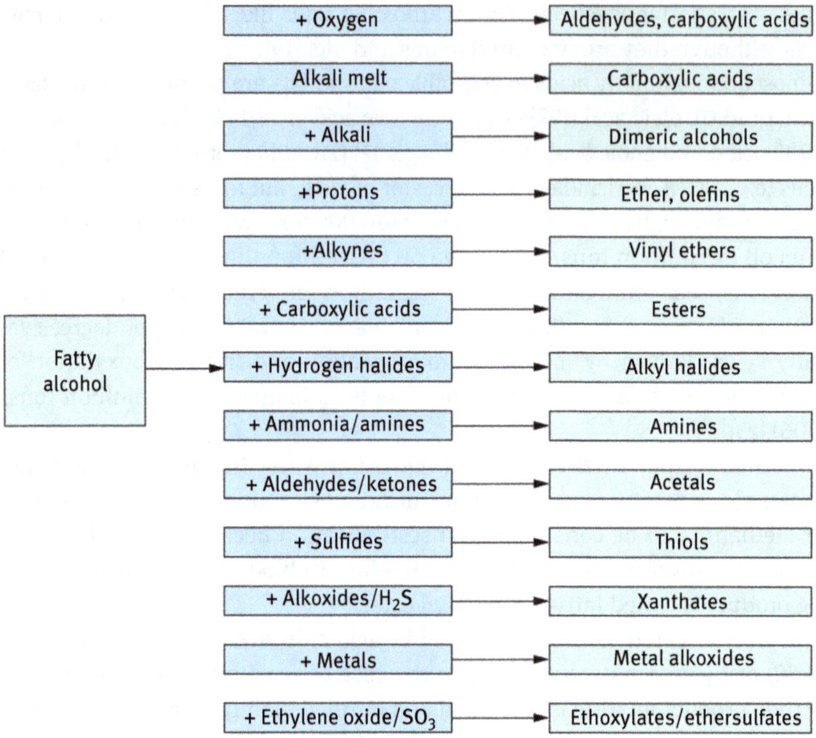

Figure 18.7: Typical reactions of fatty alcohols [26].

Because the terminal hydroxyl of the carboxylic group of fatty alcohols and fatty acids, respectively, can act as an anchor for further transformation, it can be used for a variety of reactions not yet commercialized but with great possibilities (Figure 18.7) [27].

18.2.2.3 Waxes

Waxes were classically defined as esters of a fatty acid with a fatty alcohol (e.g., beeswax), but this no longer holds as many waxes are not covered by that definition [28]. Therefore, a definition based on physical properties has been proposed [29]. The global market for waxes is relatively small (estimated to be 3.4 million tons/year), comprising mainly natural, fossil paraffin waxes (90%) and synthetic polyolefin and Fischer-Tropsch waxes [28]. So-called recent natural waxes (animal and plant waxes) make up a small percentage with only 28,000 tons/year.

Similar to all natural waxes is a hydrocarbon-like chain structure with chain lengths of approximately 20 to 60 and hence certain hydrophobic properties. As their numerous applications are driven by physical properties, only slight chemical modifications take place in the natural waxes. Additionally, due to the low production rate, waxes are not further considered as feedstock for additional processing into, for example, commodity chemicals.

18.2.3 Proteins

Proteins are found in wooden biomass (<5 wt%), cereals (~10 wt%), and herbs (up to 20 wt%) [13, 30]. They can accumulate in large-scale biomass processes like bioethanol or biodiesel production with approximately 1% of the deployed biomass [6].

As the annual world production of food protein from plants accounted for 250 million tons in 1990 [29], a huge surplus generated in biofuel plants could be further used for the production of chemicals. Proteins from plants can be directly used in animal and human nutrition, cosmetics, and health care products as well as adhesives, coatings, and further special applications [30, 31].

Amino acids can be obtained by hydrolysis of proteins, chemical synthesis, fermentation (e.g., of sugars), and enzymatic processes [32]. They serve similar purposes as proteins, but in addition, aspartic acid and glutamic acid were rated as top value-added chemicals, which can be converted to further fine chemicals [8]. Another example is the production of building blocks like ethylenediamine and butanediamine from serine and arginine, respectively [6].

18.3 Selected Key Chemicals

18.3.1 Cellulose

Among the renewable raw materials, cellulose is among the most abundant ones. In contrast to other carbohydrates, such as sucrose or starch, cellulose processing does not compete with food production. In the context of examples of the conversion of carbohydrates, the term *cellulose* may also include oligosaccharides and monosaccharides, such as glucose. This is because cellulose conversion almost always includes a depolymerization to soluble fragments, which are further converted.

As cellulose – a linear polymer of β-D-glucose with partly crystalline structure – is hardly soluble in most solvents, it is a challenge to find reaction conditions and catalysts that allow both high conversion and selectivity to value-added products. The key steps of cellulose degradation are depolymerization into glucose monomer or oligomer units via acid-catalyzed hydrolysis followed by chemical reactions such as hydrogenation or oxidation by means of catalysts. However, when using diluted acids, the product solution contains free acid, which has to be neutralized and separated.

More advantageous is the use of solids acids and especially bifunctional catalysts, exhibiting acid sites for hydrolysis in combination with a metal (Rh, Pt, Ru, and Ni) for hydrogenation, which can be used, in appropriate solvents, for the direct conversion of cellulose into value-added chemicals (e.g., sorbitol, mannitol, ethylene glycol, hydroxymethylfurfural, 2,5-dimethyltetrahydrofuran, levulinic acid, valerolactone, and isosorbide) [33–36, *Alternative monomers based on lignocellulose and their use for biopolymer production* [68]].

18.3.2 Glycerol

In the past, glycerol was produced mainly from propene via allyl chloride and epichlorohydrin, a process developed by I. G. Farben and in operation since 1943. Today, glycerol is obtained almost completely as a coproduct in oleochemistry (fat splitting) and biodiesel production (transesterification) with 110 kg crude glycerol or 100 kg pure glycerol per ton of biodiesel [37]. With the rise in biodiesel production, the availability increased while the price decreased drastically by approximately 66% within 15 years in the United States [38].

The crude glycerol obtained during biodiesel production has a purity of approximately 80–85% and includes water, salt, and further residues. This crude glycerol can be purified to pharmaceutical-grade glycerol (>99.5%) by vacuum distillation [39]. Heterogeneously catalyzed processes are known that give glycerol in higher purity (<98%) without further distillation [40].

Glycerol is a valuable C3 building block, whose modification possibilities were extensively reviewed by Behr et al. [41] and Pagliaro et al. [42]. Examples of the use of glycerol are shown in Figure 18.8.

The first plants for the biotechnological production of ethanol, acetone, and isoprene directly from glycerol with a capacity of approximately 20,000 tons/year are under construction [38].

An important intermediate from glycerol is acrolein, which has been examined in academia and industry [43] and already runs on a pilot scale but is not commercialized yet [44]. Many research reports are also focused on selective glycerol oxidation to dihydroxyacetone or glyceric acid, glycerol hydrogenolysis to 1,2- or 1,3-propanediol, and aqueous-phase reforming (APR) of glycerol to produce hydrogen. It is also possible to convert glycerol to epichlorohydrin in a commercial route via the Epicerol process.

18.3.2.1 Epichlorohydrin

Epichlorohydrin is an intermediate chemical with an annual production rate of 1.2 million tons (2007) worldwide and is used mainly for epoxy resins (80%) and elastomers [45]. It is produced in a multistep process from propene using chlorine and

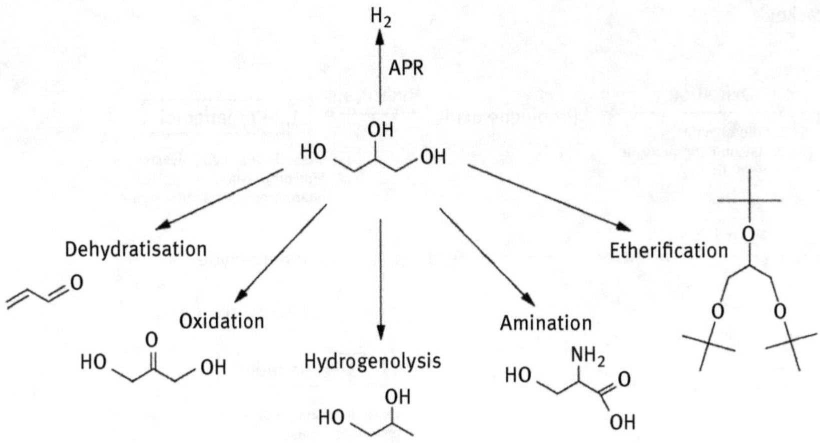

Figure 18.8: Examples for the use of glycerol.

base, releasing a major amount of chloride waste. An alternative process starting vice versa from glycerol was patented by several producers (Dow Chemical, Solvay, and Spolchemie) [46]. Plants with a total capacity of approximately 100,000 tons/year are in operation, while plants with greater capacities (~250,000 tons/year) are under construction [38]. With them, approximately 25% of the worldwide production will be produced from glycerol by the end of the decade.

This new process features the hydrochlorination of glycerol with gaseous HCl catalyzed by organic acids (Figure 18.9). First, monochlorohydrins are formed and, consequently, 1,3- and 2,3-dichlorohydrin. The reaction with base gives the desired epichlorohydrin under formation of only one equivalent of chloride salt.

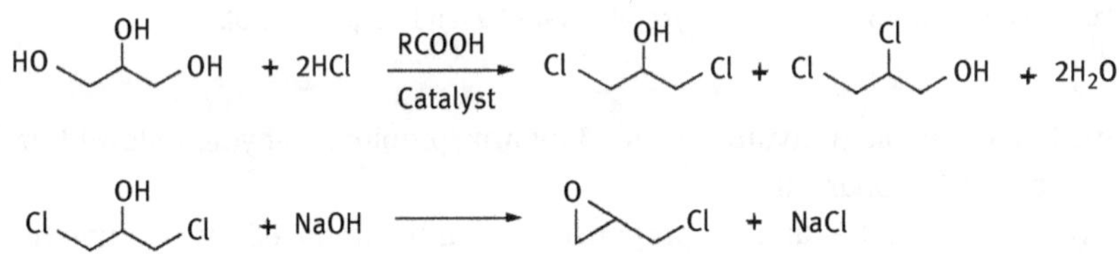

Figure 18.9: Scheme of the glycerol to epichlorohydrin process (GTE, Dow Chemical Company) [46].

18.3.2.2 Propanediols

1,2-Propanediol (propylene glycol) and 1,3-propanediol are important chemicals because of their use for the production of antifreezes and polyesters, respectively. Figure 18.10 shows the petrochemical route to both compounds based either on propylene oxide with subsequent hydrolysis (large excess of water) to

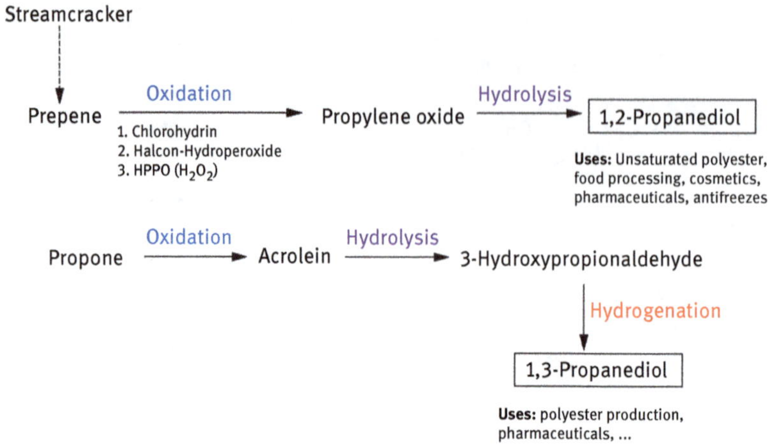

Figure 18.10: Old way – petrochemical route toward 1,2- and 1,3-propandiol.

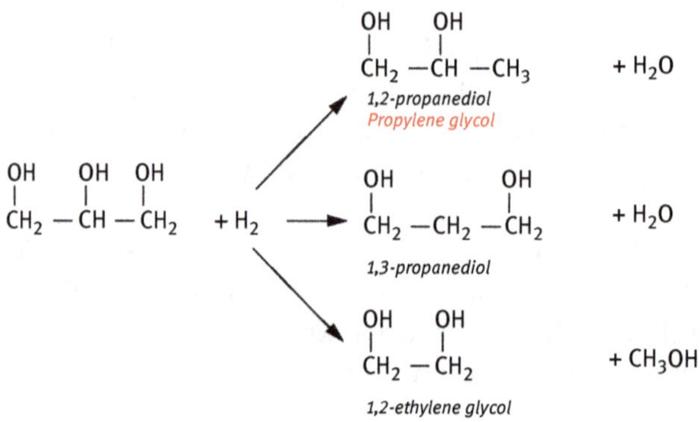

Figure 18.11: New way – hydrogenolysis of glycerol toward 1,2- and 1,3-propandiol.

1,2-propanediol or acrolein hydrolysis to 3-hydroxypropionaldehyde followed by hydrogenation to 1,3-propanediol.

However, nowadays 1,2- and 1,3-propanediols can be produced by selective hydrogenolysis with appropriate bifunctional catalysts exhibiting active sites for dehydration and hydrogenation (see Figure 18.11). Cu/ZnO-based catalysts are frequently used to produce 1,2-propanediol with high selectivity and catalyst activity; however, they suffer from strong deactivation. When Cu/ZnO is modified by Ga_2O_3, a stable catalyst can be obtained that operates even under harsh reaction conditions (e.g., 220 °C) and in the presence of water. Thus, 1,2-propanediol can be produced with a high space-time yield of 22.1 g/(g_{Cu} h) [47].

In contrast, selective hydrogenolysis of glycerol to 1,3-propanediol by means of chemo catalysis is still a challenging task. Although several attempts do exist with, for example, $Pt/WO_3/ZrO_2$ or $Ir\text{-}ReO_x/SiO_2$ catalysts [48, 49], the enzyme-catalyzed route using bacterial strains is more efficient [42] and has been commercialized (see Table 18.1).

18.3.2.3 Acrolein

Zirconium and niobium mixed oxides have been shown to catalyze the dehydration of glycerol to acrolein, at 300 °C in the presence of water with high selectivity (72%) at nearly total glycerol conversion [50]. Silica-supported niobia catalysts can also be used with similar catalytic performance [51]. Catalytic results for small-sized H-ZSM 5 zeolites showed that the high density of Brønsted acid sites favors acrolein production [52]. Acrolein production from glycerol has also been carried out in subcritical water at 360 °C and 34 MPa with catalytic quantities of $ZnSO_4$ (791 ppm [g/g]) [52].

18.3.2.4 Oxidation Products

In principle, glycerol oxidation leads, according to the complex reaction network shown in Figure 18.12, to numerous products, mainly depending on the pH of the solution and the type of catalyst (metal) used. Glyceric acid and dihydroxyacetone are the main target products because of their use for the production of amino acids and tanning components, respectively. Oxidation of the secondary OH group is more difficult and can be achieved with Pt-Bi catalysts (low pH, 50–60 °C) [54–56] or with mono- and bimetallic gold catalysts (high pH, presence of NaOH is necessary, 60 °C) also producing glyceric acid [57–60]. Pt-Bi catalysts show a high initial selectivity to dihydroxyacetone in acidic media but exhibit a strong deactivation during reaction as well. For example, in continuous experiments using a trickle-bed reactor (Liquid Hourly Space Velocity (LHSV) = 0.15–1.5 h^{-1}, 60 °C, 10 bar, O_2/glycerol = 1.4–0.14), glycerol conversion of 90% was achieved with a very high initial selectivity toward dihydroxyacetone of 80%, which decreased to 40% within 1,000 h time on stream [56]. It has been shown that glyceric acid, formed by oxidation of the primary OH group, selectively blocks those kinds of active sites that are predominantly responsible for dihydroxyacetone formation [61].

Table 18.1: Examples of the current status of industrial chemical production from biomass.

Product	Producer	Capacity, t/a	Feed	Start-up
1,3-Propandiol	Archer Daniels Midland	100,000	Glycerol	2011
	DuPont/Tate & Lyle	45,000		
	METabolic EXplorer	50,000	Crude glycerol	
3-Hydroxypropionic acid	OPX Biotechnologies	Demo scale	Sugar	
Butanediol	BioAmber	23,000		2014
	BioAmber	50,000		2014
	Genomatica	Demo scale		2012
Citric acid	Biotechconsult	Intended	Crude glycerol	
Glucaric acid	Rennovia	135,000	Glucose	2012
	Rivertop Renewables	45	Glucose	2011
Isoprene	Rivertop Renewables METabolic EXplorer	27,000 20,000	Glucose Crude glycerol	2013 2012
Itaconic acid	Biotechconsult	Intended	Crude glycerol	
Methanol (Poly) ethylene	Methanex Braskem	45,000 200,000	Waste sugar via ethanol	2011 2010
(Poly)propylene	Braskem	30,000	Sugar via ethanol	2013
Succinic acid	BASF/PURAC	75,000		2013
	BioAmber	2,000	Wheat/glucose	2011
	BioAmber	20,000		2013
	BioAmber BioAmber	35,000 65,000		2013 2014
	DSM/Roquette	300		2010
	DSM/Roquette	10,000	Starch	2012
	Myriant Technologies	13,000	Sorghum	2012
Ethylene glycol, Propylene glycol, Bio-Monoethylene glycol (BioMEG), Bio-Monopropylene glycol (BioMPG), Lignin-based Renewable Functional Fillers (RFF)	UPM	220,000	Wood (Lignocellulose)	2023

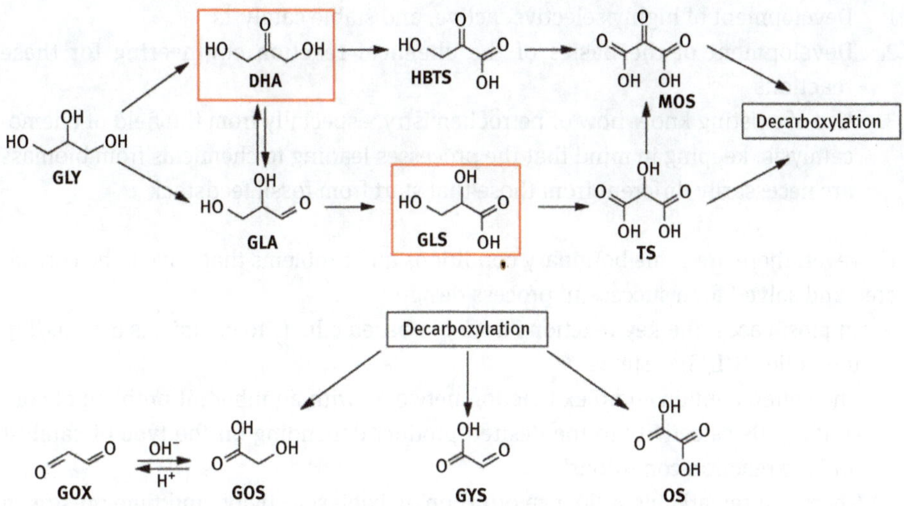

Figure 18.12: Reaction network of glycerol oxidation (GLY, glycerol; DHA, dihydroxyacetone; GLA, glyceric aldehyde; GLS, glyceric acid; HBT, hydroxypyruvic acid; MOS, mesoxalic acid; TS, tartronic acid; GOX, glyoxal; GOS, glycolic acid; GYS, glyoxylic acid; OS, oxalic acid).

18.4 Technologies and Requirements for Chemical Production from Biomass

Thermochemical processes (pyrolysis, gasification, and liquefication), biocatalysis/biotechnology, and chemocatalysis are the fundamental technologies for the production of value-added chemicals from biomass. Their application depends on the biomass feedstock and the desired products (e.g., as shown in Figure 18.4) and is aimed, in principle, at the question of whether (1) hydrodeoxygenation of the molecules (via decarboxylation, decarbonylation, or hydrogenation/reduction) is desired to reduce their high degree of oxygen functionality and improve the energy density or (2) an intelligent chemical processing/functionalization by means of biocatalysts or heterogeneous catalysts can be developed. Another strategy is the total degradation to C1 building blocks, such as CO or CH_4, followed by gradual reconstruction (e.g., via Fischer-Tropsch synthesis) in plants already established in the chemical industry.

A broad diversity of chemicals can be obtained from biomass by different chemical reactions, as exemplified previously, involving selective oxidation, hydrogenation, hydrogenolysis (of C–O or C–C bonds), hydrolysis, dehydration, C–C coupling, reforming, and so forth, known from organic chemistry. However, a precondition for the successful change of the raw material base is to develop early on the scientific and technological fundamentals for production processes of chemicals from biomass as follows:

1. Development of highly selective, active, and stable catalysts
2. Development of the basics of the chemical reaction engineering for these reactions
3. Use of existing know-how of petrochemistry, especially from the field of chemocatalysis, keeping in mind that the processes leading to chemicals from biomass are necessarily different from those that start from fossil feedstock

However, there are some boundary conditions and problems that have to be considered and solved for a successful process design:
- In most cases, the key reactions from bio-based educts to chemicals are gas/liquid/solid (G/L/S) systems.
- They often contain complex reaction networks with an inherent problem of controlling the selectivity to the desired product depending on the type of catalyst and the reaction conditions.
- There is a remarkable influence of pH on activity, selectivity, and time-on-stream behavior.
- Complete quantitative analysis of the reaction products is complicated but necessary to close material balances.
- Mass transport effects have to be considered to understand the catalyst's reactivity.
- Catalyst deactivation, for example, via leaching, dissolution of the catalyst, or decrease of metal dispersion has to be resolved.
- It is important to note that the results in terms of selectivity/yield, catalyst activity, and stability have to be verified when crude educts, instead of pure components, are used.
- Actually, most reactions are still carried out in batch reactors; however, development of continuous processes is necessary (e.g., using trickle-bed reactors).
- To understand the complex chemistry and the catalyst's mode of action in situ, catalyst characterization/operando spectroscopy is necessary, but methods for G/L/S systems are still lacking.

18.5 Economic Considerations

From the chemicals revealed in 2004 as the top 12 value-added products, only a few can be considered big sellers today.

As reported in the U.S. Department of Energy study *Top Value-Added Chemicals from Biomass.* Vol. II, *Results of Screening for Potential Candidates from Biorefinery Lignin*, the biomass production in the United States is capable of supplying all necessary raw materials for the complete chemical industry [17].

In a theoretical study, Cherubini and Strømman investigated the maximum possible yields for fuels and selected chemicals from different biomass [46]. They gave examples of three different scenarios focusing on biofuels (bioethanol and Fischer-Tropsch diesel), biochemicals (phenols, levulinic acid, formic acid, and furfural), and gasification only to Fischer-Tropsch diesel. They calculated maximum mass yields to 54%, 87%, and 41%, as well as energy yields to 88%, 81%, and 89%, respectively. The results suggest that the optimum use of biomass is dependent on the nature of biomass; while C6 carbohydrates are best converted to ethanol, C5 carbohydrates should serve as feedstock for furfurals, and lignin as feedstock for Fischer-Tropsch diesel after gasification.

Table 18.2: Forecast for the production of bulk chemicals from biomass in the Rotterdam region [43].

0–10 years	Ethylene glycol, propylene glycol, iso-propanol, acetone, butylene, methyl ethyl ketone, ethyl *tert*-butyl ether (ethylene, vinylchloride)
10–20 years	Acrylic acid, acrylonitrile, acrylamide, ε-caprolactam, ethylene, vinylchloride
20–30 years	Aromatics (benzene, toluene, xylene)

As Table 18.1 demonstrates, the chemical industry step by step has reached a point where fine chemicals are produced from biomass, especially by means of biotechnology. Beginning with the production of bulk chemicals, not only biotechnology but also chemocatalysis is needed to convert renewable feedstock into products in the high quantities characteristic of bulk chemicals.

18.6 Outlook

A case study and forecast for the explicit production of bulk chemicals from biomass in the region of Rotterdam has recently been provided by Van Haveren et al. [46]. As technologies for the conversion of, for example, ethanol, glycerol, and sugars to glycols, iso-propanol, and acetone are well known and readily available, and as the prices for feedstock are well below the selling prices for the named products, there is a clear short-term potential for these bulk chemicals (Table 18.2).

Technological barriers exist for the production of, for example, acrylic acid from biomass, while currently higher prices impede the production of N-containing chemicals (acrylonitrile, acrylamide, and ε-caprolactam) from proteins. These obstacles will be overcome according to van Haveren et al. in the next 10 to 20 years.

A clear long-term development will be the production of aromatics from biomass. As the direct extraction of aromatics from lignin still needs a breakthrough in research, the detour via pyrolysis or gasification of lignin is the current alternative for the production of aromatics.

Although these results are not transferable to all markets, the chemical industry is aware of chances to switch the value chain to feedstocks from biomass, as their prices are expected to decrease – contrary to oil prices. Based on the developments of the past decade the potential for the production of bio-based chemicals and materials can be assessed much more reliably as recently summarized in the "EU Biorefinery outlook to 2030 [64]. The potential for drop-in chemicals as well as for new bio-based value added chains is given. However, the major challenge is certainly the development of economically competitive production routes that have to address the above mentioned technological challenges regarding the catalysis, the reaction engineering and the integrated process development."

References

[1] CEFIC – The European Chemical Industry Council. Cefic review 2009–2010 – sustainability and innovation driving chemistry solutions for the future. 2010. Available from: http://asp.zone-secure.net/v2/index.jsp?id=598/765/10024&lng=en (accessed 24 January 2012).

[2] Europia. 2009 activity report. 2009. Available from: http://www.europia.com/DocShareNoFrame/Common/GetFile.asp?PortalSource=1362&DocID=25002&mfd=off&pdoc=1 (accessed 24 January 2012).

[3] Turley DB. The chemical value of biomass . In: Clark J, Deswarte F eds, Introduction to Chemicals from Biomass. Chichester: John Wiley & Sons; 2008.

[4] Fachagentur Nachwachsende Rohstoffe e. V. Daten und Fakten zu nachwachsenden Rohstoffen. 2007. Available from: http://www.fnr-server.de/ftp/pdf/literatur/pdf_303fg_dafa_071107.pdf (accessed 24 January 2012).

[5] Eurostat. Final energy consumption in Europe. 2012. Available from: http://epp.eurostat.ec.europa.eu/tgm/table.do?tab=table&tableSelection=1&labeling=labels&footnotes=yes&language=de&pcode=ten00095&plugin=1 (accessed 24 January 2012).

[6] Sanders J, Scott E, Weusthuis R, Mooibroek H. Bio-refinery as the bio-inspired process to bulk chemicals. Macromol Biosci. 2007;7:105–17.

[7] Zakzeski J, Bruijnincx PCA, Jongerius AL, Weckhuysen BM. The catalytic valorization of lignin for the production of renewable chemicals. Chem Rev. 2010;110:3552–99.

[8] Kamm B, Kamm M, Gruber PR, Kromus S. Biorefinery systems – an overview. In: Biorefineries – industrial processes and products, Vol. 1, VCH-Wiley, 2006:23.

[9] Patel M, Crank M, Dornburg V et al. BREW – Medium and Long-term Opportunities and Risks of the Biotechnological Production of Bulk Chemicals from Renewable Resources. Utrecht University; 2006.

[10] Chheda JN, Huber GW, Dumesic JA. Liquid-phase catalytic processing of biomass -derived oxygenated hydrocarbons to fuels and chemicals. Angew Chem Int Ed. 2007;46:7164–83.

[11] Lichtenthaler FW. Carbohydrates as Organic Raw Materials. Weinheim: VCH Verlagsgesellschaft, 1991.

[12] Saake B, Lehnen R. Lignin. In: Ullmann's Encyclopedia of Industrial Chemistry, Wiley-VCH Verlag GmbH & Co. KGaA, 2000.

[13] Klass DL. Biomass for renewable energy, fuels, and chemicals. San Diego: Academic Press, 1998.

[14] Van Herwijnen HWG, Fliedner E, Heep W. Verwendung nachwachsender Rohstoffe in Bindemitteln für Holzwerkstoffe. Chem Ing Tech. 2010;82:1161–68.

[15] Bugg TD, Ahmad M, Hardiman EM, Singh R. The emerging role for bacteria in lignin degradation and bio-product formation. Curr Opin Biotechnol. 2011;22:394–400.

[16] Gallezot P. Catalytic routes from renewables to fine chemicals. Catal Today. 2007;121:76–91.
[17] Holladay JE, Bozell JJ, White JF, Johnson D. Top value added chemicals from biomass. Vol. II, Results of screening for potential candidates from biorefinery lignin PNNL-16983. Richland, WA: Pacific Northwest National Laboratory; 2007.
[18] Partenheimer W. The aerobic oxidative cleavage of lignin to produce hydroxyaromatic benzaldehydes and carboxylic acids via metal/bromide catalysts in acetic acid/water mixtures. Adv Synth Catal. 2009;351:456–66.
[19] Moss GP, Smith PAS, Tavernier D. Glossary of class names of organic compounds and reactivity intermediates based on structure. Pure Appl Chem. 1995;67:1307–75.
[20] Fahy E, Subramaniam S, Brown HA et al. A comprehensive classification system for lipids. J Lipid Res. 2005;46:839–62.
[21] Thomas A. Fats and fatty oils. In: Ullmann's Encyclopedia of Industrial Chemistry, Wiley-VCH Verlag GmbH & Co. KGaA; 2000.
[22] Food and Agriculture Organization of the United Nations. Food outlook November 2011. 2011. Available from: http://www.fao.org/docrep/014/al981e/al981e00.pdf (accessed 24 January 2012).
[23] Anneken DJ, Both S, Christoph R, Fieg G, Steinberner U, Westfechtel A. Fatty acids. In: Ullmann's Encyclopedia of Industrial Chemistry, Wiley-VCH Verlag GmbH & Co. KGaA, 2000.
[24] OVID – Verband der ölsaatenverarbeitenden Industrie in Deutschland e.V. Facts and figures. 2010. Available from: http://www.ovid-verband.de/unsere-branche/daten-und-grafiken/pflanzenoele/ (accessed 24 January 2012).
[25] Norlin L-H. Tall oil. In Ullmann's Encyclopedia of Industrial Chemistry, Wiley-VCH Verlag GmbH & Co. KGaA, 2000.
[26] Noweck K, Grafahrend W. Fatty alcohols. In: Ullmann's Encyclopedia of Industrial Chemistry, Wiley-VCH Verlag GmbH & Co. KGaA, 2000.
[27] Behr A, Westfechtel A, Pérez Gomes J. Catalytic processes for the technical use of natural fats and oils. Chem Eng Technol. 2008;31:700–14.
[28] Wolfmeier U, Schmidt H, Heinrichs F-L et al. Waxes. In Ullmann's encyclopedia of industrial chemistry. Wiley-VCH Verlag GmbH & Co. KGaA, 2000.
[29] Seher A, Lange J. Gemeinschaftsarbeiten der DGF, 60. Mitteilung. Deutsche Einheitsmethoden zur Untersuchung von Fetten, Fettprodukten und verwandten Stoffen, 45. Mitt.: Analyse von Wachsen und Wachsprodukten X. Fette, Seifen, Anstrichm. 1974;76:135.
[30] Klostermeyer H, Schmandke H, Soeder CJ et al. Proteins. In: Ullmann's Encyclopedia of Industrial Chemistry. Wiley-VCH Verlag GmbH & Co. KGaA, 2000.
[31] Koutinas AA, Du C, Wang RH, Webb C. Production of chemicals from biomass. In: Clark J, Deswarte F eds, Introduction to chemicals from biomass. Chichester: John Wiley & Sons; 2008.
[32] Drauz K, Grayson I, Kleemann A, Krimmer H-P, Leuchtenberger W, Weckbecker C. Amino acids. In Ullmann's encyclopedia of industrial chemistry, Wiley-VCH Verlag GmbH & Co. KGaA, 2000.
[33] Liang G, Wu C, He L et al. Selective conversion of concentrated microcrystalline cellulose to isosorbide over Ru/C catalyst. Green Chem. 2011;13:839–42.
[34] Van De Vyver S, Geboers J, Jacobs PA, Sels BF. Recent advances in the catalytic conversion of cellulose. Chem Cat Chem. 2011;3:82–94.
[35] Kobayashi H, Ito Y, Komanoya T et al. Synthesis of sugar alcohols by hydrolytic hydrogenation of cellulose XE "cellulose" over supported metal catalysts. Green Chem. 2011;13:326–33.
[36] Cabiaca A, Guillona E, Chambonb F, Pinel C, Rataboulb F, Essayem N. Cellulose reactivity and glycosidic bond cleavage in aqueous phase by catalytic and non catalytic transformations. Appl Catal A Gen. 2011;402:1–10.
[37] Christoph R, Schmidt B, Steinberner U, Dilla W, Karinen R. Glycerol. In Ullmann's Encyclopedia of Industrial Chemistry, 7th Electronic edn, Weinheim: Wiley-VCH; 2006.

[38] Guzman DD, Wilson S, Halaln H, Lefebvre B, Taylor J, Seng S. Low glycerin price drives research. ICIS Chem Bus. 2010;278:22–23.
[39] Viinikainen TS, Karinen RS, Krause AOI. Conversion of glycerol into traffic fuels. In Centi G, Van Santen RA editors, Catalysis for renewables. Weinheim: Wiley-VCH; 2007.
[40] Bournay L, Casanave D, Delfort B, Hillion G, Chodorge JA. New heterogeneous process for biodiesel production: a way to improve the quality and the value of the crude glycerin produced by biodiesel plants. Catal Today. 2005;106:190–92.
[41] Behr A, Eilting J, Irawadi K, Leschinski J, Lindner F. Improved utilisation of renewable resources: new important derivatives of glycerol. Green Chem. 2008;10:13–30.
[42] Pagliaro M, Ciriminna R, Kimura H, Rossi M, Della Pina C. From glycerol to value-added products. Angew Chem Int Ed. 2007;46:4434–40.
[43] Coons R. Arkema finds glycerol -to-acrylic acid catalysts, demo-scale plant possible. Chem Week. 2009;171:36.
[44] Boswell C. A bigger toolbox for acrylic acid. ICIS Chem Bus. 2011;279:32–34.
[45] Sienel G, Rieth R, Rowbottom KT. Epoxides. In Ullmann's Encyclopedia of Industrial Chemistry, Wiley-VCH Verlag GmbH & Co. KGaA; 2000.
[46] Bell BM, Briggs JR, Campbell RM et al. Glycerin as a renewable feedstock for epichlorohydrin production. The GTE process. Clean Soil Air Water. 2008;36:657–61.
[47] Bienholz A, Blume R, Knop-Gericke A, Giergsdies F, Behrens M, Claus P. Prevention of catalyst deactivation in the hydrogenolysis of glycerol by Ga_2O_3-modified copper/zinc oxide catalysts XE "catalysts". J Phys Chem C. 2011;115:999–1005.
[48] Amada Y, Shinmi Y, Koso S, Kubota T, Nakagawa Y, Tomishige K. Reaction mechanism of the glycerol hydrogenolysis to 1,3-propanediol over Ir–ReOx/SiO_2 catalyst. Appl Catal B Environ. 2011;105:117–27.
[49] Qin LZ, Song MJ, Chen CL. Aqueous-phase deoxygenation of glycerol to 1,3-propanediol over Pt/WO_3/ZrO_2 catalysts in a fixed-bed reactor. Green Chem. 2010;12:1466–72.
[50] Lauriol-Garbey P, Millet JMM, Loridant S, Bellière-Baca V, Rey P. New efficient and long-life catalyst for gas-phase glycerol dehydration to acrolein. J Catal. 2011;281:362–70.
[51] Shiju NR, Brown DR, Wilson K, Rothenberg G. Glycerol valorization: dehydration to acrolein over silica-supported niobia catalysts. Top Catal. 2010;53:1217–23.
[52] Jia CJ, Liu Y, Schmidt W, Lu AH, Small-sized SF. HZSM-5 zeolite as highly active catalyst for gas phase dehydration´of glycerol to acrolein. J Catal. 2010;269:71–79.
[53] Claus P, Vogel H. The roll of chemocatalysis in the establishment of the technology platform "renewable resources.". Chem Eng Technol. 2008;31:678–99.
[54] Kimura H, Tsuto K, Wakisaka T, Kazumi Y, Inaya Y. Selective oxidation of glycerol on a platinum-bismuth catalyst. Appl Catal A Gen. 1993;96:217–28.
[55] Garcia R, Besson M, Gallezot P. Chemoselective catalytic oxidation of glycerol with air on platinum metals. Appl Catal A Gen. 1995;127:165–76.
[56] Porta F, Prati L. Selective oxidation of glycerol to sodium glycerate with gold-on-carbon catalyst: an insight into reaction selectivity. J Catal. 2004;224:397–403.
[57] Brandner A, Lehnert K, Bienholz A, Lucas M, Claus P. Production of biomass -derived chemicals and energy: chemocatalytic conversions of glycerol. Top Catal. 2009;52:278–87.
[58] Bianchi CL, Canton P, Dimitratos N, Porta F, Prati L. Selective oxidation of glycerol with oxygen using mono and bimetallic catalysts based on Au, Pd and Pt metals. Catal Today. 2005;102:203–12.
[59] Carrettin S, McMorn P, Johnston P, Griffin K, Hutchings GJ. Selective oxidation of glycerol to glyceric acid using a gold catalyst in aqueous sodium hydroxide. Chem Commun. 2002;696–97.
[60] Demirel S, Lehnert K, Lucas M, Claus P. Use of renewables for the production of chemicals: glycerol oxidation over carbon supported gold catalysts. Appl Catal B Environ. 2007;70:637–43.

[61] Wörz N, Brandner A, Claus P. Platinum-bismuth catalyzed oxidation of glycerol : kinetics and the origin of selective deactivation. J Phys Chem C. 2010;114:1164–72.
[62] Cherubini F, Strømman AH. Production of biofuels and biochemicals from lignocellulosic biomass : estimation of maximum theoretical yields and efficiencies using matrix algebra energy fuels. Energy Fuels. 2010;24:2657–66.
[63] Van Haveren J, Scott EL, Sanders J. Bulk chemicals from biomass. Biofuels Bioprod Bioref. 2008;2:41–57.
[64] EU Biorefinery Outlook to 2030, European Commission, 2021, doi:10.2777/103465
[65] Basisdaten biobasierte Produkte 2021, Fachagentur Nachwachsende Rohstoffe e.V. (FNR), https://fnr.de/fileadmin/allgemein/pdf/broschueren/basisdaten_biobasierte_produkte_2021_web.pdf
[66] Joseph J. Bozell, Gene R. Petersen, Green Chem. 2010, 12, 539–554.
[67] DOI: 10.1039/c7cs00566k, DOI: 10.1021/acs.chemrev.7b00588
[68] I. Delidovich, P. J. C. Hausoul, D. Li, M. Rose, R. Pfützenreuter, R. Palkovits, Chem. Rev. 2016, 116, 1540–1599. DOI: 10.1021/acs.chemrev.5b00354

Nikolai DeMartini, Atte Aho, Mikko Hupa, Dmitry Yu. Murzin

19 Thermal Conversion of Biomass

19.1 Introduction

Thermal conversion of biomass to energy represented almost 10% of the world's energy use in 2018 [1] (Figure 19.1). This includes traditional uses such as cooking and heating as well as the industrial production of steam and electricity. Cellulosic biomass, which is considered in this chapter, is composed of cellulose, hemicelluloses, lignin, and extractives. Cellulosic biomass has an energy content (heating value) similar to low-rank coals, such as lignite, and about 50% of that of high-rank coals such as anthracite and bituminous coals (Table 19.1). Heating value is the energy content of the solid fuel on a mass basis and is typically given as the lower heating value or higher heating value. The higher heating value includes the energy released by the combustion of the carbon and hydrogen in the fuel to carbon dioxide and water as well as the heat released by the condensation of the water formed during combustion. The lower heating value does not include the heat of condensation. Since water condenses in the flue gases outside the combustion system in most applications, the energy from the condensation of water is not captured. Biomass can have moisture contents in excess of 50%, meaning the energy density of the original feedstock can be quite low when considering transportation and storage.

Extractives and lignin have a higher energy content than cellulose and hemicelluloses due to the difference in oxygen content of these fractions. An example of an estimated energy distribution by fraction in two wood species is given in Figure 19.2. This has practical implications in current and future biorefineries as many schemes may involve fractionation of the wood to utilize the cellulose and hemicelluloses fractions for fiber products, liquid fuels, or chemicals leaving the lignin fraction to be burned to provide the energy for the Production facility. An example of this is the Kraft pulp process, where lignin, extractives, and part of the hemicelluloses are dissolved from wood in the freeing of the cellulose fibers. The dissolved organics are subsequently burned in a special boiler called a kraft recovery boiler to recover the energy in the organics and the inorganic pulping chemicals for reuse.

Combustion of solid fuels can be divided into three stages: drying, devolatilization, and char conversion (Figure 19.3). During drying, water physically present in the biomass is released. Devolatilization represents the release of part of the fuel as volatile compounds. These consist of low molecular weight compounds in

Nikolai DeMartini, Department of Chemical Engineering and Applied Chemistry, University of Toronto, Canada
Atte Aho, Mikko Hupa, Dmitry Yu. Murzin, Johan Gadolin Process Chemistry Centre, Åbo Akademi University, Henriksgatan 2, FI-20500 Turku/Åbo, Finland

https://doi.org/10.1515/9783110608458-019

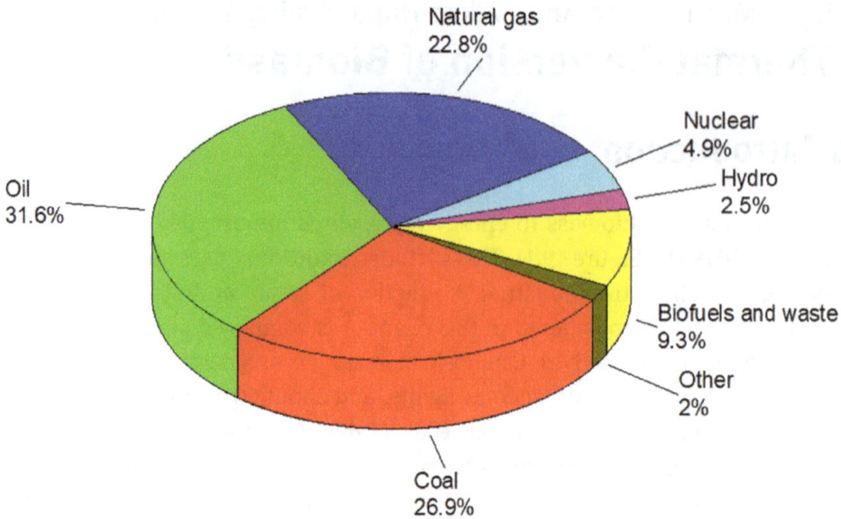

Figure 19.1: Worldwide energy distribution in 2018 by source. Based on data from [1].

Table 19.1: Lower heating value of some fuels. LHV analyzed for fuels studied at Åbo Akademi University unless noted otherwise.

Fuel	LHV ($MJ\ kg^{-1}$ d. s.)
Anthracite coal	34.9[a]
Coal, polish	28.3
Coal, brown	22.7
Peat (solid)	19.3
Wood chips	18.7
Bark, spruce	20.8
Rapeseed cake residue	20.7
Thai bagasse	17.5
Switchgrass	17.6[b]
Corn stover	16.2[b]
Straw	17.4
Sunflower seed hulls	21.0
Carbohydrates	12.3[c]
Lignin, SW	25.7[c]

Table 19.1 (continued)

Fuel	LHV (MJ kg^{-1} d. s.)
Lignin, HW	23.9[c]
Extractives	35.9[c]

Higher heating values from [a][2], [b][3], and [c][4] were converted to LHV using the concentration of hydrogen.

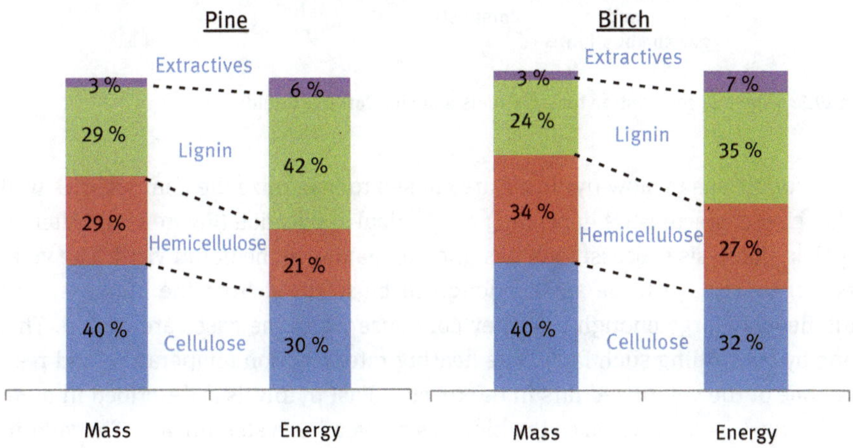

Figure 19.2: Example of calculated energy distributions for two species of wood based on the given mass compositions.

the biomass as well as higher molecular weight volatile compounds formed through the breaking of bonds in the biomass. After devolatilization, a solid carbonaceous residue called char remains. The char is then oxidized by O_2, H_2O, and CO_2, leaving the ash. In a campfire, the visible yellow flames are generated by the oxidation of the gases that are formed during devolatilization, while the glowing embers of the fire represent char oxidation where the combustion is happening at the char surface.

In this chapter, the focus will be on pyrolysis, gasification, and combustion and will not include a discussion around processes such as liquefaction and gasification in a supercritical fluid [5]. Combustion represents the complete oxidation of the biomass to carbon dioxide and water for the purpose of energy production, while pyrolysis and gasification represent partial decomposition of the biomass to form intermediate products. While pyrolysis plants, gasifiers and combustion furnaces can be technologically very different, a basic understanding of pyrolysis and gasification comes from understanding the stages of combustion. The pyrolysis process can be understood as controlled devolatilization. In slow pyrolysis, the main product is the char.

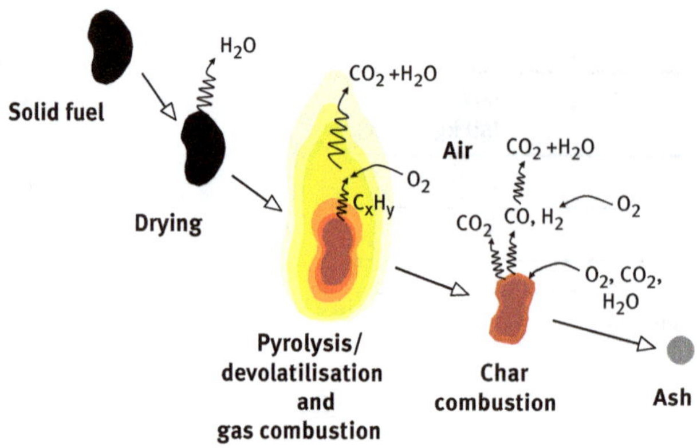

Figure 19.3: Stages of combustion for solid fuels and char forming liquids.

Process conditions in slow pyrolysis are chosen to maximize the char yield as well as other char characteristics depending on the final application (Figure 19.4). Alternatively, fast pyrolysis processes are designed to maximize the liquid yield. The volatilized organics have to be small enough to be released from the biomass, and simultaneously large enough that they condense when the gases are cooled. This is done by controlling such factors as heating rate, reaction temperature and residence time of the volatilized tars in the reactor. Fast pyrolysis is described in more detail below. Gasification takes the biomass conversion a step further. The volatilized organics and char should both be converted as much as possible to carbon monoxide and hydrogen so that only ash remains. This is done by supplying a sub-stoichiometric amount of oxygen and is covered in more detail below. In this chapter, fast pyrolysis, gasification, and combustion will be covered in that order.

19.2 Pyrolysis

19.2.1 Introduction

Pyrolysis is the thermal degradation of an organic material in the absence of oxygen, being also the initial step in combustion and gasification. The products formed in the pyrolysis are a solid residue called char, a liquid called bio-oil, and uncondensed gases. The process can be optimized for maximal bio-oil production through a process called fast pyrolysis [6]. Important requirements in this optimization are:

- moderate temperature, optimal temperature around 500 °C,
- rapid heating of the biomass particles,
- short residence time, typically under 2 s, for the pyrolysis vapors,
- fast quenching of the pyrolysis vapors to condense the bio-oil.

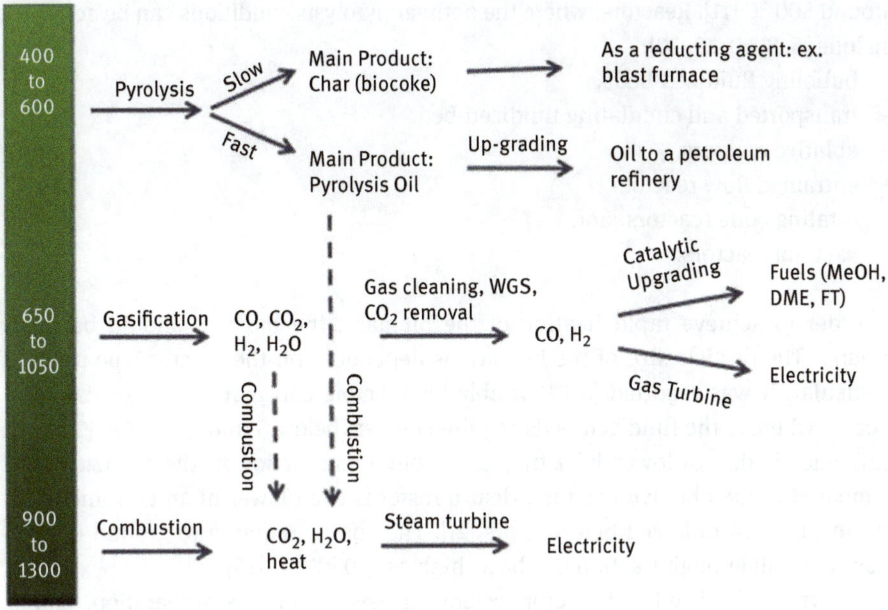

Figure 19.4: Routes for thermal conversion of biomass. Temperature is given in °C.

All these conditions can be achieved in a fluidized bed reactor system [6–10]. The yield of bio-oil through pyrolysis, which is a water containing red-brownish oxygenated organic liquid, could be as high as 75 wt% of the biomass, under the right circumstances [11]. Too rapid heating rate can have a negative effect on the bio-oil yield [12].

The by-products of pyrolysis, that is, char and gases, can be further processed, for example, combusted to provide the needed heat for the pyrolysis process [11]. The main components in the gas phase are CO_2, CO, short hydrocarbons, and hydrogen [13, 14]. Typically the char yield is around 15 wt% of the biomass, accounting for about 25% of the energy in the biomass. The pyrolysis process requires typically about 15% of the energy in the biomass [11, 15, 16], that is, pyrolysis can be considered self-sufficient in energy terms.

19.2.2 Pyrolysis Reactors

As stated above, optimal temperatures and residence times are needed for maximal bio-oil production, which can be achieved in so-called fast or flash pyrolysis, where the residence time of the pyrolysis vapors is 0.1–2 s and the optimal temperature is between 400 and 650 °C, respectively [7, 10, 16–18], with the optimum being usually

around 500 °C [11]. Reactors, where the optimal pyrolysis conditions can be achieved, include [6, 8, 10, 19, 20]:
- bubbling fluidized-beds,
- transported and circulating fluidized-bed,
- ablative reactors,
- entrained flow reactors,
- rotating cone reactors, and
- vacuum reactors.

In order to achieve rapid heating of the biomass, the biomass should be finely ground. The particle size of the biomass is dependent on the reactor type used. In particular, it was reported [21] that ablative reactors can pyrolyse large size feedstocks, whereas the fluidized beds require smaller, below 3 mm, particles [20]. The difference is due to lower heat transfer through conduction in the fluidized beds compared to the ablative reactors. Heat transfer is even lower in an entrained flow reactor, than in fluidized bed reactors [21]. The energy required to ground the biomass to small enough fraction can be as high as 100 kWh/t [16].

A typical fluidized bed reactor system consists of biomass preparation, that is, drying and grinding, biomass feeding system, a fluidized bed pyrolyzer, a char removal system and a cooling system for bio-oil collection and a place to burn the non-condensible gases. It is favorable to dry the biomass to less than 8% moisture in order to avoid phase separation of the bio-oil that might occur when the water content in the bio-oil is above 30% [16]. Continuous char removal from the pyrolysis reactor is beneficial since the char can catalyze vapor cracking, thereby decreasing the bio-oil yield [6, 8].

Valmet, previously Metso Power, has built an integrated pyrolysis pilot plant, illustrated in Figure 19.5 [22]. The integration was carried out by coupling a 2 MW_{fuel} fluidized bed pyrolyzer with a 4 MW_{th} circulating fluidized bed boiler. The pyrolyzer utilizes hot sand from the boiler as a heat source. The bio-oil is condensed with product bio-oil in a scrubber and condenser unit. The char and uncondensed gases are fed to the boiler and then combusted. In 2015 the plant was handed over to the Fortum company [23, 24]. It is reported that the integrated concept is easy and smooth to operate and has a high efficiency compared to stand-alone pyrolysis units [22].

Reports on commercial implementation of pyrolysis are scarce, while there are several pilot size units. For example, Dynamotive in Canada has built four fluidized bed units with a total capacity of 200 tons of dry wood per day [11], while Ensyn also from Canada has 8 circulating fluidized bed units with a total capacity of 100 dry tons per day [11]. Biomass Technology Group from the Netherlands has designed rotating cone pyrolyzers with a total capacity of 2 tons dry material per hour [11].

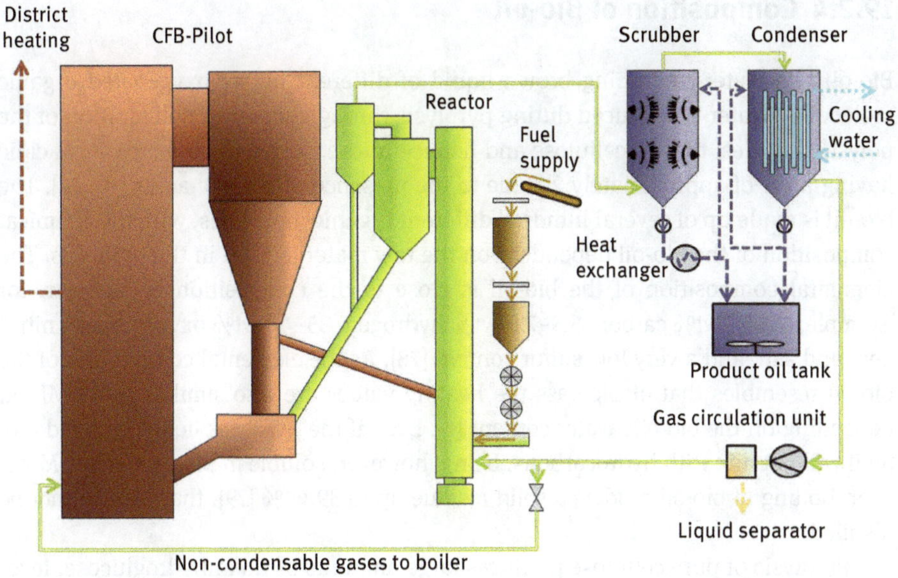

Figure 19.5: Integrated pyrolysis pilot plant [22].

19.2.3 Biomass

In principle, most types of biomass can be used as a raw material in the pyrolysis process [11]. Most of the research has been carried out using different wood as feedstock, although over 100 different types of biomass have been tested [11]. These materials include besides wood, forest residues such as bark, black liquor, and agricultural residues such as straw, olive pits, and nut shells [7, 11]. Additionally, pure biological polymers, that is, cellulose (linear polymer of D-glucose units), hemicellulose (heteropolymers of different hexoses and pentoses) and lignin (heteropolymer of *p*-coumaryl alcohol, coniferyl alcohol, and sinapyl alcohol) [25], have been tested as raw materials in the pyrolysis.

Yields of the different pyrolysis products, that is, char, bio-oil and gases are dependent on the chemical composition of biomass. Biomass with high lignin content produces more char compared to biomasses with lower lignin content [26]. Furthermore, biomass with high alkali and alkali earth metal concentrations generates less bio-oil and instead more char and gases [11, 19, 27]. Prior to pyrolysis it is beneficial to dry the biomass in order to minimize the water content in the bio-oil [11].

19.2.4 Composition of Bio-oil

Bio-oil is a water-containing brown liquid of different highly oxygenated organic molecules. Water is produced during pyrolysis through chemical dehydration of the polysaccharides, that is, cellulose and hemicellulose. Moreover, the bio-oil is acidic having a pH of approximately 2.5 due to the presence of organic acids [10, 27]. The bio-oil is made up of several hundred different organic molecules, with the chemical composition of the bio-oil dependent on the raw material used in the pyrolysis. The elemental composition of the bio-oil is close to the composition of biomass, for example, 54–58 wt% carbon, 5.5–7.0 wt% hydrogen, 35–40 wt% oxygen, some nitrogen, and ash, and a very low sulfur content [28]. As the elemental composition of the bio-oil resembles that of biomass the heating values are also similar, 16–19 MJ/kg, depending on the bio-oils water content [6]. Even if the pyrolysis liquid is called bio-oil it will not mix with hydrocarbons, being, however, soluble in polar solvents. Moreover, boiling of bio-oil renders a solid residue up to 39 wt% [29], that is, it cannot be distilled.

Pyrolysis of pure cellulose produces large amounts of 1,6-anhydroglucose, levoglucosan [30–33]. Impurities or ash-forming elements, such as alkali and metals shift the degradation of cellulose toward smaller molecules such as glycolaldehyde [7, 31, 32, 34]. Other polysaccharide-related pyrolysis degradation products are: organic acids, esters, alcohols, ketones, aldehydes, and furans [7, 10]. Thermal degradation of lignin produces a wide range of different monomeric phenols, but mostly high molar mass phenolic oligomers. The most dominating softwood lignin originating monomeric compounds are alkyl and methoxy substituted phenols, that is, cresols, guaiacols [6].

19.2.5 Utilization of Bio-oil

Some of the water-soluble chemicals in the bio-oil can be used as food flavoring, such as liquid smoke [21, 35].

Bio-oil has been tested in diesel engines [27, 36] turbines and heating boilers [6, 27]. The following conclusions were made: no char should be present in the bio-oil; the materials of the injection nozzle, needle bodies, pressure valves, and the gaskets and seals in pumps should be changed in a way that they can tolerate low pH of the bio-oil; water content and extent of thermal cracking are responsible for the ignition delay; the water in bio-oil evens out the temperature gradient and is beneficial for NO_x reduction [27]. Some of the challenges could be overcome through emulsification or by blending the bio-oil with oxygenated compounds having a high cetane number [27]. However, challenges with the material of the injection nozzles still persist.

Combustion tests of the bio-oil produced at Metso Powers pilot plant were performed in Fortum's 1.5 MW district heating plant in Masala, Finland [22]. In total, approximately 20 tons of bio-oil were combusted during 2010. Not only the burner performed well but also the bio-oil receiving, storing, and pumping system was reliable despite the fairly low outdoor temperature, −20 °C to +10 °C [22].

19.2.6 Upgrading of Bio-oil

The quality of bio-oil can be improved by applying physical or chemical, in particular, catalytic methods. This section is focused on the catalytic upgrading of bio-oil.

The quality of the highly oxygenated bio-oil can be improved by deoxygenation through various methods. The most common ones reported in the literature are catalytic hydrodeoxygenation and catalytic deoxygenation over zeolites and mesoporous materials.

Hydrogenation of the bio-oil is commonly performed over sulfided CoMo and NiMo catalysts [7, 34, 37]. Deoxygenation of the bio-oil through hydrogenation removes oxygen as water, while unsaturated C–C bonds can also be hydrogenated [7]. Typical reaction conditions are moderate temperature, 300–600 °C [7] and high hydrogen pressures, 70–200 bar [34]. Benefits with the hydrotreatment of bio-oil is, besides the lower oxygen content in the treated organic phase, also the phase separation of organics and water, which occurs when oxygen content in the organic phase is below around 2 wt% [34]. A major disadvantage with hydrogenation is, besides the saturation of aromatic rings [7], high consumption of hydrogen, which can be as high as 90 g kg^{-1} bio-oil [34]. Recently promising results on the hydrogen-assisted pyrolysis of biomass have been reported [37]. Pressures as low as 36 bars have been successfully applied in the total deoxygenation of beech wood pyrolysis oil using sulfided catalysts [37].

The required hydrogen needed for the hydrogenation of bio-oil can be produced through steam reforming of bio-oil over Ni-supported alumina catalysts [38]. Hydrogen yields as high as 73% could be produced at 950 °C without a downstream low temperature water–gas shift reactor [38].

Bio-oil can be deoxygenated over zeolites, through dehydration, decarbonylation and decarboxylation [39, 59]. Zeolites, which are crystalline aluminosilicate minerals having a three-dimensional microporous, smaller than 2 nm, framework structure, are widely used in many applications in the oil refining industry as catalysts [40]. Mesoporous materials have similar properties as zeolites, but the pore size is larger than 2 nm. Cracking, polymerization and aromatization can occur during the catalytic treatment of the bio-oil because of zeolites acidity [7]. Deoxygenation over zeolites can be performed in various ways. Zeolites can be applied as bed materials in the pyrolyzer providing simultaneously the needed heat for pyrolysis and deoxygenating the bio-oil molecules [39, 41, 42]. The pyrolysis and catalysis can be separated in a way that the biomass is first pyrolysed and then the vapors are deoxygenated without any condensation in between [43, 44, 59, 60]. One can also first produce the bio-oil and then aim to upgrade it over zeolites [45]. An advantage with the deoxygenation over zeolites compared to hydrogenation is that no hydrogen is required. A major disadvantage with the upgrading over zeolites is the high coke formation [45, 59, 60].

A summary of the parameters effecting pyrolysis is given in Table 19.2.

Table 19.2: Overview of parameters, which affect pyrolysis of biomass.

Parameter	Effects	Drawbacks
Temperature	Optimization of bio-oil yield at approximately 500 °C	High temperature favors gas formation, low temperature yields more char
Heating rate	Rapid heating rate to maximize formation of condensable pyrolysis vapors	Too high heating rate can have a negative impact on the bio-oil yield
Particle size	Small biomass particles <3 mm to increase heat transfer	Entrainment of char particles, economics of grinding
Vapor residence time	Short residence (0.1–2 s) time to minimize further degradation of the pyrolysis vapors	Longer residence times can prolong the pyrolysis reactions and result in lower bio-oil yield and higher gas yields
Vapor quenching	Rapid quenching or cooling of the vapors to maximize bio-oil yield	Risk of losing bio-oil due to insufficient condensation
Reactor type	Fluidized bed usually gives high liquid yields	High demand of inert gas for fluidization, risk of char entrainment, small biomass particle size required
Char removal	Char removal can decrease impact of vapor cracking and give a cleaner bio-oil	Cyclones or other type of char removal can make the pyrolysis set-up more complex
Biomass moisture content	Low water content in bio-oil, smaller chances of phase separation	Drying costs
Biomass type	High lignin content increases char formation	
Ash forming matter	Alkali metals influence the product distribution, more char and gases formed	Possible to remove the alkali metals by leaching in lab-scale, economically questionable in full-scale
Catalytic upgrading	Possible to decrease the oxygen content in the bio-oil through hydrodeoxygenation or upgrading over zeolites	High demand of hydrogen, rapid coking of zeolites, loss of carbon

19.3 Gasification

19.3.1 Introduction

Gasification involves adding sub-stoichiometric levels of oxygen for partial oxidation of the char and the high molecular weight volatiles. Carbon in the biomass forms carbon monoxide, carbon dioxide, non-condensable hydrocarbons such as methane, and condensable hydrocarbons (tars). Hydrogen in the biomass is converted to molecular hydrogen, water, noncondensable hydrocarbons, and tars.

While condensable hydrocarbons are a desirable product of pyrolysis, they are undesirable in gasification as they can condense, causing fouling in downstream processes. Additionally, they represent an energy loss to the system if not oxidized.

19.3.2 Gasification Reactors

Gasification can be carried out at low or high temperatures; at elevated pressures (1–32 bar); can be either directly or indirectly heated; and, can be carried out with air, oxygen, and/or steam (Figure 19.6). The boundary for high and low-temperature gasification is set by the melting of the ash. Low-temperature processes try to maximize the operating temperature while staying below the melting temperature of the ash. In high-temperature processes the temperature is sufficiently high that the ash is fully molten.

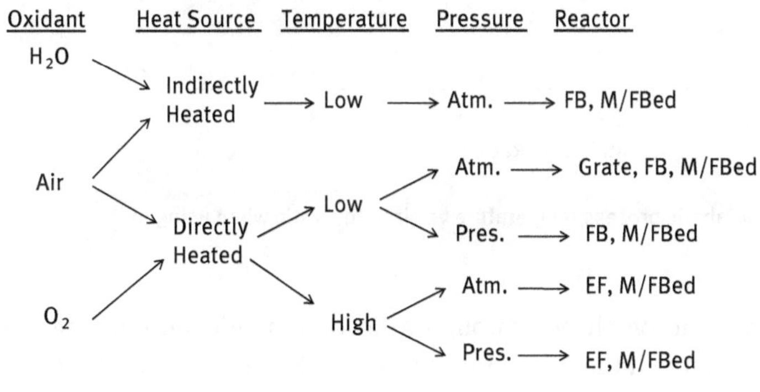

Figure 19.6: Gasification technologies/routes. (Atm., atmospheric pressure; Pres., pressurized; FB, fluidized bed (both bubbling and circulating; M/F bed, moving or fixed bed; EF, entrained flow).

Gasifier technologies include fluidized beds (bubbling and circulating), moving beds (co-current and counter current), grate (both moving and stationary), and entrained flow. Fluidized beds need to be operated below the ash melting temperature to avoid bed agglomeration and the resulting de-fluidization of the bed. Moving beds can be operated above or below the ash melting temperature, while entrained flow gasifiers are generally operated above the melting temperature of the salts.

Directly heated gasification involves the addition of air or oxygen to partially oxidize some of the fuel to carbon dioxide and water to provide the energy needed to drive the gasification reactions. The higher the amount of air or oxygen used, the higher the reactor temperature (Figure 19.7). The nitrogen in air has a diluting effect on the product gas from directly heated air blown gasification. This reduces the heating value of the product gas and increases the volume of the product gas. If the product gas is going to be burned in a boiler or kiln [46], air is used for gasification. If the product gas (also called synthesis gas) is going to be used in catalytic processes such as the production of liquid fuels, oxygen is applied. The use of oxygen minimizes the electricity needed for compression of the product gas and maximizes the concentration of the reactive components, carbon monoxide, and hydrogen gas.

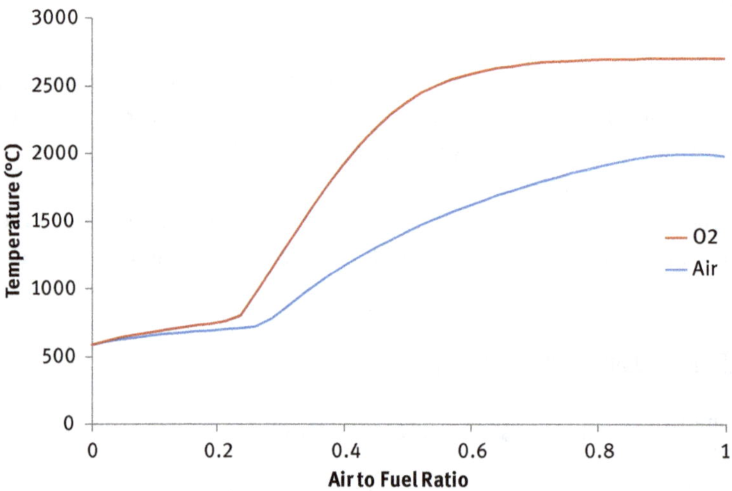

Figure 19.7: Calculated adiabatic process temperature vs air to fuel ratio when using air or oxygen as an oxidant.

Indirect heating can be accomplished through a heating mantle surrounding the gasification chamber; through the use of tubes running through the gasifier; or by recycling a heat transfer median such as sand in a fluidized bed. This is used, for example, in steam gasification.

19.3.3 Energy in Gasification

The synthesis gas from direct gasification represents about 70% of the energy in the original biomass [47]. The remaining 30% is consumed in the gasification process and in heat losses. Oxygen blown gasifiers can produce a synthesis gas with a lower heating value of 7–9.5 MJ $m^{-3}{}_n$. A calculated lower heating value (MJ/$m^{-3}{}_n$) and heat consumed or generated (deltaH) for synthesis gas generated at 900 °C and 1 bar and at adiabatic conditions is given in Figure 19.8. Air-blown gasifiers produce a syngas with a lower heating value of 2.–3.5 MJ $m^{-3}{}_n$ due to the diluting effect of nitrogen in the air [48]. Higher heating values of approximately 11–13 MJ $m^{-3}{}_n$ have been reported with steam reforming of black liquor where the heat for gasification is supplied externally [49] (Figure 19.8).

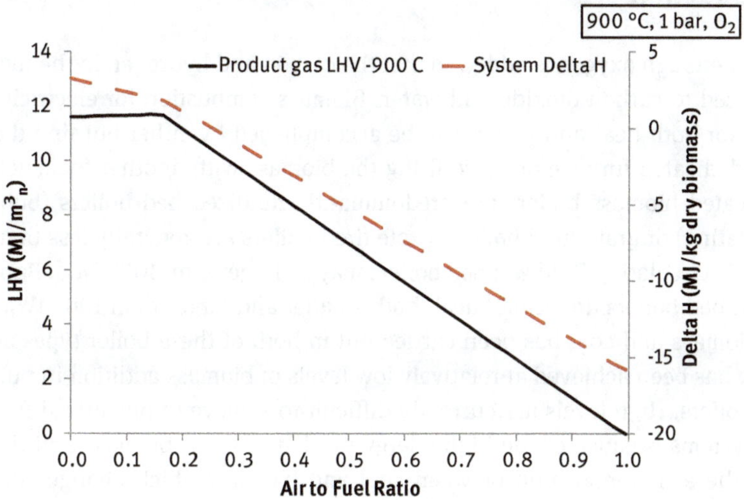

Figure 19.8: The calculated lower heating value (LHV) and heat consumed or generated (DeltaH) for biomass gasification at 900 °C, 1 barand assuming adiabatic conditions. The gas composition and heat consumed or generated at different air to fuel ratios were calculated using the Gibbs free energy minimization software FACTSage [50].

Biomass gasification combined cycle has been considered one alternative to achieving higher electricity production from biomass when compared to biomass combustion. In combined cycle, the biomass is gasified, and the syngas is cleaned to remove alkali salts and sulfur before being burned in a gas turbine. Steam is then generated with the hot flue gas from the exit of the gas turbine, and this steam in turn generates additional electricity in a steam turbine. Current coal-fired gasification combined cycle demonstration plants are attaining electricity production similar to state-of-the-art supercritical pulverized coal direct combustion boilers though they are expected to be able to convert up to 50% of the thermal energy in the coal to electricity [51].

Synthesis gas can also be used to produce fuels or chemicals by catalytic upgrading. The process involves conditioning of the synthesis gas which includes a water–gas shift reactor to adjust the H_2/CO ration; removal of reduced sulfur compounds to less than 0.1 ppmv [52] and catalytic upgrading. A study by Bain [47] estimated that gasification of wood for methanol production would result in a methanol stream equal to 56% of the heating value of the incoming wood. A study of black liquor gasification for methanol or dimethyl ether (DME) also found that 56% of the chemical energy in the black liquor could be recovered at either methanol or DME [53].

19.4 Combustion

19.4.1 Introduction

In combustion, enough oxygen is added so that the carbon and hydrogen in the fuel are fully oxidized to carbon dioxide and water. Biomass combustion for electricity generation or for both heat and power can be accomplished by either burning the biomass in a dedicated furnace or by co-firing the biomass with another fuel such as coal. Dedicated biomass boilers are predominantly fluidized bed boilers (bubbling or circulating) or grate-fired boilers. Grate fired boilers are generally less than 100 MW_{th}, while circulating fluidized bed boilers may be larger than 100 MW_{th}. Bubbling fluidized bed boilers are economical, both smaller and larger than 100 MW_{th}. Co-firing of biomass and coal has been carried out in both of these boiler types as well. Co-firing has been achieved at relatively low levels of biomass addition in pulverized coal boilers. High levels are currently difficult to achieve in pulverized fuel combustion systems because of the high energy needed to grind biomass and the difference in the ash composition between coal and biomass which changes the fouling and corrosion characteristics of the boiler ash as well as what the resulting ash can be used for.

19.4.2 Energy in Combustion

The heat released from combustion of the fuel is transferred by radiation and convection to evaporate water and create superheated steam, which is then used to create electricity in a steam turbine. Steam temperatures in state-of-the-art coal fired boilers are pushing close to 600 °C (i.e., above the critical point of water) with net electricity production reaching 45+% of the thermal energy (lower heating value) of the burned fuel [51]. Present-day sub-critical boilers are mostly just around 40% net efficiency in electricity production, but older boilers can have efficiencies as low as 30% on a lower heating value basis.

A key to extracting more electricity from biomass combustion is reaching higher steam temperatures. Steam temperatures are limited by the increased superheater tube corrosion due to the flue gas impurities (alkalies, chlorine) originating in the biomass. In co-generating plants, the steam is partly used at lower pressures for district heating or industrial processes. This results in a much higher thermal efficiency (80–90%) compared to electricity only, but at the cost of lower electricity production.

19.4.3 Co-combustion

Co-combustion involves partial replacement of a fossil fuel such as coal with biomass. Reviews of co-combustion have been published by Leckner [54] and Sami [55]. Co-combustion is successfully done at the commercial scale [56, 57]. For partial substitution of coal by biomass, the biomass can be fed with the coal through the same burner/feeder, or fed to a separate burner. Alternatively, the biomass can be first pyrolysed/gasified, and then the products can be burned in the boiler. Biomass gasification followed by combustion in a boiler has been commercially applied and is illustrated in Figure 19.9.

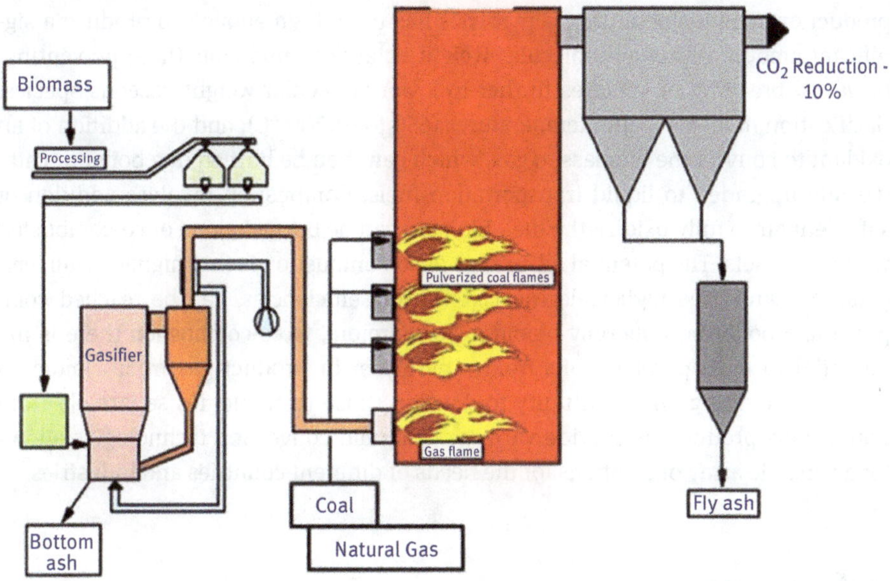

Figure 19.9: Illustration of Foster wheeler gasifier in Lahti, Finland, used for gasification of waste or biomass. The gas from the gasifier is then burned in a boiler, which can process coal or natural gas.

One factor limiting biomass use in pulverized coal-fired boilers is the difficulty in grinding biomass. It is energy-intensive and results in a larger particle size and distribution compared to pulverized coal [3, 58]. Fluidized beds have been successfully used for both co-firing biomass with coal and for firing biomass alone [56]. Fluidized beds are more fuel-flexible than pulverized fuel boilers and do not require the size reduction (i.e., grinding) of the fuel that is required for pulverized fuel units. This is the main factor limiting the use of high levels of substituting in pulverized coal boilers where substitution levels are typically less than 10% on an energy basis; though trials up to 20% have been successfully run with only slight increases in slagging [54, 57, 58]. In addition to affecting the fouling and corrosion in coal fired boilers the ash from the co-combustion of biomass with coal reduces the ability to sell the ash for industrial use such as use in concrete and cement [54, 57].

19.5 Summary

Thermal conversion is one of the ways to utilize the energy stored in biomass. Thermal conversion can be used to convert the chemical energy in biomass to liquid fuels, electricity, and heat. Pyrolysis at temperatures between 400 and 600 °C is utilized to produce a lower grade liquid fuel, which can either be burned directly in the furnace of an oil-fired boiler or catalytically upgraded for transportation fuel production. The temperature for pyrolysis has to be high enough to produce a significant amount of higher-molecular-weight volatile components (tars) preventing, however, breaking of volatiles further to lower-molecular-weight gaseous species. Gasification involves higher temperatures (650 to ~1,300 °C), and the addition of an oxidant to convert the biomass to gas, which can then be burned in a boiler or catalytically upgraded to liquid transportation fuels. Combustion involves addition of sufficient air to fully oxidize the fuel. Biomass can be burned alone or co-combusted with fossil fuels. The potential advantage of co-combustion is that higher steam temperatures, and thus higher electrical generating efficiencies, can be reached compared to a boiler burning only biomass. Furthermore, in co-combustion there is the potential to burn pyrolysis oils from a pyrolyser or product gas from a biomass gasifier, providing an opportunity to develop those technologies separately from liquid fuels production. A wide variety of thermal conversion technologies allows for a versatile array of solutions for the needs of different countries and industries.

References

[1] World Energy Outlook 2020, IEA Publications, International Energy Agency, 79.

[2] Bartock W, Sarofim AF. Fossil Fuel Combustion: A Source Book. New York: Wiley, 1991.

[3] Mani S, Tabil LG, Sokhansanj S. Grinding performance and physical properties of wheat and barley straws, corn stover and switchgrass. Biomass Bioenergy. 2004;27:339–52.

[4] Frederick WJ Black liquor properties. In: Adams TN, ed. Kraft Recovery Boilers. Atlanta, GA: TAPPI Press; 1998. 78.

[5] Peterson AA, Vogel F, Lachance RP, Fröling M, Antal MJ Jr., Tester JW. Thermochemical biofuel production in hydrothermal media: A review of sub- and supercritical water technologies. Energy Environ Sci. 2008;1:32–65.

[6] Bridgwater AV, ed. Fast Pyrolysis of Biomass: A Handbook Volume 2. Newbury, United Kingdom: CPL Press, 2002.

[7] Huber GW, Iborra S, Corma A. Synthesis of transportation fuels from biomass: chemistry, catalysts, and engineering. Chem Rev. 2006;106:4044–98.

[8] Jahirul MI, Rasul MG, Chowdhury AA, Ashwath N. Biofuels production through biomass pyrolysis – A technological review. Energies. 2012;5:4952–5001.

[9] Lappas AA, Dimitropoulos VS, Antonakou EV, Voutetakis SS, Vasalos IA. Design, construction, and operation of a transported fluid bed process development unit for biomass fast pyrolysis: effect of temperature, Ind. Eng Chem Res. 2008;47:742–47.

[10] Mohan D, Pittman CU, Steele PH. Pyrolysis of wood/biomass for bio-oil: A critical review. Energy Fuels. 2006;20:848–89.

[11] Bridgwater AV. Review of fast pyrolysis of biomass and product upgrading. Biomass Bioenergy. 2012;38:68–94.

[12] Efika CE, Onwudili JA, Williams PT. Influence of heating rates on the products of high-temperature pyrolysis of waste wood pellets and biomass model compounds. Waste Manage. 2018;76:497–506.

[13] Matta J, Bronson B, Gogolek PEG, Mazerolle D, Thibault J, Mehrani P. Comparison of multi-component kinetic relations on bubbling fluidized-bed woody biomass fast pyrolysis model performance. Fuel. 2017;210:625–38.

[14] Pereira EG, Martins MA, Perenka R, Carneiro ADCO. Pyrolysis gases burners: Sustainability for integrated production of charcoal, heat and electricity. Renewable Sustainable Energy Rev. 2017;75:592–600.

[15] Leibbrandt NH, Knoetze JH, Görgens JF. Comparing biological and thermochemical processing of sugarcane bagasse: An energy balance perspective. Biomass Bioenergy. 2011;35:2117–26.

[16] Onarheim K, Solantausta Y, Lehto J. Process simulation development of fast pyrolysis of wood using Aspen Plus. Energy Fuels. 2015;29:205–17.

[17] Akhtar A, Krepl V, Ivanova T. A combined overview of combustion, pyrolysis, and gasification of biomass. Energy Fuels. 2018;32:7294–318.

[18] Mortensen PM, Grunwaldr J-D, Jensen PA, Knudsen KG, Jensen AD. A review of catalytic upgrading of bio-oil to engine fuels. Appl Catal A Gen. 2011;407:1–19.

[19] Brown RC, ed. Thermochemical Processing of Biomass Conversion into Fuels, Chemicals and Power. Chichester, West Sussex, United Kingdom: John Wiley & Sons, Ltd, 2011.

[20] Crocker M, ed. Thermochemical Conversion of Biomass to Liquid Fuels and Chemicals. Cambridge, United Kingdom: Royal Society of Chemistry, 2010.

[21] Bridgwater AV. An overview of fast pyrolysis of biomass. Org Geochem. 1999;30:1479–93.

[22] Autio J, Lehto J, Heikkinen M, Oasmaa A, Solantausta Y, Jokela P, Alin J (2011). Production of bio-oil and chemicals in an integrated pyrolysis pilot unit. In: Proceedings of the Third Nordic Wood Biorefinery Conference, pp. 47–52.

[23] https://www.fortum.com/about-us/our-company/our-energy-production/our-power-plants/joensuu-chp-plant accessed 14.6.2018

[24] https://www.valmet.com/more-industries/bio/bio-oil/ accessed 14.6.2018.

[25] Fengel D, Wegener G. Wood, Chemistry, Ultrastructure, Reaction. Berlin, Germany: Walter de Gruyter, 1984.

[26] Nowakowski DJ, Bridgwater AV, Elliott DC, Meier D, De Wild P. Lignin fast pyrolysis: results from an international collaboration, J. Anal Appl Pyrolysis. 2010;88:53–72.

[27] Chiaramonti D, Oasmaa A, Solantausta Y. Power generation using fast pyrolysis liquids from biomass. Renewable Sustainable Energy Rev. 2007;11:1056–85.

[28] Czernik S, Bridgwater AV. Overview of applications of biomass fast pyrolysis oil. Energy Fuels. 2004;18:590–98.

[29] Branca C, Di Blasi C, Elefante R. Devolatilization and heterogeneous combustion of wood fast pyrolysis oils. Ind Eng Chem Res. 2005;44:799–810.
[30] Essig M, Richards GN, Schenck E. Mechanisms of formation of the major volatile products from the pyrolysis of cellulose. In: Schuerch C, ed. Cellulose and Wood – Chemistry and Technology. Syracuse, NY, USA: 841–62.
[31] Evans RJ, Milne TA. Molecular characterization of the pyrolysis of biomass. 1. Fundamentals. Energy Fuels. 1987;1:132–37.
[32] Richards GN. Glycolaldehyde from pyrolysis of cellulose. J Anal Appl Pyrolysis. 1987;10: 251–55.
[33] Shafizadeh F, Fu YL. Pyrolysis of cellulose. Carbohydr Res. 1973;29:113–22.
[34] Bridgwater AV. Production of high grade fuels and chemicals from catalytic pyrolysis of biomass. Catal Today. 1996;29:285–95.
[35] http://www.redarrowusa.com/natural_smoke_condensates/manufacturing_process.html (accessed 15.06.2011).
[36] Solantausta Y, Nylund N-O, Westerholm M, Koljonen T, Oasmaa A. Wood-pyrolysis oil as fuel in a diesel-power plant. Bioresour Technol. 1993;46:177–88.
[37] Stummann MZ, Høj M, Schandel CB, Hansen AB, Wiwel P, Gabrielsen J, Jensen PA, Jensen AD. Hydrogen assisted catalytic biomass pyrolysis. Effect of temperature and pressure. Biomass Bioenergy. 2018;115:97–107.
[38] Seyedeyn-Azad F, Salehi E, Abedi J, Harding T. Biomass to hydrogen via catalytic steam reforming of bio-oil over Ni-supported alumina catalysts. Fuel Process Technol. 2011;92: 563–69.
[39] Atutxa A, Aguado R, Gayubo AG, Olazar M, Bilbao J. Kinetic description of the catalytic pyrolysis of biomass in a conical spouted bed reactor. Energy Fuels. 2005;19:765–74.
[40] Čejka J, Van Bekkum H, Corma A, Schüth A, ed. Introduction to Zeolite Science and Practice. 3rd revised edition. Amsterdam, the Netherlands: Elsevier Science; 2007.
[41] Aho A, Kumar N, Eränen K, Salmi T, Hupa M, Murzin DY. Catalytic pyrolysis of woody biomass in a fluidized bed reactor: Influence of the zeolite structure. Fuel. 2008;87:2493–501.
[42] Lappas AA, Samolada MC, Iatridis DK, Voutetakis SS, Vasalos IA. Biomass pyrolysis in a circulating fluid bed reactor for the production of fuels and chemicals. Fuel. 2002;81: 2087–95.
[43] Aho A, Kumar N, Lashkul AV, Eränen K, Ziolek M, Decyk P, Salmi T, Holmbom B, Hupa M, Murzin DY. Catalytic upgrading of woody biomass derived pyrolysis vapours over iron modified zeolites in a dual-fluidized bed reactor. Fuel. 2010;89:1992–2000.
[44] Ratnasari DK, Yang W, Jönsson PG. Two-stage ex-situ catalytic pyrolysis of lignocellulose for the production of gasoline-range chemical. J Anal Appl Pyrolysis. 2018;134:454–64.
[45] Vitolo S, Seggiani M, Frediani P, Ambrosini G, Politi L. Catalytic upgrading of pyrolysis oil sto fuel over different zeolites. Fuel. 1999;78:1147–59.
[46] Francey S, Tran H, Jones A (2008) Current status of alternative fuel use in lime kilns. 2008 TAPPI Engineering, Pulping and Environmental Conf., 24.27 Aug. 2008, Portland, OR, USA.
[47] Bain RL (1992) Material and energy balances for methanol from biomass using biomass gasifiers. NREL/TP-510-17098, 128 pp.
[48] Lindblom M (2003) An overview of Chemrec process concepts, Colloquium on black liquor combustion and gasification, 13-16 May 2003, Parck City, Utah.
[49] Whitty K, Verrill CL (2004) A historical look at the development of alternative black liquor recovery technologies and the evolution of black liquor gasifier designs. 2004 International Chemical Recovery Conference. 6-10 June, Charleston, S.C., USA, Vol. 1: 13-33.

[50] Bale CW, Bélisle E, Chartrand P, Decterov SA, Eriksson G, Hack K, Jung I-H, Kang Y-B, Melançona J, Peltona AD, Robelin C, Petersen S. FactSage thermochemical software and databases – recent developments. Calphad. 2009;33:295–311.

[51] IEA (2006) Energy technology perspectives: Scenarios and strategies to 2050. 479.

[52] Korens N, Simbeck DR, Wilhelm DJ (2002) Process screening analysis of alternative gas treating and sulfur removal for gasification. Revised Final Report. U.S. Department of Energy. Task Order No. 739656-00100: 144 pp.

[53] Ekbom T, Lindblom M, Berglin N, Ahlvik P (2003) Technical and commercial feasibility study of black liquor gasification with methanol/DME production as motor fuels for automotive uses – BLGMF. (ALTENER programme of the European Union, Contract No. 4.1030/Z/01-087/2001), 200.

[54] Leckner B. Co-combustion – A summary of technology. Thermal Sci. 2007;11(4):5–40.

[55] Sami M, Annamalai K, Wooldridge M. Co-firing of coal and biomass fuel blends. Prog Energy Combust Sci. 2001;27:171–214.

[56] Kokko A, Nylund M (2005) Biomass and coal co-combustion in utility scale – Operating experience of Alholmens Kraft. 18th International Conf. on FBC, ASME, New York, USA, ISBN: 0-7918-3755-6, Paper FBC2005-78035.

[57] Overgaard P, Sander B, Junker H, Friborg K, Larsen O-H (2004) Two years' operational experience and further development of full-scale co-firing of straw. 2nd World Conf. on Biomass for Energy, Industry and Climate Protection, 10-14 May 2004, Rome, Italy, 1261–64.

[58] Pedersen L-S, Nielsen HP, Kiil S, Hansen LA, Dam-Johansen K, Kildsig F, Christensen J, Jespersen P. Fuel. 1996;2(13):1584–90.

[59] Williams PT, Horne PA. The influence of catalyst type on the composition of upgraded biomass pyrolysis oil. J Anal Appl Pyrolysis. 1995;31:39–61.

[60] Williams PT, Horne PA. Analysis of aromatic hydrocarbons in pyrolytic oil derived from biomass. J Anal Appl Pyrolysis. 1995;31:15–37.

[61] Veerapen J, Beerepoot M (2011) Co-generation and renewable: Solutions for a low-carbon energy future. OECD/IEA, 31.

Markus Antonietti

20 Biomass to Mineralized Carbon: Energy Generation and/or Carbon Sequestration

Low-value biomass (e.g., from waste streams or agricultural side products) is a potentially very valuable raw material source. For that, biomass has to be stabilized and mineralized, and a process called hydrothermal carbonization (HTC) can be used for densification of the material and energy content, but also as a unification technology. As the resulting products bind carbon from natural, regrowing resources, they potentially influence the CO_2 balance sheet in a favorable, lowering fashion. It will be discussed how such novel usage cascades based on abundant biomasses can contribute to global energy/materials generation or CO_2-sequestration schemes.

20.1 Introduction

Our industrial society currently depends on a stable support of fossil crude oil for energy generation, transportation, and the generation of chemical platform chemicals. The end of the "oil age" is, however, foreseeable, and the economic earthquakes preceding the shortage of oil are already being felt today. A further downside of the oil economy is the liberation of large amounts of CO_2. From oil alone, 12.5 billion tons (Gt) of CO_2 are generated, with the known implications for climate and weather extremes.

The replacement of smaller material streams by renewables is certainly desirable, but essentially just window dressing, as it leaves the very basic problems untouched. In particular, it cannot replace the current base load of fossil energy carriers, so humankind is essentially employing a "business as usual" approach without effective changes.

How could an actual useful instrument or technology be defined? It is very clear that it would be worthwhile to not only lower further CO_2 emissions (the current, realistic base of all climate models) but also create processes that counteract previous development by binding new atmospheric CO_2 as well as that from early industrialization. The thought of climate management instead of climate consumption is on the horizon, but currently still not the object of a public discussion. Realistically, from the chemical point of view there is no bypass around such a technology: the search for new carbon deposits is key, while part of our industrial products must be

Markus Antonietti, Department of Colloid Chemistry, Max Planck Institute of Colloids and Interfaces, Forschungscampus Golm, 14424 Potsdam, Germany, e-mail: Markus.Antonietti@mpikg.mpg.de

https://doi.org/10.1515/9783110608458-020

turned into carbon-negative deposits (i.e., products that bind carbon rather than generate it throughout their life cycle).

20.2 HTC

20.2.1 HTC of Biomass Waste for Environmentally Friendly Carbon Sequestration

The simplest path to this target is presumably indirect, but close at hand. The most effective and cost-free CO_2 collector is Mother Nature herself, and she is also the most massive sink for carbon. A rough estimate of the terrestrial biomass production (not including the contributions of oceans) amounts to 120 Gt/year as dry matter, these are approximately 60 Gt bound carbon or 220 Gt sequestered CO_2 per year [1, 2]. The natural CO_2 cycle is therefore still one order of magnitude larger than the anthropogenic one, except that nature has been in equilibrium for hundreds of millions of years. This perpetuated binding and liberation of CO_2 can indeed serve as the role model for future chemistry.

The generated biomass is, however, just a short-term, temporal sink of CO_2, as the microbial degradation after plant death liberates exactly the amount of CO_2 that was previously bound in the plant material. Plant biomass, because of its chemical structure, contains 40 wt% (carbohydrates) and 50 wt% (lignocellulosic biomass) carbon. Locking away "just" 8.5 wt% of the freshly produced biomass from the active ecosystem would indeed compensate for the complete CO_2 generation from crude oil. This nicely illustrates the overall scale of an actual effective climate operation, and global application of biomass collection and conversion schemes might indeed not be unrealistic for lowing atmospheric CO_2 by direct human operations.

There are many reasons why the "locking away" will not occur as a slag heap, and from the viewpoint of a material chemist or an economist, the carbon should instead be used for a secondary purpose while being deposited. Examples of this are adding carbon to large-scale technical products, as a component to improve concrete, as a mixing additive for street pavements, for building insulation, or even as a filling soot for mechanic or coloristic applications. For all that, we need another processing step to turn biomass into coal, coke, carbon, or soot, to accomplish densification and uniformity, and *carbonization* (i.e., the conversion of soft biomolecules into a carbon-rich powder) is the name for this process.

Relevant to the present argument, there exists not only the hot flame carbonization, as a charcoal burner is using it, but also a more effective "wet" carbonization, which takes place on the timescale of a few hundred years (throughout the formation of peat) to some millions of years (black coal). In this pathway, biomass is dehydrated under slightly acidic conditions and the exclusion of oxygen, essentially creating only water and coal. In the modern scientific literature, some experiments are described in which this coal formation is mimicked by faster technological processes. The HTC was performed by Bergius and Specht as early as 1913. In those experiments, cellulose was converted hydrothermally in coal-like substances [3, 4]. Biomass in the presence of water and potential catalysts is heated to 180–230 °C in a closed autoclave, and the reaction still takes days to weeks. More modern versions still accelerate those processes by additives and allow the generation of meaningful micro- and nanostructures with special surface chemistry. The finally described acceleration of coalification by factors of 10^6–10^9 down to the hour scale makes HTC a "polycondensation" technique that in our opinion is worth considering on a larger scale.

$$C_6H_{12}O_6 \rightarrow C_6H_4O_2 + 4\,H_2O$$

Glucose "biocoal" water

Scheme 20.1: Operating principle of HTC. Under temperature and catalysis, carbohydrates (here glucose) are converted to biocoal and water only. The sum formula of biocoal is a simplification and a matter of the reaction conditions.

In comparison to other biomass processes, HTC is not only fast but also simple and effective. On the one hand, it inherently can use wet starting products, as the reaction only effectively takes place in a water environment. Costly drying, therefore, is not demanded. On the other hand, biocoal can be easily filtered off from the wastewater stream; the product isolation is rather simple. It is also favorable that the majority of salts is kept in the water stream and can be donated back to the natural cycle. Furthermore, under weak acidic conditions, at temperatures below 220 °C, and using wastewater for processing, practically all carbon from the biomass ends up in the biocoal (i.e., the carbon efficiency [CE] is practically 1). Another beneficial aspect is that the reaction is exothermic and spontaneous (i.e., the process does not rely on excessive energy input). The real energy output is still under debate, but depending on the reaction scheme and degree of aromatization of the coal, between 5% and 30% of the original heat of combustion is liberated throughout the process. This was already observed by Bergius in his setup: without cooling, some of his reactions would have heated up to a few hundred degrees Celsius.

It is therefore my opinion that carbonization of waste biomass or of rapidly growing plants or algae (farmed just for this purpose) represents at the moment the most effective process for the removal of atmospheric CO_2. Ideally, however, this will be coupled for profitability with chemical material synthesis ("make something useful out of it") and presumably can be supported by political instruments encouraging "CO_2-negative" products.

20.2.2 HTC for "Carbon-Negative Materials"

Material chemists now look for such applied materials with a potential footprint of billions of tons. An attractive application of that scale is the use of such biocoals as ion-exchange resins or water "superabsorbers" (i.e., one has to incorporate ion- and water-binding properties into the "coal scaffolds"). Such ion-binding products could be used for the purification of drinking water or for the improvement of agricultural soils. In terms of chemistry, this technology corresponds to a replication of the natural processes of black soils, highly humic matter containing soils, which are among the most fertile soils on earth. The development of biocoals with chemically adjusted humic matter profiles could bring a "biomimetic soil improvement" and contribute to the preservation and productivity of agricultural areas, even under altered climate conditions. The need for humic matter on this planet is indeed measured in billions of tons, and the potentialities for a chemist to work with such molecules can be easily illustrated.

By choice of the biomass or adding minor amounts of special chemical comonomers, a hydrophilic surface with special functional groups and a high capillarity as a key property can be adjusted. Besides distinct chemical functionalities, a special product texture is also beneficial for such a purpose (i.e., the micro- and nanoarchitecture of the coal has to be optimized). In this way, and nevertheless using biological waste as a starting product, the whole knowledge base of modern material chemistry can be used to undertake such a project, and a "superhumus" could be synthesized, as it is rather rarely produced in nature. Figure 20.1 depicts the inner structure of this HTC product, which was synthesized from Berlin oak leaves [5]. The material combines optimal accessibility with a highly functional surface chemistry and is ideally suited for the capillary binding of water and specific ion binding, a "carbon sponge" in figurative language.

For use as a soot replacement (e.g., for printing purposes or as a filler for car tires), other disperse morphologies are regarded to be optimal (e.g., as those known from emulsion polymerization processes), and a corresponding reaction engineering of HTC including comonomers and stabilizers is providing exactly such products (Figure 20.2A). For use in gas filters, thermal insulation, mechanical strengthening, and antistatic finishing, nanofiber architectures are most optimal, and again, HTC can be directed to produce these materials (Figure 20.2B). It is important to understand that HTC is just another heterophase polymerization process that allows reapplication of known principles of material design.

Making such products is a highly attractive task, especially when biomass waste is directly used for product synthesis, without chemical deviations. The concept of CO_2-negative chemical products is new in industrial thinking, but companies could improve their CO_2 balance simply by making some of their products in this manner [6].

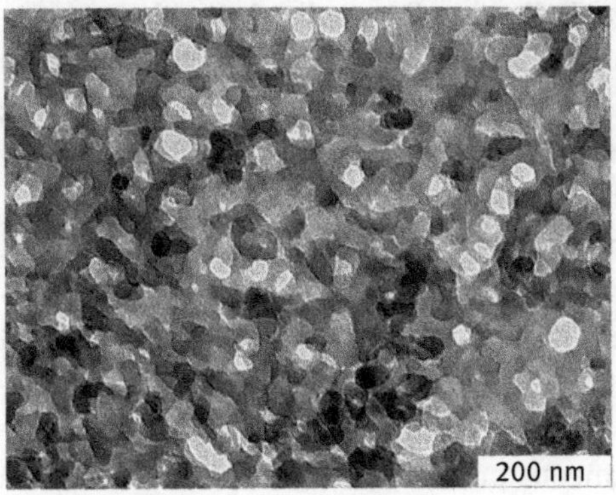

Figure 20.1: Electron microscopy picture of biocoal, which was made by hydrothermal treatment of oak leaves. The sponge-like pore structure with structural elements in the 20–50 nm region is nicely seen.

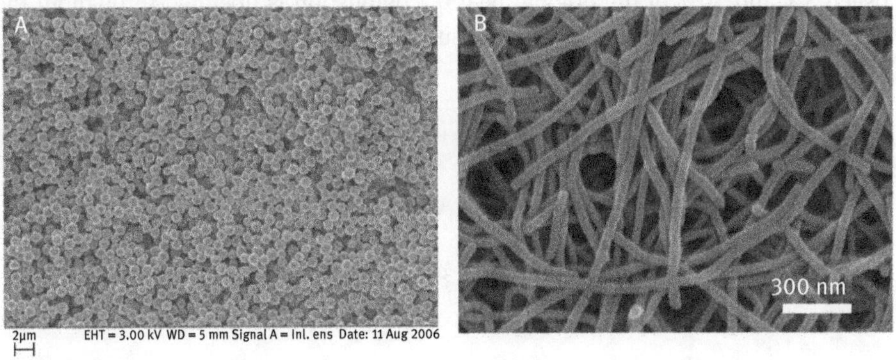

Figure 20.2: Two other relevant nanoarchitectures of hydrothermal carbons: (A) disperse spherical morphology and (B) carbon nanofibers.

20.3 Mineralized Biomass as Energy Carrier

20.3.1 "Biocoal" and Its Comparison to Other Biofuels, Biogas, and Bioethanol

Calcined biomass or biocoals can, of course, be used as materials or as carbon deposits, but they also have an energy value. As long as biocoals indeed substitute for fossil energy carriers (and are not burned just in addition), this operation is meaningful and may contribute to solving the overall problem. However, here biocoal production has to be compared with other already-existing techniques of biomass use. The generation of "biofuels" and "bioenergy" is meanwhile an established industry and comes as biogas, bioethanol (E10), or biodiesel to the public. Sadly, their generation schemes, at least those of the first and second generation of biofuels, are rather ineffective and can be questioned on the base of their overall ecobalance. In addition, they usually just convert parts of the contained biomass, with a majority essentially staying unused and turning into a waste stream.

HTC, as compared to all other current techniques of biomass/energy conversion, is not only rather simple and fast (down to 90 minutes) but also converts practically all the carbon atoms of biomass into the final coal powder (i.e., CE is close to 1). At the same time, relevant salts (e.g., K^+) stay in the aqueous phase and can be back-donated to the ecosystem.

As coalification is in nature a spontaneous process, it must be exothermic, and energy is lost going from the biomass to the calcined and densified coal. But also here, the numbers are not so bad and nicely compete with the other techniques of biofuel generation. The real energy loss or energy preservation is still under debate, but depending on reaction schemes and degree of aromatization of the coal, between 5% and 30% of the original heat of combustion is liberated throughout the process, which of course can be used to support the carbonization.

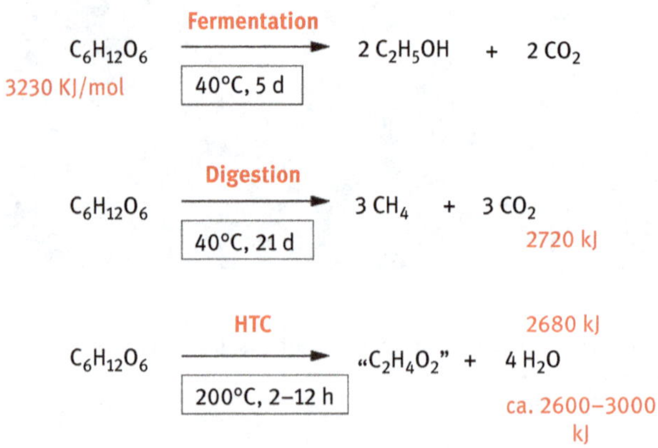

Figure 20.3: Comparison of different "renewable energy pathways" and carbon transfer schemes from carbohydrates, as well as their typical conditions. Here, preservation of combustion energy and the "carbon efficiency" (CE) of the transformation are compared. The combustion energy always concerns the complete side of the reaction equation. The "sum formula" of the coalified plant material is a schematic simplification.

A graphical comparison between HTC and "alcoholic fermentation" or "anaerobic digestion" with respect to reaction schemes, typical conditions, and mass streams is given in Figure 20.3.

All three conversions are, of course, exothermic and spontaneous and store about the same level of energy in the final products. The energy loss explains the simplicity of conversion, independent of its biological or chemical origin. HTC generates water instead of CO_2 as a side product; coalification in these terms is just the elimination of water under preservation of the carbon scaffolds.

Thus, using the HTC process to convert biomass into coal could represent a most efficient tool for valozation of the energy content [7]. The already technically realized acceleration of the coalification down to the hour range and operation of continuous processes makes it a technically attractive, realistic instrument for generating a transportable, dense, stable, and rather safe chemical energy carrier.

Current medium-scale operations for HTC calculate operating expenses on the order of €200/ton biocoal, not including a forthcoming economy of scale and further technical optimization. Even at this premature level, this corresponds to 2.7 (euro) cents/kWh heat (i.e., it can nicely compete with current energy media, being carbon neutral at the same time).

However, what needs to be considered is the societal convenience of liquid fuels, which makes them a preferred choice even when coming with a high markup. Science on energy storage also has to consider such issues (e.g., how to create convenience with solid energy carriers).

Figure 20.4 depicts a technical realization of such an HTC machine on the pilot scale (designed by Carbon Solutions Ltd.). Such a machine in principle fits into a smaller building and can run 10,000 tons of input per year (i.e., good for a communal or small industrial operation).

20.3.2 Carbon Fuel Cells

We are so used to the idea that the energy of coal has to be liberated via fire, thereby driving a heat engine. This, however, is just one option, and especially fine slurries of coal in water, as they are automatically obtained by HTC, can also be directly electrified in a fuel cell. The idea for the electrochemical conversion of the potential energy of coal into electrical energy originates from the late nineteenth century, when W. Jacques introduced the principle of the noncombustive, electrochemical oxidation of carbonaceous matter in air [8]. Carbon fuel cells are conceptually the most effective way of converting the chemical energy of carbon into electrical energy, especially when compared to classical combustion processes. Consequently, there have been many attempts in the past decades to develop and improve generally applicable carbon fuel cells [9, 10], promising higher efficiency compared to conventional coal combustion.

Besides optimizing the fuel cell device, the efficiency of carbon fuel cells can be also optimized by making special carbon colloids as a fuel, as electrochemical oxidation of a solid certainly depends on the absolute surface area and its surface structure and chemistry. Especially for the so-called indirect carbon fuel cell, which operates in water at ambient temperatures, design of carbon colloid fuels seems to be a worthy endeavor, and HTC is directly suited for that. Recently, a comparative

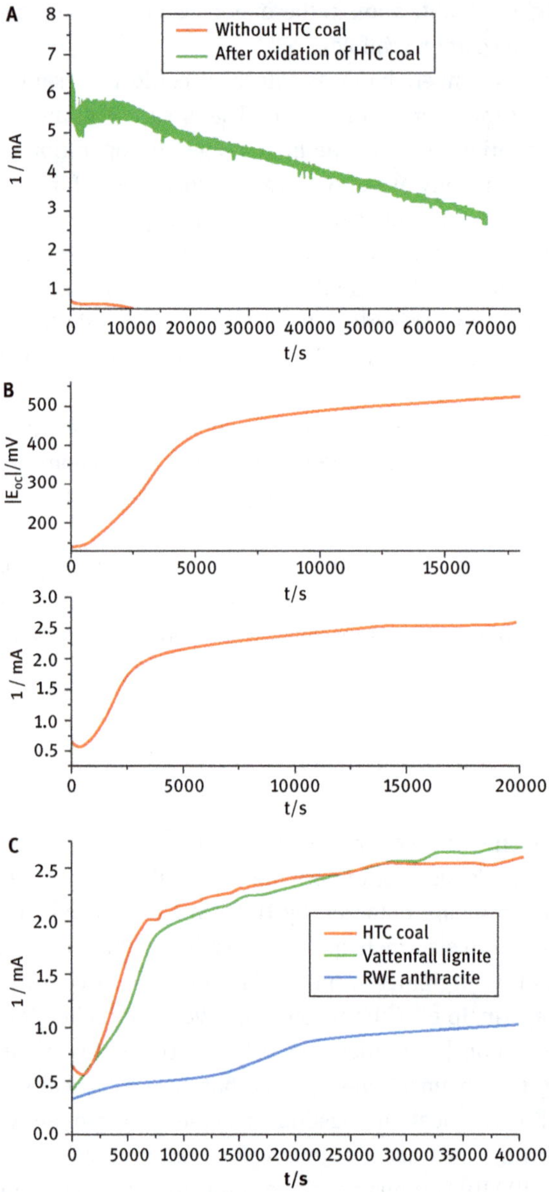

Figure 20.4: (A) Time-dependent electric current generated from the oxidation of HTC coal in an indirect carbon fuel cell. Solutions of Fe^{III} and V^{V} were prepared in 0.5 mol L^{-1} H_2SO_4.
(B) Development of open-circuit potential E_{oc} (up) and current I(down) due to Fe^{2+} formation in the anodic half-cell via oxidation of HTC coal, indicating the reducing potential of bare hydrothermal carbon (HC) dispersions. Charge equalization between the two half-cells was assured by a salt bridge containing a saturated KCl solution. Carbon felt was used as electrodes. (C) Comparison of hydrothermal and fossil carbon sources in the same setup.

study on classical carbon dusts and carbon dispersions prepared by the HTC process has been published [11]. Different samples of lignite and anthracite were applied as reference materials, using a simple Fe^{3+}-oxidation assay. The experiments revealed a significantly better capability of hydrothermal coal derived from glucose to oxidize: it highly exceeds anthracite in its oxidation rate but also shows better oxidation properties than lignite. The reason for the lower oxidation rates of anthracite and pure graphitic carbon is presumably found in their dense, ordered structure of stable aromatic graphene sheets. For lignite, the structure is more open and amorphous, thus making the carbon more accessible for oxidation. HTC coal is even more reactive, as it features a chemical structure mainly composed of aliphatic and olefinic building units, which are highly reductive and reactive, while the amount of conjugated aromatic rings can be kept remarkably low. The micrometer-sized spherical particles dispersed in water additionally offer a more accessible surface for the heterogeneous oxidation process.

The HTC coal was then employed in an anodic half-cell coupled to a model VO^{2+}/VO_2^+ cathode, making the procedure more suitable to run on a laboratory scale than using the standard oxygen cathode (note that the redox potential of VO^{2+}/VO_2^+ and oxygen/Fe^{2+} is very similar). Figure 20.4 displays the obtained results.

It is clearly seen that such cells operate over a long time and with increasing voltage, current, and efficiency. The average efficiency was estimated to be approximately 10%, which, despite the simplicity of the engineering approach, is remarkable. Such primary scientific experiments are, of course, far from any competitive technical device but already prove that carbon colloids derived from hydrothermally treated biomass can indeed act as a potential fuel for decentralized energy generation with an overall zero-emission balance of CO_2, avoiding open fires and combustion side products.

20.4 Discussion and Conclusion

In this chapter, it was delineated that calcined, densified carbon from biomass might be an energy carrier with many advantages. As carbonization stabilizes biomass against naturally occurring degradation, all carbon fixed in this way is essentially on the negative side of the CO_2 balance (i.e., atmospheric CO_2 is collected by plants and turned into solid carbon by an added technology). As approximately 10% of all biomass is already part of farming operations, treating the side products of farming is enough to "save the planet."

Whether this carbon is used as a carbon-neutral energy carrier or indeed to compensate for other CO_2 sources is not a simple question, it depends on natural resources and national development, and is in the end also a question of societal values and political agreements. If the energy content is valued on the market more than the bound CO_2 and the "boxed" climate change, it is safe to predict that we will experience a renaissance of carbon-based machines.

If, however, we can add a materials value to the (otherwise only morally and politically valued) storage operation – for example, by using the stored carbon as a soil conditioner to bring fertility and added biological value – then storing carbon might eventually become the best choice economically. In this first step, it is quite relevant and pleasing that we have this choice, and that research has the potential to contribute to both sides.

References

[1] Lieth H, Whittaker RH. Primary Productivity of the Biosphere. Berlin: Springer; 1975:205–06.

[2] Bobleter O. Hydrothermal degradation of polymers derived from plants. Prog Polym Sci. 1994;19:797–841.

[3] Bergius F, Specht H. Die Anwendung hoher Drücke bei chemischen Vorgängen, Habilitation work. Halle; 1913.

[4] Bergius F. Articles on the theory of coal formation. Naturwissenschaften. 1928;16:1–10.

[5] Titirici MM, Thomas A, YuSH, Antonietti M. A direct synthesis of mesoporous carbon with bicontinuous pore morphology from crude plant material by hydrothermal carbonization. Chem Mater. 2007;19:4205–12.

[6] Titirici MM, Thomas A, Antonietti M. Back in the black: Hydrothermal carbonization of plant material as an efficient chemical process to treat the CO_2 problem? New J Chem. 2007;31:787–89.

[7] Hu B, Wang K, Wu LH, Yu SH, Antonietti M, Titirici MM. Engineering carbon materials from the hydrothermal carbonization process of biomass. Adv Mater. 2010;22:813–28.

[8] Jacques W Method of converting potential energy of carbon into electrical energy. U.S. Patent 555511, 26.06, 1896.

[9] Horita T, Sakai N, Kawada T, Yokokawa H, Dokiya M. An investigation of anodes for direct-oxidation of carbon in solid oxide fuel-cells. J Electrochem Soc. 1995;142.

[10] Peelen WHA, Olivry M, Au SF, Fehribach JD, Hemmes K. Electrochemical oxidation of carbon in a 62/38 mol % Li/K carbonate melt. J Appl Electrochem. 2000;30.

[11] Paraknowitsch JP, Thomas A, Antonietti M. Carbon colloids prepared by hydrothermal carbonization as efficient fuel for indirect carbon fuel cells. Chem Mater. 2009;21:1170–75.

Roh Pin Lee, Bernd Meyer

21 Innovative Use of Carbon

21.1 Introduction

Carbon – the chemical element "C" in group 14 of the periodic table – is essential to life and underpins the normal functioning of our society [1, 2]. Driven by population growth and economic development [3, 4], the global demand for carbon resources has been consistently increasing over the last decades to feed the growing energy hunger [5]. In 2018, primary carbon resources in the form of oil, natural gas, and coal accounted for 84.8% of the global primary energy consumption [6]. In today's linear economy, these primary carbon resources are predominantly used energetically (i.e., combusted) as fuels by the electricity generation, heating and mobility sectors [2]. Such energetic utilization of primary carbon resources is associated with significant CO_2 emissions. In the Fifth Assessment Report of the Intergovernmental Panel on Climate Change (IPCC), the electricity, heating and mobility sectors are reported to account for 39% of the global greenhouse gas emissions in 2010 [7, 8]. In 2018, global CO_2 emissions from fuel combustion reached a historical high of 33.5 Gt CO_2, with emissions largely driven by robust population growth and economic developments in non-OECD countries, especially in China and India [9].

Besides the combustion of primary carbon resources, the incineration of secondary carbon resources (i.e., carbonaceous waste) for electricity and heat production also contributes to CO_2 emissions from carbon utilization. Today, around 2,500 thermal waste treatment plants with a total capacity of approximately 400 million tons of waste per year are in operation worldwide. This capacity is anticipated to expand to 550 million tons of waste per year with nearly 2,700 waste incineration plants by 2029 [10], with increasing tendency as growing economies such as China shift from landfilling to large-scale expansion of waste incineration in order to resolve their worsening waste disposal issue [11]. Assuming that 1 ton of municipal

Acknowledgments: This research is supported by the German Federal Ministry of Education (BMBF) through the research project grant no. 01LN1713A to the research group Global Change: STEEP-CarbonTrans. Any opinions, findings, conclusions, and recommendations in the document are those of the authors and do not necessarily reflect the view of the BMBF.

Roh Pin Lee, Institute for Energy Process Engineering and Chemical Engineering, TU Bergakademie Freiberg, Germany; Fraunhofer IMWS Branch Lab "Circular Carbon Technologies", Freiberg, Germany
Bernd Meyer, Institute for Energy Process Engineering and Chemical Engineering, TU Bergakademie Freiberg, Germany; Center for Efficient High Temperature Processes and Materials Conversion, TU Bergakademie Freiberg, Germany; State Key Laboratory of Coal Conversion, Institute of Coal Chemistry, Chinese Academy of Sciences, Taiyuan, China

https://doi.org/10.1515/9783110608458-021

waste is associated with the release of about 0.7–1.7 tons of CO_2 [12], waste incineration – at 371 to 901 million tons of CO_2 emissions (i.e., 0.37–0.9 Gt CO_2) – will account for another significant proportion of the CO_2 emissions from the energetic utilization of carbon resources.

The global focus on CO_2 emissions reduction arises from its pivotal role in contributing to global warming and increasing occurrences of extreme weather events [13, 14]. To combat the threat of climate change, the Paris Agreement aims to keep the global temperature rise this century to well below 2–1.5 °C above pre-industrial levels [15]. To achieve this target, the reduction of CO_2 emissions associated with the energetic utilization of carbon resources is an urgent priority. To date, global efforts to facilitate the transformation of our carbon-intensive society toward net zero has predominantly focused on transiting the energy-intensive electricity generation sectors from fossil to renewable energy sources [16]. In recent years, this focus has been increasingly extended to the decarbonization of the mobility sector via electric and hydrogen mobility to replace conventional combustion engines [17].

While it may eventually be possible in the long run to decarbonize to a large extent the electricity and mobility sectors, carbon-intensive and dependent industrial sectors such as the chemical industry have no alternatives to carbon resources as feedstock for their production [1, 18]. Today, 14% of crude oil and 8% of natural gas are globally consumed by the chemical industry for its production of a wide variety of products ranging from plastics, fertilizers, fine and special chemicals, pharmaceutical products to washing and personal care products. This is in turn associated with the emission of 1.5 Gt CO_2 in 2018 globally, making the chemical industry the third largest industrial CO_2 emitter [19]. Furthermore, a dependence on imported oil and natural gas by numerous countries – including for chemical production – also leads to significant carbon leakages along international supply chain, further compounding the climate impacts of the chemical sector [20, 21]. As a decarbonization of the chemical industry is not possible, the need to lower its carbon footprint is motivating the search for sustainable carbon resource alternatives for its production. However, this is not the only impetus for change in the chemical industry. Other important drivers such as supply security, structural changes as well as resource efficiency and circularity are also exerting increasing pressure on the industry to accelerate its transformation from petrochemistry via innovative use of alternative carbon resources. In this context, coal and waste as carbon feedstock alternatives are drawing increasing attention from the global community.[1] Goal is to retain the carbon contained in these resources in the system for as long as possible, rather than emitting it via combustion into the atmosphere as CO_2.

---

1 Biomass also represents another primary carbon feedstock alternative for the chemical industry. However, due to potential conflict with food production, it is not considered in this chapter.

21.2 Transformation Options for the Chemical Industry

This section briefly summarizes the key motivations and potential relevance of coal and waste as alternative carbon resources for the transformation of the carbon-intensive and dependent chemical sector.

21.2.1 Coal as Alternative Primary Carbon Feedstock

World coal reserves in 2019 are reported at 1,070 billion tons [6]. Globally, coal is mainly energetically utilized for electricity and heat production [22]. In contrast to combustion where 100% of the carbon in the coal – in the case of complete combustion – will be emitted as CO_2 into the atmosphere, the non-energetic utilization of coal as a chemical feedstock will emit much less CO_2 [23]. However, this is generally not the main motivation for such alternative utilization of coal. For countries such as China and South Africa that have considerable domestic coal reserves but are lacking in oil and gas reserves, chemical coal utilization has been intensively promoted as it not only contributes to the supply security of their chemical industries, it furthermore supports the conservation of their scarce oil and natural gas reserves for other purposes [24].

Considering the above reasons, coal as an alternative feedstock for chemical production could be a viable "bridging" development option for coal nations as they transit toward a low carbon society. In presenting a non-energetic utilization option to coal combustion, it could contribute to lowering their CO_2 emissions while reducing their dependency on imported oil and natural gas, thus contributing to their supply security as well as reduction of carbon leakages along international supply chains [16]. This is especially relevant for coal regions that are undergoing/planning a coal phaseout as they transit their energy systems toward renewable sources. Considering that coal regions in transition (e.g., in the European Union) will be facing significant adverse socio-economic impacts associated with the decline of the coal industry, chemical coal utilization could offer the affected populations in these regions a chance to adapt and develop sustainable structural change measures on a step-by-step basis – in contrast to a detrimental structural break – as they reorient themselves for a net zero society in the long term, which will be based on renewables and circularity.

21.2.2 Waste as Alternative Secondary Carbon Feedstock

Between 2012 and 2025, the amount of global solid waste generated annually is anticipated to increase exponentially from 1.3 to 2.2 billion tons [25]. Currently, only about 16% of this waste is recycled, while 46% (over 950 million tons) is disposed of unsustainably [26, 27]. This is leading to a global waste crisis, driven especially by the challenge of plastic disposal. Since 1960s, global plastic production has increased twentyfold, reaching 322 million tons in 2015. This is expected to double again over the next 20 years [28]. While plastic plays an essential role in our daily life and in our economy, vivid images of giant ocean garbage patches – made up largely of plastic waste – have elicited growing global concern that our society is drowning in waste [29–31]. The chemical industry – as producer of plastics – is being held largely accountable for the global plastic crisis [1]. Hence, besides regulatory measures in numerous countries to reduce the use of single-use plastics and promote eco-design for multiple reuse and better recyclability, there is increasing political, market, and public pressure on the global chemical industry to develop sustainable solutions for plastic waste, including its recirculation into the value chain as an alternative carbon raw material for chemical production [32–35].

The use of carbonaceous waste as a secondary carbon feedstock for the chemical industry is referred to generally as chemical recycling. Although reports about chemical recycling have focused mainly on its applicability for plastic waste, other carbonaceous waste types ranging from municipal solid waste to biogenic waste materials (e.g., waste wood, agricultural waste, and forest residues) also represent potential feedstock for the production of a wide range of chemical products [36–38].

Chemical recycling of carbonaceous waste represents an innovative use of carbon, which could contribute to the following [23, 39, 40][2]:

- solving the global waste (plastic) crisis,
- reducing emission of climate relevant gases (e.g., CO_2, CH_4) through reduction/avoidance of waste incineration and landfilling,
- diversifying feedstock for chemical production, and thus contribute to conserving primary carbon resources, increasing supply security by reducing dependency on fossil resource imports, and reducing carbon leakages along international supply chains,
- increasing resource efficiency by recirculating the carbon in waste back into the production cycle,
- promoting circularity and achievement of recycling targets, and
- creating increased value from waste (e.g., waste-to-products rather than waste-to-energy).

2 In the dominant linear economy, carbon resources go through the stages of extraction, production, distribution and consumption before ending up ultimately as waste, either physically or in the form of CO_2 through combustion [16]. While the latter – CO_2 emitted from point sources (e.g., coal power plants, cement plants, steel plants) – is also a source of carbonaceous waste, its end-of-life utilization in the form of carbon capture and utilization (CCU) is beyond the scope of this chapter.

With the prediction that 3.4 billion tons of waste would be generated per annum by 2050 [41], ways to facilitate sustainable waste management are urgently required. While waste incineration offers an option for meaningful utilization of waste compared to improper disposal and/or landfilling, the combustion of waste for the generation of electricity and heat is not only associated with significant CO_2 emissions; it also represents a waste of resources that could be otherwise channeled back into the production cycle as secondary raw materials [11]. Hence, in view of its potential applicability in addressing multiple objectives ranging from the waste crisis, the resource dilemma, CO_2 emission reduction to circularity all at the same time, the chemical utilization of waste via chemical recycling represents an attractive and relevant option for all societies. This applies especially to those nations with active chemical sectors and which are dependent on fossil resource imports.

21.2.3 Alternative Transformation Pathways Toward Sustainability

While coal and carbonaceous waste represent carbon feedstock alternatives for the global chemical industry, their potential contribution to its sustainability differs. Two aspects are of relevance in assessing their roles in the transformation of the chemical industry, namely (1) resource type and (2) product type.

- *Resource type:* Similar to the conventional feedstock crude oil and natural gas which are currently utilized for petrochemical production, coal is a fossil resource with finite reserves. The advantages of its utilization for chemical production thus lie not in the replacement of other fossil resources, but rather in the CO_2 emissions reduction potential compared to its energetic utilization (i.e., for coal power generation), the increased supply security for coal nations that are dependent on fossil imports, as well as the support of coal regions in transition as a bridge technology. However, the chemical utilization of the primary resource coal will continue to contribute to fossil resource depletion and CO_2 emissions. Hence, to support the chemical industry's decarbonization and circularity efforts, waste as a secondary carbon feedstock to replace primary fossil resources for chemical production is an attractive – and increasingly necessary – alternative. Note, however, that carbonaceous waste is distinguished according to their origins, that is, petrochemical (e.g., plastic waste) or biogenic (e.g., wood waste, agricultural waste, forest residues, and animal waste). This distinction is especially relevant when considering the emissions that will be associated with the products produced.
- *Product type:* Two major product types from the chemical utilization of carbon resources are eliciting significant market and political interest, namely (1) chemical products (e.g., plastics) and intermediate raw materials for their production (e.g., methanol, hydrogen), as well as (2) fuels (e.g., transportation fuels, fuel

gas for electricity and heat generation). While chemical recycling is generally used to refer to the use of carbonaceous waste as a secondary feedstock for the chemical industry, there is significant controversy regarding whether this term should also apply to the production of fuels from waste. This is because fuels will ultimately be emitted as CO_2 (and H_2O) following its combustion [36, 42, 43]. In this article, chemical recycling shall therefore be understood as the use of carbonaceous waste for chemical production (i.e., waste-to-products), but not for waste-to-fuels.

For chemical products, the use of coal and petrochemical waste sources as carbon feedstock for their production will facilitate the retainment of parts of the fossil carbon in the production cycle, that is, prevent their emissions as 100% fossil CO_2. However, such CO_2 reduction potential will not be realized when they are used to produce fuels (e.g., diesel, gasoline, and kerosene). Their energetic utilization as fuels in conventional mobility – similar to their combustion for electricity and heat generation – will result in the carbon contained in such resources being emitted 100% – in the case of complete combustion – as fossil CO_2 emissions. This will directly contribute to an increase in atmospheric concentrations of CO_2 and consequently, global warming [44].

In contrast, biogenic CO_2 from the combustion of biogenic waste is widely considered CO_2-neutral as it generally will not result in increases in atmospheric CO_2 concentrations [44]. Hence, the use of biogenic waste as feedstock will not only enable the production of CO_2-neutral chemical products, it also supports the production of CO_2-neutral liquid fuels. This could complement battery electric and hydrogen mobility in transforming the passenger vehicle sector toward CO_2-neutral mobility. Moreover, it represents an attractive alternative for mobility sectors (e.g., aviation, shipping, and heavy load vehicles), which have no feasible alternatives to liquid fuels in the near future [45].

Taken together, to facilitate the transformation toward a net zero society, biogenic waste would theoretically be an ideal carbon feedstock – in terms of its CO_2-neutrality – for chemical and fuels production. However, should coal and fossil-based waste be considered as potential feedstock alternatives, their utilization for chemical production followed by chemical recycling to retain the fossil carbon in the production cycle for as long as possible would be more meaningful compared to their combustion for electricity and heat production or their chemical utilization for fuels production.

Figure 21.1 illustrates the current linear economy for the chemical industry, with oil and gas at the beginning of the value chain to CO_2 at the end. In Figure 21.2, the innovative use of carbon in the form of coal and (fossil and biogenic) waste for chemical and fuel production as well as CO_2-neutral mobility for sectors that are dependent on liquid fuels is summarized. Fossil pathways are indicated in gray, while CO_2-neutral pathways are indicated in green. Fossil pathways with reduced CO_2 emissions are indicated in yellow.

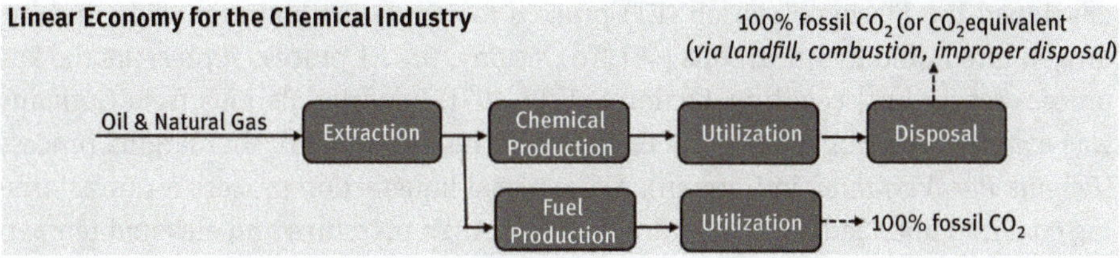

Figure 21.1: Linear economy for the chemical industry.

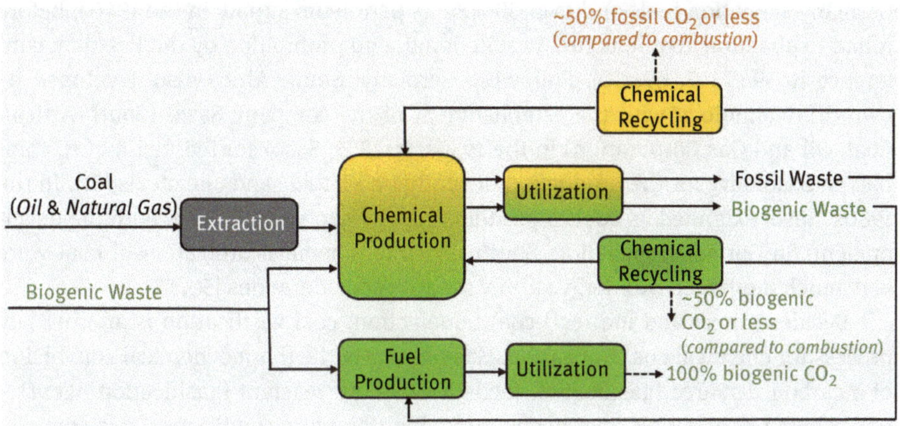

Figure 21.2: Transformation pathways for the chemical industry.

21.3 Chemical Utilization of Alternative Carbon Resources: Examples of Innovative Applications

To illustrate how the chemical utilization of primary and secondary carbon resources in the form of coal and waste respectively could support the decarbonization efforts and transformation of carbon-intensive and carbon-dependent sectors, highlights from four global developments, namely (1) coal (& waste)-to-products, (2) chemical recycling (i.e., waste-to-chemicals), (3) green hydrogen from waste, and (4) CO_2-neutral liquid fuels from waste (i.e., waste-to-fuels) will be briefly summarized in this section.

21.3.1 Coal (& Waste)-to-Products

In the beginning, industrial developments were very much driven by Germany and focused on coal-to-liquids (CtL). In 1920s, Franz Fischer and Hans Tropsch at the Kaiser-Wilhelm-Institut für Kohleforschung in Mühlheim an der Ruhr in Germany developed the Fischer–Tropsch (FT) process to convert carbon monoxide and hydrogen into liquid hydrocarbons [46]. To this day, the FT process represents the key process for indirect coal liquefaction [47]. In 1931, Friedrich Bergius from Germany was awarded the Nobel Prize in Chemistry for his invention of the Bergius process (*Bergius-Pier-Verfahren* in German), a direct coal liquefaction process for transforming coal into liquefied oil with hydrogen under high pressures and elevated temperatures for use as synthetic fuels [48]. During the mid-1930s, industrial production of synthetic fuels via CtL processes was initiated by German companies such as IG Farben and Ruhrchemie. Coal-based synthetic fuels grew to account for over 90% of Germany's aviation fuels and over 50% of its petroleum supply in the 1940s, before it had to abandon synthetic fuel production due to prohibition by the Potsdam conference in 1945 [47, 49, 50]. Following Germany, South Africa also developed its own CtL technology with the establishment of the company Sasol (South African Coal, Oil and Gas Corporation) in the 1950s. In 1955, Sasol started its first CtL complex in Sasolburg for the production of synthetic FT-fuels and chemicals [51]. In the 1980s, Sasol extended its coal-to-products activities to Secunda [52]. Both the developments in Germany as well as South Africa on chemical utilization of coal were very much motivated by energy supply security considerations [50, 53].

Besides (direct and indirect) coal liquefaction, coal gasification is another key process for chemical coal utilization. Gasification is the thermochemical conversion of a carbon resource (gasification feedstock) with a reactant (gasification agent) – under high temperatures and mostly under high pressure conditions – to a combustible gas made up of a mixture of carbon monoxide, hydrogen, carbon dioxide, methane and higher hydrocarbons [54]. It represents a key process for both direct and indirect coal liquefaction as it is used to produce hydrogen for the Bergius process as well as synthesis gas for the FT synthesis. Depending on gasification principle and operating conditions, the product gas will have different compositions and utilization potential. Generally, a low-heating value gas from air-blown gasification can be used as a fuel for heat and power generation but is not suitable for chemical production. A medium-heating value gas from oxygen (and steam) -blown gasification will compose mainly of carbon monoxide and hydrogen (i.e., syngas), which can be used for ammonia, methanol, and other synthesis. From a high-heating value, methane-rich gas, synthetic natural gas (i.e., SNG) can be synthesized to substitute natural gas as fuel or for production [55].

As with coal liquefaction, Germany also played a leading role in the development of the three basic principles of coal gasification technologies.

- *Fluidized-bed gasification process:* In 1926, the German chemist Fritz Winkler invented the fluidized-bed gasification process. This led to the commercial installation of the first Winkler gasifier (atmospheric pressure) in Leuna. The technology was then further developed in the 1980s and 1990s to the High Temperature Winkler process (HTW) by the German company Rheinische Braunkohlenwerke AG (Rheinbraun) to realize higher gasification pressures. The demonstration plant for the first pressurized HTW gasification technology was started up in 1986 in Berrenrath, Germany, and operated for over 10 years to supply syngas for methanol production [55, 56].

Besides Berrenrath, the HTW gasification technology was also implemented on a pilot and commercial scale in Japan and in Finland. In 2010, the technology was acquired by ThyssenKrupp Uhde from Rheinbraun/RWE [57]. In addition to the HTW, another fluidized-bed gasification development is the U-GAS gasification technology by the Gas Technology Institute (GTI) in the USA. Today, the U-GAS technology is commercially deployed in China for coal-to-chemicals [58].

- *Fixed-bed gasification process:* The world's first oxygen-blown fixed-bed gasification process was also developed in Germany, by the company Lurgi Kohle GmbH in Frankfurt. In 1936, the first pressurized Lurgi fixed-bed dry ash (FBDA) gasifier with 1.3-m inner diameter was implemented in Hirschfeld to supply the city of Zittau with town gas [59]. The pressurized Lurgi FBDA gasifiers – with inner diameter up to 2.6 m – were then commercialized in the 1940s in Brüx (today Most), Czech Republic. This was then the largest industrial oxygen-blown coal gasification facility in the world for syngas production [60, 61]. In the 1950s, Sasol in South Africa built the first large-scale, pressurized Lurgi gasifier – with inner diameter at 3.6 m (Mark IV) – to produce syngas of high purity for subsequent fuel and chemical production via FT synthesis. During the time of the German Democratic Republic (GDR) in the 1960's, the Lurgi FBDA gasification process was also implemented with 24 large-scale, pressurized Mark IV gasifiers in the VEB Gaskombinat Schwarze Pumpe – the largest lignite refining facility in the world till 1989 – for producing town gas from coal and later on, syngas for methanol production [62, 63].

  Today, the FBDA gasification technology is implemented especially in South Africa and China for coal-to-chemicals/fuels. In the former, Sasol has deployed over 101 Lurgi FBDA gasifiers over the last 50 years [64]. In the later, SEDIN Engineering Co., Ltd has successfully implemented the FBDA for diverse coal-to-chemicals/fuels projects [65], where it accounted for the highest amount of syngas capacity in China among domestic gasification technology providers in 2011 [66]. Furthermore, Lurgi FDBA gasifiers are also being deployed in the Great Plains Synfuels Plant in the USA [56] as well as in the Vresova IGCC Power Plant in the Czech Republic [67].

  In addition to the Lurgi FDBA, another fixed-bed gasification process is the British Gas/Lurgi fixed-bed slagging (BGL) gasification technology – a further development of the original Lurgi FBDA design. Currently, the BGL technology is owned by the Envirotherm GmbH as well as the ZEMAG Clean Energy Technology GmbH, and commercially deployed mostly in China and India [68].
- *Entrained-flow gasification process:* The world's first entrained-flow gasifier was also developed in Germany. It was a joint development by the company Heinrich Koppers GmbH in the 1940s in cooperation with Friedrich Totzek from the Koopers

Company Inc. (USA). The first commercial installment was in Finland in 1952 [55]. In the 1980s, the Koopers-Totzek entrained-flow process was further developed in the GDR time at the German Fuel Institute in Freiberg (*Deutsches Brennstoffinstitut*, abbreviated DBI). This was then implemented as the GSP process in the VEB Gaskombinat Schwarze Pumpe for producing town gas from coal, and subsequently for producing syngas for methanol production [69, 70].

The GSP process was owned by the Noell group, followed by Babcock Borsig and Future Energy before its acquisition by SIEMENS in 2006 [71]. Amongst others, the SIEMENS gasification technology is implemented by the Shenhua Ningxia Coal Group in the largest CtL facility in the world. The facility is situated in Yinchuan, China, where 20 million tons of coal are converted – via syngas generated from coal gasification – into 4 million tons of oil products annually. These include 2.7 million tons of diesel, 980,000 tons of naphtha petroleum and 340,000 tons of liquefied gas [72]. In 2016, SIEMENS gave up its gasification activities [73] and the GSP pilot plant facility (formerly SIEMENS gasification facility in Freiberg, Germany) is currently operated by the DBI-Virtuhcon GmbH in cooperation with the Institute of Energy Process Engineering and Chemical Engineering, TU Bergakademie Freiberg [74].

In addition to Germany, other entrained-flow gasification developments in the 1970s and 1980s can be traced back to GE Energy (formerly Chevron Texaco in the USA), Shell (in the Netherlands) as well as Uhde (with the Prenflo gasification technology in Germany) [75–77]. Today, the gasification technologies of GE Energy and Shell are deployed widely in China, and they are considered – besides SIEMENS – the leading overseas players in China in terms of syngas capacity [66, 78]. However, international entrained-flow gasification technology providers are facing increasing competition from domestic technology providers, in particular from the Multi-Component Slurry (MCSG) gasification technology by the Northwest Research Institute of Chemical Industry, the Opposite Multi-Burner (OMB) gasification technology by the East China University of Science and Technology and the HT-L gasification technology by the China Aerospace Science and Technology Corporation [66].

Since the 1920s, the development and deployment of gasification technologies have accelerated worldwide. Today, there are 272 gasification plants with 686 gasifiers operating globally. Additionally, 74 plants with a total of 238 gasifiers are currently under construction. While 33 gasification plants are located in the USA, China currently has the largest gasification capacity in the world [79]. Coal gasification accounts for about 5% of China's total coal consumption. It is used for the production of bulk chemicals (e.g., fertilizers, methanol, olefins, aromatics, and ethylene glycol.), synthetic fuels (e.g., diesel and SNG), as well as for power generation, metallurgy and other process industries [80]. In view of China's dominance in the use of gasification technologies, it is not surprising that coal is the dominant feedstock for gasification globally. Nevertheless, alternative feedstock such as different types of waste is gaining increasing market attention. Though interest in gasification for the production of liquid and gaseous fuels is also growing, its utilization for the production of chemicals is expected to remain the most important gasification application in the foreseeable future [79].

Figure 21.3 provides a summary overview of the diverse production possibilities and products from chemical coal utilization. Coal-to-chemicals, with its potential to retain fossil carbon in the production cycle compared to coal combustion, has significant potential in the mid-term to support coal regions in their decarbonization efforts as they shift away from energetic coal utilization (indicated in yellow). Moreover, coal-to-chemicals represents a promising transformation pathway for the chemical industry as it could support the development of the technological basis and expertise required for waste-to-chemicals (i.e., chemical recycling), and thus the first step into a circular carbon economy. In contrast, while coal-to-fuels could increase a coal's national energy supply security (as in the case of China and South Africa), its utilization is – similar to coal power and heat generation – associated with emissions of 100% fossil CO_2 (indicated in gray). Hence, the contribution of coal-to-fuels to CO_2-emissions reduction efforts is limited. Nevertheless, similar to coal-to-chemicals, its implementation could establish the technological basis for a stepwise transformation toward biogenic waste-based fuels that will be CO_2 neutral.

As with chemical coal utilization, Germany was also a pioneer in the utilization of waste as co-feedstock with coal for chemical production. This first step toward circularity was implemented on a large scale between 1990s and 2000s in Berrenrath in West Germany as well as in SVZ (i.e., Sekundärrohstoff-Verwertungszentrum – Secondary Raw Material Recovery Center) Schwarze Pumpe in East Germany [63]. Driving force for these developments was the enactment of the Packaging Ordinance and the Closed Substance and Waste Management Act in 1991 and 1996 respectively, which led Germany to becoming the first country in the world to adopt a "polluter pays" principle for municipal waste disposal [81–84].

During this early attempt at using waste as a chemical feedstock, it was necessary to use coal as a co-feedstock in the gasification process. Coal acted as an enabler for the chemical utilization of waste by stabilizing the gasification process. It also made up for the carbon losses associated with the recirculation of waste back into the production cycle. A closer look at the co-utilization of waste and coal for chemical production in Germany:

- *Berrenrath:* The HTW demonstration plant at Berrenrath near Cologne was started by Rheinbraun in early 1986 and operated until 1997. One 150 MWth HTW gasifier (10 bar) was operated initially for dried lignite (i.e., brown coal) gasification to generate syngas for subsequent methanol production. To demonstrate the process' applicability for multiple feedstock, different waste feedstock ranging from plastic waste, biomass, MSW to sewage sludge were also co-gasified with lignite (about 50% waste with 50% coal) for the production of about 300 tons of methanol per day [57, 85, 86].

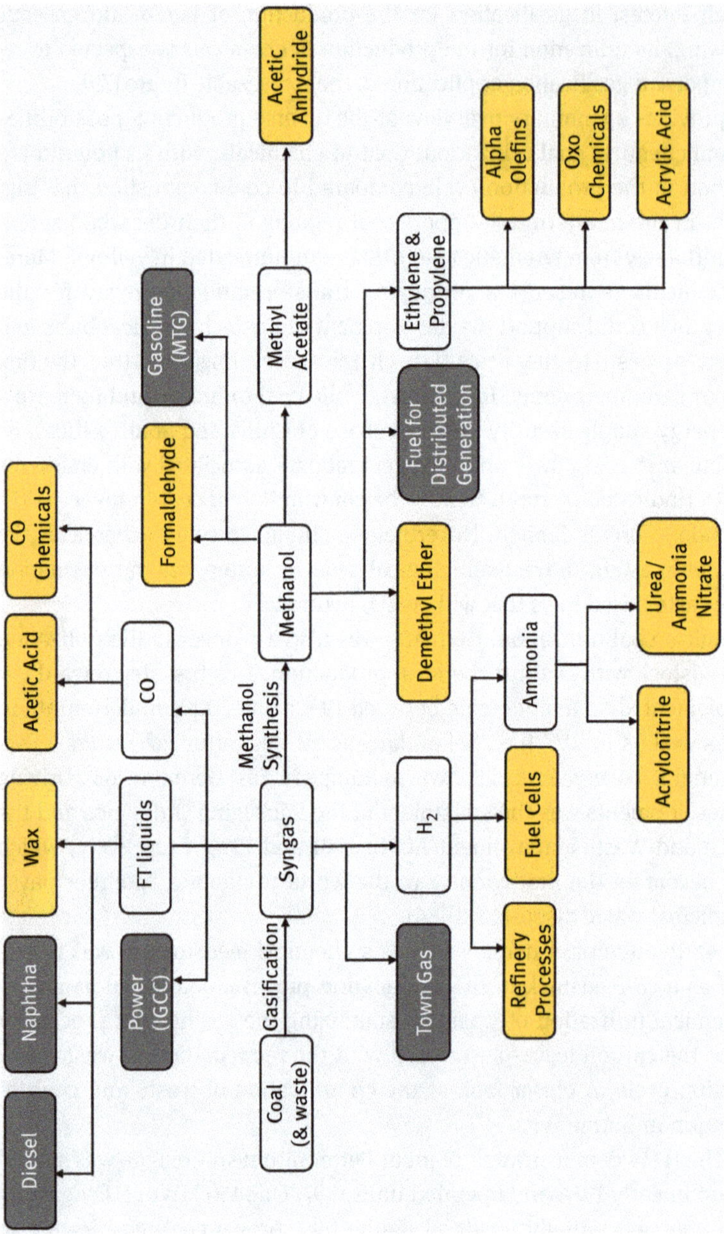

Figure 21.3: Coal-to-chemicals and coal-to-fuels routes.[3]

3 Adapted from NETL [91].

– *SVZ Schwarze Pumpe:* In contrast to Berrenrath, which focused on the demonstration of one gasification technology, the Secondary Raw Material Recovery Center in Schwarze Pumpe carried out commercial application and further development of multiple gasification technologies from 1995 to 2007. In the beginning, seven existing classical Lurgi fixed-bed dry ash (FBDA) gasifiers, which were originally operated for coal gasification, were further developed for the co-gasification of waste and coal (about 75% waste with 25% coal). At the same time, the world's first fixed-bed slagging gasifier for waste gasification also went into trial operation (i.e., British Gas/Lurgi gasification technology – abbreviated BGL gasification technology). Subsequently, the existing GSP entrained-flow gasifier was also utilized for the gasification of liquid residues. Figure 21.4 provides a simplified process illustration of the integrated gasification technology platform being operated at the SVZ Schwarze Pumpe [87, 88].

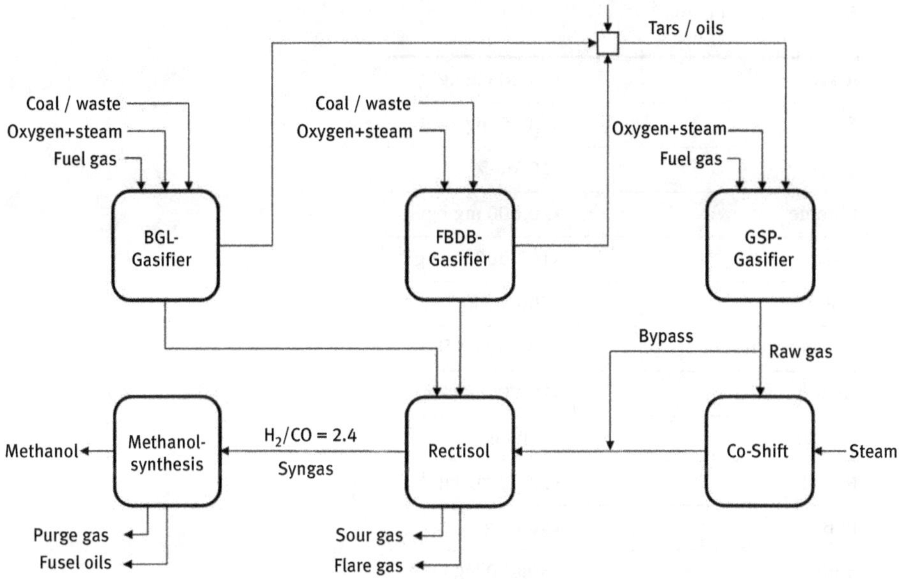

Figure 21.4: Simplified process diagram of SVZ Schwarze Pumpe.[4]

Specifically, solid waste feedstock ranging from plastic waste, sewage sludge, MSW, waste tires, shredder fractions, wood waste, electronic waste to residual waste from mechanical-biological waste treatment plants were gasified in the Lurgi FBDA and BGL gasifiers for syngas production. Liquid and pasty waste – including tars/oils that were generated as waste by-products in the BGL and

4 Adapted from Meyer et al. [86].

Lurgi FBDA gasifiers – were gasified in the GSP gasifier. Via the Rectisol, clean synthesis gas for methanol synthesis was provided while purge gases from methanol synthesis (e.g., methane) were used for power generation in a combined cycle power plant. SVZ Schwarze Pumpe thus attempted to optimize its utilization of diverse waste streams, with its key focus being on decreasing the proportion of coal as co-feedstock (i.e., increase proportion of waste as feedstock). Hence, SVZ Schwarze Pumpe could be termed as the first commercial chemical recycling center in the world.

The gasification technologies utilized in SVZ Schwarze Pumpe exhibited a relatively robust tolerance for diverse contaminants and heavy metals (see Table 21.1). Note that the input limits refer to temporary peak values – and not permanent values – for diverse contaminants.

Table 21.1: Maximum values for contaminants at SVZ Schwarze Pumpe.[5]

Arsenic	<2,000 mg kg^{-1}
Cadmium	<1,000 mg kg^{-1}
Chloride	<10 Ma-%
Chrome	<20,000 mg kg^{-1}
Copper	<100,000 mg kg^{-1}
Cyanide	<500 mg kg^{-1}
Dioxin and furan	<50,000 ng TE kg^{-1}
Lead	<10,000 mg kg^{-1}
Mercury	<200 mg kg^{-1}
Nickel	<5,000 mg kg^{-1}
PCB	<500 mg kg^{-1}
Sulfur	<30,000 mg kg^{-1}
Tin	<10,000 mg kg^{-1}
Zinc	<100,000 mg kg^{-1}
Water content	<20 Ma-%
Calorific value	No limit
Ash content	No limit

5 Adapted from Obermeier et al. [88].

During commercial operation, for example in the year 2001, the SVZ Schwarze Pumpe chemical recycling facility processed over 300,000 tons of solid waste into syngas and produced about 120,000 tons of methanol [87].

Despite its technological success in co-gasifying waste with coal for chemical production, the limited cost-effectiveness of the complex processes coupled with the low market prices for methanol and increased competition from waste-to-energy plants for waste feedstock ultimately led to the closure of the SVZ Schwarze Pumpe in 2007 [11, 89, 90]. This also signified the end of Germany's first foray into the chemical utilization of waste. Nevertheless, though 100% waste gasification was not achieved, the previous developments at Berrenrath and Schwarze Pumpe illustrated how coal gasification – in setting the technological basis as well as in providing the necessary expertise and experience – could support a step-wise transformation of the chemical industry toward waste-to-chemicals and thus its transition toward a circular carbon economy.

21.3.2 Chemical Recycling (Waste-to-Chemicals)

Chemical recycling – also known as advanced recycling or feedstock recycling – is defined generally as the breaking down of carbon-containing content in waste either via solvents or by using heat and/or reactions into basic chemical compounds that can be utilized as feedstock by the chemical industry for its production [11, 36, 91]. Especially in recent years, growing interest in chemical recycling is observable from science, industry, politics, and the civil society. However, the predominant attention is placed on the utilization of plastic waste as feedstock [33, 42, 92, 93].

In 2018, 359 million tons of plastics were produced [94]. However, while the majority of plastic waste is either incinerated or landfilled, a proportion also ends up polluting the environment due to improper disposal. As changes in this general trend of plastics production/consumption are not expected, chemical recycling is thus being seen as a potential solution to the end-of-life problem for plastic disposal, and also a key part of the solution for substituting fossil carbon via alternative carbon feedstock for production [92].

Generally, chemical recycling technologies can be characterized into three broad categories, namely (1) solvolysis, (2) pyrolysis, and (3) gasification [37, 95].[6] With the global focus on developing solutions for the plastic waste crisis, chemical recycling developments and activities in recent years have focused very much on the first two categories that represent processes suitable for small-scale and decentralized applications.

6 Please refer to Lechleitner et al. [95] and Mamani-Soliz et al. [37], for an overview of different chemical recycling processes.

- *Solvolysis:* This includes solvent extraction methods as well as chemical depolymerization processes such as alcoholysis, hydrolysis, acidolysis, aminolysis, and diverse reactions for the production of oligomers or monomers [96, 97]. Solvent extraction uses the principle of solubility to selectively separate, remove and recover plastic polymer from other materials contaminating the plastic waste. In contrast, during chemical depolymerization, the polymer chain is broken down into monomers through the use of chemicals. The recovered monomers are then purified and separated from contaminants [98]. Relatively clean and homogeneous plastic waste is required in order to achieve high yields and high-quality outputs. Hence, only a narrow spectrum of plastic waste – mostly thermoplastics and thermosets such as PET, PVC as well as polyesters, polyamides, and polyurethanes – is suitable for chemical recycling via solvolysis [96–98].

  Currently, of the 21 innovation projects forming the "Plastics Circularity Multiplier" to promote synergies among EU-funded projects for plastic circularity, six projects, namely Circular Flooring, CREATor, DeCOAT, demeto, ISOPREP, MULTI-CYCLE, and NONTOX are focusing on solvolysis development and application [99].
- *Pyrolysis:* During pyrolysis, carbonaceous waste is thermally decomposed at temperatures of around 400–600 °C in the absence of air into combustible gases, oils and solid residues. Pyrolysis processes require an external heat supply and sometimes the addition of catalysts [11]. It is a well-established process in the waste industry for producing fuel products (i.e., fuel gas or oils for electricity/heat production). However, waste-to-chemicals via the use of pyrolysis oil as a substitute feedstock for naphtha in a steam cracker is a relatively new application [98]. While it is possible to use heterogeneous and contaminated carbonaceous feedstock such as municipal solid waste, old tires, biomass, and paper sludges for pyrolysis [37], depending on technologies, the resulting pyrolysis products can be contaminated with organochlorine compounds such as PCBs, undesirable sulfur, nitrogen and oxygen compounds or even dioxins [11]. Pyrolysis oil from mixed and contaminated wastes thus need to undergo significant purification before it can consistently meet the strict specifications of steam crackers for integration into the chemical production process [98]. To ensure the quality of pyrolysis oil and reduce the need for subsequent purification, relatively pure and homogeneous waste fractions will have to be used as pyrolysis feedstock [11, 98].

  In recent years, the chemical industry has announced diverse collaborations and R&D activities focusing on the pyrolysis of plastic waste. These include BASF's investment in Quantafuel to ensure access to pyrolysis oil and purified hydrocarbons from its plant in Denmark [100], the collaboration between LyondellBasell and Karlsruhe Institute of Technology to develop a new catalyst and pyrolysis technology [101], the partnership between OMV and Borealis on the ReOil project [102] as well as the strategic collaboration between PLASTIC ENERGY with INEOS and TOTAL respectively for the construction of pyrolysis-based chemical recycling plants for plastic waste [103, 104].

- *Gasification:* Gasification refers to the thermochemical conversion of carbonaceous materials under high temperature and pressure conditions to release a synthetic gas (or syngas) that can be subsequently utilized for the production of a broad range of products ranging from chemical products (e.g., methanol, plastics, and pharmaceuticals), transportation fuels to hydrogen [105, 90, 54]. Unlike solvolysis and pyrolysis, gasification is an industrial-scale technology suitable for centralized implementation via direct integration in chemical sites. While initially developed for the chemical utilization of coal [55], it has the potential to convert large quantities of mixed-quality carbonaceous waste into a flexible intermediate feedstock for chemical production [37]. An example is the SVZ Schwarze Pumpe. In extending the range of potential feedstock from plastic waste to heterogeneous carbonaceous waste, waste gasification technologies thus hold the promise for a significant contribution not only to circularity in the chemical industry but also to reducing its dependence on (imported) fossil feedstock for its production. Moreover, compared to solvolysis and pyrolysis that require relatively pure and homogeneous waste streams as feedstock, it offers a more realistic chemical recycling solution for contaminated and mixed waste streams (e.g., unsorted municipal waste), which account for the main part of municipal solid waste.

  As mentioned in Chapter 21.3.1, Germany was a pioneer in the development of gasification technologies, and also for its implementation for waste gasification (with coal as co-feedstock). In recent years, especially in view of its potential to address the plastic challenge, a renewed interest in waste gasification is observable globally. Focus however is on mono (i.e., 100%) waste gasification. For example, Showa Denko in Japan operates a waste gasification facility since 2003 to produce hydrogen from waste plastics as raw material for ammonia synthesis [106]. ENERKEM operates a facility in Edmonton, Canada, which gasifies municipal waste to produce syngas for subsequent methanol and ethanol production [107]. In Germany, the Institute of Energy Process Engineering and Chemical Engineering at the TU Bergakademie Freiberg has concentrated on the further development of diverse gasification technologies for waste gasification. Main objective is the development of gasification technologies that are suitable for flexible feedstock and flexible products. Specifically, the FlexiSlag and FlexiEntrained gasification processes are further developments of the BGL fixed-bed slagging gasification technology and the GSP entrained-flow gasification technology that were implemented at the SVZ Schwarze Pumpe respectively. Additionally, the FlexiFluid gasification process is a further development of the HTW technology that was operated at Berrenrath [108].

Solvolysis, pyrolysis, and gasification can be integrated differently into existing chemical production pathways to increase resource efficiency and contribute to a stepwise substitution of primary carbon resources for chemical production (see Figure 21.5).

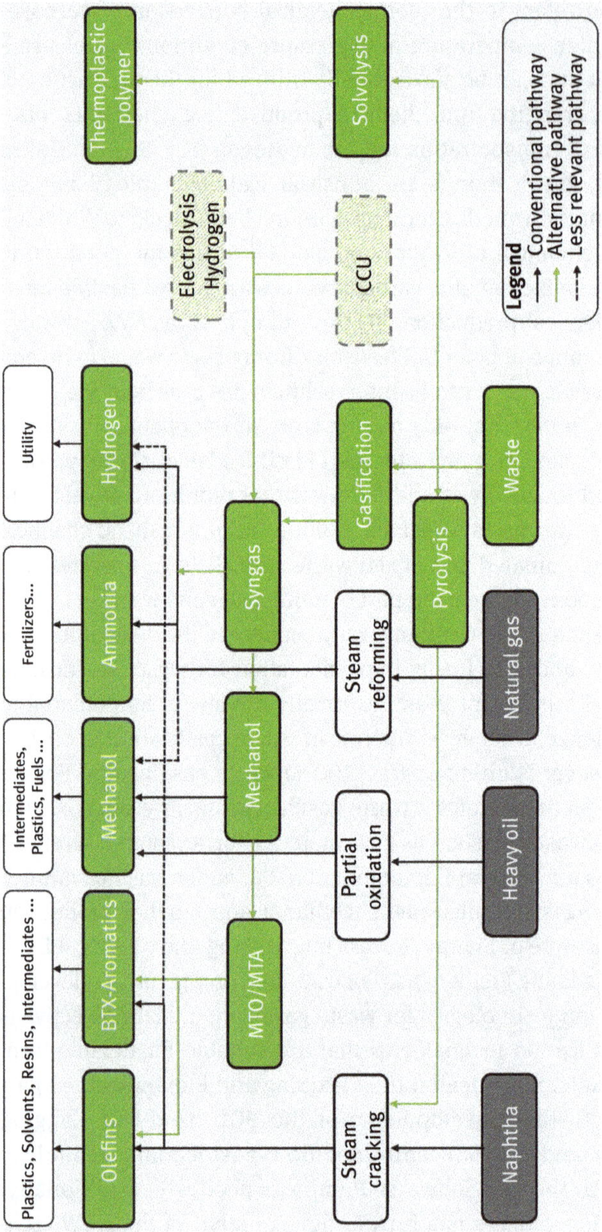

Figure 21.5: Process overview of integration opportunities for chemical recycling technologies.[7]

7 Adapted from Keller et al. [105].

Table 21.2 briefly summarizes the key pros and cons which are associated with the three different chemical recycling routes.

Table 21.2: Pros and cons of solvolysis, pyrolysis, and gasification.[8]

	Advantages	Disadvantages
Solvolysis	– Polymer recycling yield similar to mechanical recycling – Product quality similar to virgin material – Short process chain leading to high efficiency	– Require relatively pure and homogenous waste streams that limit feedstock potential – Contaminant removal potentially incomplete – Scale-up beyond pilot/demonstration still pending
Pyrolysis	– Comparatively simple process suitable for a wider range of feedstock materials compared to solvolysis – Easy integration of pyrolysis products in existing petrochemical infrastructure – Decentralized production possible	– Require relatively pure and homogenous waste streams – but of lower quality than solvolysis – which limit feedstock potential – Pyrolysis oil quality does not meet steam cracker specifications (removal of impurities and refining not proven at scale); significant catalytic hydrotreatment required – Scale-up difficulty (reactor design, heat transfer); economies-of-scale is limited by scale-up of small modules – High carbon footprint (similar to gasification) compared to solvolysis
Gasification	– Industrial-scale facilities in operation – Wide range of possible feedstock (e.g., mixed and contaminated waste) – Wide variety of products possible (e.g., syngas, SNG, and H_2) – Technologies for downstream processes (i.e., gas purification, conditioning, chemical synthesis) are commercially available – Integration of renewable hydrogen to utilize CO_2 component in the syngas as additional chemical feedstock and reduce carbon footprint	– Partial oxidation inevitably results in carbon losses, that is, carbon footprint similar to pyrolysis – Waste gasification technology for multiple waste feedstock (e.g., municipal waste, agricultural waste, and sewage sludge) still requires demonstration at scale – Complex process chain requiring high capital expenditure for large plants – Skilled personnel required for plant operation

8 Adapted from Seidl et al. [149].

21.3.3 Green Hydrogen from Waste

Hydrogen (H_2) is a gas with many advantages. It is light, energy dense, produces no direct emissions of pollutions or greenhouse gases, and – similar to natural gas – can be distributed under pressure via a pipeline network [109]. According to IEA [110, 111], industrial demand for hydrogen has more than tripled since 1975, reaching about 70 million tons in 2018. However, this demand is almost entirely met by fossil fuels, with 6% of global natural gas and 2% of global coal going to hydrogen production. Consequently, the global production of "gray" hydrogen via steam reforming of natural gas and coal gasification is associated with about 830 million tons of CO_2 emissions per year, similar to the combined emissions of Indonesia and the UK.

Today, new application opportunities for hydrogen to support decarbonization of diverse sectors are underpinning the growing global demand for hydrogen. These include applications in [111, 112]:

- the *industry sector* to replace fossil fuels in industrial processes such as oil refining, ammonia production, methanol production, and steel production;
- the *transportation sector* as a means to produce synthetic fuels or in fuels cells for hydrogen mobility;
- the *building heating sector* to reduce demand for natural gas by injection into existing natural gas networks; and
- the *power generation sector* to replace fossil fuels via utilization of hydrogen and ammonia and to increase power system flexibility.

To realize the full potential of hydrogen in supporting efforts to meet emissions reduction and circular economy goals, global attention has turned to "green" hydrogen from renewable energy sources (i.e., power-to-hydrogen). To date, R&D activities as well as project implementation have focused predominantly on green hydrogen production via water electrolysis powered by renewable energy, with installed electrolysis capacity growing from less than 1 MW in 2010 to more than 25 MW in 2019 [111]. However, green hydrogen from electrolysis faces three main challenges, namely high capital and production cost, a low conversion efficiency, as well as a lack of affordable renewable electricity.[9] This applies especially to a majority of the developed countries where there is a low temporal availability of solar and wind energy [113, 114]. In view of the dominant role green hydrogen is anticipated to play in meeting global decarbonization and circular economy targets, there is increasing interest in

9 Take Germany for example. For the production of 1 kg of hydrogen, about 60 kWh (i.e., 216 MJ) of electricity will be consumed [151]. Germany has the highest non-household electricity price in the EU at 0.18 EUR/kWh [152]. Hence, production of 1 kg of hydrogen via water electrolysis will cost 10.80 EUR. Assuming a natural gas price of 0.25 EUR/kg, this is almost 4–10 times more than hydrogen production via conventional steam reforming [153].

alternative routes for its production. Three such alternative routes involving the innovative use of biogenic carbonaceous waste sources are the production of green hydrogen via (1) biogas reforming, (2) biogas pyrolysis, and (3) gasification of biogenic waste.

- *Biogas reforming:* Biogas is a mixture of methane, CO_2 and small quantities of other gases such as hydrogen sulfide, hydrogen, nitrogen, oxygen, ammonia and water vapor [115]. It is produced via anaerobic digestion of organic waste material (e.g., agricultural waste, livestock manure, food waste, organic fraction in municipal waste, and wastewater sludges) in an oxygen-free environment (e.g. in biodigesters), in landfill gas recovery systems or in wastewater treatment plants [116]. Though it is generally combusted for energy production, biogas could be used to replace natural gas as feedstock in the reforming process, that is, in the biogas reformer to produce green – instead of gray – hydrogen [117]. The advantages of using biogas for this purpose are the reduction of greenhouse gas emissions, its reliability and its local availability [115].

  Biogas reforming processes can be categorized into conventional and non-conventional reforming processes. The former includes steam reforming, partial oxidation reforming (POR), autothermal reforming (ATR), dry reforming (DR), and dry oxidation reformation (DOR) while the later encompasses solar reforming, catalytic decomposition, and thermal plasma reforming [118].[10] During the reforming process, biogas is generally reacted with agents such as steam, air, or oxygen to produce syngas, which is made up of hydrogen and other gases [119]. As biogas contains some undesirable compounds (e.g., hydrogen sulfide) that could poison the catalysts, purification and upgrading technologies are necessary [115].

  Alves and colleagues [120] identified the ATR as an attractive path for hydrogen production from biogas due to its high energy efficiency, high hydrogen yields and lowest ratio of expense/utility compared to all other reforming processes. However, they also pointed to the need for a complex control mechanism due to the instability of the process. In contrast, Zhao et al. [121] indicated dry reforming as a promising candidate for green hydrogen production from biogas. However, not only is it more energy intensive compared to conventional steam reforming of natural gas, other challenges relating to lower catalytic performance, decreased reactant conversions and plugging of the reforming reactor also remain to be resolved.

  An innovative development is the reforming of biogas in a membrane reactor, where both chemical reactions and purification processes take place in the same compact reactor. Main advantages are the higher hydrogen yields and lower energy requirements [115]. To test the feasibility and potential of this new

10 Please refer to Nalbant and Colpan [115] for an overview of different reforming processes and a comparison of associated advantages and disadvantages.

technology, the EU funded the BIONICO project under its Horizon 2020 Program between 2015 and 2019 to develop, build, and demonstrate a biogas membrane reformer for decentralized green hydrogen production [122, 123].

- *Biogas pyrolysis*: Steam reforming is the state of the art for producing hydrogen from natural gas. To reduce the carbon footprint of hydrogen production from natural gas, attention has been increasingly focused on the production of "turquoise" hydrogen via pyrolysis of natural gas (also called methane pyrolysis). During methane pyrolysis, natural gas is heated up to temperatures of about 800–2,000 °C (depending on whether it is a catalytic process, a noncatalytic thermal process or using plasma torches) in the absence of steam, which thermally splits it into hydrogen and solid carbon. As carbon is captured in a solid, disposable form, hydrogen production via methane pyrolysis – in the case where renewable energy is used to deliver the heat required for the process – will have no CO_2 emissions. R&D activities focused on developing this alternative route of hydrogen production are currently being carried out by BASF, Karlsruhe Institute of Technology, Institute of Energy Process Engineering and Chemical Engineering at the TU Bergakademie Freiberg, Tomsk University in cooperation with GAZPROM as well as the Hazer Group [124, 125].[11]

  Similar to reforming, biogas could also be utilized to replace natural gas as feedstock for methane pyrolysis. This process is called biomethane pyrolysis. Biomethane pyrolysis is however more complicated than biogas reforming, with the technology still in the development stage. Nevertheless, industry interest is observable. For example, BASF, which is developing its own methane pyrolysis process for hydrogen production, has indicated that biogas would be an imaginable alternative to natural gas as feedstock for its process [126].
- *Gasification of biogenic waste*: As introduced in Chapter 21.3.1, gasification is a thermochemical technology which converts carbon sources – generally under high pressure and temperature conditions – into a mixture of combustible gases which includes hydrogen and carbon monoxide. The gasification of biogenic waste – often referred to as biomass gasification – could thus convert low or negative value waste feedstock into marketable and desirable products in the form of green hydrogen [127]. Gasification has thus been identified as a promising technology which is anticipated to play a key role in both the USA and the EU for green hydrogen production from renewable sources [128]. Biogenic waste which are currently gaining attention as potential carbonaceous feedstock for biomass gasification range from microalgae, crop residues, livestock waste, food waste, municipal solid waste, and sewage sludges to wood waste [127].

11 Please refer to Schneider et al. [125] for a state-of-the-art of hydrogen production via pyrolysis of natural gas.

In particular plasma gasification has been eliciting significant interest as a potentially clean technology with little environmental impact for green hydrogen production [127]. Its utilization to date has concentrated on waste-to-energy applications especially for the treatment of hazardous waste. However, it also has potential applicability for a wide range of organic wastes such as municipal solid waste, sewage sludges, biomass waste, and industrial wastes [129–133]. With the growing global interest in green hydrogen, commercial providers of plasma technologies are emphasizing the potential applicability of their technologies for such purpose. For example, together with the City of Lancaster in California, SGH2 Energy Global is planning to build a large-scale green hydrogen facility using its Solena Plasma Enhanced Gasification Technology to produce 3,800 tons of green hydrogen annually from 42,000 tons of recycled mixed paper waste [134, 135]. In Germany, Richter Recycling GmbH has announced their cooperation with PLAGAZI to produce up to 6,000 tons of green hydrogen from 45,000 tons of waste using InEnTec's plasma gasification technology [136].

Besides plasma gasification, other biomass gasification technologies which are potentially interesting for green hydrogen production include the dual fluidized-bed gasification technology from Vienna University of Technology (TU Wien), the MILENA gasification technology by the Energy Research Center of the Netherlands (ECN) as well as the Flexi-technologies by the Institute of Energy Process Engineering and Chemical Engineering, TU Bergakademie Freiberg mentioned in Chapter 21.3.2 [108, 128].

21.3.4 CO_2-Neutral Liquid Fuels (Waste-to-Fuels]

According to IEA [111], transportation is responsible for 24% of direct CO_2 emissions from the combustion of carbon resources. The transport sector thus has a crucial role to play in reducing the global carbon footprint and the transition toward a net zero society. In the last decade, the low carbon transformation in the transport sector has focused predominantly on battery electric vehicles to replace conventional petroleum-fueled internal combustion engine vehicles. Policy measures to promote electric vehicles in the passenger car segment – especially in China and other countries such as Norway, the Netherlands, Germany, Brazil, and the USA – have led to a significant growth in e-mobility worldwide [137, 138]. In recent years, this development is complemented by an expansion in hydrogen mobility – catalyzed by fuel cell vehicle developments in Asia [111]. However, while e-mobility and hydrogen-mobility could support a reduction in CO_2 emissions for road vehicles such as passenger cars and buses, other transport branches such as aviation, shipping and heavy load transportation have no real alternatives to liquid fuel for their operations in the foreseeable future [112]. Hence, there is an urgent need for a parallel development of CO_2-neutral, green liquid fuels to substitute fossil-based liquid fuels.

The chemical storage of green hydrogen in CO_2-neutral synthetic liquid fuels represents a complementary building block to developments in electric and hydrogen mobility. Synthetic fuels – assuming that they meet the appropriate standards – can be used as substitutes for diesel, gasoline and jet fuels [45]. Main advantage is that they can be used in conventional combustion engines. Thus, similar to biofuels, synthetic fuels can be distributed using the existing petrol station network as drop-in fuel or gradually blended into existing petroleum-based fuels for a stepwise transformation of conventional combustion engines toward low carbon mobility [139].

In recent years, developments have focused predominantly on electricity-based fuels, that is, E-fuels. E-fuels are defined as gaseous and liquid fuels such as hydrogen, methane, as well as synthetic petrol/diesel/kerosene fuels generated from renewable electricity [140]. The processes are commonly referred to as power-to-X (PtX), power-to-liquids (PtL) and power-to-gas (PtG). Basis is green hydrogen which is produced via water electrolysis powered by renewable energy. As such, E-fuels – unlike biofuels – will not compete with food production (i.e., no food-vs-fuel debate) [139]. To produce CO_2-neutral synthetic liquid E-fuels, green hydrogen is utilized in subsequent synthesis processes to convert CO_2 (either from point sources or from the atmosphere) to E-gasoline/E-diesel/E-kerosene/E-methanol for use in combustion engines [140]. Specifically, green hydrogen and CO_2 can be converted via FT synthesis process to E-diesel, E-kerosene, and E-gasoline. Alternatively, green hydrogen and CO_2 are first converted via methanol synthesis into E-methanol, before being catalytically converted for example into E-gasoline. Note that the costs of green hydrogen supply account for about 60–80% of the total production costs of synthetic liquid E-fuels [141]. As such, high capital and electricity costs associated with green hydrogen production (see Chapter 12) will considerably limit the economic attractiveness of E-gasoline/E-diesel/E-kerosene/E-methanol.

An alternative route to produce synthetic liquid fuels is via gasification. Its main advantage is that it requires only about one-tenth of the electricity required by water electrolysis.[12] The use of biogenic waste materials (e.g., waste wood, agricultural waste, and forest residues) as (hydro)carbon feedstock for waste gasification furthermore facilitates the chemical storage of biogenic hydrogen in CO_2-neutral synthetic liquid fuels. Note that the use of waste sources containing petrochemical waste components (e.g., municipal solid waste which contains plastic waste) as feedstock will support the production of a synthetic fuel with low carbon footprint, but which is not CO_2-neutral.

12 In a life cycle assessment of conventional and alternative production pathways for olefins, Keller and colleagues [150] determined that compared to electricity demand for olefins production via waste gasification, electricity demand increases by a factor of 12 for olefins production via carbon capture and utilization/CCU (i.e., using green hydrogen produced via renewably powered water electrolysis to react with CO_2 from point sources). A similar tendency is expected for the production of synthetic fuels.

Similar to E-fuels, following waste gasification, the syngas produced can be subsequently refined through synthesis processes to produce synthetic fuels. Through the FT process, syngas is converted via catalysts into a product mixture of synthetic naphtha, synthetic middle oil distillates (i.e., diesel and kerosene), lubricating oils and synthetic waxes [142]. Such use of waste gasification in combination with FT synthesis for the production of low carbon/CO_2-neutral synthetic liquid fuels is gaining momentum worldwide, with diverse projects in the pipeline. An example is the Sierra BioFuels Project which is using Fulcrum's gasification and FT processes to convert municipal solid waste feedstock into low carbon synthetic transportation fuels [143].

Alternatively, syngas is first converted into methanol via methanol synthesis. Following that, a second synthesis step is necessary to convert methanol to synthetic transportation fuels, for example, methanol-to-gasoline (MTG) [144]. The technical feasibility of the production of green gasoline is demonstrated for instance in the Closed-Carbon-Cycle (C3) Mobility project funded by the German Ministry for Economic Affairs and Energy. Within the project, the Institute of Energy Process Engineering and Chemical Engineering at the TU Bergakademie Freiberg in cooperation with Chemieanlagenbau Chemnitz GmbH (CAC) successfully produced 16,000 liters[13] of CO_2-neutral synthetic gasoline using CAC's MTG technology [145]. While its production cost is much lower than that of E-fuels, further development is still required to raise the economic competitiveness and energy efficiency of the MTG process for the production of CO_2-neutral synthetic liquid fuels [146].

21.4 Conclusion

The energetic utilization of primary and secondary carbon resources is associated with significant CO_2 emissions and plays a pivotal role in contributing to global warming. While it may be possible in the long run to decarbonize to a large extent the electricity and mobility sectors, other carbon-intensive and dependent sectors such as the chemical industry have no alternatives to carbon resources as feedstock for their production. Today, the chemical industry in numerous countries remains largely dependent on imported oil and gas for their production. In addition to global pressure to lower its carbon footprint, other drivers such as supply security, structural changes as well as resource efficiency and circularity are also motivating the chemical industry to accelerate its transformation from petrochemistry toward sustainable production via innovative use of alternative carbon resources.

In this article, transformation options for the chemical industry via the utilization of (1) domestic coal as an alternative primary carbon feedstock, and (2) domestic

13 A test campaign in 2020 generated an additional 45,000 L of CO_2-neutral synthetic gasoline during continuous operation (Internal report, TU Bergakademie Freiberg. Publication forthcoming).

waste as an alternative secondary carbon feedstock for chemical production are identified, and associated environmental and societal impacts discussed. To illustrate how the chemical utilization of primary and secondary carbon resources in the form of coal and waste as alternative feedstock could support the decarbonization efforts and transformation of carbon-intensive and dependent sectors, highlights from four global developments, namely (1) coal (and waste)-to-products, (2) chemical recycling (i.e., waste-to-chemicals), (3) green hydrogen from waste, and (4) CO_2-neutral liquid fuels from (i.e., waste-to-fuels) are briefly summarized, highlighting past achievements as well as recent progresses in R&D and project deployment. Insights from this article thus contribute to a better understanding of how innovative use of carbon resources via alternative transformation pathways focusing on their chemical utilization could complement global efforts toward a low carbon and net zero society.

References

[1] Lee RP, Scheibe A. The politics of a carbon transition: An analysis of political indicators for a transformation in the German chemical industry. J Clean Prod. 2020;244(22):118629. DOI: https://doi.org/10.1016/j.jclepro.2019.118629.

[2] Royal Society of Chemistry. (2021). Carbon. https://www.rsc.org/periodic-table/element/6/carbon, accessed 26 January 2021

[3] United Nations. (2021b). World population prospects 2019: Graphs/ profiles [World]. https://population.un.org/wpp/Graphs/DemographicProfiles/Line/900, accessed 26 January 2021.

[4] The World Bank Group. (2021). World Bank national accounts data, and OECD National Accounts data files: Data [GDP (current US$)]. https://data.worldbank.org/indicator/NY.GDP.MKTP.CD, accessed 26 January 2021.

[5] IEA. (2020c). World energy balances: Total energy supply by source [World 1990–2018]. International Energy Agency. https://www.iea.org/data-and-statistics?country=WORLD&fuel=Energy%20supply&indicator=TPESbySource, accessed 26 January 2021.

[6] bp. (2019). bp statistical review of world energy. bp p.l.c. https://www.bp.com/content/dam/bp/business-sites/en/global/corporate/pdfs/energy-economics/statistical-review/bp-stats-review-2020-full-report.pdf, accessed 26 January 2021.

[7] EPA. (2020). Global greenhouse gas emissions data. United States Environmental Protection Agency. https://www.epa.gov/ghgemissions/global-greenhouse-gas-emissions-data, accessed 26 January 2021.

[8] Fischedick M, Roy J, Abdel-Aziz A, Acquaye A, Allwood JM, Ceron J-P, Geng Y, Kheshgi H, Lanza A, Perczyk D, Price L, Santalla E, Sheinbaum C, Tanaka K. Industry. In: Edenhofer O, Pichs-Madruga R, Sokona Y, Farahani E, Kadner S, Seyboth K, Adler A, Baum I, Brunner S, Eickemeier P, Kriemann B, Savolainen J, Schlömer S, Von Stechow C, Zwickel T, Minx JC, editors. Climate change 2014: Mitigation of climate change: Contribution of working group III to the fifth assessment report of the intergovernmental panel on climate change. Cambridge University Press; 2014. http://pure.iiasa.ac.at/id/eprint/11116/1/ipcc_wg3_ar5_chapter10.pdf, accessed 16 March 2021.

[9] IEA. (2021a). CO2 emissions from fuel combustion: Overview: Statistics report – July 2020. International Energy Agency. https://www.iea.org/reports/co2-emissions-from-fuel-combustion-overview, accessed 26 January 2021.

[10] ecoprog. (2021). Waste to energy 2020/2021. ecoprog GmbH. https://www.ecoprog.com/publikationen/abfallwirtschaft/waste-to-energy.htm, accessed 26 January 2021.

[11] Lee RP, Meyer B, Huang Q, Voss R. Sustainable waste management for zero waste cities in China: potential, challenges and opportunities. Clean Energy. 2020;4(3):169–201. DOI: https://doi.org/10.1093/ce/zkaa013.

[12] Zero Waste Europe. (2019). The impact of waste-to-energy incineration on climate [Policy briefing]. https://zerowasteeurope.eu/wp-content/uploads/edd/2019/09/ZWE_Policy-briefing_The-impact-of-Waste-to-Energy-incineration-on-Climate.pdf, accessed 26 January 2021.

[13] Garrett CW. On global climate change, carbon dioxide, and fossil fuel combustion. Prog Energy Combust Sci. 1992;18(5):369–407. https://doi.org/10.1016/0360-1285(92)90007-N.

[14] Scientific American. (2018). CO2 can directly impact extreme weather, research suggests: Some experts are calling for a cap on CO2 emissions in addition to warming limits. E&E News. https://www.scientificamerican.com/article/co2-can-directly-impact-extreme-weather-research-suggests/#:~:text=Rising%20carbon%20dioxide%20levels%20in,the%20climate%20system%2C%20they%20say.&text=As%20a%20result%2C%20some%20experts,in%20the%20Paris%20climate%20accord, accessed 26 January 2021.

[15] United Nations. (2021a). The Paris Agreement. United Nations Framework Convention on Climate Change. https://unfccc.int/process-and-meetings/the-paris-agreement/the-paris-agreement, accessed 26 January 2021.

[16] Lee RP. Alternative carbon feedstock for the chemical industry? – Assessing the challenges posed by the human dimension in the carbon transition. J Clean Prod. 2019;219(8):786–96. DOI: https://doi.org/10.1016/j.jclepro.2019.01.316.

[17] McKinsey & Company. (2017). Hydrogen: The next wave for electric vehicles? https://www.mckinsey.com/industries/automotive-and-assembly/our-insights/hydrogen-the-next-wave-for-electric-vehicles, accessed 26 January 2021.

[18] GDCh, Dechema, DGMK, & VCI. (2010). Position paper: Change in the raw materials base. Gesellschaft Deutscher Chemiker e.V.; Gesellschaft für Chemische Technik und Biotechnologie e.V.; Deutschen Wissenschaftlichen Gesellschaft für Erdöl, Erdgas und Kohle e.V; Verband der Chemischen Industrie e.V. https://www.gdch.de/fileadmin/downloads/Service_und_Informationen/Presse_OEffentlichkeitsarbeit/PDF/pos_rohst_en.pdf, accessed 26 January 2021.

[19] dena. (2019). Feedstocks for the chemical industry [Global Alliance Powerfuels]. Deutsche Energie-Agentur GmbH. https://www.dena.de/fileadmin/dena/Publikationen/PDFs/2019/Feedstocks_for_the_chemical_industry.pdf, accessed 26 January 2021.

[20] Lee RP, Keller F, Meyer B. A concept to support the transformation from a linear to circular carbon economy: net zero emissions, resource efficiency and conservation through a coupling of the energy, chemical and waste management sectors. Clean Energy. 2017;1(1):102–13. DOI: https://doi.org/10.1093/ce/zkx004.

[21] Lee RP, Reinhardt R, Keller F, Gurtner S, Schiffer L. A raw materials transition for a low-carbon economy: challenges and opportunities for management in addressing the trilemma of competitiveness, supply security and sustainability. In: George G, Schillebeeckx S, editors. Managing Natural Resources. Edward Elgar Publishing; 2018. 61–87. https://www.elgaronline.com/view/edcoll/9781786435712/9781786435712.00012.xml.

[22] IEA. (2020a). Coal information: Overview: Statistics report – July 2020. International Energy Agency. https://www.iea.org/reports/coal-information-overview, accessed 26 January 2021.

[23] Lee RP, Laugwitz A, Keller F, Wolfersdorf C, Mehlhose F, Meyer B. Achieving a closed carbon & circular economy for the waste management sector: Net zero emissions, resource efficiency and conservation by coupling the energy, chemical and recycling sectors. In: Thomé-Kozmiensky KJ, Thiel S, Thomé-Kozmiensky E, Winter F, Juchelková D, editors. Waste managament, Volume 7: Waste to energy. TK Verlag Karl Thomé-Kozmiensky; 2017. 343–54.

[24] Minchener A (2016). Gasification based coal to chemicals in China: economic and environmental challenges. 8th International Freiberg Conference on IGCC & XtL Technologies, 12 June 2016, Cologne, Germany. https://tu-freiberg.de/fakult4/iec/evt/gasification-based-coal-to-chemicals-in-china-economic-and-environmental-challenges, accessed 26 January 2021.

[25] Hoornweg D, Bhada-Tata P (2012). What a waste: A global review of solid waste management [Urban development series knowledge papers]. World Bank. https://openknowledge.worldbank.org/handle/10986/17388 License: CC BY 3.0 IGO, accessed 26 January 2021.

[26] BBC. (2019). US top of the garbage pile in global waste crisis. BBC News. https://www.bbc.com/news/science-environment-48838699, accessed 26 January 2021.

[27] Smith N (2019). US tops list of countries fuelling the waste crisis: Waste generation and recycling indices. https://www.maplecroft.com/insights/analysis/us-tops-list-of-countries-fuelling-the-mounting-waste-crisis/, accessed 26 January 2021.

[28] European Commission. (2018). A European strategy for plastics in a circular economy. https://www.europarc.org/wp-content/uploads/2018/01/Eu-plastics-strategy-brochure.pdf, accessed 26 January 2021.

[29] Lebreton L, Slat B, Ferrari F, Sainte-Rose B, Aitken J, Marthouse R, Hajbane S, Cunsolo S, Schwarz A, Levivier A, Noble K, Debeljak P, Maral H, Schoeneich-Argent R, Brambini R, Reisser J. Evidence that the Great Pacific Garbage Patch is rapidly accumulating plastic. Sci Rep. 2018;8(1):4666. DOI: https://doi.org/10.1038/s41598-018-22939-w.

[30] Parker L (2018). We made plastic. We depend on it. Now we're drowning in it: The miracle material has made modern life possible. But more than 40 percent of it is used just once, and it's choking our waterways. Planet or Plastics. National Geographic Partners, LLC. https://www.nationalgeographic.com/magazine/2018/06/plastic-planet-waste-pollution-trash-crisis/, accessed 26 January 2021.

[31] Parker L (2019). The world agrees there's a plastic waste crisis – can it agree on a solution? Environment, planet or plastics. National Geographic Partners, LLC. https://www.nationalgeographic.com/environment/2019/03/un-environment-plastic-pollution-negotiations/, accessed 26 January 2021.

[32] BBC. (2020). Single-use plastic: China to ban bags and other items. BBC News. https://www.bbc.com/news/world-asia-china-51171491, accessed 26 January 2021.

[33] cefic. (2021a). Chemical recycling. The European Chemical Industry Council. https://cefic.org/policy-matters/innovation/chemical-recycling/, accessed 26 January 2021.

[34] European Parliament. (2019). Directive (EU) 2019/904 of the European Parliament and of the Council of 5 June 2019 on the reduction of the impact of certain plastic products on the environment: Directive (EU) 2019/904. https://eur-lex.europa.eu/legal-content/EN/TXT/PDF/?uri=CELEX:32019L0904, accessed 26 January 2021.

[35] Wiley-VCH. (2019). Chemical recycling: Racing to close the loop: The EU's circular economy package drives companies to share know-how and resources. Markets & companies. ChemManager International. https://www.chemanager-online.com/en/topics/economy-business/chemical-recycling-racing-close-loop, accessed 26 January 2021.

[36] ACC. (2021). Plastics: Advanced recycling alliance for plastics. ACC American Chemistry Council. https://plastics.americanchemistry.com/Advanced-Recycling-Alliance-for-Plastics.html, accessed 26 January 2021.

[37] Mamani-Soliz P, Seidl LG, Keller F, Lee RP, Meyer B. Chemisches Recycling – Aktueller Stand und neue Entwicklungen. In: Holm O, Thomé-Kozmiensky E, Goldmann D, Friedrich B, editors. Recycling und Sekundärrohstoffe, Band. Thomé-Kozmiensky Verlag GmbH; 2020, Vol. 13, 268–84.
[38] Recenso. (2021). Chemical recycling. Recenso GmbH. https://recenso.eu/en/chemical-recycling-en.html, accessed 26 January 2021.
[39] cefic. (2021b). Chemical recycling is essential for the EU to achieve its circular economy goals. The European Chemical Industry Council. https://cefic.org/media-corner/newsroom/chemical-recycling-is-essential-for-the-eu-to-achieve-its-circular-economy-goals/, accessed 26 January 2021.
[40] Neste. (2021). Why chemical recycling? https://www.neste.com/products/all-products/plastics/combating-plastic-pollution/why-chemical-recycling, accessed 26 January 2021.
[41] Kaza S, Yao L, Bhada-Tata P, Woerden FV (2018). What a waste 2.0: A global snapshot of solid waste management to 2050 [Urban development series knowledge papers]. World Bank. https://openknowledge.worldbank.org/handle/10986/30317 License: CC BY 3.0 IGO, accessed 26 January 2021.
[42] Zero Waste Europe. (2020a). Chemical recycling: 7 Steps to effectively legislate on chemical recycling [Joint statement]. https://zerowasteeurope.eu/wp-content/uploads/2020/07/rpa_chemical_recycling_statement.pdf, accessed 26 January 2021.
[43] Zero Waste Europe. (2020b). Chemical recycling in the US: a Trojan horse for plastic-to-fuel [Chemical recycling]. https://zerowasteeurope.eu/2020/08/chemical-recycling-in-the-us-a-trojan-horse-for-plastic-to-fuel/, accessed 26 January 2021.
[44] UCDavis. (2021). Science & climate definitions. The Regents of the University of California, Davis campus. https://climatechange.ucdavis.edu/climate-change-definitions/biogenic-carbon/, accessed 26 January 2021.
[45] European Commission. (2015). State of the art on alternative fuels transport systems in the European Union: Final report. https://ec.europa.eu/transport/sites/transport/files/themes/urban/studies/doc/2015-07-alter-fuels-transport-syst-in-eu.pdf, accessed 26 January 2021.
[46] Klerk AD. Fischer-Tropsch process. In: John Wiley & Sons, editor. Kirk-Othmer Encyclopedia of Chemical Technology. John Wiley & Sons, Inc; 2013. 1. Vol. 1084, https://doi.org/10.1002/0471238961.fiscdekl.a01.
[47] Höök M, Fantazzini D, Angelantoni A, Snowden S. Hydrocarbon liquefaction: Viability as a peak oil mitigation strategy. Philos Trans A Math Phys Eng Sci. 2014;372(2006):20120319. https://doi.org/10.1098/rsta.2012.0319.
[48] Bergius F (1932). Chemical reactions under high pressure. Nobel Lecture, 21 May 1932, Stockholm, Sweden. https://www.nobelprize.org/uploads/2018/06/bergius-lecture.pdf, accessed 26 January 2021.
[49] CAZWV. (2021). A brief overview of coal to chemicals: Opportunities and obstacles. CAZWV Chemical Alliance Zone West Virginia. https://energywv.org/assets/files/Energy-Summit-Presentations/2016/19_DIGREGORIO.pdf, accessed 26 January 2021.
[50] Stranges A. Germany's synthetic fuel industry, 1927–1945. In: Lesch JE, editor. The German Chemical Industry in the Twentieth Century. Springer Netherlands; 2000, Vol. 75, 147–216. DOI: https://doi.org/10.1007/978-94-015-9377-9_7.
[51] Sasol. (2021). Historical milestones. Sasol Technology. https://www.sasol.com/about-sasol/company-profile/historical-milestones, accessed 26 January 2021.
[52] Nierop P, Van, Erasmus HB, Van Zyl JW (2000). Sasol's achievements in the 20th century as a building block for the 21st. Sasol Technology. 2000 Gasification Technologies Council, 08 October 2000, San Francisco, California, USA. https://de.scribd.com/document/354097732/Sasol-Technology, accessed 26 January 2021.

[53] Spalding-Fecher R, Williams A, Van Horen C. Energy and environment in South Africa: charting a course to sustainability. Energy Sustainable Dev. 2000;4(4):8–17. https://doi.org/10.1016/S0973-0826(08)60259-8.

[54] Scheithauer M, Mamani Soliz PE, Lee RP, Keller F, Meyer B, Bui X-N, Huong TTT. Assessment of feasible and effective technologies for the chemical utilization of domestic coal for value-added production in Vietnam. In: Bui X-N, Lee C, Drebenstedt C, editors. Lecture Notes in Civil Engineering. Proceedings of the International Conference on Innovations for Sustainable and Responsible Mining. Springer International Publishing; 2021, Vol. 109, 364–84. DOI: https://doi.org/10.1007/978-3-030-60839-2_19.

[55] Vamvuka D. Gasification of coal. Energy Explor Exploit. 1999;17(6):515–81. https://doi.org/10.1177/014459879901700603.

[56] NETL. High Temperature Winkler (HTW) gasifier. National Energy Technology Laboratory; 2021e, https://netl.doe.gov/research/Coal/energy-systems/gasification/gasifipedia/winkler, accessed 26 January 2021.

[57] Radtke K (2011). ThyssenKrupp Uhde's PRENFLO® and HTW™ gasification technologies: Global update on technology and projects. Gasification Technologies Conference 2011, 09 October 2011, San Francisco, California, USA. https://www.netl.doe.gov/sites/default/files/netl-file/16RADTKE.pdf, accessed 26 January 2021.

[58] NETL. (2021i). U-Gas® gasifier. National Energy Technology Laboratory. https://netl.doe.gov/research/Coal/energy-systems/gasification/gasifipedia/ugas, accessed 26 January 2021.

[59] Danulat F (1941). Die Druckvergasung fester Brennstoffe mit Sauerstoff. http://www.fischer-tropsch.org/Tom%20Reels/Linked/TOM%20195%20Partial/TOM-195-0074-0083%20Part%208%20Item%203.pdf, accessed 02 February 2021.

[60] Gumz W. Verfahren unter Druck, dargelegt an Beispielen der Kohlen- und Koksverwendung. Chem Ing Tech. 1960;32(3):172–78. https://doi.org/10.1002/cite.330320308.

[61] Modde P, Krzack S. Gaserzeuger mit Drehrost/Lurgi-Druckvergasung: Verfahrensprinzip und frühere Entwicklungen, Festbettdruckvergasung von Weichbraunkohle, Festbettdruckvergasung von Salzkohle. In: Schmalfeld J, editor. Die Veredlung und Umwandlung von Kohle: Technologien und Projekte 1970 bis 2000 in Deutschland. 2008. 307–36. DGMK Deutsche Wissenschaftliche Gesellschaft für Erdöl, Erdgas und Kohle e. V.

[62] Ray GF. Der Innovationsprozeß in westeuropäischen Industrieländern. Innovation in der Energiewirtschaft. Vol. 3. Duncker & Humblot; 1979.

[63] Schmalfeld J, editor. Die Veredlung und Umwandlung von Kohle: Technologien und Projekte 1970 bis 2000 in Deutschland. 2008, DGMK Deutsche Wissenschaftliche Gesellschaft für Erdöl, Erdgas und Kohle e. V.

[64] Weiss -M-M, Turna O (2009). Lurgi´s FBDB gasification – recent developments and project up-dates. Lurgi GmbH, Air Liquide Group. 3th International Freiberg Conference on IGCC & XtL Technologies, 18 May 2009, Dresden, Germany. https://tu-freiberg.de/sites/default/files/media/professur-fuer-energieverfahrenstechnik-und-thermische-rueckstandsbehandlung-16460/publikationen/4774_593.pdf, accessed 16 March 2021.

[65] SEDIN. (2018). Coal to synthetic natural gas (SNG). Sedin Engineering Co., Ltd. CNCEC. http://en.sedin.com/index.php?m=content&c=index&a=lists&catid=30, accessed 26 January 2021.

[66] Xia L (2011). China gasification market outlook. Asiachem Consulting. Gasification Technologies Conference 2011, 10 October 2011, San Francisco, California, USA. https://docplayer.net/83902067-China-gasification-market-outlook-lei-xia-asiachem-consulting-2011-gasification-technologies-conference-san-francisco-u-s-oct.html, accessed 26 January 2021.

[67] Bucko Z, Uhelna S, Takahashi M, Vierrath H (2000). HTW Fluidized-bed gasification for 400 MW IGCC power plant, Vresova – Czech Republic. Gasification Technologies Conference 2000, 09 October 2000, San Francisco, California, USA.

[68] NETL. British Gas/ Lurgi gasifier. National Energy Technology Laboratory; 2021b, https://netl.doe.gov/research/coal/energy-systems/gasification/gasifipedia/bgl, accessed 26 January 2021.

[69] Sächsisches Staatsarchiv. (2021). 40143 Deutsches Brennstoffinstitut Freiberg [Bestand]. https://archiv.sachsen.de/archiv/bestand.jsp?oid=09.03&bestandid=40143&syg_id=235089, accessed 26.0.2021.

[70] Stark F (2013). Das war die größte Dreckschleuder der DDR. Geschichte, Schwarze Pumpe. Welt. https://www.welt.de/geschichte/article121393564/Das-war-die-groesste-Dreckschleuder-der-DDR.html, accessed 26 January 2021.

[71] NETL. (2021h). Siemens gasifiers. National Energy Technology Laboratory. https://netl.doe.gov/research/coal/energy-systems/gasification/gasifipedia/siemens, accessed 26 January 2021.

[72] China Daily. (2016). Significant coal-to-liquid project in production in Ningxia. Xinhua. China Daily Information Co. https://www.chinadaily.com.cn/china/2016-12/28/content_27805385.htm, accessed 26 January 2021.

[73] dpa. (2016). Siemens-Standort in Freiberg macht dicht. Sachsen. Deutsche Presse-Agentur, Welt. https://www.welt.de/regionales/sachsen/article151190891/Siemens-Standort-in-Freiberg-macht-dicht.html, accessed 26 January 2021.

[74] TUBAF. (2021). Ausstattung EVT – Großversuchsanlagen. TU Bergakademie Freiberg. https://tu-freiberg.de/fakult4/iec/evt/ausstattung/grossversuchsanlagen, accessed 26 January 2021.

[75] NETL. GE Energy (formerly Chevron Texaco) gasifier. National Energy Technology Laboratory; 2021d, https://netl.doe.gov/research/coal/energy-systems/gasification/gasifipedia/ge, accessed 26 January 2021.

[76] NETL. Shell gasifier. National Energy Technology Laboratory; 2021g, https://netl.doe.gov/research/coal/energy-systems/gasification/gasifipedia/shell, accessed 26 January 2021.

[77] NETL. (2021j). Uhde – Prenflo. National Energy Technology Laboratory. https://netl.doe.gov/research/coal/energy-systems/gasification/gasifipedia/uhde-prenflo, accessed 26 January 2021.

[78] NETL. China Gasification Database. National Energy Technology Laboratory; 2021c, https://netl.doe.gov/research/coal/energy-systems/gasification/gasification-plant-databases/china-gasification-database, accessed 26 January 2021.

[79] Global Syngas Technologies Council. (2021). The gasification industry. https://www.globalsyngas.org/resources/the-gasification-industry/, accessed 26 January 2021.

[80] Wang F, Guangsuo Y, Qinghua G (2017). Development of coal gasification technology in China. https://www.worldcoal.org/development-coal-gasification-technology-china, accessed 26 January 2021.

[81] BMU. (2017a). Gesetz zur Förderung der Kreislaufwirtschaft und Sicherung der umweltverträglichen Beseitigung von Abfällen: Kreislaufwirtschafts- und Abfallgesetz – KrW-/AbfG. Bundesministerium für Umwelt, Naturschutz und nukleare Sicherheit. http://www.gesetze-im-internet.de/krwg/KrWG.pdf, accessed 26 January 2021.

[82] BMU. (2017b). Verordnung über die Vermeidung und Verwertung von Verpackungsabfällen: Verpackungsverordnung (VerpackV). Bundesministerium für Umwelt, Naturschutz und nukleare Sicherheit. https://web.archive.org/web/20180613161825/http://www.gesetze-im-internet.de/verpackv_1998/VerpackV.pdf, accessed 26 January 2021.

[83] BMU. (2021). Waste policy [Background: The development of waste policy in Germany]. Bundesministerium für Umwelt, Naturschutz und nukleare Sicherheit. https://www.bmu.de/en/topics/water-waste-soil/waste-management/waste-policy/, accessed 26 January 2021.

[84] Fishbein BK. Germany, Garbage, and the Green Dot: Challenging the Throwaway Society. Der Grüne Punkt. United States Environmental Protection Agency; 1994.

[85] Adlhoch W, Sato H, Wolff J, Radtke K (2000). High-Temperature Winkler gasification of municipal solid waste. 2000 Gasification Technologies Conference, 08 October 2000, San Francisco, California, USA. https://netl.doe.gov/sites/default/files/netl-file/Gtc00320.pdf, accessed 26 January 2021.

[86] Renzenbrink W, Wischnewski R, Engelhard J, Mittelstädt A (1998). High Temperature Winkler (HTW) coal gasification: A fully developed process for methanol and electricity production. https://netl.doe.gov/sites/default/files/netl-file/GTC9808P.pdf, accessed 26 January 2021.

[87] Buttker B, Obermeier T. Methanol aus Klärschlamm: Das Konzept der SVZ Schwarze Pumpe GmbH. In: Thomé-Kozmiensky KJ, editor. Technik, Wirtschaft, Umweltschutz. Verantwortungsbewusste Klärschlammverwertung. TK Verlag Karl Thomé-Kozmiensky; 2001. https://www.tomm-c.de/fileadmin/pdf/2002/Methanol_aus_Klaerschlamm_in_der_SVZ_Schwarze_Pumpe_GmbH_2002.pdf, accessed 16 March 2021.

[88] Obermeier T, Findeisen H, Markowski J. Forderungen an die Qualität von Einsatzmaterial für die Vergasung im SVZ Schwarze Pumpe. In: Thomé-Kozmiensky KJ, editor. Ersatzbrennstoffe. TK Verlag Karl Thomé-Kozmiensky; 2002. accessed 02 February 2021.

[89] KI. (2007). Sustec: Schwarze Pumpe will keinen Abfall mehr. KI Kunststoff Information. https://www.kunststoffweb.de/branchen-news/sustec_schwarze_pumpe_will_keinen_abfall_mehr_t208261, accessed 26 January 2021.

[90] Lee RP, Wolfersdorf C, Keller F, Meyer B. Towards a closed carbon cycle and achieving a circular economy for carbonaceous resources: Net zero emissions, resource efficiency and resource conservation through coupling of the energy, chemical and recycling sectors. Erdöl, Erdgas, Kohle – Oil Gas European Magazine. 2017, 2017(6):76–80.

[91] cefic. (2020). Chemical recycling: Greenhouse gas emission reduction potential of an emerging waste management route. The European Chemical Industry Council. https://cefic.org/app/uploads/2020/12/CEFIC_Quantis_report_final.pdf?utm_campaign=Circular-economy&utm_source=Cefic-Twitter&utm_medium=post-organic&utm_content=Quantis-Report&utm_term=Europe_cefic_Anyone_Campaign_Reach_Circular_economy_uk-en__post-organic_Quantis-Report, accessed 26 January 2021.

[92] Krause L, Dietrich F, Carus M, Skoczinski P, Ruiz P, Dammer L, Raschka A (2020). Chemical recycling – status, trends and challenges.: Technologies, sustainability, policy and key players. nova Institut GmbH. http://bio-based.eu/downloads/chemical-recycling-status-trends-and-challenges/, accessed 26 January 2021.

[93] Plastics Recyclers Europe. (2018). Chemical recycling. https://www.plasticsrecyclers.eu/chemical-recycling, accessed 26 January 2021.

[94] PlasticsEurope. (2019). Plastics – the facts 2019: An analysis of European plastics production, demand and waste data. https://www.plasticseurope.org/de/resources/publications/1804-plastics-facts-2019, accessed 26 January 2021.

[95] Lechleitner A, Schwabl D, Schubert T, Bauer M, Lehner M. Chemisches Recycling von gemischten Kunststoffabfällen als ergänzender Recyclingpfad zur Erhöhung der Recyclingquote. Österreichische Wasser- Und Abfallwirtschaft. 2020;72(1-2):47–60. https://doi.org/10.1007/s00506-019-00628-w.

[96] Xanthos M, Patel SH. Solvolysis. In: Akovali G, Bernardo CA, Leidner J, Utracki LA, Xanthos M, editors. Frontiers in the Science and Technology of Polymer Recycling. Springer Netherlands; 1998, Vol. 74, 425–36. DOI: https://doi.org/10.1007/978-94-017-1626-0_20.

[97] Zhao Y-B, Lv X-D, Ni H-G. Solvent-based separation and recycling of waste plastics: A review. Chemosphere. 2018;209:707–20. DOI: https://doi.org/10.1016/j.chemosphere.2018.06.095.

[98] Hann S, Connock T (2020). Chemical recycling: state of play. Eunomia. https://www.eunomia.co.uk/reports-tools/final-report-chemical-recycling-state-of-play/#:~:text=The%20final%20report%2C%20Chemical%20Recycling,input%20and%20trouble%20dealing%20with, accessed 26 January 2021.
[99] EuPC. (2020). Plastics circularity multiplier. European Plastics Converters. https://www.plasticscircularitymultiplier.eu/, accessed 27 January 2021.
[100] BASF. (2019). BASF invests in quantafuel to jointly drive chemical recycling of mixed plastic waste. BASF SE. https://www.basf.com/dz/fr/live/News/2019/10/p-19-356.html, accessed 27 January 2021.
[101] LyondellBasell. (2018). LyondellBasell advances chemical recycling by signing agreement with the Karlsruhe Institute of Technology. LyondellBasell Industries Holdings B.V. https://www.lyondellbasell.com/en/news-events/products–technology-news/lyondellbasell-advances-chemical-recycling-by-signing-agreement-with-the-karlsruhe-institute-of-technology/, accessed 27 January 2021.
[102] Borealis. (2019). OMV reoil project: OMV and Borealis extend their partnership at the industrial site in Schwechat: OMV and Borealis are exploring synergies in order to establish a circular economy. Borealis AG. https://www.borealisgroup.com/news/omv-reoil-circular-economy-project-omv-and-borealis-extend-their-partnership, accessed 27 January 2021.
[103] INEOS. (2020). INEOS and Plastic Energy to collaborate on new advanced plastic recycling facility [News]. INEOS Group. https://www.ineos.com/news/shared-news/ineos-and-plastic-energy-to-collaborate-on-new-advanced-plastic-recycling-facility/, accessed 27 January 2021.
[104] Plastic Energy. (2020). Total and Plastic Energy announce a strategic partnership and the construction of the first chemical recycling plant in France. Plastic Energy™. https://plasticenergy.com/total-and-plastic-energy-announce-a-strategic-partnership-and-the-construction-of-the-first-chemical-recycling-plant-in-france/, accessed 27 January 2021.
[105] Keller F, Lee RP, Meyer B. Life cycle assessment of global warming potential, resource depletion and acidification potential of fossil, renewable and secondary feedstock for olefin production in Germany. J Clean Prod. 2020;250:119484. https://doi.org/10.1016/j.jclepro.2019.119484.
[106] EBARA. (2021). Fluidized-bed gasification technologies [Technologies and products]. EBARA Environmental Plant Co.,Ltd. https://www.eep.ebara.com/en/products/gas.html, accessed 27 January 2021.
[107] Enerkem. (2021). Carbon recycling: Promote a circular economy with clean fuels and chemicals made from waste. https://enerkem.com/process-technology/carbon-recycling/, accessed 27 January 2021.
[108] Lee RP, Meyer B. Synthesegas aus schwer verwertbaren Abfällen. Kreislaufwirtschaft. Umweltmagazin. 2019;2019(09), 36–39. https://www.ingenieur.de/fachmedien/umweltmagazin/abfall-und-kreislauf/synthesegas-aus-schwer-verwertbaren-abfaellen/. accessed 27 January 2021.
[109] IEA. (2019). The future of hydrogen: Seizing today's opportunities. International Energy Agency. https://www.iea.org/reports/the-future-of-hydrogen, accessed 27 January 2021.
[110] IEA. (2021b). Hydrogen. International Energy Agency. https://www.iea.org/fuels-and-technologies/hydrogen, accessed 27 January 2021.
[111] IEA. (2021c). Hydrogen: More efforts needed. International Energy Agency. https://www.iea.org/reports/hydrogen, accessed 27 January 2021.
[112] BMWi. (2020). The National Hydrogen Strategy. Federal Ministry for Economic Affairs and Energy. https://www.bmbf.de/files/bmwi_Nationale%20Wasserstoffstrategie_Eng_s01.pdf, accessed 27 January 2021.

[113] Christensen A (2020). Assessment of hydrogen production costs from electrolysis: United States and Europe [Consultant report]. International Council on Clean Transportation, Three Seas Consulting. https://theicct.org/publications/assessment-hydrogen-production-costs-electrolysis-united-states-and-europe, accessed 27 January 2021.

[114] Hydrogeit. (2020). Green hydrogen from biogas. H2-International, GeneratePress. https://www.h2-international.com/2020/09/15/green-hydrogen-from-biogas/, accessed 27 January 2021.

[115] Nalbant Y, Colpan CO. An overview of hydrogen production from biogas. In: Uyar TS, editor. Lecture Notes in Energy. Accelerating the transition to a 100% renewable energy era. Springer International Publishing; 2020, Vol. 74, 355–73. DOI: https://doi.org/10.1007/978-3-030-40738-4_16.

[116] IEA. (2020b). Outlook for biogas and biomethane: Prospects for organic growth: World energy outlook special report [Fuel report – March 2020]. International Energy Agency. https://www.iea.org/reports/outlook-for-biogas-and-biomethane-prospects-for-organic-growth/an-introduction-to-biogas-and-biomethane, accessed 27 January 2021.

[117] World Biogas Association. (2019). Global potential of biogas. https://www.worldbiogasassociation.org/wp-content/uploads/2019/09/WBA-globalreport-56ppa4_digital-Sept-2019.pdf, accessed 27 January 2021.

[118] Verma P, Samanta SK. Overview of biogas reforming technologies for hydrogen production: Advantages and challenges. In: Kumar S, Khanal SK, Yadav YK, editors. Springer Proceedings in Energy. Proceedings of the First International Conference on Recent Advances in Bioenergy Research. Springer India; 2016, Vol. 38, 227–43. DOI: https://doi.org/10.1007/978-81-322-2773-1_17.

[119] Galvagno A, Chiodo V, Urbani F, Freni F. Biogas as hydrogen source for fuel cell applications. Int J Hydrogen Energy. 2013;38(10):3913–20. https://doi.org/10.1016/j.ijhydene.2013.01.083.

[120] Alves HJ, Bley Junior C, Niklevicz RR, Frigo EP, Frigo MS, Coimbra-Araújo CH. Overview of hydrogen production technologies from biogas and the applications in fuel cells. Int J Hydrogen Energy. 2013;38(13):5215–25. https://doi.org/10.1016/j.ijhydene.2013.02.057.

[121] Zhao X, Joseph B, Kuhn J, Ozcan S. Biogas reforming to syngas: A review. IScience. 2020;23(5):101082. DOI: https://doi.org/10.1016/j.isci.2020.101082.

[122] BIONICO. (2021). Biogas membrane reformer for decentralized H2 production: Call: H2020-JTI-FCH-2014-1 Topic: FCH-02.2-2014 Decentralized hydrogen production from clean CO2-containing biogas. http://www.bionicoproject.eu/, accessed 27 January 2021.

[123] CORDIS. (2020). Biogas membrane reformer for decentrallzed hydrogen production [HORIZON 2020]. CORDIS Informationsdienst der Gemeinschaft für Forschung und Entwicklung. https://cordis.europa.eu/project/id/671459/de, accessed 27 January 2021.

[124] Drescher F (2021). Wasserstoff: Energieträger der Zukunft? Deutschlandfunk Kultur. https://www.deutschlandfunkkultur.de/wasserstoff-energietraeger-der-zukunft.976.de.html?dram:article_id=491061, accessed 27 January 2021.

[125] Schneider S, Bajohr S, Graf F, Kolb T. State of the art of hydrogen production via pyrolysis of natural gas. ChemBioEng Rev. 2020;7(5):150–58. https://doi.org/10.1002/cite.202000021.

[126] Dierks S (2019). BASF setzt in der Wasserstoffherstellung auf Pyrolyse. Neue Märkte. Energate Messenger. https://www.energate-messenger.de/news/196410/basf-setzt-in-der-wasserstoffherstellung-auf-pyrolyse, accessed 27 January 2021.

[127] Hossain MZ, Charpentier PA. Hydrogen production by gasification of biomass and opportunity fuels. In: Subramani V, Basile A, Veziroğlu TN, editors. Compendium of Hydrogen Energy: Hydrogen Production And Purification. Elsevier Ltd; 2015, Vol. 14, 137–75. https://doi.org/10.1016/B978-1-78242-361-4.00006-6.

[128] Binder M, Kraussler M, Kuba M, Luisser M (2018). Hydrogen from biomass gasification. IEA Bioenergy. https://www.ieabioenergy.com/wp-content/uploads/2019/01/Wasserstoffstudie_IEA-final.pdf, accessed 27 January 2021.

[129] Fabry F, Rehmet C, Rohani V, Fulcheri L. Waste gasification by thermal plasma: A review. Waste Biomass Valorization. 2013;4(3):421–39. https://doi.org/10.1007/s12649-013-9201-7.

[130] NETL. Plasma Gasification. National Energy Technology Laboratory; 2021f, https://netl.doe.gov/research/Coal/energy-systems/gasification/gasifipedia/westinghouse, accessed 27 January 2021.

[131] Pourali M. Application of plasma gasification technology in waste to energy – challenges and opportunities. EEE Trans Sustainable Energy. 2010;1(3):125–30. https://ieeexplore.ieee.org/document/5534883.

[132] Sanlisoy A, Carpinlioglu MO. A review on plasma gasification for solid waste disposal. Int J Hydrogen Energy. 2017;42(2):1361–65. DOI: https://doi.org/10.1016/j.ijhydene.2016.06.008.

[133] Zhovtyansky V, Valinčius V. Efficiency of plasma gasification technologies for hazardous waste treatment. In: Yun Y, editor. Gasification for Low-Grade Feedstock. InTech; 2018. DOI: https://doi.org/10.5772/intechopen.74485.

[134] Elsevier. (2020). SGH2 plant to produce green hydrogen from waste in California. https://www.sciencedirect.com/science/article/abs/pii/S1464285920302479#:~:text=In%20the%20US%2C%20SGH2%20Energy,from%20recycled%20mixed%20paper%20waste.

[135] Leiby S (2020). Green hydrogen from trash. Strategic Business Insights. http://www.strategicbusinessinsights.com/about/featured/2020/2020-09-green-hydrogen-trash.shtml, accessed 27 January 2021.

[136] ipg. (2020). Grüner Wasserstoff aus dem Industriepark Premnitz (IPP). ipg Infrastruktur- und Projektentwicklungsgesellschaft mbH. https://www.ipg-potsdam.de/gruener-wasserstoff-aus-dem-industriepark-premnitz-ipp/, accessed 27 January 2021.

[137] IEA. (2021d). Transport: Improving the sustainability of passenger and freight transport. International Energy Agency. https://www.iea.org/topics/transport, accessed 27 January 2021.

[138] Rietmann N, Lieven T. How policy measures succeeded to promote electric mobility – Worldwide review and outlook. J Clean Prod. 2019;206(1):66–75. https://doi.org/10.1016/j.jclepro.2018.09.121.

[139] VDA. (2020). Synthetic fuels – power for the future [Environment and Climate]. Verband der Automobilindustrie. https://www.vda.de/en/topics/environment-and-climate/e-fuels/synthetic-fuels.html, accessed 27 January 2021.

[140] Siegemund S, Trommler M, Kolb O, Zinnecker V (2019). The potential of electricity-based fuels for low-emission transport in the EU: An expertise by LBST and dena [E-Fuels Study]. dena Deutsche Energie-Agentur GmbH. https://www.dena.de/fileadmin/dena/Dokumente/Pdf/9219_E-FUELS-STUDY_The_potential_of_electricity_based_fuels_for_low_emission_transport_in_the_EU.pdf, accessed 26 January 2021.

[141] Schemme S, Breuer JL, Köller M, Meschede S, Walman F, Samsun RC, Peters R, Stolten D. H2-based synthetic fuels: A techno-economic comparison of alcohol, ether and hydrocarbon production. Int J Hydrogen Energy. 2020;45(8):5395–414. https://doi.org/10.1016/j.ijhydene.2019.05.028.

[142] Krylova AY. Products of the Fischer-Tropsch synthesis (a review). Solid Fuel Chem. 2014;48(1):22–35. https://doi.org/10.3103/S0361521914010030.

[143] Tiverios P, Dunlap B (2019). DOE Bioenergy technologies office (BETO): 2019 Project peer review [Bioenergy Fulcrum Sierra BioFuels, LLC]. Fulcrum™. https://www.energy.gov/sites/prod/files/2019/04/f61/Fulcrum%20Sierra%20Biofuels%20Project_EE000DPA1.pdf, accessed 27 January 2021.

[144] Dagle RA, Lizarazo-Adarme JA, Zhu Y, Jones SB, Hallen RT, White JF, Palo DR, Lebarbier VMC, King DL, Gray MJ, Biddy MJ, Wang Y, Holladay JE (2013). Single-step syngas-to-distillates (S2D) synthesis via methanol and dimethyl ether intermediates: Final Report [Conducted under the National Advanced Biomass Consortium program]. U.S. Department of Energy. https://www.pnnl.gov/main/publications/external/technical_reports/PNNL-22984.pdf, accessed 27 January 2021.
[145] Meschkat S (2020). Forscher entwickeln CO2-neutrales Benzin [Klimaschutz]. Deutschlandfunk Nova. https://www.deutschlandfunknova.de/beitrag/klimaschutz-forscher-entwickeln-co2-neutrales-benzin, accessed 27 January 2021.
[146] MDR. (2020). Grüner Strom aus CO2 aus der Luft: Treibstoff der Zukunft [Ohne Russ und Emissionen]. Mitteldeutscher Rundfunk. https://www.mdr.de/wissen/umwelt/synthetischer-kraftsstoff-aus-mitteldeutschland-100.html, accessed 27 January 2021.
[147] NETL. 12.3 Types of Coal-Derived Chemicals. National Energy Technology Laboratory; 2021a, https://netl.doe.gov/research/coal/energy-systems/gasification/gasifipedia/coal-derived-chem, accessed 26 January 2021.
[148] Meyer B, Lee RP, Reinmöller M. Chemical Recycling: A Key Building Block in the Transition Towards Zero-Waste Cities. Taiyuan, China: Taiyuan Energy Low Carbon Development Forum; 2019, 23 October 2019.
[149] Seidl LG, Mamani-Soliz PE, Lee RP, Meyer B (2020). Chemical recycling – Technology overview and current developments. Fraunhofer IMWS. NK2 Chemical Recycling Workshop, 25 February 2020, Freiberg, Germany.
[150] Keller F, Lee RP, Meyer B (2020). Consequential life cycle assessment of chemical recycling technologies in the waste-energy-chemical system. Fraunhofer IMWS. NK2 Chemical Recycling Workshop, 25 February 2020, Freiberg, Germany.
[151] Rievaj V, Gana J., Synak F. Is hydrogen the fuel of the future? Transportation Research Procedia. 2019, 40:469–474. https://doi.org/10.1016/j.trpro.2019.07.068.
[152] European Commission (2021). Electricity price statistics. https://ec.europa.eu/eurostat/statistics-explained/index.php?title=Electricity_price_statistics, accessed 26 April 2021.
[153] Dagdougui H, Sacile H, Bersani C, Ouammi A, Chapter 2 - Hydrogen Production and Current Technologies. In: Dagdougui H. Sacile R, Bersani C, Ouammi A, (Eds), Hydrogen Infrastructure for Energy Applications, Academic Press, 2018: 7–21. https://doi.org/10.1016/B978-0-12-812036-1.00002-0.

Index

https://doi.org/10.1515/9783110608458-022

www.ingramcontent.com/pod-product-compliance
Lightning Source LLC
LaVergne TN
LVHW080333021026
849160LV00028B/1065
9783110608434